Multi-Omics in Biomedical Sciences and Environmental Sustainability

Minu Kesheri • Swarna Kanchan
Donat-P. Häder • Rajeshwar P. Sinha
Editors

Multi-Omics in Biomedical Sciences and Environmental Sustainability

Applications and Recent Advances

Springer

Editors
Minu Kesheri
Department of Biomedical Sciences
Joan C. Edwards School of Medicine
Marshall University
Huntington, WV, USA

Swarna Kanchan
Department of Biomedical Sciences
Joan C. Edwards School of Medicine
Marshall University
Huntington, WV, USA

Donat-P. Häder
Department of Botany
Friedrich-Alexander University
Möhrendorf, Bayern, Germany

Rajeshwar P. Sinha
Department of Botany
Banaras Hindu University
Varanasi, Uttar Pradesh, India

ISBN 978-981-96-7066-6 ISBN 978-981-96-7067-3 (eBook)
https://doi.org/10.1007/978-981-96-7067-3

This Springer imprint is published by the registered company Springer Nature Singapore Pte Ltd.
The registered company address is: 152 Beach Road, #21-01/04 Gateway East, Singapore 189721, Singapore

Foreword

It's a great opportunity and pleasure to pen down the foreword for this edited book "Multi-omics in Biomedical Sciences and Environmental Sustainability—Applications and Recent Advances" edited by *Drs. Minu Kesheri, Swarna Kanchan, Donat-P. Häder*, and *Rajeshwar P. Sinha*. This reminds me of the memories and golden days spent together with my friend, Prof. Dr. Rajeshwar P. Sinha. Prof. Dr. Donat-P. Häder is a beacon in the field of environmental sciences and a veteran who has been an appointed member of the Environmental Effects Panel of the United Nations on the effects of ozone depletion and climate change apart from being an Emeritus Professor, and former Director of the Botanical Institute and the Chair of Ecophysiology of Plants at the Friedrich-Alexander University, Erlangen-Nürnberg, Germany. My friend Prof. Dr. Rajeshwar P. Sinha has been a DAAD Fellow awardee and has a vast scientific experience owing to his scientific stays and travels around the globe visiting countries of Europe, Japan, Republic of Korea, the UK, and the USA. He is among the top 2% most cited researchers in the world as showcased by Stanford University, USA. It's a pleasure sharing that I have known Dr. Minu Kesheri who is an exuberant researcher and academician being inspired and guided by great mentors like Donat-P. Häder and Rajeshwar P. Sinha. It's an honor to testify Dr. Minu Kesheri as a potential torchbearer who would aid in imparting knowledge she gained during her doctoral research with Prof. Sinha and her postdoctoral research experience and teaching during her international scientific stays in Europe and the USA. I am confident that Dr. Swarna Kanchan's excellent international scientific research experiences gained from several European countries and renowned Universities in the USA would synergize this group of editors to deliver an edited book that is polished and highlights the recent advances in the field of biomedical sciences and environmental sustainability.

Multi-omics is the buzzword in the arena of omics that fuels the research findings by bestowing a top angle view of the biological systems by painting a holistic panorama of integration of various biological processes, their environment and interactions among various biomolecules. Moreover, in this era of artificial intelligence, machine learning, bioinformatics, and data science are key tools to unlock the treasure of information obtained from various omics technologies comprising of genomics, metagenomics, transcriptomics, meta-transcriptomics, proteomics, metabolomics, integromics, phenomics, lipidomics, etc.

Owing to my decades of experience in scientific research and looking at the topics covered in this book regarding biomedical sciences and environmental sustainability, I am highly positive that this edited book will be an exuberant and comprehensive yet broad collection of topics illustrating state of the art in a lucid style. The chapters shedding light on the computational codes, scripts, and bioinformatic tools and pipelines that are in demand nowadays empower it with the potential to revolutionize the findings and arouse interests of both academicians and researchers. All the chapters in this book describe the basic concepts, delve deep into recent advances in relevant technologies, and highlight various applications providing fruitful scientific insights. Last but not the least, I convey my heartiest congratulations to the authors from all over the world who have contributed the chapters and to the editors for their relentless work. I wish this book a great readership and fantastic success!

Rakesh Pandey
Professor Emeritus AcSIR & CSIR-Emeritus Scientist,
Crop Protection & Production Division, CSIR-Central
Institute of Medicinal and Aromatic Plants (CIMAP)
Lucknow, India

Preface

Multi-omics alludes to a cutting-edge approach that serves as a conduit of data incorporating various biomolecular platforms encompassing genomics, transcriptomics, proteomics, metabolomics, epigenomics, neuromics, phenomics, etc. Various omics data when fueled by analyses comprising bioinformatics, statistics, data science, machine learning, and artificial intelligence paints a holistic panorama of interactions occurring in biological systems and their niches, enabling better understanding of numerous aspects pertaining to morphological, physiological, biomedical, novel drug discovery and development, and resilience strategies in response to climate change, advances in bioremediation approaches, futuristic prediction models of pandemics to alert healthcare systems, and paving ways for achieving environmental sustainability.

This edited book is a compilation of various topics pertaining to applications and recent advances in the field of biomedical sciences and environmental sustainability. The topics under the two broad horizons, biomedical sciences and environmental sustainability, discuss the state of the art in the respective arena. Chapter 1 describes the coming of age of translational omics in precision medicine. Chapter 2 explores the multi-omics landscape describing its scope crossing the horizons of genomes. Chapter 3 presents insights in disease dynamics from microbiome and multi-omics perspectives. Chapter 4 illuminates multiresolution insights into single-cell landscapes while integrating genomics, epigenomics, and proteomics for brain studies. Chapter 5 enumerates multi-omics applications in drug discovery and development. Chapter 6 revolves around the buzzwords of the era elucidating machine learning approaches for the diagnosis of pre- and post-Covid-19 phases. Chapter 7 enlightens human brain disorders through multi-omics approaches. Chapter 8 discusses various applications of multi-omics in human genetic disorders. Chapter 9 encompasses metabolomics approaches for microbiome analysis. Chapter 10 presents a multi-omics perspective on brain microbiome in health and disease. Chapter 11 delegates microbiome dynamics in colorectal cancer describing its implications for personalized treatment and multi-omics analysis. Chapter 12 showcases human–environment interactions through a lens comprising multi-omics and interactome perspective. Chapter 13 opens the second broad horizon of environmental sustainability by advocating for bioremediation as a boon for a sustainable environment. Chapter 14 describes multi-omics approaches for environmental sustainability. Chapter 15 reviews the innovative applications of phenomics in understanding plant root systems. Chapter 16 describes the industrial production of biomedically significant alginates by harnessing omics with machine learning technologies. Chapter 17 elucidates proteomics and bioinformatics approaches for exploring resilience strategies in cyanobacteria highlighting the state of the art of AlphaFold modeling skills. Chapter 18 describes advancements and trends in omics of cyanobacterial ultraviolet-screening compounds, scytonemin and mycosporine-like amino acids. Chapter 19 posits metabolomics as an approach to improve production of cyanobacterial natural products. Chapter 20 discusses the significance of omics in cyanobacterial biofuel production. Chapter 21 summarizes lichen symbiosis from a system biology perspective.

This book bestows its readers a comprehensive overview touching base with the basic concepts to the trailblazing advances regarding various facets of omics technologies and its applications in the field of biomedical sciences and environmental sustainability through illustrative figures and a lucid style. The incorporation of various updated bioinformatics tools, computational algorithms and codes aids in deciphering complex data providing multifaceted as well as a top angle overview of experimental analyses. This book would be advantageous to bibliophiles spanning from the research fraternity, academics, and doyens in the arena of biomedical and environmental sciences as well as professionals related to the development of novel drug design and biotechnological applications. We wish the readers happy reading!

Huntington, WV, USA — Minu Kesheri
Huntington, WV, USA — Swarna Kanchan
Möhrendorf, Bayern, Germany — Donat-P. Häder
Varanasi, Uttar Pradesh, India — Rajeshwar P. Sinha

Acknowledgments

The editors express their heartfelt thanks to the Almighty God for everything he has bestowed to us. We convey our sincere gratitude to the most reputed scientists and researchers who contributed as authors leading to successful culmination of various chapters in the form of an edited book. Dr. Minu Kesheri and Dr. Swarna Kanchan are grateful to Dr. Brad D. Smith, President, Marshall University, Huntington, WV, USA; Dr. David Gozal, Vice President for health affairs and the dean of the Joan C. Edwards School of Medicine, Marshall University; Dr. David A Dampier, Dean of the College of Engineering and Computer Sciences at Marshall University Huntington, WV, USA; Dr. Gary O. Rankin, Professor and Chair, Department of Biomedical Sciences, and Vice Dean for Basic Sciences, Marshall University, Huntington, WV, USA; Dr. James Denvir, Professor and Director, Data Science Core Facility, Department of Biomedical Sciences, Marshall University, Huntington, WV, USA; Dr. Krista L. Denning, Professor and Chair, Department of Pathology, Marshall University, Huntington, WV, USA, for their consistent encouragement moral support and mentorship. We devote our heartfelt gratitude for the encouragement and guidance from Prof. Greg Hampikian and Dr. Eric Hayden, Chair of the Department of Biological Sciences at Boise State University, Boise, ID, USA; thanks, are due to Dr. Pascal Durrens, INRIA, France and Donald A. Primerano, Emeritus Professor and former Director of the genomics core at the Department of Biomedical Sciences, Joan C. Edwards School of Medicine, Marshall University, Huntington, WV, USA, for their motivation and blessings that keeps our spirits high. It would not be extravagant to thankfully count upon the kind support from Prof. Dr. Rakesh Pandey, Emeritus Professor AcSIR and CSIR-Emeritus Scientist at the Crop Protection & Production Division, CSIR-Central Institute of Medicinal and Aromatic Plants (CIMAP), Lucknow, India, for his shout-out in the form of "Foreword" for this edited book.

The editors extend their heartfelt gratitude for the industriousness and support from the publisher's team at Springer Nature, whose conscientious guidance, patience, proficiency, professionalism, and devotion deserve plaudits. We are grateful to our panel of reviewers for their assessment, constructive feedback and insightful comments that helped to present excellent content conferring state of art to various topics discussed in this book in a reader-friendly way. We express our sincere gratitude to all the contributors. Thanks are also due to Varsha K. Singh for her excellent technical assistance during the editing of this book. We would like to acknowledge Ashish P. Singh, Sapana Jha, Palak Rana, and Riya Tripathi for their cooperation while editing of the book. We also wish to offer our indebtedness and gratitude to our family, friends, and mentors for their unwavering support, encouragement, belief, and patience. Last but not least, Dr. Minu Kesheri and Dr. Swarna Kanchan are grateful to their beloved son Adarsh Keshari for his patience and supportive motivation throughout the compilation of this project. We appreciate the research fraternity for their assiduous research and wish that this book draws interests of our readers engendering ideas for delving deep to unravel the mysteries of scientific world.

Minu Kesheri
Swarna Kanchan
Donat-P. Häder
Rajeshwar P. Sinha

Contents

Editors and Contributors

About the Editors

Minu Kesheri earned her Ph.D. degree from Banaras Hindu University, Varanasi, Uttar Pradesh, India. She was a Postdoctoral Fellow at Boise State University, Boise, Idaho, USA. During her scientific stays at the University of Bordeaux campus in France, Poland, Switzerland, Germany, Spain, and Belgium, she delivered invited talks, presented posters, attended workshops, and served as a judge and chaired sessions at conferences. She served as Assistant Professor and Head of the Biological and Environmental Sciences Department in India at Shridhar University, Presidency University, and Amity University. She procured fundings as Principal Investigator and was awarded as young scientist. She served as Research Associate at Department of Biomedical Sciences, Joan C. Edwards School of Medicine, Marshall University, Huntington, West Virginia, USA. She has published high impact peer-reviewed research articles, book chapters, edited and authored books. She is a reviewer and editorial board member of international scientific journals and committees.

Swarna Kanchan obtained his Ph.D. degree from Birla Institute of Technology and Science, Pilani, Rajasthan, India. He served as a Postdoctoral Fellow at INRIA Bordeaux in collaboration with the University of Bordeaux, France, Nencki Institute of Experimental Biology, Warsaw, Poland; and Boise State University, Boise, Idaho, USA. He worked as an Assistant Professor at the Department of Life Sciences, Presidency University, India. He served as a Staff Scientist-Bioinformatics at Joan C. Edwards School of Medicine,

Marshall University, Huntington, WV, USA. He has published several peer-reviewed research articles, book chapters, and books. He is an active reviewer and editorial board member of international scientific journals and committees. He has delivered invited lectures, served at technical program committee member in various conferences, poster session judge, and presented posters and participated in workshops and conferences at national and international institutions in India, France, Germany, Switzerland, Poland, and the USA.

Donat-P. Häder, prior to his retirement, served as the director of the Botanical Institute, held the chair of Ecophysiology of Plants at the Friedrich-Alexander University, Erlangen-Nürnberg, Germany, and was the director of the Botanical Garden. He has authored over 800 scientific publications, has produced 32 books, and holds seven national and international patents. He is the editor of several international journals. His research interests allude to cyanobacteria, gravitational biology of microorganisms and ecophysiology and ecotoxicology of aquatic ecosystems, and he is a founder and an appointed member of the Environmental Effects Panel of the United Nations on the Effects of Ozone Depletion and Climate Change. He is vice president of a German photobiological society. He ranks among the top 1% most cited scientists in the world.

Rajeshwar P. Sinha, DAAD Fellowship Awardee and Fellow, Society for Applied Biotechnology, India, serves as a Senior Professor of Molecular Biology, Department of Botany, Banaras Hindu University (BHU), Varanasi, India. He earned his Ph.D. in Biotechnology from BHU, Varanasi, India, and visited countries such as Argentina, Austria, Belgium, Canada, China, Germany, Greece, France, Italy, Japan, Luxembourg, Norway, Poland, the Republic of Korea, Spain, Switzerland, the Netherlands, the UK, and the USA in the field of academics/research. His research aims at UV-B radiation impacts on cyanobacteria. He has published over 550 research papers/reviews/book chapters/conference proceedings and edited/authored 14 books. He is a lifetime member of several national and international scientific societies and editorial board member of various national and international journals. He has over 17,690 citations with an h-index of 60 and i10-index of 169. He stands among the top 2% most cited scientists in the world.

Contributors

Haseen Ahmed Department of Botany, Rajkiya Mahila Snatkottar Mahavidyalay (Affiliated to Veer Bahadur Singh Purvanchal University, Jaunpur, Uttar Pradesh), Ghazipur, Uttar Pradesh, India

Tanzeel Ahmed School of Biotechnology, IFTM University, Moradabad, Uttar Pradesh, India

Anupama Avasthi Amity Institute of Biotechnology, Amity University, Noida, Uttar Pradesh, India

Jyothi Belurappa Department of Biotechnology, REVA University, School of Applied Sciences, Yelahanka, Bengaluru, Karnataka, India

Sajad Bhat Department of Adult Neurosurgery, Stanford School of Medicine, Stanford, CA, USA

Ashutosh Singh Chauhan Air Pollution and Stone Conservation Laboratory, Archaeological Survey of India, Agra Fort, Agra, Uttar Pradesh, India

Bhaskar Chandra Srinivas Chittoori Department of Civil Engineering, Boise State University, Boise, ID, USA

Poornima Dubey Department of Nanotechnology, Indian Institute of Technology, Uttarakhand, India, Roorkee

Department of Biological Engineering, Massachusetts Institute of Technology, Cambridge, MA, USA

Wolfgang Gärtner Institute for Analytical Chemistry, University of Leipzig, Leipzig, Germany

Amit Gupta Laboratory of Photobiology and Molecular Microbiology, Department of Botany, Institute of Science, Banaras Hindu University, Varanasi, Uttar Pradesh, India

Manish Kumar Gupta Department of Biotechnology, Veer Bahadur Singh Purvanchal, University, Jaunpur, Uttar Pradesh, India

Donat-P. Häder Emeritus from Friedrich-Alexander University, Department of Biology, Möhrendorf, Germany

Sapana Jha Laboratory of Photobiology and Molecular Microbiology, Department of Botany, Institute of Science, Banaras Hindu University, Varanasi, Uttar Pradesh, India

Poonam Kaithal Department of Molecular and Cellular Engineering, Jacob Institute of Biotechnology and Bioengineering SHUATS, Prayagraj, Uttar Pradesh, India

Swarna Kanchan Department of Biological Sciences, Boise State University, Huntington, WV, USA

Department of Biomedical Sciences, Joan C. Edwards School of Medicine, Marshall University, Huntington, WV, USA

Pramod Katara Computational Omics Lab, Centre of Bioinformatics, IIDS, University of Allahabad, Prayagraj, Uttar Pradesh, India

Chandandeep Kaur Happiest Health, Bangalore, Karnataka, India

Harleen Kaur CSIR—National Physical Laboratory, New Delhi, India

Minu Kesheri Department of Biological Sciences, Boise State University, Boise, ID, USA

Department of Biomedical Sciences, Joan C. Edwards School of Medicine, Marshall University, Huntington, WV, USA

Vivek Keshri Department of Biotechnology, REVA University, School of Applied Sciences, Bengaluru, Karnataka, India

Neha Kumari Laboratory of Photobiology and Molecular Microbiology, Department of Botany, Institute of Science, Banaras Hindu University, Varanasi, Uttar Pradesh, India

Bhagaban Mallik Department of Biological Sciences, IISER, Bhopal, Madhya Pradesh, India

Department of Anatomy and Cell Biology, University of Iowa, Iowa City, IA, USA

Tarun Mishra Department of Biological Sciences, IISER, Bhopal, Madhya Pradesh, India

Department of Microbiology and Immunology, University of Iowa, Iowa City, IA, USA

Jainendra Pathak Department of Botany, Pt. Jawaharlal Nehru College (Affiliated to Bundelkhand University, Jhansi), Banda, Uttar Pradesh, India

Shraddha Piparia Department of Pediatrics, University of California, San Diego, CA, USA

Palak Rana Laboratory of Photobiology and Molecular Microbiology, Department of Botany, Institute of Science, Banaras Hindu University, Varanasi, Uttar Pradesh, India

Bhawna Rathi Amity Institute of Biotechnology, Amity University, Noida, Uttar Pradesh, India

Ratnaprabha Ratna-Raj Department of Civil Engineering, Boise State University, Boise, ID, USA

Ahmad Reza Rezaei Medical University of Bialystok, Bialystok, Poland

Susanta Roy Bio Explore, Dimapur, Nagaland, India

Shivani Sharda Amity Institute of Biotechnology, Amity University, Noida, Uttar Pradesh, India

Ashok Kumar Sharma Research Bioinformatician II, Cedars Sinai Medical Center, Los Angeles, CA, USA

Ashish P. Singh Laboratory of Photobiology and Molecular Microbiology, Department of Botany, Institute of Science, Banaras Hindu University, Varanasi, Uttar Pradesh, India

Lavanya Singh Amity Institute of Biotechnology, Amity University, Noida, Uttar Pradesh, India

Satendra Singh Department of Computational Biology and Bioinformatics, Sam Higginbottom, University of Agriculture, Technology and Sciences, Allahabad, Uttar Pradesh, India

Smrita Singh School of Biotechnology, IFTM University, Moradabad, Uttar Pradesh, India

Varsha K. Singh Laboratory of Photobiology and Molecular Microbiology, Department of Botany, Institute of Science, Banaras Hindu University, Varanasi, Uttar Pradesh, India

Pallavi Singhal Meerut Institute of Technology, Meerut, Uttar Pradesh, India

Rajeshwar P. Sinha Laboratory of Photobiology and Molecular Microbiology, Department of Botany, Institute of Science, Banaras Hindu University, Varanasi, Uttar Pradesh, India

Shambhavee Srivastav Amity Institute of Biotechnology, Amity University, Noida, Uttar Pradesh, India

Upasna Srivastava Yale University, New Haven, CT, USA

Department of Cell Biology, Cancer Institute, Japanese Foundation for Cancer Research, Tokyo, Japan

Sonal Sukreet Alzheimer's Therapeutic Research Institute-UCS, San Diego, CA, USA

Pankaj Bharat Tiwari Department of Biological Sciences, IISER, Bhopal, Madhya Pradesh, India

Riya Tripathi Laboratory of Photobiology and Molecular Microbiology, Department of Botany, Institute of Science, Banaras Hindu University, Varanasi, Uttar Pradesh, India

1 Translational Omics: Precision and All "Coming of Age"

Lavanya Singh, Shambhavee Srivastav, Anupama Avasthi, Bhawna Rathi, Wolfgang Gärtner, Minu Kesheri, Swarna Kanchan, and Shivani Sharda

Abstract

Omics technologies have grown at a very rapid pace, offering unprecedented opportunities to deal with complex biological and environmental challenges. Translational omics connects high-throughput data generation to practical applications in precision medicine, environmental management, and industrial biotechnology. In this regard, the intricate molecular profiling of genomic, transcriptomic, proteomic, and metabolomic data offers a holistic view of biological systems, opening new avenues in understanding and solving complex phenomena. Integration of multi-omics data into sophisticated computational tools will pave the way toward biomarker identification and the enhancement of drug responses and improved environmental interventions. We discuss methodological improvement, computational models, and metadata integration approaches underlying the field of omics in unveiling how they impact both clinical, sustainable agriculture, and environmental management applications. This chapter explains the need for translational omics: from the laboratory to real applications.

L. Singh · S. Srivastav · A. Avasthi · B. Rathi · S. Sharda (✉)
Amity Institute of Biotechnology, Amity University, Noida, India
e-mail: ssharda@amity.edu

W. Gärtner
Institute for Analytical Chemistry, University of Leipzig, Leipzig, Germany

M. Kesheri · S. Kanchan
Biology Department, Boise State University, Boise, ID, USA

Joan C. Edwards School of Medicine, Marshall University, Huntington, WV, USA

M. Kesheri et al. (eds.), *Multi-Omics in Biomedical Sciences and Environmental Sustainability*, https://doi.org/10.1007/978-981-96-7067-3_1

Keywords

EcoMicrobiology · Environmental sustainability · Multi-omics · Pharmaco-precision omics · Translational omics

Objectives

- Conceptual learning of omics data analysis and pre-requisites of multi-omics and translational omics.
- Inclusion of meta-omics and application-driven multi-modal omics profiling for biological challenges.
- Converging the aspects of translational omics for emerging disciplines: Pharmaco-EcoMicrobiology, Pharmacomicrobiomics ensuing precision omics strategies.
- Potential designs of interdisciplinary studies toward biological sustainability and global health.

1.1 Introduction—Concepts of Translational Omics

The biological sciences, in this millennia, have witnessed an unprecedented interest and surge in complex high-dimensional data generation using omics research. The research into high-throughput data hitherto has impacted not just human biology or the biomedical field but extended to environmental and industrial applications. With meticulously designed experiments and discovery-based strategies, such data of biomolecular sequence, structure, function, interactions, drug responses, cellular imaging, and population studies have customized the generation of high-throughput complex knowledge base and generative biology (Mock et al. 2024). The enticing prospects of untying the complexities while spanning precision experiments toward useful outcomes due to the colossal unactionable data and challenges of customization. The cutting-edge multidisciplinary technologies grouped as "multi-omics" aims to close the existing interlude of lab explored analyses while the unequivocal "Translational Omics" realizes the potential data collected by multiple omics disciplines toward the environment's, medical, industrial, and societal benefits. This chapter unfolds as a beacon of illumination in the vast sea of biological data, shedding light on how Translation Omics offers a path to translate raw data into actionable insights, revolutionizing our comprehension of the intricate molecular machinery that underpins life itself. It is evidence of the value of multidisciplinary cooperation and the blending of data science, biology, clinical, and ecological practices in the pursuit of enhanced human well-being and a deeper understanding of biology.

We envisage the transitions and transformations of the knowledge base created by the advanced techniques as these negotiate the complex terrain of omics technology and its translation (Havas et al. 2023). Altogether, translation omics is more than simply a trendy term in science; rather, it is a ray of hope and opportunity that holds the key for sustainable and pertinent decisions commanding holistic life. The translational omics ethos emerging to emanate the need for precision studies comprises of the heterogenous details available for various bio-systems and bio-niches, which needs structural compilation of multi-omics and related studies that is presented in this scientific excursion to disentangle and critically evaluate rhetoric labels of translation. We present the panoramic view of the multi-omics data and meta-data formulating the approaches, processes, and developments in the aspects of medical, environmental, and industrial relevance. This chapter transitions from the learning of omics sciences, their multifarious profiling, and translation to sustainable scientific logics.

1.2 Biosystems—Multi-Omics Profiling

Translational Research with a major segment devotes its references in biomedical sciences, also known as "Translational medicine." The translation process though integrates the research initiatives not just in medical sciences but also in environmental aspects and innovation in artificial intelligence based on biological big data. The concepts of translational sciences have submerged the ever-expanding need for up-close biological necessities that have direct impact on the research benefits toward society (Pinu et al. 2019a).

The basic cores of biosystems studies are hereby layered to connect and draw conclusions while reconstructing the various pathways to molecular studies, which are part and parcel of these translational sciences. For this, major impactful demarcation has come from omics explorations in the varied yet interlocked fields including genomes, transcriptomics, proteomics, and metabolomics, with the profound insights necessary for expanding our understanding of complex biological systems. The huge diversity of information serves as the basis of biological big data compilation, a domain of innovative profiling networks to pave the way for precision medicine and similarly precision environmental health, agricultural health, and biodiversity measures (Zheng et al. 2024).

The tailoring of the computational and molecular technologies in omics, while exporting the precision and accuracy-driven translational omics have proven to be the cornerstone of innovative and positively disruptive solutions globally in the fields of healthcare, agriculture, and environmental health for impacting scientific, humanitarian, and economic sustainability (Denholm and Martin 2008). We outline the vital types of biological data and metadata as key ingredients for successful implementation of translational omics anticipated toward potential solutions to existing and unexpected future bioinformatics needs (Kumari et al. 2016, 2018). The rigorous integration of signature omics data includes the genomic wide associations, meta-data unique features, various regulatory omics as well as proteomic/metabolomic analyses, interaction profiles of host—microbiomes, socio-population, and environmental profiles for creating useful biological intelligence (Wishart 2016) and making proactive predictions to alert the health care systems during

Covid-19 led recent pandemic (Kanchan et al. 2024). The high-dimensional complex data spawn a web of experimental and computational models, which are pitched to proficiency while removing over-representation and over-fitting as these data are generated by measuring multi-variables per sample than total number of biological samples. The involvement of a multitude of biomarkers and variations in sample analyses needs a relook while harmonizing the translation of discovery driven technical data to biological outputs (Kumar et al. 2024). With the ever-growing need for further assimilation of biological big data and machine learning models, utilize the pattern learning processes embedded in artificial intelligence for accurate identification of complex associations required for managing effective translational omics (Blum et al. 2018; Fig. 1.1). The landscape of factors that contour translational omics formulate with some of the following targets:

- Systematic evidence of multi-omics studies with execution and interpretation of complementary omics profiling.
- Study information extracted to include experimental study designs, sampling design, size calculations, variant characterization, and regulatory methodologies.
- Algorithm improvements and in silico modeling precision.

1.2.1 Translation Medicine—Converging the Biomedical Profiles

In the pursuit of biomedical advancements, the convergence afforded by multi-omics profiling extends beyond a mere amalgamation of data as it serves to be the

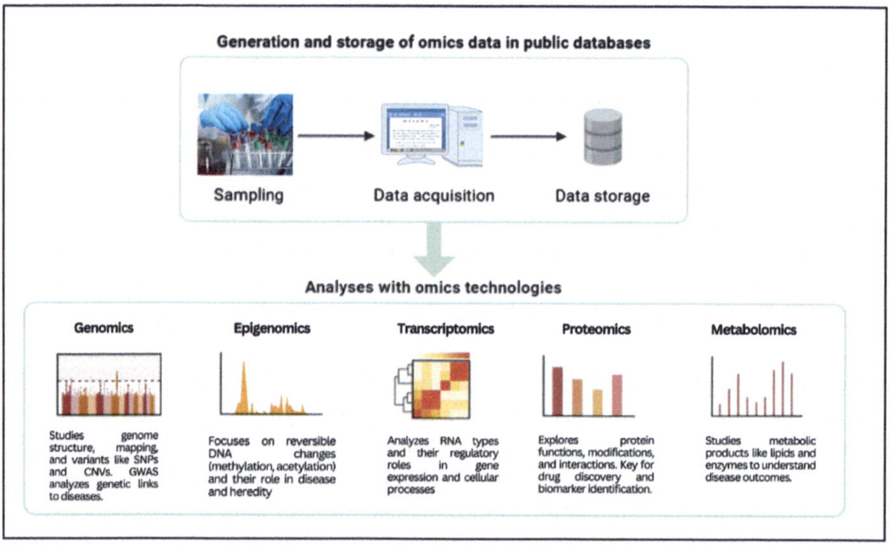

Fig. 1.1 Multi-omics data profiles—individual technologies toward integration

key unlocking precision medicine frontiers (Hu and Wang 2021). As genomics, transcriptomics, proteomics, and metabolomics intertwine, researchers gain a nuanced understanding of the molecular intricacies governing health and disease. This integrated approach is particularly transformative in identifying subtle variations in individual genomes, deciphering intricate signaling pathways, and uncovering regulatory networks that underlie diverse health conditions (Babu and Snyder 2023).

Precision medicine, propelled by insights from multi-omics studies, ensues a new era where treatment strategies are tailored to the unique genetic makeup of individuals (Lau and Wu 2018). Biomarkers identified through this convergence become the linchpin for early disease detection, prognosis, and the development of targeted therapies. The biosystems perspective thus not only reshapes the landscape of biomedical research but also steers clinical applications toward a future where healthcare is personalized, predictive, and profoundly impactful (Table 1.1).

1.2.2 Environmental Translation Omics—Stabilizing the Environmental and Toxicology Profiles

Metagenomics also known as environmental omics are vital in advancing multi-level investigations of organisms, populations, and communities within natural ecosystems (Pathak et al. 2009; Kanchan et al. 2020). It integrates information from genomes, metagenomes, transcriptomes, metatranscriptomes, proteomes, and metaproteomes to explore the taxonomy, diversity, phylogeny, and functional adaptations of these systems, providing a comprehensive understanding of ecosystem dynamics (Denholm and Martin 2008). This bionomic discipline integrates biotic and abiotic aspects and therefore directly influences ecosystem functions and genetic speciation. It explores the structure, diversity, and relationships of organisms comprising a biocomplex, showing synergetic, commensal, and pathogenetic associations, elucidates genomic plasticity and adaptive processes through functional annotation, assesses interactions between ecosystems and socio-economic or cultural developments, forecasts anthropogenic-factor-influenced evolutionary constraints, and resolves multi-omics data into actionable insights (Srivastav et al. 2024; Kesheri et al. 2024a).

Environmental omics processes the characterization of sample sources as bacterial, fungal, archaeal, microscopic eukaryotes, and viruses, and descriptors as eco-geographic location, environmental factors, and phenotypic relationships, ecophysiological variations (Kesheri et al. 2014), temporal variations (Kesheri et al. 2021, 2022), and climate change. Its domain extends across water, soil, forest, agricultural, phyllosphere, and host–microbial ecosystems, emphasizing the importance of interdisciplinary collaboration among biologists, bioinformaticians, and geoscientists (Crawford et al. 2017). This collaboration facilitates biodiversity assessments and functional trophic network mapping within ecosystems. Quality, annotation, and storage of large-scale genetic and molecular data were ensured by various techniques like shotgun sequencing, DNA barcoding, gene expression

Table 1.1 Biologically relevant omics metadata—outcome targeting translational research

Sr. No.	Metadata type	Description	Applications	Impact
1	Patient metadata	Patient demographics, genetic profiles, and lifestyle information	Personalized medicine, epidemiological studies	Tailored healthcare and treatment strategies
2	Clinical metadata	Diagnostics profiling, treatment protocols, medical history, and outcomes	Drug efficacy studies, health data analytics	Reduces diagnostic errors, improves treatment regimes
3	Biological metadata	Profiling samples as per tissue type, collection methods, and biological characteristics	Biomarker discovery, tissue-specific research	Improves sample relevance, data accuracy, and targeted therapies
4	Temporal metadata	Captures time-series data like treatment duration and disease progression	Longitudinal studies, chronic disease management	Understanding temporal health patterns and disease management
5	Instrumentation Metadata	Document calibration settings, measurement units, and experimental conditions	Reproducibility of experiments, inter-study comparisons	Ensures reliability and standardization of data
6	Environmental Metadata	Records geospatial, temporal, and ecological data related to the sample origin	Climate studies, biodiversity conservation	Enhances understanding of ecosystem dynamics and impacts
7	Ethical Metadata	Includes patient consent, compliance with regulations, and privacy details	Clinical trials, public health studies	Ensures ethical data use and enhances trust in research

profiling, and proteome analysis using computational and statistical tools (Kaur et al. 2021). The field touches on some significant dimensions: how to resolve the population abundance of rare or cryptic species; accurate population sampling; societal impacts; computational annotation; and, most importantly, the integration of interdisciplinary data. It aims to capture taxonomic distributions, molecular physiology, phylogenetic relationships, interactions between organisms and their environments, and societal interfaces. These insights are essential for creating robust predictive and experimental models that quantify anthropogenic influences, trophic relations, and eco-computing niches. By adopting standardized workflows, environmental omics manages meta-omics data for ecological biodiversity translation, addressing eco-toxicological variables and sustainability challenges. It bridges molecular patterns with complex ecosystem dynamics and offers a pathway for long-term environmental monitoring and impact assessment through the assessment of risk factors and individual variations.

1.2.3 Sourcing the Biological Metadata Toward Translation Omics

Metadata is an indispensable tool for translational research, serving as the backbone that transforms raw omics data into actionable insights of lasting significance to clinical, environmental, and industrial applications. Metadata embeds critical context and structure, ensuring that the collected data remains interpretable and applicable across various domains, thus solidifying its contribution as an invariable cornerstone of translational research. With the advancement of omics technologies, open-access data repositories support secondary analyses, biological computations, and integrative studies (Cernava et al. 2022). However, this wealth of heterogeneous data requires a robust framework for compatibility, reusability, and global scalability. Metadata underlies the success of omics research: this success is based on proper documentation and curation of sampling methods, experimental conditions, gene annotations, software versions, and data-processing techniques, which are known to directly support translational applications by allowing exact interpretation, integration, and the finding of novel phenomena that can be translated into new clinical, environmental, and industrial innovations. These elements are necessary for the interpretation and integration of data in an accurate manner to help researchers find new phenomena and translate their findings into real applications. Metadata ties together unprocessed omics data with practical results, enabling clinicians and researchers to work through the complexity of datasets to bring discoveries to the clinical, environmental, or industrial context. Metadata in translational omics encompasses the most critical information regarding samples, such as tissue types, collection techniques, and patient demographics. It provides information about experimental conditions, treatment regimens, and environmental factors, as well as the details of steps in data processing such as normalization and quality control. Clinical annotations associate omics data with the patient's diagnosis, disease stage, and history of treatment for biomarker discovery and therapeutic progress. Longitudinal metadata captures temporal patterns, which help in understanding disease progression or treatment responses over time (Stvilia et al. 2007). Valuable biological metadata is branched to include health and environmental aspects scoring most the bio-ecosystems that are presented further.

1.2.3.1 Health Metadata

Metadata in health and medicine is considered one of the most critical areas for data management, sharing, and analysis. In the health and medical sphere, metadata serves as a central part that catalogues, provides context, and makes available data sets, clinical studies, and other health materials. Metadata is the connecting link between raw data and actionable information that gives thorough insights into the structure and function of datasets, bringing immense benefits to academics, doctors, and policymakers alike. In the following sections, we delve into the diverse types of health metadata and their crucial role in translational research:

(a) *Patient Metadata*: The metadata of the patients are quite heterogeneous, starting from simple demographics up to detailed genetic information. This type of metadata helps in tailored care by better understanding the specific profiles of each patient. A doctor can utilize the patient metadata for treatment personalization, while a researcher can apply this to research on the distribution of diseases within various populations that would facilitate more advanced epidemiological studies.
(b) *Clinical Metadata*: Clinical metadata are used to record all the healthcare encounters, such as diagnostic tests, treatment regimens, and histories of medications. Such metadata help in the clinical decision-making process, thereby reducing the probability of misdiagnosis, which can be traced back to investigate the outcome of interventions.
(c) *Biological Metadata*: In genomic science and molecular biology, biological metadata provides important contextual information to the biological samples from tissue type, collection method, and genetic variation. Linking genomic data with specific samples helps keep those data more accurate and less error-prone, thus enriching precision medicine and genetic research fields.
(d) *Temporal Metadata*: Temporal metadata capture changes in patient health over time and aids in the assessment of treatment efficacy and disease progression. In chronic conditions such as diabetes, metadata will enable dynamic adjustment of plans for varying treatment responses based on changing health trends.
(e) *Instrumentation Metadata*: The metadata of instrumentation encompasses calibration settings as well as units of measurement utilized by medical devices during diagnostics and treatment. Therefore, the reliability and consistency of data brought about by various studies can be assured, thereby supporting clinical care and research as well.

1.2.3.2 Environmental Metadata

Although the concept of metadata may seem abstract, it is highly critical in organizing and contextualizing diverse datasets for environmental research. Here, metadata provides a comprehensive description of environmental datasets and study projects and resources. As the foundational layer, metadata makes accurate interpretation and application of environmental knowledge at all phases in the research process possible. In what follows, we discuss some of the most important types of environmental metadata and how they have contributed to the progress of environmental research (Hirschman et al. 2008).

Types of Environmental Metadata:

(a) *Geospatial Metadata*: Geospatial metadata are more than just geographical coordinates; it also includes CRS, map projections, and spatial resolutions. This type of metadata provides insight into the spatial context of environmental data (Durante and Hardy 2015). For instance, in climate change research, geospatial metadata can identify areas with temperature anomalies or habitat shifts, thus allowing for focused conservation.

(b) *Temporal Metadata*: Temporal metadata are quite crucial in environmental monitoring with regard to trends and changes with time. They include data collection frequency and coverage extent that provide researchers with information about long-term trends, seasonal changes, and influences of ephemeral events like storms or hurricanes (Desaules 2011). In this way, the dynamic nature of the environmental process is captured with high accuracy.
(c) *Sampling Metadata*: Sampling metadata hold critical information on the methods of collection of environmental data, including sampling equipment, collection depths, and protocols. For example, in oceanography, sampling metadata shows the equipment used to collect water samples at different depths. This ensures comparability between different datasets and methodologies.
(d) *Biological Metadata*: Biological metadata, which are relevant to environmental studies, enrich biological samples with information regarding species, habitats, and collection methods. These essential metadata protect the integrity of ecological studies and supports biodiversity assessments by linking samples to their corresponding environmental contexts.
(e) *Instrumentation Metadata*: Instrumentation metadata in environmental research refer to the kinds of tools and technologies used in gathering and analysing data. This includes calibration information, units of measurement, and technical specifications which allow for standardization and reliability from study to study but also support teamwork among scientists.

Therefore, environmental metadata allow for a systematic integration of datasets, thereby accurately modelling and deciding on conservation, sustainability, and policy development. Standardized workflows and emphasis on interoperability enable the translation of environmental research into actionable solutions to global challenges.

1.2.3.3 Impact of Metadata in Translation Biology

Metadata are an important element in environmental and health research in enhancing data discovery, integration, and access. It makes it easy for researchers to identify relevant datasets within a short period and enables integration of data across diverse sources to provide a holistic picture of ecosystems or health concerns. More importantly, metadata ensure reproducibility and validity of research because of meticulous documentation that makes experiments and research even more reliable. It also informs policy decision-making in environmental protection and in clinical settings among healthcare providers. Moreover, metadata are crucial in epidemiological studies since they help in the identification of risk factors, tracking of disease outbreaks, and assessment of public health interventions.

1.3 Status of Translation of High-Throughput Data in Varied Biological Fields

The translation of high-throughput data across diverse biological fields is a dynamic landscape, facilitated by pioneering technologies such as next-generation sequencing and omics techniques. In genomics, the focus on genetic markers, especially the post-Human Genome Project, enhances risk assessment and personalized therapy (Hasin et al. 2017; Marais 2018). Proteomics translation illuminates protein interactions crucial for personalized medicine and drug discovery (Al-Amrani et al. 2021; Sahu et al. 2023). Transcriptomics data revolutionize our understanding of gene expression, offering insights into disease causes and treatment targets. Metabolomics translation advances our knowledge of metabolic underpinnings, paving the way for personalized healthcare strategies. Systems biology integration provides holistic insights, predicting system behavior and identifying regulatory nodes. In cancer research, high-throughput technologies unveil molecular subtypes and personalized therapies, shifting from generic to customized strategies (Ringborg 2019; Sharda et al. 2023). Microbiome research, fueled by high-throughput sequencing, uncovers microbial populations' health effects, guiding therapeutic interventions. Challenges persist, including data analysis, reproducibility, and ethical considerations, requiring collaborative efforts for the responsible translation of high-throughput data, transforming healthcare, and advancing biological understanding (Table 1.2).

1.3.1 Multi-Omics and Systems Research Facets

Advancing Integration Multi-omics modalities are thriving through the continuous evolution of high-throughput research involving refining methods for integrating diverse datasets of omics, while providing a more nuanced understanding of biological systems. An inherent capacity building is required to allow the collation of large heterogenous datasets with corroborative biological knowledge for establishing standardized methodologies that improve the synergistic analysis of multi-omics data, so contributing to the predictive potential of systems biology (Veenstra 2021).

Predictive Modeling As challenges to the real world medical and environmental health mount, the requirements of creation of prediction models based on multi-omics data is a critical frontier. Predictive capacities are relying on advancing algorithms that predict system behavior under a variety of scenarios, opening the path for preventive measures, probable individualized treatments, and management of biological crisis. Researchers here aim to uncover critical regulatory nodes of multi-omics by analyzing the intricate interplay of bio-molecular interactions, creating the possibilities toward therapeutic targets and pathways for precision omics.

Table 1.2 Status of high-throughput data translation—key advances, applications, and challenges in translating high-throughput data across various biological omics fields

Sr. No.	Field	Key advances	Applications	Future challenges
1	Genomics	Whole-genome sequencing, GWAS, CRISPR technologies	Risk prediction, genetic disease therapy	Ethical concerns, high costs of sequencing
2	Transcriptomics	Advances in single-cell RNA sequencing (scRNA-seq)	Understanding gene expression and regulation	Data complexity, need for advanced tools
3	Proteomics	High-resolution mass spectrometry, protein interaction networks	Drug discovery, protein-targeted therapies	Protein diversity and dynamic range issues
4	Metabolomics	Comprehensive metabolic pathway mapping	Biomarker identification, disease diagnostics	Standardization and reproducibility (Pinu et al. 2019b)
5	Epigenomics	Development of methylation arrays	Cancer epigenetics, heritable disease studies	Data interpretation complexity
6	Microbiomics	Advances in microbiome sequencing and functional analysis	Gut health studies, probiotics development	Variability among individuals, host–microbiome interactions
7	Multi-Omics Integration	Tools for integrating genomics, transcriptomics, and proteomics data	Systems biology, precision medicine	Computational power, harmonization of datasets

Appending Emerging Technologies Revolutionary technological advances (e.g., single-cell omics and spatial omics) are expanding the landscape of multi-omics investigations. These tools enable the dissection of biological systems at a finer resolution than previously possible, permitting the study of cellular heterogeneity and spatial architecture. The incorporation of developing technologies into multi-omics research has the potential to uncover additional levels of complexity in biological systems (Wimmers and Pulendran 2020; Liang et al. 2023).

1.3.2 Environmental Impact of Omics Translation—Handling AMR and Meta Studies

Environmental translational omics is a new frontier that stands at the crossroads of molecular biology and ecological challenges, coalescing multi-omics to reveal profound insights into the intricate workings of ecosystems. Translational environmental processes bridge existing gaps of biochemical foundations, molecular data, and organism–environment interactions for unravelling ecosystem responses to natural and anthropogenic changes, thus endeavoring the world's most pressing concerns, involving biodiversity loss, pollution, habitat degradation, climate change, and adaptations, driving environmental processes. Integrating high-dimensional datasets to unfold the intricate interactions between organisms with their habitats, through high-throughput sequencing technologies and bioinformatics algorithms, conceivable details of the mapping of microbial communities at structural and functional strata with unprecedented fidelity can technically be achieved. Such insights are crucial in the identification of keystone species, understanding microbial dynamics (Mallik et al. 2024), and assessing resilience in the ecosystem to stress conditions. For instance, soil omics has helped in understanding how microbial communities play significant roles in nutrient cycling, carbon sequestration, pathogen suppression, dysbiosis, and gut microbiome composition (Mishra et al. 2024) thus presenting valuable input for sustainable agriculture and land management.

The heterogeneous datasets of omics studies from different ecological contexts need to be analyzed for holistic scale dynamics of an ecosystem, to further concoct the processes for global ecological balance. Such capabilities are potentiated to identify important signature markers, ecological interactions, and intervention points that nurture resilience against ecological disruptions and improve the robustness of environmental omics translation. Prodigious emphasis is put on translating effective omics strategies for tackling antimicrobial resistance, or AMR to unlock the complex intricacies of resistant genes within microbial communities. The findings open avenues for the design of tailored approaches for mitigating the environmental impacts of AMR and preserving the efficacy of antibiotics currently in use. The novel mechanisms of resistance unveiled through integrative omics provide for designing targeted antimicrobial agents and strategies to combat horizontal gene transfer within microbial ecosystems (Pandey 2017). In the era of unknown microbes/viruses, significant focus for tackling AMR hotspots such as hospitals, agricultural sites, and wastewater treatment facilities is pioneered to complement meta-studies in environmental omics.

Translational studies in environmental omics also bear ethical considerations, whereby balancing the preservation of ecological components needs to be addressed with possible unforeseen impacts of human interventions needs immediate measures. This requires creating an ethical framework for scientists based on clear stakeholder communication, adherence to conservation principles, and proactive long-term evaluation of ecological impacts. Such frameworks would allow omics applications not to harm biodiversity, ecosystem integrity, or social equity but to harmonize scientific progress with ecological stewardship.

1.3.2.1 Environmental Translational Omics

Modern environmental science in multi-omics paradigms is vital for unravelling the molecular complexities that regulate ecosystems with transformative narratives in environmental matrices, inclusive of temporal responses of organisms to dynamic environmental stimuli. Some of these details include adaptive strategies of aquatic organisms with identified molecular stress biomarkers, which are directed interventions toward rehabilitation in response to human pollution degrading aquatic ecosystems (Kesheri et al. 2022). Environmental omics further improve the efficiency of agricultural production through the discovery of genes involved in the traits for plant resilience and optimization of yields toward sustainability in agricultural systems (Kesheri et al. 2024b). Research studies comprising proteomic analysis of soil microbial communities gives thorough understanding of nutrient cycling and carbon sequestration, thus postulating strategies for reducing greenhouse gas emissions. In the face of climate change, translational omics serve as an environmental forecaster, revealing adaptive genetic signatures in organisms responding to climatic changes while offering preventive strategies to preserve biodiversity and maintain ecological balance. Translational omics reveal the intricate molecular interplay between genetics and environmental factors, thereby specifying targeted public health interventions and paving the way for personalized medicine. Emerging tools, such as single-cell omics and spatial transcriptomics, are refining our comprehension of how microenvironments affect organismal responses at the cellular level. Translational omics takes raw molecular data and translates it into action to restore ecosystems, maintain sustainable agriculture, and ensure ecological conservation. It serves as a kind of connective tissue between scientific exploration and practical solutions, clearing the path toward a resilient and sustainable future.

The ability to trace the origin of contaminants, determine the fate of pollutants, and assess the health status of aquatic systems makes water systems a matter of great interest in environmental omics. Using translational approaches, scientists have identified genetic and enzymatic traits of organisms that facilitate bioremediation of toxins, thus providing innovative solutions aimed at restoring polluted water bodies. These insights also involve marine and freshwater biodiversity exploration, which provides critical information necessary for saving endangered species and maintaining ecological equilibrium (Schully and Khoury 2014). Pollution mitigation is another fundamental component of environmental omics. Characterization of the molecular properties of pollutants and their interaction with biological systems can be done to design specific interventions for effective remediation. Omics-based bioremediation takes the impressive potential of microorganisms and plants in degrading or immobilizing contaminants. Environmental challenges of heavy metal pollution, oil spills, and industrial waste, need propelling green technologies through phytoremediation and microbial fuel cell technologies, well aligned with the sustainability agenda and preservation of biodiversity. Multi-omics processing of study outcomes entails the power to discover adaptive genetic variants enabling species to survive in an ever-changing world. These genetic variants are an important source of information for restoration ecology, focusing efforts on degraded habitats and ensuring that ecosystems are conserved against global environmental change's impacts. With the molecular details of how an organism responds to environmental stressors, translational omics will help conservationists anticipate and mitigate effects of global environmental change. Advances in technology, especially artificial intelligence and machine learning, have changed the face of processing and interpretation of omics data. Predictive models derived multi-omics and AI-driven tools effectuate early detection of ecological disturbances, exemplified by unravelling rise of invasive species, spread of harmful algal blooms. Such models offer valuable insights for policymakers and environmental managers to implement proactive measures that mitigate ecological harm while fostering long-term sustainability.

Nevertheless, there exist crucial challenges for translational efforts, predominantly for inexact standardized protocols for the integration and interpretation of data. Interdisciplinary strengths are needed to overcome the complexity of environmental datasets at individualistic and interaction scales as well as to attune ethical issues arising from ecological manipulations to ensure translational applications are responsible and effective (Zhang et al. 2019). This demands an imperative need for collaboration between biologists, data scientists, ecologists, and policymakers to overcome the challenges and translate omics potential fully.

1.3.2.2 Breathing in Future with Translational Omics

Air pollution accumulating due to vehicular fossil fuel combustion, energy sources in multiple industrial units add up as major contributors to release pollutants likewise of nitrogen oxides (NO_2), sulphur dioxides, carbon monoxide (CO), and respirable particulates, marring the urban and rural population with inflammatory pulmonary conditions. Translational omics is poised to have a pivotal role in advancing strategies to combat air pollution and mitigate the impact of airborne pathogens. While creating the biomolecular hub information targeted and efficient interventions beyond traditional methods, translational omics contributes to the environment-friendly air pollution removal and feasibility of curbing further pollutant generation through latest discoveries, including the production of biofuels like biodiesel, biogas, and bioethanol. Tackling the multiple environmental issues through such sustainable alternatives where microbial potential and less polluting renewable processes are presented needs high-throughput translational efforts to coordinate efficient and viable outputs.

Biodiesel, derived from renewable resources such as plant oils, animal fats, or algae, presents itself as a promising alternative to traditional diesel, offering a sustainable and eco-friendly energy solution. A liquid fuel produced through the transesterification of vegetable oils, animal fats, or waste cooking oil presents a potential solution as an alternative to fossil fuels. The technical process of biodiesel blended with fossil diesels (up to 20% volume), can efficiently power modern compression-ignition engines without requiring modifications to transport or storage. Chemically, while possessing a higher oxygen content than fossil diesel, that promotes complete fuel combustion, it lacks aromatic and sulphurous compounds, resulting in emissions containing significantly fewer unburned hydrocarbons, polycyclic aromatic hydrocarbons (PAHs), carbon monoxide, and particulate matter compared to fossil diesels. One of its primary advantages lies in the reduction of greenhouse gas emissions, contributing to mitigating climate change impacts. Additionally, biodiesel's biodegradability fosters environmental safety, as it breaks down more readily than conventional diesel in the event of spills or leaks.

Alternative energy sources need translational, multi-modal omics so as to acclimatise putative large-scale cultivation contributing to deforestation, habitat loss, and biodiversity decline. Furthermore, the commercial/beneficial offsetting of the potential benefits of alternate fuels needs creating a network of bio-systems omics and chemical studies to ensure its positive impact on the environment and overall sustainability (Wafi and Mirnezami 2018).

Biodiesel, as a part of green technology is exemplifying the need and transition of translational omics to the exploration and development of sustainable fuel sources, with multi-modal studies percolating crucial processes elucidating the intricate molecular mechanisms involved in biodiesel production. Genomic analyses have facilitated the identification of microbial strains with optimal traits for lipid accumulation, while proteomic studies have unveiled enzymes crucial to the biodiesel synthesis pathways. Metabolomic insights have further guided the optimization of enzymatic and metabolic processes, enhancing the overall efficiency of biodiesel production by offering a comprehensive molecular understanding for formulating bioprospecting expansion in synthetic biology applications (Pinu et al. 2019b).

1.3.2.3 The Aqueous Revolution of Translational Omics

An aspect of aquatic life forms is the most crucial to balance the eco-biosystem ecology. Translational omics, as a comprehensive tool for investigating the molecular intricacies of the aqueous geochemical cycle, conveys profound insights into water chemistry and the dynamic interactions within aquatic environments. Integrating the omics datasets facilitating a holistic understanding of the composition and functional capabilities of microbial communities in aqueous systems, influencing geochemical cycling through activities such as nitrogen fixation, denitrification, and sulphate reduction.

In the realm of contaminant fate, omics approaches contribute to understanding the molecular mechanisms underlying pollutant degradation. As due to excessive industrialization, urbanization, and rising level of pollutants, the water bodies are continuously exposed to diverse group of toxic pollutants whose concentration, chemical forms and bioavailability in the aqueous system are governed by natural geochemical cycles as well as human activities. The chemical form or species of these toxic compounds or heavy metals is governed by the prevailing specific physical and chemical parameters like pH, salinity, dissolved organics, hardness and sedimentary load, which impacts their mobility, bio-availability, and hence toxicity. To overcome the challenging situation, an increased emphasis has been given on in-situ innovative technologies like phytoremediation, capable of detoxification of contaminated water bodies. The high-throughput sciences accommodate the potential aspects of alternate metagenomic achievements that can examine a multitude of energy sourcing, phytoremediation, remediation of water pollutants for sustaining green and ecological strategies. Translational omics establish the integrative framework for comprehending functional stratification and interaction of phyto- and zooplanktons, microbial world in aqueous bodies with genetic traits conducive to pollutant uptake, tolerance, and transformation, facilitating the selection or engineering of plants optimized for aqueous phytoremediation (Kussmann and Kaput 2014; Schully and Khoury 2014). Surpassing traditional analyses and contributing to a nuanced understanding, translational omics targets the broader ecological implications of geochemical processes in aquatic ecosystems.

1.3.2.4 Soil Craft: Tomorrow's Harvest

Soil dynamics accord for utmost decisive characteristic of agricultural practices becoming the backbone of nutrition and life sustenance. Translational omics imperatively furnish genetic, proteinaceous, and metabolic constituents governing soil geological cycles whereby composition/functional capabilities, molecular intricacies, and microbial community dynamics, illuminates factors governing the eco-biosystems. This encompasses a multitude of soil ecological features toward nutritional complementation, nitrogen fixation, denitrification, and carbon cycling, toxicology, and detoxification processes highlighting impact of proteomics, metabolomics, and microbiomics interactions.

Recent translational geological studies recorded the presence of toxic action of a widely applied fungicide formulation (chlorothalonil), following a time series approach and using a soil model arthropod, *Folsomia candida*. Results of the above mentioned studies showed a mechanistic mode of action for chlorothalonil, affecting pathways of detoxification and excretion, immune response, cellular respiration, protein metabolism, and oxidative stress defence, causing irregular cell signalling (JNK and NOD ½ pathways), DNA damage and abnormal cell proliferation, leading to impairment in developmental features such as molting cycle and reproduction in invertebrates. This integrative and holistic approach, facilitated by translational omics, advances our comprehension of the molecular mechanisms governing toxin interactions in soil, thereby laying the groundwork for more efficacious remediation strategies and the implementation of sustainable land management practices.

1.3.2.5 Host Harbor: Health Horizons Through Translational Omics

Transcending into the world of personalized medicine, multi-modal omics is an emerging powerful force in health care, shaping precision medicine, with the exhaustive layering of genomic, proteomic, metabolomic, and microbiomics interactome profiles. Meta-omics appended to the host molecular profiles garners toward technology personifying medical interventions. Some of the remarkable achievements are the development of targeted cancer treatments such as Herceptin targeting bio-markers for breast cancer and imatinib for chronic myeloid leukemia. Postulated as pharmacogenomics, network of omics outcomes fine-tuned drug therapies based on genetic differences, such as the role of CYP2D6 in deciding which medication to use, such as codeine. Translational omics also facilitates the early detection of diseases by discovering early disease onset biomarkers, such as PSA in prostate cancer, and facilitates drug development, such as monoclonal antibodies like rituximab. Metabolomics is another important component that helps identify unique metabolic signatures for diseases such as diabetes, thus enhancing the accuracy of diagnosis (Deidda et al. 2015).

The mounting colossal data for biological interpretations in the research of both non-communicable and infectious diseases, translational omics accelerates putative progressive outcomes, e.g., vaccine development for precipitous pandemic breakouts like SARS-CoV-2019 (Kariyawasam et al. 2022), conscripted evidences to tackle antimicrobial resistance (AMR), thereby guiding strategies against resistant pathogens and promoting novel therapeutic development. Meticulous enmeshing of multi-omics datasets explicates immune responses and host–pathogen interactions, opening next-generation therapeutics and improved public health measures (Kedaigle and Fraenkel 2018). Research paradigms of translational studies are extending to predicting treatment responses and precision diagnostics, fostering personalized therapies and real-time AMR surveillance, with advanced algorithms and inclusive approaches tracing susceptibility and resistance patterns, microbiome diversity and dynamics fostering the health versus disease debacle, strategizing infection controls, antimicrobial treatments, emergence of resistance, ultimately contributing to the development of sustainable interventions and public health advancements (Tang et al. 2022; Zhang et al. 2022).

1.3.3 Transitioning the Omics Studies to Biotechnology Industry

The conceptual need of translational sciences is required at the industrial spectrum culminating the journey of bench to bedside, spurring innovation in a variety of fields, from pharmaceuticals to agriculture, while optimizing production processes, develop novel products, and enhance quality control (Amer and Baidoo 2021). The long road toward precision processes at lab front to be delivered through bioprocess optimization in areas like biopharmaceuticals and biofuels. Biotechnologists may fine-tune conditions to optimize productivity, decrease waste, and expedite manufacturing processes with a global aim to adhere to the regulatory landscape of commercial biotechnology. Researchers and industry leaders are coordinating toward developing guidelines for omics driven technologies data validation, quality assurance, and ethical issues, coagulating the business translation while also encouraging responsible and sustainable industrial practices.

1.4 Emerging Disciplines of Pharmaco-EcoMicrobiology

1.4.1 Pharmaco-EcoMicrobiology

Pharmaco-EcoMicrobiology is the annexation of pharmacology, microbial ecosystems and environment for health outcomes with aspects of antibiotic resistance, biodiversity preservation, and sustainable environmental practices. The research advancements in the field foster this inter-connectedness for augmenting revolutionary approaches to pharmaceutical–microbial pollution, fostering interdisciplinary and innovative solutions to protect ecosystems and human health (Nassar et al. 2017).

Pharmaco-EcoMicrobiology is an interdisciplinary area of research that is rapidly growing to address very important concerns about pharmaceutical pollution in natural ecosystems. Environmental entry of pharmaceutical agents, including antibiotics and antimicrobials, mostly occurs through human and animal excretion processes, mismanagement of waste disposals, and hospital and pharmaceutical manufacturing discharging wastes. The released compounds stay in the soil, groundwater, and surface water for a long time, thereby causing severe environmental contamination (Hignite and Azarnoff 1977). This type of contamination disturbs the diverse nature of microbial populations at the same time as providing selective pressure that fosters the development and spread of antibiotic-resistant bacteria (Shahid et al. 2010). This resistance has the potential to disrupt both ecological stability and public health on a global scale. This form of resistance has terrible implications for environmental and public health in that it will allow the horizontal transfer of natural resistance genes from environmental sources to clinical pathogens and thus exacerbates the global issue of antimicrobial resistance. More ecological damage besides their impact on microbial populations comes from pharmaceuticals. The presence of antibiotics in environmental settings disrupts nutrient cycling, negatively affects ecosystem services such as nitrogen fixation, and destabilizes food webs by altering key microbial interactions. This problem is further complicated by the presence of antimicrobial chemicals in household products (such as triclosan and quaternary ammonium compounds), agricultural sprays, and sites contaminated near pharmaceutical manufacturing facilities (Shahid 2011).

Microbial ecotoxicology forms a cornerstone of Pharmaco-EcoMicrobiology by investigating how sublethal concentrations of pharmaceuticals impact microbial populations, metabolic pathways, and the selection of resistant strains. Microbes also hold good promise as excellent bioremediators in natural ecosystems—the agents capable of degrading or transforming pharmaceutical contaminants (Shahid et al. 2010). For microbial bioremediation, there have been possibilities of alleviating the effects of pollution in diverse environmental settings. Culture-dependent sequencing techniques, in which about 80% of gut microbiota can be grown, have greatly improved the functional validation of microbial processes, including drug metabolism (Wu et al. 2020). Organ-on-chip and culture-based methods provide tools for the detailed study of the interactions between pharmaceuticals and microbial communities in the context of human health. These techniques enable extended periods of metabolic study and better understanding of pharmaceutical biodegradation (Ingber 2022; Caracciolo et al. 2015).

The prioritization of regulation and remedial practices for water, soil, agriculture output, and livestock through concerted efforts of biological and physical technologies may enervate the diverse levels of contamination, while ensuring the adoption of sustainable pharmaceutical management practices (Fig. 1.2).

1.4.2 Pharmacomicrobiomics: Unveiling the Interplay Between Gut Microbiome and Drug Response

In the realm of human health, a novel frontier has emerged at the intersection of pharmacology, microbiology, and personalized well-being. This frontier is known as PharmacoMicrobiomics, a field poised to illuminate the intricate relationship between pharmaceuticals, the human microbiome, and overall health. In an era marked by unprecedented advances in genomics and microbiome research, understanding the profound impact of drugs on our microbial companions has become a pivotal pursuit (Chen et al. 2022).

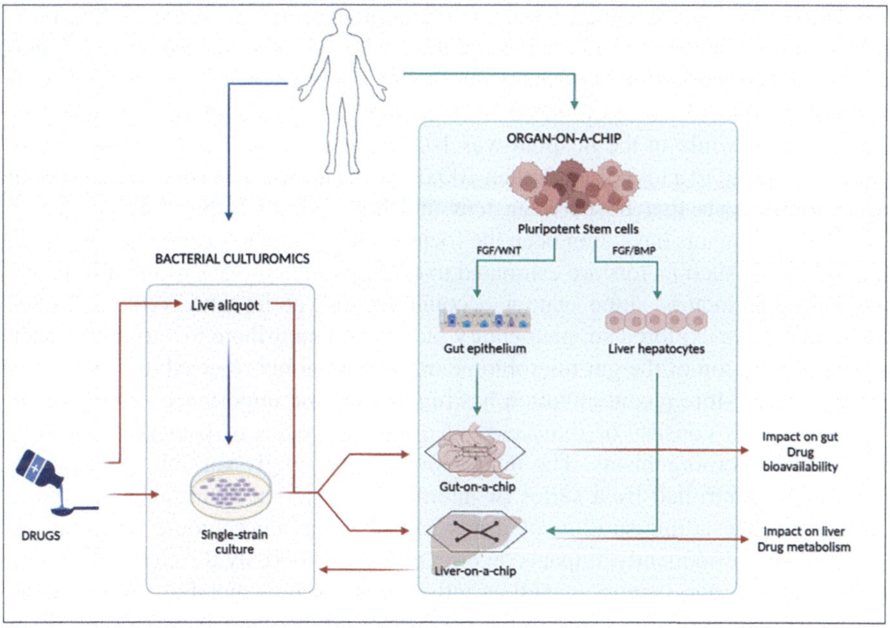

Fig. 1.2 Integrating multi-omics for processing high throughput biological and translational research - a paradigm towards translational omics.

Pharmacomicrobiomics delves deep into the world of the human microbiome, particularly within the gastrointestinal tract. These microbial communities, collectively known as the microbiome, wield significant influence over digestion, metabolism, immune function, and even mental health (ElRakaiby et al. 2014). Yet, until recently, the intricate interactions between pharmaceutical compounds and these microorganisms remained largely uncharted. At the heart of pharmacomicrobiomics lies the recognition that drugs, whether antibiotics, immunosuppressants, or common medications, can wield substantial influence over the composition and function of our gut microbiome. Remarkably, these microbial communities reciprocate, altering the pharmacokinetics and efficacy of drugs. This phenomenon underscores the bidirectional relationship between our health and the microbiome (Dikeocha et al. 2022).

Our journey into pharmacomicrobiomics holds the promise of transformative discoveries and innovations. It offers the potential to personalize medications based on individual microbiome profiles, enhancing treatment efficacy while minimizing side effects. Furthermore, it highlights the critical role of the microbiome in mediating drug responses, paving the way for personalized medicine and the development of novel therapeutics (Sharma et al. 2019).

It is very important to understand that inter-individual medication reactions are complex before delving into pharmacomicrobiomics. The response rates for the common drugs usually lie in the range of 50–75%, thus leaving out the major population who are not helped by standard treatment protocols. Adverse drug reactions also contribute to the added load on the health systems. According to 22 studies, the median proportion of hospital admissions in Europe were caused by an ADR was 3.5%, while 13 studies found that the median percentage of patients who suffered an ADR while in the hospital was 10.1% (Bouvy et al. 2015). These severe toxic effects lead to a loss of more than 100,000 lives annually in America and result in economic costs that lie between tens and hundreds of billions. Even though genetic determinants have long been the focus of discussion to explain variability in drug response, such factors are estimated to explain only 20–95% of the difference. Since genetic factors alone cannot account for the entire observed variability, researchers have explored supplementary factors that contribute to variability, such as the contribution of the gut microbiome, more commonly referred to as the "second genome." More recent attention has highlighted the importance of human gut microbiota, that consists of thousands of different species of bacteria as well as many other microorganisms. The microbiota is surprisingly variable among individuals and controlled by a series of internal and external determinants. Studies have shown the administration of drugs can alter gut microbiome and that gut microbiome, subsequently, impacts on drug efficacy or toxicity through modulation of the host immune system as well as influencing the host metabolism and direct biotransformation of drugs due to the production of enzymes in microbes (Pryor et al. 2020).

The emerging field of pharmacomicrobiomics aims at understanding how drug action and drug toxicity are intricately related with the composition of the microbiome. It seems that the gut microbiome does not just wait passively for a medication response (Doestzada et al. 2018). The functional and mechanistic allying of microbiome influences drug responses as per host drug metabolism, modulation of the host's metabolic processes, and immune system interactions impact treatment efficacy and side effects. Thus, a vital cue for the development of personalized medicine is appended through pharmacomicrobiomics, so as to make tailored therapeutic interventions by analyzing and modifying an individual's gut microbiome. Some of the most recent developments that fall in this line include the culturomics and organ-on-chip technology. Techniques mimic the gut–liver complex environment and hence allow for a very accurate simulation of in vivo pharmacological conditions. Organ-on-chip models derived from patient-specific induced pluripotent stem cells (iPSCs) offer a great insight into drug metabolism, toxicity, and microbial interactions (Palasantzas et al. 2023). Recent breakthroughs in bacterial culture techniques have now enabled researchers to culture approximately 80% of gut microorganisms, which can be functionally validated for microbial processes like drug metabolism.

These targeted in vitro models are likely to find a great deal of application in the future to include genomes and metagenomes in pharmacological studies, allowing for the concurrent analysis of patient-specific microbiota, intestinal epithelial cells, and hepatocytes. With a detailed process of analyzing microbial metabolites and pharmaceutical degradation products affecting host cellular components, the pharmacomicrobiomics domain, is challenging horizon potentiating treatment regimes, manage the hazardous multi-level drug–microbiota effects and improving patient outcomes.

Community medicine as the aspect of psycho-social phenomics represents a much-required interpretation of network medicine as it deals with the interplay between genotype, psychological characteristics, and psychosocial environments. With the recent pandemic outbreak, this interdependent translation of bench to bedside need to integrate the factors determining behavior, mental health, and social interactions (Zeggini et al. 2019). The omics perspectives of psycho-social phenomics investigates differential genetic markers, such as SNPs and gene–environment interactions, gene expression modulations due to impact of environmental factors, including psychological traits—personality, family dynamics, life stress events affecting neurological disorders like depression, schizophrenia, and social environment. Research methods that are being used for such research purposes include genome-wide association studies, twin and family studies, and epigenetic analyses (Ebstein et al. 2010; Slavich and Cole 2013). This discipline has become more complete through the incorporation of genomics, psychology, and sociology in studying how sociocultural factors such as support from family, socioeconomic position, and cultural norms impact mental health and behavior (Tanksley et al. 2019). With the deeper understanding and knowledge of how life experience influences genetic predispositions, and psychological profiles, the design of public health policies to reduce systemic inequalities and changing mental health care, reducing stigma, and informing interventions that maximize individual well-being and social justice (Sanchez-Roige et al. 2017). Applications of psycho-social phenomics range from personalized medicine to social policy (Fig. 1.3). Understanding a patient's genetic predispositions, psychological traits, and social context can elucidate the therapeutic regimes for targeted medical interventions based on the research insights for formulating policies aimed at balancing the therapy as per psycho-social health status among vulnerable populations (Houle et al. 2010).

1.5 Fields of Translational Studies

1.5.1 Integrative Systems and Synthetic Biology

Attenuation of the translational sciences in biological phenomenon is poised to be realized through integrative systems and synthetic biology outcomes. Having achieved the data enplaned biological exploration, connecting the machine learning algorithms to the nodal aspects of experimental validation will stand out for sustaining the essence of life. Correlating the systems knowledge to the engineering applications of biology will upturn the pivotal avenues of deep learning in novel or de novo designs and generative biology (Janes et al. 2017). Impacting the diversity of life forms, systems analysis plays a multifaceted role recording the engineering of bio-parts/modules that reverberate traditional boundaries between microbiology, ecology, medicine, computer science, mathematics, and engineering. Applications of translational sciences have borne the fruit of many such state of art scientific

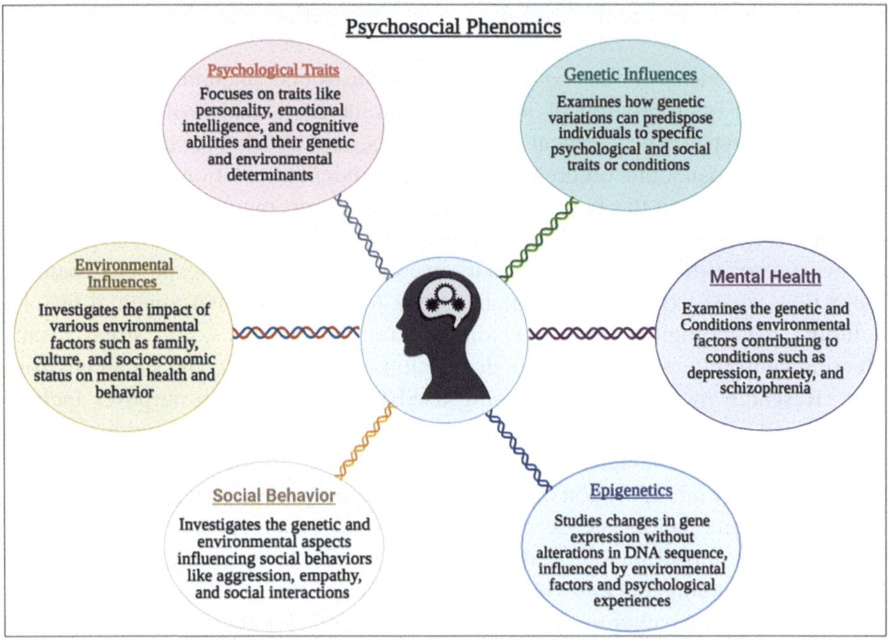

Fig. 1.3 Psychosocial phenomics

endeavors like biophotonics, optogenetics, point of care biosensors, bio-fabricated textiles, synthetic green constructions, leather alternatives, aviation fuels, space explorations, sustainable/novel fragrances, biomining, biofertilisers, bioremediation, bio-surfactants, biofuels, designer chemicals, theragnostic devices (Abernethy and Wheeler 2011). The diverse exploration across the design by intent processes in synthetic biology is comprehensible with deeper knowledge and insights of multi-omics inroads and realization of its translations in designer biology. As we navigate this complex landscape, the imperative to improve our understanding of bio-ecosystem biome dynamics and enhance our ability to engineer them for societal benefit remains paramount. The integration of integrative systems and synthetic biology offers a promising access unlocking opportunities for the functional convergence of these disciplines as key to widespread deployment and transformative applications in bioenergy, agriculture, and human health (Leggieri et al. 2021).

Synthetic biology pioneers to organize, assemble, and modulate the face of modern biology to intercept the cross-scale systems biology and multi-modal omics data so as to improvise designs from single modal level to the multi-scale biology. This aspect of translational sciences is the realization of molecular scale aspects of engineering biology, for designing de novo pathways, and networks based developing novel biological components, with a goal of fine-tuning integrative biological research and reprogramming organisms (Yu 2011). The outputs as genetically reprogrammed organisms (GROs) will impact the broader goals of biological, medical and environmental sustainability (Sharma et al. 2022). These would target microbiome engineering and biocontainment of engineered communities with potential applications in bioenergy, agriculture, and human health (Leggieri et al. 2021). Systems and synthetic biology research effectively attenuate the multi-faceted engineering of gene parts/modules as well as novel feedback/metabolite networks for processes and products of translational omics. The culmination of bridging the challenging terrain of translational knowledge gaps will be achieved through integration of diverse experimental and computational methodologies, fostering collaboration across various scientific disciplines (Zamir 2016).

1.5.2 Molecular–Clinical Translation and Precision Medicine

The clinico–molecular transformation of conventional frameworks governing drug discovery and development will restructure the dynamic arena of precision medicine, as out-of-the-box mix-omics processing followed through incorporation of digital technologies, and artificial intelligence creates the niches of translational precision medicine. Rooted in the principles of multi-omics profiling (Gao et al. 2022), digital biomarkers, model-based data integration, artificial intelligence, biomarker-guided trial designs, and patient-centric companion diagnostics, Translational Precision Medicine represents a pivotal shift from serendipity-driven to data-driven medicine (Hartl et al. 2021). The ever-changing paradigm of evolving healthcare ecosystem pivots through the conceptual backdrop of transcending through the conservative and risk-averse nature of traditional drug development, the translation calls for innovative expeditions in mitigating risks, augmenting designed clinical trials, for sustainable pharmaceutical productivity (Lill et al. 2021). This transformation, embodied by the aspiration to deliver the right medicine to the right patient at the right dose and time, establishes the foundation for addressing the challenges and potentials of precision medicine implementation. Identifying a critical bottleneck in drug development—the Translational Gap, particularly at the interface of drug discovery and early clinical development—the narrative draws from the establishment of the National Center for Advancing Translational Sciences by the National Institutes of Health. The exploration examines how translational medicine, as defined by the European Society for Translational Medicine, integrates various R&D tools to bridge this gap (Maggiolini 2021). The process mitigates the end-to-end biomarker-guided drug development cycle, with meticulous translation of omics signatures undertaking data-driven mechanism pairing for forward translation from bench-to-bedside, back-translation from bedside-to-bench, into clinically relevant biomarkers, advancing healthcare and pharmaceutical innovation (Petrovski and Goldstein 2014).

1.5.3 Nanotechnologies and Omics

"Nano-omics" exemplars leverage enrichment of nanotechnology tools for scoring the enrichment and subsequent omics analysis of circulating biomolecule, hence refining the sensitivity and specificity of diagnostic assays. The transformative potential of nano-omics have a spectrum of applications in avenues of biomedicine and environmental management as the versatility of nanotechnologies in synergy with omics, reshape the landscape of biological research (Luke et al. 2018). Sensitive tracing of biochemical and molecular processes generates the multi-marker systems as part of multimodal panel implies the need of discrete profiling through nano-omics, thereby revealing the theragnostic translation of research processes as point of care devices (Lee et al. 2020; Jiang et al. 2022).

1.6 Directed Sustainability and Development Through Translational Studies

Translational studies have developed from the mere role of translation of scientific information for medical and environmental applications to a strategic tool in translating knowledge and innovation into practical solutions customized to the sustainability targets. This purposeful alignment represents a paradigm change in which translational activities are directed not only at scientific achievements but also toward tackling specific concerns in the environmental, social, and economic realms (Sanford and Smolka 2022). Translational research applies data science in a deliberate way to address defined issues toward achieving sustainability. For example, climate modelling becomes a focused project at the interface of technology and environmental sustainability. It is a data-driven approach to solving issues and achieving the goal targeted and measured in nature; thus, translational efforts capitalize on the precision and efficiency of data science. Similarly, clean energy innovations in renewable energy technology, are regarded as the integration of both technology and translational research. These innovations become focused efforts that contribute to the global shift toward sustainable energy by intentionally transforming scientific understanding into practical applications. Focused technology development is aimed at solving specific issues related to energy sustainability. This intentional approach ensures that translational efforts in technology align with sustainability objectives and yield tangible outcomes. In this context, putative sustainability can be achieved in the following fields:

(a) Sustainable Agriculture and Food Security
(b) Biodiversity Conservation and Ecosystem Restoration
(c) Healthcare Accessibility and Equity
(d) Education and Knowledge Transfer
(e) Economic Resilience through Innovations and Entrepreneurs.

Through the sustainable processes, challenges in aligning translational research with sustainability are met with hurdles of ethical issues and stakeholder involvement with prospects of global collaboration and knowledge sharing. Such practices will lay the foundation for novel educational aspects and capacity development. Hence, the concept of targeted sustainability and development through translational research indicates the urgency of purposeful and intentional endeavors in solving specific concerns within the environmental, social, and economic spectra. The linkage of translational efforts with the definition of specific sustainability goals maximizes the possibility for transformational outcomes and unlocks the door to a future in which translational research is used in attaining measurable, yet targeted, sustainable development goals.

1.7 Toolkits and Ways Ahead in Translational Omics—Big Data and AI

The fusion of artificial intelligence serves as a vital orchestrator in the ever-changing landscape of translational omics, meticulously weaving the symphony of each omics modality acting as a separate musical instrument, adding nuanced data points to the overall composition. Artificial intelligence carefully assesses the molecular parts like deconstruction of musical notes, mastering the discovery of nuanced notes, while identifying variants and mutations with potential repercussions. Protein expression, interactions, and changes, proteomic analysis adds dynamic harmonics to this symphony, and metabolomics provides the throbbing foundation for this musical encounter, complementing the composition. Artificial intelligence delves into metabolic markers, decoding biological metabolic pathways, and enhancing the musical score's contextual richness. AI's integration of various multi-omics layers goes beyond data aggregation to create a unified and all-encompassing soundscape.

The translational result, like the final piece of a musical performance, arises as a thorough account that includes not just raw data but also a network of interconnected patterns and relationships. This complicated structure emerges in clinical settings, exposing prospective biomarkers, therapeutic targets, and individualized treatment techniques. These insights guide choices and facilitate tailored interventions in a manner similar to how practitioners and scholars analyze a complex musical score. The convergence of omics data via AI integration transforms the molecular landscape into a symphony of precision medicine, distinguished by sophistication, nuance, and constant evolution. AI acts as a conductor, promoting a harmonious confluence of biological insights and technological advances.

Toolkits and Way Ahead in Translational Omics—Big Data and AI:

(a) *Harnessing Big Data in Translational Omics*: Integrating the expansive realm of big data, deployment of sophisticated toolkits for seamless data storage, efficient processing, and comprehensive analysis.
(b) *AI-Driven Data Analysis Platforms*: Navigating the future of translational omics for charting the development of advanced AI-driven platforms interfacing with current omics datasets, empower researchers with robust tools for unravelling intricate patterns and establishing nuanced relationships within complex biological data and discovery of subtle associations that might escape conventional methodologies (Terranova et al. 2021).

(c) *Navigating Challenges and Seizing Opportunities in Data Standardization*: The cornerstone of effective interpretation in translational omics lies in the imperative standardization addressing concerns related to data heterogeneity to ensure the reliability and repeatability of translational omics research (Sweet and Moulaison 2013).
(d) *Ethical Deliberations in the Era of AI-powered Translational Omics*: Pertinent issues surrounding data privacy, consent mechanisms, and the potential for biases can be steered through AI powered translational omics for navigating this ethical terrain, harnessing the transformative potential of AI while upholding the principles of integrity, privacy, and societal benefit (Ahmed et al. 2024).

1.8 Future Prospects of Translational Sciences—Assaying the Tipping Point of Success and Failure in Translational Studies

The panorama of translational sciences is on the cusp of a transformative evolution, promising unprecedented advancements and reshaping the future of healthcare and scientific inquiry. Scientific translation of translational omics conjoins innovation, technology, and concerted consortia efforts of high-throughput biology (Jazieh 2011). Translational omics, at the tight juncture of scientific discovery, need to surpass the threshold toward practical realization of laboratory research, data integration, and discovery (D'Adamo et al. 2020). Potential inefficiencies in experimental designs, research practices, and effective communication often hinder the translation of scientific discoveries into applicable solutions (Austin 2021). Such barriers contribute toward what is referred to as the "research paradox," culminating in gaps of targets to expected achievements. The way forward can be sensitive predictability and systematic methodologies to correct these inefficiencies by shifting paradigms from empirical methods toward more predictable, data-driven strategies. Artificial intelligence, machine learning, and patient-centric models are identified as transformative forces that will enhance the speed and efficiency of translating research into clinical practice. The more one learns about the foundational principles governing the translational process, the easier it is to get beyond the trial-and-error processes and more systematically and predictably move toward practical application from discovery (Van De Wall et al. 2023).

Global collaboration is also a crucial component in the resolution of issues that plague efficient translational research. World economies and policy makers should now focus on international collaboration sharing resources and advanced capabilities, thereby fast-tracking the conversion of scientific breakthroughs into transforming leading-edge research into more feasible biological modalities. In this pursuit, the optimization of the inherent fields of translational sciences, as depicted below, should be adhered for the major fields of biological interventions in:

(a) Technological Innovations and Precision Medicine
(b) Interdisciplinary Collaborations and Convergence

(c) Data Integration and Artificial Intelligence
(d) Patient-Centric Approaches and Ethical Considerations
(e) Global Perspectives and Accessible Biological Interventions

In conclusion, translational sciences draw a picture of a dynamic and collaborative field, driven by technological innovation, data integration, and a commitment to a sustainable ethos in biology. The future of translational sciences is characterized by the breaking down of traditional silos, fostering interdisciplinary collaborations that span across scientific disciplines, biological specialties, and technological domains (Wafi and Mirnezami 2018). The convergence of expertise from diverse fields will amplify the translational process, leading to more holistic solutions. As collaborative networks expand, we can anticipate accelerated translation of scientific discoveries into practical applications, bridging the gap between bench and bedside.

1.9 Conclusions

The compilation of this chapter discusses conversant aspects of translational omics, with a focus on bridging high-throughput data generation and real-world applications. From the integration of multi-omics data to the addressment of environmental and societal challenges, translational omics provides the basis for precision medicine, sustainable development, and industrial innovation. The most important areas that have been discussed in the process are those of metadata utilization, AI-driven tools, and interdisciplinary collaborations that can take this field forward. Moving forward, ongoing evolution in technologies and methodologies has been appended that shape the future of translational sciences and drive impactful solutions for global health, environmental sustainability, and societal progress.

Short Chapter Summary

The anthology of the ever-evolving transformative field of translational omics is processed with potential inputs from the futuristic research designs while challenging the hurdles of integration of multi-omics heterogeneous data. Key topics include the role of metadata in ensuring data reproducibility and interpretability, advancements in high-throughput data translation across biological fields, and the emergence of innovative disciplines such as Pharmaco-EcoMicrobiology and Pharmacomicrobiomics. This chapter also highlights the integration of artificial intelligence and big data analytics, which are revolutionizing omics research by enabling predictive modeling and personalized applications. Special attention is given to the application of translational omics in addressing global challenges, including antimicrobial resistance, environmental degradation, and socio-economic disparities. With the potential to align with global sustainability goals and foster interdisciplinary collaborations, translational omics paves the way to novel solutions for some of the most pressing issues of impactful advances in health, environment, and industry affecting our society and science.

Acknowledgments Concept and writing, Lavanya Singh, Shambhavee Srivastav, and Shivani Sharda; Chapter compilation, Editing, and Review—Anupama Avasthi, Bhawna Rathi, Wolfgang Gaertner, Shivani Sharda, Swarna Kanchan, and Minu Kesheri.

References

Abernethy AP, Wheeler JL (2011) True translational research: bridging the three phases of translation through data and behavior. Transl Behav Med 1(1):26–30

Ahmed Z, Wan S, Zhang F, Zhong W (2024) Artificial intelligence for omics data analysis. BMC Methods 1(1):4

Al-Amrani S, Al-Jabri Z, Al-Zaabi A, Alshekaili J, Al-Khabori M (2021) Proteomics: concepts and applications in human medicine. World J Biol Chem 12(5):57

Amer B, Baidoo EEK (2021) Omics-driven biotechnology for industrial applications. Front Bioeng Biotechnol 9:613307

Austin CP (2021) Opportunities and challenges in translational science. Clin Transl Sci 14(5):1629–1647

Babu MM, Snyder M (2023) Multi-omics profiling for health. Mol Cell Proteomics 22(6):100561

Blum B, Mousavi F, Emili A (2018) Single-platform 'multi-omic' profiling: unified mass spectrometry and computational workflows for integrative proteomics-metabolomics analysis. Mol Omics 14(5):307–319

Bouvy JC, De Bruin ML, Koopmanschap MA (2015) Epidemiology of adverse drug reactions in Europe: a review of recent observational studies. Drug Saf 38(5):437–453

Caracciolo AB, Topp E, Grenni P (2015) Pharmaceuticals in the environment: biodegradation and effects on natural microbial communities. A review. J Pharm Biomed Anal 106:25–36

Cernava T, Rybakova D, Buscot F, Clavel T, McHardy AC, Meyer F, Meyer F, Overmann J, Stecher B, Sessitsch A, Schloter M (2022) Metadata harmonization-standards are the key for a better usage of omics data for integrative microbiome analysis. Environ Microbiome 17(1):33

Chen HQ, Gong JY, Xing K, Liu MZ, Ren H, Luo JQ (2022) Pharmacomicrobiomics: exploiting the drug-microbiota interactions in antihypertensive treatment. Front Med 8:742394

Crawford SE, Hartung T, Hollert H, Mathes B, van Ravenzwaay B, Steger-Hartmann T, Studer C, Krug HF (2017) Green toxicology: a strategy for sustainable chemical and material development. Environ Sci Eur 29:1–6

D'Adamo GL, Widdop JT, Giles EM (2020) The future is now? Clinical and translational aspects of "Omics" technologies. Immunol Cell Biol 99(2):168–176

Deidda M, Piras C, Bassareo PP, Dessalvi CC, Mercuro G (2015) Metabolomics, a promising approach to translational research in cardiology. IJC Metab Endocr 9:31–38

Denholm EM, Martin WJ (2008) Translational research in environmental health sciences. Transl Res 151(2):57–58

Desaules A (2011) The role of metadata and strategies to detect and control temporal data bias in environmental monitoring of soil contamination. Environ Monit Assess 184(11):7023–7039

Dikeocha IJ, Al-Kabsi AM, Miftahussurur M, Alshawsh MA (2022) Pharmacomicrobiomics: influence of gut microbiota on drug and xenobiotic metabolism. FASEB J 36(6):e22350

Doestzada M, Vila AV, Zhernakova A, Koonen DPY, Weersma RK, Touw D, Kuipers F, Wijmenga C, Fu J (2018) Pharmacomicrobiomics: a novel route towards personalized medicine? Protein Cell 9(5):432–445

Durante K, Hardy D (2015) Discovery, management, and preservation of geospatial data using Hydra. J Map Geogr Libr 11(2):123–154

Ebstein RP, Israel S, Chew SH, Zhong S, Knafo A (2010) Genetics of human social behavior. Neuron 65(6):831–844

ElRakaiby MT, Dutilh BE, Rizkallah MR, Boleij A, Cole JN, Aziz RK (2014) Pharmacomicrobiomics: the impact of human microbiome variations on systems pharmacology and personalized therapeutics. OMICS 18(7):402–414

Gao P, Shen X, Zhang X, Jiang C, Zhang S, Zhou X, Rose SMS, Snyder M (2022) Precision environmental health monitoring by longitudinal exposome and multi-omics profiling. Genome Res 32(6):1199–1214

Hartl D, De Luca V, Kostikova A, Laramie JM, Kennedy SR, Ferrero E, Siegel RW, Fink M, Ahmed SS, Millholland JM, Schuhmacher A, Hinder M, Piali L, Roth A (2021) Translational precision medicine: an industry perspective. J Transl Med 19(1):245

Hasin Y, Seldin MM, Lusis AJ (2017) Multi-omics approaches to disease. Genome Biol 18(1):1–5

Havas A, Schartinger D, Weber M (2023) Innovation studies, social innovation, and sustainability transitions research: from mutual ignorance towards an integrative perspective? Environ Innov Soc Tr 48:100754

Hignite CE, Azarnoff DL (1977) Drugs and drug metabolites as environmental contaminants: chlorophenoxyisobutyrate and salicylic acid in sewage water effluent. Life Sci 20(2):337–341

Hirschman L, Clark C, Cohen KB, Mardis S, Luciano JS, Kottmann R, Cole JR, Markowitz V, Kyrpides NC, Morrison N, Schriml LM, Field D (2008) Habitat-lite: a GSC case study based on free text terms for environmental metadata. OMICS 12(2):129–136

Houle D, Govindaraju DR, Omholt SW (2010) Phenomics: the next challenge. Nat Rev Genet 11(12):855–866

Hu C, Wang J (2021) Multi-omics profiling: the way toward precision medicine in metabolic diseases. J Mol Cell Biol 13(8):576–593

Ingber DE (2022) Human organs-on-chips for disease modelling, drug development and personalized medicine. Nat Rev Genet 23(8):467–491

Janes KA, Chandran PL, Ford RM, Lazzara MJ, Papin JA, Peirce SM, Saucerman JJ, Lauffenburger DA (2017) An engineering design approach to systems biology. Integr Biol 9(7):574–583

Jazieh AR (2011) Future of translational research: why go pragmatic? Pragmat Obs Res 2:1–4

Jiang P, Sinha S, Aldape K, Hannenhalli S, Sahinalp SC, Ruppin E (2022) Big data in basic and translational cancer research. Nat Rev Cancer 22(11):625–639

Kanchan S, Sinha RP, Chaudière J, Kesheri M (2020) Computational metagenomics: current status and challenges. In: Katara P (ed) Recent trends in 'computational omics': concepts and methodology. Nova Science Publishers, USA, pp 371–395

Kanchan S, Ogden E, Kesheri M, Skinner A, Miliken E, Lyman D, Armstrong J, Sciglitano L, Hampikian G (2024) COVID-19 hospitalizations and deaths predicted by SARS-CoV-2 levels in Boise, Idaho wastewater. Sci Total Environ 907:167742

Kariyawasam RM, Julien DA, Jelinski DC, Larose SL, Rennert-May E, Conly JM, Dingle TC, Chen JZ, Tyrrell GJ, Ronksley PE, Barkema HW (2022) Antimicrobial resistance (AMR) in COVID-19 patients: a systematic review and meta-analysis (November 2019-June 2021). Antimicrob Resist Infect Control 11(1):45

Kaur P, Singh A, Chana I (2021) Computational techniques and tools for omics data analysis: state-of-the-art, challenges, and future directions. Arch Comput Methods Eng 28(7):4595–4631

Kedaigle A, Fraenkel E (2018) Turning omics data into therapeutic insights. Curr Opin Pharmacol 42:95–101

Kesheri M, Kanchan S, Richa, Sinha RP (2014) Isolation and in-silico analysis of Fe- superoxide dismutase in *Nostoc commune*. Gene 553(2):117–125

Kesheri M, Kanchan S, Sinha RP (2021) Isolation and *in silico* analysis of antioxidants in response to temporal variations in the cyanobacterium *Oscillatoria* sp. Gene Rep 23:101023

Kesheri M, Kanchan S, Sinha RP (2022) Responses of antioxidants for resilience to temporal variations in the cyanobacterium *Microcystis aeruginosa*. S Afr J Bot 148:190–199

Kesheri M, Kanchan S, Salisbury TB, Sinha RP (2024a) Microbial omics in environment and health. Springer Nature, Singapore. ISBN: 978-981-97-1768-2

Kesheri M, Kanchan S, Srivastava U, Chittoori B, Ratna-Raj R, Sinha RP, Vaishampayan A, Rastogi RP, Primerano DA (2024b) Ecology and environmental omics. In: Gupta MK, Katara P, Mondal S, Singh RL (eds) Integrative omics. Academic Press, London, pp 315–331

Kumar P, Kanchan S, Kesheri M (2024) Multi-omics in human disease biomarker discovery. In: Kesheri M, Kanchan S, Salisbury TB, Sinha RP (eds) Microbial omics in environment and health. Springer, Singapore, pp 205–239

Kumari A, Kanchan S, Kesheri M (2016) Applications of bio-molecular databases in bioinformatics. In: Dey N, Bhateja V, Hassanien AE (eds) Advancements in bio-medical sensing, imaging, measurements and instrumentation, vol 651. Springer-Verlag, Germany, pp 329–351
Kumari A, Kesheri M, Sinha RP, Kanchan S (2018) Integration of soft computing approach in plant biology and its applications in agriculture. In: Purohit HJ, Raje DV, More RP, Kalia VC (eds) Soft computing in biological sciences. Springer Nature, Singapore, pp 265–328
Kussmann M, Kaput J (2014) Translational genomics. Appl Transl Genom 3(3):43–47
Lau E, Wu JC (2018) Omics, big data, and precision medicine in cardiovascular sciences. Circ Res 122(9):1165–1168
Lee MW, Miljanic M, Triplett T, Ramírez C, Aung KL, Eckhardt SG, Capasso A (2020) Current methods in translational cancer research. Cancer Metastasis Rev 40(1):7–30
Leggieri PA, Liu Y, Hayes MM, Connors B, Seppälä S, O'Malley M, Venturelli OS (2021) Integrating systems and synthetic biology to understand and engineer microbiomes. Annu Rev Biomed Eng 23(1):169–201
Liang A, Kong Y, Chen Z, Qiu Y, Wu Y, Zhu X, Li Z (2023) Advancements and applications of single-cell multi-omics techniques in cancer research: unveiling heterogeneity and paving the way for precision therapeutics. Biochem Biophys Rep 37:101589
Lill JR, Mathews WR, Rose CM, Schirle M (2021) Proteomics in the pharmaceutical and biotechnology industry: a look to the next decade. Expert Rev Proteomics 18(7):503–526
Luke DA, Sarli C, Suiter AM, Carothers BJ, Combs T, Allen JL, Beers C, Evanoff B (2018) The translational science benefits model: a new framework for assessing the health and societal benefits of clinical and translational sciences. Clin Transl Sci 11(1):77–84
Maggiolini M (2021) Translational signaling and systems biology. J Transl Med 19(1):1–2
Mallik B, Mishra T, Dubey P, Kesheri M, Kanchan S (2024) Exploring the secrets of microbes: unveiling the hidden world through microbial omics in environment and health. In: Kesheri M, Kanchan S, Salisbury TB, Sinha RP (eds) Microbial omics in environment and health. Springer, Singapore, pp 269–294
Marais K (2018) Introduction: translation and development. Translator 24(4):295–300
Mishra T, Mallik B, Kesheri M, Kanchan S (2024) The interplay of gut microbiome in health and diseases. In: Kesheri M, Kanchan S, Salisbury TB, Sinha RP (eds) Microbial omics in environment and health. Springer, Singapore, pp 1–34
Mock M, Langmead CJ, Grandsard P, Edavettal S, Russell A (2024) Recent advances in generative biology for biotherapeutic discovery. Trends Pharmacol Sci 45(3):255–267
Nassar AF, Wu T, Nassar SF, Wisnewski AV (2017) UPLC-MS for metabolomics: a giant step forward in support of pharmaceutical research. Drug Discov Today 22(2):463–470
Palasantzas VE, Tamargo-Rubio I, Le K, Slager J, Wijmenga C, Jonkers IH, Kumar V, Fu J, Withoff S (2023) iPSC-derived organ-on-a-chip models for personalized human genetics and pharmacogenomics studies. Trends Genet 39(4):268–284
Pandey A (2017) Moving to translational proteomics. Proteomics 17(6):1770042
Pathak GP, Ehrenreich A, Losi A, Streit WR, Gärtner W (2009) Novel blue light-sensitive proteins from a metagenomic approach. Environ Microbiol 11(9):2388–2399
Petrovski S, Goldstein DB (2014) Phenomics and the interpretation of personal genomes. Sci Transl Med 6(254):254
Pinu FR, Beale DJ, Paten AM, Kouremenos KA, Swarup S, Schirra HJ, Wishart DS (2019a) Systems biology and multi-omics integration: viewpoints from the metabolomics research community. Metabolites 9(4):76
Pinu FR, Goldansaz SA, Jaine J (2019b) Translational metabolomics: current challenges and future opportunities. Metabolites 9(6):108
Pryor R, Martínez-Martínez D, Quintaneiro LM, Cabreiro F (2020) The role of the microbiome in drug response. Annu Rev Pharmacol Toxicol 60(1):417–435
Ringborg U (2019) Translational cancer research – a coherent cancer research continuum. Mol Oncol 13(3):517–520

Sahu N, Mishra S, Kesheri M, Kanchan S, Sinha RP (2023) Identification of cyanobacteria- based natural inhibitors against SARS-CoV-2 druggable target ACE2 using molecular docking study, ADME and toxicity analysis. Indian J Clin Biochem 38(3):361–373

Sanchez-Roige S, Gray JC, MacKillop J, Chen C, Palmer AA (2017) The genetics of human personality. Genes Brain Behav 17(3):e12439

Sanford EJ, Smolka MB (2022) A field guide to the proteomics of post-translational modifications in DNA repair. Proteomics 22(15–16):2200064

Schully SD, Khoury MJ (2014) What is translational genomics? An expanded research agenda for improving individual and population health. Appl Transl Genom 3(4):82–83

Shahid M (2011) Environmental dissemination of NDM-1: time to act sensibly. Lancet Infect Dis 11(5):334–335

Shahid M, Khardori N, Tripathi T, Bergman S (2010) Pharmaco-ecomicrobiology: a newer component of medical sciences bridging pharmacovigilance, ecology, and environmental microbiology. J Infect Public Health 3(1):1–4

Sharda S, Avasthi A, Bose S, Kaur N (2023) Cellular interactions network in cancer: integrative disease models. In: Sobti RC, Ganguly NK, Kumar R (eds) Handbook of oncobiology: from basic to clinical sciences. Springer Nature, Singapore, pp 905–924

Sharma A, Buschmann MM, Gilbert JA (2019) Pharmacomicrobiomics: the holy grail to variability in drug response? Clin Pharmacol Ther 106(2):317–328

Sharma P, Singh SP, Iqbal HM, Parra-Saldivar R, Varjani S, Tong YW (2022) Genetic modifications associated with sustainability aspects for sustainable developments. Bioengineered 13(4):9509–9521

Slavich GM, Cole SW (2013) The emerging field of human social genomics. Clin Psychol Sci 1(3):331–348

Srivastav S, Lavanya S, Avasthi A, Kaur N, Gaertner W, Kesheri M, Kanchan S, Sharda S (2024) Integromics: tracking the multi-omic expanse in theragnostics. In: Kesheri M, Kanchan S, Salisbury TB, Sinha RP (eds) Microbial omics in environment and health. Springer, Singapore, pp 135–173

Stvilia, B., Gasser, L., & Twidale, M. B. (2007). Metadata Quality Problems in Federated Collections. In L. Al-Hakim (Ed.), Challenges of Managing Information Quality in Service Organizations (pp. 154-186). IGI Global Scientific Publishing. https://doi.org/10.4018/978-1-59904-420-0.ch008 https://www.igi-global.com/gateway/chapter/6547

Sweet LE, Moulaison HL (2013) Electronic health records data and metadata: challenges for big data in the United States. Big Data 1(4):245–251

Tang R, Luo R, Tang S, Song H, Chen X (2022) Machine learning in predicting antimicrobial resistance: a systematic review and meta-analysis. Int J Antimicrob Agents 60(5–6):106684

Tanksley PT, Motz RT, Kail RM, Barnes JC, Liu H (2019) The genome-wide study of human social behavior and its application in sociology. Front Sociol 4:53

Terranova N, Venkatakrishnan K, Benincosa LJ (2021) Application of machine learning in translational medicine: current status and future opportunities. AAPS J 23(4):74

Van De Wall G, Van Hattem A, Timmermans J, Ritskes-Hoitinga M, Bleich A, Leenaars C (2023) Comparing translational success rates across medical research fields. ALTEX 40(4):584–594

Veenstra TD (2021) Systems biology and multi-omics. Proteomics 21(3–4):e2000306

Wafi A, Mirnezami R (2018) Translational-omics: future potential and current challenges in precision medicine. Methods 151:3–11

Wimmers F, Pulendran B, (2020) Emerging technologies for systems vaccinology — multi-omics integration and single-cell (epi)genomic profiling. Curr Opin Immunol 65:57–64. https://doi.org/10.1016/j.coi.2020.05.001

Wishart DS (2016) Emerging applications of metabolomics in drug discovery and precision medicine. Nat Rev Drug Discov 15(7):473–484

Wu Q, Liu J, Wang X, Feng L, Wu J, Zhu X, Wen W, Gong X (2020) Organ-on-a-chip: recent breakthroughs and future prospects. Biomed Eng Online 19(1):1–19

Yu D (2011) Translational research: current status, challenges and future strategies. Am J Transl Res 3(5):422–433

Zamir E (2016) Integrative systems and synthetic biology of cell-matrix adhesion sites. Cell Adhes Migr 10(5):451–460

Zeggini E, Gloyn AL, Barton A, Wain LV (2019) Translational genomics and precision medicine: moving from the lab to the clinic. Science 365(6460):1409–1413

Zhang X, Li L, Butcher J, Stintzi A, Figeys D (2019) Advancing functional and translational microbiome research using meta-omics approaches. Microbiome 7(1):1–12

Zhang ZJ, Lehmann C, Cole CG, Pamer EG (2022) Translating microbiome research from and to the clinic. Ann Rev Microbiol 76(1):435–460

Zheng Y, Liu Y, Yang J, Dong L, Zhang R, Tian S, Yu Y, Ren L, Hou W, Zhu F, Mai Y (2024) Multi-omics data integration using ratio-based quantitative profiling with quartet reference materials. Nat Biotechnol 42(7):1133–1149

Beyond the Genome: Exploring the Multi-Omics Landscape

2

Smrita Singh, Tanzeel Ahmed, and Ashutosh Singh Chauhan

Abstract

While the genome has long been considered the instruction manual for life, revealing crucial insights into health and disease, relying solely on genomic information paints an incomplete picture of biological complexity. Multi-omics has emerged as a powerful solution, integrating data from various biological layers—including the genome, transcriptome, proteome, and metabolome—to provide a more comprehensive understanding. This chapter explores the exciting realm of multi-omics, examining its fundamental principles, methodologies, and significant potential for advancing biomedicine. The requirement for a multi-layered approach is justified by our discussion of the shortcomings of single-omics techniques in representing the complex nature of biological systems. Computational techniques for data integration and interpretation are examined, as well as important technologies for producing and analyzing multi-omics data. We give strong instances of how multi-omics is being applied to identify novel biomarkers, understand disease mechanisms, and create individualized treatments for a range of illnesses. Finally, we address existing challenges and future directions for the field, stressing the importance of developing robust analytical tools, standardized protocols, and fostering collaborative efforts to fully unlock the power of multi-omics for improving human health.

Keywords

Bioinformatics · Metabolomics · Multi-omics · System biology · Transcriptomics

S. Singh (✉) · T. Ahmed
School of Biotechnology, IFTM University, Moradabad, Uttar Pradesh, India

A. S. Chauhan
Air Pollution and Stone Conservation Laboratory, Archaeological Survey of India, Agra Fort, Agra, Uttar Pradesh, India

M. Kesheri et al. (eds.), *Multi-Omics in Biomedical Sciences and Environmental Sustainability*, https://doi.org/10.1007/978-981-96-7067-3_2

Objectives

- To understand genomics to further study other omics disciplines
- To understand epigenetic modifications that play a pivotal role in diverse biological processes
- To understand how multi-omics is being used to uncover disease mechanisms, pinpoint new biomarkers, and develop personalized treatments

2.1 Introduction

Genomics provides a foundational understanding of an organism's genetic blueprint. However, a comprehensive understanding of biological systems requires going beyond the genome and exploring the intricate interplay among various biomolecules and their interactions. This is where the field of multi-omics comes into play. Multi-omics research integrates different high-throughput "omics" technologies, such as genomics, transcriptomics, proteomics, and metabolomics, to offer a holistic view of biological systems (Sun and Hu 2016; Amer and Baidoo 2021). By capturing the complex relationships between genes, transcripts, proteins, and metabolites, multi-omics studies can uncover novel insights into the underlying mechanisms of biological processes and diseases. Multi-omics data integration utilizes empirical, functional, and other techniques to combine information from multiple omics domains, enabling a robust characterization of biochemical signatures that are reflective of organism phenotypes. Recent population studies have started incorporating high-throughput omics data beyond a single type of molecule, providing a combined view of multiple functional layers at a system level and paving the way for precision medicine, which emphasizes tailored medical practice to optimize clinical outcomes given the unique individual profile compared to the population average. Genome-wide profiling of genetic variants and RNA transcripts has been used to study human diseases in populations for over a decade. The next frontier is to harness the power of integrated multi-omics approaches to gain a deeper understanding of human biology and disease mechanisms. As we summarize in the following sections, the wealth of genomic, transcriptomic, and epigenomic data obtained through high-throughput sequencing technologies has enriched the scope of multi-omics studies, enabling researchers to explore the complex interplay of various biomolecules and their roles in biological processes and diseases. The development of high-throughput sequencing technology is closely related to the advancement of multi-omics research, as it has become efficient and fast to sequence and analyze the genomes, transcriptomes, and DNA/RNA samples of organisms, thereby generating a large amount of multi-omics data. Therefore, the development of attention-based approaches, such as PACS, contributes to future research on multi-omics data (Pan et al. 2023).

2.2 Core Omics Disciplines

Genomics, the study of an organism's entire genetic blueprint, has been transformative in our understanding of biology. High-throughput DNA sequencing has enabled rapid and comprehensive analysis of genomes, providing valuable insights into genetic factors associated with diseases, evolutionary relationships, and new diagnostic and therapeutic approaches. However, solely focusing on the genome has its limitations. While the genome outlines the fundamental building blocks of life, it does not reveal the dynamic processes that shape an organism's observable characteristics (phenotype). It is akin to having a complete list of car parts without understanding how those parts interact to make the vehicle function. To fully comprehend the complexity of biological systems, we must go beyond the genome and explore other layers of information.

Genomics serves as a crucial starting point for other omics investigations. The genome serves as the foundation for studying other omics disciplines. For instance, transcriptomics examines the RNA transcripts produced from the genome, providing insights into gene expression patterns. Similarly, epigenomics investigates modifications to the genome that affect gene expression without altering the DNA sequence itself. Additionally, proteomics and metabolomics study the proteins and metabolites encoded by the genome, revealing how genes are translated into functional molecules. Genomics can identify genes associated with specific diseases. However, understanding the contribution of these genes to disease development often requires integrating information from other omics disciplines. Transcriptomics can reveal altered gene expression patterns in diseased tissues, while proteomics can identify abnormal protein levels or modifications. Although genomics can help predict an individual's response to certain drugs based on their genetic makeup, other omics data, such as metabolomics, can provide additional insights into drug metabolism and potential side effects. Furthermore, genomics plays a crucial role in personalized medicine by identifying genetic variations that influence disease risk and treatment response. Integrating data from other omics disciplines can further refine personalized treatment strategies. By integrating genomics with other omics disciplines, researchers can gain a more comprehensive understanding of biological systems and develop more effective strategies for diagnosing, treating, and preventing diseases. Transcriptomics encompasses various types of RNA, and specific techniques are used for the analysis having advantages and disadvantages also (Table 2.1).

2.3 Applications of Transcriptomics

Transcriptomics highlights how genomic information is translated into functional molecules, influencing cellular function and contributing to health and disease. It adds depth to genomics by revealing gene expression patterns and molecular functionality. It has several applications (Table 2.2).

Table 2.1 Types of RNA, their function, analysis techniques, and their advantages and disadvantages

Type of RNA	**Function**	**Notes**	
mRNA	Carries the genetic code from DNA to ribosomes for translation into proteins	mRNA levels are often used as a proxy for gene expression	
tRNA	Delivers amino acids to ribosomes during protein synthesis	Essential for the accurate translation of mRNA into protein	
rRNA	A structural component of ribosomes	Plays a critical role in ribosome function and protein synthesis	
ncRNA	A diverse group of RNA molecules that regulate gene expression without encoding proteins	Includes microRNAs, long noncoding RNAs, and small interfering RNAs	
Technique	**Description**	**Advantages**	**Disadvantages**
RNA sequencing	Converts RNA into cDNA and sequences it to determine RNA transcript abundance	Comprehensive view of the transcriptome, novel transcript discovery, and alternative splicing identification	Requires more resources and computational analysis
Microarrays	Uses probes to measure the abundance of specific RNA transcripts	Useful for targeted gene expression studies	Less comprehensive than RNA-Seq

Table 2.2 Applications of transcriptomics in various fields

Application	Description
Understanding gene regulation	Reveals patterns of gene expression under different conditions, providing insights into gene regulation mechanisms
Identifying disease biomarkers	Detects changes in gene expression profiles that can serve as biomarkers for various diseases
Developing new drug targets	Pinpoints dysregulated genes in diseases, suggesting potential targets for drug development
Studying developmental processes	Tracks shifts in gene expression during development, shedding light on cellular differentiation and tissue formation
Investigating environmental responses	Explains how organisms respond to environmental changes at the molecular level through gene expression analysis

Epigenomics is the study of heritable changes in gene expression that do not involve alterations to the underlying DNA sequence. Epigenetic modifications, such as DNA methylation and histone modifications, can influence which genes are expressed, ultimately affecting the phenotype of an organism. The epigenome can be thought of as the software that directs the hardware of the genome. Epigenomics involves the following mechanisms that drive heritable changes in gene expression without altering the underlying DNA sequence.

DNA Methylation: The addition of a methyl group to cytosine bases, often at CpG dinucleotides, typically resulting in gene silencing (Delage and Dashwood 2008; Costa 2010; Handy et al. 2011; Tammen et al. 2012; Choi et al. 2013; Moosavi and Ardekani 2016).

Histone Modifications: Chemical alterations, such as acetylation, methylation, phosphorylation, and ubiquitination, to the histone proteins around which DNA is wrapped. These modifications can either activate or repress gene expression depending on the specific change and its location (Delage and Dashwood 2008; Costa 2010; Handy et al. 2011).

Noncoding RNAs: Certain RNA molecules, like microRNAs and long noncoding RNAs, can regulate gene expression by interacting with mRNA or DNA (Costa 2010; Moosavi and Ardekani 2016).

Epigenetic modifications play a pivotal role in diverse biological processes, including development, disease, and environmental responses. During development, epigenetic mechanisms ensure that the correct genes are expressed at the right time and place, driving cell differentiation and tissue formation (Delage and Dashwood 2008; Handy et al. 2011; Tammen et al. 2012; Choi et al. 2013; Moosavi and Ardekani 2016). Conversely, dysregulation of epigenetic pathways can contribute to various diseases, such as cancer, autoimmune disorders, and neurological conditions, often marked by aberrant DNA methylation patterns (Costa 2010; Moosavi and Ardekani 2016). Furthermore, environmental factors, including diet, stress, and toxin exposure, can influence epigenetic modifications and alter gene expression, with long-term consequences for health and disease risk (Handy et al. 2011; Tammen et al. 2012; Choi et al. 2013; Moosavi and Ardekani 2016).

To study these epigenetic phenomena, researchers employ a range of techniques, such as bisulfite sequencing to map DNA methylation patterns, chromatin immunoprecipitation to identify histone modifications, and RNA sequencing to elucidate the role of noncoding RNAs in gene regulation. By understanding how epigenetic modifications influence gene expression, we can gain deeper insights into the intricate interplay between the environment and the genome in shaping phenotypes and disease susceptibility.

2.4 Proteomics

Proteomics is the large-scale study of proteins, provides a comprehensive view of the proteome, the entire set of proteins expressed in a cell, tissue, or organism. Proteins serve as the functional workhorses of the cell, catalyzing biochemical reactions, transporting molecules, and carrying out a wide array of essential functions. Deciphering the complexities of protein structure, function, and interactions is crucial for advancing our understanding of biological systems and for developing effective diagnostic techniques and disease treatments (Gligorijević et al. 2015; Sun and Hu 2016). Proteomics, the comprehensive study of proteins, provides valuable insights into the structure, function, and interactions of these essential cellular components.

Protein Structure and Function: Proteomics investigates the relationship between the three-dimensional structure of proteins and their biological activities. Alterations in protein structure can lead to dysfunction and disease (Yates 2019).

Protein–Protein Interactions: Proteins rarely operate in isolation; they form intricate networks and interact with one another to regulate cellular processes. Proteomics research focuses on elucidating these protein–protein interactions and understanding how they coordinate various cellular functions (Bruce et al. 2013; Yates 2019; Kesheri et al. 2022, 2024). Posttranslational Modifications: Proteins can undergo chemical modifications after their synthesis, and these modifications can significantly alter their activity and function. Proteomics identifies and characterizes these posttranslational modifications, such as phosphorylation, glycosylation, and ubiquitination (Bruce et al. 2013; Yates 2019; Srivastav et al. 2024). Techniques in Proteomics: Proteomics research employs a variety of analytical techniques, including Mass Spectrometry: A powerful technique that determines the mass-to-charge ratio of ions, enabling the identification and quantification of proteins, as well as the characterization of PTMs (Bruce et al. 2013). Protein Arrays: Miniaturized platforms that allow for the simultaneous analysis of thousands of proteins, facilitating the study of protein–protein interactions, biomarker discovery, and drug target identification. 2D Gel Electrophoresis: A technique that separates proteins based on their size and charge, enabling the visualization of changes in protein expression levels (Bruce et al. 2013).

2.4.1 Applications of Proteomics

There are several applications of proteomics in different areas (Fig. 2.1).

Biomarker Discovery: Proteomic analysis can identify proteins that exhibit differential expression between healthy and diseased tissues or biological fluids, which can serve as potential biomarkers for disease diagnosis and prognosis (Fredolini et al. 2010).

Drug Target Identification: Proteomic studies can pinpoint proteins involved in disease pathways, providing insights into potential therapeutic targets (Fredolini et al. 2010; Sahu et al. 2023).

Fig. 2.1 Proteomics application in various fields

Understanding Disease Mechanisms: Proteomics can elucidate the molecular mechanisms underlying diseases by detecting proteins that are dysregulated in disease states (Fredolini et al. 2010).

By offering a comprehensive view of the proteome, proteomics provides valuable insights into the intricate molecular processes that govern cellular function and contribute to health and disease.

2.5 Metabolomics

Metabolomics is the comprehensive study of small molecules, known as metabolites, within a biological system. These metabolites are the intermediate and end products of metabolism, and their levels can offer valuable insights into the physiological state of an organism (Dunn et al. 2010; Baker 2011; Bingol and Brüschweiler 2016; Quintás-Cardama et al. 2021). Metabolomics encompasses the comprehensive study of diverse metabolites, which are the intermediate and end products of cellular processes. These small molecules, including amino acids, sugars, lipids, nucleotides, and organic acids, serve as indicators of the dynamic physiological state of an organism (Clish 2015). Metabolomics provides insights into metabolic pathways, the interconnected series of chemical reactions occurring within cells. By measuring metabolite levels, researchers can gain an understanding of pathway activity and identify potential bottlenecks or dysregulation (Dunn et al. 2010; Quintás-Cardama et al. 2021). Various analytical techniques are employed in metabolomics research, such as Nuclear Magnetic Resonance Spectroscopy, which measures the magnetic properties of atomic nuclei to identify and quantify metabolites, and Mass Spectrometry, which determines the mass-to-charge ratio of ions to detect and quantify metabolites, often coupled with chromatography for complex mixture separation (Dunn et al. 2010; Bingol and Brüschweiler 2016).

2.5.1 Applications of Metabolomics

- Disease Biomarker Discovery: Metabolomics can detect changes in metabolite levels that are indicative of disease, allowing for the identification of potential biomarkers for diagnosis, prognosis, and monitoring treatment response (Baker 2011; Clish 2015; Quintás-Cardama et al. 2021).
- Drug Discovery and Development: Metabolomics can be employed to study drug metabolism, pinpoint drug targets, and evaluate drug efficacy and toxicity (Shyur and Yang 2008; Quintás-Cardama et al. 2021).
- Personalized Medicine: Metabolomics can provide insights into an individual's unique metabolic profile, enabling the tailoring of treatments and optimization of drug therapies (Clish 2015).
- Nutritional Studies: Metabolomics can be used to investigate the effects of diet on metabolism and identify relevant nutritional biomarkers (Shyur and Yang 2008).

- Systems Biology: Metabolomics is a valuable tool for systems biology research, allowing researchers to study the complex interactions between genes, proteins, and metabolites (Dunn et al. 2010). By offering a comprehensive view of the metabolome, metabolomics provides a powerful approach for understanding the dynamic interplay of metabolic processes in health and disease. It complements other omics disciplines by offering a functional readout of cellular activity.

2.6 Expanding Omics Horizons

By concentrating on metabolomics, genomes, proteomics, transcriptomics, and other areas, we may broaden our omics horizons and find more intriguing subjects to investigate further (Fig. 2.2).

- *Interactomics*: Interactomics is the study of protein–protein interactions and how these interactions form complex networks that regulate cellular processes. These networks, known as interactomes, provide a systems-level understanding of biological processes and disease mechanisms. Key aspects of interactomics:
- *Mapping Protein–Protein Interactions*: Various experimental techniques are used to identify and map PPIs, including yeast two-hybrid assays, co-immunoprecipitation, and affinity purification coupled with mass spectrometry. Computational methods also play a role in predicting and analyzing PPIs.
- *Interactome Networks*: PPIs form complex networks where proteins interact with each other in specific ways. These networks can be represented graphically, with proteins as nodes and interactions as edges. Analyzing these networks reveals important information about cellular organization and function.

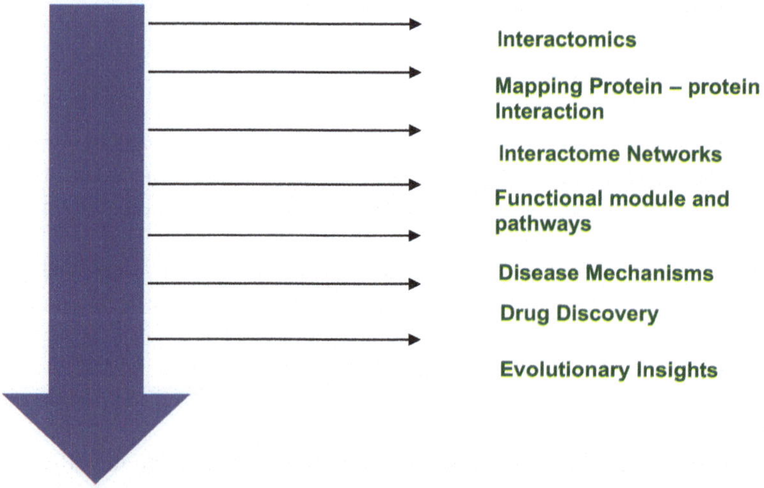

Fig. 2.2 Expanding horizons of Omics deep down in the domain

- *Functional Modules and Pathways*: Interactomes are organized into functional modules and pathways, where groups of proteins work together to carry out specific cellular processes. Identifying these modules helps understand how different cellular components coordinate their activities.
- *Disease Mechanisms*: Disruptions in PPIs can lead to diseases. Interactomics can identify disease-associated PPIs and provide insights into disease mechanisms, leading to the development of new therapeutic targets.
- *Drug Discovery*: Interactomics plays a role in drug discovery by identifying potential drug targets and predicting drug efficacy and side effects. By understanding how drugs affect PPI networks, researchers can develop more targeted and effective therapies.
- *Evolutionary Insights*: Comparing interactomes across different species provides insights into the evolution of protein interactions and cellular processes.

2.7 Interactomics Challenges

Interactomics research faces several challenges:

- *Data Completeness and Accuracy*: Current interactome maps are incomplete and may contain false positives or false negatives. Improving the accuracy and completeness of interactome data is an ongoing effort.
- *Dynamic Nature of Interactomes*: PPIs are not static but change dynamically in response to cellular signals and environmental cues. Capturing these dynamic interactions is a challenge.
- *Complexity of Interactome Networks*: Interactome networks are highly complex, making it challenging to analyze and interpret them. Advanced computational tools and systems biology approaches are needed to understand these complex networks.

Despite these challenges, interactomics is a rapidly growing field that provides valuable insights into cellular processes, disease mechanisms, and drug discovery. As technologies and analytical methods continue to improve, interactomics will play an increasingly important role in advancing our understanding of biology and medicine.

2.8 Lipidomics

Lipidomics is the comprehensive study of lipids within biological systems, encompassing their diverse roles in cellular function and their implications for health and disease (Eyster 2007; Dunn et al. 2010; Xu et al. 2017; Yuan et al. 2024). It involves identifying, quantifying, and characterizing the complete set of lipids (the lipidome) in a given biological sample, as well as studying their interactions with other biomolecules like proteins and carbohydrates.

Key aspects of lipidomics:

- *Lipid Diversity and Function*: Lipids are a diverse group of biomolecules with various functions, including energy storage, structural components of cell membranes, signaling molecules, and enzyme cofactors (Eyster 2007; Xu et al. 2017; Yuan et al. 2024). Understanding the specific roles of different lipid classes is crucial for comprehending cellular processes.
- *Analytical Techniques*: Lipidomics relies heavily on advanced analytical techniques, such as mass spectrometry and nuclear magnetic resonance spectroscopy, to identify and quantify lipids in complex biological samples (Dunn et al. 2010; Bingol and Brüschweiler 2016). These techniques enable researchers to analyze hundreds or even thousands of lipid species simultaneously.
- *Lipid Metabolism and Homeostasis*: Lipidomics investigates the pathways involved in lipid synthesis, modification, and degradation (Xu et al. 2017). Understanding lipid metabolism is essential for comprehending how cells maintain lipid homeostasis and how disruptions in these pathways contribute to disease.
- *Lipid Signaling*: Lipids play crucial roles in cellular signaling pathways, mediating various cellular processes such as growth, differentiation, and inflammation (Eyster 2007; Wasserman et al. 2020). Lipidomics helps unravel the complex interplay of lipids in these signaling networks.
- *Health and Disease*: Alterations in lipid metabolism and composition are implicated in various diseases, including cardiovascular disease, diabetes, cancer, and neurological disorders (Xu et al. 2017; Wasserman et al. 2020; Yuan et al. 2024). Lipidomics can identify lipid biomarkers associated with these diseases, aiding in diagnosis, prognosis, and treatment development.
- *Drug Discovery and Development*: Lipidomics plays a role in drug discovery by identifying potential drug targets and evaluating the efficacy and safety of lipid-modifying therapies.

Lipidomics research faces several challenges:

- *Lipid Complexity*: The lipidome is incredibly complex, with thousands of different lipid species. Characterizing and quantifying all these species is a significant analytical challenge.
- *Dynamic Nature of Lipids*: Lipid composition and levels can change rapidly in response to various stimuli. Capturing these dynamic changes requires specialized experimental approaches.
- *Data Analysis and Interpretation*: Analyzing and interpreting large lipidomics datasets requires sophisticated bioinformatics tools and expertise.

Despite these challenges, lipidomics is a rapidly evolving field that provides valuable insights into cellular processes, disease mechanisms, and drug discovery. As technologies and analytical methods continue to improve, lipidomics will play an increasingly important role in advancing our understanding of biology and medicine.

2.9 Glycomics

Glycomics is the comprehensive study of glycans, which are complex carbohydrates, and their diverse roles in biological systems (Rademacher et al. 1988). It encompasses the characterization of all glycan structures in a given cell type or organism, as well as their functions in various biological processes, including cell signaling, immunity, and disease (Kölbl et al. 2015; Haas et al. 2018).

Key aspects of glycomics:

- *Glycan Structure and Diversity*: Glycans exhibit remarkable structural diversity due to variations in monosaccharide composition, branching patterns, and glycosidic linkages. This structural complexity contributes to their diverse biological functions (Winchester and Fleet 2000; Vuorio et al. 2020).
- *Biosynthesis and Modification*: Glycans are synthesized and modified by a complex network of enzymes, including glycosyltransferases and glycosidases. Understanding these biosynthetic pathways is crucial for comprehending how glycan structures are generated and regulated (Winchester and Fleet 2000; Haas et al. 2018).
- *Analytical Techniques*: Glycomics utilizes various analytical techniques, such as mass spectrometry, chromatography, and nuclear magnetic resonance spectroscopy, to characterize glycan structures and quantify their abundance in biological samples. These techniques enable researchers to analyze complex mixtures of glycans and identify specific glycan modifications.
- *Cellular Functions*: Glycans play essential roles in various cellular processes, including cell–cell recognition, adhesion, signaling, and immune responses (Winchester and Fleet 2000; Haas et al. 2018). They also participate in protein folding, stability, and trafficking.
- *Glycans in Immunity*: Glycans are crucial components of the immune system, mediating interactions between immune cells and pathogens (Haas et al. 2018). They also play a role in immune recognition and regulation.
- *Glycans in Disease*: Alterations in glycan structures are associated with various diseases, including cancer, inflammatory disorders, and infectious diseases (Winchester and Fleet 2000; Kölbl et al. 2015; Haas et al. 2018). Glycomics can identify glycan biomarkers associated with these diseases, aiding in diagnosis, prognosis, and treatment development.
- *Therapeutic Applications*: Glycomics research has led to the development of glycan-based therapeutics, such as carbohydrate-based vaccines and enzyme inhibitors targeting glycosylation pathways (Winchester and Fleet 2000; Haas et al. 2018).

Glycomics research faces several challenges:

- *Glycan Complexity*: The structural complexity of glycans makes their analysis and characterization challenging.
- *Dynamic Nature of Glycans*: Glycan structures can change dynamically in response to cellular signals and environmental cues. Capturing these dynamic changes requires specialized experimental approaches.
- *Data Analysis and Interpretation*: Analyzing and interpreting large glycomics datasets requires sophisticated bioinformatics tools and expertise.

Despite these challenges, glycomics is a rapidly evolving field that provides valuable insights into cellular processes, disease mechanisms, and therapeutic development. As technologies and analytical methods continue to improve, glycomics will play an increasingly important role in advancing our understanding of biology and medicine.

2.10 Exposomics

Exposomics is the study of the totality of environmental exposures an individual experiences throughout their lifetime, and how these exposures impact health and disease. It considers all nongenetic exposures, from conception onward, including chemical, biological, physical, and social factors (DeBord et al. 2016).

Key aspects of exposomics:

- *Scope of Exposures*: Exposomics encompasses a wide range of exposures, including pollutants, diet, lifestyle factors, medications, infections, and socioeconomic factors (DeBord et al. 2016). It aims to capture the complex interplay of these exposures over time.
- *Lifetime Exposure Assessment*: A key challenge in exposomics is accurately assessing cumulative exposures throughout an individual's life. This requires integrating data from various sources, including environmental monitoring, personal sensors, biomarkers, and questionnaires (DeBord et al. 2016).
- *Health Impacts*: Exposomics investigates how environmental exposures influence disease risk, progression, and treatment response (DeBord et al. 2016). It aims to identify specific exposures or combinations of exposures that contribute to various health outcomes.
- *Omics Technologies*: Exposomics utilizes various "omics" technologies, such as genomics, transcriptomics, proteomics, and metabolomics, to understand how environmental exposures affect biological processes at the molecular level (DeBord et al. 2016). These technologies provide insights into the mechanisms by which exposures influence health and disease.
- *Personalized Medicine*: Exposomics data can be integrated with other individual-level information, such as genetics and lifestyle factors, to develop personalized risk assessments and prevention strategies.

- *Public Health Implications*: Exposomics research can inform public health policies and interventions aimed at reducing harmful environmental exposures and promoting healthier environments (DeBord et al. 2016).

Exposomics research faces several challenges:

- *Exposure Complexity*: Accurately measuring and characterizing all relevant exposures is a significant challenge.
- *Data Integration*: Integrating data from diverse sources, including environmental monitoring, personal sensors, and biological samples, requires sophisticated computational tools and methods.
- *Longitudinal Studies*: Understanding the long-term effects of cumulative exposures requires large-scale, longitudinal studies that follow individuals over many years.

Despite these challenges, exposomics is a rapidly evolving field with the potential to revolutionize our understanding of how environmental factors influence human health. As technologies and analytical methods continue to improve, exposomics will play an increasingly important role in developing personalized prevention strategies and improving public health outcomes.

2.11 Microbiomics

Microbiomics is the study of microbial communities (microbiota), including bacteria, archaea, fungi, protists, and viruses, and their collective genomes (microbiome), in various environments, including the human body, soil, oceans, and the built environment (Statnikov et al. 2013; Kashyap et al. 2017; Ray et al. 2020; Hu et al. 2020; Olm et al. 2021). It investigates the composition, diversity, function, and interactions of these microbial communities and their influence on human health, disease, and the environment. Microbiomics is a field that explores the diverse world of microorganisms, including bacteria, archaea, fungi, protists, and viruses, and their collective genomes across various environments (Statnikov et al. 2013; Wang et al. 2015; Kashyap et al. 2017; Ray et al. 2020; Olm et al. 2021).

Key aspects of microbiomics :

- Microbial Diversity and Composition: Researchers investigate the vast diversity of microorganisms present in different environments and how their composition varies among individuals, populations, and ecosystems. Advanced sequencing technologies enable the identification and quantification of the diverse microbial species in a sample.
- Microbial Function and Metabolism: Microbiomics examines the metabolic activities and functions of microbial communities, including how microbes break down complex molecules, produce essential nutrients, and interact with their environment (Wang et al. 2015; Kashyap et al. 2017; Ray et al. 2020).

- Host–Microbe Interactions: Microbiomics explores the complex interactions between microbes and their hosts, such as humans, animals, and plants. These interactions can be beneficial, like the gut microbiota aiding in digestion and nutrient absorption, or detrimental, as in the case of pathogenic microbes causing disease (Kashyap et al. 2017; Ray et al. 2020; Olm et al. 2021).
- Microbiome and Human Health: Microbiomics research has revealed the profound influence of the human microbiome on various aspects of health, including digestion, immunity, metabolism, and even behavior. Disruptions in the microbiome, known as dysbiosis, are linked to various diseases, such as inflammatory bowel disease, obesity, and autoimmune disorders (Kashyap et al. 2017; Olm et al. 2021).
- Microbiome and Disease: Microbiomics investigates the role of the microbiome in the development and progression of various diseases, including the identification of microbial biomarkers and the development of microbiome-based diagnostics and therapeutics (Statnikov et al. 2013; Olm et al. 2021).
- Environmental Microbiomics: Microbiomics studies microbial communities in various environments, such as soil, oceans, and the built environment, to understand the ecological roles of microbes in nutrient cycling, bioremediation and assessing potential to cope with environmental stress for instance at extreme environmental conditions such as cold stress at Antarctica (Kesheri et al. 2014).

Microbiomics research faces several key challenges:

- Complexity of Microbial Communities: Microbial communities are highly intricate and dynamic, making their comprehensive analysis and characterization a significant challenge.
- Need for Standardized Methodologies: Establishing standardized protocols for sample collection, processing, and data analysis is crucial to ensure reproducibility and comparability across different studies.
- Interpreting Complex Data: Translating the complex microbiomics data into meaningful biological insights requires advanced bioinformatics tools and specialized expertise.

Despite these obstacles, microbiomics is a rapidly evolving field with immense potential to revolutionize our understanding of the microbial world and its profound impact on human health, disease, and environmental dynamics. As technologies and analytical approaches continue to advance, microbiomics is poised to play an increasingly pivotal role in developing novel diagnostics, therapeutics, and strategies for environmental management.

2.12 Advanced Omics Approaches

2.12.1 Single-Cell Omics

Single-cell omics is the analysis of individual cells within a population, enabling a high-resolution examination of cellular heterogeneity and function (Schier 2020; Zhu et al. 2020; Vandereyken et al. 2023). Unlike traditional bulk omics approaches that average measurements across millions of cells, single-cell omics allows researchers to analyze the unique molecular profiles of individual cells, revealing subtle differences that would otherwise be obscured (Schier 2020; Zhu et al. 2020; Vandereyken et al. 2023).

Key aspects of single-cell omics:

- Cellular Heterogeneity: Single-cell omics unveils the substantial variations in gene expression, protein levels, and metabolic activity within seemingly homogenous cell populations, highlighting the extent of cellular heterogeneity in tissues and organs (Schier 2020; Zhu et al. 2020; Vandereyken et al. 2023).
- Multi-omics Approaches: Single-cell omics can be applied to various molecular layers, including genomics, transcriptomics, proteomics, epigenomics, and metabolomics. Combining multiple omics modalities provides a more comprehensive understanding of cellular function and regulation (Zhu et al. 2020).
- Technological Advancements: Recent technological advancements, such as microfluidics, droplet-based sequencing, and mass cytometry, have enabled high-throughput single-cell analysis, allowing researchers to analyze thousands or even millions of individual cells in a single experiment (Zhu et al. 2020).
- Single-cell omics has widespread applications across various scientific fields. In developmental biology, it enables the study of cellular differentiation and lineage tracing during development (Schier 2020; Zhu et al. 2020). In immunology, it allows for the characterization of immune cell populations and their responses to infection or disease (Zhu et al. 2020). Within cancer biology, single-cell omics provides insights into tumor heterogeneity and the identification of rare cancer stem cells (Zhu et al. 2020). In neurobiology, it facilitates the investigation of neuronal diversity and function in the brain (Zhu et al. 2020). Furthermore, in microbiome research, single-cell omics enables the analysis of the diversity and function of microbial communities at the individual cell level (Olm et al. 2021).
- Analyzing and interpreting the large datasets generated by single-cell omics requires specialized bioinformatics tools and expertise (Zhu et al. 2020; Vandereyken et al. 2023).
- Commonly used techniques include dimensionality reduction, clustering, and trajectory inference methods, which help extract meaningful insights from these complex single-cell data.

Single-cell omics faces several challenges, such as technical noise due to the small amount of starting material, which necessitates the use of specialized data analysis methods. Additionally, the cost and scalability of single-cell omics technologies can present obstacles, and the integration of data from different single-cell omics modalities requires sophisticated computational tools and methods. Despite these challenges, single-cell omics is transforming biological research by offering unprecedented insights into cellular heterogeneity and function. As technologies and analytical methods continue to improve, single-cell omics is poised to play an increasingly important role in advancing our understanding of complex biological systems and disease mechanisms.

2.12.2 Spatial Omics

Spatial omics is a rapidly evolving field that combines high-throughput omics technologies with spatial information, allowing researchers to analyze the molecular profiles of cells and tissues while preserving their spatial context (Bressan et al. 2023). This approach provides a deeper understanding of how cells interact and function within their native environment, revealing insights into tissue organization, cell–cell communication, and disease development.

Key aspects of spatial omics:

- Preserving Spatial Context: Traditional omics techniques often require tissue dissociation, which can result in the loss of valuable spatial information. Spatial omics overcomes this limitation by analyzing molecules directly within tissues, preserving their spatial location and relationships (Bressan et al. 2023).
- Multi-omics Capabilities: Spatial omics can be applied to various molecular layers, including genomics, transcriptomics, proteomics, and metabolomics (Bressan et al. 2023). This allows researchers to study multiple molecular modalities simultaneously within the same tissue section, providing a more comprehensive view of cellular function and interactions.
- Technological Approaches: Several technologies enable spatial omics analysis, including imaging-based methods that use techniques like microscopy and mass spectrometry imaging to visualize and quantify the distribution of molecules within tissues, (Bressan et al. 2023) and sequencing-based methods that employ spatially barcoded probes or in situ sequencing to capture and analyze nucleic acids directly within tissues (Bressan et al. 2023).

Applications of Spatial omics have diverse applications across various scientific domains:

- Cancer Research: Investigating tumor heterogeneity, interactions within the tumor microenvironment, and mechanisms of metastasis (Bressan et al. 2023).
- Developmental Biology: Analyzing tissue development, cellular differentiation processes, and morphogenesis.
- Immunology: Studying immune cell infiltration, activation, and interactions within tissues (Bressan et al. 2023).

- Neuroscience: Mapping brain regions, neuronal circuits, and cell–cell communication patterns (Bressan et al. 2023).

Analyzing and visualizing spatial omics data require specialized bioinformatics tools and methods. These tools integrate spatial information with molecular profiles, enabling researchers to visualize the distribution of molecules within tissues and identify spatial patterns and correlations (Bressan et al. 2023).

Spatial omics faces several key challenges, including the following:

Spatial Resolution: Achieving high spatial resolution while maintaining molecular sensitivity remains an ongoing challenge.

Data Integration: Integrating data from different spatial omics modalities and with other omics datasets requires sophisticated computational tools and methods.

Cost and Scalability: Spatial omics technologies can be expensive, and scaling up experiments to analyze large tissue sections or multiple samples can be logistically demanding.

Despite these challenges, spatial omics is revolutionizing biological research by providing a spatially resolved view of cellular and tissue function. As technologies and analytical methods continue to improve, spatial omics will play an increasingly important role in advancing our understanding of complex biological systems and disease mechanisms.

2.13 Temporal Omics

Temporal omics, also known as longitudinal omics or time-course omics, is the study of how biological systems change over time at a molecular level (Liang and Kelemen 2017; Bodein et al. 2021). It involves collecting and analyzing omics data from the same individual or system at multiple time points, allowing researchers to track dynamic changes in molecular profiles and understand the temporal relationships between different biological processes. Key aspects of temporal omics:

Longitudinal Data Collection: Temporal omics studies involve collecting samples and generating omics data at multiple time points, ranging from hours to years, to capture dynamic changes in molecular profiles that would be missed with single-time-point analyses (Liang and Kelemen 2017; Bodein et al. 2021).

Studying Dynamic Processes: Temporal omics is crucial for investigating dynamic biological processes, such as development, aging, disease progression, and responses to environmental stimuli, as tracking changes in molecular profiles over time can provide insights into the underlying mechanisms driving these processes (Liang and Kelemen 2017; Bodein et al. 2021).

Data Analysis: Analyzing temporal omics data necessitates specialized bioinformatics tools and methods, which can identify temporal patterns, trends, and correlations in molecular data, revealing how different biological processes unfold over time. Common techniques include time-series analysis, trajectory inference, and dynamic network modeling (Liang and Kelemen 2017; Bodein et al. 2021).

2.13.1 Applications of Temporal Omics

- Disease Monitoring: Tracking temporal changes in molecular profiles can enable the early detection of disease warning signs, predict disease trajectories, and monitor the efficacy of therapeutic interventions.
- Pharmaceutical Research: Temporal omics data can inform the study of drug mechanisms of action, the identification of potential drug targets, and the optimization of personalized treatment strategies.
- Personalized Healthcare: Longitudinal omics data can be utilized to develop customized treatment plans based on an individual's unique molecular profile and its dynamic changes over time.
- Systems Biology: Temporal omics provides valuable datasets for constructing dynamic models of biological systems, enabling researchers to simulate and forecast system behavior under diverse conditions.

2.13.2 Key Challenges of Temporal Omics

Designing appropriate sampling strategies, particularly for long-term studies, can be complex. Factors such as sample frequency, duration, and storage require careful consideration.

- Temporal omics datasets can be inherently complex and heterogeneous, necessitating advanced data integration and analysis methods.
- Analyzing large-scale temporal omics datasets often requires substantial computational resources and specialized expertise.

Despite these hurdles, temporal omics remains a powerful approach for investigating dynamic biological processes. As technologies and analytical techniques continue to evolve, this field holds great potential to transform our understanding of health, disease, and environmental interactions. As a result, temporal omics is poised to play an increasingly pivotal role in advancing biological research and personalized medicine.

2.14 Integrative and Applied Omics

Integrative Omics: Integrating data from multiple omics disciplines, such as genomics, transcriptomics, proteomics, and metabolomics, provides a more holistic and comprehensive understanding of biological systems (Sun and Hu 2016; Huang et al. 2017; Grapov et al. 2018; Amer and Baidoo 2021). Instead of examining each "ome" in isolation, integrative omics seeks to unravel the complex interrelationships between different biological layers and how they contribute to phenotype, health, and disease. Key aspects of integrative omics include the following:

- The Need for Integration: Biological systems are intricate and interconnected. Genes influence RNA expression, which in turn affects protein levels and metabolic activity. Studying each "ome" individually offers valuable insights, but it fails to capture the full picture of these intricate interactions. Integrative omics addresses this limitation by combining data from multiple sources (Sun and Hu 2016).
- Data Integration Challenges: Integrating data from diverse omics technologies presents significant computational challenges. Each "ome" has unique characteristics, data formats, and analytical methods. Combining these heterogeneous datasets requires sophisticated computational tools and algorithms (Huang et al. 2017; Grapov et al. 2018).
- Computational Approaches: Various computational methods are employed for integrative omics, including machine learning algorithms that can uncover patterns and relationships from complex datasets (Gligorijević et al. 2015; Huang et al. 2017) network analysis methods for constructing and analyzing biological networks, (Huang et al. 2017), and statistical modeling techniques for identifying correlations and causal relationships between different omics datasets (Gligorijević et al. 2015).

2.14.1 Applications of Integrative Omics

Integrative omics is crucial for systems biology, which aims to understand the behavior of complex biological systems holistically. By combining data from multiple omics layers, researchers can identify more robust and accurate disease biomarkers. Additionally, integrative omics can aid in drug discovery and development by providing a more comprehensive understanding of disease mechanisms, allowing for the identification of new targets and the optimization of therapeutic strategies. Furthermore, integrative omics can be leveraged to develop personalized treatments tailored to an individual's unique molecular profile. This holistic approach is essential for advancing our knowledge of biology and translating these insights into clinical applications.

2.14.2 Computational Omics

Computational omics focuses on developing and applying computational approaches to analyze and interpret the large, complex datasets generated by various omics technologies (Gligorijević et al. 2015; Sun and Hu 2016; Huang et al. 2017; Markowetz 2017; de Anda-Jáuregui and Hernández-Lemus 2020). Key aspects of computational omics include the following:

- Data Processing and Analysis: Computational tools are essential for processing and analyzing raw omics data, including tasks such as quality control, normalization, feature extraction, and statistical analysis (de Anda-Jáuregui and Hernández-Lemus 2020).
- Data Integration and Interpretation: Integrating data from multiple omics platforms is a significant challenge. Computational methods are employed to combine data from diverse sources and interpret the intricate relationships between different biological layers (Gligorijević et al. 2015; Sun and Hu 2016; Huang et al. 2017; de Anda-Jáuregui and Hernández-Lemus 2020).
- Machine Learning and Artificial Intelligence: Machine learning algorithms are increasingly utilized in computational omics to identify patterns, predict outcomes, and construct predictive models. These algorithms can learn from complex datasets and make predictions about disease risk, treatment response, and other important biological phenomena (de Anda-Jáuregui and Hernández-Lemus 2020; Grapov et al. 2018; Kanchan et al. 2024).
- Network Analysis: Computational methods are used to construct and analyze biological networks, such as gene regulatory networks, protein–protein interaction networks, and metabolic networks, providing insights into the complex interactions between various biological molecules and pathways (de Anda-Jáuregui and Hernández-Lemus 2020).
- Bioinformatics and Data Visualization: Bioinformatics tools and techniques are essential for managing, analyzing, and visualizing omics data. Data visualization methods help researchers explore and interpret complex datasets, identify patterns, and communicate their findings effectively (McEntyre and Ostell 2018).
- Development of New Algorithms and Software: Computational omics is a rapidly evolving field, and new algorithms and software tools are continuously being developed to address the challenges of analyzing and interpreting large omics datasets. Computational omics plays a crucial role in extracting meaningful biological insights from the vast amounts of data generated by various omics technologies, advancing our understanding of complex biological systems, and translating these findings into clinical applications.

2.15 Experimental Omics

Experimental omics encompasses the diverse experimental designs and data acquisition strategies employed across various omics disciplines. Careful experimental design is paramount for generating high-quality, reliable data that can be effectively analyzed and interpreted.

Careful experimental design is crucial for generating high-quality, reliable omics data that can be effectively analyzed and interpreted. Key considerations in the study design phase include the following:

- Defining a clear and focused research question to guide the experimental approach.

- Selecting appropriate samples, such as cell lines, animal models, or human samples, to ensure the generalizability of the findings.
- Incorporating appropriate controls to account for experimental variability and isolate the effects of the experimental manipulation.
- Incorporating sufficient biological and technical replicates to assess the reproducibility of the results and increase the statistical power of the analysis.
- In the data acquisition phase, researchers must choose the appropriate omics technology based on the research question and the type of data being collected, such as gene expression, protein levels, or metabolite concentrations. Proper sample preparation and optimization of data acquisition parameters are also essential for ensuring the quality and integrity of the omics data.
- Finally, quality control and data preprocessing steps, such as quality assessment, data normalization, and data transformation, are necessary to prepare the omics data for downstream analysis and interpretation. By carefully considering these experimental designs and data acquisition strategies, researchers can generate high-quality omics datasets that can be effectively leveraged to gain valuable biological insights.

2.15.1 Key Areas of Translational Omics

- Biomarker Discovery: Omics technologies can identify molecular markers associated with disease risk, diagnosis, prognosis, or treatment response. These biomarkers can be leveraged to develop noninvasive diagnostic tests, predict disease progression, and monitor treatment effectiveness (Clish 2015; Quintás-Cardama et al. 2021).
- Drug Discovery and Development: Omics data can be utilized to identify novel drug targets and optimize drug development pipelines. By elucidating the molecular mechanisms underlying disease, researchers can design more targeted and effective therapies (Altman 2012; Roychowdhury and Chinnaiyan 2015).
- Personalized Medicine: Translational omics is crucial for personalized medicine, as it enables the tailoring of treatments to individual patients based on their unique molecular profiles. This approach can improve treatment outcomes by selecting the most effective therapy for each patient and minimizing adverse effects (Altman 2012; Clish 2015; Roychowdhury and Chinnaiyan 2015).
- Diagnostics Development: Omics-based diagnostics are being developed for a wide range of diseases, including cancer, infectious diseases, and genetic disorders. These tests can provide earlier and more accurate diagnoses, leading to improved patient outcomes (Fredolini et al. 2010).
- Theranostics: This emerging field combines diagnostics and therapeutics by leveraging the same molecular information for both diagnosis and treatment. For instance, a diagnostic test could identify a specific genetic mutation that makes a patient eligible for a targeted therapy (Srivastav et al. 2024).

2.15.2 Challenges Translational Omics

- Data Complexity and Integration: Integrating data from multiple omics platforms and clinical sources can be a computationally complex task (Clish 2015; Grapov et al. 2018).
- Validation and Clinical Trials: Omics-based findings must undergo rigorous validation in clinical trials before they can be translated into clinical practice (Boja et al. 2014; Roychowdhury and Chinnaiyan 2015; Grapov et al. 2018).
- Regulatory Hurdles: The regulatory landscape for omics-based diagnostics and therapeutics is still evolving, presenting challenges for their adoption (Roychowdhury and Chinnaiyan 2015; Grapov et al. 2018).
- Cost and Accessibility: Omics technologies can be expensive, potentially limiting their accessibility in certain healthcare settings (Boja et al. 2014).

Despite these obstacles, translational omics holds immense potential for improving human health. As omics technologies continue to advance and our understanding of complex biological systems grows, translational omics will play an increasingly crucial role in developing new and more effective approaches to disease diagnosis, treatment, and prevention (Clish 2015).

2.16 Clinical Omics

Clinical omics applies omics technologies and principles within clinical settings to enhance patient care. It focuses on using omics data to diagnose and manage diseases, predict treatment responses, and personalize treatment strategies (Clish 2015; Grapov et al. 2018).

2.16.1 Key Applications of Clinical Omics

- Disease Diagnostics: Omics-based tests can detect diseases earlier and more accurately than traditional methods (Fredolini et al. 2010; Clish 2015).
- Treatment Selection and Monitoring: Omics data can guide treatment decisions by predicting which patients are most likely to respond to specific therapies, improving treatment outcomes, and minimizing adverse effects (Roychowdhury and Chinnaiyan 2015; Grapov et al. 2018).
- Disease Prognosis and Risk Stratification: Omics can identify biomarkers that predict disease progression and patient outcomes, enabling clinicians to tailor treatment strategies accordingly (Fredolini et al. 2010; Quintás-Cardama et al. 2021).
- Pharmacogenomics examines how an individual's genetic characteristics affect their response to drugs. This knowledge can assist clinicians in selecting the appropriate medication and dosage for each patient, minimizing adverse drug

reactions and maximizing treatment efficacy (Roychowdhury and Chinnaiyan 2015).
- Omics data can also monitor disease progression and treatment response over time. This information can help clinicians adjust treatment strategies as needed and improve long-term patient outcomes.

2.16.2 Clinical Omics Challenges

- Data Interpretation and Clinical Validity: Interpreting complex omics data and translating it into clinically actionable information can be challenging. Rigorous validation is essential to ensure the clinical validity and utility of omics-based tests.
- Integration with Electronic Health Records: Integrating omics data with electronic health records is crucial for seamless data management and clinical decision-making. Standardized data formats and interoperable systems are needed to facilitate data sharing and integration.
- Ethical and Privacy Concerns: The use of omics data in clinical settings raises ethical and privacy concerns. Protecting patient privacy and ensuring responsible use of omics data are paramount.
- Cost and Reimbursement: Omics technologies can be expensive, which can limit their accessibility in some healthcare settings. Clear reimbursement policies are needed to ensure that patients have access to these potentially life-saving technologies.

Despite these challenges, clinical omics holds immense potential to transform healthcare. As omics technologies continue to advance and become more affordable, clinical omics will play an increasingly important role in improving patient care and advancing precision medicine. Precision medicine, also known as personalized medicine, tailors medical treatment to individual patients based on their unique genetic, molecular, and environmental profiles (König et al. 2017; Ginsburg and Phillips 2018; Wiedermann 2023). It moves away from a "one-size-fits-all" approach to healthcare and aims to optimize treatment efficacy and minimize adverse effects by considering individual variability.

- Omics data contributes to precision medicine in the following ways: Identifying Genetic Predispositions: Genomics identifies genetic variations that influence disease risk. This information can guide preventative measures, early detection strategies, and treatment decisions (König et al. 2017; Ginsburg and Phillips 2018; Uddin et al. 2019; Wiedermann 2023).
- Pharmacogenomics leverages genetic information to predict individual responses to specific drugs, enabling clinicians to select the most effective medication and dosage while minimizing adverse effects (Vogenberg et al. 2010; König et al. 2017; Uddin et al. 2019). Omics data can identify molecular targets for the development of more effective therapies that address the underlying causes of disease

in each patient (Vogenberg et al. 2010; Roychowdhury and Chinnaiyan 2015). Additionally, omics can be used to stratify patients for clinical trials, improving the efficiency of these studies and accelerating the introduction of new treatments (Wiedermann 2023). Omics data can also track changes in a patient's molecular profile during treatment, providing real-time feedback on the efficacy of the therapy and allowing for adjustments as needed. Furthermore, omics can identify individuals at high risk for specific diseases, enabling the implementation of personalized prevention strategies, such as lifestyle modifications or targeted screening programs. While precision medicine holds immense potential to enhance healthcare, it faces several significant challenges. Protecting patient privacy and securing sensitive omics data are crucial considerations. Interpreting complex omics data and translating it into clinically actionable information can be challenging, underscoring the need for rigorous validation to ensure the clinical validity and utility of omics-based tests. The cost of omics technologies can also limit accessibility for some patients, though efforts are underway to improve affordability. Lastly, the use of omics data in healthcare raises ethical considerations, such as informed consent, genetic discrimination, and equitable access to precision medicine.

2.17 Multi-Omics

Multi-omics approaches involve the integrated analysis of data from various omics disciplines, such as genomics, transcriptomics, proteomics, and metabolomics. By combining these diverse datasets, researchers can gain a more comprehensive understanding of complex biological systems and processes (Sun and Hu 2016; Huang et al. 2017; Grapov et al. 2018). Here are several specific examples of how multi-omics is being applied:

- Cancer Research: Multi-omics is utilized to identify cancer subtypes, predict treatment response, and discover new therapeutic targets. For instance, integrating genomic data with transcriptomic and proteomic data can reveal how genetic mutations affect gene expression and protein function, leading to a better understanding of cancer development and progression (Grapov et al. 2018; Lu and Zhan 2018).
- Infectious Disease Research: Multi-omics is employed to study host–pathogen interactions, identify biomarkers of infection, and develop new diagnostic and therapeutic strategies. By integrating host transcriptomic data with proteomic and metabolomic data, researchers can gain insights into how pathogens manipulate host cellular processes and how the host immune system responds to infection.
- Drug Discovery and Development: Multi-omics is utilized to identify new drug targets, predict drug efficacy and toxicity, and personalize drug therapies. Integrating genomic data with transcriptomic, proteomic, and metabolomic data

can help identify potential drug targets and predict how individuals will respond to different drugs based on their unique molecular profiles.

- Systems Biology: Multi-omics is essential for systems biology research, which aims to understand the behavior of complex biological systems as a whole. By integrating data from multiple omics levels, researchers can construct comprehensive models of biological systems and predict how they will respond to different perturbations.
- Multi-omics is being applied to study complex diseases such as diabetes, cardiovascular disease, and neurodegenerative disorders. By integrating data from multiple omics levels, researchers can gain a deeper understanding of the complex interplay of genetic and environmental factors that contribute to these diseases (Sun and Hu 2016; Grapov et al. 2018). These examples illustrate the diverse ways in which multi-omics approaches are being applied to study complex biological questions.

2.18 Conclusions

Multi-omics represents a significant advancement in biological research, shifting from single-cell omics studies to a more holistic understanding of complex biological systems. By combining data from various molecular levels (genomics, transcriptomics, proteomics, metabolomics, etc.), multi-omics offers a complete view of the interactions between biomolecules. This approach has the potential to transform our understanding of cellular function, development, disease, and treatment. Multi-omics has led to progress in disease diagnostics, drug discovery, and personalized medicine. Identifying new biomarkers, predicting disease risk, and personalizing treatment strategies based on individual molecular profiles offer great promise for improving health. It also facilitates the development of new drug targets and the faster introduction of effective treatments. Integrating and interpreting large amounts of diverse data requires advanced computational tools and skills. Data standardization, validation, and accessibility are key obstacles. Ethical concerns about data privacy, informed consent, and equitable access to these technologies must be addressed. Despite these challenges, multi-omics has a promising future. As technologies improve and our understanding of biological systems deepens, multi-omics will play a vital role in transforming healthcare and scientific discovery. Continued investment in research, development, and education will lead to a new era of precision medicine, with treatments tailored to individual molecular profiles, improving health outcomes for everyone.

Short Chapter Summary
This chapter discusses the growing field of multi-omics, which goes beyond just genomics to incorporate data from other biological areas such as transcriptomics, proteomics, and metabolomics. This broader perspective allows for a more complete understanding of complex biological systems, leading to better insights into how cells function, how diseases develop, and how treatments can be personalized.

This chapter also covers the potential benefits and difficulties of using multi-omics in research and clinical settings, including issues with data integration, validation, ethics, and cost. Lastly, it emphasizes how multi-omics can advance precision medicine by tailoring treatments to each patient's unique molecular profile.

Acknowledgments I am grateful to Prof. M. P. Pandey, Dr. Sanjeev Agrawal, Prof. Vaibhav Trivedi, and Prof. Tanzeel Ahmad for their unwavering encouragement in writing this chapter.

Authors' Contributions Chapter concept and design, Smrita Singh and Ashutosh Singh Chauhan; chapter writing, Smrita Singh, Tanzeel Ahmed and Ashutosh Singh Chauhan; chapter editing and review, Tanzeel Ahmed and Ashutosh Singh Chauhan.

References

Altman RB (2012) Translational bioinformatics: linking the molecular world to the clinical world. Clin Pharmacol Ther 91:994. https://doi.org/10.1038/clpt.2012.49
Amer B, Baidoo EEK (2021) Omics-driven biotechnology for industrial applications. Front Bioeng Biotechnol 9:613307. https://doi.org/10.3389/fbioe.2021.613307
Baker M (2011) Metabolomics: from small molecules to big ideas. Nat Methods 8:117. https://doi.org/10.1038/nmeth0211-117
Bingol K, Brüschweiler R (2016) Knowns and unknowns in metabolomics identified by multidimensional NMR and hybrid MS/NMR methods. Curr Opin Biotechnol 43:17. https://doi.org/10.1016/j.copbio.2016.07.006
Bodein A, Scott-Boyer M, Périn O, Cao KL, Droit A (2021) timeOmics: an R package for longitudinal multi-omics data integration. Bioinformatics 38:577. https://doi.org/10.1093/bioinformatics/btab664
Boja ES, Kinsinger CR, Rodriguez H, Srinivas PR (2014) Integration of omics sciences to advance biology and medicine. Clin Proteomics 11:45. https://doi.org/10.1186/1559-0275-11-45
Bressan D, Battistoni G, Hannon GJ (2023) The dawn of spatial omics. Science 381:abq4964. https://doi.org/10.1126/science.abq4964
Bruce C, Stone KL, Gulcicek EE, Williams KR (2013) Proteomics and the analysis of proteomic data: overview of current protein-profiling technologies. Curr Protoc Bioinformatics 41:1321s41. https://doi.org/10.1002/0471250953.bi1321s41
Choi S, Claycombe K, Martínéz JA, Friso S, Schalinske KL (2013) Nutritional epigenomics: a portal to disease prevention. Adv Nutr 4:530. https://doi.org/10.3945/an.113.004168
Clish CB (2015) Metabolomics: an emerging but powerful tool for precision medicine. Cold Spring Harb Mol Case Stud 1:588. https://doi.org/10.1101/mcs.a000588
Costa FF (2010) Epigenomics in cancer management. Cancer Manag Res 2:255. https://doi.org/10.2147/cmr.s7280
de Anda-Jáuregui G, Hernández-Lemus E (2020) Computational oncology in the multi-omics era: state of the art. Front Oncol 10:423. https://doi.org/10.3389/fonc.2020.00423
DeBord DG, Carreón T, Lentz TJ, Middendorf PJ, Hoover MD, Schulte PA (2016) Use of the "exposome" in the practice of epidemiology: a primer on -omictechnologies. Am J Epidemiol 184:302. https://doi.org/10.1093/aje/kwv325
Delage B, Dashwood RH (2008) Dietary manipulation of histone structure and function. Annu Rev Nutr 28:347. https://doi.org/10.1146/annurev.nutr.28.061807.155354
Dunn WB, Broadhurst D, Atherton HJ, Goodacre R, Griffin JL (2010) Systems level studies of mammalian metabolomes: the roles of mass spectrometry and nuclear magnetic resonance spectroscopy. Chem Soc Rev 40:387. https://doi.org/10.1039/b906712b

Eyster KM (2007) The membrane and lipids as integral participants in signal transduction: lipid signal transduction for the non-lipid biochemist. Adv Physiol Educ 31:5. https://doi.org/10.1152/advan.00088.2006

Fredolini C, Liotta LA, Petricoin EF (2010) Application of proteomic technologies for prostate cancer detection, prognosis, and tailored therapy. Crit Rev Clin Lab Sci 47:125. https://doi.org/10.3109/10408363.2010.503558

Ginsburg GS, Phillips KA (2018) Precision medicine: from science to value. Health Aff 37:694. https://doi.org/10.1377/hlthaff.2017.1624

Gligorijević V, Malod-Dognin N, Pržulj N (2015) Integrative methods for analyzing big data in precision medicine. Proteomics 16(5):741. https://doi.org/10.1002/pmic.201500396

Grapov D, Fahrmann JF, Wanichthanarak K, Khoomrung S (2018) Rise of deep learning for genomic, proteomic, and metabolomic data integration in precision medicine. OMICS 22(10):630. Mary Ann Liebert, Inc

Haas Q, Simillion C, Gunten S (2018) A cartography of Siglecs and sialyltransferases in gynecologic malignancies: is there a road towards a sweet future? Front Oncol 8. https://doi.org/10.3389/fonc.2018.00068

Handy DE, Castro R, Loscalzo J (2011) Epigenetic modifications. Circulation 123(19):2145. https://doi.org/10.1161/circulationaha.110.956839

Hu Y, Fang L, Nicholson C, Wang K (2020) Implications of error-prone long-read whole-genome shotgun sequencing on characterizing reference microbiomes. iScience 23(6):101223. https://doi.org/10.1016/j.isci.2020.101223

Huang S, Chaudhary K, Garmire LX (2017) More is better: recent progress in multi-omics data integration methods. Front Genet 8:84. https://doi.org/10.3389/fgene.2017.00084

Kanchan S, Ogden E, Kesheri M, Skinner A, Miliken E, Lyman D, Armstrong J, Sciglitano L, Hampikian G (2024) COVID-19 hospitalizations and deaths predicted by SARS-CoV-2 levels in Boise, Idaho wastewater. Sci Total Environ 907:167742. https://doi.org/10.1016/j.scitotenv.2023.167742. PMID: 37852488

Kashyap P, Chia N, Nelson H, Segal E, Elinav E (2017) Microbiome at the frontier of personalized medicine. Mayo Clin Proc 92(12):1855. https://doi.org/10.1016/j.mayocp.2017.10.004

Kesheri M, Kanchan S, Richa, Sinha RP (2014) Isolation and in-silico analysis of Fe- superoxide dismutase in Nostoc commune. Gene 553(2):117–125. https://doi.org/10.1016/j.gene.2014.10.010

Kesheri M, Kanchan S, Sinha RP (2022) Responses of antioxidants for resilience to temporal variations in the cyanobacterium Microcystis aeruginosa. S Afr J Bot 148:190–199. https://doi.org/10.1016/j.sajb.2022.04.017

Kesheri M, Kanchan S, Kumar A, Srivastava U, Sharda S, Mallik B, Mishra T, Kaithal P, Narayan J, Kumar P, Priya P, Sinha RP (2024) Peculiar endosymbiosis in the cyanobiont Nostoc azollae 0708: an in silico approach. In: Kesheri M, Kanchan S, Salisbury TB, Sinha RP (eds) Microbial omics in environment and health. Springer, Singapore. https://doi.org/10.1007/978-981-97-1769-9_13

Kölbl AC, Andergassen U, Jeschke U (2015) The role of glycosylation in breast cancer metastasis and cancer control. Front Oncol 5. https://doi.org/10.3389/fonc.2015.00219

König IR, Fuchs O, Hansen G, Mutius E, Kopp M (2017) What is precision medicine? Eur Respir J 50(4):1700391. https://doi.org/10.1183/13993003.00391-2017

Liang Y, Kelemen A (2017) Computational dynamic approaches for temporal omics data with applications to systems medicine. BioData Min 10(1). https://doi.org/10.1186/s13040-017-0140-x

Lu M, Zhan X (2018) The crucial role of multiomic approach in cancer research and clinically relevant outcomes. EPMA J 9(1):77. https://doi.org/10.1007/s13167-018-0128-8

Markowetz F (2017) All biology is computational biology. PLoS Biol 15(3). https://doi.org/10.1371/journal.pbio.2002050

Moosavi A, Ardekani AM (2016) Role of epigenetics in biology and human diseases. Iran Biomed J 20(5):246. https://doi.org/10.22045/ibj.2016.01

Olm MR, Crits-Christoph A, Bouma-Gregson K, Firek B, Morowitz MJ, Banfield JF (2021) inStrain profiles population microdiversity from metagenomic data and sensitively detects shared microbial strains. Nat Biotechnol 39(6):727. https://doi.org/10.1038/s41587-020-00797-0

Pan L, Liu D, Feng Z, Liu W, Peng S (2023) PACS: prediction and analysis of cancer subtypes from multi-omics data based on a multi-head attention mechanism model. arXiv (Cornell University). https://doi.org/10.48550/arxiv.2308.10917

Quintás-Cardama A, Martínez-Sena T, Conde I, Pareja E, Kleinjans J, Castell JV (2021) Metabolomic analysis to discriminate drug-induced liver injury (DILI) phenotypes. Arch Toxicol 95(9):3049. https://doi.org/10.1007/s00204-021-03114-z

Rademacher TW, Parekh RB, Dwek RA (1988) Glycobiology. Annu Rev Biochem 57(1):785. https://doi.org/10.1146/annurev.bi.57.070188.004033

Ray P, Lakshmanan V, Labbé J, Craven KD (2020) Microbe to microbiome: a paradigm shift in the application of microorganisms for sustainable agriculture. Front Microbiol 11. https://doi.org/10.3389/fmicb.2020.622926

Roychowdhury S, Chinnaiyan AM (2015) Translating cancer genomes and transcriptomes for precision oncology. CA Cancer J Clin 66(1):75. https://doi.org/10.3322/caac.21329

Sahu N, Mishra S, Kesheri M, Kanchan S, Sinha RP (2023) Identification of cyanobacteria- based natural inhibitors against SARS-CoV-2 druggable target ACE2 using molecular docking study, ADME and toxicity analysis. Indian J Clin Biochem 38(3):361–373. https://doi.org/10.1007/s12291-022-01056-6

Schier AF (2020) Single-cell biology: beyond the sum of its parts. Nat Methods 17(1):17. https://doi.org/10.1038/s41592-019-0693-3

Shyur L, Yang N (2008) Metabolomics for phytomedicine research and drug development. Curr Opin Chem Biol 12(1):66. https://doi.org/10.1016/j.cbpa.2008.01.032

Srivastav S, Lavanya, Avasthi A, Kaur N, Gaertner W, Kesheri M, Kanchan S, Sharda S (2024) Integromics: tracking the multi-omics expanse in theragnostics. In: Kesheri M, Kanchan S, Salisbury TB, Sinha RP (eds) Microbial omics in environment and health. Springer, Singapore. https://doi.org/10.1007/978-981-97-1769-9_6

Statnikov A, Henaff M, Narendra V, Konganti K, Li Z, Yang L, Pei Z, Blaser MJ, Aliferis CF, Alekseyenko AV (2013) A comprehensive evaluation of multicategory classification methods for microbiomic data. Microbiome 1(1). https://doi.org/10.1186/2049-2618-1-11

Sun YV, Hu Y (2016) Integrative analysis of multi-omics data for discovery and functional studies of complex human diseases. Adv Genet 147. https://doi.org/10.1016/bs.adgen.2015.11.004

Tammen SA, Friso S, Choi S (2012) Epigenetics: the link between nature and nurture. Mol Asp Med 34(4):753. https://doi.org/10.1016/j.mam.2012.07.018

Uddin M, Wang Y, Woodbury-Smith M (2019) Artificial intelligence for precision medicine in neurodevelopmental disorders. NPJ Digit Med 2(1). https://doi.org/10.1038/s41746-019-0191-0

Vandereyken K, Sifrim A, Thienpont B, Voet T (2023) Methods and applications for single-cell and spatial multi-omics. Nat Rev Genet 24(8):494. https://doi.org/10.1038/s41576-023-00580-2

Vogenberg FR, Barash CI, Pursel M (2010) Personalized medicine: part 1: evolution and development into theranostics. Available at: https://pubmed.ncbi.nlm.nih.gov/21037908. Accessed 13 Jan 2025

Vuorio J, Škerlová J, Fábry M, Veverka V, Vattulainen I, Řezáčová P, Martinez-Seara H (2020) N-glycosylation blocks and simultaneously fosters different receptor-ligand binding sites: the chameleonic CD44–hyaluronan interaction. bioRxiv. https://doi.org/10.1101/2020.03.23.004002

Wang W, Xu S, Ren Z, Tao L, Jiang J, Zheng S (2015) Application of metagenomics in the human gut microbiome. World J Gastroenterol 21(3):803. https://doi.org/10.3748/wjg.v21.i3.803

Wasserman AH, Venkatesan M, Aguirre A (2020) Bioactive lipid signaling in cardiovascular disease, development, and regeneration. Cells 9(6):1391. https://doi.org/10.3390/cells9061391

Wiedermann CJ (2023) Advancing precision medicine in South Tyrol, Italy: a public health development proposal for a bilingual, autonomous province. J Pers Med 13(6):972. https://doi.org/10.3390/jpm13060972

Winchester B, Fleet GWJ (2000) Modification of glycosylation as a therapeutic strategy. J Carbohydr Chem 19(4):471. https://doi.org/10.1080/07328300008544094

Xu S, Zhang X, Liu P (2017) Lipid droplet proteins and metabolic diseases. Biochim Biophys Acta (BBA) – Mol Basis Dis 1864(5):1968. https://doi.org/10.1016/j.bbadis.2017.07.019

Yates JR (2019) Recent technical advances in proteomics. F1000Res 8:351. https://doi.org/10.12688/f1000research.16987.1

Yuan Y, Li P, Li J, Zhao Q, Chang Y, He X (2024) Protein lipidation in health and disease: molecular basis, physiological function and pathological implication. Signal Transduct Target Ther 9(1). https://doi.org/10.1038/s41392-024-01759-7

Zhu C, Preißl S, Ren B (2020) Single-cell multimodal omics: the power of many. Nat Methods 17(1):11. https://doi.org/10.1038/s41592-019-0691-5

Disease Dynamics: Insights from Microbiome and Multi-Omics Analysis

3

Tarun Mishra, Pankaj Bharat Tiwari, Ahmad Reza Rezaei, Bhagaban Mallik, Swarna Kanchan, and Minu Kesheri

Abstract

The study of disease dynamics has traditionally focused on pathogen–host interactions, yet recent advances in microbiome research and multi-omics analysis have unveiled a more complex picture of disease progression. This chapter discusses the critical insights emerging from integrating microbiome data with multi-omics approaches, including genomics, transcriptomics, proteomics, and metabolomics. Emerging fields of single-celled and spatial multi-omics offer unprecedented resolution in studying cellular and tissue-specific interactions within the microbiome and host, further advancing our understanding of disease dynamics and providing insights crucial for the development of targeted therapeutic interventions. By leveraging these data, we can better understand the intricate relationships between host systems and microbial communities, revealing how dysbiosis and microbial signaling contribute to disease onset, progression,

T. Mishra (✉)
Department of Biological Sciences, IISER, Bhopal, Madhya Pradesh, India

Department of Microbiology and Immunology, University of Iowa, Iowa City, IA, USA

P. B. Tiwari
Department of Biological Sciences, IISER, Bhopal, Madhya Pradesh, India

A. R. Rezaei
Medical University of Bialystok, Bialystok, Poland

B. Mallik
Department of Biological Sciences, IISER, Bhopal, Madhya Pradesh, India

Department of Anatomy and Cell Biology, University of Iowa, Iowa City, IA, USA

S. Kanchan · M. Kesheri
Department of Biomedical Sciences, Marshall University, Huntington, WV, USA

Biology Department, Boise State University, Boise, ID, USA

M. Kesheri et al. (eds.), *Multi-Omics in Biomedical Sciences and Environmental Sustainability*, https://doi.org/10.1007/978-981-96-7067-3_3

and recovery. The integration of multi-omics data further allows for the identification of biomarkers, personalized therapeutic targets, and a deeper comprehension of host–microbe interactions across diverse diseases, from autoimmune disorders to cancer and infectious diseases. This convergence of microbiome and multi-omics data holds transformative potential for predictive modeling of disease dynamics and the development of precision medicine strategies, marking a new era in our understanding of health and disease.

Keywords

Gut microbiome · Human diseases · Microbiome · Multi-omics · Proteomics · Transcriptomics

Objectives

- The interaction between host and microbiome contributes toward the well-being of organisms.
- Dysbiosis in the microbiome leads to development of multiple diseases.
- The understanding of microorganism diversity in health and diseased individual will help researchers to generate personalized medicine.
- We aim to compile the knowledge of how multi-omics can be used to understand the role of microbiome in the development, prevention, and control of diseases.

3.1 Introduction

The study of disease dynamics has traditionally focused on pathogen–host interactions, providing valuable insights into disease mechanisms. However, this approach overlooks the critical role of the human microbiome—a vast, and intricate community of microorganisms that inhabit various parts of the body (Gilbert et al. 2018). Advances in microbiome research have demonstrated the pivotal role of these microbial ecosystems in maintaining physiological balance, regulating immune responses, and influencing metabolism. Dysbiosis, or the disruption of microbial communities, has been increasingly implicated in the pathogenesis of diverse diseases, including metabolic disorders, autoimmune diseases, cancer, and infectious diseases (Ramdas et al. 2020; Fulcher et al. 2022; Giri et al. 2022).

The advancement of multi-omics technologies, including genomics, transcriptomics, proteomics, metabolomics, and other specialized approaches, have revolutionized our ability to study these complex host–microbiome interactions (Mallik et al. 2022; Mishra et al. 2024). These high-throughput methods provide a comprehensive view of molecular networks underpinning health and disease (Uehara et al. 2024). For example, genomics identifies genetic susceptibilities; transcriptomics and proteomics reveal gene expression patterns and protein modifications; and metabolomics highlights dynamic biochemical changes, particularly the metabolic interplay between the host and its microbiota (Kuehne et al. 2017; Deek et al. 2024).

The integration of multi-omics data has been transformative in microbiome research, enabling a systems-level understanding of the intricate cross-talk between microbial communities and their host (Srivastava et al. 2024). This approach has facilitated the identification of novel disease biomarkers, critical microbial pathways, and molecular mechanisms that drive disease onset and progression (Chetty and Blekhman 2024). Moreover, advancement in single-cell and spatial omics provide cell-specific and localized insights into host–microbe interactions, enabling the study of tissue-specific processes that are otherwise difficult to capture (Yuan et al. 2024).

These advances hold profound implications for precision medicine, enabling therapeutic interventions tailored to an individual's microbiome composition and molecular profile. For example, gut dysbiosis has been linked to metabolic disorders, neurological conditions, and immune dysregulation. Multi-omics approaches have identified microbial signatures associated with these diseases, paving the way for personalized treatments targeting specific microbial pathways or restoring microbial balance. Similarly, studies integrating host genomics with microbial metabolomics have revealed how microbial metabolites influence host–gene regulation and immune responses, leading to innovative therapeutic strategies (Fig. 3.1) (Ahmed 2022).

The convergence of microbiome research and multi-omics technologies marks a paradigm shift in the study of disease dynamics. It not only advances our understanding of the biological complexity underlying health and disease but also provides powerful tools for predictive modeling and the development of targeted, effective treatments. The purpose of this chapter is to develop a theoretical framework for integrating multi-omics to deepen our understanding of microbiome composition and its influence on human health. We provide examples demonstrating how these insights improve our understanding of health and discuss the potential for microbiome-based interventions to mitigate diseases. Finally, we outline the future directions to advance this field.

3.2 Characterizing the Human Microbiome

The human microbiome is a diverse and dynamic ecosystem of microorganisms that inhabit various parts of the human body, performing metabolic functions since millions of years. The first evidence of microbes as part of the normal human system emerged in the mid-1880s when Theodor Escherich observed a bacterium (now known as *Escherichia coli*) in the gut of both healthy and diarrhea-affected children (Ruiz and Silhavy 2022). Later studies revealed the presence of other microbes, such as *Veillonella parvula*, in multiple body sites, including the oral cavity, digestive system, urinary tract, and upper respiratory tract (Kennedy and Chang 2020).

In healthy adults, microbiomes are found in five major body regions: the airways, skin, oral cavity, gastrointestinal tract, and urinogenital tract (Proctor 2011). Among these, the largest microbiota population resides in the gut, where individuals host 10–100 trillion microbial cells, including bacteria, viruses, fungi, archaea, and

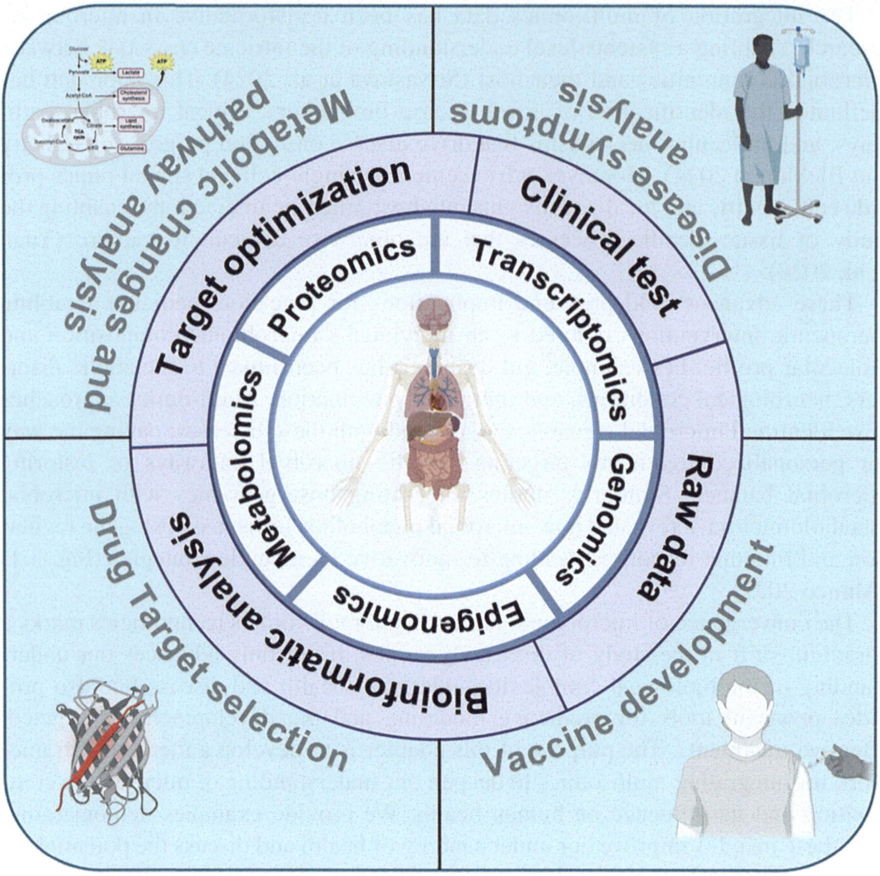

Fig. 3.1 Integrative role of multi-omics approaches in advancing our understanding of disease mechanisms and enhancing therapeutic approaches

protozoans (Proctor 2011). Initial estimation showed that these organisms alone outnumber human cells by a 10:1 ratio, but recent studies revised this to 1:1 (Sender et al. 2016). Despite these vast numbers, microbes account for only 2–6 pounds of body weight in a 200-pound adult (Sender et al. 2016). Far from being harmful, the microbiome is essential for health. Microbes synthesize vitamins that humans cannot produce, break down complex dietary compounds to extract nutrients, train the immune system to recognize harmful pathogens, and produce anti-inflammatory compounds that protect against infections (Mishra et al. 2024). These microorganisms form a symbiotic relationship with the host, playing crucial roles in development, nutrition, immune regulation, and pathogen resistance throughout life (Chiu et al. 2017; Panwar et al. 2021).

The European Metagenomics of the Human Intestinal Tract (Meta-HIT) consortium reported in 2010 that the human gut microbiome contains approximately 3.3 million nonredundant genes—vastly outnumbering the roughly 22,000 genes in the human genome (Zhang et al. 2022). This immense genetic diversity within the microbiome is striking. While humans share about 99.9% genetic similarity, the composition of an individual's gut microbiome can differ by as much as 80–90% from that of others. These findings suggest that, as technology and knowledge advance, tailoring medical treatments to target microbiome variations could potentially be more effective than focusing solely on the relatively uniform human genome. Below, we describe the efforts to categorize the composition and complex dynamics of the microbiota.

3.2.1 Historical Studies and Cultivation Challenges

The exploration of the human microbiome dates back to the 1680s when Antonie van Leeuwenhoek, often regarded as the father of microbiology, conducted pioneering studies. Using his self-crafted microscopes, he compared the bacteria present in his fecal and oral samples, meticulously documenting the differences between these sites in both healthy and diseased states (Berg et al. 2020). His observations laid the groundwork for microbiome research, but the tools of the time limited further exploration.

Early microbiome studies frequently emphasized *Escherichia coli* as a representative gut microbe. This focus was largely influenced by the cultivation techniques available at the time, which relied heavily on isolating and studying individual microbial species. These methods often overlooked many microbes, as many require specific, intricate environmental conditions for growth that could not be replicated in laboratory settings. This led to a significant gap in understanding the complexity and diversity of the human gut microbiome. The introduction of anaerobic cultivation techniques in the 1970s marked a turning point in microbiome research. These methods facilitated the growth of microbes that thrive in oxygen-free environments, leading to the identification of over 300 new bacterial species from the human gut (Hanisakova et al. 2022). Quantitative techniques, such as viable cell counting on selective media, provided deeper insights into microbial populations. For instance, one study analyzing stool samples from 141 Americans revealed *Bacteroides* and anaerobic cocci as the dominant genera, while *Clostridium* species were comparatively less prevalent. Other common, though less abundant, bacteria included members of the genera *Eubacterium*, *Lactobacillus*, *Bifidobacterium*, and *Streptococcus*, alongside facultative anaerobes like *Escherichia* (Lloyd-Price et al. 2016).

Despite these advancements, the limitations of traditional techniques became increasingly apparent. Selective media, while useful, struggled to differentiate closely related species, such as the high-abundance *Bacteroides* strains. Additionally, many microbial species remained uncultivable due to their reliance on complex symbiotic relationships or environmental factors that were difficult to replicate in vitro. These challenges underscored the necessity for innovative molecular approaches, such as DNA-sequencing-based techniques, to more effectively uncover the vast microbial diversity within the human gut.

The evolution of microbiome research—from the rudimentary observations of Leeuwenhoek to the breakthroughs in anaerobic cultivation—highlighted both the progress made and the significant gaps that remained, setting the stage for the genomic revolution in microbiome study.

3.2.2 Techniques to Study Microbiome Diversity

The advent of advanced DNA sequencing technologies has revolutionized microbiome research, giving rise to metagenomics—a powerful approach for studying entire microbial communities without requiring cultivation (Kanchan et al. 2020). Unlike traditional techniques that focus on the genome of a single bacterial strain, metagenomics investigates the combined genomes of microorganisms directly sampled from natural environments (Kim et al. 2024).

Early efforts using 16S RNA sequencing revealed that 85–90% of gut bacteria belonged to three groups: *Bacteroides*, *Clostridium* cluster IV, and *Clostridium* cluster XIVa. These studies showed that many sequences did not match any documented species, highlighting the presence of previously uncharacterized microbes (Yen and Johnson 2021).

The emergence of various sequencing techniques has significantly clarified the taxonomic composition of microbes. However, a substantial portion of functional diversity remains unexplored. Early studies revealed significant interpersonal variation, even among twins, while suggesting the existence of a shared set of microbial genes across all individuals. This led to the concept of a "core microbiome," defined by conserved functional traits rather than taxonomic composition (Neu et al. 2021).

Modern studies go beyond merely describing site-specific microbial differences. They employ cutting-edge molecular tools to investigate the underlying causes of microbial variations and their physiological implications, focusing on the roles microbial communities play in health and disease. One key approach is to assign unidentified sequences to databases such as KEGG (Kyoto Encyclopedia of Genes and Genomes), COG (Clusters of Orthologous Groups), MetaCyc, EggNOG, BioCyc, PATRIC (Pathosystems Resource Integration Center), and HMP (Human Microbiome Project) (Kanehisa and Goto 2000; Karp et al. 2002). These resources facilitate the identification of microbial interactions and networks, linking specific microbes or microbial functions to host phenotypes and advancing our understanding of the microbiome's role in health and disease.

3.3 Microbial Diversities Among Anatomical Sites

Microbial diversity not only varies across different body sites and between individuals but also undergoes significant changes throughout a person's life. Factors such as mode of delivery, lifestyle, health status, and genetic makeup influence the establishment and evolution of these microbial communities.

3.3.1 Inheritance of Microbiota

A fetus is generally considered sterile or nearly sterile until the rupture of the amniotic sac. After vaginal delivery, the newborn's microbiota closely resembles that of the mother's vaginal microbiome, predominantly comprising *Lactobacillus*. Similarly, breastfed infants have gut microbiota dominated by *Lactobacillus* and *Bifidobacteria*, supported by the presence of these bacteria in human milk (Reyman et al. 2019).

Modern lifestyles, however, can disrupt the maternal-to-infant transfer of microbiota. For instance, cesarean delivery significantly alters the initial microbial community. Infants born via cesarean harbor a distinct microbiota compared to those born vaginally, often requiring months to develop a stable microbiome. Notably, shortly after birth, the infant's microbiota remains largely undifferentiated across body sites, with the mode of delivery influencing which microbial populations colonize first (Hoang et al. 2021) (Fig. 3.2).

3.3.2 Postnatal Development and Microbiota Stabilization

An infant's microbiome evolves as it grows, becoming increasingly diverse and densely populated. For example, the development of teeth and dietary changes introduce complex food materials, fostering the colonization of oral bacteria such as *Streptococcus oralis*, *Streptococcus mutans*, *Streptococcus salivarius*, and *Actinomyces naeslundii*.

The gut microbiota undergoes three distinct developmental phases:

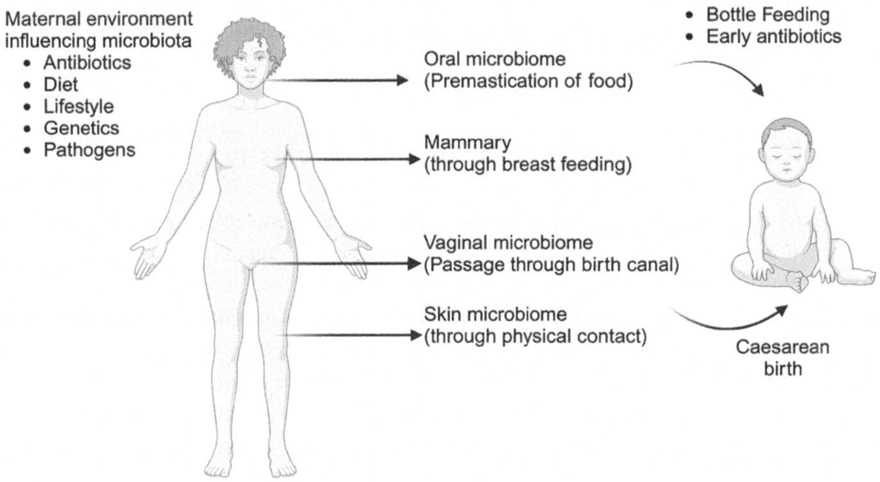

Fig. 3.2 Factors contributing to vertical transmission of microbiome in early life (mother to child)

1. Developmental Phase (3–14 months): Early microbial colonization.
2. Transitional Phase (15–30 months): Microbial diversity increases with dietary and environmental changes.
3. Stable Phase (31–46 months): A more resilient and balanced microbiota emerges.

These observations were drawn from a study analyzing over 12,500 fecal samples from 903 children across three European countries (Germany, Sweden, Finland) and three US states (Colorado, Georgia, Washington).

The skin microbiota also develops over time, influenced by factors such as hormonal changes during puberty. For example, *Lactobacillus*, *Prevotella*, and *Sneathia* are found on the skin of vaginally delivered newborns, while *Staphylococcus*, *Corynebacterium*, and *Cutibacterium* dominate in those delivered via cesarean section (Reyman et al. 2019; Hoang et al. 2021). Over time, *Staphylococcus* and *Streptococcus* populations decline, while *Actinobacteria* and *Proteobacteria* increase. Additionally, prepubertal children have a broader diversity of fungal species compared to adolescents and adults.

In infants up to 6 weeks old, the vaginal microbiota mainly includes *Lactobacilli*, *Streptococci*, and *Enterococci*. As children grow and approach puberty, the pH of the genitourinary tract remains neutral, supporting a slightly different microbial composition. However, after puberty, hormonal changes, particularly elevated estrogen levels, lower the pH, fostering an acidic environment dominated by *Lactobacilli*. This shift is critical for reproductive health but also makes women of reproductive age more susceptible to urogenital infections (Song et al. 2020).

Throughout life, the human microbiota remains dynamic, stabilizing during middle age. However, in old age, a decline in microbial diversity and functionality can lead to various health issues. This has led to the hypothesis that co-evolved microbiota play dual roles: supporting essential host functions during the active phases of life while potentially contributing to age-related decline and eventual host demise (Sun et al. 2023).

3.3.3 Gastrointestinal Tract Microbiome Diversity

The gastrointestinal tract is the most densely inhabited microbial habitat. Most of the gut microbiota studies make use of fecal samples because of the easy and non-invasive nature of collection. In some studies gut microbiota are studied using intestinal tissues and biopsy samples (Zreloff et al. 2023). The gut microbiome varies with in healthy individuals due to variability in diet, genetic makeup, lifestyle, and age (Sun et al. 2023). The healthy human gut microbiota is predominantly (>99%) composed of the phyla *Firmicutes*, *Bacteroidetes*, *Actinobacteria*, and *Proteobacteria*. Key bacterial families include *Lachnospiraceae*, *Ruminococcaceae*, *Prevotellaceae*, *Bacteroidaceae*, and *Rikenellaceae*. In the small intestine, *Lactobacillaceae* and *Enterobacteriaceae* are more prevalent (Zhang et al. 2022; Mishra et al. 2024).

Common bacterial species in a healthy gut include *Faecalibacterium prausnitzii*, *Roseburia intestinalis*, *Ruminococcus intestinalis*, *Blautia obeum*, *Dorea formicigenerans*, *Clostridium cluster XIVa*, *Eubacterium hallii*, *Prevotella copri*, *Bacteroides uniformis*, *Collinsella aerofaciens*, *Bifidobacterium bifidum*, and *Escherichia coli*. Additionally, the gut microbiome includes methanogenic archaea like *Methanobrevibacter smithii*, eukaryotic microbes such as *Saccharomyces* and *Candida*, and a diverse array of viruses, including bacteriophages. These diverse communities contribute to maintaining gut health (Fan and Pedersen 2021).

A study focusing on the Indian population revealed that the gut microbiome of vegetarians was predominantly composed of *Firmicutes* (62%), followed by *Bacteroidetes* (24%). However, in nonvegetarians, the proportion of *Bacteroidetes* increased significantly. A comparative analysis between Indian and Danish populations showed that the Danish cohort exhibited greater microbiota diversity than the Indian cohort. In both groups, *Firmicutes* was the most abundant phylum, followed by *Bacteroidetes*. However, *Actinobacteria*, *Proteobacteria*, and *Elusimicrobia* were more abundant in the Indian cohort, whereas *Tenericutes*, *Bacteroidetes*, *Verrucomicrobia*, and *Synergistetes* were prevalent in the Danish cohort (Dhakan et al. 2019). In the Japanese cohort, *Prevotella* and *Candida* were dominant at the genus level, while *Bacteroides* and *Saccharomyces* were the most abundant microbes overall. Similarly, a comparative study of African and European children highlighted distinct population-specific microbial signatures. Rural African populations displayed remarkable microbial diversity and richness compared to European populations. Dominant taxa in the African gut microbiome included *Prevotella*, *Treponema*, *Sutterella*, *Anaerovibrio*, members of the *Clostridiaceae* family, *Faecalibacterium*, and *Dialister*. Many of these microbes were enriched with carbohydrate-active enzyme (CAZyme) functions and exhibited fibrinolytic properties. Interestingly, *Bifidobacteria* and *Lactobacillus* species were rarely detected in the gut microbiota of African children (Gupta et al. 2017; Costea et al. 2018; Syromyatnikov et al. 2022).

3.3.4 Oral Microbiome

The human oral cavity, hosting over 700 bacterial species across 12 phyla and 185 genera, is the second most diverse microbial site in the body. Essential for digestion and oral-systemic health, the oral microbiota begins developing at birth, influenced by delivery mode and feeding practices. The microbiome includes bacteria, protozoa (*Entamoeba gingivalis*, *Trichomonas tenax*), fungi (primarily *Candida*), and viruses. Despite dense microbial colonization, acute oral infections are rare due to the balance between host immune defenses and microbial symbionts. However, immunocompromised individuals highlight the importance of this balance, as they are prone to severe viral, fungal, and nonoral microbial infections (Sampaio-Maia and Monteiro-Silva 2014; Guo et al. 2022).

3.3.5 Urinary Tract Microbiome

The vaginal microbiota is dynamic, influenced by hormonal fluctuations, life stages, and individual factors. Dominated by *Lactobacillus* species, it maintains an acidic environment through lactic acid production, providing stability and infection resistance. Ethnicity impacts microbiota composition: *Lactobacillus* species dominate in Asian and white American women, while anaerobic bacteria such as *Gardnerella* and *Prevotella* are more prevalent in Hispanic and black women. During pregnancy, microbial diversity decreases, with a higher prevalence of *Lactobacillus*, *Clostridiales*, and *Bacteroidales*. Vaginal microbiota varies with sexual activity, douching, stress, geography, and race. Research categorizes microbiomes into five community state types (CSTs), with different *Lactobacillus* species dominating CSTs I–V. CST IV, rich in anaerobic bacteria, is more common in black and Hispanic women and associated with higher Nugent scores (a Gram stain-based scoring system used to diagnose bacterial vaginosis (BV). Variations are likely driven more by host factors, such as immune responses and epithelial interactions, than by cultural or behavioral differences (Onywera et al. 2019; Song et al. 2020).

3.3.6 Microbiome on Skin

Skin, the largest human organ, serves as a physical barrier against disease and hosts a diverse microbiota crucial for pathogen defense, immune education, and nutrient metabolism. Microbial colonization is influenced by skin physiology, with sebaceous areas dominated by *Propionibacterium* species and humid areas by *Staphylococcus* and *Corynebacterium*. Skin microbiota is relatively stable over time but varies with factors like delivery method and hormonal changes, particularly during puberty. Prepubescent individuals have a more diverse microbial and fungal composition, while postpubescent microbiota favors lipophilic organisms like *Propionibacterium* and *Malassezia* (Schneider et al. 2023; Townsend and Kalan 2023).

Microbial interactions prevent harmful colonization through colonization resistance, though dysbiosis and imbalanced microbial communities can lead to conditions like acne (*Propionibacterium acnes*), eczema (*Staphylococcus aureus*), and chronic wounds (*Staphylococcus epidermidis*). Research shows that *S. epidermidis* enhances immune responses and protects against external pathogens in mice (Pastar et al. 2020). Scalp microbiota is shaped by host factors like sebum, moisture, and pH, with imbalances linked to dandruff, seborrheic dermatitis, and atopic dermatitis. Studies highlight population-specific variations in dominant taxa, with *Malassezia*, *Propionibacterium*, and *Staphylococcus* most prevalent (Ganju et al. 2016) (Table 3.1).

Advancements in our understanding of the human microbiome have highlighted the significant role microorganisms play in maintaining individual health and well-being. Numerous studies have shown that imbalances in microbiome composition can lead to the onset of autoimmune diseases and a weakened immune response,

Table 3.1 The microbiome diversity at various anatomical sites of the human body in healthy and diseased individuals

Body location	Microbiome in healthy individuals	Microbiome alteration in disease	Disease	References
Oral cavity	*Streptococcus, Gemella, Prevotella, Neisseria, Abiotrophia,* and *Granulicatella*	↑ *Porphyromonas, Tannerella, Prevotella, Filifactor*	Dental cavities, gingivitis, periodontitis, oral cancer	Sampaio-Maia and Monteiro-Silva (2014) and Guo et al. (2022)
Respiratory tract	*Staphylococcaceae, Peptoniphilaceae, Carnobacteriacea, Staphylococcus, Corynebacterium, Alloicoccus, Granulicatella,* and *Moraxella*	↑ *Streptococcus, Haemphillus, Moraxella, Proteobacteria, Escherichia, Roseateles,* and *Pseudomonas*	Bronchiolitis, rhinosinusitis, asthma and acute respiratory illness (ARI)	Teo et al. (2015), Wen et al. (2018), and Kang and Kang (2021)
Gastro-intestinal tract	*Akkermansia muciniphila, Arthrobacter, Bacillus, Bacteriodes, Bifidobacterium, Clostridium, Coprobacillus, Enterococcus, Eschrichia/Shigella, Eubacterium rectable, Faecalibacterium, Helicobacter pylori, Lactobacillus, Lactococcus, Lysinibacillus, prausnitzii, Prevotella, Psedomonas, Solibacillus, Stenotrophomonas, Streptococcus, Veillonella*	↑ *Fusobacterium, Gemella, Parvimonas, Prevotella,* and *Proteobacteria* ↓ *Bacteroides, Eubactium rectale, Faecalibacterium prausnitzii, Akkermania muciniphila,* and *Spirochaetes*	Gastroesophageal reflux disease (GERD), Barrett's esophagus, or esophageal carcinoma, appendicitis	Sinha et al. (2016) and Yeoh et al. (2021)
Skin	*Propionibacterium, Staphylococcus, Streptococcus,* and *Corynebacterium*	↑ *S. aureus, S. epidermidis, P. acnes, Proteobacteria* ↓ *Propionibacterium, Acinetobacter Cutibacterium, Corynebacterium,* and *Staphylococcus*	Psoriasis, atopic dermatitis, alopecia, and systemic lupus erythematosus	Ganju et al. (2016), Clausen et al. (2018), and Emmert et al. (2021)
Vagina	*Lactobacillus crispatus, L. iners, L. gasseri, L. jensenii, Bifidobacterium* Very low abundance of anaerobes, *Prevotell, Atopobium, Sneathia Gardnerella*	↑ *Sneathia, Atopobium, Gardnerella* ↓ *Lactobacilli*	Vulvovaginal infection (RVVI), HPV infection, bacterial vaginosis	Ilhan et al. (2019), Onywera et al. (2019), Chorna et al. (2020), and Liu et al. (2023)

increasing susceptibility to viral and fungal infections (Clausen et al. 2018). Disease progression within the human body is often multifactorial, driven by complex interactions among genetic, environmental, microbial, immunological, and other yet-to-be-identified factors. While individual studies offer valuable insights, their findings can sometimes be inconsistent or contradictory. Traditional research methods that rely on survey and administrative data are often limited by issues such as selection bias, confounding variables, misclassification, and other forms of bias. Moreover, these approaches frequently lack detailed and quantitative information on key variables like infection duration, antibiotic usage, and lifestyle factors. Multi-omics platforms provide a way to overcome many of these limitations. By integrating data across various omics layers, such as genomics, transcriptomics, proteomics, and metabolomics, these technologies offer a multidimensional perspective on disease pathways (Kanchan et al. 2024). They enable the identification of complex risk factors and interactions that drive disease progression, from initial exposure to advanced stages.

In this context, we explore the diverse applications of omics platforms and the opportunities they offer for understanding the molecular pathways involved in disease progression.

3.4 Omics in Disease Research

As medical technology advances rapidly, understanding the pathogenesis of diseases requires in-depth exploration. Omics technologies, including genomics, transcriptomics, proteomics, metabolomics, single-cell transcriptomics, single-cell multi-omics, and spatial transcriptomics, offer high-throughput methods to efficiently uncover detailed biological insights into human diseases (de Tayrac et al. 2009). For example, combination of multi-omics approaches has been used to profile primary tissues, assess biological aging of organs and systems, and uncover systems biology mechanisms in cardiovascular diseases (Sousa et al. 2024). Furthermore, environmental factors, genetic backgrounds, and gene regulation are integral to disease development, reinforcing the need for multi-omics research to build systems biology frameworks.

Recent advances in single-cell omics and spatial omics provide even finer resolution of molecular interactions within and between cells, offering insights into development, physiology, and pathology. For instance, integrating single-cell and spatial transcriptomics has clarified spatial immune-malignant cell networks in colorectal cancer (Liu et al. 2024a). Despite these advancements, systematic guidelines on conducting multi-omics research—covering experimental design, data integration, and analysis tools—remain scarce, posing challenges for new researchers in the field.

3.4.1 Genomics

Genomics focuses on analyzing entire genomes to identify, characterize, and quantify all genes, uncover their interrelationships, and understand their influence on the organism. It is one of the earliest and most widely applied omics fields in medicine, exemplified by the Human Genome Project. Core components of genomics include high-throughput DNA sequencing, sequence assembly, and genome annotation.

A key application of genomics is the genome-wide association study (GWAS), which identifies sequence variations across the human genome, such as single nucleotide polymorphisms (SNPs), and screens for disease-associated SNPs. Techniques used in genomics include genotyping arrays, third-generation sequencing for whole-genome sequencing, and exome sequencing. GWAS analyzes genetic variants across multiple genomes to identify genotype-phenotype associations (Witsch-Baumgartner et al. 2022).

3.4.2 Transcriptomics

Transcriptomics studies the expression of all RNAs in a given cell population, offering insights into molecular changes induced by environmental factors or pathogens. It encompasses the analysis of both protein-coding RNAs (mRNAs) and various noncoding RNAs, including long noncoding RNAs, microRNAs, small-interfering RNAs, small nuclear RNAs, piwi-interacting RNAs, enhancer RNAs, and circular RNAs. Noncoding RNAs have demonstrated significant associations with diseases such as diabetes, cancer, cardiovascular conditions, and central nervous system disorders. Given that mRNAs constitute only 1–4% of the overall transcriptome, exploring noncoding RNAs is a growing focus in disease research.

RNA sequencing (RNA-seq) is the primary technology used in transcriptomics, enabling the quantification and characterization of RNA transcripts even from minimal RNA samples. With advancements in technology, single-cell RNA sequencing (scRNA-seq) has emerged as a powerful tool to analyze transcripts from specific cell types, providing valuable insights into diseases such as cancer and Alzheimer's disease (Li and Wang 2021; Jovic et al. 2022).

3.4.3 Proteomics

Proteomics focuses on the comprehensive identification and quantification of all proteins within cells or tissues. Since gene transcription levels can be influenced by posttranscriptional modifications, RNA analysis often shows limited correlation with actual protein expression. Proteomics overcomes this limitation by directly quantifying protein expression, providing insights into changes driven by environmental factors or disease progression.

Proteins are analyzed using techniques such as mass spectrometry, affinity proteomics, protein chips, and reverse-phase protein microarrays. Mass spectrometry, widely applied in medical research, includes stable isotope labeling proteomics and label-free proteomics. Labeled proteomics notably reduces detection time and minimizes batch effects across samples. Protein–protein interactions are identified by combining immunoprecipitation with mass spectrometry, such as purifying target proteins with antibodies and detecting interacting partners (Sutandy et al. 2013).

Posttranslational modifications (PTMs)—such as phosphorylation, glycosylation, ubiquitination, acetylation, and nitrosylation—play critical roles in intracellular signaling, protein transport, and enzymatic activity. Analyzing modified proteins provides deeper insights into disease progression. For instance, phosphoproteomics has revealed novel mechanisms underlying type 2 diabetes, lung adenocarcinoma, nonalcoholic steatohepatitis, and Alzheimer's disease (Zhong et al. 2023).

3.4.4 Metabolomics

Metabolomics studies small molecule metabolites produced during cellular metabolism, including substrates, intermediates, and end products such as carbohydrates, fatty acids, and amino acids. Metabolite analysis captures real-time physiological dynamics, with abnormal metabolite levels or ratios often linked to disease. This field is instrumental in elucidating the mechanisms of disease progression (Yachida et al. 2019).

Metabolomics integrates well with other omics layers (e.g., genomics, transcriptomics, proteomics). For instance, mRNA levels can predict metabolite concentrations, gut microbiota correlate with amino acid levels in fibromyalgia patients, and creatine kinase expression aligns with acylcarnitine and acetyl-CoA levels in hypertrophic cardiomyopathy (Deek et al. 2024).

The Human Metabolome Database (HMDB) provides a free resource for detailed information on human metabolites, while MetaboAnalyst 5.0 offers a platform for metabolomics analysis. Analytical technologies include nuclear magnetic resonance (NMR) and mass spectrometry-based methods such as gas chromatography-mass spectrometry (GC-MS), liquid chromatography-mass spectrometry (LC-MS), and capillary electrophoresis-mass spectrometry (CE-MS) (Wishart et al. 2022).

3.4.5 Single-Cell Omics

The rapid advancement of single-cell sequencing technology has revolutionized medical research, enabling the study of cellular heterogeneity in genomics, transcriptomics, and epigenomics. Applications include identifying transcriptional responses linked to drug resistance in lung cancer, mutations associated with resistance in acute myeloid leukemia, and epigenetic patterns in breast cancer resistance. Single-cell proteomics has enabled detailed protein analysis at the single-cell level to study tumor heterogeneity, stem cell differentiation, and circulating tumor cells. The field is moving toward multi-omics, integrating data on mutations, mRNAs, and proteins for a comprehensive understanding of cellular processes. Tools like 10× Chromium and BD Rhapsody™ facilitate such analyses (Vandereyken et al. 2023).

3.4.6 Microbiomics

Microbiomics focuses on studying the community of microorganisms, including bacteria, archaea, protozoa, fungi, and viruses, that live symbiotically or pathologically on plants and animals. The human microbiome is unique to each individual and is influenced more by the environment than by genetics. This makes microbiomics effective for explaining clinical variations in phenotypes beyond genetic factors. The gut microbiome, in particular, is linked to conditions such as neurological disorders, renal failure, and cancer (Kumar 2021).

3.5 Advanced Multi-Omics Approaches

Advancements in multi-omics technologies, such as next-generation sequencing (NGS) (Tsui et al. 2025) and shotgun metagenomics (Jovel et al. 2016), have revolutionized biomedical research by enabling the simultaneous analysis of diverse omics layers, including genomics, proteomics, transcriptomics, and metabolomics. This integration provides a holistic understanding of biological systems and disease mechanisms, facilitating the identification of genetic variants, gene expression patterns, and epigenetic modifications (Hasin et al. 2017; Martinez Arbas et al. 2021; Marabita et al. 2022).

Shotgun metagenomics, a powerful tool for characterizing entire microbial communities, involves fragmenting DNA into smaller pieces for sequencing and reassembling the genome using computational algorithms (Hunter et al. 2012; Sharpton 2014). This technique identifies diverse microbial species, including bacteria, viruses, and other nonbacterial microflora, along with their functional roles (Chen et al. 2017; Mitchell et al. 2018). By broadening the scope of microbiome research, shotgun metagenomics uncovers complex interactions within microbial communities and their environments (Greenman et al. 2024; Uehara et al. 2024). Integrating metagenomic data with host genomic information further enhances our understanding of how microbial communities influence host health and disease states (Hunter et al. 2012; Sharpton 2014; Chen et al. 2017; Mitchell et al. 2018).

Multi-omics integration provides a paradigm shift in research by linking data from different omics layers through computational pipelines, machine learning models, and mathematical frameworks (Jiang et al. 2019). Analytical approaches for multi-omics can be categorized into three main types: genome-first, which focuses on genetic loci involved in disease mechanisms; phenotype-first, which examines metabolic pathway dysregulation linked to disease prognosis; and environment-first, which investigates how environmental stimuli perturb genetic or signaling pathways (Hasin et al. 2017). The choice of categorization of various multi-omics integration approaches is invariably complex and many authors proposes their own classification (Chetty and Blekhman 2024). However, in this section we have tried to cover most of the integration approaches and presented a conceptual overview of each approach.

3.5.1 Dimensionality Reduction and Clustering Methods

Dimensionality reduction method mostly operates in the forefront of any omics analysis presenting an easy way of visualizing the dataset structure by reducing the dimensions of a set of variables while preserving the information contained in it. The most exploited method in this category is Principal Component Analysis (PCA), which presents a two-dimensional plot comparing two features under analysis establishing an association index based on high variance. Another method called Principal coordinate analysis relies on a similar approach of two-dimensional embedding of the datasets preserving the distances between the data points. Other strategies included in this category are canonical Co-inertia Analysis, t-SNE, Partial Least Square regression, UMAP, Canonical Correlation Analysis, and Isomap (Daniel et al. 2012; Hasin et al. 2017). Each strategy employs a different statistical approach but summarizes the data in a two-dimensional scale. The methods have been used at varied scales to analysis different kinds of datasets originating from SNP and gene expression or co-expression network, Mrna, and microRNA data, microbiome and metabolomics data and identification of biomarker association to specific cellular responses. The major limitation of this method is that the scale is limited to two datasets comparison which were individually grouped based on linear combinations (Meng et al. 2016b).

Clustering methods rely on clustering of datasets based on covariance analysis to identify significant associative patterns between the individual omics datasets. This strategy help in ruling out datasets in the further analysis that demonstrate weak associations allowing establishment of more complex relationship pattern between the multi-omics data (Rappoport and Shamir 2018). Cluster analysis approaches have been widely used to identify microbiome composition of clinical relevance and their association pattern with the various disease states, identify cancer subtypes, gut enterotype classifications, protein abundance and co-expression patterns (Costea et al. 2018). The similarity network in a cluster analysis can be established based on approaches such as Bray–Curtis dissimilarity method, Kullback–Leibler divergence, Manhattan distance, or Euclidean distance. Several network analysis approaches integrate the clustering methods for complex pattern analysis. Included among those are iCluster, PARADIGM, Patient-specific Data Fusion (PSDF), Similarity network fusion (SNF), and LRACluster (Vaske et al. 2010; Labory et al. 2020).

3.5.2 Correlation-Based Methods

Correlation-based analysis approach a widely used statistical tool packet to construct complex biological relationship network based on the quantifiable measures to compute pairwise association between two independent covariates (Heintz-Buschart et al. 2016). The quantitative computational tool used in mapping the correlation network encompasses Pearson's correlation coefficient, Kendall's tau Spearman's rank coefficient, and SparCC (Liu et al. 2016). Unlike Kendall's tau and Spearman's rank coefficient, Pearson's correlation cannot identify association between nonlinear and statistically significant datasets and thus is used in conjunction with the earlier ones for compositional data analysis. However, all of them tend to introduce spurious associations in the data processing. Modified methods such as SparCC and centered-log-ratio covariate analysis were introduced to counter the spurious associations occurring during compositional microbiome analysis. Canonical Correlation Analysis is another method in this category that sorts paired datasets to identify highly associative features. False discovery rate (FDR) or Familywise error rate (FWER), a thresholding procedure for statistically related p values to construct a biological network of more meaningful datasets with high statistically robust correlation. Weighted Gene Co-expression Network Analysis (WGCNA) was a modified marginal correlation analysis approach that considers the weighted edge in the biological networks which was missing in earlier thresholding approaches (Langfelder and Horvath 2008). Other types weighted network approaches include Network modules, Scale-free topology, Node connectivity, and Clustering coefficient (Amano et al. 2018). These weighted network analysis methods provide an edge over unweighted network with a more meaningful interpretation driven from connection strengths. Correlation based approaches have been actively used to analyze biological datasets for miRNA-target predictions, oncogene expression in cancer progression and metastasis, microarray data, microbiome data, and host–microbe association (Jiang et al. 2019).

3.5.3 Regression-Based Methods

Regression-based network analyzes data where the associative relationship of one set of variables is predicted from the other set. This method of analysis can be applied to datasets with direct relationship and one of the datasets (response) is invariably dependent on the other (predictor). Regression analysis approaches are superior to above discussed data integration methods due to its ability to incorporate significant covariate information in the biological network. Two modalities of operation, namely, logistic regression, and linear regression are the most widely used ones in this category (Jiang et al. 2019). The regression analysis can operate on two frameworks; generalized and reduced/regularized. The generalized regression model is limited by the inability to perform variable data selection for computed analysis which is overcome by a more exhaustive regularized regression framework. Regularized regression simultaneously performs parameter estimation and variable selection in datasets with multiple predictors enabling data reduction to small and interpretable covariates in a single model (Wu et al. 2019). The most widely used regularized regression approaches are Lasso, Elastic net regularization, Ridge and Linear discriminant analysis. Multiple frameworks such as PLINK, HOMINID, PLATO, BOOST, epiGPU, interSNP, and PLATO are available that integrate and analyze multi-omics data from biological networks using regression-based models (Fang et al. 2015; Yan et al. 2018). Datasets from multiple single omics layers can be integrated and computed to establish significant association and simplified interpretation expanding our knowledge about disease pathology. Regression analysis has been implemented in many cases to identify gene regulatory network, bimolecular interaction pattern, DNA methylation pattern, microbiome composition to disease progression, and microbial metabolome (Li et al. 2019).

3.5.4 Fusion Methods

Fusion methods analyze post preprocessing of the data from dimensionality reduction and combines to generate a single dataset while preserving the property of individual dataset. The pattern analysis is based on computed alignment and bias correction of the individual dataset. The method is more specifically used in the analysis of genomics and transcriptomics data. Most popularly used methods in this category include Pattern Fusion Analysis (PFA), Semidefinite Programming/Support Vector Machine (SDP/SVM), Multiple co-inertia analysis (MCIA), Penalized Multivariate Analysis (PMA), and non-negative matrix factorization (NMF) (Meng et al. 2014; Shi et al. 2017; Labory et al. 2020). These methods are applicable in gene and microRNA expression analysis and cancer subtype identification but cannot provide information on biomarker identification and tumor progression mechanism (Meng et al. 2014; Labory et al. 2020).

3.5.5 Modeling Method

Gaussian distribution-based models are extensively used in bivariate and multivariate network analysis of multilayer omics data. The Gaussian graphical models analyze the data network based on the calculation of inverse covariance matrix framework but remain largely inapplicable for multidimensional variables. To overcome the limitation of multidimensional analysis, Gaussian based graphical models have incorporated two statistical strategies: Graphical lasso and neighborhood selection method (Drton and Maathuis 2017). Extended version of the graphical model is the conditional graphical model that involves additional steps of data conditioning to rule out external factor-dependent spurious discrepancy in the analysis. The conditioning of the datasets is dependent on the existing information available for each individual dataset to construct a defined biological network. This method has been widely used for biomarker analysis, gene expression and regulatory network identification (Gao and Cui 2015; Jiang et al. 2019).

3.5.6 Bayesian Networks

Dynamic Bayesian network (DBN) is a type of graphical model popularly used for data integration and network analysis. It is a probability-based model that categorizes variables with respect to their joint probability distribution and conditional patterning framework (Ruiz-Perez et al. 2021). This type of modeling-based approach has been widely used in genomic and microbiome compositional studies. Network analysis tools such as CONEXIC, BNP, QTLnet, and PALM integrate dynamic Bayesian network algorithms for multi-omics data analysis (Ruiz-Perez et al. 2021). DBN integrated network framework has been exploited to understand AD prognosis, host–microbiome interaction, and genomic interaction. DBN is a competitively advantageous analytical tool due to its probabilistic model approach, existing knowledge of the relationship among the variables, allows parallel analysis of dataset employing different statistical tools, and can compute disparate data features (Jiang et al. 2019). However, integration of a large number of variables limits the applicability of DBN due to high computational cost, data complexity, and interpretability.

3.5.7 Spatial Multi-Omics

Spatial multi-omics has emerged as a transformative tool advancing the analysis of single cell sequencing data filling the missing gaps of spatial information about cellular heterogeneity, cellular location in a tissue system, cellular differentiation, and quantitative gene expression pattern (Liu et al. 2024b). Computational advancements in spatial analysis have led to development of pipeline models to effectively integrate the multilayers of omics data in microbiome and metabolome analysis and establish the mechanism of disease pathology. It was recognized and listed as a revolutionary transformative technology by Nature methods in 2022 (Vandereyken et al. 2023). Various spatial multi-omics analysis pipelines at various resolution range have been developed to uncover more detailed insights into the cellular organization and differentiation at a single-cell level. Included among these tools are MERFISH, HiPR-FISH, seq-FISH, OligoFISSEQ, DBiT-seq, Visium, GeoMx, and DSP. More recently, many more analysis tools had been added to the list that includes ExSeq, Spatial-CITE-seq, CUT&Tag-RNA-seq, SPOTS, SM-OMICS, and GeoMX DSP and has been continuously updated and modified for more efficient and detailed cellular analysis (Vandereyken et al. 2023).

These methods collectively address the complexities of multi-omics data, enabling a comprehensive understanding of biological systems and advancing precision medicine efforts (Table 3.2).

Table 3.2 Various multi-omics tools for data integration and analysis

Omics data integration tools	Category	Method	Integrated omic layers	Platform	References
PARADIGM	Unsupervised	Bayesian network	Metagenomics, Metatransciptomics, Metaproteomics	R	Vaske et al. (2010)
iCluster	Unsupervised	Matrix factorization	Metagenomics, Metatransciptomics	R	Pollack et al. (2002)
iClusterPlus	Unsupervised	Matrix factorization	Metagenomics, Metatransciptomics	R	Mo et al. (2013)
iClusterBayes	Unsupervised	Bayesian network	Metagenomics, Metatransciptomics	R	Mo et al. (2018)
LRAcluster	Unsupervised	Matrix factorization	Metagenomics, Metatransciptomics, Metaproteomics	R	Wu et al. (2015)
PSDF	Unsupervised	Bayesian network	Metagenomics, Metatransciptomics, Metaproteomics	MATLAB	Yuan et al. (2011)
BCC	Unsupervised	Bayesian network	Metagenomics, Metatransciptomics, Metaproteomics	R	Lock and Dunson (2013)
MDI	Unsupervised	Bayesian network	Metagenomics, Metatransciptomics	MATLAB	Kirk et al. (2012)
SNF	Unsupervised	Network-based	Metagenomics, Metatransciptomics	MATLAB/R	Wang et al. (2014)
PFA	Unsupervised	Fusion-based	Metagenomics, Metatransciptomics	MATLAB	Shi et al. (2017)
CCA	Unsupervised	Correlation based	Metagenomics, Metatransciptomics	R	Moon et al. (2022)
CNAmet	Unsupervised	Correlation based	Metagenomics, Metatransciptomics	R	Louhimo and Hautaniemi (2011)
PINSPlus	Unsupervised	Similarity-based	Metagenomics, Metatransciptomics, Metaproteomics	R	Nguyen et al. (2019)
NEMO	Unsupervised	Similarity-based	Metagenomics, Metatransciptomics	R	Rappoport and Shamir (2019)
moCluster	Unsupervised	Matrix factorization	Metagenomics, Metatransciptomics, Metaproteomics	R	Meng et al. (2016a)
JIVE	Unsupervised	Matrix factorization	Metagenomics, Metatransciptomics	MATLAB	Lock et al. (2013)
MFA	Unsupervised	Matrix factorization	Metagenomics, Metatransciptomics, Metaproteomics	PYTHON/R	de Tayrac et al. (2009)

rMKL-LPP	Unsupervised	Similarity-based	Metagenomics, Metatransciptomics, Metaproteomics	WEB	Speicher and Pfeifer (2015)
iNMF	Unsupervised	Matrix factorization	Metagenomics, Metatransciptomics, Metaproteomics	R	Yang and Michailidis (2016)
MOFA	Unsupervised	Bayesian network	Metagenomics, Metatransciptomics, Metaproteomics, Metabolomics	PYTHON/R	Argelaguet et al. (2018)
NetICS	Unsupervised	Network-based	Metagenomics, Metatransciptomics, Metaproteomics	MATLAB	Dimitrakopoulos et al. (2018)
Joint Bayesian factor	Unsupervised	Bayesian network	Metagenomics, Metatransciptomics, Metaproteomics	R	Ray et al. (2014)
MCIA	Unsupervised	Matrix factorization	Metagenomics, Metatransciptomics, Metaproteomics	PYTHON/R	Meng et al. (2014)
CONEXIC	Unsupervised	Network-based/ Bayesian	Metagenomics, Metatransciptomics, Metaproteomics	R	Akavia et al. (2010)
IntegrOmics	Unsupervised	Regression-based	Metagenomics, Metatransciptomics, Metaproteomics	R	Le Cao et al. (2009)
iPAC	Unsupervised	Multi-step analysis	Metagenomics, Metatransciptomics, Metaproteomics	Python	Aure et al. (2013)
MOO	Unsupervised	Network-based	Metagenomics, Metatransciptomics	R	Mosca and Milanesi (2013)
Multiplex	Unsupervised	Network-based	Metagenomics, Metatransciptomics	Python	Menichetti et al. (2014)
NuChart	Unsupervised	Network-based	Metagenomics, Metatransciptomics	R	Merelli et al. (2013)
SteinerNet	Unsupervised	Network-based	Metatransciptomics, Metaproteomics, Metabolomics	WEB	Tuncbag et al. (2012)
COALESCE	Unsupervised	Bayesian network	Metagenomics, Metatransciptomics, Metaproteomics	R	Huttenhower et al. (2009)
TMD	Unsupervised	Bayesian network	Metagenomics, Metatransciptomics	MATLAB	Savage et al. (2010)
Kernel Fusion	Unsupervised	Network-based	Metagenomics, Metatransciptomics	MATLAB	De Bie et al. (2007)
Endeavour	Unsupervised	Network-based	Metagenomics, Metatransciptomics	WEB	Aerts et al. (2006)
CAMELOT	Unsupervised	Regression-based	Metagenomics, Metatransciptomics, Metaproteomics, Metabolomics	Python	Chen et al. (2009)

(continued)

Table 3.2 (continued)

Omics data integration tools	Category	Method	Integrated omic layers	Platform	References
stSVM	Unsupervised	Network-based	Metagenomics, Metatransciptomics	R	Cun and Frohlich (2014)
3Omics	Unsupervised	Matrix factorization	Metatransciptomics, Metaproteomics, Metabolomics	WEB	Kuo et al. (2013)
IMP	Unsupervised	Network-based	Metagenomics, Metatransciptomics, Metaproteomics, Metabolomics	WEB	Narayanasamy et al. (2016)
PALM	Unsupervised	Bayesian network	Metagenomics, Metatransciptomics, Metabolomics		Ruiz-Perez et al. (2021)
WGCNA	Unsupervised	Correlation based	Metagenomics, Metatransciptomics	R	Langfelder and Horvath (2008)
MCD	Unsupervised	Multi-step analysis	Metagenomics, Metatransciptomics, Metaproteomics	Python	Chari et al. (2010)
iBAG	Supervised	Multi-step analysis	Metagenomics, Metatransciptomics	R	Wang et al. (2013)
FSMKL	Supervised	Multivariate analysis	Metagenomics, Metatransciptomics, Metaproteomics	MATLAB	Seoane et al. (2014)
iOmicsPASS	Supervised	Network-based	Metagenomics, Metatransciptomics, Metaproteomics	R	Koh et al. (2019)
sMBPLS	Supervised	Multivariate analysis	Metagenomics, Metatransciptomics, Metaproteomics, Metabolomics	MATLAB	Li et al. (2012)
T-SVD	Supervised	Multivariate analysis	Metagenomics, Metatransciptomics, Metaproteomics	R	Ma et al. (2014)
iGC	Supervised	Multi-step analysis	Metagenomics, Metatransciptomics	R	Lai et al. (2017)
SVM/SDP	Supervised	Similarity-based	Metagenomics, Metatransciptomics	R	Lanckriet et al. (2004)
OncoIMPACT	Supervised	Multi-step analysis	Metagenomics, Metatransciptomics	R	Bertrand et al. (2015)
PaintOmics	Supervised	Multivariate analysis	Metatransciptomics, Metaproteomics, Metabolomics	WEB	Garcia-Alcalde et al. (2011)
MONGKIE	Supervised	Multivariate analysis	Metagenomics, Metatransciptomics, Metaproteomics, Metabolomics	Python	Jang et al. (2016)

PMA	Unsupervised/ Supervised	Correlation based	Metagenomics, Metatransciptomics	R	Parkhomenko et al. (2009)
mixOmics	Unsupervised/ Supervised	Matrix factorization	Metagenomics, Metatransciptomics, Metabolomics	R	Rohart et al. (2017)
MiBiOmics	Unsupervised/ Supervised	Matrix factorization	Metagenomics, Metatransciptomics, Metaproteomics	WEB	Zoppi et al. (2021)
Joint NMF	Semisupervised	Matrix factorization	Metagenomics, Metatransciptomics, Metaproteomics	Python	Zhang et al. (2012)
FALDA	NA	NA	Metagenomics, Metatransciptomics, Metaproteomics	Python	
PhenoLink	NA	NA	Metagenomics, Metatransciptomics, Metabolomics	WEB	Bayjanov et al. (2012)
gNOMO	NA	NA	Metagenomics, Metatransciptomics, Metaproteomics	R	Munoz-Benavent et al. (2020)
COMBI	NA	NA	Metagenomics, Metatransciptomics, Metaproteomics, Metabolomics	R	Munoz-Benavent et al. (2020)
MapMan	NA	NA	Metatransciptomics, Metabolomics	R	Thimm et al. (2004)

PARADIGM Pathway Recognition Algorithm using Data Integration on Genomic Models, *LRAcluster* Low rank approximation-based clustering, *PSDF* Patient-Specific Data Fusion, *BCC* Bayesian consensus clustering, *MDI* Multiple dataset integration, *SNF* Similarity network fusion, *PFA* Pattern fusion analysis, *CCA* Canonical correlation analysis, *PINSPlus* Perturbation clustering for data INtegration and disease Subtyping, *NEMO* Neighborhood based multi-omics clustering, *JIVE* Joint and individual variation explained, *MFA* Multiple factor analysis, *rMKL-LPP* Regularized multiple kernel learning-locality preserving projections, *iNMF* Integrative non-negative matrix factorization, *MOFA* Multi-omics factor analysis, *NetICS* Network-based integration of multi-omics data, *CNAmet* Copy number alteration and DNA methylation analysis, *MCIA* Multiple co-inertia analysis, *CONEXIC* Copy number and expression in cancer, *iPAC* In-trans Process Associated and cis-Correlated, *MCD* Multiple Concerted Disruption, *iBAG* Integrative Bayesian Analysis of Genomics data, *FSMKL* Feature selection multiple kernel learning, *sMBPLS* Sparse multi-block partial least squares, *T-SVD* Thresholding-Singular value decomposition, *iGC* Integrative gene expression and copy number alteration analysis, *SDP/SVM* Semidefinite Programming/Support Vector Machine, *PMA* Penalized multivariate analysis, *CAMELOT* CAusal Modelling with Expression Linkage for cOmplex Traits, *FALDA* Factor Analysis and linear discriminant analysis, *COALESCE* Combinatorial Algorithm for Expression and Sequence-based Cluster Extraction, *TMD* Transcriptional Modules Discovery, *MONGKIE* Modular Network Generation and Visualization with Knowledge Integration Environments, *MOO* Multi-objective optimization, *st-SVM* Network smoothed T-statistic SVMs, *IMP* Integrated meta-omic pipeline, *COMBI* Compositional omics model-based integration, *PALM* Pipeline for the analysis of longitudinal multi-omics data, *WGCNA* Weighted gene correlation network analysis

3.6 Microbiome-Multi-Omics Convergence

Integrating multi-omics data into microbiome research has provided valuable insights into the interactions between microbial communities and host biology. For example, dysbiosis, or an imbalance in gut microbiota, has been linked to the development of metabolic and inflammatory diseases (Lloyd-Price et al. 2019). Functional metagenomics can be used to discover novel host genomics and metabolomics and facilitates the precise identification of microbial signatures associated with disease states, opening new avenues for targeted enzymes that benefit industrial processes and potentially enhance gut health by promoting fiber digestion or mitigating dysbiosis. Combining microbiome analysis with interventions is particularly promising for conditions like Parkinson's disease, where gut microbiome alterations have been linked to disease progression and treatment efficacy (Kwon et al. 2024).

Moreover, specific microbial signatures have been identified in diseases such as Crohn's disease and breast cancer, highlighting the potential of microbiome-targeted therapies (Rivera-Pinto et al. 2018; Banerjee et al. 2022) (Fig. 3.3).

3.6.1 Multi-Omics and Metabolic Modeling

Microbiome metabolic modeling offers a sophisticated approach to understand the complex biochemical reactions within microbial ecosystems by constructing genome-scale metabolic models. These models enable predictions about microbial abundances, metabolite production and consumption rates, and inter-microbial and microbe-environment interactions (Ankrah et al. 2021). One notable resource, AGORA2 (Assembly of Gut Organisms through Reconstruction and Analysis, version 2), includes genome-scale metabolic reconstructions for 7302 strains (Heinken et al. 2023). Initially applied to identify a growth medium for *Bacteroides caccae*, AGORA has since been used to uncover metabolic differences between inflammatory bowel disease (IBD) patients and healthy individuals, providing valuable insights into disease-associated microbial metabolism (Heinken et al. 2021).

Integrating multi-omics data refines these models by incorporating experimental and environmental constrain (Heinken et al. 2021). For instance, IOMA (Integrative Omics-Metabolic Analysis) combines metabolic reconstructions with proteomics and metabolomics measurements to identify metabolic fluxes consistent with known pathways and experimental results, enhancing the accuracy of predictions (Yizhak et al. 2010).

The convergence of metabolomics and metagenomics has also shed light on the role of microbes in drug metabolism and host gene regulation. Gut microbes can metabolize pharmaceutical compounds, altering their pharmacokinetics and therapeutic effects, thus influencing individualized drug responses. Furthermore, microbial metabolites like short-chain fatty acids (SCFAs), such as butyrate, modulate host gene expression, supporting anti-inflammatory pathways and maintaining intestinal barrier integrity (O'Riordan et al. 2022; Chetty and Blekhman 2024).

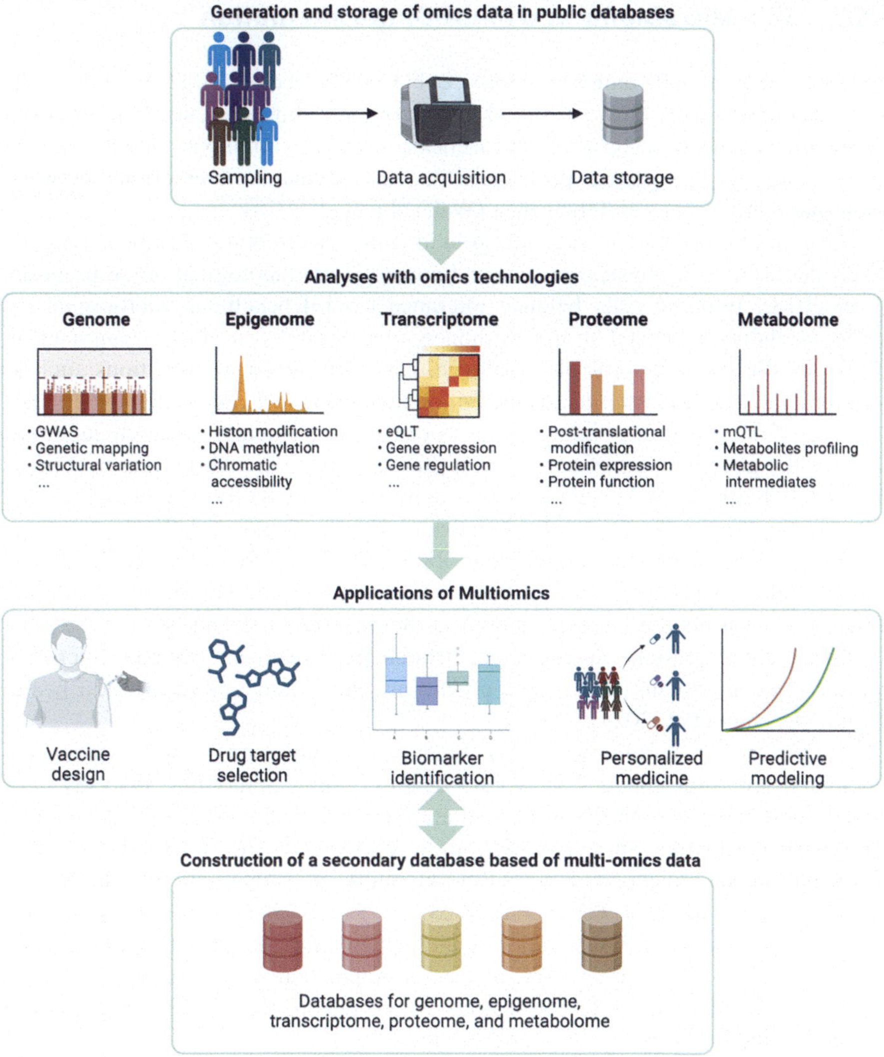

Fig. 3.3 General workflow for applying multi-omics in clinical settings. The integration of different omics data provides a broader picture of how complex biological system impacts an individual's health

These advancements, coupled with computational approaches like canonical correlation analysis (CCA) and Bayesian networks, provide a systems-level understanding of microbiome–host interactions. By standardizing methodologies and overcoming data limitations, multi-omics integration drives innovation in microbiome research, paving the way for precision medicine and personalized therapeutic strategies (Jiang et al. 2019; Chetty and Blekhman 2024; Kesheri et al. 2024).

3.6.2 Skin Microbiome and Multi-Omics Integration

Advancements in multi-omics technologies have enhanced our understanding of the intricate relationships between the skin microbiome, host genetics, and environmental factors. These integrative approaches unravel the complexity of the skin ecosystem, shaped by genetic predispositions and dynamic interactions between microbial communities and their host (Watanabe et al. 2018).

The skin microbiome plays a vital role in cutaneous immunity, emphasizing the skin's dual role as a physical barrier and an active immunological organ (Clausen et al. 2018). Immune cells balance tolerance toward beneficial microorganisms while defending against harmful pathogens and their by-products (Ganju et al. 2016). Dysbiosis, or microbial imbalance, has been linked to conditions such as psoriasis and atopic dermatitis. Multi-omics studies provide a systems-level understanding of microbial–host interactions. For instance, genomics and metagenomics uncover the genetic composition of microbial communities, while transcriptomics and metabolomics reveal changes in gene expression and microbial metabolites that influence immune responses and skin barrier integrity (Dessi et al. 2024).

Recent findings demonstrate the microbiome's role in disease-specific contexts. For example, microbiome profiling of melanoma patients revealed compositional differences that may serve as diagnostic biomarkers (Mekadim et al. 2022). Therapeutic interventions, such as near-infrared laser treatment, can alter microbial communities in patients with atopic conditions, potentially enhancing skin health (Munoz Declara et al. 2024).

Additionally, metabolomics and metagenomics studies on wound tissues have elucidated temporal changes in wound healing, while lipidomics analyses have linked microbial communities to lipid alterations in patients with atopic dermatitis. Environmental factors, such as pollution, have also been shown to influence skin pigmentation and molecular changes through multi-omics integration (Ashrafi et al. 2020; Emmert et al. 2021; Chen et al. 2023).

By integrating multi-omics data, researchers gain reproducible insights into the interplay between host, microbial, and environmental factors, advancing clinical applications in dermatology. These findings are paving the way for personalized treatments and innovative therapeutic strategies.

3.6.3 Cervicovaginal Microbiota Study

Application of multi-omics methodologies in studying the cervicovaginal microbiota has provided critical insights into the intricate relationships between microbial populations and host biological processes. By integrating data from genomics, transcriptomics, proteomics, and metabolomics, this approach enables a comprehensive analysis of the cervicovaginal environment and its role in reproductive health and disease. For example, lactic acid produced by *Lactobacillus* species has been shown to enhance cervicovaginal epithelial barrier integrity by upregulating tight junction protein expression, illustrating the metabolic contributions of specific microbial taxa to host health maintenance (Delgado-Diaz et al. 2022). Multi-omics approaches have also uncovered links between the cervicovaginal microbiota, human papillomavirus (HPV) infection, and inflammatory responses, identifying key metabolic signatures linked to cervical dysplasia and oncogenic mechanisms (Ilhan et al. 2019). These findings emphasize the potential of metabolomics and microbiome analysis in identifying biomarkers for predicting disease progression.

Advanced computational methods that integrate microbiome data with host genomic and metabolomics profiles have further elucidated mechanisms underlying host–microbiome interactions. This integration is particularly relevant for understanding the cervicovaginal microbiota's role in conditions such as bacterial vaginosis and HPV-associated diseases, paving the way for novel therapeutic strategies (Liu et al. 2023).

Alterations in the cervicovaginal microbiome are linked to conditions such as bacterial vaginosis and cervical intraepithelial neoplasia, underscoring the role of dysbiosis in the etiology of these diseases (Onywera et al. 2019). Moreover, studies combining cervicovaginal microbiota analysis with urine metabolomics have revealed potential associations with HPV infections, suggesting that shifts in microbial populations are closely tied to changes in host metabolism (Chorna et al. 2020).

3.6.4 Autoimmune Disorder and Microbiome

Autoimmune diseases (AIDs) occur when the immune system attacks the body's own tissues, affecting approximately 3–5% of the global population. Although the exact causes remain unclear, genetic factors, environmental influences (such as infections, diet, and drugs), and changes in the human microbiome are implicated. Microbial shifts can disrupt immune tolerance, potentially triggering autoimmune responses. Dysbiosis of the gut microbiota, characterized by reduced diversity and the overgrowth of *Proteobacteria* and *Enterobacteriaceae*, has been linked to inflammatory bowel diseases (IBD) like CD and ulcerative colitis (UC) (Lavelle and Sokol 2020; Gardiner et al. 2022; Fiocchi 2023; Haraoui and Blaser 2023). Multi-omics data can highlight microbial genes affecting host immunity or metabolism. CRISPR can target these genes to study their roles. Precise editing of microbial communities using CRISPR can restore balance in dysbiotic states, offering therapeutic potential for diseases like IBD or metabolic disorders (Mishra et al. 2022a).

Beyond the gut, the microbiome influences the brain through the "gut–brain axis," affecting behavioral and cognitive functions via microbial metabolites like short-chain fatty acids (SCFAs) and neurotransmitters such as GABA and serotonin. In multiple sclerosis (MS), dysbiosis includes reduced SCFA-producing bacteria and increased *Akkermansia* and *Methanobrevibacter*, with *Clostridium perfringens* toxins implicated in relapses. Similarly, type 1 diabetes mellitus (T1DM) has been linked to microbiome changes, including reduced butyrate-producing bacteria like *Intestinimonas* and an increase in *Blautia* (Qin et al. 2012; Pedersen et al. 2016; Fan and Pedersen 2021). These findings underscore the microbiome's role in autoimmune and inflammatory diseases and its potential as a target for therapeutic strategies.

3.6.5 COVID-19 and Multi-Omics

In early 2020, the SARS-CoV-2 virus began to spread globally, causing a devastating pandemic that resulted in the deaths of millions worldwide. Vaccination emerged as the most effective method to prevent infectious diseases, including COVID-19 (Mishra et al. 2021, 2022b). Developing vaccines requires a deep understanding of pathogen-specific markers and metabolites, as well as the ability of an antigen to stimulate a robust and long-lasting immune response. Recent advances in genomics, proteomics, and metabolomics have significantly contributed to identifying SARS-CoV-2-associated markers, enabling the development of vaccines and therapeutic interventions aimed at preventing the spread of the virus and reducing the severity of infections.

Researchers have leveraged these advancements by collecting plasma samples from SARS-CoV-2-infected individuals at various stages of COVID-19, alongside samples from healthy donors. Multi-omics approaches, including metabolomic profiling, proteomic analysis, and single-cell RNA sequencing for transcriptomics, were used to gain a comprehensive understanding of the disease. These studies identified viral RNA in multiple host cell types and revealed significant changes in the host cell transcriptome, offering insights into the molecular mechanisms driving the infection. Such data have been instrumental in understanding the interplay between the virus and host cells, as well as in identifying potential therapeutic targets.

Based on this enhanced understanding of COVID-19 pathogenesis and the application of multi-omics techniques, researchers designed and computationally tested over 80 potentially active compounds for their efficacy and safety. After rigorous computational and toxicity studies, six compounds—amsacrine, bosutinib, tretinoin, crizotinib, nidanib, and sunitinib—were finalized as promising candidates for further investigation. These compounds represent a significant step forward in antiviral drug development and highlight the power of multi-omics in accelerating drug discovery (Chen et al. 2021). In recent years, metallomics, a field focusing on the role of metals and metal-containing biomolecules in biological systems, has also become integrated with other omics technologies. This integration has further enriched our understanding of COVID-19 by uncovering the molecular processes and host responses involved in the virus's initiation, progression, and transmission. Multi-omics technologies, therefore, not only enhance our ability to respond to emerging infectious diseases but also pave the way for innovative, data-driven approaches to disease prevention and treatment.

3.6.6 The Impact of Gut Microbiome on Central Nervous System (CNS) Development

The gut microbiome profoundly influences CNS development and function through neural, endocrine, metabolic, and immunological mechanisms (Greenwood et al. 2017; Osadchiy et al. 2019; Mallik et al. 2022). Multi-omics approaches, such as metabolomics, transcriptomics, and proteomics, have been pivotal in uncovering these complex interactions. For instance, metabolomics identifies microbial metabolites like short-chain fatty acids (SCFAs) and tryptophan derivatives that regulate neurodevelopmental pathways, while transcriptomics and proteomics reveal gene expression changes and protein interactions critical for processes like synaptogenesis and myelination. Studies on germ-free animals demonstrate that postnatal microbiome colonization sets a functional baseline for the hypothalamic-pituitary-adrenal (HPA) axis, regulating stress responses (Mlynarska et al. 2022).

The microbiome exerts neural influence via the enteric nervous system (ENS) and vagal afferent nerves, which serve as communication conduits between the gut and CNS (Mayer et al. 2022). Microbial metabolites, such as short-chain fatty acids (SCFAs), and neuroactive molecules, including serotonin and gamma-aminobutyric acid (GABA), further mediate gut–brain signaling. Certain probiotics enhance excitability in colonic neurons, demonstrating a direct interaction between gut microbes and the ENS.

Immunologically, the microbiome modulates CNS immunity by shaping microglial maturation and systemic inflammation, as shown by transcriptomics and single-cell analyses (Hou et al. 2022; Mallik and Frank 2022; Mallik et al. 2022). Multi-omics studies indicate that microbiota regulates tryptophan metabolism, influencing serotonin and kynurenine pathways implicated in neurological conditions like Alzheimer's and Huntington's disease (Quigley 2017; Raut et al. 2017). Together, these pathways highlight the gut microbiome's integral role in CNS development, with implications for therapeutic interventions targeting neurodevelopmental and neurodegenerative disorders.

3.6.7 Cancer Microbiome

Emerging evidence highlights the profound association between microbiota and cancer, underscoring their role in tumor initiation and progression (Sepich-Poore et al. 2021; Singh et al. 2024). Studies utilizing germ-free (GF) and gnotobiotic animals—reared in aseptic environments and colonized with specific microbes—have provided critical insights into this relationship. Comparisons with conventionally raised animals have demonstrated microbiota's involvement in tumor induction and modulation of immune responses (Gopalakrishnan et al. 2018). Interestingly, these investigations revealed diverse outcomes, such as increased liver cancer prevalence, reduced small intestinal polyposis, and minimal changes in mammary tumor incidences in GF mice. Contradictory results from carcinogen-induced tumor studies in GF and conventional animals further suggest strain-specific, protocol-dependent, and organ-specific variations in carcinogenesis.

Microbiota in the oral cavity also play a pivotal role in cancer biology, as certain oral bacteria can translocate to other body sites, contributing to gut dysbiosis and gastrointestinal cancers. For instance, genomic analyses of colorectal adenocarcinoma tissues have shown enrichment of *Fusobacterium nucleatum* and other *Fusobacterium* species (Bashir et al. 2016). Meta-analyses have corroborated these findings, identifying elevated levels of oral microbiota such as *Peptostreptococcus stomatis*, *Parvimonas micra*, and *F. nucleatum* in colorectal cancer tissues. Consequently, oral microbiota analysis is being explored as a potential noninvasive biomarker for cancer diagnosis (Huh and Roh 2020).

The microbiota's role in colorectal cancer (CRC) has been categorized into three models: individual microbes, microbial populations, and interactions between specific microbes and the microbial community. Individual bacteria, such as *Bacteroides fragilis*, *Escherichia coli*, and *F. nucleatum*, contribute to CRC pathogenesis by inducing inflammation, causing DNA damage, or impairing repair mechanisms. On the other hand, altered gut microbiota can promote tumorigenesis by inducing pro-inflammatory pathways, as shown in studies where fecal transplants from CRC patients triggered immune responses and oncogenic pathways in experimental animals (Dzakula et al. 2024).

Microbial contributions to cancer can be direct or indirect. For example, *F. nucleatum* promotes colorectal tumor growth through Fusobacterium adhesin A (FadA) binding to E-cadherin, activating β-catenin signaling and cellular proliferation (Jiang et al. 2024). Indirectly, chronic immune activation by microbiota can sustain a pro-tumor environment. Recognizing the dual roles of microbiota in cancer—either as causative agents or accelerators of tumor growth—highlights the potential of early microbial detection and modulation as strategies for cancer prevention and treatment.

3.7 Conclusions and Future Perspectives

As multi-omics research continues to advance, particularly in the study of the cervicovaginal microbiota, several promising yet challenging approaches are emerging. The growing volume and complexity of multi-omics data demand sophisticated analytical tools capable of extracting meaningful insights from diverse, heterogeneous datasets. Technologies such as artificial intelligence (AI) and machine learning (ML) have demonstrated immense potential in integrating various omics layers—including genomics, transcriptomics, proteomics, and metabolomics—to unravel the intricate relationships between microbial communities and host biology (Khorraminezhad et al. 2020). These computational tools are increasingly recognized as essential for translating multi-omics data into actionable knowledge.

One of the major challenges in multi-omics research is the integration of diverse data types, which often differ in quality, resolution, and accuracy. Combining demographic and population-level information with multi-omics datasets can provide a more holistic understanding of disease. For instance, including variables such as age, ethnicity, and lifestyle factors alongside multi-omics data could refine insights into how the cervicovaginal microbiota influences health and disease (Gardiner et al. 2022). However, the lack of standardized methods for integrating such diverse datasets remains a significant barrier. The establishment of universally accepted protocols is crucial for improving data interoperability, enabling collaborative research efforts, and fostering a more unified approach to multi-omics analysis (Khorraminezhad et al. 2020).

AI's role in analyzing multi-omics datasets is rapidly expanding. Machine learning algorithms are increasingly used to classify microbial subgroups, predict disease outcomes, and identify potential biomarkers based on complex multi-omics profiles (Marcos-Zambrano et al. 2021; Gardiner et al. 2022). These capabilities are particularly valuable in the study of HPV infections, where integrating microbiome data with host gene expression and metabolic profiles has provided insights into disease progression and therapeutic opportunities (Ilhan et al. 2019). For example, ML models can identify key microbial taxa and metabolic pathways associated with cervical dysplasia, aiding in early diagnosis and targeted treatment strategies.

The integration of multi-omics data also holds significant promise for identifying biomarkers that facilitate early disease diagnosis and support personalized medicine. Specific metabolic signatures linked to the cervicovaginal microbiota have been associated with an increased risk of conditions such as cervical cancer, underscoring the potential of multi-omics approaches to uncover therapeutic targets and improve patient outcomes (Lloyd-Price et al. 2019). Beyond HPV-related conditions, these approaches may also be applied to other cervicovaginal diseases, such as bacterial vaginosis and pelvic inflammatory disease, further highlighting the breadth of their clinical utility.

Despite these advancements, significant challenges remain. Ensuring the reproducibility and generalizability of findings is critical, particularly as multi-omics research transitions into clinical applications. Variability in sample collection, processing, and analysis can lead to inconsistent results, emphasizing the need for rigorous quality control measures and robust experimental designs. The establishment of standardized methodologies is essential for validating results and ensuring reliability across studies.

Moreover, ethical considerations surrounding data sharing and patient privacy are increasingly important as multi-omics datasets grow in size and clinical relevance. Developing secure frameworks for data storage and sharing will be essential to address concerns about confidentiality and to build public trust in the application of multi-omics technologies (Fiocchi 2023). Balancing the need for open science with the protection of sensitive patient data will be critical for the future success of this field. As the field evolves, the integration of multi-omics research with cutting-edge computational tools promises to unlock new avenues for understanding and treating cervicovaginal diseases. By addressing current challenges and leveraging technological advancements, researchers can pave the way for transformative discoveries in reproductive health and precision medicine. Despite these advancements, challenges remain in ensuring the reproducibility and reliability of multi-omics studies. As highlighted by Fiocchi, the integration of multi-omics data requires robust quality control measures and standardized methodologies to ensure the validity of findings (Fiocchi 2023). Furthermore, ethical considerations surrounding data sharing and patient privacy must be addressed as multi-omics research becomes more prevalent in clinical settings.

3.8 Short Chapter Summary

This chapter delves into the integration of multi-omics methodologies to unravel complex interactions between microbial communities, host genetics, and environmental factors, emphasizing their significance in health and disease. This chapter underscores the transformative potential of multi-omics technologies in microbiome research. The methodologies discussed highlight how multi-omics tools are advancing clinical research and personalized medicine, offering innovative solutions to complex health challenges.

Author Contributions Conceptualization, Tarun Mishra, Pankaj Tiwari and Ahmad Reza Rezaei; methodology and investigation, Tarun Mishra, Pankaj Tiwari, and Ahmad Reza Rezaei; writing—original draft, Tarun Mishra, Pankaj Tiwari, Ahmad Reza Rezaei and Bhagaban Mallik; writing—review and editing, Tarun Mishra, Pankaj Tiwari, Swarna Kanchan, and Minu Kesheri.

References

Aerts S, Lambrechts D, Maity S, Van Loo P, Coessens B, De Smet F, Tranchevent LC, De Moor B, Marynen P, Hassan B, Carmeliet P, Moreau Y (2006) Gene prioritization through genomic data fusion. Nat Biotechnol 24(5):537–544. https://doi.org/10.1038/nbt1203

Ahmed Z (2022) Multi-omics strategies for personalized and predictive medicine: past, current, and future translational opportunities. Emerg Top Life Sci 6(2):215–225. https://doi.org/10.1042/ETLS20210244

Akavia UD, Litvin O, Kim J, Sanchez-Garcia F, Kotliar D, Causton HC, Pochanard P, Mozes E, Garraway LA, Pe'er D (2010) An integrated approach to uncover drivers of cancer. Cell 143(6):1005–1017. https://doi.org/10.1016/j.cell.2010.11.013

Amano SI, Ogawa KI, Miyake Y (2018) Node property of weighted networks considering connectability to nodes within two degrees of separation. Sci Rep 8(1):8464. https://doi.org/10.1038/s41598-018-26781-y

Ankrah NYD, Bernstein DB, Biggs M, Carey M, Engevik M, Garcia-Jimenez B, Lakshmanan M, Pacheco AR, Sulheim S, Medlock GL (2021) Enhancing microbiome research through genome-scale metabolic modeling. mSystems 6(6):e0059921. https://doi.org/10.1128/mSystems.00599-21

Argelaguet R, Velten B, Arnol D, Dietrich S, Zenz T, Marioni JC, Buettner F, Huber W, Stegle O (2018) Multi-omics factor analysis-a framework for unsupervised integration of multi-omics data sets. Mol Syst Biol 14(6):e8124. https://doi.org/10.15252/msb.20178124

Ashrafi M, Xu Y, Muhamadali H, White I, Wilkinson M, Hollywood K, Baguneid M, Goodacre R, Bayat A (2020) A microbiome and metabolomic signature of phases of cutaneous healing identified by profiling sequential acute wounds of human skin: An exploratory study. PLoS One 15(2):e0229545. https://doi.org/10.1371/journal.pone.0229545

Aure MR, Steinfeld I, Baumbusch LO, Liestol K, Lipson D, Nyberg S, Naume B, Sahlberg KK, Kristensen VN, Borresen-Dale AL, Lingjaerde OC, Yakhini Z (2013) Identifying in-trans process associated genes in breast cancer by integrated analysis of copy number and expression data. PLoS One 8(1):e53014. https://doi.org/10.1371/journal.pone.0053014

Banerjee S, Maiti TK, Roy RN (2022) Enzyme producing insect gut microbes: an unexplored biotechnological aspect. Crit Rev Biotechnol 42(3):384–402. https://doi.org/10.1080/07388551.2021.1942777

Bashir A, Miskeen AY, Hazari YM, Asrafuzzaman S, Fazili KM (2016) *Fusobacterium nucleatum*, inflammation, and immunity: the fire within human gut. Tumour Biol 37(3):2805–2810. https://doi.org/10.1007/s13277-015-4724-0

Bayjanov JR, Molenaar D, Tzeneva V, Siezen RJ, van Hijum SA (2012) PhenoLink--a web-tool for linking phenotype to ~omics data for bacteria: application to gene-trait matching for *Lactobacillus plantarum* strains. BMC Genomics 13:170. https://doi.org/10.1186/1471-2164-13-170

Berg G, Rybakova D, Fischer D, Cernava T, Verges MC, Charles T, Chen X, Cocolin L, Eversole K, Corral GH, Kazou M, Kinkel L, Lange L, Lima N, Loy A, Macklin JA, Maguin E, Mauchline T, McClure R, Mitter B, Ryan M, Sarand I, Smidt H, Schelkle B, Roume H, Kiran GS, Selvin J, Souza RSC, van Overbeek L, Singh BK, Wagner M, Walsh A, Sessitsch A, Schloter M (2020) Microbiome definition re-visited: old concepts and new challenges. Microbiome 8(1):103. https://doi.org/10.1186/s40168-020-00875-0

Bertrand D, Chng KR, Sherbaf FG, Kiesel A, Chia BK, Sia YY, Huang SK, Hoon DS, Liu ET, Hillmer A, Nagarajan N (2015) Patient-specific driver gene prediction and risk assessment through integrated network analysis of cancer omics profiles. Nucleic Acids Res 43(7):e44. https://doi.org/10.1093/nar/gku1393

Chari R, Coe BP, Vucic EA, Lockwood WW, Lam WL (2010) An integrative multi-dimensional genetic and epigenetic strategy to identify aberrant genes and pathways in cancer. BMC Syst Biol 4:67. https://doi.org/10.1186/1752-0509-4-67

Chen BJ, Causton HC, Mancenido D, Goddard NL, Perlstein EO, Pe'er D (2009) Harnessing gene expression to identify the genetic basis of drug resistance. Mol Syst Biol 5:310. https://doi.org/10.1038/msb.2009.69

Chen IA, Markowitz VM, Chu K, Palaniappan K, Szeto E, Pillay M, Ratner A, Huang J, Andersen E, Huntemann M, Varghese N, Hadjithomas M, Tennessen K, Nielsen T, Ivanova NN, Kyrpides NC (2017) IMG/M: integrated genome and metagenome comparative data analysis system. Nucleic Acids Res 45(D1):D507–D516. https://doi.org/10.1093/nar/gkw929

Chen T, Philip M, Le Cao KA, Tyagi S (2021) A multi-modal data harmonisation approach for discovery of COVID-19 drug targets. Brief Bioinform 22(6). https://doi.org/10.1093/bib/bbab185

Chen YE, Bousbaine D, Veinbachs A, Atabakhsh K, Dimas A, Yu VK, Zhao A, Enright NJ, Nagashima K, Belkaid Y, Fischbach MA (2023) Engineered skin bacteria induce antitumor T cell responses against melanoma. Science 380(6641):203–210. https://doi.org/10.1126/science.abp9563

Chetty A, Blekhman R (2024) Multi-omic approaches for host-microbiome data integration. Gut Microbes 16(1):2297860. https://doi.org/10.1080/19490976.2023.2297860

Chiu L, Bazin T, Truchetet ME, Schaeverbeke T, Delhaes L, Pradeu T (2017) Protective microbiota: from localized to long-reaching co-immunity. Front Immunol 8:1678. https://doi.org/10.3389/fimmu.2017.01678

Chorna N, Romaguera J, Godoy-Vitorino F (2020) Cervicovaginal microbiome and urine metabolome paired analysis reveals niche partitioning of the microbiota in patients with human papilloma virus infections. Metabolites 10(1). https://doi.org/10.3390/metabo10010036

Clausen ML, Agner T, Lilje B, Edslev SM, Johannesen TB, Andersen PS (2018) Association of disease severity with skin microbiome and filaggrin gene mutations in adult atopic dermatitis. JAMA Dermatol 154(3):293–300. https://doi.org/10.1001/jamadermatol.2017.5440

Costea PI, Hildebrand F, Arumugam M, Backhed F, Blaser MJ, Bushman FD, de Vos WM, Ehrlich SD, Fraser CM, Hattori M, Huttenhower C, Jeffery IB, Knights D, Lewis JD, Ley RE, Ochman H, O'Toole PW, Quince C, Relman DA, Shanahan F, Sunagawa S, Wang J, Weinstock GM, Wu GD, Zeller G, Zhao L, Raes J, Knight R, Bork P (2018) Enterotypes in the landscape of gut microbial community composition. Nat Microbiol 3(1):8–16. https://doi.org/10.1038/s41564-017-0072-8

Cun Y, Frohlich H (2014) netClass: an R-package for network based, integrative biomarker signature discovery. Bioinformatics 30(9):1325–1326. https://doi.org/10.1093/bioinformatics/btu025

Daniel E, Hüttenberger L, Hamann B (2012) A survey of dimension reduction methods for high-dimensional data analysis and visualization. OpenAccess Ser Inform 27:135–149. https://doi.org/10.4230/OASIcs.VLUDS.2011.135

De Bie T, Tranchevent LC, van Oeffelen LM, Moreau Y (2007) Kernel-based data fusion for gene prioritization. Bioinformatics 23(13):i125–i132. https://doi.org/10.1093/bioinformatics/btm187
de Tayrac M, Le S, Aubry M, Mosser J, Husson F (2009) Simultaneous analysis of distinct Omics data sets with integration of biological knowledge: multiple factor analysis approach. BMC Genomics 10:32. https://doi.org/10.1186/1471-2164-10-32
Deek RA, Ma S, Lewis J, Li H (2024) Statistical and computational methods for integrating microbiome, host genomics, and metabolomics data. Elife 13. https://doi.org/10.7554/eLife.88956
Delgado-Diaz DJ, Jesaveluk B, Hayward JA, Tyssen D, Alisoltani A, Potgieter M, Bell L, Ross E, Iranzadeh A, Allali I, Dabee S, Barnabas S, Gamieldien H, Blackburn JM, Mulder N, Smith SB, Edwards VL, Burgener AD, Bekker LG, Ravel J, Passmore JS, Masson L, Hearps AC, Tachedjian G (2022) Lactic acid from vaginal microbiota enhances cervicovaginal epithelial barrier integrity by promoting tight junction protein expression. Microbiome 10(1):141. https://doi.org/10.1186/s40168-022-01337-5
Dessi A, Pintus R, Fanos V, Bosco A (2024) Integrative multiomics approach to skin: the sinergy between individualised medicine and futuristic precision skin care? Meta 14(3). https://doi.org/10.3390/metabo14030157
Dhakan DB, Maji A, Sharma AK, Saxena R, Pulikkan J, Grace T, Gomez A, Scaria J, Amato KR, Sharma VK (2019) The unique composition of Indian gut microbiome, gene catalogue, and associated fecal metabolome deciphered using multi-omics approaches. Gigascience 8(3). https://doi.org/10.1093/gigascience/giz004
Dimitrakopoulos C, Hindupur SK, Hafliger L, Behr J, Montazeri H, Hall MN, Beerenwinkel N (2018) Network-based integration of multi-omics data for prioritizing cancer genes. Bioinformatics 34(14):2441–2448. https://doi.org/10.1093/bioinformatics/bty148
Drton M, Maathuis MH (2017) Structure learning in graphical modeling. Annu Rev Stat Appl 4:365–393
Dzakula A, Loncarek K, Vocanec D (2024) Palliative care – too complex to make it simple. Croat Med J 65(2):165–166. https://doi.org/10.3325/cmj.2024.65.165
Emmert H, Baurecht H, Thielking F, Stolzl D, Rodriguez E, Harder I, Proksch E, Weidinger S (2021) Stratum corneum lipidomics analysis reveals altered ceramide profile in atopic dermatitis patients across body sites with correlated changes in skin microbiome. Exp Dermatol 30(10):1398–1408. https://doi.org/10.1111/exd.14185
Fan Y, Pedersen O (2021) Gut microbiota in human metabolic health and disease. Nat Rev Microbiol 19(1):55–71. https://doi.org/10.1038/s41579-020-0433-9
Fang H, Huang C, Zhao H, Deng M (2015) CCLasso: correlation inference for compositional data through Lasso. Bioinformatics 31(19):3172–3180. https://doi.org/10.1093/bioinformatics/btv349
Fiocchi C (2023) Omics and multi-omics in IBD: no integration, no breakthroughs. Int J Mol Sci 24(19). https://doi.org/10.3390/ijms241914912
Fulcher JA, Li F, Tobin NH, Zabih S, Elliott J, Clark JL, D'Aquila R, Mustanski B, Kipke MD, Shoptaw S, Gorbach PM, Aldrovandi GM (2022) Gut dysbiosis and inflammatory blood markers precede HIV with limited changes after early seroconversion. EBioMedicine 84:104286. https://doi.org/10.1016/j.ebiom.2022.104286
Ganju P, Nagpal S, Mohammed MH, Nishal Kumar P, Pandey R, Natarajan VT, Mande SS, Gokhale RS (2016) Microbial community profiling shows dysbiosis in the lesional skin of Vitiligo subjects. Sci Rep 6:18761. https://doi.org/10.1038/srep18761
Gao B, Cui Y (2015) Learning directed acyclic graphical structures with genetical genomics data. Bioinformatics 31(24):3953–3960. https://doi.org/10.1093/bioinformatics/btv513
Garcia-Alcalde F, Garcia-Lopez F, Dopazo J, Conesa A (2011) Paintomics: a web based tool for the joint visualization of transcriptomics and metabolomics data. Bioinformatics 27(1):137–139. https://doi.org/10.1093/bioinformatics/btq594
Gardiner LJ, Carrieri AP, Bingham K, Macluskie G, Bunton D, McNeil M, Pyzer-Knapp EO (2022) Combining explainable machine learning, demographic and multi-omic data to inform precision medicine strategies for inflammatory bowel disease. PLoS One 17(2):e0263248. https://doi.org/10.1371/journal.pone.0263248

Gilbert JA, Blaser MJ, Caporaso JG, Jansson JK, Lynch SV, Knight R (2018) Current understanding of the human microbiome. Nat Med 24(4):392–400. https://doi.org/10.1038/nm.4517
Giri J, Basu M, Roy S, Mishra T, Jana K, Chande A, Ukil A (2022) Translationally controlled tumor protein-mediated stabilization of host antiapoptotic protein MCL-1 is critical for establishment of infection by intramacrophage parasite *Leishmania donovani*. J Immunol 208(11):2540–2548. https://doi.org/10.4049/jimmunol.2100748
Gopalakrishnan V, Helmink BA, Spencer CN, Reuben A, Wargo JA (2018) The influence of the gut microbiome on cancer, immunity, and cancer immunotherapy. Cancer Cell 33(4):570–580. https://doi.org/10.1016/j.ccell.2018.03.015
Greenman N, Hassouneh SA, Abdelli LS, Johnston C, Azarian T (2024) Improving bacterial metagenomic research through long-read sequencing. Microorganisms 12(5). https://doi.org/10.3390/microorganisms12050935
Greenwood JR, Finnegan EJ, Watanabe N, Trevaskis B, Swain SM (2017) New alleles of the wheat domestication gene Q reveal multiple roles in growth and reproductive development. Development 144(11):1959–1965. https://doi.org/10.1242/dev.146407
Guo H, Li J, Yao H, Liu Y, Ji Y, Zhang J, Zhao Y, Du M (2022) The dynamic communities of oral microbiome in neonates. Front Microbiol 13:1052525. https://doi.org/10.3389/fmicb.2022.1052525
Gupta VK, Paul S, Dutta C (2017) Geography, ethnicity or subsistence-specific variations in human microbiome composition and diversity. Front Microbiol 8:1162. https://doi.org/10.3389/fmicb.2017.01162
Hanisakova N, Vitezova M, Rittmann SKR (2022) The historical development of cultivation techniques for methanogens and other strict anaerobes and their application in modern microbiology. Microorganisms 10(2). https://doi.org/10.3390/microorganisms10020412
Haraoui LP, Blaser MJ (2023) The microbiome and infectious diseases. Clin Infect Dis 77(Suppl 6):S441–S446. https://doi.org/10.1093/cid/ciad577
Hasin Y, Seldin M, Lusis A (2017) Multi-omics approaches to disease. Genome Biol 18(1):83. https://doi.org/10.1186/s13059-017-1215-1
Heinken A, Hertel J, Thiele I (2021) Metabolic modelling reveals broad changes in gut microbial metabolism in inflammatory bowel disease patients with dysbiosis. NPJ Syst Biol Appl 7(1):19. https://doi.org/10.1038/s41540-021-00178-6
Heinken A, Hertel J, Acharya G, Ravcheev DA, Nyga M, Okpala OE, Hogan M, Magnusdottir S, Martinelli F, Nap B, Preciat G, Edirisinghe JN, Henry CS, Fleming RMT, Thiele I (2023) Genome-scale metabolic reconstruction of 7,302 human microorganisms for personalized medicine. Nat Biotechnol 41(9):1320–1331. https://doi.org/10.1038/s41587-022-01628-0
Heintz-Buschart A, May P, Laczny CC, Lebrun LA, Bellora C, Krishna A, Wampach L, Schneider JG, Hogan A, de Beaufort C, Wilmes P (2016) Integrated multi-omics of the human gut microbiome in a case study of familial type 1 diabetes. Nat Microbiol 2:16180. https://doi.org/10.1038/nmicrobiol.2016.180
Hoang DM, Levy EI, Vandenplas Y (2021) The impact of Caesarean section on the infant gut microbiome. Acta Paediatr 110(1):60–67. https://doi.org/10.1111/apa.15501
Hou K, Wu ZX, Chen XY, Wang JQ, Zhang D, Xiao C, Zhu D, Koya JB, Wei L, Li J, Chen ZS (2022) Microbiota in health and diseases. Signal Transduct Target Ther 7(1):135. https://doi.org/10.1038/s41392-022-00974-4
Huh JW, Roh TY (2020) Opportunistic detection of *Fusobacterium nucleatum* as a marker for the early gut microbial dysbiosis. BMC Microbiol 20(1):208. https://doi.org/10.1186/s12866-020-01887-4
Hunter CI, Mitchell A, Jones P, McAnulla C, Pesseat S, Scheremetjew M, Hunter S (2012) Metagenomic analysis: the challenge of the data bonanza. Brief Bioinform 13(6):743–746. https://doi.org/10.1093/bib/bbs020
Huttenhower C, Mutungu KT, Indik N, Yang W, Schroeder M, Forman JJ, Troyanskaya OG, Coller HA (2009) Detailing regulatory networks through large scale data integration. Bioinformatics 25(24):3267–3274. https://doi.org/10.1093/bioinformatics/btp588

Ilhan ZE, Laniewski P, Thomas N, Roe DJ, Chase DM, Herbst-Kralovetz MM (2019) Deciphering the complex interplay between microbiota, HPV, inflammation and cancer through cervicovaginal metabolic profiling. EBioMedicine 44:675–690. https://doi.org/10.1016/j.ebiom.2019.04.028

Jang Y, Yu N, Seo J, Kim S, Lee S (2016) MONGKIE: an integrated tool for network analysis and visualization for multi-omics data. Biol Direct 11(1):10. https://doi.org/10.1186/s13062-016-0112-y

Jiang D, Armour CR, Hu C, Mei M, Tian C, Sharpton TJ, Jiang Y (2019) Microbiome multi-omics network analysis: statistical considerations, limitations, and opportunities. Front Genet 10:995. https://doi.org/10.3389/fgene.2019.00995

Jiang Y, Huang Y, Hu Y, Yang Y, You F, Hu Q, Li X, Zhao Z (2024) Banxia xiexin decoction delays colitis-to-cancer transition by inhibiting E-cadherin/beta-catenin pathway via *Fusobacterium nucleatum* FadA. J Ethnopharmacol 328:117932. https://doi.org/10.1016/j.jep.2024.117932

Jovel J, Patterson J, Wang W, Hotte N, O'Keefe S, Mitchel T, Perry T, Kao D, Mason AL, Madsen KL, Wong GK (2016) Characterization of the gut microbiome using 16S or shotgun metagenomics. Front Microbiol 7:459. https://doi.org/10.3389/fmicb.2016.00459

Jovic D, Liang X, Zeng H, Lin L, Xu F, Luo Y (2022) Single-cell RNA sequencing technologies and applications: a brief overview. Clin Transl Med 12(3):e694. https://doi.org/10.1002/ctm2.694

Kanchan S, Sinha RP, Chaudière J, Kesheri M (2020) Computational metagenomics: current status and challenges. In: Katara P (ed) Recent trends in 'computational omics': concepts and methodology. Nova Science Publishers, USA, pp 371–395

Kanchan S, Kesheri M, Srivastava U, Katharia H, Ratna-Raj R, Chittoori B, Bogomolnaya L, Sinha RP, Denvir J (2024) Integrative omics data mining: challenges and opportunities. In: Gupta MK, Katara P, Mondal S, Singh RL (eds) Integrative omics: concepts, methodology, and application. Academic Press (Elsevier), United States, pp 237–255. https://doi.org/10.1016/B978-0-443-16092-9.00015-1

Kanehisa M, Goto S (2000) KEGG: Kyoto encyclopedia of genes and genomes. Nucleic Acids Res 28(1):27–30. https://doi.org/10.1093/nar/28.1.27

Kang HM, Kang JH (2021) Effects of nasopharyngeal microbiota in respiratory infections and allergies. Clin Exp Pediatr 64(11):543–551. https://doi.org/10.3345/cep.2020.01452

Karp PD, Riley M, Paley SM, Pellegrini-Toole A (2002) The MetaCyc database. Nucleic Acids Res 30(1):59–61. https://doi.org/10.1093/nar/30.1.59

Kennedy MS, Chang EB (2020) The microbiome: composition and locations. Prog Mol Biol Transl Sci 176:1–42. https://doi.org/10.1016/bs.pmbts.2020.08.013

Kesheri M, Kanchan S, Srivastava S, Ratna-Raj R, Chittoori B, Vaishampayan A, Rastogi RP, Sinha RP, Primerano D (2024) Ecology and environmental omics. In: Gupta MK, Katara P, Mondal S, Singh RL (eds) Integrative omics: concepts, methodology, and application. Academic Press (Elsevier), United States, pp 315–331

Khorraminezhad L, Leclercq M, Droit A, Bilodeau JF, Rudkowska I (2020) Statistical and machine-learning analyses in nutritional genomics studies. Nutrients 12(10). https://doi.org/10.3390/nu12103140

Kim N, Ma J, Kim W, Kim J, Belenky P, Lee I (2024) Genome-resolved metagenomics: a game changer for microbiome medicine. Exp Mol Med 56(7):1501–1512. https://doi.org/10.1038/s12276-024-01262-7

Kirk P, Griffin JE, Savage RS, Ghahramani Z, Wild DL (2012) Bayesian correlated clustering to integrate multiple datasets. Bioinformatics 28(24):3290–3297. https://doi.org/10.1093/bioinformatics/bts595

Koh HWL, Fermin D, Vogel C, Choi KP, Ewing RM, Choi H (2019) iOmicsPASS: network-based integration of multiomics data for predictive subnetwork discovery. NPJ Syst Biol Appl 5:22. https://doi.org/10.1038/s41540-019-0099-y

Kuehne A, Hildebrand J, Soehle J, Wenck H, Terstegen L, Gallinat S, Knott A, Winnefeld M, Zamboni N (2017) An integrative metabolomics and transcriptomics study to identify metabolic alterations in aged skin of humans in vivo. BMC Genomics 18(1):169. https://doi.org/10.1186/s12864-017-3547-3

Kumar PS (2021) Microbiomics: were we all wrong before? Periodontology 2000 85(1):8–11. https://doi.org/10.1111/prd.12373

Kuo TC, Tian TF, Tseng YJ (2013) 3Omics: a web-based systems biology tool for analysis, integration and visualization of human transcriptomic, proteomic and metabolomic data. BMC Syst Biol 7:64. https://doi.org/10.1186/1752-0509-7-64

Kwon D, Zhang K, Paul KC, Folle AD, Del Rosario I, Jacobs JP, Keener AM, Bronstein JM, Ritz B (2024) Diet and the gut microbiome in patients with Parkinson's disease. NPJ Parkinsons Dis 10(1):89. https://doi.org/10.1038/s41531-024-00681-7

Labory J, Fierville M, Ait-El-Mkadem S, Bannwarth S, Paquis-Flucklinger V, Bottini S (2020) Multi-omics approaches to improve mitochondrial disease diagnosis: challenges, advances, and perspectives. Front Mol Biosci 7:590842. https://doi.org/10.3389/fmolb.2020.590842

Lai YP, Wang LB, Wang WA, Lai LC, Tsai MH, Lu TP, Chuang EY (2017) iGC-an integrated analysis package of gene expression and copy number alteration. BMC Bioinf 18(1):35. https://doi.org/10.1186/s12859-016-1438-2

Lanckriet GR, De Bie T, Cristianini N, Jordan MI, Noble WS (2004) A statistical framework for genomic data fusion. Bioinformatics 20(16):2626–2635. https://doi.org/10.1093/bioinformatics/bth294

Langfelder P, Horvath S (2008) WGCNA: an R package for weighted correlation network analysis. BMC Bioinf 9:559. https://doi.org/10.1186/1471-2105-9-559

Lavelle A, Sokol H (2020) Gut microbiota-derived metabolites as key actors in inflammatory bowel disease. Nat Rev Gastroenterol Hepatol 17(4):223–237. https://doi.org/10.1038/s41575-019-0258-z

Le Cao KA, Gonzalez I, Dejean S (2009) integrOmics: an R package to unravel relationships between two omics datasets. Bioinformatics 25(21):2855–2856. https://doi.org/10.1093/bioinformatics/btp515

Li X, Wang CY (2021) From bulk, single-cell to spatial RNA sequencing. Int J Oral Sci 13(1):36. https://doi.org/10.1038/s41368-021-00146-0

Li W, Zhang S, Liu CC, Zhou XJ (2012) Identifying multi-layer gene regulatory modules from multi-dimensional genomic data. Bioinformatics 28(19):2458–2466. https://doi.org/10.1093/bioinformatics/bts476

Li H, Wang Y, Jiang J, Zhao H, Feng X, Zhao B, Wang L (2019) A novel human microbe-disease association prediction method based on the bidirectional weighted network. Front Microbiol 10:676. https://doi.org/10.3389/fmicb.2019.00676

Liu J, Tang W, Chen G, Lu Y, Feng C, Tu XM (2016) Correlation and agreement: overview and clarification of competing concepts and measures. Shanghai Arch Psychiatry 28(2):115–120. https://doi.org/10.11919/j.issn.1002-0829.216045

Liu Y, Li T, Guo R, Chen T, Wang S, Wu D, Li J, Liu Z, Zhao Y, Yin J, Qin J, Sun L, Chen W (2023) The vaginal microbiota among the different status of human papillomavirus infection and bacterial vaginosis. J Med Virol 95(3):e28595. https://doi.org/10.1002/jmv.28595

Liu C, Zou H, Ruan Y, Fang L, Wang B, Cui L, Wu T, Chen Z, Dang T, Lan Y, Zhao W, Zhang C, Meng H, Zhang Y (2024a) Multi-omics reveals the immunological features and the immune checkpoint blockade potential of colorectal medullary carcinoma. Clin Cancer Res 31:773. https://doi.org/10.1158/1078-0432.CCR-24-2505

Liu X, Peng T, Xu M, Lin S, Hu B, Chu T, Liu B, Xu Y, Ding W, Li L, Cao C, Wu P (2024b) Spatial multi-omics: deciphering technological landscape of integration of multi-omics and its applications. J Hematol Oncol 17(1):72. https://doi.org/10.1186/s13045-024-01596-9

Lloyd-Price J, Abu-Ali G, Huttenhower C (2016) The healthy human microbiome. Genome Med 8(1):51. https://doi.org/10.1186/s13073-016-0307-y

Lloyd-Price J, Arze C, Ananthakrishnan AN, Schirmer M, Avila-Pacheco J, Poon TW, Andrews E, Ajami NJ, Bonham KS, Brislawn CJ, Casero D, Courtney H, Gonzalez A, Graeber TG, Hall AB, Lake K, Landers CJ, Mallick H, Plichta DR, Prasad M, Rahnavard G, Sauk J, Shungin D, Vazquez-Baeza Y, White RA 3rd, Investigators I, Braun J, Denson LA, Jansson JK, Knight R,

Kugathasan S, McGovern DPB, Petrosino JF, Stappenbeck TS, Winter HS, Clish CB, Franzosa EA, Vlamakis H, Xavier RJ, Huttenhower C (2019) Multi-omics of the gut microbial ecosystem in inflammatory bowel diseases. Nature 569(7758):655–662. https://doi.org/10.1038/s41586-019-1237-9

Lock EF, Dunson DB (2013) Bayesian consensus clustering. Bioinformatics 29(20):2610–2616. https://doi.org/10.1093/bioinformatics/btt425

Lock EF, Hoadley KA, Marron JS, Nobel AB (2013) Joint and individual variation explained (Jive) for integrated analysis of multiple data types. Ann Appl Stat 7(1):523–542. https://doi.org/10.1214/12-AOAS597

Louhimo R, Hautaniemi S (2011) CNAmet: an R package for integrating copy number, methylation and expression data. Bioinformatics 27(6):887–888. https://doi.org/10.1093/bioinformatics/btr019

Ma X, Xiao L, Wong WH (2014) Learning regulatory programs by threshold SVD regression. Proc Natl Acad Sci USA 111(44):15675–15680. https://doi.org/10.1073/pnas.1417808111

Mallik B, Frank CA (2022) Roles for mitochondrial complex I subunits in regulating synaptic transmission and growth. Front Neurosci 16:846425. https://doi.org/10.3389/fnins.2022.846425

Mallik B, Bhat S, Kumar V (2022) Role of Bin-Amphiphysin-Rvs (BAR) domain proteins in mediating neuronal signaling and disease. Synapse 76(11–12):e22248. https://doi.org/10.1002/syn.22248

Marabita F, James T, Karhu A, Virtanen H, Kettunen K, Stenlund H, Boulund F, Hellstrom C, Neiman M, Mills R, Perheentupa T, Laivuori H, Helkkula P, Byrne M, Jokinen I, Honko H, Kallonen A, Ermes M, Simila H, Lindholm M, Widen E, Ripatti S, Perala-Heape M, Engstrand L, Nilsson P, Moritz T, Miettinen T, Sallinen R, Kallioniemi O (2022) Multiomics and digital monitoring during lifestyle changes reveal independent dimensions of human biology and health. Cell Syst 13(3):241–255 e247. https://doi.org/10.1016/j.cels.2021.11.001

Marcos-Zambrano LJ, Karaduzovic-Hadziabdic K, Loncar Turukalo T, Przymus P, Trajkovik V, Aasmets O, Berland M, Gruca A, Hasic J, Hron K, Klammsteiner T, Kolev M, Lahti L, Lopes MB, Moreno V, Naskinova I, Org E, Paciencia I, Papoutsoglou G, Shigdel R, Stres B, Vilne B, Yousef M, Zdravevski E, Tsamardinos I, Carrillo de Santa Pau E, Claesson MJ, Moreno-Indias I, Truu J (2021) Applications of machine learning in human microbiome studies: a review on feature selection, biomarker identification, disease prediction and treatment. Front Microbiol 12:634511. https://doi.org/10.3389/fmicb.2021.634511

Martinez Arbas S, Busi SB, Queiros P, de Nies L, Herold M, May P, Wilmes P, Muller EEL, Narayanasamy S (2021) Challenges, strategies, and perspectives for reference-independent longitudinal multi-omics microbiome studies. Front Genet 12:666244. https://doi.org/10.3389/fgene.2021.666244

Mayer EA, Nance K, Chen S (2022) The gut-brain axis. Annu Rev Med 73:439–453. https://doi.org/10.1146/annurev-med-042320-014032

Mekadim C, Skalnikova HK, Cizkova J, Cizkova V, Palanova A, Horak V, Mrazek J (2022) Dysbiosis of skin microbiome and gut microbiome in melanoma progression. BMC Microbiol 22(1):63. https://doi.org/10.1186/s12866-022-02458-5

Meng C, Kuster B, Culhane AC, Gholami AM (2014) A multivariate approach to the integration of multi-omics datasets. BMC Bioinf 15:162. https://doi.org/10.1186/1471-2105-15-162

Meng C, Helm D, Frejno M, Kuster B (2016a) moCluster: identifying joint patterns across multiple omics data sets. J Proteome Res 15(3):755–765. https://doi.org/10.1021/acs.jproteome.5b00824

Meng C, Zeleznik OA, Thallinger GG, Kuster B, Gholami AM, Culhane AC (2016b) Dimension reduction techniques for the integrative analysis of multi-omics data. Brief Bioinform 17(4):628–641. https://doi.org/10.1093/bib/bbv108

Menichetti G, Remondini D, Panzarasa P, Mondragon RJ, Bianconi G (2014) Weighted multiplex networks. PLoS One 9(6):e97857. https://doi.org/10.1371/journal.pone.0097857

Merelli I, Lio P, Milanesi L (2013) NuChart: an R package to study gene spatial neighbourhoods with multi-omics annotations. PLoS One 8(9):e75146. https://doi.org/10.1371/journal.pone.0075146

Mishra T, Sreepadmanabh M, Ramdas P, Sahu AK, Kumar A, Chande A (2021) SARS CoV-2 nucleoprotein enhances the infectivity of lentiviral spike particles. Front Cell Infect Microbiol 11:663688. https://doi.org/10.3389/fcimb.2021.663688

Mishra T, Bhardwaj V, Ahuja N, Gadgil P, Ramdas P, Shukla S, Chande A (2022a) Improved loss-of-function CRISPR-Cas9 genome editing in human cells concomitant with inhibition of TGF-beta signaling. Mol Ther Nucleic Acids 28:202–218. https://doi.org/10.1016/j.omtn.2022.03.003

Mishra T, Dalavi R, Joshi G, Kumar A, Pandey P, Shukla S, Mishra RK, Chande A (2022b) SARS-CoV-2 spike E156G/Delta157-158 mutations contribute to increased infectivity and immune escape. Life Sci Alliance 5(7):e202201415. https://doi.org/10.26508/lsa.202201415

Mishra T, Mallik B, Kesheri M, Kanchan S (2024) The interplay of gut microbiome in health and diseases. Springer Nature Singapore, pp 1–34. https://doi.org/10.1007/978-981-97-1769-9_1

Mitchell AL, Scheremetjew M, Denise H, Potter S, Tarkowska A, Qureshi M, Salazar GA, Pesseat S, Boland MA, Hunter FMI, Ten Hoopen P, Alako B, Amid C, Wilkinson DJ, Curtis TP, Cochrane G, Finn RD (2018) EBI metagenomics in 2017: enriching the analysis of microbial communities, from sequence reads to assemblies. Nucleic Acids Res 46(D1):D726–D735. https://doi.org/10.1093/nar/gkx967

Mlynarska E, Gadzinowska J, Tokarek J, Forycka J, Szuman A, Franczyk B, Rysz J (2022) The role of the microbiome-brain-gut axis in the pathogenesis of depressive disorder. Nutrients 14(9). https://doi.org/10.3390/nu14091921

Mo Q, Wang S, Seshan VE, Olshen AB, Schultz N, Sander C, Powers RS, Ladanyi M, Shen R (2013) Pattern discovery and cancer gene identification in integrated cancer genomic data. Proc Natl Acad Sci USA 110(11):4245–4250. https://doi.org/10.1073/pnas.1208949110

Mo Q, Shen R, Guo C, Vannucci M, Chan KS, Hilsenbeck SG (2018) A fully Bayesian latent variable model for integrative clustering analysis of multi-type omics data. Biostatistics 19(1):71–86. https://doi.org/10.1093/biostatistics/kxx017

Moon S, Hwang J, Lee H (2022) SDGCCA: supervised deep generalized canonical correlation analysis for multi-omics integration. J Comput Biol 29(8):892–907. https://doi.org/10.1089/cmb.2021.0598

Mosca E, Milanesi L (2013) Network-based analysis of omics with multi-objective optimization. Mol BioSyst 9(12):2971–2980. https://doi.org/10.1039/c3mb70327d

Munoz Declara S, D'Alessandro A, Gori A, Cerasuolo B, Renzi S, Berlanda M, Zini E, Monici M, Cavalieri D, Zanna G (2024) Evaluation of the impact of near-infrared multiwavelength locked system laser therapy on skin microbiome in atopic dogs. Animals (Basel) 14(6). https://doi.org/10.3390/ani14060906

Munoz-Benavent M, Hartkopf F, Van Den Bossche T, Piro VC, Garcia-Ferris C, Latorre A, Renard BY, Muth T (2020) gNOMO: a multi-omics pipeline for integrated host and microbiome analysis of non-model organisms. NAR Genomics Bioinf 2(3):lqaa058. https://doi.org/10.1093/nargab/lqaa058

Narayanasamy S, Jarosz Y, Muller EE, Heintz-Buschart A, Herold M, Kaysen A, Laczny CC, Pinel N, May P, Wilmes P (2016) IMP: a pipeline for reproducible reference-independent integrated metagenomic and metatranscriptomic analyses. Genome Biol 17(1):260. https://doi.org/10.1186/s13059-016-1116-8

Neu AT, Allen EE, Roy K (2021) Defining and quantifying the core microbiome: challenges and prospects. Proc Natl Acad Sci USA 118(51). https://doi.org/10.1073/pnas.2104429118

Nguyen H, Shrestha S, Draghici S, Nguyen T (2019) PINSPlus: a tool for tumor subtype discovery in integrated genomic data. Bioinformatics 35(16):2843–2846. https://doi.org/10.1093/bioinformatics/bty1049

O'Riordan KJ, Collins MK, Moloney GM, Knox EG, Aburto MR, Fulling C, Morley SJ, Clarke G, Schellekens H, Cryan JF (2022) Short chain fatty acids: microbial metabolites for gut-brain axis signalling. Mol Cell Endocrinol 546:111572. https://doi.org/10.1016/j.mce.2022.111572

Onywera H, Williamson AL, Mbulawa ZZA, Coetzee D, Meiring TL (2019) Factors associated with the composition and diversity of the cervical microbiota of reproductive-age Black South African women: a retrospective cross-sectional study. PeerJ 7:e7488. https://doi.org/10.7717/peerj.7488

Osadchiy V, Martin CR, Mayer EA (2019) The gut-brain axis and the microbiome: mechanisms and clinical implications. Clin Gastroenterol Hepatol 17(2):322–332. https://doi.org/10.1016/j.cgh.2018.10.002

Panwar RB, Sequeira RP, Clarke TB (2021) Microbiota-mediated protection against antibiotic-resistant pathogens. Genes Immun 22(5–6):255–267. https://doi.org/10.1038/s41435-021-00129-5

Parkhomenko E, Tritchler D, Beyene J (2009) Sparse canonical correlation analysis with application to genomic data integration. Stat Appl Genet Mol Biol 8:Article 1. https://doi.org/10.2202/1544-6115.1406

Pastar I, O'Neill K, Padula L, Head CR, Burgess JL, Chen V, Garcia D, Stojadinovic O, Hower S, Plano GV, Thaller SR, Tomic-Canic M, Strbo N (2020) *Staphylococcus epidermidis* boosts innate immune response by activation of gamma delta T cells and induction of perforin-2 in human skin. Front Immunol 11:550946. https://doi.org/10.3389/fimmu.2020.550946

Pedersen HK, Gudmundsdottir V, Nielsen HB, Hyotylainen T, Nielsen T, Jensen BA, Forslund K, Hildebrand F, Prifti E, Falony G, Le Chatelier E, Levenez F, Dore J, Mattila I, Plichta DR, Poho P, Hellgren LI, Arumugam M, Sunagawa S, Vieira-Silva S, Jorgensen T, Holm JB, Trost K, Meta HITC, Kristiansen K, Brix S, Raes J, Wang J, Hansen T, Bork P, Brunak S, Oresic M, Ehrlich SD, Pedersen O (2016) Human gut microbes impact host serum metabolome and insulin sensitivity. Nature 535(7612):376–381. https://doi.org/10.1038/nature18646

Pollack JR, Sorlie T, Perou CM, Rees CA, Jeffrey SS, Lonning PE, Tibshirani R, Botstein D, Borresen-Dale AL, Brown PO (2002) Microarray analysis reveals a major direct role of DNA copy number alteration in the transcriptional program of human breast tumors. Proc Natl Acad Sci USA 99(20):12963–12968. https://doi.org/10.1073/pnas.162471999

Proctor LM (2011) The human microbiome project in 2011 and beyond. Cell Host Microbe 10(4):287–291. https://doi.org/10.1016/j.chom.2011.10.001

Qin J, Li Y, Cai Z, Li S, Zhu J, Zhang F, Liang S, Zhang W, Guan Y, Shen D, Peng Y, Zhang D, Jie Z, Wu W, Qin Y, Xue W, Li J, Han L, Lu D, Wu P, Dai Y, Sun X, Li Z, Tang A, Zhong S, Li X, Chen W, Xu R, Wang M, Feng Q, Gong M, Yu J, Zhang Y, Zhang M, Hansen T, Sanchez G, Raes J, Falony G, Okuda S, Almeida M, LeChatelier E, Renault P, Pons N, Batto JM, Zhang Z, Chen H, Yang R, Zheng W, Li S, Yang H, Wang J, Ehrlich SD, Nielsen R, Pedersen O, Kristiansen K, Wang J (2012) A metagenome-wide association study of gut microbiota in type 2 diabetes. Nature 490(7418):55–60. https://doi.org/10.1038/nature11450

Quigley EMM (2017) Microbiota-brain-gut axis and neurodegenerative diseases. Curr Neurol Neurosci Rep 17(12):94. https://doi.org/10.1007/s11910-017-0802-6

Ramdas P, Sahu AK, Mishra T, Bhardwaj V, Chande A (2020) From entry to egress: strategic exploitation of the cellular processes by HIV-1. Front Microbiol 11:559792. https://doi.org/10.3389/fmicb.2020.559792

Rappoport N, Shamir R (2018) Multi-omic and multi-view clustering algorithms: review and cancer benchmark. Nucleic Acids Res 46(20):10546–10562. https://doi.org/10.1093/nar/gky889

Rappoport N, Shamir R (2019) NEMO: cancer subtyping by integration of partial multi-omic data. Bioinformatics 35(18):3348–3356. https://doi.org/10.1093/bioinformatics/btz058

Raut S, Mallik B, Parichha A, Amrutha V, Sahi C, Kumar V (2017) RNAi-mediated reverse genetic screen identified *Drosophila* chaperones regulating eye and neuromuscular junction morphology. G3 (Bethesda) 7(7):2023–2038. https://doi.org/10.1534/g3.117.041632

Ray P, Zheng L, Lucas J, Carin L (2014) Bayesian joint analysis of heterogeneous genomics data. Bioinformatics 30(10):1370–1376. https://doi.org/10.1093/bioinformatics/btu064

Reyman M, van Houten MA, van Baarle D, Bosch A, Man WH, Chu M, Arp K, Watson RL, Sanders EAM, Fuentes S, Bogaert D (2019) Impact of delivery mode-associated gut microbiota dynamics on health in the first year of life. Nat Commun 10(1):4997. https://doi.org/10.1038/s41467-019-13014-7

Rivera-Pinto J, Egozcue JJ, Pawlowsky-Glahn V, Paredes R, Noguera-Julian M, Calle ML (2018) Balances: a new perspective for microbiome analysis. mSystems 3(4). https://doi.org/10.1128/mSystems.00053-18

Rohart F, Gautier B, Singh A, Le Cao KA (2017) mixOmics: an R package for 'omics' feature selection and multiple data integration. PLoS Comput Biol 13(11):e1005752. https://doi.org/10.1371/journal.pcbi.1005752

Ruiz N, Silhavy TJ (2022) How *Escherichia coli* became the flagship bacterium of molecular biology. J Bacteriol 204(9):e0023022. https://doi.org/10.1128/jb.00230-22

Ruiz-Perez D, Lugo-Martinez J, Bourguignon N, Mathee K, Lerner B, Bar-Joseph Z, Narasimhan G (2021) Dynamic bayesian networks for integrating multi-omics time series microbiome data. mSystems 6(2). https://doi.org/10.1128/mSystems.01105-20

Sampaio-Maia B, Monteiro-Silva F (2014) Acquisition and maturation of oral microbiome throughout childhood: An update. Dent Res J (Isfahan) 11(3):291–301

Savage RS, Ghahramani Z, Griffin JE, de la Cruz BJ, Wild DL (2010) Discovering transcriptional modules by Bayesian data integration. Bioinformatics 26(12):i158–i167. https://doi.org/10.1093/bioinformatics/btq210

Schneider AM, Nolan ZT, Banerjee K, Paine AR, Cong Z, Gettle SL, Longenecker AL, Zhan X, Agak GW, Nelson AM (2023) Evolution of the facial skin microbiome during puberty in normal and acne skin. J Eur Acad Dermatol Venereol 37(1):166–175. https://doi.org/10.1111/jdv.18616

Sender R, Fuchs S, Milo R (2016) Revised estimates for the number of human and bacteria cells in the body. PLoS Biol 14(8):e1002533. https://doi.org/10.1371/journal.pbio.1002533

Seoane JA, Day IN, Gaunt TR, Campbell C (2014) A pathway-based data integration framework for prediction of disease progression. Bioinformatics 30(6):838–845. https://doi.org/10.1093/bioinformatics/btt610

Sepich-Poore GD, Zitvogel L, Straussman R, Hasty J, Wargo JA, Knight R (2021) The microbiome and human cancer. Science 371(6536). https://doi.org/10.1126/science.abc4552

Sharpton TJ (2014) An introduction to the analysis of shotgun metagenomic data. Front Plant Sci 5:209. https://doi.org/10.3389/fpls.2014.00209

Shi Q, Zhang C, Peng M, Yu X, Zeng T, Liu J, Chen L (2017) Pattern fusion analysis by adaptive alignment of multiple heterogeneous omics data. Bioinformatics 33(17):2706–2714. https://doi.org/10.1093/bioinformatics/btx176

Singh RK, Mallik B, Ranjan A, Tripathi R, Verma SS, Sharma V, Gupta SC, Singh AK (2024) *Semecarpus anacardium* L.f. leaf extract exhibits activities against breast cancer and prolongs the survival of tumor-bearing mice. Nat Prod Res 38(6):1080–1084. https://doi.org/10.1080/14786419.2023.2208719

Sinha R, Ahn J, Sampson JN, Shi J, Yu G, Xiong X, Hayes RB, Goedert JJ (2016) Fecal microbiota, fecal metabolome, and colorectal cancer interrelations. PLoS One 11(3):e0152126. https://doi.org/10.1371/journal.pone.0152126

Song SD, Acharya KD, Zhu JE, Deveney CM, Walther-Antonio MRS, Tetel MJ, Chia N (2020) Daily vaginal microbiota fluctuations associated with natural hormonal cycle, contraceptives, diet, and exercise. mSphere 5(4). https://doi.org/10.1128/mSphere.00593-20

Sousa P, Silva L, Camara JS, Guedes de Pinho P, Perestrelo R (2024) Integrating OMICS-based platforms and analytical tools for diagnosis and management of pancreatic cancer: a review. Mol Omics 21:108. https://doi.org/10.1039/d4mo00187g

Speicher NK, Pfeifer N (2015) Integrating different data types by regularized unsupervised multiple kernel learning with application to cancer subtype discovery. Bioinformatics 31(12):i268–i275. https://doi.org/10.1093/bioinformatics/btv244

Srivastava U, Kesheri M, Kanchan S, Singh S (2024) Computational omics protocol for the comparative study of microbiome analysis. In: Kesheri M, Kanchan S, Salisbury TB, Sinha RP (eds) Microbial omics in environment and health. Springer, Singapore. https://doi.org/10.1007/978-981-97-1769-9_5

Sun L, Li Z, Hu C, Ding J, Zhou Q, Pang G, Wu Z, Yang R, Li S, Li J, Cai J, Sun Y, Li R, Zhen H, Sun S, Zhang J, Fang M, Chen Z, Lv Y, Cao Q, Sun Y, Gong R, Huang Z, Duan Y, Liu H, Dong J, Li J, Ruan J, Lu H, He B, Li N, Li T, Xue W, Li Y, Shen J, Yang F, Zhao C, Liang Q, Zhang M, Chen C, Gong H, Hou Y, Wang J, Zhang Y, Yang H, Zhu S, Xiao L, Jin Z, Guo H, Zhao P, Brix S, Xu X, Jia H, Kristiansen K, Yang Z, Nie C (2023) Age-dependent changes in the gut microbiota and serum metabolome correlate with renal function and human aging. Aging Cell 22(12):e14028. https://doi.org/10.1111/acel.14028

Sutandy FX, Qian J, Chen CS, Zhu H (2013) Overview of protein microarrays. Curr Protoc Protein Sci Chapter 27(1):2721. https://doi.org/10.1002/0471140864.ps2701s72

Syromyatnikov M, Nesterova E, Gladkikh M, Smirnova Y, Gryaznova M, Popov V (2022) Characteristics of the gut bacterial composition in people of different nationalities and religions. Microorganisms 10(9). https://doi.org/10.3390/microorganisms10091866

Teo SM, Mok D, Pham K, Kusel M, Serralha M, Troy N, Holt BJ, Hales BJ, Walker ML, Hollams E, Bochkov YA, Grindle K, Johnston SL, Gern JE, Sly PD, Holt PG, Holt KE, Inouye M (2015) The infant nasopharyngeal microbiome impacts severity of lower respiratory infection and risk of asthma development. Cell Host Microbe 17(5):704–715. https://doi.org/10.1016/j.chom.2015.03.008

Thimm O, Blasing O, Gibon Y, Nagel A, Meyer S, Kruger P, Selbig J, Muller LA, Rhee SY, Stitt M (2004) MAPMAN: a user-driven tool to display genomics data sets onto diagrams of metabolic pathways and other biological processes. Plant J 37(6):914–939. https://doi.org/10.1111/j.1365-313x.2004.02016.x

Townsend EC, Kalan LR (2023) The dynamic balance of the skin microbiome across the lifespan. Biochem Soc Trans 51(1):71–86. https://doi.org/10.1042/BST20220216

Tsui DCC, Lee JK, Tambaoan CFB, Hughes J, Fendler B, Decker B, Frampton GM, Schrock AB, Camidge DR (2025) Genomic analysis of comprehensive next generation sequencing data to explore the criteria for MET amplification as an actionable biomarker in NSCLC. Lung Cancer 199:108081. https://doi.org/10.1016/j.lungcan.2025.108081

Tuncbag N, McCallum S, Huang SS, Fraenkel E (2012) SteinerNet: a web server for integrating 'omic' data to discover hidden components of response pathways. Nucleic Acids Res 40(Web Server issue):W505–W509. https://doi.org/10.1093/nar/gks445

Uehara M, Inoue T, Hase S, Sasaki E, Toyoda A, Sakakibara Y (2024) Decoding host-microbiome interactions through co-expression network analysis within the non-human primate intestine. mSystems 9(5):e0140523. https://doi.org/10.1128/msystems.01405-23

Vandereyken K, Sifrim A, Thienpont B, Voet T (2023) Methods and applications for single-cell and spatial multi-omics. Nat Rev Genet 24(8):494–515. https://doi.org/10.1038/s41576-023-00580-2

Vaske CJ, Benz SC, Sanborn JZ, Earl D, Szeto C, Zhu J, Haussler D, Stuart JM (2010) Inference of patient-specific pathway activities from multi-dimensional cancer genomics data using PARADIGM. Bioinformatics 26(12):i237–i245. https://doi.org/10.1093/bioinformatics/btq182

Wang W, Baladandayuthapani V, Morris JS, Broom BM, Manyam G, Do KA (2013) iBAG: integrative Bayesian analysis of high-dimensional multiplatform genomics data. Bioinformatics 29(2):149–159. https://doi.org/10.1093/bioinformatics/bts655

Wang B, Mezlini AM, Demir F, Fiume M, Tu Z, Brudno M, Haibe-Kains B, Goldenberg A (2014) Similarity network fusion for aggregating data types on a genomic scale. Nat Methods 11(3):333–337. https://doi.org/10.1038/nmeth.2810

Watanabe H, Nakamura I, Mizutani S, Kurokawa Y, Mori H, Kurokawa K, Yamada T (2018) Minor taxa in human skin microbiome contribute to the personal identification. PLoS One 13(7):e0199947. https://doi.org/10.1371/journal.pone.0199947

Wen Z, Xie G, Zhou Q, Qiu C, Li J, Hu Q, Dai W, Li D, Zheng Y, Wen F (2018) Distinct nasopharyngeal and oropharyngeal microbiota of children with influenza A virus compared with healthy children. Biomed Res Int 2018:6362716. https://doi.org/10.1155/2018/6362716

Wishart DS, Guo A, Oler E, Wang F, Anjum A, Peters H, Dizon R, Sayeeda Z, Tian S, Lee BL, Berjanskii M, Mah R, Yamamoto M, Jovel J, Torres-Calzada C, Hiebert-Giesbrecht M, Lui VW, Varshavi D, Varshavi D, Allen D, Arndt D, Khetarpal N, Sivakumaran A, Harford K, Sanford S, Yee K, Cao X, Budinski Z, Liigand J, Zhang L, Zheng J, Mandal R, Karu N, Dambrova M, Schioth HB, Greiner R, Gautam V (2022) HMDB 5.0: the human metabolome database for 2022. Nucleic Acids Res 50(D1):D622–D631. https://doi.org/10.1093/nar/gkab1062

Witsch-Baumgartner M, Schwaninger G, Schnaiter S, Kollmann F, Burkhard S, Grobner R, Muhlegger B, Schamschula E, Kirchmeier P, Zschocke J (2022) Array genotyping as diagnostic approach in medical genetics. Mol Genet Genomic Med 10(9):e2016. https://doi.org/10.1002/mgg3.2016

Wu D, Wang D, Zhang MQ, Gu J (2015) Fast dimension reduction and integrative clustering of multi-omics data using low-rank approximation: application to cancer molecular classification. BMC Genomics 16:1022. https://doi.org/10.1186/s12864-015-2223-8

Wu C, Zhou F, Ren J, Li X, Jiang Y, Ma S (2019) A selective review of multi-level omics data integration using variable selection. High Throughput 8(1). https://doi.org/10.3390/ht8010004

Yachida S, Mizutani S, Shiroma H, Shiba S, Nakajima T, Sakamoto T, Watanabe H, Masuda K, Nishimoto Y, Kubo M, Hosoda F, Rokutan H, Matsumoto M, Takamaru H, Yamada M, Matsuda T, Iwasaki M, Yamaji T, Yachida T, Soga T, Kurokawa K, Toyoda A, Ogura Y, Hayashi T, Hatakeyama M, Nakagama H, Saito Y, Fukuda S, Shibata T, Yamada T (2019) Metagenomic and metabolomic analyses reveal distinct stage-specific phenotypes of the gut microbiota in colorectal cancer. Nat Med 25(6):968–976. https://doi.org/10.1038/s41591-019-0458-7

Yan J, Risacher SL, Shen L, Saykin AJ (2018) Network approaches to systems biology analysis of complex disease: integrative methods for multi-omics data. Brief Bioinform 19(6):1370–1381. https://doi.org/10.1093/bib/bbx066

Yang Z, Michailidis G (2016) A non-negative matrix factorization method for detecting modules in heterogeneous omics multi-modal data. Bioinformatics 32(1):1–8. https://doi.org/10.1093/bioinformatics/btv544

Yen S, Johnson JS (2021) Metagenomics: a path to understanding the gut microbiome. Mamm Genome 32(4):282–296. https://doi.org/10.1007/s00335-021-09889-x

Yeoh YK, Zuo T, Lui GC, Zhang F, Liu Q, Li AY, Chung AC, Cheung CP, Tso EY, Fung KS, Chan V, Ling L, Joynt G, Hui DS, Chow KM, Ng SSS, Li TC, Ng RW, Yip TC, Wong GL, Chan FK, Wong CK, Chan PK, Ng SC (2021) Gut microbiota composition reflects disease severity and dysfunctional immune responses in patients with COVID-19. Gut 70(4):698–706. https://doi.org/10.1136/gutjnl-2020-323020

Yizhak K, Benyamini T, Liebermeister W, Ruppin E, Shlomi T (2010) Integrating quantitative proteomics and metabolomics with a genome-scale metabolic network model. Bioinformatics 26(12):i255–i260. https://doi.org/10.1093/bioinformatics/btq183

Yuan Y, Savage RS, Markowetz F (2011) Patient-specific data fusion defines prognostic cancer subtypes. PLoS Comput Biol 7(10):e1002227. https://doi.org/10.1371/journal.pcbi.1002227

Yuan CU, Quah FX, Hemberg M (2024) Single-cell and spatial transcriptomics: bridging current technologies with long-read sequencing. Mol Asp Med 96:101255. https://doi.org/10.1016/j.mam.2024.101255

Zhang S, Liu CC, Li W, Shen H, Laird PW, Zhou XJ (2012) Discovery of multi-dimensional modules by integrative analysis of cancer genomic data. Nucleic Acids Res 40(19):9379–9391. https://doi.org/10.1093/nar/gks725

Zhang F, Aschenbrenner D, Yoo JY, Zuo T (2022) The gut mycobiome in health, disease, and clinical applications in association with the gut bacterial microbiome assembly. Lancet Microbe 3(12):e969–e983. https://doi.org/10.1016/S2666-5247(22)00203-8

Zhong Q, Xiao X, Qiu Y, Xu Z, Chen C, Chong B, Zhao X, Hai S, Li S, An Z, Dai L (2023) Protein posttranslational modifications in health and diseases: functions, regulatory mechanisms, and therapeutic implications. MedComm 4(3):e261. https://doi.org/10.1002/mco2.261

Zoppi J, Guillaume JF, Neunlist M, Chaffron S (2021) MiBiOmics: an interactive web application for multi-omics data exploration and integration. BMC Bioinf 22(1):6. https://doi.org/10.1186/s12859-020-03921-8

Zreloff ZJ, Lange D, Vernon SD, Carlin MR, Cano RJ (2023) Accelerating gut microbiome research with robust sample collection. Res Rev J Microbiol Biotechnol 12(1):33–47

Multiresolution Insights into Single-Cell Landscapes: Integrating Genomics, Epigenomics, and Proteomics for Brain Studies

4

Upasna Srivastava, Sonal Sukreet, Swarna Kanchan, Minu Kesheri, and Manish Kumar Gupta

Abstract

A multiresolution framework to characterize single-cell state landscapes enhances our understanding of the regulatory genomic circuitry. Single-cell technologies have revolutionized cellular heterogeneity studies, allowing for the dissection of complex tissue compositions and identification of distinct cell states and types at unprecedented resolution. This framework integrates single-cell RNA sequencing (scRNA-seq) with other assays such as single-cell ATAC-seq (scATAC-seq), single-cell DNA methylation sequencing, single-nuclei sequencing, single-cell proteomics, and spatial analysis. By combining these datasets, researchers construct comprehensive maps of cellular states and transitions, revealing how individual cells contribute to disease progression and treatment response. Advanced computational tools and machine learning algorithms analyze the high-dimensional data generated by single-cell assays. Techniques like dimensionality reduction, clustering, and trajectory inference identify distinct cell populations and their dynamic changes over time. This approach highlights tissue heterogeneity and uncovers the regulatory mechanisms driving cell state transitions and their impact on disease. Integrating epigenomic, single cell, and

U. Srivastava (✉)
Yale School of Medicine, New Haven, CT, USA
e-mail: upasna.srivastava@yale.edu

S. Sukreet
Alzheimer's Therapeutic Research Institute-UCS, San Diego, CA, USA

S. Kanchan · M. Kesheri
Marshall University, Huntington, WV, USA

Boise State University, Boise, ID, USA

M. K. Gupta
Department of Biotechnology, Veer Bahadur Singh Purvanchal, University, Jaunpur, UP, India

M. Kesheri et al. (eds.), *Multi-Omics in Biomedical Sciences and Environmental Sustainability*, https://doi.org/10.1007/978-981-96-7067-3_4

spatial data decodes the regulatory circuitry underlying diseases. Mapping interactions between genomic loci, epigenomic marks, and protein expression at single-cell resolution pinpoints critical regulatory elements and pathways involved in disease onset and progression. This chapter unravels various integrative strategies of omics technologies in general while emphasizing its significance in precision medicine, enabling the development of targeted therapies pertaining to neurosciences. Moreover, incorporation of case studies, pipelines, and codes for bioinformatic analysis with relevant illustrations emphasize recent advances of multi-omics technologies presenting the state of art.

Keywords

Bioinformatics · Multi-omics · scRNA-seq · Single-cell proteomics · Spatial data

Objectives

- To leverage single-cell technologies for dissecting cellular heterogeneity and to highlight the advantages and limitations of snRNA-seq in analyzing complex tissues
- To integrate multi-omics approaches that combine genomics, epigenomics, and proteomics to create comprehensive maps of cellular states, utilizing advanced computational pipelines and tools
- To investigate molecular and cellular mechanisms underlying neurodegenerative diseases and psychiatric disorders by using single-cell and spatial transcriptomics facilitating identification of vulnerable cell populations and pathways for potential therapeutic targets
- To explore the synergistic role of proteomics fueled by bioinformatic analysis and the application of single-cell proteomics

4.1 Introduction

Omics technologies play a crucial role by providing a comprehensive view of biological systems. By integrating transcriptomics, proteomics, and epigenomics, researchers can decipher complex biological processes at unprecedented resolution. Single-cell methods enable detailed studies of gene expression, protein dynamics, and epigenetic regulation in crops, facilitating the identification of stress responses and developmental pathways.

For example, genomics enabled the identification of fructose 1,6-bisphosphate aldolase (FBA) as a vital enzyme in the metabolic pathways of Community Associated Methicillin-resistant *Staphylococcus aureus* (CA-MRSA), while proteomics facilitated the structural modeling and functional analysis of the protein. Metabolomics connects FBA's role in glycolysis and gluconeogenesis to the pathogen's survival, highlighting it as a drug target. Bioinformatics tools, integrating omics data, supported structural validation and virtual screening to identify potential inhibitors. Together, omics technologies bridge molecular insights and therapeutic discovery to combat antimicrobial resistance (Yadav et al. 2013). Identification of genomic islands can also aid in the search for drug targets. Studies have focused on predicting genomic islands in *Burkholderia cenocepacia* J2315, along with sequence and structural analysis of proteins encoded by identified genomic islands for functional annotation (Jaiswal et al. 2023). Integrating omics data, such as transcriptomics for gene expression profiles and genomics for identifying viral genome-host interactions, predicting Covid-19 pandemic by combining machine learning and genomics data obtained from wastewater surveillance aided in proactive decision making and preparedness of health care systems (Kanchan et al. 2024; Meadows et al. 2025).

Nutrimetabolomics leverages metabolomics to investigate the intricate relationship between diet and metabolism by identifying and quantifying metabolites that reflect nutritional intake and physiological responses. When integrated with other omics approaches like genomics, proteomics, and microbiomics, nutrimetabolomics provides deeper insights into nutrient-gene interactions, protein activity, and the gut microbiome's contribution to metabolic processes (Mishra et al. 2024). This comprehensive understanding paves the way for precision nutrition strategies aimed at improving health and preventing diseases through dietary interventions (Srivastava et al. 2023).

High-throughput omics technologies have also transformed ecological and environmental research, enabling insights into chemical toxicity, biomarkers, and environmental adaptation (Mallik et al. 2024; Shukla et al. 2024). Genomics, transcriptomics, proteomics, and metabolomics are used to study pollutant impacts, assess health risks, and monitor ecosystems. While transcriptomics dominates (43%), integrating multiomics (13%) is crucial for a deeper understanding of complex environmental interactions and revising environmental protection laws. Multi-omics enhances studies on toxicity mechanisms, environmental monitoring, and genome-diet-environment interactions, supporting sustainability, and risk assessment efforts (Kesheri et al. 2024a; Bujak et al. 2024).Okamoto et al. (2021) used patient-derived organoids and single-cell analysis to uncover a reduction in OLFM4-associated clusters in metastatic colorectal cancer. This highlights how cell-state transitions can be tracked with high resolution. Although focused on cancer, such approaches could be adapted to study cellular heterogeneity and progression in brain tissues. Integrating these methods holds promise for advancing neurobiological research. Building on this, Sakahara et al. (2024) demonstrated that $OLFM4^+$ stem cells can give rise to Paneth-like cells, which in turn support the maintenance of the $OLFM4^+$ population in advanced colorectal cancer. This reciprocal interaction within the tumor niche underscores the role of supportive niche cells in regulating stem cell maintenance. Leveraging organoid and single-cell approaches in brain research could therefore elucidate comparable stem cell–niche dynamics underlying neurodevelopment and disease.

Finally, computational omics protocols provide robust frameworks for comparative microbiome analysis. Metagenomics powered by bioinformatic analysis unveils the microbial communities present in the environmental samples as well as human tissues from various organs such as brain, lungs, gut (Kanchan et al. 2020; Kim et al. 2024), and most recently in plasma (Graf et al. 2025). A study emphasizes the use of computational omics protocols to analyze and compare microbiomes, enabling deeper insights into microbial communities and their roles in various ecosystems. These methods encompassing various multi-omics technologies that facilitate efficient data integration, visualization (Srivastava and Singh 2013) and interpretation, strengthening microbiome research's contributions to health (Kumar et al. 2024), agriculture (Srivastava and Singh 2022; Chaturvedi et al. 2024; Srivastava et al. 2025), and environmental studies (Kesheri et al. 2024b, c).

4.2 General Overview of the Human Brain

The brain is an incredibly intricate organ, playing a central role in regulating nearly all bodily functions. It consists of different cell types that collaborate to interpret information, manage body processes, and support cognitive abilities such as thinking, memory, emotions, and awareness. The two main categories of brain cells are neurons and glial cells. Neurons are the fundamental units of the brain and nervous system, specialized for transmitting information through electrical impulses. Glial cells provide support, protection, and nutrition to neurons. Although they do not transmit electrical impulses like neurons, they are essential for maintaining the environment in which neurons function. There are several types of glial cells. Astrocytes are star-shaped cells that support and nourish neurons by regulating blood flow, maintaining the blood-brain barrier, and ensuring the proper chemical environment for signal transmission. Specialized brain cell types play critical roles in development and regeneration. These cells play a vital role in brain plasticity and repair, especially in areas like the hippocampus, which is involved in memory and learning. The interaction between neurons and glia is crucial for brain function and maintenance. These glia-neuron interactions ensure neurons operate efficiently by providing structural and metabolic support, maintaining proper signaling conditions, and protecting the brain from damage and disease. Together, they create an environment conducive to effective neural communication and brain health. Percentages of brain cell types are shown in Table 4.1. Recent advancements in single-cell genomic techniques have greatly enhanced our understanding of the brain's cellular diversity, identifying hundreds of distinct transcriptomic cell types across different regions of the mouse brain.

4.3 Brain Cell Quantification

Cell quantification in the human brain can be represented using the following three key measures:

Table 4.1 Percentage of brain cell types

Cell type	Approx. proportion (%)	Function	References
Neurons	~10%	Signal transmission and processing	Azevedo et al. (2009)
Astrocytes	~36%	Support neurons, maintain homeostasis, regulate synapses, and blood-brain barrier	Nedergaard et al. (2003) and Verkhratsky and Butt (2013)
Oligodendrocytes	~36%	Myelinate axons, enhance conduction velocity, and provide metabolic support	Fields and Stevens-Graham (2002)
Microglia	~13%	Immune surveillance, phagocytosis, and inflammatory response	Ransohoff and Perry (2009) and Barres (2008)
Ependymal cells	~5%	Line ventricles produce and circulate cerebrospinal fluid (CSF)	Del Bigio (2010) and Purves et al. (2001)

(i) *Total Neuron Count (N):* The sum of all neurons in a given brain region.
(ii) *Total Glial Cell Count (G):* The sum of all glial cells, including astrocytes, oligodendrocytes, and microglia, in the same region.
(iii) *Glia-to-Neuron Ratio (GNR):* This ratio is calculated using the formula GNR = G/N, where G is the total number of glial cells and N is the total number of neurons, and GNR is their ratio value.

To determine the total glial cell, count G, most studies combine the numbers of astrocytes, oligodendrocytes, and microglia. Therefore, the GNR reflects the ratio of all glial cells to neurons in a given structure, regardless of the individual sizes of these cells (Von Bartheld et al. 2016).

4.4 Principal Approaches to Estimate the Number of Cells in the Brain

1. *Model-Based or Design-based Counting:* This involves counting stained cells, nuclei, nucleoli, or their fragments in histological sections.
2. *DNA Content Measurement:* Cell numbers are calculated by extracting and measuring total DNA content.
3. *Direct Enumeration:* This includes counting cell nuclei in homogenized brain tissue (a precursor to the isotropic fractionator) and the isotropic fractionator method itself.

These methods, while distinct, ultimately contribute to the determination of neuron and glial cell numbers and the calculation of GNR in brain structures.

4.5 Omics Approaches for Brain Studies

4.5.1 Single-Cell Sequencing of Brain Cells

Single-cell sequencing has become a powerful tool in neuroscience, enabling the analysis of individual cells to uncover the complex cellular architecture of the brain. Given the brain's diverse and intricate network of cell types, including neurons and glia, scRNA-seq allows researchers to dissect the specific gene expression profiles of individual cells, providing deep insights into both normal brain function and the molecular underpinnings of various neurological diseases. Recent advancements in scRNA-seq have provided unprecedented insights into the cellular diversity, molecular states, and functional specialization of brain cells. These studies have enabled the identification of distinct neuronal and glial subtypes, the characterization of developmental trajectories, and the discovery of region-specific gene expression patterns in both health and disease. Furthermore, scRNA-seq has highlighted interspecies differences in cellular composition and transcriptional profiles, particularly between human and mouse brains. The table below provides an overview of key recent scRNA-seq studies, summarizing their species, brain regions analyzed, methodologies, and major findings. Table 4.2 summarizes an overview of recent brain scRNA-seq studies.

4.5.2 Advancements in Single-Cell Analysis Technologies

Advancements in single-cell analysis technologies, particularly scRNA-seq and spatial transcriptomics, have revolutionized our understanding of neurobiology. These tools allow researchers to examine gene expression at the resolution of individual cells, shedding light on the dynamic processes that occur during brain development, aging, and disease. Single-cell technologies have facilitated the identification of previously unknown cell types and states, contributing to a comprehensive map of the brain's cellular diversity.

A. *Neurobiology:* Single-cell technologies have enabled the mapping of cellular diversity in the brain, revealing the distinct roles that various cell types play in neural circuits and how these roles contribute to overall brain function.
B. *Disease Understanding:* In the context of neurological disorders, single-cell analysis has provided insights into the specific cell populations affected by diseases such as Alzheimer's, Parkinson's, and schizophrenia. This understanding is critical for the development of targeted therapies aimed at specific cell types.

4.5.3 Advantages of Single-Cell RNA-Seq

I. *High Resolution:* Single-cell RNA-seq offers high-resolution data, capturing the gene expression profiles of individual cells. This level of detail is particu-

Table 4.2 A list of recent brain single-cell RNA sequencing (scRNA-seq) studies

Study	Species	Brain region	Method	Key findings
Zeisel et al. (2015)	Mouse	Cortex, Hippocampus	Microfluidics-based scRNA-seq	Identified 47 distinct cell types, including excitatory and inhibitory neurons, glia, and vascular cells
Tasic et al. (2016)	Mouse	Visual Cortex	SMART-Seq2	Classified neuronal subtypes based on gene expression; found substantial heterogeneity in excitatory neurons
Lake et al. (2016)	Human	Primary Visual Cortex	Droplet-based scRNA-seq	Identified conserved cell types between human and mouse, and highlighted differences in astrocytes and neurons
Darmanis et al. (2015)	Human	Cortex (Adult and Fetal)	SMART-Seq	Identified developmental differences in neurons and glia; highlighted transcriptional changes in aging
Saunders et al. (2018)	Mouse	Whole Brain	Drop-seq	Mapped >565,000 single cells and identified rare cell populations across multiple brain regions
Zhou et al. (2020)	Human	Prefrontal Cortex	10x Genomics Chromium	Characterized neuronal and non-neuronal diversity; found distinct transcriptomic signatures for cell types
Yao et al. (2021)	Human	Brain (Multi-regional)	Droplet-based scRNA-seq	Constructed a single-cell atlas of the human brain, showing region-specific gene expression patterns
Hodge et al. (2019)	Human	Cortex	10x Genomics Chromium	Revealed extensive diversity in interneurons and their spatial organization
Tabula Muris Consortium (2018)	Mouse	Whole Brain	Multiple platforms (10x, SmartSeq2)	Comprehensive atlas of cell types across the mouse brain and body, including neurons, glia, and vascular cells

larly important in the brain, where even closely related cells can exhibit distinct functional profiles.

II. *Discovery of Rare Cell Types:* The technology enables the identification of rare cell populations that might be overlooked in bulk RNA-seq, revealing new cell types and states that are critical for brain function.

III. *Insights into Cellular Heterogeneity:* Single-cell RNA-seq is instrumental in understanding the cellular heterogeneity within brain tissue, which is essential for deciphering the complex functions and dysfunctions of the brain.

4.5.4 Challenges in Single-Cell RNA-Seq

The intricate network of dendrites, axons, and glial processes in brain tissues presents multiple challenges for scRNA-seq. Distal cellular compartments, such as dendrites and axons, contain substantial populations of mRNAs crucial for local protein synthesis, essential for synaptic function and plasticity (Cajigas et al. 2012; Tushev et al. 2018). However, these transcripts are often lost during tissue dissection and cell dissociation. Because all transcripts originate in the nucleus, scRNA-seq can still capture newly synthesized dendritic and axonal mRNAs in the nucleus or soma before transport to these distal compartments.

scRNA-seq cannot be applied to frozen human post-mortem brain tissue, as cell membranes rupture during freezing. An alternative approach is to isolate and sequence RNA from single nuclei using single-nucleus RNA sequencing (snRNA-seq) (Grindberg et al. 2013). Nuclear transcripts comprise approximately 20%–50% of all RNA in the cell, including immature and unspliced RNA molecules containing introns (Bakken et al. 2018). Despite these limitations, snRNA-seq provides robust markers of brain cell types (Lake et al. 2017; Ding et al. 2020) with resolution comparable to scRNA-seq (Bakken et al. 2018). Dendritic transcripts can be individually identified in single cells through manual microdissection techniques (Middleton et al. 2019).

4.6 Single Nucleus RNAseq for Brain Diseases

4.6.1 Single Nucleus RNA Sequencing (snRNA-seq) Pipeline

snRNA-seq Pipeline is used for studying brain diseases, designed to account for the complexities of brain tissue, such as cellular diversity and sensitivity to degradation. It can be used to investigate neurodegenerative diseases like Alzheimer's, Parkinson's (ALS), and other brain disorders.

A. Sample Preparation: *Tissue Collection*
   I. Obtain fresh-frozen or fixed postmortem brain tissue (human or animal models).
   II. Fresh-frozen tissue is ideal for preserving RNA integrity, while fixation (e.g., with paraformaldehyde) allows for longer-term storage but may reduce RNA quality.
B. *Nuclei Isolation:*
   I. *Mechanical Disruption:* Homogenize tissue gently with a Dounce homogenizer to disrupt cell membranes and release nuclei while preserving nuclear membranes.
   II. *Nuclei Lysis:* Use mild detergents (e.g., NP-40) in a lysis buffer to separate nuclei from the cytoplasmic material.

    III. *Filtration and Purification:* Filter the homogenate using a 30–40 μm filter to remove debris. Purify nuclei using density gradient centrifugation (e.g., with sucrose or OptiPrep).

C. *Nuclei Staining and Sorting:*
    I. Stain nuclei with 4′,6-diamidino-2-phenylindole (DAPI) or similar nuclear stain for DNA content.
    II. Use Fluorescence-Activated Nuclei Sorting (FANS) to isolate single nuclei populations and remove debris or doublets.

D. Library Preparation for snRNAseq
    I. *Reverse Transcription*: To perform reverse transcription to convert RNA from single nuclei into complementary DNA (cDNA). Use primers specific to polyadenylated mRNA transcripts to ensure the conversion of mRNA into cDNA.
    II. *cDNA Amplification:* To amplify the cDNA using polymerase chain reaction (PCR)-based methods. This step increases the quantity of the cDNA library while maintaining the original RNA species' representation.
    III. Library Construction
    Step a: Fragment the cDNA and prepare it for sequencing by using a platform-compatible method (e.g., Illumina or 10x Genomics).
    Step b: Add unique molecular identifiers (UMIs) and barcodes to each nucleus's RNA. This enables identification of individual nuclei after pooling them for sequencing.

E. Sequencing
    Step a: Perform high-throughput sequencing using platforms such as Illumina NovaSeq, HiSeq, or NextSeq for paired-end sequencing.
    Step b: Ensure sufficient sequencing depth, typically aiming for 50,000 to 100,000 reads per nucleus, to capture the full transcriptome of each nucleus.

4.6.2 Computational Pipeline (Step-by-Step)

A. *Loading 10x CellRanger Data:* Data generated from the 10x Genomics CellRanger pipeline, which processes raw sequencing data, aligns it to a reference genome, and generates gene expression matrices for each nucleus.
B. *Integration Based on Common Anchors:* The snRNA-seq data from multiple samples are integrated by identifying common cellular states across datasets using "anchors." This ensures that shared cellular populations are aligned across different datasets.
C. *Quality Control:* Quality control measures remove low-quality nuclei, such as those with high mitochondrial gene content, and potential doublets (two or more nuclei captured in a single droplet).

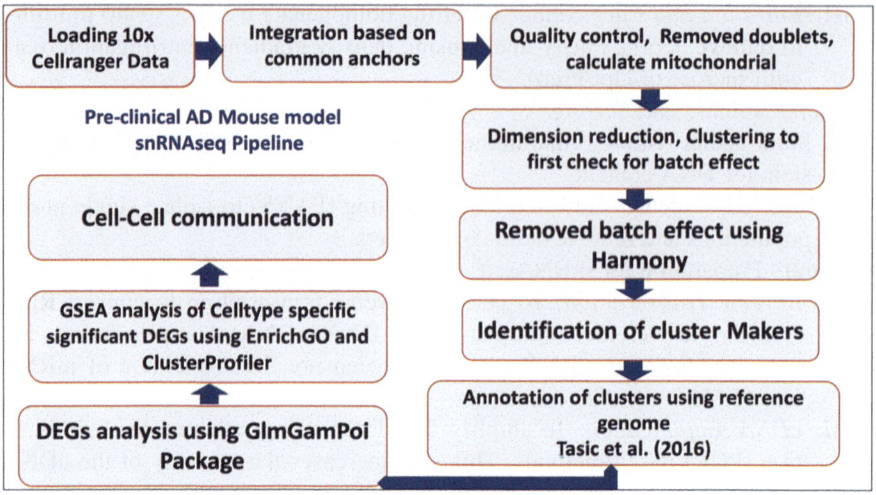

Fig. 4.1 Computational pipeline for snRNAseq data analysis

D. *Dimension Reduction, Clustering, and Batch:* Dimensionality reduction is applied (e.g., PCA, UMAP, or t-SNE) to reduce the complexity of the dataset, followed by clustering to group similar nuclei. Initial clustering is also checked to determine if batch effects (unwanted technical variation) are present.
E. *Removed Batch Effect Using Harmony:* Harmony is a tool that corrects for batch effects, aligning the clusters across different samples while preserving the biological variability of the dataset.
F. *Identification of Cluster Markers:* Cluster-specific marker genes are identified, helping to characterize the distinct cell populations present in the dataset.
G. *Annotation of Clusters Using Reference Genome:* Clusters are annotated using a reference, such as the Tasic et al. (2016) brain atlas, to identify specific cell types or states based on known gene expression profiles.
H. *Differential Expression Gene (DEG) Analysis Using GlmGamPoi Package:* GlmGamPoi, a statistical package, is used for identifying differentially expressed genes (DEGs) between conditions or across cell populations.
I. *Gene Set Enrichment Analysis (GSEA):* Gene set enrichment analysis is performed on cell-type-specific DEGs using tools like EnrichGO and ClusterProfiler to understand biological processes and pathways enriched in the DEGs.
J. *Cell-Cell Communication:* Finally, the analysis investigates potential cell-cell communication, using computational tools as Cell Chat v2 to predict ligand-receptor interactions between the identified cell types.

Pipeline (as shown in Fig. 4.1) is designed to uncover cell-type-specific gene expression changes and interactions in the brain of a preclinical AD mouse model, providing insights into disease mechanisms.

4.6.3 Advantages of Single Nucleus RNA-Seq

A. *Bias Reduction:* snRNA-seq may reduce some of the technical artifacts associated with scRNA-seq. Since nuclei are more uniform in size and morphology, snRNA-seq is more likely to capture an unbiased sample of all cell types, including those that might be more prone to cell death during isolation for scRNA-seq, such as cortical layer 5 pyramidal tract neurons (Tasic et al. 2018).
B. *Minimized Spurious Gene Activation:* Tissue dissection and cell isolation for scRNA-seq can spuriously activate gene expression. Although this can be mitigated with a transcription inhibitor (Wu et al. 2014), snRNA-seq is naturally less susceptible to this issue, offering a more accurate representation of the cells' in vivo state (Lacar et al. 2016).

4.6.4 Technical Differences and Data Analysis Considerations

The technical differences between scRNA-seq and snRNA-seq require careful consideration in data analysis and interpretation. Differential expression observed between scRNA-seq and snRNA-seq data may reflect the localization of transcripts within the cell, such as the highly abundant nuclear-localized noncoding RNA **Malat1**, which regulates RNA splicing among other functions (Bakken et al. 2018). While scRNA-seq data are typically analyzed by mapping sequenced fragments to exons, snRNA-seq data analysis should also consider the abundance of intronic reads from pre-mRNA, which may account for more than 75% of fragments (Bakken et al. 2018; Gaidatzis et al. 2015). The unique data from exonic and intronic reads can help track variations in mRNA expression over time (Bergen et al. 2020; La et al. 2018).

4.7 Single-Cell and Spatial Transcriptomics: Deciphering Brain Complexity

Single-cell and spatial transcriptomics provide complementary insights into brain complexity. While scRNA-seq focuses on the gene expression of individual cells, spatial transcriptomics preserves the spatial context of gene expression, allowing for the mapping of where specific genes are active within brain tissue. This spatial information is crucial for understanding how different regions of the brain contribute to overall function and how these regions are altered in disease.

4.8 Impact on Neuroscience Research

A. *Brain Development and Plasticity:* Single-cell transcriptomics have been pivotal in understanding brain development, revealing how distinct cell types arise and interact during neurogenesis. Additionally, by capturing the local translation of mRNAs in dendrites, this technology has illuminated key mechanisms underlying synaptic plasticity, which are essential for learning and memory.

B. *Neurodegenerative Diseases:* The ability to profile individual cell types has provided critical insights into the specific cellular changes that occur in neurodegenerative diseases such as Alzheimer's, Parkinson's, and ALS. Single-cell transcriptomics helps identify vulnerable cell populations and the molecular changes that precede cell death, offering potential for earlier diagnostic markers and targeted therapies.
C. *Psychiatric Disorders:* In psychiatric conditions like schizophrenia and depression, single-cell transcriptomics has uncovered the cellular and molecular underpinnings of these disorders. By identifying specific cell types and pathways affected, this technology paves the way for more precise and effective treatments.

4.9 Proteomics: The Functional Output of Brain Genomes

4.9.1 What Is Proteomics?

Proteomics refers to the large-scale study and analysis of proteins within a specific cell, tissue, or organism at a particular time. Unlike the relatively static nature of the genome, the proteome, the complete set of proteins expressed by a cell or tissue is highly dynamic and complex. This complexity arises from various factors, including alternative splicing of gene transcripts and post-translational modifications (PTMs) such as glycosylation, phosphorylation, and myristylation. These processes result in multiple protein isoforms from a single gene as seen in cyanobacteria (Kesheri et al. 2014, 2021). They thereby significantly increase the number of functional proteins compared to approximately 30,000 genes in the human genome (Venter et al. 2001). The proteome is not only larger but also more variable across different conditions, reflecting changes in the cellular environment, developmental stage, or disease state (Wilkins et al. 1999). As a natural extension of functional genomics, proteomics enables the study of protein expression, interactions, and modifications, offering insights into cellular mechanisms that are not fully captured by genomic data alone. However, proteomic studies generate vast amounts of data, requiring advanced computational tools and bioinformatics resources for effective analysis and interpretation (Aebersold and Mann 2003), analyzing protein-protein interactions using string (Kesheri et al. 2022). Thus, proteomics provides a powerful approach to understanding cellular function, bridging the gap between genetic information and phenotypic expression.

4.9.2 Single-Cell Proteomics of the Mouse Brain

The mouse brain is a highly complex organ composed of numerous cell types, each with distinct molecular and functional characteristics. Understanding this cellular heterogeneity is key to unraveling the brain's normal function and its dysfunction in diseases. Single-cell proteomics, a powerful approach that allows for the quantification and analysis of proteins at the individual cell level, provides unprecedented insights into cellular processes. A computational pipeline for single-cell proteomics in the mouse brain, describing each step from sample preparation to data analysis, while highlighting the significance of this method in neuroscientific research, has been discussed.

Step 1: Sample Preparation and Isolation of Single Cells: The first critical step in single-cell proteomics involves isolating individual cells from the complex brain tissue. This requires careful tissue dissociation followed by cell isolation using techniques such as fluorescence-activated cell sorting (FACS) or magnetic-activated cell sorting (MACS). In the context of the mouse brain, tissue is typically dissociated enzymatically and mechanically to yield a single-cell suspension. Each cell must be carefully handled to ensure the retention of its proteome for downstream analysis.

Step 2: Protein Extraction and Enzymatic Digestion: Once single cells are isolated, protein extraction is performed. This step involves lysing the cells to release proteins while minimizing losses. Given the small amount of protein in each cell, efficient lysis is crucial. After extraction, proteins are digested into peptides using enzymes such as trypsin. This is a standard step in proteomics workflows, as peptides are more easily analyzed by mass spectrometry than intact proteins. Two main challenges in Single-cell Proteomics occur as limited protein quantities necessitate highly sensitive methods and minimizing protein loss during extraction.

Step 3: Peptide Labeling and Quantification: In single-cell proteomics, multiplexing strategies are often employed to enhance throughput and sensitivity. Techniques such as tandem mass tag (TMT) labeling allow for the simultaneous quantification of proteins from multiple samples in a single experiment. TMT or other similar techniques use chemical tags to label peptides from different samples, enabling their pooled analysis without losing information about individual sample origins. Alternatively, label-free quantification methods can be used for direct protein quantification.

Step 4: Mass Spectrometry Acquisition: Mass spectrometry (MS) is the backbone of proteomics, providing both identification and quantification of peptides. For single-cell proteomics, high-resolution mass spectrometers, such as Orbitrap or time-of-flight (TOF) instruments, are typically used. These instruments separate and identify peptides based on their mass-to-charge ratio. In the context of single-cell proteomics of the mouse brain, it is critical to ensure that the mass spectrometry platform can detect low-abundance proteins, which is a common challenge in single-cell analysis.
Step 5: Data Preprocessing and Identification: Once raw mass spectrometry data is acquired, it must be processed to extract meaningful information. Tools such as MaxQuant and Proteome Discoverer are widely used for processing mass spectrometry data. These tools convert raw spectral data into peptide spectral matches (PSMs) and protein identifications. This process involves spectral matching against reference databases, such as UniProt, to identify the peptides based on their mass spectra.

Step 6: Quantification and Normalization: After identifying the peptides, their abundance must be quantified. TMT-labeled peptides or label-free methods can be used to quantify protein levels. Given the variability inherent in single-cell proteomics, normalization is a crucial step to correct for technical biases. Techniques such as quantile normalization or median scaling are employed to standardize the data, ensuring that the differences observed between cells or experimental groups reflect true biological variation.

Normalization Methods:

(a) *Quantile Normalization:* Ensures uniform distributions of peptide intensities. Here is R code.

```
library(preprocessCore)
data <- as.matrix(data)
# Input data matrix
normalized_data <- normalize.quantiles(data)
```

(b) *Median Scaling:* Adjusts data to a common scale to minimize technical variation. Here is R code for Median Scaling:

```
 # Sample data matrix (e.g., rows are proteins/genes, columns
are samples)
 data <- matrix(c(100, 200, 300, 400, 500, 600, 700,
800, 900),
 nrow = 3, ncol = 3, byrow = TRUE)
 colnames(data) <- c("Sample_1", "Sample_2", "Sample_3")
 rownames(data) <- c("Protein_1", "Protein_2", "Protein_3")
 # View the raw data
 print("Raw Data:")
 print(data)
 # Step 1: Compute the median for each column (sample)
 medians <- apply(data, 2, median)
 print("Sample Medians:")
 print(medians)
 # Step 2: Subtract the median from each value (Median
Centering)
 median_centered_data <- sweep(data, 2, medians, "-")
 # Step 3 (Optional): Divide by the Median Absolute Deviation
(MAD) for scaling
 mad_values <- apply(data, 2, mad)
 scaled_data <- sweep(median_centered_data, 2, mad_
values, "/")
 # View the Median Scaled Data
 print("Median Centered Data (Median Subtracted):")
 print(median_centered_data)
```

```
    print("Median Scaled Data (Median Centered + Divided
by MAD):")
    print(scaled_data)
```

Step 7: Statistical Analysis and Differential Expression
Quantified proteomic data are subjected to statistical analysis to identify proteins that are differentially expressed between conditions. Common methods for differential expression analysis include t-tests, ANOVA, or the limma package in R, which is designed for high-dimensional data analysis. These tools allow researchers to pinpoint proteins that are significantly altered in response to experimental conditions, providing insights into brain function and pathology.

Statistical Tools:

(a) *Limma:* A widely used R package for analyzing proteomics data.
(b) *ANOVA and t-tests:* Traditional statistical tests for differential expression.

Step 8: Functional Annotation and Pathway Enrichment

Once differentially expressed proteins are identified, the next step is to determine their biological significance. Functional annotation using tools like Gene Ontology (GO), KEGG, or Reactome allows researchers to assign biological functions to proteins. Enrichment analysis can identify biological pathways or processes that are over-represented in the dataset, offering insights into how specific cell populations in the brain may respond to various stimuli or disease states.

Tools for Functional Analysis:

I. *Gene Ontology (GO):* Provides functional classifications for proteins.
II. *KEGG and Reactome:* Databases for pathway mapping and enrichment analysis.

Step 9: Data Visualization and Reporting

Data visualization is crucial for communicating findings. Heatmaps, volcano plots, and pathway diagrams are commonly used to display the results of single-cell proteomics experiments (Fig. 4.2). Tools such as R, Python (Matplotlib, Seaborn), or Perseus offer flexible platforms for creating publication-quality figures. Effective visualizations are essential for identifying key patterns in the data and conveying these findings in reports and publications, as shown using Table 4.3. Here are a few data visualization techniques as:

(a) *Heatmaps:* Display protein expression across different cells or conditions.
(b) *Volcano Plots:* Highlight differentially expressed proteins.
(c) *Pathway Diagrams:* Illustrate enriched biological pathways.

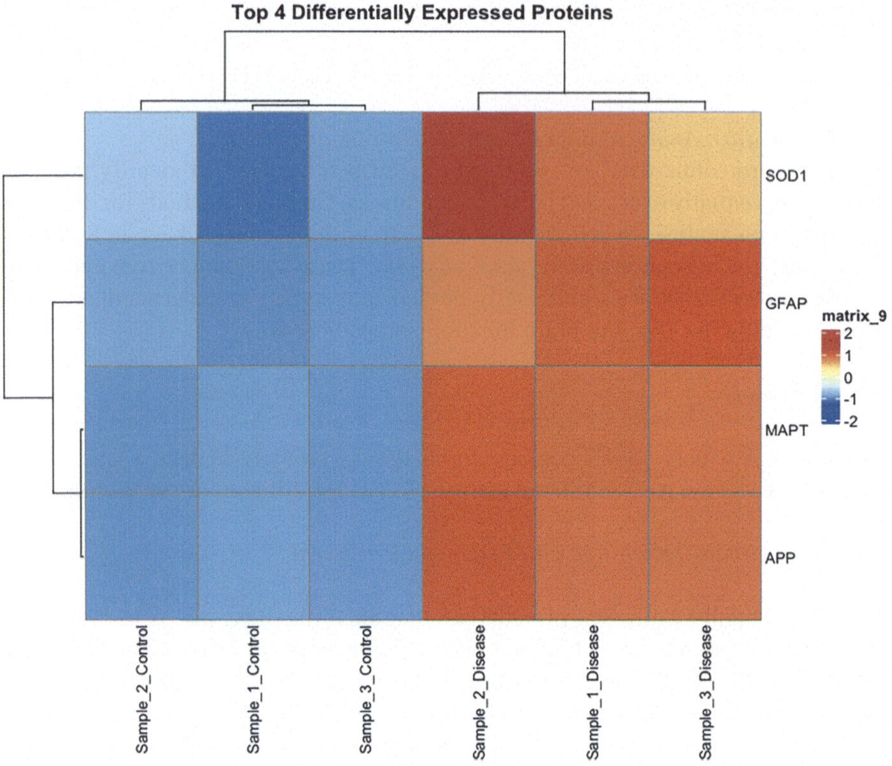

Fig. 4.2 Heatmap for top 4 differentially expressed proteins

Step 10: Comparative Computational Proteomics Analysis using R.

4.10 Neuroscience-Based Specific Markers

In neuroscience, marker guides typically refer to tools or methodologies used to identify specific cell types, proteins, or other biological structures within the brain or nervous system. These markers can be used in various imaging techniques, such as immunohistochemistry, fluorescence microscopy, or in situ hybridization, to study the organization, function, and pathology of the nervous system. Here are some common categories of markers used in neuroscience as shown in Table 4.4.

4.11 The Importance of Studying Proteins in Understanding Brain Cell Function

Highlights the importance of studying proteins in understanding brain cell function, we will outline a few key studies.

Table 4.3 Example dataset for comparative computational proteomics analysis using R pipeline. (See appendix at the end of the chapter)

Protein_ID	Gene_Symbol	Sample_1_Control	Sample_2_Control	Sample_3_Control	Sample_1_Disease	Sample_2_Disease	Sample_3_Disease
P00001	APP	120.5	115.8	118.2	300.2	310.5	298.6
P00002	GFAP	45.2	50.1	47.8	150.8	140.5	155.4
P00003	MAPT	80.6	78.2	79.4	200.5	205.3	198.9
P00004	GAPDH	200.1	198.5	199.8	205.6	210.3	202.5
P00005	SOD1	150.2	155.6	152.8	175.4	180.3	170.2

Table 4.4 List of the most famous markers used in neuroscience research

Category	Marker	Target	Application	References
Neuronal markers	DCX (Doublecortin)	Immature neurons	Marker for neuronal precursor and migration	Francis et al. (1999)
	TBR1	Cortical neurons	Early-born neuron marker	Hevner et al. (2001)
	Calbindin	Calcium-binding protein	Marker for specific interneuron subtypes	Celio (1990)
	Parvalbumin	Calcium-binding protein	Marker for fast-spiking interneurons	Celio (1986)
	Ctip2 (Bcl11b)	Layer V cortical neurons	Marker for subcortical projection neurons	Arlotta et al. (2005)
	ChAT(Choline Acetyltransferase)	Cholinergic neurons	Acetylcholine synthesis enzyme	Eckenstein and Sofroniew, (1983)
Glial cell markers	S100β	Astrocytes	Astrocyte and Schwann cell identification	Arlotta et al. (2005)
	MBP (Myelin Basic Protein)	Myelin	Oligodendrocyte marker and myelination	Campagnoni and Macklin, (1988)
	AQP4 (Aquaporin 4)	Astrocytes	Astrocyte-specific water channel protein	Nielsen et al. (1997)
	NG2 (Neural/ Glial Antigen 2)	Oligodendrocyte precursor cells	Marker for OPCs and pericytes	Nielsen et al. (1997)
Stem cell/ progenitor markers	SOX9	Glial progenitors	Marker for gliogenesis and neural crest cells	Kaneko et al. (2000)
	Musashi-1	Neural stem/ progenitor cells	RNA-binding protein in neural progenitors	Kaneko et al. (2000)
Neurotransmitter-specific markers	VGAT (Vesicular GABA Transporter)	GABAergic neurons	GABA vesicle transporter	McIntire et al. (1997)
	VMAT2 (Vesicular Monoamine Transporter 2)	Monoaminergic neurons	Monoamine vesicle transporter	Erickson et al. (1996)
	DAT (Dopamine Transporter)	Dopaminergic neurons	Dopamine reuptake transporter	Giros et al. (1996)
	SERT (Serotonin Transporter)	Serotonergic neurons	Serotonin reuptake transporter	Blakely et al. (1991)

(continued)

Table 4.4 (continued)

Category	Marker	Target	Application	References
Synaptic markers	Synapsin	Synaptic vesicles	Synaptic vesicle-associated protein	De Camilli et al. (1983)
	Bassoon	Active zone protein	Presynaptic active zone marker	tom Dieck et al. (1998)
	Homer1	Postsynaptic density	Postsynaptic scaffolding protein	Brakeman et al. (1997)
Disease-related markers	FUS (Fused in Sarcoma)	ALS and frontotemporal dementia	RNA-binding protein involved in neurodegeneration	Kwiatkowski et al. (2009)
	PrP (Prion Protein)	Prion diseases	Prion disease marker	Prusiner (1982)
	Huntingtin	Huntington's disease	Huntington's disease protein marker	The Huntington's Disease Collaborative Research Group (1993)
Vascular markers	Collagen IV	Basement membrane	Marker for blood vessels and basal lamina	Timpl (1989)
	Von Willebrand Factor (vWF)	Endothelial cells	Marker for blood vessel endothelium	Wagner et al. (1982)
	Glut1	Endothelial cells	Glucose transporter in the blood-brain barrier	Morgello et al. (1995)

4.11.1 Synaptic Plasticity and Memory Formation

Study: Molecular Mechanisms of Synaptic Function and Plasticity in Memory Formation (Kandel et al. 2001).

Summary: This seminal study by Nobel laureate Eric Kandel and colleagues demonstrated the role of proteins in synaptic plasticity, particularly focusing on the protein cAMP response element-binding protein (CREB). CREB is essential for converting short-term memories to long-term memories by regulating gene expression that strengthens synaptic connections.

Data Example: Kandel's work showed that increasing CREB activity enhances long-term potentiation (LTP) in hippocampal neurons, a key process in memory formation.

4.11.2 Proteins in Neurodegenerative Diseases

Study: Amyloid-beta Protein and Alzheimer's Disease: Molecular and Cellular Mechanisms (Hardy and Selkoe 2002).

Summary: This study explored how amyloid-beta (Aβ) protein aggregation in the brain leads to the pathology of Alzheimer's disease. The accumulation of Aβ proteins forms plaques that disrupt neuronal function and leads to neurodegeneration.

Data Example: Hardy and Selkoe provided data showing that the presence of Aβ oligomers correlate with synaptic dysfunction and cognitive decline in Alzheimer's patients, thus highlighting the importance of protein studies in understanding and potentially treating neurodegenerative diseases.

4.11.3 Protein Biomarkers in Neurological Disorders

Study: CSF Biomarkers for Alzheimer's Disease: Levels of Tau, Amyloid-beta, and Phosphorylated Tau Relate to Disease Severity (Blennow et al. 2010).

Summary: This study identified cerebrospinal fluid (CSF) levels of tau, phosphorylated tau, and amyloid-beta as biomarkers for Alzheimer's disease. It showed how these proteins can be used to diagnose the disease and assess its progression.

Data Example: Blennow et al. (2010) presented data indicating that elevated levels of phosphorylated tau and reduced levels of amyloid-beta in CSF are associated with increased disease severity, demonstrating the practical application of protein studies in clinical diagnostics.

4.11.4 Molecular Basis of Schizophrenia

Study: Dysbindin-1 and Schizophrenia: Association of a Synaptic Protein with a Complex Psychiatric Disorder (Weickert et al. 2004).

Summary: This study investigated the role of the dysbindin-1 protein, a component of the dystrophin-associated glycoprotein complex, in schizophrenia. Dysbindin-1 is involved in synaptic transmission, and mutations in the gene encoding this protein have been linked to schizophrenia.

Data Example: Weickert and colleagues found reduced levels of dysbindin-1 in the brains of schizophrenia patients, correlating with synaptic deficits and supporting the hypothesis that dysbindin-1 dysfunction contributes to the pathophysiology of schizophrenia.

4.11.5 Neurodevelopmental Disorders and Protein Dysfunction

Study: The role of FMRP in Fragile X Syndrome and the Regulation of Synaptic Function (Bear et al. 2004).

Summary: Fragile X syndrome is caused by the loss of the FMRP (Fragile X Mental Retardation Protein), leading to intellectual disability and autism. This study elucidated how the absence of FMRP disrupts synaptic protein synthesis, affecting synaptic function and plasticity.

Data Example: Bear et al. (2004) provided evidence that the lack of FMRP results in excessive protein synthesis at synapses, impairing synaptic development and plasticity, which underlies the cognitive deficits observed in Fragile X syndrome.

These examples demonstrate how studying specific proteins provides deep insights into brain cell function and the mechanisms underlying various neurological conditions. These studies have not only advanced our understanding of brain function but also opened new avenues for therapeutic interventions.

4.12 Comparative Genomics in Brain Studies

Genome comparison employs various techniques to align and analyze DNA sequences from different organisms, enabling researchers to explore evolutionary relationships and genetic diversity. Sequence alignment is a fundamental approach, with tools like BLAST (Altschul et al. 1990) facilitating pairwise comparisons to identify homologous regions across species. ClustalW and MAFFT (Sievers et al. 2011) are commonly used for multiple sequence alignment, allowing researchers to align sequences from several organisms simultaneously. Whole-genome alignment, conducted using tools such as MUMmer and LASTZ, helps detect large-scale structural variations, including insertions, deletions, and inversions. Additionally, synteny analysis examines the conservation of gene order across different species, revealing evolutionary events such as chromosomal rearrangements (Hiller et al. 2004).

4.13 Conclusions

The chapter underscores the transformative potential of single-cell technologies and multi-omics integration emphasizing neuroscience. By bridging the gaps between genomics, epigenomics, and proteomics, these methodologies offer a comprehensive understanding of cellular mechanisms underlying brain function and disease. As the field continues to evolve, such approaches promise to uncover novel therapeutic targets and revolutionize the landscape of precision medicine in neuroscience. The conclusion of this chapter emphasizes that the significant advancements in Single-cell approaches, including transcriptomics and proteomics, have been instrumental in mapping complex cellular networks and identifying rare or disease-specific cell types. These techniques have provided critical insights into how individual cells influence brain function and contribute to disorders such as Alzheimer's, Parkinson's, and schizophrenia. The chapter underscores the value of integrating multi-omics data as genomic, epigenomic, and proteomics, to create comprehensive models of cellular states, which can guide the development of precision medicine strategies, whereas single-cell proteomics of the mouse brain offers unparalleled insights into the molecular underpinnings of brain function and disease. This computational pipeline, from sample preparation to data analysis, provides a comprehensive framework for researchers interested in exploring proteomic heterogeneity at the single-cell level. As this field continues to evolve, new techniques and computational tools will further enhance our ability to understand the complex proteomic landscapes of individual brain cells, enabling novel discoveries in neuroscience. The computational pipeline described for single-cell proteomics, from sample preparation to data analysis, provides a robust framework for advancing neuroscience research. As the field continues to evolve, emerging methodologies and computational tools will further enhance our understanding of the brain's complexity, paving the way for the identification of new therapeutic targets. In conclusion, the chapter highlights the transformative potential of single-cell research in neuroscience, offering more targeted approaches to studying brain diseases and their molecular foundations. This ongoing evolution promises to deliver breakthroughs in diagnosing and treating neurological disorders.

Short Chapter Summary

Single-cell technologies, such as scRNA-seq and snRNA-seq, provide insights into cellular diversity, enabling the identification of rare or previously unknown cell types and states in the brain. The integration of multi-omics approaches, combining genomics, epigenomics, and proteomics, allows researchers to decode intricate molecular mechanisms underlying brain function and dysfunction, neurodegenerative diseases like Alzheimer's and Parkinson's, as well as psychiatric disorders like schizophrenia and depression. The chapter also highlights the importance of spatial transcriptomics for preserving the spatial organization of cells, which is critical for understanding regional brain functions. Advanced computational tools, such as Harmony and CellChat, are detailed, providing researchers with robust pipelines for analyzing large-scale single-cell datasets. Furthermore, the integration of multi-omics data facilitates a systems-level understanding of molecular disruptions, driving innovations in precision medicine by enabling the development of targeted therapies. Through these insights, the chapter empowers researchers to uncover novel biomarkers, therapeutic targets, and regulatory circuits, advancing both basic neuroscience and clinical applications.

Author Contributions Chapter concept, computational design, writing, and editing. Upasna Srivastava, Swarna Kanchan, Minu Kesheri; review, Sonal Sukreet, Manish Kumar Gupta.

Appendix

```
# 1. Install and Load Required Libraries
# Install required packages if not already installed
if (!requireNamespace("tidyverse", quietly = TRUE)) install.
packages("tidyverse")
if (!requireNamespace("limma", quietly = TRUE)) install.
packages("limma")
if (!requireNamespace("pheatmap", quietly = TRUE)) install.
packages("pheatmap")
if (!requireNamespace("BiocManager", quietly = TRUE)) install.
packages("BiocManager")
if (!requireNamespace("clusterProfiler", quietly = TRUE)) BiocMana
ger::install("clusterProfiler")
if (!requireNamespace("org.Hs.eg.db", quietly = TRUE))
BiocManager::install("org.Hs.eg.db")
if (!requireNamespace("enrichplot", quietly = TRUE)) install.
packages("enrichplot")
if (!requireNamespace("ggplot2", quietly = TRUE)) install.
packages("ggplot2")

# Load required libraries
library(tidyverse)
library(limma)
library(pheatmap)
library(clusterProfiler)
library(org.Hs.eg.db) # Human annotation database
library(enrichplot)
library(ggplot2)

#2. Load Example Proteomics Data
# Replace "brain_proteomics_example.csv" with your file path
protein_data <- read.csv("brain_proteomics_example.csv")

# View the data structure
head(protein_data)
protein_data
# A tibble: 5 × 8
 Protein_ID Gene_Symbol Sample_1_Control Sample_2_Control
Sample_3_Control Sample_1_Disease
```

```
 <chr> <chr> <dbl> <dbl> <dbl> <dbl>
1 P00001 APP 120. 116. 118. 300.
2 P00002 GFAP 45.2 50.1 47.8 151.
3 P00003 MAPT 80.6 78.2 79.4 200.
4 P00004 GAPDH 200. 198. 200. 206.
5 P00005 SOD1 150. 156. 153. 175.
# i 2 more variables: Sample_2_Disease <dbl>, Sample_3_
Disease <dbl>

# 3. Preprocess Data
# Ensure intensities are numeric and handle missing values
# Remove non-intensity columns (Protein_ID, Gene_Symbol assumed
as metadata)
protein_data <- protein_data %>% select(-Protein_ID,
-Gene_Symbol)

# Separate control and disease columns
control_cols <- protein_data %>% select(contains("Control"))
disease_cols <- protein_data %>% select(contains("Disease"))

# Check for missing values
sum(is.na(protein_data)) # Count missing values

normalized_data <- as.data.frame(normalized_data )
normalized_data
Sample_1_Control Sample_2_Control Sample_3_Control Sample_1_
Disease Sample_2_Disease Sample_3_Disease
1 120.5 115.8 118.2 300.2 310.5 298.6
2 45.2 50.1 47.8 150.8 140.5 155.4
3 80.6 78.2 79.4 200.5 205.3 198.9
4 200.1 198.5 199.8 205.6 210.3 202.5
5 150.2 155.6 152.8 175.4 180.3 170.2
row.names(normalized_data ) <- example$Gene_Symbol
normalized_data
 Sample_1_Control Sample_2_Control Sample_3_Control Sample_1_
Disease Sample_2_Disease Sample_3_Disease
APP 120.5 115.8 118.2 300.2 310.5 298.6
GFAP 45.2 50.1 47.8 150.8 140.5 155.4
MAPT 80.6 78.2 79.4 200.5 205.3 198.9
GAPDH 200.1 198.5 199.8 205.6 210.3 202.5
SOD1 150.2 155.6 152.8 175.4 180.3 170.2
```

```
# 5. Differential Expression Analysis using Limma
# Create a design matrix for control and disease groups
> group <- factor(c(rep("Control", ncol(control_cols)),
rep("Disease", ncol(disease_cols))))
> design <- model.matrix(~ group)
> # Apply limma pipeline
> fit <- lmFit(normalized_data , design)
> fit <- eBayes(fit)
>
>
> fit
An object of class "MArrayLM"
$coefficients
 (Intercept) groupDisease
APP 118.1667 184.933333
GFAP 47.7000 101.200000
MAPT 79.4000 122.166667
GAPDH 199.4667 6.666667
SOD1 152.8667 22.433333

$rank
[1] 2
$assign
[1] 0 1

$qr
$qr
 (Intercept) groupDisease
1 -2.4494897 -1.2247449
2 0.4082483 1.2247449
3 0.4082483 0.2898979
4 0.4082483 -0.5265986
5 0.4082483 -0.5265986
6 0.4082483 -0.5265986
attr(,"assign")
[1] 0 1
attr(,"contrasts")
attr(,"contrasts")$group
[1] "contr.treatment"
```

```
$qraux
[1] 1.408248 1.289898

$pivot
[1] 1 2

$tol
[1] 1e-07

$rank
[1] 2

$df.residual
[1] 4 4 4 4 4

$sigma
 APP GFAP MAPT GAPDH SOD1
4.859698 5.666569 2.503331 2.841361 4.049897

$cov.coefficients
 (Intercept) groupDisease
(Intercept) 0.3333333 -0.3333333
groupDisease -0.3333333 0.6666667

$stdev.unscaled
 (Intercept) groupDisease
APP 0.5773503 0.8164966
GFAP 0.5773503 0.8164966
MAPT 0.5773503 0.8164966
GAPDH 0.5773503 0.8164966
SOD1 0.5773503 0.8164966

$pivot
[1] 1 2

$Amean
 APP GFAP MAPT GAPDH SOD1
210.6333 98.3000 140.4833 202.8000 164.0833

$method
[1] "ls"
```

```
$design
 (Intercept) groupDisease
1 1 0
2 1 0
3 1 0
4 1 1
5 1 1
6 1 1
attr(,"assign")
[1] 0 1
attr(,"contrasts")
attr(,"contrasts")$group
[1] "contr.treatment"

$df.prior
[1] Inf

$s2.prior
[1] 17.29367

$var.prior
[1] 0.9251942 0.9251942

$proportion
[1] 0.01

$s2.post
[1] 17.29367 17.29367 17.29367 17.29367 17.29367

$t
 (Intercept) groupDisease
APP 49.21665 54.464972
GFAP 19.86715 29.804552
MAPT 33.07026 35.979474
GAPDH 83.07826 1.963409
SOD1 63.66927 6.606872

$df.total
[1] 20 20 20 20 20

$p.value
 (Intercept) groupDisease
APP 2.398660e-22 3.206983e-23
GFAP 1.226479e-14 4.768070e-18
MAPT 6.200368e-19 1.179350e-19
```

```
GAPDH 7.154100e-27 6.365492e-02
SOD1 1.436476e-24 1.956659e-06

$lods
 (Intercept) groupDisease
APP 885.0979 857.019523
GFAP 139.8218 253.114268
MAPT 396.7307 371.159848
GAPDH 2531.7088 -3.910043
SOD1 1484.7875 7.654670

$F
[1] 9179.630 2120.415 4070.867 7136.535 4692.323

$F.p.value
[1] 0 0 0 0 0
# Extract results
> results <- topTable(fit, coef = 2, adjust = "BH", number =
nrow(normalized_data))
> # View significant proteins (adjusted p-value < 0.05)
> significant_proteins <- results %>%
+ filter(adj.P.Val < 0.05) %>%
+ arrange(logFC) # Sort by log fold change
> print(significant_proteins)
 logFC AveExpr t P.Value adj.P.Val B
SOD1 22.43333 164.0833 6.606872 1.956659e-06 2.445824e-06 7.65467
GFAP 101.20000 98.3000 29.804552 4.768070e-18 7.946783e-18
253.11427
MAPT 122.16667 140.4833 35.979474 1.179350e-19 2.948376e-19
371.15985
APP 184.93333 210.6333 54.464972 3.206983e-23 1.603492e-22
857.01952
```


```
> print(significant_proteins)
 logFC AveExpr t P.Value adj.P.Val B
SOD1 22.43333 164.0833 6.606872 1.956659e-06 2.445824e-06 7.65467
GFAP 101.20000 98.3000 29.804552 4.768070e-18 7.946783e-18
253.11427
MAPT 122.16667 140.4833 35.979474 1.179350e-19 2.948376e-19
371.15985
APP 184.93333 210.6333 54.464972 3.206983e-23 1.603492e-22
857.01952
```

```
> logFC AveExpr t P.Value adj.P.Val B
1 22.43333 164.0833 6.606872 1.956659e-06 2.445824e-06 7.65467
2 101.20000 98.3000 29.804552 4.768070e-18 7.946783e-18 253.11427
3 122.16667 140.4833 35.979474 1.179350e-19 2.948376e-19
371.15985
4 184.93333 210.6333 54.464972 3.206983e-23 1.603492e-22
857.01952

# 6. Visualize Results
# Heatmap of Differentially Expressed Proteins
# Select the top 4 most significant proteins
top_proteins <- rownames(significant_proteins[1:4, ])

# Subset normalized data for heatmap
heatmap_data <- normalized_data[top_proteins, ]

# Create a heatmap
pheatmap(
 heatmap_data,
 cluster_rows = TRUE,
 cluster_cols = TRUE,
 scale = "row",
 main = "Top 4 Differentially Expressed Proteins"
)

# Volcano Plot
results <- results %>%
 mutate(Significant = ifelse(adj.P.Val < 0.05 & abs(logFC) > 1,
"Yes", "No"))

ggplot(results, aes(x = logFC, y = -log10(P.Value), color =
Significant)) +
 geom_point(alpha = 0.7) +
 scale_color_manual(values = c("No" = "gray", "Yes" = "red")) +
 labs(title = "Volcano Plot", x = "Log Fold Change", y =
"-log10(P-value)") +
 theme_minimal()

# 7. Functional Analysis
# Example significant gene symbols (replace with your actual
gene list)
library(org.Hs.eg.db)
# 7. Functional Analysis
# Example significant gene symbols (replace with your actual
gene list)
```

```
significant_genes <- c("APP", "GFAP", "MAPT", "GAPDH", "SOD1")

# Convert gene symbols to ENTREZ IDs
entrez_ids <- bitr(significant_genes, fromType = "SYMBOL", toType
= "ENTREZID", OrgDb = org.Hs.eg.db)

# Check converted IDs
print(entrez_ids)

# (A) GO Enrichment Analysis
go_enrichment <- enrichGO (
 gene = entrez_ids$ENTREZID,
 OrgDb = org.Hs.eg.db,
 ont = "BP", # Biological Process
 pAdjustMethod = "BH",
 pvalueCutoff = 0.05,
 qvalueCutoff = 0.05,
 readable = TRUE
)

# View the top enriched GO terms
# View the top enriched GO terms
> head(go_enrichment, 2)
 ID Description GeneRatio BgRatio pvalue p.adjust qvalue
geneID Count
GO:0021782 GO:0021782 glial cell development 4/5 122/18870
8.273997e-09 4.815466e-06 1.289002e-06 APP/GFAP/MAPT/SOD1 4
GO:0014002 GO:0014002 astrocyte development 3/5 43/18870
1.098684e-07 2.981567e-05 7.981043e-06 APP/GFAP/MAPT 3

# (B) KEGG Pathway Analysis
> kegg_enrichment <- enrichKEGG(
+ gene = entrez_ids$ENTREZID,
+ organism = "hsa", # 'hsa' for Homo sapiens
+ pvalueCutoff = 0.05,
+ qvalueCutoff = 0.05
+ )

> kegg_enrichment
# over-representation test
#...@organism hsa
#...@ontology KEGG
#...@keytype kegg
#...@gene chr [1:5] "351" "2670" "4137" "2597" "6647"
#...pvalues adjusted by 'BH' with cutoff <0.05
```

```
#...2 enriched terms found
'data.frame': 2 obs. of 9 variables:
 $ ID : chr "hsa05010" "hsa05022"
 $ Description: chr "Alzheimer disease" "Pathways
of neurodegeneration - multiple diseases"
 $ GeneRatio : chr "3/5" "3/5"
 $ BgRatio : chr "391/8541" "483/8541"
 $ pvalue : num 0.000889 0.00165
 $ p.adjust : num 0.0157 0.0157
 $ qvalue : num 0.0104 0.0104
 $ geneID : chr "351/4137/2597" "351/4137/6647"
 $ Count : int 3 3
#...Citation
 T Wu, E Hu, S Xu, M Chen, P Guo, Z Dai, T Feng, L Zhou, W Tang,
L Zhan, X Fu, S Liu, X Bo, and G Yu.
 clusterProfiler 4.0: A universal enrichment tool for interpreting
omics data.
 The Innovation. 2021, 2(3):100141 # View the top enriched
pathways

# 8. Bar Plot for Enriched Pathways
# Bar plot for the top 10 GO terms
barplot(go_enrichment, showCategory = 10, title = "Top 10 GO
Enriched Terms (Biological Process)")

# Bar plot for the top 10 KEGG pathways
barplot(kegg_enrichment, showCategory = 10, title = "Top 10 KEGG
Enriched Pathways")

# ggplot2-based Bar Plot for KEGG
kegg_df <- as.data.frame(kegg_enrichment)

# Select top 10 pathways for plotting
top_kegg <- kegg_df %>% top_n(-10, p.adjust)

# Create ggplot bar plot
ggplot(top_kegg, aes(x = reorder(Description, -p.adjust), y =
-log10(p.adjust))) +
 geom_bar(stat = "identity", fill = "steelblue") +
 coord_flip() +
 labs(title = "Top 10 KEGG Enriched Pathways", x = "Pathway", y =
"-log10(Adjusted P-value)") +
 theme_minimal()
```

References

Aebersold R, Mann M (2003) Mass spectrometry-based proteomics. Nature 422(6928):198–207

Altschul SF, Gish W, Miller W, Myers EW, Lipman DJ (1990) Basic local alignment search tool. J Mol Biol 215(3):403–410

Azevedo, Frederico AC et al. "Equal numbers of neuronal and nonneuronal cells make the human brain an isometrically scaled-up primate brain." J Comp Neurol vol. 513(5) (2009):532–41. https://doi.org/10.1002/cne.21974

Bakken TE, Hodge RD, Miller JA, Yao Z, Nguyen TN, Aevermann B, Lein ES (2018) Single-nucleus and single-cell transcriptomes compared in matched cortical cell types. PLoS One 13(2):e0209648

Barres BA (2008) The mystery and magic of glia: a perspective on their roles in health and disease. Neuron 60(3):430–440

Bartheld V, Christopher S, Bahney J, Herculano-Houzel S (2016) The search for true numbers of neurons and glial cells in the human brain: a review of 150 years of cell counting. J Comp Neurol 524(18):3865–3895

Bergen V, Lange M, Peidli S, Wolf FA, Theis FJ (2020) Generalizing RNA velocity to transient cell states with dynamical modeling. Nat Biotechnol 38(12):1408–1414

Bujak JP, Pereira AL, Azevedo J, Bujak AA, Leshyk V, Pham Gia M, Stadtlander T, Vasconcelos V, Winstead DJ (2024) *Azolla* as a safe food: suppression of cyanotoxin-related genes and cyanotoxin production in its symbiont, *Nostoc azollae*. Plants 27(19):2707

Cajigas IJ, Tushev G, Will TJ, tom Dieck S, Fuerst N, Schuman EM (2012) The local transcriptome in the synaptic neuropil revealed by deep sequencing and high-resolution imaging. Neuron 74(3):453–466

Chaturvedi P, Pierides I, Zhang S, Schwarzerova J, Ghatak A, Weckwerth W (2024) Multiomics for crop improvement. In: Pandey MK, Bentley A, Desmae H, Roorkiwal M, Varshney RK (eds) Frontier technologies for crop improvement. Sustainability sciences in Asia and Africa. Springer, Singapore, pp 107–141

Darmanis S, Sloan SA, Zhang Y, Enge M, Caneda C, Shuer LM, Quake SR (2015) A survey of human brain transcriptome diversity at the single-cell level. Proc Natl Acad Sci USA 112(23):7285–7290

Del Bigio, Marc R. "Ependymal cells: biology and pathology." Acta Neuropathol vol. 119(1) (2010):55–73. https://doi.org/10.1007/s00401-009-0624-y

Ding SL, Royall JJ, Sunkin SM, Ng L, Facer BA, Lesnar P, Lein ES (2020) Comprehensive cellular-resolution atlas of the adult human brain. J Comp Neurol 528(1):25–47

Fields RD, Stevens-Graham B (2002) New insights into neuron-glia communication. Science 298(5593):556–562

Gaidatzis D, Burger L, Florescu M, Stadler MB (2015) Analysis of intronic and exonic reads in RNA-seq data characterizes transcriptional and post-transcriptional regulation. Nat Biotechnol 33(7):722–729

Graf EH, Bryan A, Bowers M, Grys TE (2025) One size fits small: the narrow utility for plasma metagenomics. J Appl Lab Med 10(1):171–183

Grindberg RV, Yee-Greenbaum JL, McConnell MJ, Novotny M, O'Shaughnessy AL, Lambert GM, Quake SR (2013) RNA-sequencing from single nuclei. Proc Natl Acad Sci 110(49):19802–19807

Hiller M, Schaar BT, Bejerano G, Pachter L, Haussler D (2004) Genome alignment and phylogenetic footprinting. Nat Rev Genet 5(9):644–655

Hodge RD, Bakken TE, Miller JA, Smith KA, Barkan ER, Graybuck LT, Close JL, Long B, Johansen N, Penn O, Yao Z (2019) Conserved cell types with divergent features in human versus mouse cortex. Nature 573(7772):61–68

Jaiswal D, Paul P, Srivastava U, Singh AK, Singh S (2023) Identification and analysis of genomic islands in *Burkholderia cenocepacia* genome by using computational approaches. Int J Life Sci Appl Sci 5(3):25–24

Kanchan S, Sinha RP, Chaudière J, Kesheri M (2020) Computational metagenomics: current status and challenges. In: Katara P (ed) Recent trends in 'computational omics': concepts and methodology. Nova Science Publishers, New York, pp 371–395

Kanchan S, Ogden E, Kesheri M, Skinner A, Miliken E, Lyman D, Armstrong J, Sciglitano L, Hampikian G (2024) COVID-19 hospitalizations and deaths predicted by SARS-CoV-2 levels in Boise, Idaho wastewater. Sci Total Environ 907:167742

Kesheri M, Kanchan S, Richa, Sinha RP (2014) Isolation and in-silico analysis of Fe- superoxide dismutase in *Nostoc commune*. Gene 553(2):117–125

Kesheri M, Kanchan S, Sinha RP (2021) Isolation and in-silico analysis of antioxidants in response to temporal variations in the cyanobacterium *Oscillatoria* sp. Gene Rep 23:101023

Kesheri M, Kanchan S, Sinha RP (2022) Responses of antioxidants for resilience to temporal variations in the cyanobacterium *Microcystis aeruginosa*. Afr J Bot 148:190–199

Kesheri M, Kanchan S, Srivastava S, Ratna-Raj R, Chittoori B, Vaishampayan A, Rastogi RP, Sinha RP, Primerano D (2024a) Ecology and environmental omics. In: Gupta MK, Katara P, Mondal S, Singh RL (eds) Integrative omics: concepts, methodology, and application. Academic Press, Elsevier, Cambridge, pp 315–331

Kesheri M, Kanchan S, Kumar A, Srivastava U, Sharda S, Malik B, Mishra T, Kaithal P, Narayan J, Kumar P, Priya P, Sinha RP (2024b) Peculiar endosymbiosis in the cyanobiont *Nostoc azollae* 0708: an in-silico approach. In: Kesheri M, Kanchan S, Salisbury TB, Sinha RP (eds) Microbial omics in environment and health. Springer Nature, Singapore, pp 343–365

Kesheri M, Kanchan S, Salisbury TB, Sinha RP (2024c) Microbial omics in environment and health. Springer Nature, Singapore

Kim N, Junyeong M, Wonjong K, Jungyeon K, Peter B, Insuk L (2024) Genome-resolved metagenomics: a game changer for microbiome medicine. Exp Mol Med 56(7):1501–1512

Kumar P, Kanchan S, Kesheri M (2024) Multi-omics in human disease biomarker discovery. In: Kesheri M, Kanchan S, Salisbury TB, Sinha RP (eds) Microbial omics in environment and health. Springer, Singapore, pp 205–239

La MG, Soldatov R, Zeisel A, Braun E, Hochgerner H, Petukhov V, Linnarsson S (2018) RNA velocity of single cells. Nature 560(7719):494–498

Lacar B, Linker SB, Jaeger BN, Krishnaswami S, Barron J, Kelder M, Lein ES (2016) Nuclear RNA-seq of single neurons reveals molecular signatures of activation. Nat Commun 7(1):1–10

Lake BB, Ai R, Kaeser GE, Salathia NS, Yung YC, Liu R, Zhang K (2016) Neuronal subtypes and diversity revealed by single-nucleus RNA sequencing of the human brain. Science 352(6293):1586–1590

Lake BB, Codeluppi S, Yung YC, Gao D, Chun J, Kharchenko PV, Linnarsson S (2017) A comparative strategy for single-nucleus and single-cell transcriptomes confirms accuracy in predicted cell-type expression from nuclear RNA. Sci Rep 7(1):1–12

Mallik B, Mishra T, Dubey P, Kesheri M, Kanchan S (2024) Exploring the secrets of microbes: unveiling the hidden world through microbial omics in environment and health. In: Kesheri M, Kanchan S, Salisbury TB, Sinha RP (eds) Microbial omics in environment and health. Springer, Singapore, pp 269–294

Meadows T, Coats ER, Narum S, Top EM, Ridenhour BJ, Stalder T (2025) Epidemiological model can forecast COVID-19 outbreaks from wastewater-based surveillance in rural communities. Water Res 26:122671

Middleton SA, Eberwine J, Kim J (2019) A practical approach to single-cell RNA-sequencing of human post-mortem tissue. J Neurosci Methods 311:172–179

Mishra T, Mallik B, Kesheri M, Kanchan S (2024) The interplay of gut microbiome in health and diseases. In: Kesheri M, Kanchan S, Salisbury TB, Sinha RP (eds) Microbial omics in environment and health. Springer, Singapore, pp 1–34

Nedergaard, Maiken et al. "New roles for astrocytes: redefining the functional architecture of the brain." Trends Neurosci vol. 26(10) (2003):523–30. https://doi.org/10.1016/j.tins.2003.08.008

Okamoto T, Yaginuma K, Natsume Y, Yamanaka H, Kusama D, Fukuda M, Yao R (2021) Comparative analysis of patient-matched PDOs revealed a reduction in OLFM4-associated clusters in metastatic lesions in colorectal cancer. Stem Cell Rep 16(4):954–967. https://doi.org/10.1016/j.stemcr.2021.02.012

Purves D, Augustine GJ, Fitzpatrick D, Katz LC, Lamantia AS, McNamara JO, Williams SM (2001) Neuroscience, 2nd edn. Sinauer Associates, Sunderland

Ransohoff, Richard M, and V Hugh Perry. "Microglial physiology: unique stimuli, specialized responses." Annu Rev Immunol vol. 27 (2009):119–45. https://doi.org/10.1146/annurev.immunol.021908.132528

Sakahara M, Okamoto T, Srivastava U, Natsume Y, Yamanaka H, Suzuki Y, Obama K, Nagayama S, Yao R (2024) Paneth-like cells produced from OLFM4$^+$ stem cells support OLFM4$^+$ stem cell growth in advanced colorectal cancer. Commun Biol 7(1):27. https://doi.org/10.1038/s42003-023-05504-8

Saunders A, Macosko EZ, Wysoker A et al (2018) A single-cell atlas of cell types in the mouse brain. Nat Neurosci 21(2):276–287

Shukla N, Bhatia A, Kesheri M, Kanchan S, Narayan J (2024) Planet in peril: unveiling the microbial consequences of environmental degradation on human health through omics approaches. In: Kesheri M, Kanchan S, Salisbury TB, Sinha RP (eds) Microbial omics in environment and health. Springer, Singapore. (pp. 35–56)

Sievers F, Wilm A, Dineen D, Gibson TJ, Karplus K, Li W, Higgins DG (2011) Fast, scalable generation of high-quality protein multiple sequence alignments using Clustal Omega. Mol Syst Biol 7(1):539

Srivastava U, Singh G (2013) Comparative homology modelling for HPV type 16 E 7 proteins by using MODELLER and its validations with SAVS and ProSA Web server. J Comput Intell Bioinform 6(27):27–33

Srivastava U, Singh S (2022) Approaches of single-cell analysis in crop improvement. In: Wani SH, Kumar A (eds) Genomics of cereal crops. Springer Protocols Handbooks, Humana, New York, NY, pp 271–294

Srivastava U, Kanchan S, Kesheri M, Singh S (2023) Nutrimetabolomics: metabolomics in nutrition research. In: Soni V, Hartman TE (eds) Metabolomics. Springer, Cham, pp 241–268

Srivastava U, Kanchan S, Kesheri M, Sharma A (2025) Single-cell technology for improvement in significant crops. In: Chen J-T (ed) Guide to plant single-cell technology. Academic, pp 299–320

Tabula Muris Consortium (2018) Single-cell transcriptomics of 20 mouse organs creates a tabula Muris. Nature 562(7727):367–372

Tasic B, Menon V, Nguyen TN, Kim TK, Jarsky T, Yao Z, Zeng H (2016) Adult mouse cortical cell taxonomy revealed by single cell transcriptomics. Nat Neurosci 19(2):335–346

Tasic B, Yao Z, Graybuck LT, Smith KA, Nguyen TN, Bertagnolli D, Zeng H (2018) Shared and distinct transcriptomic cell types across neocortical areas. Nature 563(7729):72–78

Tushev G, Glock C, Heumüller M, Biever A, Jovanovic M, Schuman EM (2018) Alternative 3′ UTRs modify the localization, regulatory potential, stability, and plasticity of mRNAs in neuronal compartments. Neuron 98(3):495–511

Venter JC, Adams MD, Myers EW, Li PW, Mural RJ, Sutton GG, Kalush F (2001) The sequence of the human genome. Science 291(5507):304–1351

Verkhratsky A, Butt A (2013) Glial physiology and pathophysiology. Wiley-Blackwell

Wilkins MR, Gasteiger E, Bairoch A, Sanchez JC, Williams KL, Appel RD, Hochstrasser DF (1999) Protein identification and analysis tools in the ExPASy server. Humana Press 112:531–552

Wu AR, Neff NF, Kalisky T, Dalerba P, Treutlein B, Rothenberg ME, Quake SR (2014) Quantitative assessment of single-cell RNA-sequencing methods. Nat Methods 11(1):41–46

Yadav PK, Singh G, Gautam B, Singh S, Yadav M, Srivastav U, Singh B (2013) Molecular modeling, dynamics studies and virtual screening of Fructose 1, 6 biphosphate aldolase-II in community acquired- methicillin resistant *Staphylococcus aureus* (CA-MRSA). Bioinformation 9(3):158–164

Yao Z, Liu H, Xie F, Fischer S, Adkins RS, Aldridge AI, Mukamel EA (2021) A transcriptomic and epigenomic cell atlas of the mouse primary motor cortex. Nature 598(7879):103–110

Zeisel A, Muñoz-Manchado AB, Codeluppi S, Lönnerberg P, La Manno G, Juréus A, Linnarsson S (2015) Cell types in the mouse cortex and hippocampus revealed by single-cell RNA-seq. Science 347(6226):1138–1142

Zhou Y, Song WM, Andhey PS (2020) Large-scale spatial transcriptomics reveals the molecular, cellular, and spatial organization of the human dorsolateral prefrontal cortex. Cell Rep 32(5):108511

Multi-omics Applications in Drug Discovery and Development

5

Vivek Keshri and Jyothi Belurappa

Abstract

In the last decade, significant strides have been made in the field of drug discovery and development, driven by the rapid advancements in multi-omics technologies. Multi-omics, encompassing genomics, transcriptomics, proteomics, metabolomics, and epigenomics, has emerged as a powerful approach to unravel the complexity of diseases and identify novel therapeutic targets. Genomic studies have unveiled genetic variations associated with drug responses, leading to the development of pharmacogenomics-based strategies for personalized drug prescriptions. Additionally, transcriptomic analysis has shed light on disease-related pathways and gene expression profiles, guiding the discovery of new drug targets and repurposing existing drugs for novel indications. Furthermore, proteomics has played a pivotal role in characterizing protein interactions, post-translational modifications, and drug-protein interactions, aiding in the development of targeted therapies and the optimization of drug candidates. Metabolomics, on the other hand, has offered insights into the dynamic metabolic changes associated with disease states. Moreover, the application of epigenomics has provided crucial information on epigenetic modifications, uncovering potential druggable epigenetic targets that could influence disease progression. In conclusion, the integration of multi-omics approaches has revolutionized drug discovery and development, allowing for the identification of personalized therapeutic strategies and the acceleration of precision medicine initiatives. This comprehensive review aims at providing an overview of the various multi-omics applications in drug development and highlights their impact.

V. Keshri (✉) · J. Belurappa
Department of Biotechnology, REVA University, School of Applied Sciences, Bengaluru, Karnataka, India

M. Kesheri et al. (eds.), *Multi-Omics in Biomedical Sciences and Environmental Sustainability*, https://doi.org/10.1007/978-981-96-7067-3_5

Keywords

Epigenomics · Genomics · Metabolomics · Proteomics · Transcriptomics

Objectives

- Explore the advancements in genomics, transcriptomics, proteomics, metabolomics, and epigenomics that have driven drug discovery and development over the past decade.
- Highlight how multi-omics approaches unravel disease complexity, identify novel therapeutic targets, and inform pharmacogenomics-based strategies for personalized medicine.
- Discuss the role of transcriptomics, proteomics, and metabolomics in understanding drug-protein interactions, metabolic changes, and disease pathways to optimize drug candidates and repurpose existing drugs.
- Examine the application of epigenomics in identifying druggable epigenetic targets, advancing precision medicine, and accelerating the development of tailored therapeutic strategies.

5.1 Introduction

Multi-omics has bestowed a cutting-edge approach that incorporates data at the biomolecular level pertaining to genomics, transcriptomics, proteomics, metabolomics, and epigenomics to drive drug discovery and development over the past decade. Integration of various omics technologies facilitates the creation of a holistic view of the functioning of various physiological processes in living systems and gives an overview of cross-talks and interactions at the microbial and metabolite levels (Kesheri et al. 2024). Multi-omics has been used to explore disease complexity, identification, and repurposing of therapeutic targets and pharmacogenomics-based precision medicine (Du et al. 2024; Srivastav et al. 2024). Bioinformatics has opened avenues for molecular docking, simulation and suitability tests for various natural compounds as druggable targets (Sahu et al. 2023; Graf et al. 2025) leading to advancements in simulation profiling of multi-omic disease models with unique pipelines unraveling the intricacies of complex disorders.

Multi-omics has leveraged biomarker discovery and identification of novel drug molecules (Ivanisevic and Sewduth 2023; Kumar et al. 2024). Theragnostics-mediated identification of new molecular biomarkers has been eased by the utilization of high-throughput detection, generation of molecular imaging probes along with nano-devices that help in diagnosis as well as therapy. The onset of data-sharing initiatives as well as cloud computing platforms has increased data accessibility, boosting accelerated discoveries globally (Kaithal et al. 2024; Wang et al. 2024). Thus, bridging the force of big data through computational techniques, and integration of various omics technologies has aided in unveiling the complexities of hazardous diseases and engendered customized medical care through novel drug discovery and developments. This chapter aims at providing an overview of the latest progress and challenges of multi-omics for deciphering treatments for human diseases, emphasizing the synergistic effects of integrating omics technologies leveraging novel drug discovery and development.

5.2 Genomics

The study of an organism's entire genetic material has revolutionized the field of drug discovery and development. Genomics plays an important role in drug discovery by giving crucial insights into the genetic basis of diseases and helping the creation of more tailored and effective therapies. It enables researchers to identify and validate prospective therapeutic targets by researching genetic variations associated with diseases. Researchers can uncover genes or genomic areas linked to certain diseases or treatment responses by analyzing large-scale genomic data, such as genome-wide association studies (GWAS) (Kesheri et al. 2016; Visscher et al. 2017). This information aids in prioritizing prospective targets for further inquiry.

5.2.1 Genomic Biomarkers for Disease Diagnosis, Prognosis, and Treatment Response

Genomic biomarkers offer a promising approach to disease diagnosis (Manolio et al. 2013). Through genomic analysis, researchers can identify specific genetic variants or gene expression patterns that are associated with various diseases. These biomarkers can be utilized to develop diagnostic tests that enable early and accurate disease detection. The availability of genomic biomarkers for disease diagnosis facilitates timely interventions and improves patient outcomes. Accurate prognosis assessment is crucial for determining disease progression and tailoring treatment strategies. Genomic biomarkers provide valuable information about the expected course of a disease based on an individual's genetic profile. By examining genomic biomarkers, clinicians can assess the likelihood of disease recurrence, disease severity, or progression (Muzny et al. 2012; Prensner et al. 2012; Tanase et al. 2017; Rebouissou and Nault 2020). This enables personalized treatment plans and proactive interventions to optimize patient outcomes. One of the key advantages of genomic biomarkers is their ability to predict an individual's response to specific treatments. By analyzing genetic variants or gene expression patterns associated with treatment response, researchers can identify patients who are more likely to benefit from a particular therapy. This information aids in treatment selection, minimizing ineffective treatments, and reducing potential adverse effects. Genomic biomarkers also support the development of targeted therapies for specific patient subgroups, enhancing treatment efficacy (Flaherty et al. 2010; Swanton 2012; Collins and Varmus 2015).

5.2.2 Pharmacogenomics Enhancing Drug Discovery and Development

Pharmacogenomics is the study of how a person's genetic composition affects their reaction to medications (Evans and Relling 1999; Johnson et al. 2011; Caudle et al. 2013). Researchers can predict how individuals will respond to specific pharmaceuticals by researching genetic differences that alter drug metabolism, efficacy, or toxicity. This knowledge aids in the customization of drug therapies to individual patients, the optimization of drug selection and dosing regimes, and the reduction of adverse drug reactions. By identifying genetic variants associated with increased susceptibility to adverse effects, healthcare providers can proactively avoid prescribing medications that may cause harm to certain individuals (Wang et al. 2011). This knowledge can also aid in the development of guidelines and recommendations for medication use based on a person's genetic profile. Overall, pharmacogenomics plays a crucial role in advancing precision medicine by considering an individual's genetic composition in drug selection and dosing. This field holds promise for improving therapeutic outcomes, minimizing adverse reactions, and optimizing treatment strategies for patients.

5.2.3 Medication Repurposing

Researchers can uncover new targets or pathways associated with certain diseases and connect them with existing medications that affect those targets. When compared to beginning from scratch, this method can dramatically reduce the medication development timeline and costs. The process involves analyzing genetic data to identify novel targets or pathways related to specific diseases and then matching them with already available medications that can affect those targets. By leveraging genomic data, researchers can uncover previously unknown connections between genetic factors and diseases, leading to the identification of potential new uses for existing drugs (Sirota et al. 2011; Iorio et al. 2016; Emilsson et al. 2018; Sweeney et al. 2018; Musa et al. 2018). This approach offers several advantages compared to starting the medication development process from scratch. Firstly, it can significantly reduce the timeline required for medication development. Since the pharmaceuticals being repurposed already have established safety profiles and have undergone previous clinical trials, the necessary steps for regulatory approval may be streamlined. This can expedite the availability of new treatment options to patients. Secondly, medication repurposing through genomics can potentially save substantial expenditures. Developing a new medication from the beginning involves extensive research, preclinical and clinical trials, and regulatory processes, which are time-consuming and costly. By repurposing existing medications, researchers can bypass some of these stages, resulting in cost savings and improved resource allocation. In summary, utilizing genomics in medication repurposing efforts allows researchers to harness genetic data to identify new therapeutic targets and match them with existing drugs, leading to a faster and more cost-effective approach to developing new treatments for various diseases.

5.2.4 Bioinformatics Tools and Genomics

The introduction of high-throughput sequencing technologies and bioinformatics tools has resulted in a massive amount of genetic data. Integrating genomics with other omics data (such as transcriptomics, proteomics, and metabolomics) and utilizing advanced data analysis tools, such as artificial intelligence and machine learning, enables researchers to extract valuable insights from these complex datasets. These discoveries aid in target prioritization, disease mechanism knowledge, and the development of novel therapeutic methods. Researchers integrate genomics data with other omics data, including transcriptomics, proteomics, and metabolomics. By combining multiple layers of biological information, a more comprehensive understanding of complex biological systems can be achieved. The discoveries resulting from the integration of genomics with other omics data and the application of advanced data analysis tools have several important implications. Firstly, they aid in target prioritization, helping researchers identify specific genes, proteins, or metabolic pathways that play a crucial role in diseases. This knowledge is valuable in the development of targeted therapies and personalized medicine approaches.

To summarize, genomics is a valuable tool for understanding the genetic basis of diseases and optimizing drug research and development processes. By deciphering the complexity of the genome, genomics has the potential to provide patients with more precise, effective, and personalized medicines.

5.3 Transcriptomics

The science of drug discovery and development has undergone a revolution thanks to transcriptomics, which has shed light on the complex connections between genes, gene expression, and the development of disease. It has revolutionized drug discovery and development by providing a systems-level understanding of gene expression and its relevance to diseases. It has transformed the way researchers identify drug targets, validate their therapeutic potential, and develop personalized treatment strategies, leading to more effective and targeted therapies. Drug discovery employs a range of methodologies and platforms, including Microarray and RNA-Seq, to evaluate global gene expression profiles, providing essential insights into the transcriptional landscape in various biological samples. Each method has its advantages and limitations, and the technique used is determined by the individual research question, sample availability, required throughput, and budget constraints.

Using transcriptomics data in drug development faces several challenges and limitations. Firstly, interpreting and integrating the vast amount of data generated requires sophisticated bioinformatics tools and expertise. Technical variability and batch effects can introduce noise and confound analysis. The limited understanding of noncoding RNAs, such as microRNAs and long noncoding RNAs, hinders comprehensive analysis. Additionally, transcriptomics data may not fully capture protein-level changes or post-translational modifications. Validation and reproducibility of findings across different platforms and datasets are crucial. Finally, the complexity of translating transcriptomics findings into clinically relevant applications and overcoming regulatory hurdles presents further challenges. Overall, while transcriptomics delivers useful insights, careful consideration of these constraints is required for its successful implementation in drug development. There are several emerging trends and advancements in transcriptomics for drug discovery.

The integration of transcriptomics with other omics technologies, such as proteomics and metabolomics, enables a more comprehensive understanding of biological processes and drug responses. Single-cell transcriptomics is gaining momentum, allowing researchers to investigate cellular heterogeneity and identify rare cell populations relevant to drug discovery. Moreover, the application of machine learning and artificial intelligence algorithms to analyze data is improving the identification of novel drug targets and biomarkers. Additionally, the development of miniaturized and cost-effective platforms, such as droplet-based technologies, is facilitating high-throughput transcriptomic profiling. There are various computational methods and tools used by the scientific community for analyzing transcriptomics data in drug discovery. Differential gene expression analysis is commonly performed using tools such as DESeq2, edgeR, and limma, which identify genes that are differentially expressed between experimental conditions. Gene set enrichment analysis (GSEA) is used to determine if predefined gene sets or pathways are significantly enriched in the data. Network-based analysis tools, like WGCNA (Weighted Gene Co-expression Network Analysis) and STRING (Search Tool for the Retrieval of Interacting Genes/Proteins),, enable the identification of co-expressed gene modules and functional interactions between genes. Integrated analysis tools, such as ToppGene Suite and DAVID, facilitate the interpretation of transcriptomics data by incorporating functional annotation and pathway enrichment analysis.

Current transcriptomics research in drug discovery has several gaps and future directions to explore. There is a need for improved integration of multi-omics data to gain a more comprehensive understanding of disease mechanisms and drug responses. Additionally, the functional annotation of noncoding RNAs and the development of standardized pipelines for their analysis are areas requiring further exploration. Addressing technical challenges related to data reproducibility, standardization, and scalability will enhance the utility of transcriptomics in large-scale drug discovery efforts. Moreover, the application of artificial intelligence and machine learning algorithms holds promise for advanced data analysis and prediction of drug efficacy and safety. Finally, the translation of transcriptomics findings into clinical practice and the implementation of regulatory frameworks for utilizing transcriptomics in drug development are crucial areas for future advancement in this field.

5.4 Proteomics

Proteomics is the study of the complete set of proteins present in a particular cell, tissue, or organism, and it encompasses the identification, characterization, and quantification of these proteins. In the context of drug discovery and development, proteomics plays a crucial role in understanding the complex molecular mechanisms underlying diseases and identifying potential targets for drug intervention. By analyzing protein expression patterns, modifications, interactions, and pathways involved in disease states, proteomics provides valuable insights into disease biology and facilitates the discovery and development of new drugs. Proteomics also aids in the evaluation of drug safety and efficacy, the identification of biomarkers for disease diagnosis and prognosis, and the optimization of treatment regimens, ultimately contributing to personalized medicine and the development of more effective and targeted therapies.

5.4.1 Techniques Used in Proteomics for Drug Discovery

One commonly used approach is mass spectrometry, which facilitates protein identification, quantification, and characterization. It can be coupled with various separation techniques, such as liquid chromatography or gel electrophoresis, to enhance protein analysis. Another key technology is protein microarrays, which allow high-throughput screening of protein-protein interactions, protein-small molecule interactions, and antibody profiling. Additionally, techniques like two-dimensional gel electrophoresis, shotgun proteomics, and label-free quantification methods are used for protein profiling and differential expression analysis. Furthermore, advanced methodologies like protein-protein interaction networks, structural proteomics, and chemical proteomics enable the exploration of protein functions, protein complex dynamics, and drug-target interactions. The integration of multiple proteomic techniques and bioinformatics tools provides a comprehensive understanding of the proteome and its implications in drug discovery and development.

5.4.2 Challenges and Limitations

Applying proteomics in drug discovery comes with several challenges and limitations. One major challenge is the complex nature of the proteome, as proteins can undergo various posttranslational modifications and exhibit different isoforms, making the comprehensive analysis and interpretation of proteomic data challenging. Another limitation is the wide dynamic range of protein expression levels, with highly abundant proteins potentially overshadowing the detection of low-abundance proteins of interest. Additionally, the sensitivity and throughput of proteomic technologies can be limiting factors, especially when analyzing large sample sets. Standardization and reproducibility of proteomic workflows and data analysis also pose challenges, requiring careful experimental design and rigorous quality control measures. Despite these challenges, advancements in technology, data analysis methods, and collaborative efforts are continuously addressing these limitations, making proteomics an invaluable tool in drug discovery.

5.5 Metabolomics

Metabolomics is the comprehensive study of small molecules, known as metabolites, within a biological system. It involves the identification, quantification, and analysis of metabolites to understand the metabolic processes occurring in cells, tissues, or organisms. Metabolomics contributes to drug discovery and development by providing insights into drug metabolism, pharmacokinetics, and the biochemical pathways affected by drug interventions. It enables the identification of potential biomarkers for disease diagnosis, prognosis, and therapeutic response prediction. Metabolomics plays a key role in refining potential drug candidates and selecting the most effective ones by analyzing their impact on cellular metabolism and metabolic pathways. This helps ensure that the drugs work as intended while minimizing unwanted effects. Furthermore, metabolomics plays a crucial role in understanding drug toxicity and safety assessment, facilitating the identification of potential adverse effects and aiding in drug optimization. Metabolomics contributes to the study of drug resistance mechanisms by providing insights into the metabolic alterations and adaptive responses that occur in resistant cells or organisms. Metabolomic profiling allows for the identification and quantification of metabolic changes associated with drug resistance, such as alterations in energy metabolism, activation of alternative metabolic pathways, and changes in the levels of specific metabolites involved in drug response. By comparing the metabolic profiles of drug-resistant and drug-sensitive cells or organisms, metabolomics can identify specific metabolic pathways or metabolites that are associated with resistance. This information helps uncover the underlying mechanisms of drug resistance, including the activation of detoxification pathways, altered drug metabolism, changes in drug uptake or efflux, and compensatory metabolic rewiring. Understanding the metabolic adaptations in drug-resistant cells can guide the development of strategies to overcome resistance, such as combination therapies targeting key metabolic vulnerabilities or the development of drugs that specifically inhibit the altered metabolic pathways identified through metabolomics analysis (Srivastava et al. 2023).

Computational and statistical approaches play a crucial role in metabolomics data analysis for drug discovery. One common approach is univariate analysis, which involves statistical tests, such as t-tests or ANOVA, to identify individual metabolites that show significant differences between experimental groups. Multivariate analysis methods, including principal component analysis (PCA), partial least squares-discriminant analysis (PLS-DA), and hierarchical clustering, are used to explore the overall patterns and relationships in metabolomics data. Pathway analysis tools, such as MetaboAnalyst and MetScape, are employed to identify affected metabolic pathways and understand their functional implications. Additionally, machine learning algorithms, such as random forests and support vector machines, are utilized for classification and prediction tasks. Furthermore, network-based approaches and correlation analyses help uncover metabolite interactions and identify potential biomarkers or key metabolites associated with drug response or disease progression. These computational and statistical techniques enable the extraction of meaningful information from complex metabolomics data, aiding in the identification of potential drug targets, biomarkers, and therapeutic strategies in drug discovery (Kesheri et al. 2024).

5.6 Epigenomics

Epigenomics, which is the study of changes in gene expression or cellular phenotype without involving changes to the DNA sequence itself, has become a promising field with important applications in drug discovery and development. Epigenetic changes have been linked to a number of disorders and are critical in controlling gene expression. The creation of new therapeutic approaches can be facilitated by comprehending and focusing on these epigenetic modifications. Here are a few significant uses of epigenomics in the development and discovery of pharmaceuticals. By examining the epigenetic changes linked to certain diseases, epigenomics aids in the identification and validation of prospective therapeutic targets. The analysis of epigenetic markers, including DNA methylation, histone modifications, and noncoding RNAs, enables researchers to pinpoint the genes or genomic areas that are epigenetically dysregulated in pathological situations. Further research into the therapeutic potential of these dysregulated targets is possible.

Epigenomic profiling can find epigenetic biomarkers that are useful for diagnosing, forecasting, or managing disease. Epigenetic biomarkers can be used to identify diseases early, track their development, evaluate treatment effectiveness, and stratify patients for individualized medicines. It may assist inform therapy choices and shed light on disease subgroups. Because epigenetic changes can be undone, they are appealing therapeutic targets. Small compounds have been created as possible medications that specifically affect epigenetic enzymes such as DNA methyltransferases, histone deacetylases, or histone methyltransferases. These epigenetic modifiers work to reverse abnormal epigenetic marks linked to disease to restore normal gene expression patterns. To increase the effectiveness of other therapies, epigenetic modulators can be used in conjunction with them. Drugs targeting the epigenome can make tumor cells more susceptible to immunotherapy, radiation treatment, or conventional chemotherapy. These combination medicines can overcome drug resistance mechanisms, inhibit tumor growth, and enhance therapeutic results by changing the epigenetic landscape. By highlighting epigenetic dependencies or vulnerabilities in illnesses, epigenomic data might direct therapeutic repurposing efforts. It is possible to target particular epigenetic abnormalities with already available medications that have epigenetic effects or interact with epigenetic modifiers. Given that repurposed medications have previously undergone safety testing, this strategy offers prospects for accelerated drug development. Epigenomic profiling can be used to evaluate how possible drug candidates affect epigenetic markers. It helps check if a drug works on the intended epigenetic changes or causes unwanted effects on the epigenome. The most effective drug candidates can be chosen using this information, and treatment plans can be improved. Epigenomic profiling makes it possible to identify patient-specific epigenetic patterns that can direct the development of personalized therapy strategies. Therapeutic techniques can be adapted to target particular epigenetic abnormalities by identifying an individual's epigenetic landscape, enhancing therapeutic effectiveness and reducing side effects.

Epigenomics, in summary, has significant ramifications for the development and identification of new medicines. Researchers can uncover therapeutic targets, create epigenetic modulators, repurpose existing medications, and direct approaches to precision medicine by clarifying epigenetic changes and their involvement in diseases. Developing innovative medicines that target epigenetic dysregulation and enhance patient outcomes is made possible by epigenomics, which offers considerable potential (Fig. 5.1).

Here are some multi-omics bioinformatics applications and tools that can be useful for drug developments:

1. Bioconductor: A collection of R packages for analyzing and visualizing genomic data. It includes tools for microarray analysis, RNA-seq, ChIP-seq, and more. https://www.bioconductor.org/.
2. Omics Integrator: A tool that integrates multiple omics datasets to infer functional relationships between genes, proteins, and metabolites. http://fraenkel.mit.edu/omicsintegrator.
3. Cytoscape: A popular tool for visualizing and analyzing biological networks, including omics data, to identify key nodes and pathways relevant to drug development (Su et al. 2014).

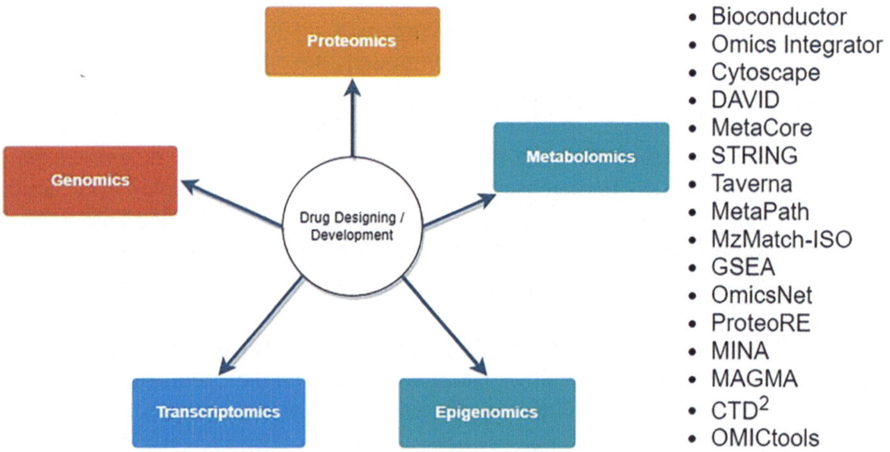

Fig. 5.1 Illustration of the concept of utilizing multi-omics approaches in drug designing, where omics technologies with available bioinformatics tools/applications are integrated to provide diseases and enable the development of effective therapeutic interventions

4. DAVID (Database for Annotation, Visualization, and Integrated Discovery): Useful for functional annotation and enrichment analysis of omics data, helping identify biological themes and pathways (Dennis et al. 2003).
5. MetaCore (Clarivate Analytics): A platform for pathway analysis, network visualization, and data integration across multiple omics data types. https://clarivate.com/lp/metacore-integrated-pathway-analysis-for-multi-omics-data/.
6. STRING (Search Tool for the Retrieval of Interacting Genes/Proteins): A database and web resource for protein-protein interaction networks, which can aid in drug target identification (Szklarczyk et al. 2019).
7. Taverna: A workflow management system that can integrate and analyze large-scale omics datasets (Oinn et al. 2004).
8. MetaPath: A tool for pathway-based analysis of multi-omics data, which helps in understanding disease mechanisms and identifying potential drug targets (Swanton 2012).
9. MzMatch-ISO: Specifically designed for metabolomics data, this tool helps in the identification and quantification of metabolites, which can be valuable in drug development studies. https://mzmatch.sourceforge.net/.
10. GSEA (Gene Set Enrichment Analysis): Useful for determining whether a priori-defined set of genes shows statistically significant differences between two biological states, providing insights into drug response and mechanisms (Subramanian et al. 2005).
11. OmicsNet: A tool for integrative analysis and visualization of multi-omics data to identify novel biomarkers and drug targets (Zhou et al. 2022).
12. ProteoRE: Focused on proteomics data, this tool helps in the analysis and interpretation of quantitative proteomics experiments to identify potential drug targets. https://proteore.org/
13. Molecular Interaction Network Analysis (MINA): Elucidate candidate disease gene (Kara et al. 2019).
14. MAGMA (Multi-marker Analysis of GenoMic Annotation): A tool for gene-based analysis of omics data, helpful in identifying genetic associations with diseases and potential drug targets.
15. Cancer Target Discovery and Development (CTD^2) Network Tools: A collection of tools developed by the CTD^2 Network for analyzing and interpreting cancer multi-omics data for drug discovery (Aksoy et al. 2017).
16. OMICtools: Although not a tool itself, it is a valuable resource for finding various bioinformatics tools relevant to multi-omics data analysis (Henry et al. 2014).

The choice of tools and applications may depend on the specific type of multi-omics data being analyzed and the goals of drug development research (Table 5.1).

Table 5.1 Concise overview of key omics disciplines, their descriptions, and significant applications in understanding disease mechanisms, biomarker discovery, drug development, and precision medicine

Section	Description	Key points
5.2 Genomics	Study of the entire genetic material to gain insights into disease mechanisms and enable the development of tailored therapies	Enables target identification using GWAS Facilitates genomic biomarkers for disease diagnosis and personalized treatments Supports pharmacogenomics to predict drug responses Utilizes medication repurposing for cost-effective drug development Integrates bioinformatics tools for data analysis and multi-omics integration
5.2.1 Genomic biomarkers	Biomarkers aid in early diagnosis, prognosis, and treatment response	Identify specific genetic variants or expression patterns Assist in early detection and personalized treatment plans Predict treatment response and reduce adverse effects
5.2.2 Pharmacogenomics	Study of genetic factors influencing drug metabolism and response	Tailors drug therapies to genetic profiles Optimizes drug selection and dosing Minimizes adverse drug reactions
5.2.3 Medication Repurposing	Use of genomics data to identify new applications for existing drugs	Reduces drug development time and costs Leverages genetic insights for identifying novel targets
5.2.4 Bioinformatics Tools	Integration of genomics with omics data for deeper insights into disease mechanisms	Employs AI and machine learning for advanced data analysis Aids in target prioritization and therapy development
5.3 Transcriptomics	Systems-level study of gene expression for drug target discovery and validation	Uses microarray and RNA-Seq for gene expression profiling Integrates multi-omics data for comprehensive insights Employs computational tools like DESeq2, GSEA, and WGCNA for data analysis

(continued)

Table 5.1 (continued)

Section	Description	Key points
5.4 Proteomics	Study of the entire set of proteins in a system to understand disease mechanisms and drug targets	Techniques include mass spectrometry, protein microarrays, and 2D gel electrophoresis Addresses challenges like dynamic protein range and post-translational modifications Integrates advanced tools for detailed protein analysis
5.5 Metabolomics	Study of metabolites to understand biochemical pathways and drug metabolism	Provides insights into pharmacokinetics, drug toxicity, and resistance mechanisms Utilizes computational tools like PCA and PLS for data analysis Supports biomarker identification and lead optimization
5.6 Epigenomics	Study of changes in gene expression or phenotype without altering DNA sequence	Controls gene expression

5.7 Conclusions

Multi-omics approaches support the stratification of patient populations, enabling personalized medicine strategies that maximize efficacy while minimizing adverse effects. As technological advancements continue to enhance the capabilities of high-throughput sequencing, mass spectrometry, and bioinformatics, the potential of multi-omics to reshape therapeutic development becomes even more evident. Interdisciplinary collaborations between scientists, clinicians, and computational experts are further driving the integration of multi-omics into routine drug development pipelines. These efforts are laying the foundation for the next generation of therapeutics—ones that are not only more effective but also safer and tailored to meet the unique needs of individual patients. Consequently, multi-omics is set to remain a cornerstone of innovative research and a transformative force in addressing the challenges of complex diseases across diverse populations.

Chapter Summary

This chapter delves into the synergistic combination of diverse omics studies that provide an unparalleled opportunity to investigate into the complex biological networks underlying diseases, allowing for a more precise characterization of molecular mechanisms and pathways. The advent of multi-omics approaches has revolutionized the field of drug discovery and development by offering a holistic and in-depth understanding of diseases and the factors influencing individual variability in drug responses. By integrating data from genomics, transcriptomics, proteomics, metabolomics, and epigenomics, researchers have been able to unlock new possibilities for identifying therapeutic targets, discovering reliable biomarkers, and predicting patient-specific drug responses. These approaches have also significantly contributed to the acceleration of drug repurposing initiatives, making it possible to find novel uses for existing medications in a more efficient and cost-effective manner. This precision facilitates the development of targeted therapies tailored to address specific molecular abnormalities, ultimately improving therapeutic outcomes.

Acknowledgments The authors would like to acknowledge the support of REVA University throughout the preparation of this chapter. I would like to extend my sincere appreciation to my colleagues and fellow researchers at REVA University, Bengaluru (Karnataka), India. Their insightful discussions and collaborative spirit enriched my understanding of the subject matter and enhanced the quality of this work. I want to express my gratitude to the editorial team for their assistance and professionalism during the publication process. Their expertise and attention to detail helped refine this chapter into its final form.

Author Contributions Chapter concept and design, Vivek Keshri; chapter writing, Vivek Keshri; chapter editing and review, Vivek Keshri, and Jyothi B.

References

Aksoy BA, Dancík V, Smith K, Mazerik JN, Ji Z, Gross B, Nikolova O, Jaber N, Califano A, Schreiber SL, Gerhard DS, Hermida LC, Jagu S, Sander C, Floratos A, Clemons PA (2017) CTD2 dashboard: a searchable web interface to connect validated results from the cancer target discovery and development network. Database (Oxford) 2017:54. https://doi.org/10.1093/database/bax054

Caudle KE, Thorn CF, Klein TE, Swen JJ, McLeod HL, Diasio RB, Schwab M (2013) Clinical pharmacogenetics implementation consortium guidelines for dihydropyrimidine dehydrogenase genotype and fluoropyrimidine dosing. Clin Pharmacol Ther 94:640–645. https://doi.org/10.1038/clpt.2013.172

Collins FS, Varmus H (2015) A new initiative on precision medicine. N Engl J Med 372:793–795. https://doi.org/10.1056/nejmp1500523

Dennis G, Sherman BT, Hosack DA, Yang J, Gao W, Lane HC, Lempicki RA (2003) DAVID: database for annotation, visualization, and integrated discovery. Genome Biol 4:1–11. https://doi.org/10.1186/gb-2003-4-9-r60

Du P, Fan R, Zhang N, Wu C, Zhang Y (2024) Advances in integrated multi-omics analysis for drug-target identification. Biomol Ther 14(6):692. https://doi.org/10.3390/biom14060692

Emilsson V, Ilkov M, Lamb JR, Finkel N, Gudmundsson EF, Pitts R, Hoover H, Gudmundsdottir V, Horman SR, Aspelund T, Shu L, Trifonov V, Sigurdsson S, Manolescu A, Zhu J, Olafsson O, Jakobsdottir J, Lesley SA, To J, Zhang J, Harris TB, Launer LJ, Zhang B, Eiriksdottir G, Yang X, Orth AP, Jennings LL, Gudnason V (2018) Co-regulatory networks of human serum proteins link genetics to disease. Science 361(6404):769–773. https://doi.org/10.1126/science.aaq1327

Evans WE, Relling MV (1999) Pharmacogenomics: translating functional genomics into rational therapeutics. Science 286(5439):487–491. https://doi.org/10.1126/science.286.5439.487

Flaherty KT, Puzanov I, Kim KB, Ribas A, McArthur GA, Sosman JA, O'Dwyer PJ, Lee RJ, Grippo JF, Nolop K, Chapman PB (2010) Inhibition of mutated, activated BRAF in metastatic melanoma. N Engl J Med 363(9):809–819. https://doi.org/10.1056/nejmoa1002011

Graf EH, Bryan A, Bowers M, Grys TE (2025) One size fits small: the narrow utility for plasma metagenomics. J Appl Lab Med 10(1):171–183. https://doi.org/10.1093/jalm/jfae122

Henry VJ, Bandrowski AE, Pepin AS, Gonzalez BJ, Desfeux A (2014) OMICtools: an informative directory for multi-omic data analysis. Database (Oxford) 2014:bau069. https://doi.org/10.1093/database/bau069

Iorio F, Knijnenburg TA, Vis DJ, Bignell GR, Menden MP, Schubert M, Aben N, Gonçalves E, Barthorpe S, Lightfoot H, Cokelaer T (2016) A landscape of pharmacogenomic interactions in cancer. Cell 166:740–754. https://doi.org/10.1016/j.cell.2016.06.017

Ivanisevic T, Sewduth RN (2023) Multi-omics integration for the design of novel therapies and the identification of novel biomarkers. Proteomes 11(4):34. https://doi.org/10.3390/proteomes11040034

Johnson JA, Gong L, Whirl-Carrillo M, Gage BF, Scott SA, Stein CM, Anderson JL, Kimmel SE, Lee MTM, Pirmohamed M, Wadelius M, Klein TE, Altman RB (2011) Clinical pharmacogenetics implementation consortium guidelines for CYP2C9 and VKORC1 genotypes and warfarin dosing. Clin Pharmacol Ther 90:625–629. https://doi.org/10.1038/clpt.2011.185

Kaithal P, Kanchan S, Kesheri M (2024) Recent advances in biological omics databases and tools in human health. In: Kesheri M, Kanchan S, Salisbury TB, Sinha RP (eds) Microbial omics in environment and health. Springer, Singapore. https://doi.org/10.1007/978-981-97-1769-9_12

Kara S, Pirela-Morillo GA, Gilliam CT, Wilson GD (2019) Molecular interaction network approach (MINA) identifies association of novel candidate disease genes. J Autoimmun 97:48–58. https://doi.org/10.1016/j.jaut.2018.10.002

Kesheri M, Sinha RP, Kanchan S (2016) Advances in soft computing approaches for gene prediction: a bioinformatics approach. In: Dey N, Bhateja V, Hassanien AE (eds) Advancements in bio- medical sensing, imaging, measurements and instrumentation, vol 651. Springer-Verlag, Cham, pp 383–405. https://doi.org/10.1007/978-3-319-33793-7_17

Kesheri M, Kanchan S, Srivastava S, Ratna-Raj R, Chittoori B, Vaishampayan A, Rastogi RP, Sinha RP, Primerano D (2024) Ecology and environmental omics. In: Gupta MK, Katara P, Mondal S, Singh RL (eds) Integrative omics: concepts, methodology, and application. Academic Press (Elsevier), pp 315–331. https://doi.org/10.1016/B978-0-443-16092-9.00019-9

Kumar P, Kanchan S, Kesheri M (2024) Multi-omics in human disease biomarker discovery. In: Kesheri M, Kanchan S, Salisbury TB, Sinha RP (eds) Microbial omics in environment and health. Springer, Singapore. https://doi.org/10.1007/978-981-97-1769-9_8

Manolio TA, Chisholm RL, Ozenberger B, Roden DM, Williams MS, Wilson R, Bick D, Bottinger EP, Brilliant MH, Eng C, Frazer KA (2013) Implementing genomic medicine in the clinic: the future is here. Genet Med 15(4):258–267. https://doi.org/10.1038/gim.2012.157

Musa A, Ghoraie LS, Zhang SD, Glazko G, Yli-Harja O, Dehmer M, Haibe-Kains B, Emmert-Streib F (2018) A review of connectivity map and computational approaches in pharmacogenomics. Brief Bioinform 19:506–523. https://doi.org/10.1093/bib/bbw112

Muzny DM, Bainbridge MN, Chang K, Dinh HH, Drummond J, Fowler G (2012) Comprehensive molecular characterization of human colon and rectal cancer. Nature 487:330–337. https://doi.org/10.1038/nature11252

Oinn T, Addis M, Ferris J, Marvin D, Senger M, Greenwood M, Carver T, Glover K, Pocock MR, Wipat A, Li P (2004) Taverna: a tool for the composition and enactment of bioinformatics workflows. Bioinformatics 20:3045–3054. https://doi.org/10.1093/bioinformatics/bth361

Prensner JR, Rubin MA, Wei JT, Chinnaiyan AM (2012) Beyond PSA: the next generation of prostate cancer biomarkers. Sci Transl Med 4:127rv3. https://doi.org/10.1126/scitranslmed.3003180

Rebouissou S, Nault JC (2020) Advances in molecular classification and precision oncology in hepatocellular carcinoma. J Hepatol 72:215–229. https://doi.org/10.1016/j.jhep.2019.08.017

Sahu N, Mishra S, Kesheri M, Kanchan S, Sinha RP (2023) Identification of cyanobacteria- based natural inhibitors against SARS-CoV-2 druggable target ACE2 using molecular docking study, ADME and toxicity analysis. Ind J Clin Biochem 38(3):361–373. https://doi.org/10.1007/s12291-022-01056-65.8

Sirota M, Dudley JT, Kim J, Chiang AP, Morgan AA, Sweet-Cordero A, Sage J, Butte AJ (2011) Discovery and preclinical validation of drug indications using compendia of public gene expression data. Sci Transl Med 3:96ra77. https://doi.org/10.1126/scitranslmed.3001318

Srivastav S, Lavanya AA, Kaur N, Gaertner W, Kesheri M, Kanchan S, Sharda S (2024) Integromics: tracking the multi-omics expanse in theragnostics. In: Kesheri M, Kanchan S, Salisbury TB, Sinha RP (eds) Microbial omics in environment and health. Springer, Singapore. https://doi.org/10.1007/978-981-97-1769-9_6

Srivastava U, Kanchan S, Kesheri M, Singh S (2023) Nutrimetabolomics: metabolomics in nutrition research. In: Soni V, Hartman TE (eds) Metabolomics. Springer, Cham, pp 241–268. https://doi.org/10.1007/978-3-031-39094-4_8

Su G, Morris JH, Demchak B, Bader GD (2014) Biological network exploration with cytoscape 3. Curr Protoc Bioinforma 47:8.13.1–8.13.24. https://doi.org/10.1002/0471250953.bi0813s47

Subramanian A, Tamayo P, Mootha VK, Mukherjee S, Ebert BL, Gillette MA, Paulovich A, Pomeroy SL, Golub TR, Lander ES, Mesirov JP (2005) Gene set enrichment analysis: a knowledge-based approach for interpreting genome-wide expression profiles. Proc Natl Acad Sci USA 102:15545–15550. https://doi.org/10.1073/pnas.0506580102

Swanton C (2012) Intratumor heterogeneity: evolution through space and time. Cancer Res 72:4875–4882. https://doi.org/10.1158/0008-5472.CAN-12-2217

Sweeney TE, Azad TD, Donato M, Haynes WA, Perumal TM, Henao R, Bermejo-Martin JF, Almansa R, Tamayo E, Howrylak JA, Choi A (2018) Unsupervised analysis of transcriptomics in bacterial sepsis across multiple datasets reveals three robust clusters. Crit Care Med 46:915–925. https://doi.org/10.1097/CCM.0000000000003084

Szklarczyk D, Gable AL, Lyon D, Junge A, Wyder S, Huerta-Cepas J, Simonovic M, Doncheva NT, Morris JH, Bork P, Jensen LJ (2019) STRING v11: protein-protein association networks with increased coverage, supporting functional discovery in genome-wide experimental datasets. Nucleic Acids Res 47:D607–D613. https://doi.org/10.1093/nar/gky1131

Tanase CP, Codrici E, Popescu ID, Mihai S, Enciu AM, Necula LG, Preda A, Ismail G, Albulescu R (2017) Prostate cancer proteomics: current trends and future perspectives for biomarker discovery. Oncotarget 8:18497–18512. https://doi.org/10.18632/oncotarget.14501

Visscher PM, Wray NR, Zhang Q, Sklar P, McCarthy MI, Brown MA, Yang J (2017) 10 years of GWAS discovery: biology, function, and translation. Am J Hum Genet 101:5–22. https://doi.org/10.1016/j.ajhg.2017.06.005

Wang L, McLeod HL, Weinshilboum RM (2011) Genomics and drug response. N Engl J Med 364:1144–1153. https://doi.org/10.1056/nejmra1010600

Wang Z, Zhao Y, Zhang L (2024) Emerging trends and hot topics in the application of multi-omics in drug discovery: a bibliometric and visualized study. Curr Pharm Anal 21(1):20–32. https://doi.org/10.1016/j.cpan.2024.12.001

Zhou G, Pang Z, Lu Y, Ewald J, Xia J (2022) OmicsNet 2.0: a web-based platform for multi-omics integration and network visual analytics. Nucleic Acids Res 50:W527–W533. https://doi.org/10.1093/nar/gkac376

Machine Learning Approaches for the Diagnosis of Pre- and Post-COVID-19 Phases

6

Upasna Srivastava, Swarna Kanchan, Minu Kesheri, Shraddha Piparia, and Satendra Singh

Abstract

The global pandemic scenario of pre- and post-sequelae infections due to highly evolving SARS-CoV-2 mutants warrants an urgent need for its rapid and accurate analysis and prediction, which might be crucial in combating and addressing the ensuing medical challenges. Detailed descriptions of various machine learning approaches as significant tools encompassing models classifying COVID-19 disease severity, predicting the mortality of the patients, accuracy, precision, sensitivity, and specificity rates using supervised and unsupervised learning-based machine learning approaches have been provided. A comprehensive review of modern approaches to machine learning models for diagnosis, analysis and prediction of pre- and post-effects is integrated with the effects of COVID-19. This chapter also focuses on the studies on diagnostic analyses based on machine learning approaches using available pediatric and adult COVID-19 public domain data comprising of X-ray images, CT scan images, CXR images, lung

U. Srivastava (✉)
Yale School of Medicine, New Haven, CT, USA
e-mail: upasna.srivastava@yale.edu

S. Kanchan · M. Kesheri
Joan C. Edwards School of Medicine, Marshall University, Huntington, WV, USA

Boise State University, Boise, ID, USA

S. Piparia
Department of Pediatrics, University of California, San Diego, CA, USA

S. Singh
Department of Computational Biology and Bioinformatics, Sam Higginbottom University of Agriculture, Technology and Sciences, Allahabad, Uttar Pradesh, India

M. Kesheri et al. (eds.), *Multi-Omics in Biomedical Sciences and Environmental Sustainability*, https://doi.org/10.1007/978-981-96-7067-3_6

ultrasound (LUS), etc. The diagrammatic illustrations, computational models, exclusive examples, and lucid style used in this chapter will aid in an easy understanding of the complex topics discussed. This chapter will benefit those involved in healthcare and scientific brotherhood in developing COVID-19 resilience by improving quality of care and increasing patient survival.

Keywords

CT scan images · CXR images · Lung ultrasound (LUS) · Machine learning approach (ML) · ROC (Receiver · operating characteristic) · SARS-CoV-2

Objectives

- To understand how artificial intelligence and machine learning have helped in preventing the spread of COVID-19 and what roles they have played and how effective they have been.
- To understand the different machine learning techniques that have been used in COVID-19 research as how they work, why researchers chose them, and what kind of results they got.
- Medicine and healthcare industriesAnother goal is to dive into how machine learning has been used for diagnosing COVID-19 including the kinds of data these systems rely on, and how accurate or useful they have been in real-world situations.
- Finally, to explore the bigger picture as all the different ways AI and machine learning have been applied to help with COVID-19, especially in terms of prevention and public health strategies.

6.1 Introduction

COVID-19 has led to one of the most challenging pandemics of the current century, caused by severe acute respiratory syndrome coronavirus 2 (SARS-COV-2). COVID-19 was first discovered in Wuhan, China in December 2019 and quickly infected millions of people around the world. According to World Health Organization (WHO) data, 651,918,402 confirmed cases of COVID-19, including 6,656,601 on December 23, 2022. Coronavirus is an RNA virus of the coronaviridae family that is responsible for demanding disease. According to the Centers for Disease Control and Prevention (CDC 2022), infected people exhibit many symptoms, including fever, dry cough, shortness of breath, fatigue, loss of taste and odor, diarrhea, and hiding places. The Coronaviridae family consists of a group of large, individual, and positive RNA viruses isolated from several species, and is known to cause cold and diarrhea diseases in humans (Dey et al. 2020).

On May 20, 2022, approximately 521,920,560 COVID-199 cases and approximately 6,274,323 deaths were reported to the World Health Organization. The

SARS-COV-2 virus is constantly altered by mutations. Therefore, various virus tribes of SARS-COV-2 have been reported, including the United Kingdom, South Africa, Brazil, and Nigeria, and are now spreading around the world. Although we have put so much effort into understanding COVID-19, some developing countries face major challenges, vaccinations, and complex import guidelines and strategies due to lack of funding. This requires antioxidant-rich ethnopharmacological approaches and study of natural products (Kumari et al. 2018), and features stress relief and increased immunity as a global response to the fight against COVID-19 (Sahu et al. 2023). Bioinformatics has served as a boon in the arena of medicine (Priya et al. 2017) healthcare (Kumari et al. 2016) and environmental aspects (Kesheri et al. 2024a; b) encompassing healthcare and drug design (Sahu et al. 2023). The application of machine learning in the fight against COVID-19 has proven valuable for diagnostic purposes, patient triage, and predictive modeling. Among the algorithms most frequently used are Linear Regression (LR), Support Vector Machines (SVM), Multi-Layer Perceptron (MLP), and Vector Auto-Regression models. Deep learning-based predictive modeling has become very popular (Gudigar et al. 2021). Besides this, numerous variables including the accessibility of healthcare resources, environmental conditions, and socioeconomic variables could be connected to the death rate of COVID-19.

The research in this area is focused on five different facets—detection, bibliometric analyses, textual analyses, topic modeling, and comparative analysis of COVID-19 with other coronaviruses. Initial analysis was focused on adults since children seem to be less affected by COVID-19, but later studies revealed the risks and possible side effects of COVID-19 in pediatric patients as well. Some studies have focused on the Post-Acute Sequela of COVID-19 (PASC) (Munipalli et al. 2022). Several data augmentation approaches have been studied in the context of X-ray and CT image analysis (Yoo et al. 2020). Predictive modeling based on COVID-19 data has been carried out using AI and machine learning, and noisy chest X-rays have also been analyzed with machine learning and deep learning tools to detect the presence of the virus. Deep learning is described as the most common term associated with COVID-19 according to one study with word cloud on research conducted in COVID-19 (Boukobza et al. 2022).

As a result, machine learning can be used as a useful tool for screening, tracking, prediction characteristics and distribution/infection/hospitalization/death of COVID-19. Machine learning methods include heart disease (Ayon et al. 2020), diabetes (Islam Ayon and Milon Islam 2019), liver disorders, breast cancer (Hasan et al. 2016; Islam et al. 2017), COVID-19 (Rahman et al. 2021a, 2021b). This chapter of this book focuses on the importance and influence of machine learning between people and COVID-19. Using machine learning algorithms provide detailed overview of all technologies that seem to detect and process new corona viruses. Machine training can be an important tool for decision-making practices and can work as a help for early detection of COVID-19.

6.2 The Contribution of Artificial Intelligence and Machine Learning to COVID-19 Prevention

While artificial intelligence (AI) can reproduce human behavior with consciousness and sensitivity, machine learning (ML) is a branch of artificial intelligence, and computers can test the data set by providing the ability to learn about learning data. Mashechka Training requires effective algorithms to create predictive models as well as educational sets and test sets. Machine training is divided into two major categories: control and uncontrollable. The following method is a machine learning approach and is determined by using an example of the expected results. This example is used to teach machine learning algorithms and classifies test data for accurate prediction results. The following methods of teaching a mechanical learning student set a parameter for learning in each example and getting close to expected results. Nevertheless, unearthed machine learning methods use algorithms to analyze non-aged data sets. This method of machine learning shows the patterns hidden in raw data without the need for human intervention. Therefore, this method is called a machine learning method. In other words, the controlled training uses the indicated input, while the uncontrolled learning algorithm uses a non-emotional data set. In general, the ML method is widely used in the medical field, so it can be used for diagnosis and treatment of COVID-19.

6.3 Machine Learning Approaches Applied in COVID-19 Research

6.3.1 Support Vector Machine (SVM)

The support vector machine (SVM) is used for clear classifications of data points by searching overdue in N-daily space (Wang et al. 2017; Rahman et al. 2021a, b). Vector support is more data point to Hyperflocon. These data points not only affect the location, but also affect the direction of the hyper plus. The SVM selects the distance between the hyperfrabos and the field, which can be separated between the data points of the two classes. The purpose of SVM is to maximize the value of this margin to obtain a clear boundary that can be used for future data points. The classifier of the support vector predicts the seriousness of the COVID-19 infection and reaches 87% of the accuracy in the prediction case (Guhathakurata et al. 2021).

Benefits of Support Vector Machines Include

I. Their robustness, as SVMs effectively reduce the impact of outliers by focusing on the data points that matter most—those closest to the hyperplane.
II. Strong performance in distinguishing between different categories or classes.
III. High effectiveness in dealing with high-dimensional data.
IV. Efficient memory usage compared to some other machine learning algorithms.

However, SVMs Also Have Certain Limitations

I. They are generally not well-suited for handling large datasets due to computational complexity.
II. SVM performance tends to decline with noisy datasets, which may affect accuracy.
III. SVM does not inherently provide probability estimates for classifications, as it operates based on margin-based separation rather than probabilistic modeling.

6.3.2 Artificial Neural Network (ANN) in COVID-19

Artificial neural networks (ANNs) are a well-established machine learning technique modeled after the structure and functioning of the human brain. ANNs are considered a type of soft computing system that can efficiently handle and interpret large datasets through a series of input nodes. These networks are composed of artificial neurons that are interconnected to form layers, allowing them to process complex information in a structured manner (Mollalo et al. 2020). Due to their adaptability, ANNs have been effectively utilized in various biological domains, including gene prediction, genetic identification, and protein structure forecasting. The standard ANN architecture is organized into three main layers: the input layer, one or more hidden layers, and the output layer. Each layer passes processed data to the next, enabling a flow of information from input to output (Meraihi et al. 2022; Kanchan et al. 2024).

In an ANN, each input is assigned an initial weight, which is often randomly generated. These weights are refined using the backpropagation algorithm, coupled with a gradient descent optimization technique, to minimize error and improve model accuracy. In a feedforward neural network, the data flows unidirectionally from the input layer through the hidden layers to the output layer. Alternatively, recurrent neural networks (RNNs) introduce loops that allow information to be fed back into the network, creating dynamic behavior based on temporal sequences. The network is final output is computed through a combination of weighted inputs and activation functions. This process involves aggregating the inputs at each node, applying weights and activation functions across layers, and summing the results to produce the final output value.

6.3.3 Deep Learning Approaches for COVID-19 Detection and Evaluation

Deep learning, as described by Goodfellow et al. (2016), is a subfield of machine learning that emphasizes the development of complex neural network architectures composed of multiple hidden layers. These deep architectures enable models to learn intricate patterns and make more accurate decisions. The depth or number of hidden layers significantly influences the model's predictive performance. Given their strong learning capabilities, deep learning models are considered reliable tools for the effective diagnosis of COVID-19 infections (Thepade and Jadhav 2020). Several research efforts have demonstrated that the application of deep learning techniques has significantly improved the accuracy of COVID-19 detection from CT scan images (Zhang et al. 2020). As a result, deep learning has become a widely accepted method in the field of medical diagnostics due to its sensitivity, precision, and diagnostic efficiency (Zhang et al. 2020). Beyond Artificial Neural Networks (ANNs), various deep learning algorithms have been employed in COVID-19-related studies, showcasing their adaptability and effectiveness in clinical applications.

6.3.4 Convolutional Neural Network (CNN) for COVID-19 Prediction

Convolutional Neural Networks (CNNs), often referred to as ConvNets, are a type of deep learning architecture that has gained considerable attention in the medical field, especially for interpreting CT and chest X-ray images. These networks are a special kind of artificial neural network where each layer contains multiple neurons. The architecture of a CNN can vary based on the number of layers, the type of layers used, and the number of neurons per layer all of which significantly influence how well the network performs. CNNs process data through a series of layers that apply mathematical operations to extract meaningful features. What makes CNNs unique is the presence of convolutional layers usually the first one to three that are designed to capture essential, abstract features such as edges or textures from the input image. These initial layers help the model recognize foundational patterns, and the subsequent deeper layers build on this information to differentiate between complex objects or abnormalities in the image. Due to this layered approach and feature extraction capability, CNNs have become one of the most effective tools for detecting abnormalities, disease patterns, and irregularities in chest radiographs, including both CT and X-ray images. Their high accuracy, flexibility in architecture design, and proven performance make them a preferred choice in diagnostic imaging tasks.

CNN architectures come in different forms, depending on how layers like convolutional, pooling, fully connected, and dropout layers are arranged. One well-known architecture is Inceptionv3, a deep CNN model with 159 layers. It was initially developed as part of GoogleNet and is widely used for image classification and object detection tasks (Szegedy et al. 2016). Another important CNN model is Xception, proposed by François Chollet. This architecture includes approximately 22,910,480 parameters and has a total model size of about 88 megabytes, making it one of the larger but highly capable CNN frameworks available for image analysis. In contrast, MobileNet is designed to be more lightweight and efficient. It utilizes a method called depth-wise separable convolution to reduce computational complexity. MobileNet has around 4,253,864 parameters and a compact model size of roughly 16 megabytes, making it suitable for deployment on mobile devices and in real-time diagnostic applications.

6.3.5 Recurrent Neural Network (RNN) for COVID-19

Recurrent neural networks (Schuster and Paliwal 1997) are a specific type of ANN used in deep learning, allowing information to flow in both directions between deep layers and the initial layers, aiding in memory retention. As a result, RNN is well suited for handling sequential data, such as learning and generating signals. Unfortunately, RNN is not ideal for tasks that require analyzing video sequences with significant time gaps (such as video classification). Long short-term memory is the widely utilized recurrent neural network across all types of recurrent neural networks. When dealing with extensive datasets, long short-term memory (LSTM) tends to outperform gated recurrent unit (GRU), whereas on smaller datasets, GRU exhibits comparable or even superior performance. Chimmula and Zhang employed long short-term memory (LSTM) neural networks for predicting the transmission of COVID-19 in Canada (Chimmula and Zhang 2020).

6.3.6 Generative Adversarial Network (GAN)

GAN (Creswell et al. 2018) is a sophisticated machine learning method. Generative adversarial networks (GANs) are a type of deep learning-based machine learning algorithm that employs two neural networks to create new, synthetic instances of data. These are commonly employed in image and video applications, particularly in voice recognition. In summary, generative adversarial networks (GANs) are built upon the rivalry between two networks, known as the "generator" and "discriminator," within a specific framework. The generator neural network is in charge of generating new instances of an object, while the discriminator neural network evaluates the authenticity of the object.

6.3.7 Decision Tree (DT)

A decision tree (DT) is a widely recognized supervised learning machine learning algorithm that utilizes a tree-shaped graph, making it highly effective in classifying categorical variables (bishop 2008). The study of decision trees is centered around three key nodes: the root node, the interior node, and the leaf node. It establishes a training model that can forecast the category or class of the dependent variable by applying a set of decision rules. Decision tree ensembles were utilized to forecast the spread of coronavirus disease infections and also in comparative studies.

6.3.8 Regression Analysis

Regression analysis is a machine learning statistical modeling technique that enables the prediction of a continuous outcome variable (y). It contains one or multiple explanatory variables (x). In summary, the regression model establishes a relationship between y and the x variables in an equation that can be utilized to forecast the outcome (y) for a given set of variables (x).

6.3.9 Linear Regression

Linear regression is a widely used method for forecasting a continuous variable, assuming a linear connection between the outcome and the predictor variables. The linear regression equation can be written as: Y = b0 + bx, where: b0 is the intercept and b are the regression coefficient associated with the predictor variable x. The coefficients in linear regression are used to estimate the outcome value with the least amount of error. In the ordinary least squares method, beta coefficients are computed, and then a t-test is conducted to determine if these coefficients are significantly different from zero. A nonzero beta coefficient signifies a substantial connection between the predictors (x) and the outcome variable (y), which can be visually represented using a scatter plot in Python. When dealing with multiple predictor variables, linear regression can be expressed as: Y = b0 + b1x1 + b2x2. By raising the value of predictor variable x1, we could enhance the explanatory power of predictor variable x2 in accounting for the variation. Linear regression models can also include both continuous and categorical predictor variables.

```
# R-Packages to determine the linear regression
The R function lm () is used to compute the linear
regression model.
#Build the model
 model <- lm (predictor ~., data = Genomic.data)
#Summarize the model
Summary (model)
#Make predictions
predictions <- model %>% predict (Genomic.data)
#Model performance
#(a) Prediction error, RMSE
RMSE (predictions, Genomic.data$predictor)
 #(b) r-square
 R2(predictions, Genomic.data$predictor).
```

Nevertheless, in certain instances, the connection between the result and the variables being studied is not a straight line. In these circumstances, you need to construct a nonlinear regression model, such as the polynomial and spline regression model. In addition to this, when dealing with complex datasets involving multiple omics (Srivastava et al. 2024a), we can utilize various strategies such as:

6.3.10 Principal Component Regression

Principal component regression (PCR) entails creating the initial principal components z1,..,zqz1,..,zq, and subsequently employing these components as predictors in a linear regression model that is fitted using the least squares method. The number of principal components helps us understand how much of the data's variability is explained and how it is related to each other. In simpler terms, we believe that the directions x1,xpx1,..,xp indicate the greatest variation in the directions that are linked to y. While this assumption may not always hold true, it is frequently a sufficient basis for achieving positive outcomes.

PCR consists of two phases:

A. Principal component analysis is used on the training data to reduce its dimensionality.
B. Linear regression-based training on the transformed samples.

In principal component analysis, the transformation is purely unsupervised, without any use of information about the targets. Consequently, principal component regression may not yield accurate results in datasets where the target variable is highly correlated with directions that have low variability. In fact, the principal component analysis technique reduces the dimensionality of the data by projecting it into a lower-dimensional space, where the variance of the projected data is maximized along each axis. Partial least squares (PLS) are a versatile tool that can both transform and regress data, and it shares similarities with principal component regression. The samples are first reduced in dimensionality before being used as input for a linear regressor. The primary distinction between PCR and partial least squares (PLS) lies in the supervised nature of the PLS transformation.

6.4 Partial Least Squares Regression

Partial least squares (PLS) regression simplifies the predictors to a smaller number of components. The least squares regression analysis is conducted on the components of the data, rather than the original data itself. PLS regression is a solution for: That is, when the variables are highly correlated—when there are more predictors than observations, it encourages overfitting due to a lack of the degree of freedom—when ordinary least-squares regression either generates coefficients with high standard errors or fails to produce accurate results. The predictors can be measured with error, which makes them more robust to dealing with uncertainty (Kuhn and Johnson 2013).

6.5 Penalized Regression-Ridge Regression

Penalized regression-ridge regression shrinks the coefficients in the model toward 0, but never to 0. This shrinkage leads to a substantial reduction in the variance of the predictions, with a penalty of a slightly increased bias. The number of variables that minimize noise while retaining the predictive power of the model will be chosen based on log lambda on the graph. We believe that this and LASSO could be highly accurate as they shrink coefficients and may be able to reduce noise better than other models.

```
# R code for using of Ridge regression model on data
 library (glmnet)
 set. seed (101)
 X.train = model. matrix (new_Patients ~. -date, train.
dat) [, -1]
 Y. train <- train.dat$ new_Patients
 x_test <- model. matrix (new_Patients ~. -date, test.dat) [, -1]
 y_test <- test.dat$new_Patients
 # Finding the best Lambda Using Cross-Validation
 # train = sample (6720, 5377) # 80% training
 # test = (-train)
 ridge. modell <- glmnet (x = x_train, y = y_train, alpha = 0,
standardize = F)
 cv.out <- cv. glmnet (x = x_train, y = y_train, alpha = 0,
standardize = F)
 plot (ridge. modell, xvar = "lambda")
Note: The impact of different "λ" parameters and select the best
one for our ridge regression model.
 bestlam_ridge <- cv.out$lambda.min
 ridge_train_preds <- predict (cv.out, x_train, s = "lambda.min")
 ridge_test_preds <- predict (ridge.modell, x_test, s =
bestlam_ridge)
 sprintf("Test MSE: %.4f", mean((ridge_test_preds - y_test)^2))
 sprintf("Train MSE: %.4f",
mean((ridge_train_preds - y_train)^2))
 ridge_mse <- mean ((ridge_test_preds - y_test) ^2)
```

6.6 LASSO Regression

LASSO, unlike the ridge regression model, the coefficients are actually reduced to 0. As a result, the lasso regression and ridge regression models are known for their high accuracy, as they shrink coefficients and effectively reduce noise compared to other models.

```
# R code for Lasso Regression
 library (glmnet)
 set. seed (100)
 lasso. modell <- glmnet (x = x_train, y = y_train, alpha = 1,
standardize = F)
 plot (lasso. modell, xvar = "lambda")
 set. seed (100)
 lasso.cv <- cv.glmnet(x = x_train, y = y_train, alpha = 1,
standardize = F)
 plot(lasso.cv)
```

```
best_lasso <- lasso.cv$lambda.min
# Comparing LASSO training and testing MSE
lasso_train_preds <- predict(lasso.cv, x_train, s =
"lambda.min")
lasso_test_preds <- predict (lasso. model1, x_test, s =
best_lasso)
sprintf("Test MSE: %.4f", mean((lasso_test_preds - y_test)^2))
sprintf("Train MSE: %.4f",
mean((lasso_train_preds - y_train)^2))
lasso_mse <- mean ((lasso_test_preds - y_test) ^2)
# LASSO Regression Most Influential Features
library (broom)
coef (lasso.cv, s = "lambda.min") %>% tidy () %>% filter (row! =
"(Intercept)") %>%
ggplot (aes (value, reorder (row, value))) + geom_point() +
xlab("Coefficient") + ylab(NULL)
```

6.7 Regression Model Accuracy Metrics

In this chapter, we discuss various statistical metrics used to evaluate the effectiveness of a regression model.

(a) *R-squared*, also known as the coefficient of determination, quantifies the proportion of variability in a dependent variable that can be explained by an independent variable. In multiple regression models, r2 represents the squared correlation between the actual outcome values and the predicted values generated by the model. R2 ranges from 0 to 1, with a higher value indicating a better model.

(b) *Root mean squared error (RMSE)* is a metric used to assess the average discrepancy between predicted values and the actual variable, indicating the effectiveness of a model. Multiple studies have demonstrated the effectiveness of RMSE as a validation method for the generated model. Residual standard error (RSE): RSE is also referred to as model sigma, a different form of the RMSE. RMSE and RSE are distinct from each other when it comes to handling large multivariate data. The lower value of RSE indicates a more accurate model.

(c) *Mean absolute error (MAE):* MAE calculates the average magnitude of the differences between the actual and predicted outcomes. Mae is more sensitive to outliers than RMSE. In addition to the above, there are other metrics like AIC (Akaike's information criteria) and BIC (Bayesian information criteria) that can be utilized for evaluating and choosing models. AIC is primarily employed to penalize the increase in error when incorporating additional variables into a model. The lower the AIC value, the higher the quality of the model. BIC is a more stringent version of AIC that imposes a stronger penalty for including additional variables in the model.

Some packages are assisting in making this model tidy, for data manipulation and visualization, model, functions for evaluating regression model performance metrics and broom use for tidy data frame containing the model statistical metrics.

```
Note: # R code for making regression models:
library(tidyverse)
library(modelr)
library(broom)
# Load the data data("Covid19")
 #Creating model
 model <- lm (Hospitalization ~., data = Covid19)
 summary(model)
 AIC (model)
 BIC (model)
 library(modelr)
 data. frame (R2 = rsquare (model, data = Covid19),
 RMSE = rmse (model, data = Covid19),
 glance(model)
```

6.8 Multilayer Perceptron (MLP) Models

We can model the spread of COVID-19 infection using a multilayer perceptron model. This could help to model the spread of the disease, which predicts the maximal number of people who contracted the disease per location in each time unit based on the maximal number of people who recovered per location in each time unit, and the maximal number of deaths per location in each time unit. MLP is a type of fully connected, feed-forward artificial neural network (ANN), consisting of neurons arranged in layers as an input layer, an output layer, and one or more hidden layers (Hui et al. 2020). Cross-validation is performed to validate the results. The cross-validation method used in this research is the K-fold algorithm (Goodfellow et al. 2016; Lorencin et al. 2019).

6.9 Vector Auto-Regression Models

Vector autoregression (var) is a statistical model that helps us understand how different quantities interact and change over time. Var is a type of random process model. By including, multivariate time series, var models are generalized.

6.10 Diagnosis of COVID-19 Using Tomography Scans: MT-nCov-Net

Mt-nCov-net is a novel hybrid model that was created to extract visual features from volumetric thoracic CT scans for the detection of COVID-19. We could also refer to this as deep CNN architecture for classifying CT images. Advanced deep neural networks for the identification of COVID-19 from chest CT images will offer a wide range of learning opportunities. This approach is also known as a hybrid methodology for diagnosing COVID-19 pneumonia, combining the best features from multiple networks (Ding et al. 2023).

6.11 Analysis of COVID-19 Using the XG Boost Model

The XGBoost model is a decision tree-based machine learning model, by which we can uncover the nonlinearity in the time series of COVID-19 cases, and it can help improve the prediction of COVID-19 cases in the United States over the ARIMA model because in the ARIMA model we can predict the time-series data, and it can show autocorrelations in data while in the XGBoost model (decision tree-based machine learning model) by which we can uncover the nonlinearity in the time series of COVID-19 cases. We can predict the data of vaccination in the USA using this model. Also, the XG Boost model was built based on pre-vaccination-induced immunity. Moreover, XGBoost acts as a stronger classifier than other classifiers due to having the advantage of avoiding overfitting effectively. This model also deals with missing values, and reducing the running time by parallel and distributed calculation effectively. We could use the xts, TSstudio, and tseries R-bioconductor packages to analyze data and the ggplots2 and dygraphs packages to draw diagrams. The proposed models were established via forecast and XGBoost packages. In the preparation of data for using XGBoost, we need to convert all other forms of data into numeric vectors and make a sparse matrix using flags on every possible value of that variable. Conversely, a dense matrix is a matrix where most of the values are nonzeros).

```
# R codes for using of XG Boost model
library (Matrix)
sparse. Matrix <- sparse. model. matrix (response ~. -1,
data = Data)
# Load all the libraries
library(xgboost)
library(readr)
library(stringr)
library(caret)
library(car)
# Load the dataset
set. seed (100)
df. train = read.csv ("train.csv")
df.test = read.csv("test.csv")
# Load labels of the train data
labels = df. train['Patient_Status']
df. train = df. train [-grep (' Patient_Status ', colnames (df.
train))]
```

```
# Combine train and test data
Df. All = rbind (df. train, df.test)

# Data Cleaning & Features loading
There were some missing data (NA) in the dataset. In this step,
we mainly deal with all the missing variables using imputation
methods in the data and create variables for the categorical
features.
Df. All$sex [is.na (Df. All$sex)] = "male"
Df. All $Hospitalized [is.na (Df. All $ Hospitalized)] = "no"
Df. All $Salary [is.na (Df. All $Salary)] = mean (Df. All
$Salary, na.rm = TRUE)
Df. All $Salary_Term [is.na (Df. All $Salary_Term)] = 360
Df. All $patient _History [is.na (Df. All $patient _
History)] = "yes"
Df. All $Dependents [is.na (Df. All$Dependents)] = 0
# Split train and test data set
X = Df. All [Df. All $patient_id %in df. train$patient_id,]
y <- recode (labels$Patient_Status," 'Y'=1; 'N'=0")
X_test = Df. All [Df. All $patient_id %in% df.test$ patient_id,]
# Run the model
set. seed (100)
xgb <- xgboost (data = data.matrix(X[,-1]),
 label = y,
 eta = 0.1,
 max_depth = 15,
 nround=25,
 subsample = 0.5,
 colsample_bytree = 0.5,
 eval_metric = "merror",
 objective = "multi: softprob",
 num_class = 12,
 nthread = 3
)
# Score the Test
y_pred <- predict (xgb, data. matrix (X_test [, -1]))
#Check the model (check top 10 models)
model <- xgb. dump (xgb, with. stats = T)
model [1:10]
names <- dimnames(data. matrix (X [, -1])) [[2]]
importance_matrix <- xgb.importance(names, model = xgb)
importance_matrix
#Graph generation plot
xgb. plot. importance (importance_matrix [1:10,])
xgb. plot.importance(importance_matrix[1:10,])
```

6.12 CXR and CT Medical Imaging Methods

The image processing and machine learning techniques can expedite and provide definite coronavirus detection using medical imaging, which can aid in early detection, diagnosis, and potential cure for the eradication of COVID-19. We can examine the COVID-19 image data by first preprocessing the images through normalization to improve their quality and eliminate noise. Secondly, the process of dividing images into segments is achieved through fuzzy c-means clustering. The CT images are significantly more advantageous than the X-ray images because CT images can provide a three-dimensional view of the lungs. These images can be transformed into two-dimensional images, providing a sagittal axial coronal view, which aids in the diagnosis of the condition (Benmalek et al. 2021).

6.13 Novel Corona Virus Datasets for Study

To give deep insights into COVID-19-related research, we obtain a large-scale scholarly dataset from the COVID-19 open research dataset. We can also get the publication metadata for COVID-19 studies from PubMed central (PMC), BioRxiv preprint servers, medRxiv preprint servers, and who COVID-19 databases using the code below to extract a csv files. Also here is few links. Some well-known sources for COVID-19 datasets include the following:

I. *Johns Hopkins University COVID-19 Data Repository.*
   (a) GitHub: https://github.com/CSSEGISandData/COVID-19
II. *Our World in Data—COVID-19 Dataset.*
   (a) Website: https://ourworldindata.org/coronavirus
III. *COVID-19 Data Hub.*
   (a) Website: https://covid19datahub.io/
IV. *Google COVID-19 Mobility Reports.*
   (a) Website: https://www.google.com/covid19/mobility/
V. *Kaggle COVID-19 Datasets.*
   (a) Website: https://www.kaggle.com/datasets

6.14 Applications of Machine Learning in COVID-19

The various advanced developments in machine learning and artificial intelligence have significantly reduced human intervention in medical practices and in the improvement of treatment, medication, screening, prediction, forecasting, contact tracing, and drug or vaccine development process for COVID-19. The support vector machine (SVM) has also been utilized in the detection and classification of COVID-19 cases. The clinical data and test results were utilized in their research to assess the effectiveness of SVM's performance. The simulation outcomes showcased the efficiency of the SVM model, attaining an accuracy of 81.48%, sensitivity of 83.33%, and specificity of 100%. Deep learning integrated with support vector machine (SVM) uses x-ray images for predicting COVID-19 (Srivastava et al. 2024b). Instead of using a deep learning-based classifier, also employed for classification, as it does not require a large dataset for training and validation. A fully connected layer of the CNN model from deep learning is extracted to feed into SVM for classification, which uses three categories of X-ray images from COVID-19 patients, pneumonia patients, and healthy people. The method is advantageous for the medical practitioner to categorize among the COVID-19 patients, pneumonia patients, and healthy individuals. Wang and his co-workers proposed a deep CNN model, called residual network34 (resnet34), for COVID-19 diagnosis in CT scan images (wang et al. 2020). Wang et al.'s CNN-based covid-net12 achieved a 93.3% test accuracy for multi-class classification on a test cohort of 100 normal, 100 pneumonia, and 100 COVID-19 images from the COVIDx dataset with the rest of the images of each class being used to train their model. The efficacy of resnet34 was confirmed through the analysis of CT scan images obtained from 99 patients, including 55 individuals with typical viral pneumonia and 44 patients with COVID-19. The simulation results demonstrated that resnet34 attains an overall accuracy of 73.10%, specificity of 67%, and sensitivity of 74%.

6.15 Conclusions and Future Perspectives

Representation is used in machine learning-based approaches to address intricate issues. Deep learning-based models, such as CNN, proposed custom CNN, DCNN, and other methods have lately been used for COVID-19 classification, diagnostics, and detection, by researchers to combat the COVID-19 outbreak. Therefore, the combination of different omics technologies with machine learning, deep learning, and artificial intelligence methods is essential for gaining a deeper understanding and effectively managing the COVID-19 situation (Srivastava et al. 2024c). The utilization of these methods is rapidly increasing due to their ability to open up new possibilities for diagnosis, sentiment analysis, public surveillance, and illness prevention. Various diagnostic methods for COVID-19 that utilize images and advanced techniques like deep learning and artificial intelligence have been developed, and their correlation with the traditional RT–PCR test has been examined. By combining machine learning, deep learning, and artificial intelligence techniques, researchers are exploring innovative approaches to address the challenges posed by the COVID-19 pandemic.

Chapter Summary

This chapter provides an in-depth overview of various machine learning methodologies currently employed to address critical aspects of the COVID-19 pandemic. These include evaluating disease severity, forecasting patient mortality, and assessing model performance through key metrics such as accuracy, precision, sensitivity, and specificity. Both supervised and unsupervised learning approaches are discussed to highlight their respective contributions to these tasks. Additionally, the chapter presents a curated selection of large-scale datasets sourced from scholarly research and public repositories, which serve as foundational resources for training and validating machine learning models. Overall, this chapter delivers a thorough and up-to-date review of cutting-edge machine learning techniques developed for the diagnosis, monitoring, and prediction of COVID-19, including its short- and long-term impacts on affected populations.

References

Ayon SI, Islam MM, Hossain MR (2020) Coronary artery heart disease prediction: a comparative study of computational intelligence techniques. IETE J Res 68(4):2488–2507

Benmalek E, Elmhamdi J, Jilbab A (2021) Comparing CT scan and chest X-ray imaging for COVID-19 diagnosis. Biomed Eng Adv 1:100003

Bishop CM (2008) Pattern recognition and machine learning. Springer-Verlag, New York

Boukobza A, Burgun A, Roudier B, Tsopra R (2022) Deep neural networks for simultaneously capturing public topics and sentiments during a pandemic: application on a COVID-19 tweet data set. JMIR Med Inform 10(5):e34306

Chimmula VKR, Zhang L (2020) Time series forecasting of COVID-19 transmission in Canada using LSTM networks. Chaos, Solitons Fractals 135:109864

Creswell A, White T, Dumoulin V, Arulkumaran K, Sengupta B, Bharath AA (2018) Generative adversarial networks: an overview. IEEE Signal Process Mag 35(1):53–65

Dey SK, Rahman MM, Siddiqi UR, Howlader A (2020) Analyzing the epidemiological outbreak of COVID-19: a visual exploratory data analysis approach. J Med Virol 92(6):632–638

Ding W, Abdel-Basset M, Hawash H, Elkomy OM (2023) MT-nCov-net: a multitask deep-learning framework for efficient diagnosis of COVID-19 using tomography scans. IEEE Trans Cybern 53(2):1285–1298. https://doi.org/10.1109/TCYB.2021.3123173

Goodfellow I, Bengio Y, Courville A (2016) Deep learning, volume 1. MIT Press, Cambridge

Gudigar A, Raghavendra U, Nayak S, Ooi CP, Chan WY, Gangavarapu MR, Dharmik C, Samanth J, Kadri NA, Hasikin K, Barua PD, Chakraborty S, Ciaccio EJ, Acharya UR (2021) Role of artificial intelligence in COVID-19 detection. Sensors 21:8045

Guhathakurata S, Kundu S, Chakraborty A, Banerjee JS (2021) A novel approach to predict COVID-19 using support vector machine. Data Sci COVID-19:351–364

Haque MR, Islam MM, Iqbal H, Reza MS, Hasan MK (2018) Performance evaluation of random forests and artificial neural networks for the classification of liver disorder. In: International conference on computer, communication, chemical, material and electronic engineering (IC4ME2), pp 1–5

Hasan MK, Islam MM, Hashem MMA (2016) Mathematical model development to detect breast cancer using multigene genetic programming informatics, electronics & vision, (ICIEV). IEEE, pp 574–579

Islam Ayon S, Milon Islam M (2019) Diabetes prediction: a deep learning approach. Int J Inf Eng Electron Bus 11:21–27

Islam MM, Iqbal H, Haque MR, Hasan MK (2017) Prediction of breast cancer using support vector machine and K-nearest neighbors. In: IEEE region 10 humanitarian technology conference (R10-HTC), 2018, pp 226–229

Kanchan S, Ogden E, Kesheri M, Skinner A, Miliken E, Lyman D, Armstrong J, Sciglitano L, Hampikian G (2024) COVID-19 hospitalizations and deaths predicted by SARS-CoV-2 levels in Boise, Idaho wastewater. Sci Total Environ 907:167742

Kesheri M, Kanchan S, Srivastava S, Ratna-Raj R, Chittoori B, Vaishampayan A, Rastogi RP, Sinha RP, Primerano D (2024a) Ecology and environmental omics. In: Gupta MK, Katara P, Mondal S, Singh RL (eds) Integrative omics: concepts, methodology, and application. Academic Press (Elsevier), pp 315–331

Kesheri M, Kanchan S, Salisbury TB, Sinha RP (2024b). Microbial omics in environment and health. Springer Nature, Singapore. [ISBN:978-981-97-1768-2]

Kuhn M, Johnson K (2013) Applied predictive modeling. Springer, New York

Kumari A, Kanchan S, Kesheri M (2016) Applications of bio-molecular databases in bioinformatics. In: Dey N, Bhateja V, Hassanien AE (eds) Advancements in bio-medical sensing, imaging, measurements and instrumentation, vol 651. Springer-Verlag, Heidelberg, pp 329–351

Kumari A, Kesheri M, Sinha RP, Kanchan S (2018) Integration of soft computing approach in plant biology and its applications in agriculture. In: Purohit HJ, Raje DV, More RP, Kalia VC (eds) Soft computing in biological sciences. Springer Nature, Singapore, pp 265–278

Lorencin I, Anđelić N, Mrzljak V, Car Z (2019) Genetic algorithm approach to design of multilayer perceptron for combined cycle power plant electrical power output estimation. Energies 12(22):4352

Meraihi Y, Gabis AB, Mirjalili S, Ramdane-Cherif A, Alsaadi FE (2022) Machine learning-based research for COVID-19 detection, diagnosis, and prediction: a survey. SN Comput Sci 3:286

Mollalo A, Rivera KM, Vahedi B (2020) Artificial neural network modeling of novel coronavirus (COVID-19) incidence rates across the continental United States. Int J Environ Res Public Health 17(12):4204

Munipalli B, Seim L, Dawson NL, Knight D, Dabrh AMA (2022) Post-acute sequelae of COVID-19 (PASC): a meta-narrative review of pathophysiology, prevalence, and management. SN Compr Clin Med 4(1):90

Priya P, Kesheri M, Sinha RP, Kanchan S (2017) Molecular dynamics simulations for biological systems. In: Karâa WBA, Dey N (eds) Biomedical image analysis and mining techniques for improved health outcomes. Medical Information Science Reference, pp 286–313

Rahman MA, Islam SMS, Tungpunkom P, Sultana F, Alif SM, Banik B, Cross WM (2021a) COVID-19: factors associated with psychological distress, fear, and coping strategies among community members across 17 countries. Glob Health 17:1–19

Rahman MM, Nooruddin S, Hasan KA, Dey NK (2021b) HOG+ CNN net: diagnosing COVID-19 and pneumonia by deep neural network from chest X-ray images. SN Comput Sci 2(5):371

Sahu N, Mishra S, Kesheri M, Kanchan S, Sinha RP (2023) Identification of cyanobacteria-based natural inhibitors against SARS-CoV-2 druggable target ACE2 using molecular docking study, ADME, and toxicity analysis. Indian J Clin Biochem 38(3):361–373

Schuster M, Paliwal KK (1997) Bidirectional recurrent neural networks. IEEE Trans Signal Process 45(11):2673–2681

Srivastava U, Kanchan S, Kesheri M, Gupta MK, Singh S (2024a) Types of omics data: genomics, metagenomics, epigenomics, transcriptomics, proteomics, metabolomics, and phenomics. In: Gupta MK, Katara P, Mondal S, Singh RL (eds) Integrative omics: concepts, methodology, and application. Academic Press (Elsevier), pp 13–34

Srivastava U, Kesheri M, Kanchan S, Singh S (2024b) Computational omics protocol for the comparative study of microbiome analysis. In: Kesheri M, Kanchan S, Salisbury TB, Sinha RP (eds) Microbial omics in environment and health. Springer, Singapore

Srivastava U, Kanchan S, Kesheri M, Gupta MK, Singh S (2024c) Integrative omics approaches for identification of biomarkers. In: Gupta MK, Katara P, Mondal S, Singh RL (eds) Integrative omics: concepts, methodology, and application. Academic Press (Elsevier), pp 157–173

Szegedy C, Vanhoucke V, Ioffe S, Shlens J (2016) Rethinking the inception architecture for computer vision. In: The IEEE conference on computer vision and pattern recognition (CVPR), pp 2818–2826

Thepade SD, Jadhav K (2020) COVID-19 identification from chest X-ray images using local binary patterns with assorted machine learning classifiers. In: 2020 IEEE Bombay section signature conference (IBSSC), pp 46–51

Wang H, Gu J, Wang S (2017) An effective intrusion detection framework based on SVM with feature augmentation. Knowl Based Syst 136:130–139

Wang LL, Lo K, Chandrasekhar Y, Reas R, Yang J, Burdick D, Kohlmeier S (2020) Cord-19: the covid-19 open research dataset. ArXiv:arXiv-2004

Yoo SH, Geng H, Chiu TL, Yu SK, Cho DC, Heo J, Lee H (2020) Deep learning-based decision-tree classifier for COVID-19 diagnosis from chest X-ray imaging. Front Med 7:427

Zhang HT, Zhang JS, Zhang HH, Nan YD, Zhao Y, Fu EQ, Zhang T (2020) Automated detection and quantification of COVID-19 pneumonia: CT imaging analysis by a deep learning-based software. Eur J Nucl Med Mol Imaging 47:2525–2532

Exploring Human Brain Disorders Through Multi-Omics Approaches

7

Swarna Kanchan, Minu Kesheri, Upasna Srivastava, Bhagaban Mallik, Pramod Katara, Shivani Sharda, Chandandeep Kaur, and Rajeshwar P. Sinha

Abstract

Among all the organs, brain is considered as most complex organ in humans. Brain includes various disorders called as neurodegenerative diseases, such as Alzheimer's disease and Parkinson's disease, which represent one of the top healthcare problems in developed countries. In recent years, the omics technologies have contributed a lot to the identification of novel biomarkers, such as

S. Kanchan (✉) · M. Kesheri
Boise State University, Boise, ID, USA

Joan C. Edwards School of Medicine, Marshall University, Huntington, WV, USA
e-mail: kanchan@marshall.edu

U. Srivastava
Yale School of Medicine, New Haven, CT, USA

B. Mallik
Department of Biological Sciences, IISER, Bhopal, Madhya Pradesh, India

Department of Anatomy and Cell Biology, University of Iowa, Iowa City, IA, USA

P. Katara
Computational Omics Lab, Centre of Bioinformatics, IIDS, University of Allahabad, Prayagraj, Uttar Pradesh, India

S. Sharda
Amity Institute of Biotechnology, Amity University, Noida, Uttar Pradesh, India

C. Kaur
Happiest Health, Bangalore, Karnataka, India

R. P. Sinha
Laboratory of Photobiology and Molecular Microbiology, Department of Botany, Institute of Science, Banaras Hindu University, Varanasi, Uttar Pradesh, India

M. Kesheri et al. (eds.), *Multi-Omics in Biomedical Sciences and Environmental Sustainability*, https://doi.org/10.1007/978-981-96-7067-3_7

genes and biological pathways, to understand the mechanisms of various brain disorders. These omics approaches help in identifying the molecular signatures of various brain disorders and their interactions at various levels. In this chapter, we describe the applications of omics, single-cell omics, and spatial transcriptomics in brain research. Various omics technologies individually as well as their integration to elucidate various stages of neurological disorders, disease progression and identification of biomarkers have been highlighted. An extensive overview illustrating the state of the art in omics methodologies for exploring brain diseases in lucid style makes this chapter informative and reader friendly.

Keywords

Alzheimer's disease · Human brain diseases · Multi-omics · Omics technologies · Parkinson's disease

Objectives

- To explore various mechanisms of neurodegeneration to understand the mechanisms of neurodegenerative diseases such as Alzheimer's and Parkinson's diseases
- To highlight various omics approaches in studying the various stages of neurological disorders
- To identify genetic variants and novel biomarkers for neurodegenerative diseases using multi-omics technologies and their integration
- To highlight single-cell transcriptomics and spatial omics applications in studying various neurological disorders at single-cell resolution

7.1 Introduction

The brain is a complex organ that is composed of many specialized cells. These specialized cells control various physical and cognitive functions such as decision-making, learning, and memory. Brain cells differ in morphology and functions. They are classified into multiple parts based on their anatomical locations. Neurons are the structural and functional unit of our brain which participates in communication. Therefore, progressive loss of neurons, in terms of structure, and functions are responsible for neurodegeneration. Dysfunction of synapses and neural networks also leads to neurodegeneration. Deposition of mutated proteins in the brain is also associated with neurodegeneration. The blood–brain barrier (BBB) plays a vital role in and acts as a real barrier to various brain disorders (Lamptey et al. 2022). Disruption of the blood-brain barrier is responsible for the early stages of neurodegenerative diseases. The most common neurodegenerative disorders are dementias, which include Alzheimer's and Parkinson's diseases.

Various omics technologies (Srivastava et al. 2024a) have been applied in neuroscience studies, to identify the alterations in genes, proteins, and metabolites and identifying novel biomarkers (Kumar et al. 2024; Srivastava et al. 2024b). Various spatial omics studies for various regions of the brain help further construct a brain atlas. Various single-cell/single-nucleus studies further help in exploring the biological processes by monitoring changes at single cell resolution (Geschwind and Konopka 2009; Hasin et al. 2017; Dong et al. 2021).

7.2 Types of Neurological Disorders

7.2.1 Dementias

Dementia is a group of diseases that is popular globally due to its increasing incidence in the aging population. Out of 55 million, 12 million individuals in Europe have been diagnosed with dementia. Neurodegenerative dementias are due to the accumulation of misfolded proteins inside the neurons, in extracellular matrix and other cell types of the brain. People with neurogenerative dementia spend 10–20% of their total income on treatment. Among all forms of dementias, Alzheimer's disease (AD) is considered as most prevalent (45–60%) dementia.

7.2.2 Alzheimer's Disease

Alzheimer's is a major health concern in several countries including the United States. So far, approximately 5.4 million people in the United States are affected with this disease (Suehs et al. 2013). Alzheimer's disease (AD), is characterized by extensive memory loss and cognitive dysfunction and faulty synaptic connections. Amyloid-β accumulation and neurodegeneration are mainly responsible for Alzheimer's disease (Strooper and Karran 2016; Long and Holtzman 2019). Damage to the essential neurotransmitter systems and microglia illustrated the AD disease progression (Keren-Shaul et al. 2017).

7.2.3 Parkinson's Disease

Parkinson's disease (PD) is considered the second most common, progressive, neurodegenerative disease affecting 2–3% of the population having aged more than 65 (Poewe et al. 2017). Parkinson's disease symptoms include muscle stiffness, unsteady walking, and problems in coordination and balance. Parkinson's disease (PD) is characterized by a loss of dopaminergic neurons, accumulation of protein aggregates called Lewy bodies (LBs) in the midbrain (Mahul-Mellier et al. 2020) resulting in impaired movement (Demaagd et al. 2015).

7.3 Omics Technologies

7.3.1 Epigenomics

Epigenetics studies various changes that are not encoded by DNA. Thus, epigenetics influence gene expression as well as and protein expression without directly involving DNA. Epigenomic changes are associated with various pathological conditions of the brain and help in revealing the mechanisms of brain disorders (O'Connor et al. 2023). Various epigenetic changes are dynamic and reversible, which are due to various factors such as diet, and exposure to various chemicals (Ghosh and Saadat 2023).

Various epigenetic changes are due to various biological phenomenon such as DNA methylation, histone modifications, chromatin remodeling, and noncoding RNA regulation. Histone acetylation and deacetylation regulate gene transcription by altering the chromatin structure. Acetylation/deacetylation of histones leads the accessibility to transcription factors to DNA at transcription start site (Lu et al. 2015). Dysregulation of histone acetylation affects various signal transduction pathways such as cell differentiation as well as cell apoptosis. Histone acetylation dysregulation also leads to vascular remodeling, as well as inflammation reactions/responses neuronal plasticity, and metabolic reprogramming (Mihaylova and Shaw 2013). Dysregulation of histone acetylation is considered an important epigenetic marker that plays a significant role in Alzheimer's disease (AD) etiology. In a recent study, an increase in histone acetylation was observed associated with AD pathogenesis (Nativio et al. 2020). In another study, histone deacetylase inhibitors were observed to act as neuroprotective in AD based animal models. Genome-wide analysis based on histone-associated epigenetic changes utilizing AD brains samples (Nativio et al. 2018) illustrated the loss of an age-related protective epigenetic pathway in AD (Nativio et al. 2018) when compared to the normal aging profile.

Ben-Hattar and coworkers for the first time studied the epigenetic changes associated with AD involving CpG methylation (Ben-Hattar and Jiricny 1988). A data mining (Kanchan et al. 2024) study based on GEO datasets that implemented a new computational framework identified 21 differentially methylated genes for AD (Mallik and Zhao 2020). DNA methylation changes could also affect beta-amyloid (Ab) and tau deposition patterns in Alzheimer's disease. On the other side, DNA methylation variations also affect alpha-synuclein expression in Parkinson's disease (Jowaed et al. 2010; Matsumoto et al. 2010).

Epigenomics-based datasets and a few data mining studies are available online at the National Cell Repository for Alzheimer's Disease (NCRAD), Gene Expression Omnibus (GEO), and Roadmap Epigenomics (Consortium et al. 2015), and (Barrett et al. 2013).

7.3.2 Transcriptomics (Bulk RNA Sequencing)

Transcriptomics refers to the study of the transcriptome. Transcriptomics analysis has been used to investigate the mRNA expression levels and identify differentially expressed genes (DEGs) in various diseases.

A standard transcriptomics data analysis workflow (as shown in Fig. 7.1) starts with quality checks of sequence reads using FastQC (Andrews 2010). Adapter trimming and quality-based preprocessing of RNA-seq data are performed further by Trimmomatic (Bolger et al. 2014). These sequence reads are subjected to alignment with reference genomes using Bowtie2 (Langmead and Salzberg 2012), Hisat2 (Kim et al. 2019), STAR (Dobin et al. 2013), etc. Samtools (Li et al. 2009) are further used for file (sam to bam and vice versa) conversion, sorting, and indexing of

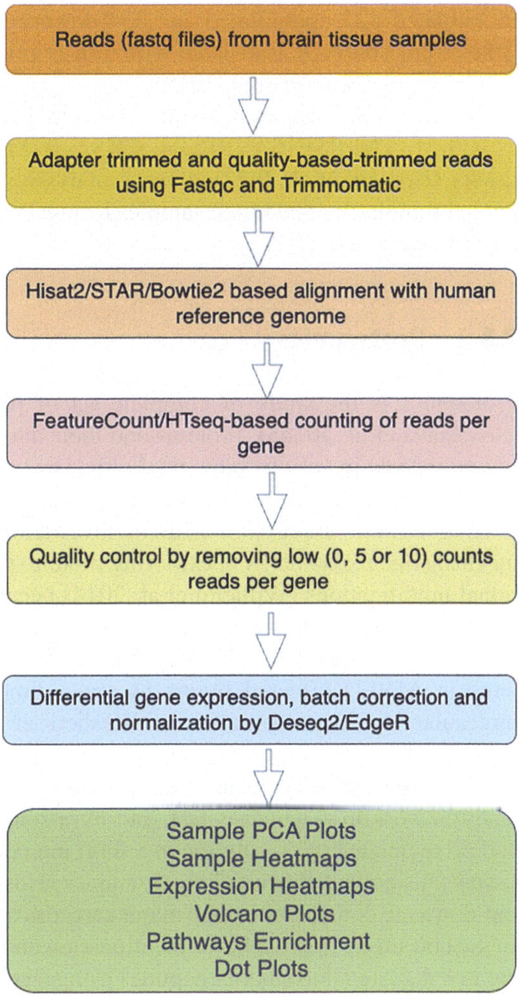

Fig. 7.1 A standard pipeline for transcriptomics bulk data analysis from brain tissue samples

aligned files. FeatureCounts (Liao et al. 2014) are used further for counting reads per gene. Tools like DESeq2 (Love et al. 2014) and EdgeR (Robinson et al. 2009), are commonly used for analyzing differential gene expression between control and treatment groups. Gene set enrichment analysis is subjected to investigate gene ontology terms and various affected biological pathways using GSEA (Subramanian et al. 2005), DAVID and Gprofiler (Reimand et al. 2007).

Transcriptomics studies also help in identifying gene expression signatures associated with AD and PD. Such signatures could be beneficial to address the mechanisms associated with AD and PD progression. A study by Nativio and coworkers was conducted to investigate the changes in gene expression and identify the molecular mechanisms underlying AD. In this study, AD (AD; $n = 12$; mean age = 68), healthy older (Old; $n = 10$; mean age = 68) and healthy younger (Younger; $n = 8$, mean age = 52) brains from the lateral temporal lobe region were studied (Nativio et al. 2018). Transcriptomics study comparing AD and healthy old (as control), investigated 421 upregulated and 434 downregulated genes. Most downregulated genes were observed associated with transcription factors and chromatin structure-related genes (Nativio et al. 2018).

A recent RNA-seq study identified several genes, such as upregulated genes such as RPL21, TNFRSF10A, RPL34, CKS2, B2M, DTX2, and HLA-B, in PD brain tissues (Hossain et al. 2022). A meta-analysis of 26 different RNA-seq studies utilizing brain tissues and blood samples from AD and PD patients was also performed (Hooshmand et al. 2022).

7.3.3 Proteomics

Proteomics is the study of complete set of proteins produced by an individual (Srivastava et al. 2024a). Proteins and their interactions with other molecules play important role in various biological processes; any problem in such biological processes leads to various neurodegenerative diseases. RNA quantification and characterizing protein expression help in investigating disease states and underlying mechanisms. It also helps in exploring protein–protein interactions and posttranslational modifications (Wilhelm et al. 2014) because these parameters further affect the pathogenesis and disease progression of neurodegenerative diseases (Verrou et al. 2020). In brief, proteomics aims in identifying the abnormality in such proteins in AD/PD affected brains by three-dimensional structural prediction and molecular dynamics simulations (Kesheri et al. 2015; Priya et al. 2017; Li et al. 2019).

Mass spectrometry technology provides the opportunity for proteomics data analysis workflow, which is fast, and ultra-sensitive. It allows detection of tissue/cell proteome at high resolution in a short interval of time utilizing a low number of tissue/cells collected from clinical samples (Hosp and Mann 2017). Bottom-up and top-down are considered as two major categories defined for proteomics approaches. In the bottom-up approach, for identification and quantification of peptides proteins are first digested into smaller peptides using trypsin utilizing liquid chromatography coupled with tandem mass spectrometry (LC–MS/MS). Bottom-up proteomics analyzed more than 2000 proteins and identified affected and altered biological pathways in the disease-affected brain (Hondius et al. 2016, 2018). On the other hand, using "top-down" approaches proteins are analyzed directly which provides more detailed information about protein modifications, etc.

Lachen-Montes and colleagues examined proteome changes in AD-affected olfactory bulbs compared to healthy controls (Lachen-Montes et al. 2017). In this study, 1311 proteins from 110 (initial AD cases), 125 (intermediate AD cases), and 159 (advanced AD cases) samples were studied for differentially expressed genes by comparing with the healthy controls. More than 50% of the proteins were observed associated with progressive synaptic degeneration during AD disease progression. Xu and colleagues also performed proteomics analysis considering six brain regions. Heavily AD-affected regions samples were collected from hippocampus, entorhinal cortex, and cingulate gyrus, lightly AD-affected regions samples were collected from sensory cortex and motor cortex, and the least affected AD sample was collected from the cerebellum (Xu et al. 2019). Among a total of 4825 proteins, around 30% of proteins were observed as altered in the heavily affected regions, whereas only 11–13% proteins were altered.

Orbitrap HF-X with 40 Hz scanning speed or Tims TOF pro could be used to analyze thousands of proteins and their isoforms from the samples collected from affected brain samples (Kelstrup et al. 2018) or (Meier et al. 2015; Bekker-Jensen et al. 2017).

Nowadays, mass spectrometry-based proteomics datasets are compiled into databases (Kaithal et al. 2024) such as MassIVE (Choi et al. 2020), PRoteomics IDEntifications Database (PRIDE) (Perez-Riverol et al. 2022), and ProteomicsDB (Samaras et al. 2020), iProteomicsDB (Samaras et al. 2020), which could help in quick access to various types of proteomics datasets. The availability of such databases will encourage data mining-based studies in various proteomics.

7.3.4 Metabolomics

Metabolomics refers to the identification and analysis of all metabolites in each organism/sample (Srivastava et al. 2023). Metabolomics studies various metabolites that are considered as end products of various complex biochemical processes such as organic acids, sugars, nucleotides, and amino acids. Metabolomics is further categorized into targeted and untargeted metabolomics. Targeted metabolomics focuses on analyzing pre-selected metabolites of interest. On the other hand, untargeted metabolomics analyzes all the metabolites present within a sample. Metabolomics analysis utilizes cerebrospinal fluid (CSF) collected from AD patients because cerebrospinal fluid transports various metabolites. The composition of metabolites in CSF changes in AD patients during the disease progression (Quintero et al. 2021). Glucose hypometabolism is observed frequently in AD patients at clinical stages which is characterized by insulin resistance. In another study, Van der Velpen et al. illustrated a higher concentration of tryptophan, quinolinic acid and other intermediates of tryptophan pathway associated with AD pathogenesis and disease progression (van der Velpen et al. 2019). Metabolome studies also illustrated the differences in metabolic concentrations present in different body fluids and tissues between PD patients and controls (Luan et al. 2015; Hatano et al. 2016; Kori et al. 2016). Various metabolic studies as Kori et al. (2016) discussed 54 metabolite markers for PD retrieved from 14 metabolomic studies. In a study by Luan et al. (2015), out of 46 metabolic signatures, 18 metabolite signatures were observed to have a high discrimination ability for early-stage PD disease, compared to middle-/advanced-stage PD disease (Luan et al. 2015). For various metabolomics resources, Cerebrospinal Fluid (CSF) Metabolome Database (Wishart et al. 2008), Human Metabolome Database (HMDB) (Wishart et al. 2022) are available nowadays in public domain.

7.3.5 Lipidomics

Lipidomics studies the identification and quantification of various lipid molecules. Lipidomics studies are considered as important in brain research. Lipid dysregulations are observed in AD and PD pathogenesis due to the presence of lipids in myelination of neurons and signal transduction via various lipid mediators and intermediates (Castellanos et al. 2021). Phosphatidylcholines and phosphocholines were also observed associated with various AD and PD studies (Stephenson et al. 2017; Proitsi et al. 2017).

With the generation of large-scale metabolomics and lipidomics-based high-throughput experimental data, various open accessible metabolomics and lipidomics databases such as the Alzheimer's Disease Neuroimaging Initiative (ADNI) database (Dimitriadis and Liparas 2018), LIPID MAPS website provides various open-access lipidomics database resources which is available at https://www.lipidmaps.org/.

7.3.6 Single-Cell and Single-Nucleus RNA Sequencing

Nowadays, single-cell RNA-seq (scRNA-seq) and single-nucleus RNA-seq (snRNA-seq) techniques have become popular (Poirion et al. 2016; Hwang et al. 2018), and these technologies provide gene expression information in specific cell types at a single-cell resolution (Ma and Lim 2021; Slovin et al. 2021). scRNA-seq and snRNA-seq techniques provide information about the functional changes associated with highly heterogeneous cells in the brain and thus elucidating the role of cell types in neurodegenerative diseases such as AD and PD (Jiang et al. 2020). scRNA-seq and snRNA-seq study from AD patients illustrated the complexity of human microglia (Masuda et al. 2019; Alsema et al. 2020) associated with AD and PD development (Zhou et al. 2020). Moreover, single-cell high-throughput technologies-based studies using AD and PD samples have been applied to other glial cells like astrocytes and oligodendrocytes (Lian et al. 2015; Liddelow et al. 2017).

A study based on snRNA-seq datasets illustrated the upregulation of LINGO1 in both excitatory neurons and oligodendrocytes leading to mitochondrial and estrogen signaling dysfunction in AD (Grubman et al. 2019; Mathys et al. 2019; Lau et al. 2020). Dopamine neurons were identified in scRNA-seq studies, which illustrated the abnormalities in mice-based Parkinson's disease models (Poulin et al. 2014; Saunders et al. 2018). Furthermore, scRNA-seq was performed on human embryonic stem cells (hESCs) and human induced pluripotent stem cells (hiPSCs) also illustrate their role in Parkinson's disease (Manno et al. 2016).

GEO (Barrett et al. 2013), Sequence Read Archive, Mount Sinai Brain Bank (Wang et al. 2018), and Synapse database (Lüscher and Isaac 2009) include scRNA-seq and snRNA-seq datasets.

7.3.7 Spatial Omics

Nowadays, spatial omics technologies have become popular and are used to visualize the anatomical location of various biological molecules. These technologies enable us to investigate the structural organization of complex tissues. Such technologies also help in the visualization of the cell-cell interactions and tissue microenvironments (Liao et al. 2021). Various tools such as SpatialLIBD (Pardo et al. 2022), Bento (Mah et al. 2023), SpatialExperiment (Righelli et al. 2022), Squidpy (Palla et al. 2022), SM-Omics (Vickovic et al. 2022), and ATHENA (Martinelli and Rapsomaniki 2022) are available for analysis of spatial omics datasets. These tools compute spatial statistics for diseased states compared to controls (Mah et al. 2023). Databases such as spatial TranscriptOmics DataBase (STOmicsDB) and SpatialDB could be used for data mining-based studies and comparison of spatial omics data with other experimental data (Fan et al. 2020).

7.4 Multi-Omics for Exploring Brain Disorders

Multi-omics data refer to the integration of multiple omics to provide a holistic panorama of biological functions (Kumar et al. 2024). Identifying genetic variants and discovering novel biomarkers of brain disorders are the major goals of multi-omics analysis as illustrated in Fig. 7.2. In last 15 years, various multi-omics datasets were generated from both patients and healthy controls tissue samples, muti-omics datasets were also generated from few cultured cells, as well from population cohorts.

The Genotype-Tissue Expression (GTEx) database aims to identify gene expression changes across individuals and multiple tissues including brain (GTEx Consortium 2017). Omics Discovery Index (OmicsDI) is considered as other multi-omics database from various resources (Perez-Riverol et al. 2017).

Accelerating Medicines Partnership (AMP)-AD is considered the largest database, which utilizes cutting-edge systems as well as network biology approaches to human brain-based AD (all stages) clinical and pathological datasets. ROSMAP,

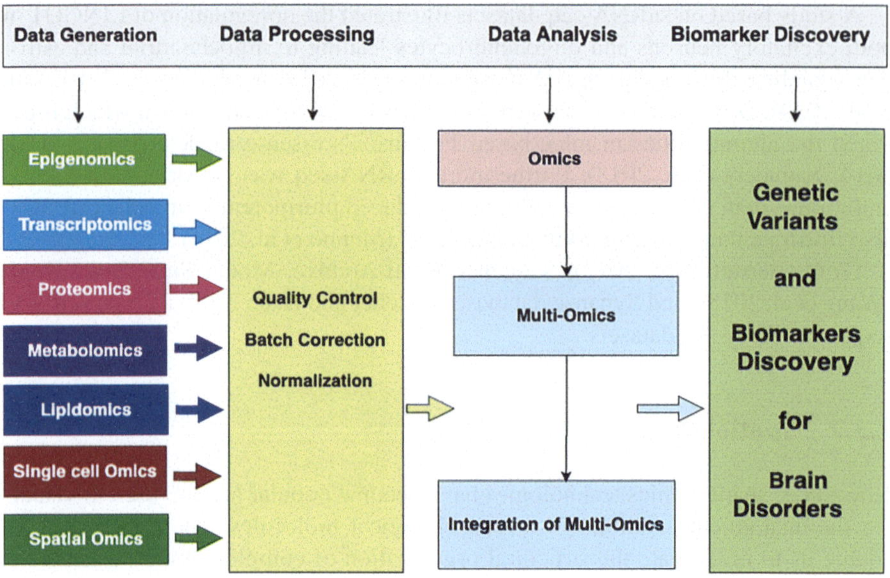

Fig. 7.2 A systematic workflow of multi-omics analysis for exploring brain disorders

Mount Sinai Brain Bank (MSBB), and MayoRNAseq studies in (AMP)-AD contributed most of the multi-omics datasets by integrating multidimensional human genomic, epigenomic, RNA, proteomic-based molecular datasets. MSBB alone provided over 1700 brain samples-based datasets from four brain regions (superior temporal gyrus, frontal pole, para hippocampal gyrus, and inferior frontal gyrus). ROSMAP in AMP-AD provided multi-omics datasets of 2214 brain samples which includes multi-omics datasets.

AMP-PD is another one of the largest multi-omics datasets, which includes PD data (clinical and pathological) from patients and healthy controls (Iwaki et al. 2021). In its first release in 2018, AMP-PD incorporated 4298 contributors in total from four cohorts, but till October 27, 2023, AMP-PD included 3566 individuals from PD cases and 4376 individuals from healthy control from various cohorts.

Multi-omics data integration for investigating brain disorders leverages identification of disease subtyping using multi-omics profiles and discovering of novel biomarkers for new biological insights of brain disorders studies (Dong et al. 2021). Various multi-omics integration methods such as Bayesian methods, Matrix Factorization, multiple kernel learning, correlation-based, network-based methods network-based methods, are available nowadays (Kanchan et al. 2024). Several multi-omics dataset integration tools are available which are called as mixOmics (Rohart et al. 2017), Paintomics (Hernández-de-Diego et al. 2018), Multi-Omics Factor Analysis or MOFA+ (Argelaguet et al. 2020), and miodin (Ulfenborg 2019).

7.5 Conclusions

This chapter summarizes various mechanisms of neurodegeneration and Alzheimer's and Parkinson's disease. This chapter also discusses various omics approaches and their applications in identifying the biomarkers/signatures of neurological disorders such as AD and PD. Various omics databases are also discussed for quick access of omics datasets related to brain disorders facilitating data mining studies in brain disorders. Efforts have been made to envision the power of integration of multi-omics analysis to paint holistic and elaborate aspects of neurogenerative diseases.

Short Chapter Summary
This chapter summarizes various mechanisms of neurodegeneration and Alzheimer's and Parkinson's disease through the lens of various omics approaches. The applications of various omics technologies and their integrative capacity as multi-omics to elucidate various stages of neurological disorders, disease progression, and identification of biomarkers have been highlighted.

Author Contributions Chapter concept and design, Swarna Kanchan and Minu Kesheri; chapter writing, Swarna Kanchan, Minu Kesheri, Bhagaban Mallik, Shivani Sharda, Chandandeep Kaur, Pramod Katara; chapter editing and review, Rajeshwar P. Sinha. Swarna Kanchan and Minu Kesheri contributed equally as first authors.

References

Alsema AM, Jiang Q, Kracht L, Gerrits E, Eggen BJL (2020) Profiling microglia from Alzheimer's disease donors and non-demented elderly in acute human postmortem cortical tissue. Front Mol Neurosci 13:1–14

Andrews S (2010) FastQC: a quality control tool for high throughput sequence data

Argelaguet R, Arnol D, Bredikhin D, Deloro Y, Velten B, Marioni JC, Stegle O (2020) MOFA+: a statistical framework for comprehensive integration of multi-modal single-cell data. Genome Biol 21(1):111

Barrett T, Wilhite SE, Ledoux P (2013) NCBI GEO: archive for functional genomics data sets: update. Nucleic Acids Res 41:D991–D995

Bekker-Jensen DB, Kelstrup CD, Batth TS (2017) An optimized shotgun strategy for the rapid generation of comprehensive human proteomes. Cell Syst 4:587–599.e584

Ben-Hattar J, Jiricny J (1988) Methylation of single CpG dinucleotides within a promoter element of the *Herpes simplex* virus tk gene reduces its transcription in vivo. Gene 65:219–227

Bolger AM, Lohse M, Usadel B (2014) Trimmomatic: a flexible trimmer for Illumina sequence data. Bioinformatics 30(15):2114–2120

Castellanos DB, Martín-Jiménez CA, Rojas-Rodríguez F, Barreto GE, González J (2021) Brain lipidomics as a rising field in neurodegenerative contexts: Perspectives with Machine Learning approaches. Front Neuroendocrinol. 61:100899. https://doi.org/10.1016/j.yfrne.2021

Choi M, Carver J, Chiva C (2020) MassIVE.quant: a community resource of quantitative mass spectrometry-based proteomics datasets. Nat Methods 17:981–984

Consortium RE, Kundaje A, Meuleman W (2015) Integrative analysis of 111 reference human epigenomes. Nature 518:317–330

DeMaagd G, Philip A (2015) Parkinson's Disease and Its Management: Part 1: Disease Entity, Risk Factors, Pathophysiology, Clinical Presentation, and Diagnosis. P T. 40(8):504–32

Dimitriadis SI, Liparas D (2018) Alzheimer's Disease Neuroimaging Initiative. How random is the random forest? Random forest algorithm on the service of structural imaging biomarkers for Alzheimer's disease: from Alzheimer's disease neuroimaging initiative (ADNI) database. Neural Regen Res 13:962–970. https://doi.org/10.4103/1673-5374.233433

Dobin A, Davis CA, Schlesinger F, Drenkow J, Zaleski C, Jha S, Batut P, Chaisson M, Gingeras TR (2013) STAR: ultrafast universal RNA-seq aligner. Bioinformatics 29(1):15–21

Dong X, Liu C, Dozmorov M (2021) Review of multi-omics data resources and integrative analysis for human brain disorders. Brief Funct Genomics 20:223–234

Fan Z, Chen R, Chen X (2020) SpatialDB: a database for spatially resolved transcriptomes. Nucleic Acids Res 48:D233–D237

Geschwind DH, Konopka G (2009) Neuroscience in the era of functional genomics and systems biology. Nature 461:908–915

Ghosh P, Saadat A (2023) Neurodegeneration and epigenetics: a review. Neurologia (Engl Ed) 38:e62–e68

Grubman A, Chew G, Ouyang JF, Sun G, Choo XY, McLean C, Simmons RK, Buckberry S, Vargas-Landin DB, Poppe D, Pflueger J (2019) A single-cell atlas of entorhinal cortex from individuals with Alzheimer's disease reveals cell-type-specific gene expression regulation. Nat Neurosci 22:2087–2097

GTEx Consortium (2017) Data Analysis & Coordinating Center (LDACC)—Analysis Working Group, Statistical Methods Groups—Analysis Working Group, genetic effects on gene expression across human tissues. Nature 550:204–213

Hasin Y, Seldin M, Lusis A (2017) Multi-omics approaches to disease. Genome Biol 18:1–5

Hatano T, Saiki S, Okuzumi A, Mohney RP, Hattori N (2016) Identification of novel biomarkers for Parkinson's disease by metabolomic technologies. J Neurol Neurosurg Psychiatry 87:295–301

Hernández-de-Diego R, Tarazona S, Martínez-Mira C, Balzano-Nogueira L, Furió-Tarí P, Pappas GJ Jr, Conesa A (2018) PaintOmics 3: a web resource for the pathway analysis and visualization of multi-omics data. Nucleic Acids Res 46(W1):W503–W509. https://doi.org/10.1093/nar/gky466

Hooshmand K, Halliday GM, Pineda SS, Sutherland GT, Guennewig B (2022) Overlap between Central and Peripheral Transcriptomes in Parkinson's Disease but Not Alzheimer's Disease. Int J Mol Sci. 23(9):5200. https://doi.org/10.3390/ijms23095200

Hondius DC, van Nierop P, Li KW, Hoozemans JJ, van der Schors RC, van Haastert ES, van der Vies SM, Rozemuller AJ, Smit AB (2016) Profiling the human hippocampal proteome at all pathologic stages of Alzheimer's disease. Alzheimers Dement 12:654–668

Hondius DC, Eigenhuis KN, Morrema THJ, van der Schors RC, van Nierop P, Bugiani M, Li KW, Hoozemans JJ, Smit AB, Rozemuller AJ (2018) Proteomics analysis identifies new markers associated with capillary cerebral amyloid angiopathy in Alzheimer's disease. Acta Neuropathol Commun 6:46

Hosp F, Mann M (2017) A primer on concepts and applications of proteomics in neuroscience. Neuron 96:558–571

Hossain MB, Islam MK, Adhikary A, Rahaman A, Islam MZ (2022) Bioinformatics Approach to Identify Significant Biomarkers, Drug Targets Shared Between Parkinson's Disease and Bipolar Disorder: A Pilot Study. Bioinform Biol Insights. 16:11779322221079232. https://doi.org/10.1177/11779322221079232

Hwang B, Lee JH, Bang D (2018) Single-cell RNA sequencing technologies and bioinformatics pipelines. Exp Mol Med 50:1–14

Iwaki H, Leonard HL, Makarious MB, Bookman M, Landin B, Vismer D, Casey B, Gibbs JR, Hernandez DG, Blauwendraat C, Vitale D (2021) Accelerating medicines partnership: Parkinson's disease. Genetic resource. Mov Disord 36(8):1795–1804. https://doi.org/10.1002/mds.28549

Jiang J, Wang C, Qi R, Fu H, Ma Q (2020) scREAD: a single-cell RNA-seq database for Alzheimer's disease. iScience 23:101769

Jowaed A, Schmitt I, Kaut O, Wüllner U (2010) Methylation regulates alpha-synuclein expression and is decreased in Parkinson's disease patients' brains. J Neurosci 30:6355–6359

Kaithal P, Kanchan S, Kesheri M (2024) Recent advances in biological omics databases and tools in human health. In: Kesheri M, Kanchan S, Salisbury TB, Sinha RP (eds) Microbial omics in environment and health. Springer, Singapore, pp 311–341. https://doi.org/10.1007/978-981-97-1769-9_12

Kanchan S, Kesheri M, Srivastava U, Katharia H, Ratna-Raj R, Chittoori B, Bogomolnaya L, Sinha RP, Denvir J (2024) Integrative omics data mining: challenges and opportunities. In: Gupta MK, Katara P, Mondal S, Singh RL (eds) Integrative omics: concepts, methodology, and application. Academic Press, London, pp 237–255. https://doi.org/10.1016/B978-0-443-16092-9.00015-1

Kelstrup CD, Bekker-Jensen DB, Arrey TN, Hogrebe A, Harder A, Olsen JV (2018) Performance evaluation of the Q exactive HF-X for shotgun proteomics. J Proteome Res 17:727–738

Keren-Shaul H, Spinrad A, Weiner A, Matcovitch-Natan O, Dvir-Szternfeld R, Ulland TK, David E, Baruch K, Lara-Astaiso D, Toth B, Itzkovitz S (2017) A unique microglia type associated with restricting the development of Alzheimer's disease. Cell 169:1276–1290

Kesheri M, Kanchan S, Chowdhury S, Sinha RP (2015) Secondary and tertiary structure prediction of proteins: a bioinformatic approach. In: Zhu Q, Azar AT (eds) Complex system modelling and control through intelligent soft computations, vol 319. Springer-Verlag, Berlin, Heidelberg, pp 541–569. https://doi.org/10.1007/978-3-319-12883-2_19

Kim D, Paggi JM, Park C, Bennett C, Salzberg SL (2019) Graph-based genome alignment and genotyping with HISAT2 and HISAT-genotype. Nat Biotechnol 37(8):907–915

Kori M, Aydin B, Unal S, Arga KY, Kazan D (2016) Metabolic biomarkers and neurodegeneration: a pathway enrichment analysis of Alzheimer's disease, Parkinson's disease, and amyotrophic lateral sclerosis. Omics 20:645–661

Kumar P, Kanchan S, Kesheri M (2024) Multi-omics in human disease biomarker discovery. In: Kesheri M, Kanchan S, Salisbury TB, Sinha RP (eds) Microbial omics in environment and health. Springer, Singapore, pp 205–239. https://doi.org/10.1007/978-981-97-1769-9_8

Lachen-Montes M, Gonzalez-Morales A, Zelaya MV, Pérez-Valderrama E, Ausín K, Ferrer I, Fernández-Irigoyen J, Santamaría E (2017) Olfactory bulb neuroproteomics reveals a chronological perturbation of survival routes and a disruption of prohibitin complex during Alzheimer's disease progression. Sci Rep 7:9115

Lamptey RNL, Chaulagain B, Trivedi R, Gothwal A, Layek B, Singh J (2022) A review of the common neurodegenerative disorders: current therapeutic approaches and the potential role of nanotherapeutics. Int J Mol Sci 23:1851. https://doi.org/10.3390/ijms23031851

Langmead B, Salzberg SL (2012) Fast gapped-read alignment with Bowtie2. Nat Methods 9(4):357–359

Lau SF, Cao H, Fu AK, Ip NY (2020) Single-nucleus transcriptome analysis reveals dysregulation of angiogenic endothelial cells and neuroprotective glia in Alzheimer's disease. Proc Natl Acad Sci USA 117:25800–25809

Li H, Handsaker B, Wysoker A, Fennell T, Ruan J, Homer N, Marth G, Abecasis G, Durbin R (2009) The sequence alignment/map format and SAMtools. Bioinformatics 25(16):2078–2079

Li KW, Ganz AB, Smit AB (2019) Proteomics of neurodegenerative diseases: analysis of human post-mortem brain. J Neurochem 151:435–445

Lian H, Yang L, Cole A, Sun L, Chiang AC, Fowler SW, Shim DJ, Rodriguez-Rivera J, Taglialatela G, Jankowsky JL, Lu HC (2015) NFκB-activated astroglial release of complement C3 compromises neuronal morphology and function associated with Alzheimer's disease. Neuron 85:101–115

Liao Y, Smyth GK, Shi W (2014) FeatureCounts: an efficient general-purpose program for assigning sequence reads to genomic features. Bioinformatics 30(7):923–930

Liao J, Lu X, Shao X, Zhu L, Fan X (2021) Uncovering an organ's molecular architecture at single-cell resolution by spatially resolved transcriptomics. Trends Biotechnol 39:43–58

Liddelow SA, Guttenplan KA, Clarke LE, Bennett FC, Bohlen CJ, Schirmer L, Bennett ML, Münch AE, Chung WS, Peterson TC, Wilton DK (2017) Neurotoxic reactive astrocytes are induced by activated microglia. Nature 541:481–487

Long JM, Holtzman DM (2019) Alzheimer disease: an update on pathobiology and treatment strategies. Cell 179(2):312–339

Love MI, Huber W, Anders S (2014) Moderated estimation of fold change and dispersion for RNA-seq data with DESeq2. Genome Biol 15:1–21

Lu X, Wang L, Yu C, Yu D, Yu G (2015) Histone acetylation modifiers in the pathogenesis of Alzheimer's disease. Front Cell Neurosci 9:226. https://doi.org/10.3389/fncel.2015.00226

Luan H, Liu LF, Tang Z, Zhang M, Chua KK, Song JX, Mok VCT, Li M, Cai Z (2015) Comprehensive urinary metabolomic profiling and identification of potential noninvasive marker for idiopathic Parkinson's disease. Sci Rep 5:1–11

Lüscher C, Isaac JT (2009) The synapse: center stage for many brain diseases. J Physiol 587(Pt 4):727–729. https://doi.org/10.1113/jphysiol.2008.167742

Ma S, Lim SB (2021) Single-cell RNA sequencing in Parkinson's disease. Biomedicines 9:368

Mah CK, Ahmed N, Lopez N, Lam DC, Pong A, Monell A, Kern C, Han Y, Prasad G, Cesnik AJ, Lundberg E (2023) Bento: a toolkit for subcellular analysis of spatial transcriptomics data. Genome Biol 25(1):1–25. https://doi.org/10.1101/2022.06.10.495510

Mahul-Mellier AL, Burtscher J, Maharjan N, Weerens L, Croisier M, Kuttler F, Leleu M, Knott GW, Lashuel HA (2020) The process of Lewy body formation, rather than simply α-synuclein fibrillization, is one of the major drivers of neurodegeneration. Proc Natl Acad Sci U S A. 117(9):4971-4982. https://doi.org/10.1073/pnas.1913904117

Mallik S, Zhao Z (2020) Detecting methylation signatures in neurodegenerative disease by density-based clustering of applications with reducing noise. Sci Rep 10:22164

La Manno G, Gyllborg D, Codeluppi S, Nishimura K, Salto C, Zeisel A, Borm LE, Stott SRW, Toledo EM, Villaescusa JC, Lönnerberg P, Ryge J, Barker RA, Arenas E, Linnarsson S (2016) Molecular Diversity of Midbrain Development in Mouse, Human, and Stem Cells. Cell. 167(2):566-580.e19. https://doi.org/10.1016/j.cell.2016.09.027

Martinelli A, Rapsomaniki M (2022) ATHENA: analysis of tumor heterogeneity from spatial omics measurements. Bioinformatics 38:3151–3153

Masuda T, Sankowski R, Staszewski O, Böttcher C, Amann L, Sagar N, Scheiwe C, Nessler S, Kunz P, van Loo G, Coenen VA (2019) Spatial and temporal heterogeneity of mouse and human microglia at single-cell resolution. Nature 566:388–392

Mathys H, Davila-Velderrain J, Peng Z, Gao F, Mohammadi S, Young JZ, Menon M, He L, Abdurrob F, Jiang X, Martorell AJ (2019) Single-cell transcriptomic analysis of Alzheimer's disease. Nature 570:332–337

Matsumoto L, Takuma H, Tamaoka A, Kurisaki H, Date H, Tsuji S, Iwata A (2010) CpG demethylation enhances alpha-synuclein expression and affects the pathogenesis of Parkinson's disease. PLoS One 5:e15522

Meier F, Beck S, Grassl N, Lubeck M, Park MA, Raether O, Mann M (2015) Parallel Accumulation-Serial Fragmentation (PASEF): multiplying sequencing speed and sensitivity by synchronized scans in a trapped ion mobility device. J Proteome Res 14:5378–5387

Mihaylova MM, Shaw RJ (2013) Metabolic reprogramming by class I and II histone deacetylases. Trends Endocrinol Metab. 24(1):48–57. https://doi.org/10.1016/j.tem.2012.09.003

Nativio R, Donahue G, Berson A, Lan Y, Amlie-Wolf A, Tuzer F, Toledo JB, Gosai SJ, Gregory BD, Torres C, Trojanowski JQ (2018) Dysregulation of the epigenetic landscape of normal aging in Alzheimer's disease. Nat Neurosci 21:497–505

Nativio R, Lan Y, Donahue G, Sidoli S, Berson A, Srinivasan AR, Shcherbakova O, Amlie-Wolf A, Nie J, Cui X, He C, Wang LS, Garcia BA, Trojanowski JQ, Bonini NM, Berger SL (2020) An integrated multi-omics approach identifies epigenetic alterations associated with Alzheimer's disease. Nat Genet 52:1024–1035. https://doi.org/10.1038/s41588-020-0696-0

O'Connor LM, O'Connor BA, Lim SB, Zeng J, Lo CH (2023) Integrative multi-omics and systems bioinformatics in translational neuroscience: a data mining perspective. J Pharm Anal 13:836–850

Palla G, Spitzer H, Klein M, Fischer D, Schaar AC, Kuemmerle LB, Rybakov S, Ibarra IL, Holmberg O, Virshup I, Lotfollahi M (2022) Squidpy: a scalable framework for spatial omics analysis. Nat Methods 19:171–178

Pardo B, Spangler A, Weber LM, Page SC, Hicks SC, Jaffe AE, Martinowich K, Maynard KR, Collado-Torres L (2022) spatialLIBD: an R/Bioconductor package to visualize spatially-resolved transcriptomics data. BMC Genomics 23:434

Perez-Riverol Y, Bai M, da Veiga Leprevost F, Squizzato S, Park YM, Haug K, Carroll AJ, Spalding D, Paschall J, Wang M, Del-Toro N (2017) Discovering and linking public omics data sets using the omics discovery index. Nat Biotechnol 35:406–409

Perez-Riverol Y, Bai J, Bandla C, García-Seisdedos D, Hewapathirana S, Kamatchinathan S, Kundu DJ, Prakash A, Frericks-Zipper A, Eisenacher M, Walzer M (2022) The PRIDE database resources in 2022: a hub for mass spectrometry-based proteomics evidences. Nucleic Acids Res 50:D543–D552

Poewe W, Seppi K, Tanner CM, Halliday GM, Brundin P, Volkmann J, Schrag AE, Lang AE (2017) Parkinson disease. Nat Rev Dis Primers 3:17013

Poirion OB, Zhu X, Ching T, Garmire L (2016) Single-cell transcriptomics bioinformatics and computational challenges. Front Genet 7:163

Poulin JF, Zou J, Drouin-Ouellet J, Kim KY, Cicchetti F, Awatramani RB (2014) Defining midbrain dopaminergic neuron diversity by single-cell gene expression profiling. Cell Rep. Nov 6;9(3):930-43. https://doi.org/10.1016/j.celrep.2014.10.008

Priya P, Kesheri M, Sinha RP, Kanchan S (2017) Molecular dynamics simulations for biological systems. In: Karâa WBA, Dey N (eds) Biomedical image analysis and mining techniques for improved health outcomes. IGI Global, Hershey, pp 286–313. https://doi.org/10.4018/978-1-4666-8811-7.ch014

Proitsi P, Kim M, Whiley L, Simmons A, Sattlecker M, Velayudhan L, Lupton MK, Soininen H, Kloszewska I, Mecocci P, Tsolaki M, Vellas B, Lovestone S, Powell JF, Dobson RJB, Legido-Quigley C (2017) Association of blood lipids with Alzheimer's disease: a comprehensive lipidomics analysis. Alzheimers Dement 13:140–151. https://doi.org/10.1016/j.jalz.2016.08.003

Quintero ME, Pontes JGM, Tasic L (2021) Metabolomics in degenerative brain diseases. Brain Res 1773:147704. https://doi.org/10.1016/j.brainres.2021.147704

Reimand J, Kull M, Peterson H, Hansen J, Vilo J (2007) g:Profiler—a web-based toolset for functional profiling of gene lists from large-scale experiments. Nucleic Acids Res 35:W193–W200

Righelli D, Weber LM, Crowell HL, Pardo B, Collado-Torres L, Ghazanfar S, Lun AT, Hicks SC, Risso D (2022) Spatial experiment: infrastructure for spatially-resolved transcriptomics data in R using bioconductor. Bioinformatics 38:3128–3131

Robinson MD, McCarthy DJ, Smyth GK (2009) edgeR: a Bioconductor package for differential expression analysis of digital gene expression data. Bioinformatics 26(1):139–140

Rohart F, Gautier B, Singh A, Lê Cao KA (2017) mixOmics: an R package for 'omics feature selection and multiple data integration. PLoS Comput Biol 3(11):e1005752

Samaras P, Schmidt T, Frejno M, Gessulat S, Reinecke M, Jarzab A, Zecha J, Mergner J, Giansanti P, Ehrlich HC, Aiche S (2020) ProteomicsDB: a multi-omics and multi-organism resource for life science research. Nucleic Acids Res 48:D1153–D1163

Saunders A, Macosko EZ, Wysoker A, Goldman M, Krienen FM, de Rivera H, Bien E, Baum M, Bortolin L, Wang S, Goeva A (2018) Molecular diversity and specializations among the cells of the adult mouse brain. Cell 174:1015–1030

Slovin S, Carissimo A, Panariello F, Grimaldi A, Bouché V, Gambardella G, Cacchiarelli D (2021) Single-cell RNA sequencing analysis: a step-by-step overview. Methods Mol Biol 2284:343–365

Srivastava U, Kanchan S, Kesheri M, Singh S (2023) Nutrimetabolomics: metabolomics in nutrition research. In: Soni V, Hartman TE (eds) Metabolomics. Springer, Cham, pp 241–268. https://doi.org/10.1007/978-3-031-39094-4_8

Srivastava U, Kanchan S, Kesheri M, Gupta MK, Singh S (2024a) Types of omics data: genomics, metagenomics, epigenomics, transcriptomics, proteomics, metabolomics, and phenomics. In: Gupta MK, Katara P, Mondal S, Singh RL (eds) Integrative omics: concepts, methodology, and application. Academic Press, London, pp 13–34. https://doi.org/10.1016/B978-0-443-16092-9.00002-3

Srivastava U, Kanchan S, Kesheri M, Gupta MK, Singh S (2024b) Integrative omics approaches for identification of biomarkers. In: Gupta MK, Katara P, Mondal S, Singh RL (eds) Integrative

omics: concepts, methodology, and application. Academic Press, London, pp 157–173. https://doi.org/10.1016/B978-0-443-16092-9.00010-2
Stephenson DJ, Hoeferlin LA, Chalfant CE (2017) Lipidomics in translational research and the clinical significance of lipid-based biomarkers. Transl Res 189:13–29. https://doi.org/10.1016/j.trsl.2017.06.006
Strooper BD, Karran E (2016) The cellular phase of Alzheimer's disease. Cell 164(4):603–615
Subramanian A, Tamayo P, Mootha VK, Mukherjee S, Ebert BL, Gillette MA, Paulovich A, Pomeroy SL, Golub TR, Lander ES, Mesirov JP (2005) Gene set enrichment analysis: a knowledge-based approach for interpreting genome-wide expression profiles. Proc Natl Acad Sci USA 102(43):15545–15550
Suehs BT, Davis CD, Alvir J, van Amerongen D, Pharmd NC, Joshi AV, Faison WE, Shah SN (2013) The clinical and economic burden of newly diagnosed Alzheimer's disease in a medicare advantage population. Am J Alzheimers Dis Other Dement 28:384–392
Ulfenborg B (2019) Vertical and horizontal integration of multi-omics data with miodin. BMC Bioinformatics 20(1):649. https://doi.org/10.1186/s12859-019-3224-4
van der Velpen V, Teav T, Gallart-Ayala H, Mehl F, Konz I, Clark C, Oikonomidi A, Peyratout G, Henry H, Delorenzi M, Ivanisevic J, Popp J (2019) Systemic and central nervous system metabolic alterations in Alzheimer's disease. Alzheimers Res Ther 11:93
Verrou KM, Tsamardinos I, Papoutsoglou G (2020) Learning pathway dynamics from single-cell proteomic data: a comparative study. Cytometry A 97:241–252
Vickovic S, Lötstedt B, Klughammer J, Mages S, Segerstolpe Å, Rozenblatt-Rosen O, Regev A (2022) SM-Omics is an automated platform for high-throughput spatial multi-omics. Nat Commun 13:795
Wang M, Beckmann ND, Roussos P, Wang E, Zhou X, Wang Q, Ming C, Neff R, Ma W, Fullard JF, Hauberg ME (2018) The Mount Sinai cohort of large-scale genomic, transcriptomic and proteomic data in Alzheimer's disease. Sci Data 5:180–185
Wilhelm M, Schlegl J, Hahne H, Gholami AM, Lieberenz M, Savitski MM, Ziegler E, Butzmann L, Gessulat S, Marx H, Mathieson T, Lemeer S, Schnatbaum K, Reimer U, Wenschuh H, Mollenhauer M, Slotta-Huspenina J, Boese JH, Bantscheff M, Gerstmair A, Faerber F, Kuster B (2014) Mass-spectrometry-based draft of the human proteome. Nature. May 29;509(7502):582–7. https://doi.org/10.1038/nature13319
Wishart DS, Lewis MJ, Morrissey JA, Flegel MD, Jeroncic K, Xiong Y, Cheng D, Eisner R, Gautam B, Tzur D, Sawhney S (2008) The human cerebrospinal fluid metabolome. J Chromatogr B Analyt Technol Biomed Life Sci 871(2):164–173
Wishart DS, Guo AC, Oler E, Wang F, Anjum A, Peters H, Dizon R, Sayeeda Z, Tian S, Lee BL, Berjanskii M (2022) HMDB 5.0: the human metabolome database for 2022. Nucleic Acids Res 50(D1):D622–D631
Xu J, Patassini S, Rustogi N, Riba-Garcia I, Hale BD, Phillips AM, Waldvogel H, Haines R, Bradbury P, Stevens A, Faull RL (2019) Regional protein expression in human Alzheimer's brain correlates with disease severity. Commun Biol 2:43
Zhou Y, Song WM, Andhey PS, Swain A, Levy T, Miller KR, Poliani PL, Cominelli M, Grover S, Gilfillan S, Cella M (2020) Human and mouse single-nucleus transcriptomics reveal TREM2-dependent and TREM2-independent cellular responses in Alzheimer's disease. Nat Med 26:131–142

Applications of Multi-omics in Human Genetic Disorders

8

Poonam Kaithal, Swarna Kanchan, and Minu Kesheri

Abstract

The rise of multi-omics approaches has changed how we might interpret human hereditary issues by giving thorough bits of knowledge into the intricate interplay of genetic, molecular, and ecological factors underlying disease pathogenesis. This chapter features ongoing advances in multi-omics applications, enveloping genomics, transcriptomics, proteomics, metabolomics, and epigenomics in unraveling the molecular mechanisms of various genetic disorders. Integrative analysis of multi-omics data helped in investigating novel disease-associated genes, biomarkers and pathways to empower customized diagnostics, remedial interactions and prognostics. Moreover, in this chapter, we also discussed the challenges in the coordination of multi-omics data, including data integration strategies, moral contemplations and computational strategies. By utilizing the power of multi-omics advancements, scientists are ready to take huge steps towards elucidating the genetic design of human diseases and making an interpretation of these findings into clinically noteworthy insight for improved patient care and accurate medication strategies.

Keywords

Data integration · Genetic disorders · Multi-omics · Omics · Personalized medicines

P. Kaithal
Department of Molecular and Cellular Engineering, Jacob Institute of Biotechnology and Bioengineering SHUATS, Prayagraj, India

S. Kanchan · M. Kesheri (✉)
Joan C. Edwards School of Medicine, Marshall University, Huntington, WV, USA

Boise State University, Boise, ID, USA
e-mail: minukesheri@boisestate.edu

M. Kesheri et al. (eds.), *Multi-Omics in Biomedical Sciences and Environmental Sustainability*, https://doi.org/10.1007/978-981-96-7067-3_8

Objectives

- To explore the advances in multi-omics approaches in human genetic diseases
- To highlight the integrative analysis of multi-omics data
- To address challenges in multi-omics data integration
- To emphasize the clinical impact of multi-omics research

8.1 Introduction

In recent years, the appearance of high-throughput technologies has revolutionized biomedical research, permitting scientists to investigate biological systems at various molecular levels simultaneously. Multi-omics, an interdisciplinary approach, includes the integration of data from multiple omics fields, involving genomics, epigenomics, transcriptomics, proteomics, and metabolomics. This reconciliation offers a holistic view of biological processes, providing insights into the complex interplay of molecular components underlying pathological and physiological conditions (HajYasien 2023).

8.1.1 Importance of Multi-omics Integration

Combining multi-omics data allows researchers to delve into complex molecular systems and pinpoint the critical factors influencing disease development. By integrating insights from various omics layers, scientists can uncover cause-and-effect relationships, identify biomarkers, and explore new therapeutic possibilities for a range of genetic disorders. Additionally, multi-omics techniques help reveal molecular patterns linked to disease risk, progression, and treatment response, laying the groundwork for tailored approaches in personalized medicine (Subramanian et al. 2020).

8.1.2 Evolution and Historical Context of Multi-omics

The idea of multi-omics arose to address the shortcomings of single-omics studies, which typically offer only a limited view of complex biological systems. Initially, efforts to combine omics data relied on basic correlation analyses. However, with progress in computational biology and bioinformatics, more advanced techniques, such as network-based methods and machine learning, have been developed for data integration. Today, multi-omics has grown into a robust analytical framework, fueling significant advancements in biomedical research and transforming clinical practices (Ganguly 2024).

8.2 Genomics and Human Genetic Disorders

Genomics involves the exploration of an organism's complete genome, including its DNA sequence, structure, functions, and variations (Srivastava et al. 2024a). The introduction of high-throughput sequencing technologies has revolutionized genomic research, allowing for detailed studies of genetic variations and their impact on human health and disease. Metagenomics serves as a pivotal tool in the holistic identification of genomes present in any sample (Kanchan et al. 2020). Metagenomic sequencing of plasma has paved a way for diagnosing numerous infectious syndromes although significant development is warranted in this regard (Graf et al. 2025).

8.2.1 Methodologies and Techniques in Genomic Studies

Genomic studies utilize a range of techniques, including whole-genome sequencing (WGS), whole-exome sequencing (WES), and genome-wide association studies (GWAS). WGS offers a comprehensive view of an individual's entire genome, while WES focuses specifically on sequencing the protein-coding regions. GWAS, on the other hand, examines genetic differences by comparing the genomes of individuals with a particular trait or disease to those without, helping to identify variants linked to specific conditions (Ginsberg and Mirnics 2006).

8.2.2 Role of Genomics in Identifying Genetic Variants

Genomics is essential for uncovering the genetic variations responsible for human genetic disorders. Through genome analysis of individuals with these conditions, scientists can identify mutations such as single-nucleotide polymorphisms (SNPs), insertions, deletions, and structural changes. These variations can interfere with protein function, disrupt regulatory regions, or alter chromosomal structure, ultimately contributing to the development of disease traits (Duzkale et al. 2013).

Genomic approaches have significantly advanced our understanding of the genetic basis of various human disorders. For instance, WGS has been instrumental in uncovering previously unknown genes and mutations linked to diseases like cystic fibrosis, Duchenne muscular dystrophy, and Huntington's disease. In another example, GWAS have pinpointed common genetic variants that contribute to complex conditions, such as diabetes, cardiovascular diseases, and autoimmune disorders. These breakthroughs demonstrate the power of genomic tools in revealing the genetic underpinnings of diverse medical conditions (Ng and Henikoff 2006).

8.2.3 Challenges and Limitations in Genomic Studies

While genomic research holds great promise, it also faces several challenges. These include difficulties in interpreting noncoding regions of the genome, identifying rare or unique genetic variants, and overcoming the need for large-scale collaborations to address sample size constraints. Additionally, ethical concerns surrounding genetic testing and data sharing need careful consideration. On top of these, integrating genomic data with other omics technologies presents its own set of computational and analytical hurdles, necessitating the development of advanced bioinformatics tools and methods (Tam et al. 2019).

8.3 Transcriptomics in Human Genetic Disorders

Transcriptomics focuses on studying an organism's transcriptome, which encompasses all the RNA transcripts generated by its genome. By analyzing these transcripts, scientists gain a deeper understanding of gene expression patterns, alternative splicing, and RNA modifications. This information sheds light on the cellular processes and regulatory mechanisms that contribute to human genetic disorders (Papassotiropoulos et al. 2006).

8.3.1 Transcriptomic Technologies and Methodologies

A variety of techniques are being used in transcriptomic research, with RNA sequencing (RNA-seq) being the most applied method. RNA-seq allows researchers to measure gene expression levels, identify alternative splicing forms, and detect noncoding RNAs. Other methods, like microarrays and quantitative polymerase chain reaction (qPCR), are also used in transcriptomic studies, each offering distinct advantages and limitations depending on the specific research needs (Chen et al. 2023).

8.3.2 Application of Transcriptomics in Identification of Gene Expression Patterns

Transcriptomics is vital for uncovering disrupted gene expression patterns linked to human genetic disorders. By comparing the transcriptomes of diseased and healthy tissues, researchers can pinpoint genes that are differentially expressed and may play a role in disease development. Through transcriptomic profiling, abnormal expression of genes involved in key cellular processes like cell growth, programmed cell death, and immune response has been identified, offering valuable insights into the molecular mechanisms behind genetic disorders (Raghavachari and Garcia-Reyero 2018).

8.3.3 Regulation of Gene Expression in Genetic Disorders

Gene expression dysregulation is a key feature of many genetic disorders, often caused by mutations in regulatory regions, changes in chromatin structure, or abnormal signaling pathways. Transcriptomic research has uncovered the intricate regulatory networks that control gene expression in disease conditions, highlighting the contributions of transcription factors, epigenetic regulators, and noncoding RNAs to the development and progression of these diseases (Matharu and Ahituv 2020).

Several case studies highlight the power of transcriptomic analysis in uncovering the molecular foundations of human genetic disorders. For instance, RNA-seq studies have led to the discovery of new genes and pathways linked to conditions like autism spectrum disorders, neurodegenerative diseases, and cancer. Additionally, transcriptomic profiling of patient samples has played a key role in identifying diagnostic biomarkers and therapeutic targets, advancing the field of precision medicine (Yang et al. 2020).

Transcriptomics is a rapidly advancing field, fueled by improvements in sequencing technologies, computational tools, and data analysis techniques (Kumar et al. 2024). Looking ahead, emerging areas such as single-cell transcriptomics, spatial transcriptomics, and longitudinal studies aim to track gene expression changes over time. Additionally, combining transcriptomic data with other omics layers offers exciting potential for gaining deeper insights into complex genetic disorders and developing more personalized treatment approaches (Gleason 2019).

8.4 Epigenomics and Human Genetic Disorders

Epigenomics focuses on the changes to the genome that influence gene expression without altering the underlying DNA sequence. These changes involve processes such as DNA methylation, histone modifications, chromatin remodeling, and the regulation by noncoding RNAs. Epigenetic mechanisms are vital for development, cell differentiation, and the progression of diseases, making epigenomics a key area of study in understanding human genetic disorders (Portela and Esteller 2010).

8.4.1 Epigenomic Technologies and Methodologies

Epigenomic studies use a range of technologies to examine and map epigenetic changes (Priya et al. 2024). These methods include bisulfite sequencing for DNA methylation analysis, chromatin immunoprecipitation followed by sequencing (ChIP-seq) for profiling histone modifications, and RNA sequencing (RNA-seq) for exploring noncoding RNAs. Newer techniques, such as single-cell epigenomics and genome-wide DNA methylation arrays, are broadening the horizons of epigenomic research (Mensaert et al. 2014).

8.4.2 Role of Epigenetics in the Pathogenesis of Genetic Disorders

Epigenetic changes play a key role in the development of many genetic disorders by altering gene expression and cellular functions. Abnormal DNA methylation, histone modifications, and chromatin remodeling have been linked to conditions like cancer, neurodevelopmental disorders, and metabolic diseases. These changes can disrupt gene regulation, genomic stability, and cellular identity, ultimately contributing to the emergence of disease characteristics (Moosavi and Ardekani 2016).

8.4.3 Techniques and Tools for Studying Epigenetic Alterations

Researchers employ a range of methods to study the epigenetic changes linked to genetic disorders. These include bioinformatics tools for analyzing epigenomic data, CRISPR-based genome editing to explore the functional impact of epigenetic modifications, and high-throughput screening techniques to identify potential therapeutic targets in epigenetic modifiers. Additionally, integrating epigenomic data with other omics technologies provides deeper insights into the molecular processes that drive disease development (Li 2021).

8.4.4 Epigenomic Signatures as Biomarkers for Genetic Disorders

Epigenomic markers, such as DNA methylation patterns and histone modification profiles, show great potential as tools for diagnosing and predicting the course of genetic disorders. These markers can help differentiate between healthy and diseased states, forecast disease progression, and tailor treatment plans based on individual patient responses. As noninvasive indicators, epigenetic biomarkers offer a way to detect and monitor diseases, supporting early intervention and the development of personalized treatment strategies (Ladd-Acosta 2015).

8.4.5 Ethical Considerations in Epigenomic Research

Ethical issues are a critical aspect of epigenomic research, especially when it comes to consent, privacy, and data sharing. Researchers must follow strict ethical guidelines and regulations to conduct epigenomic studies responsibly, particularly those involving human participants. Moreover, promoting transparency, ensuring access to data, and engaging with communities are key to building trust and upholding ethical standards in the field of epigenomics (Dupras et al. 2022).

8.5 Proteomics and Human Genetic Disorders

Proteomics involves studying all the proteins expressed by an organism, tissue, or cell at a specific point in time. Since proteins are central to biological functions (Kesheri et al. 2015), changes in their expression, abundance, or posttranslational modifications can play a role in the development of genetic disorders. Proteomic techniques allow researchers to identify and measure proteins, helping to uncover their involvement in disease processes and identify potential targets for therapy (Gloerich et al. 2007).

8.5.1 Proteomic Technologies and Methodologies

Proteomic studies use a range of advanced technologies and methods to thoroughly examine the proteome. Key techniques include mass spectrometry approaches like liquid chromatography-mass spectrometry (LC-MS) and tandem mass spectrometry (MS/MS). Other methods involve protein microarrays, gel-based techniques such as 2D gel electrophoresis, and assays for studying protein-protein interactions, such as the yeast two-hybrid system. Each method has distinct strengths and can be adapted to explore specific research questions, particularly in understanding human genetic disorders (Cho 2007).

8.5.2 Application of Proteomics in Identifying Disease-Associated Protein Profiles

Proteomics is essential for uncovering protein profiles and biomarkers linked to human genetic disorders. By analyzing the proteomes of healthy and diseased tissues or cells, researchers can pinpoint proteins with altered expression, identify posttranslational modifications, and study protein-protein interactions involved in disease development. This approach helps reveal disease-specific patterns, which can aid in diagnosing conditions, predicting outcomes, and tracking responses to treatments (Xiao et al. 2005).

8.5.3 Protein-Protein Interactions and Signaling Pathways in Genetic Disorders

Protein-protein interactions and signaling pathways are critical for normal cellular functions, but they often become disrupted in genetic disorders. Proteomics techniques, including interactome mapping and pathway analysis, offer valuable insights into the intricate networks of protein interactions and signaling processes involved in disease. By shedding light on the molecular mechanisms driving these conditions, proteomics helps identify potential therapeutic targets and supports the development of more precise, targeted treatments (Lage 2014).

8.5.4 Integration of Proteomic Data with Other Omics Approaches

Combining proteomic data with genomic, transcriptomic, and other omics information significantly deepens our understanding of the molecular foundations of human genetic disorders. Multi-omics approaches allow for a more thorough exploration of disease mechanisms, uncovering links between genetic variations and observable traits while identifying new pathways associated with diseases. By taking a systems-level approach to analyzing disease processes, integrative studies pave the way for discovering biomarkers and therapeutic targets that hold promise for clinical applications (Misra et al. 2019).

8.5.5 Challenges and Future Prospects in Proteomic Studies

Proteomic research has made remarkable progress, but it still encounters challenges like the complexity of samples, limitations in dynamic range, and the intricacies of data analysis. Looking ahead, the field aims to develop more high-throughput and sensitive technologies, explore single-cell proteomics, and integrate spatial proteomics with advanced imaging methods. Additionally, fostering collaboration and promoting data-sharing initiatives will be crucial for driving advancements in proteomics and transforming discoveries into clinical solutions for human genetic disorders (Chiou and Wu 2011).

8.6 Metabolomics and Human Genetic Disorders

Metabolomics involves the in-depth study of small molecules, or metabolites, found in biological systems (Srivastava et al. 2023). These metabolites are the final products of cellular processes and metabolic pathways, offering a snapshot of the functional state of cells and tissues. By analyzing metabolic patterns, metabolomics sheds light on the unique metabolic profiles linked to human genetic disorders, providing a comprehensive understanding of disease mechanisms and uncovering opportunities for diagnosis and treatment (Troisi et al. 2020).

8.6.1 Metabolomic Technologies and Methodologies

Metabolomic analysis focuses on detecting, identifying, and measuring metabolites using advanced techniques like mass spectrometry (MS), nuclear magnetic resonance (NMR) spectroscopy, and chromatography methods such as gas and liquid chromatography. These tools allow researchers to profile metabolites in various biological samples, including blood, urine, tissues, and cells. This approach helps uncover metabolic changes specific to certain diseases, offering valuable insights into their underlying mechanisms (Johanningsmeier et al. 2016).

8.6.2 Metabolic Pathways Implicated in Genetic Disorders

Metabolic pathways are essential for maintaining cellular balance, but can become disrupted in genetic disorders. Metabolomic research has uncovered changes in key pathways, such as those involved in energy production, amino acid processing, lipid metabolism, and nucleotide synthesis, in conditions like inborn errors of metabolism, metabolic syndromes, and mitochondrial disorders. These altered metabolites can serve as biomarkers, aiding in disease diagnosis, predicting outcomes, and monitoring treatment effectiveness (Boneh 2015).

8.6.3 Metabolomic Biomarkers for Disease Diagnosis and Prognosis

Metabolomic biomarkers offer significant potential for diagnosing genetic disorders and evaluating disease severity and progression. By analyzing metabolite patterns in patient samples, researchers can identify unique biomarkers that differentiate healthy individuals from those with specific conditions. These biomarkers provide noninvasive, sensitive tools for early detection, patient classification, and personalized treatment approaches, ultimately improving clinical care and patient outcomes (Kelly et al. 2016).

8.6.4 Metabolomics in Drug Discovery and Personalized Medicine

Metabolomics is essential in drug discovery and development, helping to understand how drugs are metabolized, their pharmacokinetics, and their potential toxicity. By analyzing the metabolic responses of patients, metabolomics can identify individuals who will benefit from a drug and those who will not, supporting more personalized treatment plans while reducing side effects. Additionally, metabolomic studies can uncover new uses for existing medications and reveal potential therapeutic targets for genetic diseases (Wishart 2016).

8.6.5 Data Integration Strategies for Multi-omics Analyses

Combining metabolomic data with other omics fields, such as genomics, transcriptomics, and proteomics, offers a deeper understanding of the biological processes and disease mechanisms behind genetic disorders. Multi-omics strategies help uncover connections between genotype and phenotype, clarify disease networks, and identify new therapeutic targets using bioinformatics analysis (Sahu et al. 2023). By providing a comprehensive, systems-level perspective on disease development, these integrated analyses support the advancement of precision medicine and personalized healthcare (Picard et al. 2021).

8.7 Integrative Multi-omics Analysis in Human Genetic Disorders

Integrative multi-omics analysis combines data from various omics layers, such as genomics, transcriptomics, epigenomics, proteomics, and metabolomics, to provide a holistic view of complex biological systems (Kanchan et al. 2024). By merging insights from these molecular levels, this approach enhances our understanding of the genetic foundations, molecular processes, and clinical features of human genetic disorders. This chapter delves into the principles, techniques, and practical applications of integrative multi-omics in studying genetic diseases (Subramanian et al. 2020).

8.7.1 Approaches for Integrating Multi-omics Data

A variety of methods are used to integrate multi-omics data, ranging from basic correlation analyses to advanced computational techniques. Network-based approaches, such as protein-protein interaction networks and gene regulatory networks, help visualize and interpret complex molecular interactions. Machine learning techniques, including random forests, support vector machines, and neural networks, play a key role in predicting disease phenotypes and identifying biomarkers from multi-omics datasets (Kumar et al. 2024; Srivastava et al. 2024b). Additionally, tools like pathway enrichment analysis and functional annotation help pinpoint biological pathways disrupted in genetic disorders (Jendoubi 2021).

Integrative multi-omics analysis has proven to be a powerful tool in understanding the molecular basis of human genetic disorders. By combining insights from genomic, transcriptomic, and epigenomic data, researchers have uncovered new genes and regulatory elements linked to conditions like autism spectrum disorders, intellectual disabilities, and neurodegenerative diseases. These approaches have also helped clarify the connections between genetic variations and observable traits, distinguish disease subtypes, and anticipate how patients might respond to treatments (Jendoubi 2021).

Application of various omics when combined as integrative omics after applying various statistical and machine learning methods could be utilized for biomarker discovery, identification of human genetic variants and their disease associations, etc., as shown in Fig. 8.1

8.7.2 Challenges and Considerations in Multi-omics Data Integration

Integrating data from multiple omics sources comes with its own set of challenges, such as differences in data types, batch-to-batch variability, missing information, and the complexities of reducing high-dimensional data. To ensure that analyses are reliable and reproducible, it is crucial to standardize data formats, apply consistent

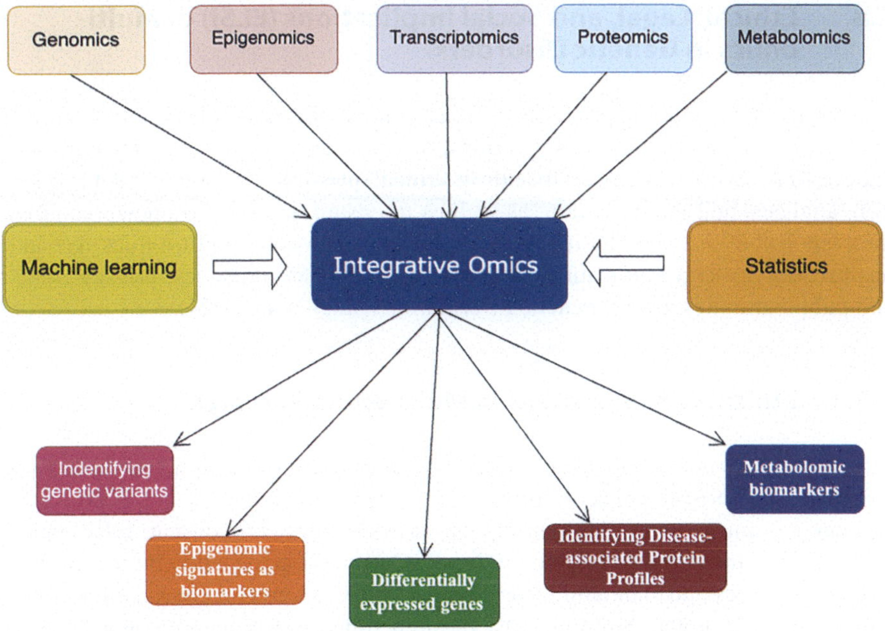

Fig. 8.1 Various applications of integrative omics in identifying human genetic disorders

normalization methods, and implement thorough quality control measures. Collaboration across disciplines and open data-sharing initiatives also play a vital role in addressing technical obstacles and fostering transparency in multi-omics research (Canzler et al. 2020).

The future of integrative multi-omics analysis is full of potential for deepening our understanding of human genetic disorders and transforming clinical care. New technologies like single-cell omics, spatial omics, and long-read sequencing provide innovative ways to explore cellular diversity and spatial organization within diseased tissues. Additionally, combining multi-omics data with clinical information, environmental influences, and electronic health records opens exciting possibilities for predicting disease risk, tailoring treatments to individual patients, and ultimately enhancing health outcomes (Chengoden et al. 2023; Shukla et al. 2024).

Integrative multi-omics analysis offers a transformative way to explore the complexities of human genetic disorders. By combining data from various molecular layers, scientists can uncover disease mechanisms, discover biomarkers, and create targeted therapies that advance precision medicine. While there are still challenges to overcome, ongoing progress in multi-omics technologies and techniques promises to revolutionize how we understand and treat genetic diseases (Chengoden et al. 2023).

8.8 Ethical, Legal, and Social Implications (ELSI) of Multi-omics in Genetic Disorders

The use of multi-omics approaches to study genetic disorders brings with it a range of ethical, legal, and social considerations. As scientists explore the molecular foundations of diseases, it is crucial to address ethical questions surrounding data collection, analysis, and interpretation, as well as to ensure that legal frameworks guide the responsible use and sharing of these data. Additionally, multi-omics research has broader societal implications, influencing healthcare practices, privacy rights, and public attitudes toward genetic information (Singh et al. 2024).

8.8.1 Ethical Considerations in Multi-omics Research

Ethical considerations are a key aspect of multi-omics research, encompassing issues like informed consent, privacy safeguards, and data sharing practices. Researchers must ensure that participants provide informed consent, fully understanding the study's purpose, potential risks, and possible benefits. Protecting participants' privacy and maintaining the confidentiality of sensitive genetic information are equally important. Building trust through transparency and accountability is vital, not just with participants but also with the broader community impacted by the research (Mohr et al. 2024).

8.8.2 Legal Frameworks Governing Multi-omics Data

Multi-omics research operates within a complex landscape of legal frameworks that address data privacy, security, and intellectual property rights. Regulations like the General Data Protection Regulation (GDPR) in the European Union and the Health Insurance Portability and Accountability Act (HIPAA) in the United States establish guidelines for handling personal health information, including its collection, storage, and sharing. Furthermore, intellectual property laws influence the ownership and commercialization of multi-omics data and related discoveries, highlighting the need to address issues of equitable access and fair benefit-sharing (Mohr et al. 2024).

8.8.3 Social Implications of Multi-omics in Healthcare

The integration of multi-omics approaches into healthcare carries significant social implications, particularly regarding healthcare disparities, access to genetic testing, and insurance coverage. Genetic insights gained from multi-omics research can reveal an individual's susceptibility to certain diseases, raising concerns about potential discrimination in areas like employment and insurance. Additionally, unequal access to genetic testing and personalized treatments could worsen existing healthcare inequalities, further impacting health outcomes for underserved populations (Mohr et al. 2024).

8.8.4 Addressing ELSI Concerns in Multi-omics Research

Addressing ethical, legal, and social issues (ELSI) in multi-omics research involves creating ethical guidelines, fostering community engagement, and implementing policy measures. Ethical guidelines establish clear principles to guide researchers in conducting responsible studies while protecting participants' rights. Community engagement efforts bring together stakeholders—such as patients, advocacy groups, and policymakers—to ensure their perspectives and concerns are incorporated into the research process. Meanwhile, policy interventions aim to introduce laws and regulations that safeguard privacy, encourage data sharing, and work toward reducing healthcare access and delivery disparities (Haga 2024).

In summary, multi-omics research offers incredible potential to deepen our understanding of genetic disorders and enhance patient care. However, it is crucial to acknowledge and address the ethical, legal, and social challenges that come with it. By upholding ethical standards, adhering to legal requirements, and actively involving stakeholders, researchers can overcome these obstacles and ensure that the benefits of multi-omics research extend to individuals, communities, and society at large.

8.9 Advancements in Multi-omics Research

Multi-omics research has transformed our understanding of human genetic disorders by offering a detailed, holistic view of the molecular processes that drive disease. By combining data from genomics, transcriptomics, epigenomics, proteomics, and metabolomics, scientists have discovered intricate interactions between genes, proteins, and metabolites that play a role in disease development and progression. Innovations in high-throughput technologies, computational techniques, and bioinformatics tools have made it possible to integrate and analyze multi-omics data, allowing researchers to explore the complexity of genetic disorders with an unparalleled level of detail (Chakraborty et al. 2018).

The knowledge gathered through multi-omics research has the potential to transform precision medicine and personalized healthcare. By uncovering unique disease markers, molecular subtypes, and specific therapeutic targets, these methods pave the way for treatment plans customized to individual differences in disease risk, progression, and treatment response. Whether it is for early diagnosis, predicting health risks, crafting targeted treatments, or monitoring therapies, multi-omics has the power to reshape medical practice and enhance outcomes for patients with a variety of genetic conditions.

8.10 Challenges and Opportunities of Multi-omics Research

While multi-omics research has made remarkable strides, several hurdles still need to be overcome. Technical issues, such as integrating, standardizing, and validating data, remain significant barriers to adopting multi-omics in clinical settings.

Additionally, ethical, legal, and social challenges—like ensuring data privacy, obtaining informed consent, and addressing equity concerns—must be thoughtfully managed to use multi-omics responsibly and fairly in healthcare. However, these challenges also open doors for collaboration across disciplines, technological advancements, and the development of policies that can drive progress in multi-omics and its application in clinical practice (Wang et al. 2023).

8.11 Future Directions

The future of multi-omics research in understanding human genetic disorders is filled with exciting possibilities. Innovative technologies like single-cell omics, spatial omics, and long-read sequencing are unlocking new ways to study cellular diversity, spatial organization, and the structure of genomes in diseased tissues. Alongside these advancements, tools such as artificial intelligence, machine learning, and network analysis are transforming how we interpret complex multi-omics data, enabling more precise and actionable insights for clinical care. Collaborative efforts, including large-scale multi-omics consortia and data-sharing platforms, will play a crucial role in speeding up discoveries, ensuring research is reproducible, and amplifying the impact of multi-omics on improving human health (Seneviratne et al. 2020; Kaithal et al. 2024).

8.12 Conclusions

In summary, multi-omics research is revolutionizing our understanding of the molecular foundations of human genetic disorders and shaping the future of precision medicine. By combining data from various omics layers, researchers are uncovering new disease mechanisms, discovering biomarkers, and creating personalized therapies tailored to everyone. Although challenges persist, ongoing innovation, collaboration, and interdisciplinary efforts in this field promise to transform healthcare and bring meaningful improvements to the lives of those affected by genetic disorders.

As we move closer to the goal of personalized medicine, multi-omics research serves as a guiding light, offering hope for a future where the intricate complexities of genetic disorders are met with precise and compassionate care.

Short Chapter Summary
This chapter explores how multi-omics approaches have revolutionized the study of human genetic disorders by providing a comprehensive understanding of the genetic, molecular, and environmental factors involved in disease development. It highlights recent advancements in genomics, transcriptomics, proteomics, metabolomics, and epigenomics that have uncovered the molecular mechanisms underlying various genetic conditions.

This chapter discusses how integrative multi-omics analyses have led to the identification of novel disease-associated genes, biomarkers, and pathways, enabling personalized diagnostics, treatments, and prognostics. It also addresses the opportunities and challenges of integrating multi-omics data, focusing on data analysis strategies, ethical considerations, and computational methods. By harnessing the potential of multi-omics, researchers are making significant strides in decoding the genetic basis of diseases and translating these discoveries into actionable insights for precision medicine and enhanced patient care.

References

Boneh A (2015) Signal transduction in inherited metabolic disorders: a model for a possible pathogenetic mechanism. J Inherit Metab Dis 38:729–740

Canzler S, Schor J, Busch W, Schubert K, Rolle-Kampczyk UE, Seitz H, Kamp H, von Bergen M, Buesen R, Hackermüller J (2020) Prospects and challenges of multi-omics data integration in toxicology. Arch Toxicol 94:371–388

Chakraborty S, Hosen MI, Ahmed M, Shekhar HU (2018) Onco-multi-OMICS approach: a new frontier in cancer research. Biomed Res Int 2018:9836256

Chen TY, You L, Hardillo JA, Chien MP (2023) Spatial transcriptomic technologies. Cells 12:2042

Chengoden R, Victor N, Huynh-The T, Yenduri G, Jhaveri RH, Alazab M, Bhattacharya S, Hegde P, Maddikunta PK, Gadekallu TR (2023) Metaverse for healthcare: a survey on potential applications, challenges and future directions. IEEE Access 11:12765–12795

Chiou SH, Wu CY (2011) Clinical proteomics: current status, challenges, and future perspectives. Kaohsiung J Med Sci 27:1–4

Cho WC (2007) Proteomics technologies and challenges. Genom Proteom Bioinform 5:77–85

Dupras C, Knoppers T, Palmour N, Beauchamp E, Liosi S, Siebert R, Berner AM, Beck S, Charest I, Joly Y (2022) Researcher perspectives on ethics considerations in epigenetics: an international survey. Clin Epigenetics 14:110

Duzkale H, Shen J, McLaughlin H, Alfares A, Kelly MA, Pugh TJ, Funke BH, Rehm HL, Lebo M (2013) A systematic approach to assessing the clinical significance of genetic variants. Clin Genet 84:453–463

Ganguly D (2024) Multi-omics studies in interpreting the evolving standard model for immune functions. Brief Funct Genomics 23:75–81

Ginsberg SD, Mirnics K (2006) Functional genomic methodologies. Prog Brain Res 158:15–40

Gleason LU (2019) Applications and future directions for population transcriptomics in marine invertebrates. Curr Mol Biol Rep 5:116–127

Gloerich J, Wevers RA, Smeitink JA, Van Engelen BG, Van Den Heuvel LP (2007) Proteomics approaches to study genetic and metabolic disorders. J Proteome Res 6:506–512

Graf EH, Bryan A, Bowers M, Grys TE (2025) One size fits small: The narrow utility for plasma metagenomics. J Appl Lab Med 10:171–183

Haga SB (2024) Artificial intelligence, medications, pharmacogenomics, and ethics. Pharmacogenomics 25:611–622

HajYasien A (2023) Introduction to multi-omics technology. In: Alkhateeb A, Rueda L (eds) Machine learning methods for multi-omics data integration. Springer International Publishing, Cham, pp 1–11

Jendoubi T (2021) Approaches to integrating metabolomics and multi-omics data: a primer. Meta 11:184

Johanningsmeier SD, Harris GK, Klevorn CM (2016) Metabolomic technologies for improving the quality of food: practice and promise. Annu Rev Food Sci Technol 7:413–438

Kaithal P, Kanchan S, Kesheri M (2024) Recent advances in biological omics databases and tools in human health. In: Kesheri M, Kanchan S, Salisbury TB, Sinha RP (eds) Microbial omics in environment and health. Springer, Singapore, pp 311–341
Kanchan S, Sinha RP, Chaudière J, Kesheri M (2020) Computational metagenomics: current status and challenges. In: Katara P (ed) Recent trends in 'computational omics': concepts and methodology. Nova Science Publishers, New York, pp 371–395
Kanchan S, Kesheri M, Srivastava U, Katharia H, Ratna-Raj R, Chittoori B, Bogomolnaya L, Sinha RP, Denvir J (2024) Integrative omics data mining: challenges and opportunities. In: Gupta MK, Katara P, Mondal S, Singh RL (eds) Integrative omics: concepts, methodology, and application. Academic, pp 237–255
Kelly RS, Vander Heiden MG, Giovannucci E, Mucci LA (2016) Metabolomic biomarkers of prostate cancer: prediction, diagnosis, progression, prognosis, and recurrence. Cancer Epidemiol Biomarkers Prev 25:887–906
Kesheri M, Kanchan S, Chowdhury S, Sinha RP (2015) Secondary and tertiary structure prediction of proteins: a bioinformatic approach. In: Zhu Q, Azar AT (eds) Complex system modelling and control through intelligent soft computations. Springer-Verlag, pp 541–569
Kumar P, Kanchan S, Kesheri M (2024) Multi-omics in human disease biomarker discovery. In: Kesheri M, Kanchan S, Salisbury TB, Sinha RP (eds) Microbial omics in environment and health. Springer, Singapore, pp 205–239
Ladd-Acosta C (2015) Epigenetic signatures as biomarkers of exposure. Curr Environ Health Rep 2:117–125
Lage K (2014) Protein–protein interactions and genetic diseases: The interactome. Biochim Biophys Acta Mol basis Dis 1842(10):1971–1980
Li Y (2021) Modern epigenetics methods in biological research. Methods 187:104–113
Matharu N, Ahituv N (2020) Modulating gene regulation to treat genetic disorders. Nat Rev Drug Discov 19(11):757–775
Mensaert K, Denil S, Trooskens G, Van Criekinge W, Thas O, De Meyer T (2014) Next-generation technologies and data analytical approaches for epigenomics. Environ Mol Mutagen 55(3):155–170
Misra BB, Langefeld C, Olivier M, Cox LA (2019) Integrated omics: tools, advances and future approaches. J Mol Endocrinol 62(1):R21–R45
Mohr AE, Ortega-Santos CP, Whisner CM, Klein-Seetharaman J, Jasbi P (2024) Navigating challenges and opportunities in multi-omics integration for personalized healthcare. Biomedicines 12(7):1496
Moosavi A, Ardekani AM (2016) Role of epigenetics in biology and human diseases. Iran Biomed J 20(5):246
Ng PC, Henikoff S (2006) Predicting the effects of amino acid substitutions on protein function. Annu Rev Genomics Hum Genet 7(1):61–80
Papassotiropoulos A, Fountoulakis M, Dunckley T, Stephan DA, Reiman EM (2006) Genetics, transcriptomics, and proteomics of Alzheimer's disease. J Clin Psychiatry 67(4):652
Picard M, Scott-Boyer MP, Bodein A, Périn O, Droit A (2021) Integration strategies of multi-omics data for machine learning analysis. Comput Struct Biotechnol J 19:3735–3746
Portela A, Esteller M (2010) Epigenetic modifications and human disease. Nat Biotechnol 28(10):1057–1068
Priya P, Sardar RK, Kanchan S, Kesheri M (2024) Current status and future strategy in crop improvement using epigenomics. In: Kesheri M, Kanchan S, Salisbury TB, Sinha RP (eds) Microbial omics in environment and health. Springer, Singapore, pp 241–267
Raghavachari N, Garcia-Reyero N (2018) Overview of gene expression analysis: Transcriptomics. In: Raghavachari N, Garcia-Reyero N (eds) Gene expression analysis: methods in molecular biology. Humana Press, New York, pp 1–6
Sahu N, Mishra S, Kesheri M, Kanchan S, Sinha RP (2023) Identification of cyanobacteria-based natural inhibitors against SARS-CoV-2 druggable target ACE2 using molecular docking study, ADME and toxicity analysis. Indian J Clin Biochem 38(3):361–373

Seneviratne CJ, Suriyanarayanan T, Widyarman AS, Lee LS, Lau M, Ching J, Delaney C, Ramage G (2020) Multi-omics tools for studying microbial biofilms: current perspectives and future directions. Crit Rev Microbiol 46(6):759–778

Shukla N, Bhatia A, Kesheri M, Kanchan S, Narayan J (2024) Planet in peril: unveiling the microbial consequences of environmental degradation on human health through omics approaches. In: Kesheri M, Kanchan S, Salisbury TB, Sinha RP (eds) Microbial omics in environment and health. Springer, Singapore, pp 35–56

Singh M, Sukunathan A, Jain S, Gupta SK, Singh RL, Gupta MK (2024) Omics technology policy and society research. In: Gupta MK, Katara P, Mondal S, Singh RL (eds) Integrative omics: concepts, methodology, and application. Academic, pp 379–400

Srivastava U, Kanchan S, Kesheri M, Singh S (2023) Nutrimetabolomics: metabolomics in nutrition research. In: Soni V, Hartman TE (eds) Metabolomics. Springer, Cham, pp 241–268

Srivastava U, Kanchan S, Kesheri M, Gupta MK, Singh S (2024a) Types of omics data: genomics, metagenomics, epigenomics, transcriptomics, proteomics, metabolomics, and phenomics. In: Gupta MK, Katara P, Mondal S, Singh RL (eds) Integrative omics: concepts, methodology, and application. Academic, pp 13–34

Srivastava U, Kanchan S, Kesheri M, Gupta MK, Singh S (2024b) Integrative omics approaches for identification of biomarkers. In: Gupta MK, Katara P, Mondal S, Singh RL (eds) Integrative omics: concepts, methodology, and application. Academic, pp 157–173

Subramanian I, Verma S, Kumar S, Jere A, Anamika K (2020) Multi-omics data integration, interpretation, and its application. Bioinform Biol Insig 14:1177932219899051

Tam V, Patel N, Turcotte M, Bossé Y, Paré G, Meyre D (2019) Benefits and limitations of genome-wide association studies. Nat Rev Genet 20(8):467–484

Troisi J, Cavallo P, Colucci A, Pierri L, Scala G, Symes S, Jones C, Richards S (2020) Metabolomics in genetic testing. Adv Clin Chem 94:85–153

Wang X, Fan D, Yang Y, Gimple RC, Zhou S (2023) Integrative multi-omics approaches to explore immune cell functions: Challenges and opportunities. iScience 26(4):21

Wishart DS (2016) Emerging applications of metabolomics in drug discovery and precision medicine. Nat Rev Drug Discov 15(7):473–484

Xiao Z, Prieto D, Conrads TP, Veenstra TD, Issaq HJ (2005) Proteomic patterns: their potential for disease diagnosis. Mol Cell Endocrinol 230(1–2):95–106

Yang X, Kui L, Tang M, Li D, Wei K, Chen W, Miao J, Dong Y (2020) High-throughput transcriptome profiling in drug and biomarker discovery. Front Genet 11:19

Metabolomics Approaches for Microbiome Analysis

9

Susanta Roy and Smrita Singh

Abstract

Microbiome, the complex community of microorganisms residing in various ecosystems, is pivotal in influencing host health and disease. Metabolomics, the comprehensive study of metabolites within biological systems, has emerged as a powerful tool for elucidating the functional interactions between the microbiome and its host. This chapter traces the evolution of metabolomics approaches in microbiome analysis, from their initial applications to their current state-of-the-art methodologies. Early metabolomics studies focused on cataloging and linking metabolites to specific microbiome functions. With advancements in analytical technologies, such as mass spectrometry and nuclear magnetic resonance, coupled with sophisticated bioinformatics tools, metabolomics now provides deeper insights into the dynamic metabolic exchanges between host and microbiome. These insights have led to the identification of disease biomarkers, the discovery of novel therapeutic targets, and a better understanding of the microbiome's role in health and disease. As we explore the journey from the inception of metabolomics in microbiome research to today's cutting-edge approaches, this chapter highlights the challenges and opportunities in the field and emphasizes the potential for future discoveries in personalized medicine and microbiome-targeted therapies.

Keywords

Analytical technologies · Disease biomarkers · Metabolomics · Microbiome analysis

S. Roy
Bio Explore, Dimapur, Nagaland, India

S. Singh (✉)
School of Biotechnology, IFTM University, Moradabad, UP, India

M. Kesheri et al. (eds.), *Multi-Omics in Biomedical Sciences and Environmental Sustainability*, https://doi.org/10.1007/978-981-96-7067-3_9

Objectives

- Microbiome's role in human health and diseases
- Overview of omics technologies (metagenomics, metatranscriptomics, metabolomics) in microbiome research
- Challenges in metabolite identification and the role of AI in overcoming them
- Future prospects of microbiome-targeted therapies and ethical considerations in precision medicine

9.1 Introduction

The human microbiome, the collective term encompassing the trillions of microorganisms that reside within the human body, has become a topic of increasing scientific interest due to its significant influence on various aspects of human health and disease. Metabolomics, the comprehensive analysis of small molecules or metabolites within a biological system, has emerged as a powerful approach to elucidate the intricate interactions between the microbiome and the host (Xie et al. 2013; Ursell et al. 2014). The human microbiome, a complex ecosystem composed of trillions of microorganisms including bacteria, archaea, fungi, and viruses, inhabits various sites within the human body, such as the skin, oral cavity, and gut. The gut microbiome, in particular, harbors the largest and most diverse microbial community and plays a crucial role in human health and disease (Hattori and Taylor 2009; Cho and Blaser 2012; Grice and Segre 2012; Kim et al. 2013; Plotnikoff and Riley 2014; Proctor et al. 2019).

The gut microbiome plays a crucial role in human health and disease through various mechanisms. It is essential for the breakdown of complex dietary components that are indigestible by the human host, producing short-chain fatty acids that serve as energy sources and contribute to gut health (Grice and Segre 2012; Kim et al. 2013; Mishra et al. 2024). Additionally, the microbiome is pivotal in the development and regulation of the immune system, promoting tolerance to beneficial microbes and responsiveness to pathogens, as well as influencing the production of immune cells and signaling molecules (Grice and Segre 2012; Proctor et al. 2019). Moreover, the microbiome provides a protective barrier against the colonization and proliferation of harmful pathogens by competing for resources and space, and producing antimicrobial substances (Grice and Segre 2012). Beyond digestion and immunity, the gut microbiome has been shown to influence various physiological processes, including brain function, behavior, and metabolism, highlighting the intricate connection between the microbiome and the central nervous system (Cho and Blaser 2012). Alterations in the composition and function of the microbiome, or dysbiosis, have been linked to a wide range of diseases, such as inflammatory bowel disease, obesity, type 2 diabetes, cardiovascular disease, and certain neurological disorders, underscoring the importance of understanding the specific microbial changes associated with these conditions for the development of targeted therapies and diagnostic tools (Cho and Blaser 2012; Kim et al. 2013; Proctor et al. 2019).

Early research into the human microbiome revealed that healthy individuals have distinct microbiota, with microbial communities varying significantly across body sites. Humans host approximately 1.1 times more microbial cells than human cells, underscoring the microbiota's crucial role in health. Disruptions to the microbiota can have substantial health impacts, a phenomenon observed not only in humans but also in animals, insects, plants, and various environments like soil and oceans (Bauermeister et al. 2021; Shukla et al. 2024).

Microbiomes are communities of microorganisms, including viruses, bacteria, archaea, unicellular eukaryotes, and fungi, specific to various habitats and closely interacting with their hosts or environments. Present across all environments—humans, animals, plants, insects, soil, water, and air—microbiomes play essential roles in maintaining life and ensuring the health of Eukaryotes, including humans. Their diverse functions are integral to the ecosystem's stability and organismal health (Lange et al. 2022).

Current research in microbiome science has significantly enhanced our understanding of their critical roles across various domains, including human health, food systems, and planetary health. In human health, microbiomes are known to influence essential processes like development, growth, digestion, and immunity, as well as their involvement in diseases such as metabolic syndrome, obesity, diabetes, inflammatory bowel disease, and mental disorders (Lange et al. 2022).

Metagenomics addresses the question, "What is the composition of a microbial community under different conditions?" Metatranscriptomics explores, "What genes are collectively expressed under different conditions?" Meanwhile, metabolomics focuses on, "What by-products are produced under different conditions?" The metabolites released by microbial communities play a crucial role in maintaining the health of their environmental niches. Whether microbiome studies are biomedical or environmental, it is clear that each omics approach provides valuable insights. However, the most comprehensive understanding comes from integrative studies that combine all available omics datasets. While these efforts hold great promise, careful integration is essential to ensure meaningful and accurate results (Aguiar-Pulido et al. 2016). Traditionally, research on microorganisms is based on cultured organisms, which have several disadvantages.

Metabolomics is a comprehensive study of low-molecular-weight molecules or metabolites in various organisms, including biofluids, cells, tissues, organs, and organisms. These metabolites are the final products of genomic, transcriptomic, and/or proteomic perturbations. Metabolomics studies have shown biological changes resulting from disease or environmental interactions and revealed novel mechanisms not previously identified by other biochemical experiments. Both nuclear magnetic resonance (NMR) spectroscopy and mass spectrometry (MS)-based metabolomics have been applied to numerous in vitro and in vivo studies. Metabolomics has revealed metabolic pathway disruption in various disease types, including cardiovascular disease, diabetes, inborn errors of metabolism, alcohol-induced liver injury, and various cancers. Exposure to environmental influences such as ionizing radiation and xenobiotic use in rodents and nonhuman primates has also been explored by metabolomics. The influence of individual gene alterations

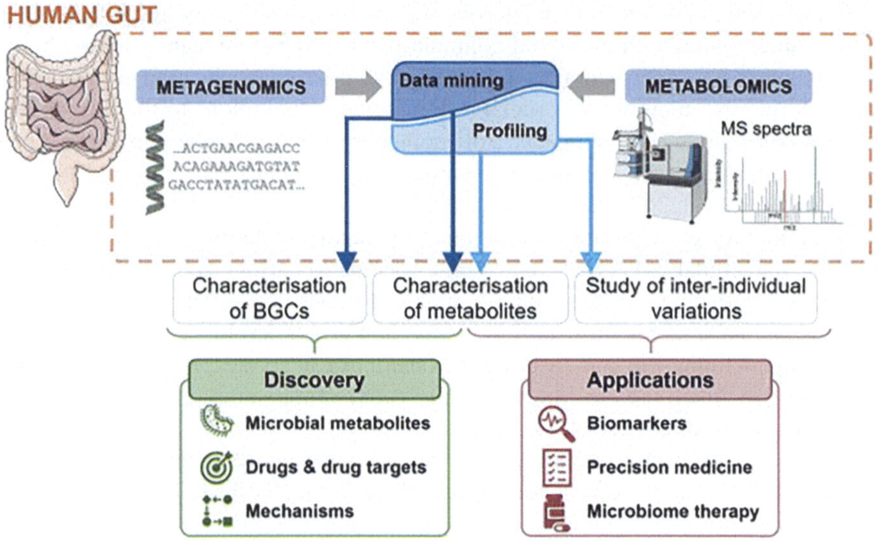

Fig. 9.1 Metagenome–metabolome crosstalk for human gut microbiome study and their clinical applications. Both data can be mined to discover new metabolites, drugs, drug targets and biological mechanisms. A profiling strategy can also be applied to seek microbial biomarkers of certain health conditions or treatment efficacy, and to identify how gut microbes modify disease prognosis and use this knowledge to define personalized treatments based on precision medicine or microbiome therapies to counteract the effects of these microbes. (Source: https://ars.els-cdn.com/content/image/1-s2.0-S0165993623003357-gr1_lrg.jpg)

on the metabolome has also been demonstrated by metabolomics. Although metabolomics can identify metabolic signatures and mechanisms of disease, challenges need to be considered when implementing a metabolomic study (Johnson and Gonzalez 2011).

Recent advancements in metabolomics and metagenomics data analysis have significantly aided in the development of modern medicine, natural products, and drug development (Fig. 9.1). This review explores the progress of metabolomics in microbiome research, highlighting the field's problems and opportunities, as well as the promise for future breakthroughs in personalized medicine and microbiome-targeted therapeutics.

9.2 Milestones in Metagenomics

The journey of metagenomics began with Leeuwenhoek's 1676 observations of oral microbiota and advanced in 1888 with Robert Koch's isolation of microorganisms using solid cultures. In 1931, Winogradsky's microbial ecology experiments provided foundational insights, followed by Carl Woese's 1977 proposal of rRNA as a taxonomic marker and Fred Sanger's development of DNA sequencing the same year. Kary Mullis revolutionized molecular biology with PCR in 1980, and

Giovannoni et al. conducted the first microbial community study using 16S rRNA libraries in 1990. The term "metagenomics" was coined in 1998 by Handelsman et al., marking a new era. Technological advancements accelerated with the release of the first NGS machine by Roche in 2005, Solexa's GA sequencer in 2006, and the publication of the Human Microbiome Project in 2008. The 2011 release of the PacBio RS sequencer further expanded metagenomic research capabilities, enabling deeper exploration of microbial diversity and functions (Escobar-Zepeda et al. 2015). Recent milestones have further expanded the field's applications and impact. The 2015 Ocean Sampling Day created the largest standardized microbial dataset from marine environments, while research in the same year linked commonly used drugs to shifts in gut microbial gene expression. Studies in 2016 on the human skin microbiome highlighted site-specific microbial signatures, and in 2018, the gut microbiome was found to influence cancer immunotherapy outcomes. By 2019, metagenomic analyses revealed novel uncultured bacterial candidates, expanding the phylogenetic scope of human microbiota studies. These achievements demonstrate the growing potential of metagenomics, from improving healthcare to advancing agricultural sustainability (Navgire et al. 2022). The early 2000s witnessed the rise of high-throughput sequencing technology, which enabled the rapid sequencing of entire microbial populations. To deal with the expanding number of metagenomic data, various specialized databases and bioinformatics tools were developed in the mid-2000s. Projects such as CAMERA (Community Cyberinfrastructure for Advanced Microbial Ecology Research and Analysis) and IMG/M (Integrated Microbial Genomes with Microbiomes) offered systems for storing and sharing metagenomic data. Currently, metagenomics is advancing alongside other omics technologies. Metagenomics is fast evolving as sequencing technologies and bioinformatics tools increase (Fig. 9.2) (Kanchan et al. 2020).

9.2.1 Metabolomics and the Microbiome

Metabolomics approaches, such as nuclear magnetic resonance spectroscopy and mass spectrometry, enable the comprehensive characterization of host metabolic profiles that are correlated with the intestinal microbial community (Xie et al. 2013). These techniques provide a holistic overview of the metabolites produced through the co-metabolic activities of the gut microbiota and the host, offering insights into the complex molecular mechanisms underlying host–microbe interactions (Xie et al. 2013). The study of these metabolic interactions has led to a better understanding of how the gut microbiome can influence various physiological processes, including the metabolism of bile acids, short-chain fatty acids, and choline (Xie et al. 2013).

Advancements in metabolomics have also empowered researchers to identify specific microbial species or groups responsible for the biotransformation of certain metabolites. For instance, the Cgr2 protein from the gut bacterium *Eggerthellalenta* has been found to inactivate digoxin, a plant toxin and a cardiac drug (Couvillion et al. 2020). Additionally, metabolomics studies have revealed the impact of diet,

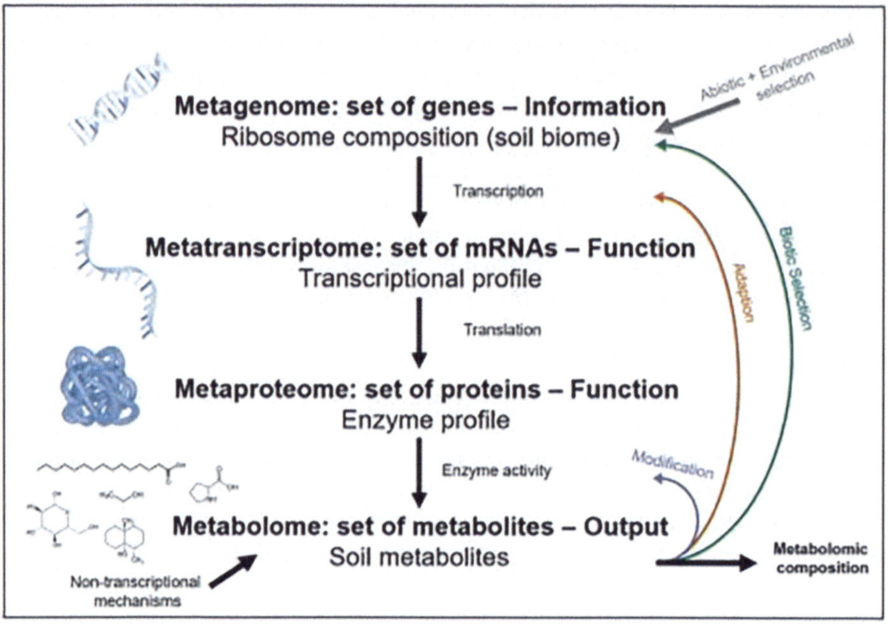

Fig. 9.2 Representation of the biological hierarchy, starting with the genome and culminating with the metabolome. (Adapted from Takahashi et al. 2012)

antibiotics, and other environmental factors on the composition and function of the gut microbiome, providing a more holistic understanding of the dynamic relationship between the microbiome and host health.

9.3 Metabolomics as a Tool for Microbiome Analysis

Metabolomics, the comprehensive analysis of small molecules or metabolites within a biological system, has emerged as a powerful approach to elucidate the intricate interactions between the microbiome and the host (Kashyap et al. 2017). This analytical technique offers a unique perspective by capturing the functional readout of the microbial community's activity, providing insights into the actual metabolic processes occurring within the microbiome and their subsequent impact on the host (Srivastava et al. 2024).

Why metabolomics is valuable in microbiome research:

- *Functional Characterization:* In contrast to other omics techniques such as *16S rRNA* gene sequencing or metagenomics that predominantly focus on microbial composition and genetic potential, metabolomics provides a functional characterization of the microbiome. This analytical approach enables researchers to identify the specific metabolites produced by the microbial community, offering insights into the metabolic pathways employed by these microorganisms and their subsequent impact on host physiology (Kashyap et al. 2017).

- *Host–Microbe Interactions:* Metabolomics enables researchers to unravel the complex metabolic interactions between the host and its microbiome. By analyzing the metabolite profiles in host tissues and biofluids, researchers can gain deeper insights into the ways the microbiome influences the host's metabolism, immune function, and overall health (Xie et al. 2013; Lee-Sarwar et al. 2020).
- *Disease Biomarker Discovery:* Alterations in the microbial community's metabolic activity are frequently linked to disease conditions. Metabolomics analysis can be leveraged to pinpoint specific metabolites that may act as biomarkers for diverse pathologies, potentially enabling earlier detection and more personalized treatment approaches (Ganesan et al. 2021).
- *Dietary and Environmental Influences:* Metabolomics can be used to evaluate the effects of dietary, environmental, and other external factors on the metabolic activity of the microbiome. This knowledge can inform tailored dietary recommendations or the development of strategies to therapeutically modulate the microbiome (Couvillion et al. 2020; Kesheri et al. 2024a).
- *Drug Metabolism and Efficacy:* Metabolomics analysis can elucidate the impact of the microbiome on drug metabolism, influencing the efficacy and potential side effects of certain medications. Computational analytical approach enables the study of complex interactions between the gut microbial community and pharmacological agents, paving the way for personalized treatment strategies that optimize drug effectiveness while minimizing adverse reactions (Couvillion et al. 2020; Sahu et al. 2023; Bujak et al. 2024; Kesheri et al. 2024b).
- *Integration with Other Omics Data:* Integrating metabolomics data with other omics approaches, such as metagenomics, metatranscriptomics, and metaproteomics, provides a more comprehensive understanding of the microbiome's composition, function, and interactions with the host. This multi-omics perspective can lead to deeper insights into the complex mechanisms underlying microbiome-related health and disease states (Petriz and Franco 2017; Peters et al. 2019; Kanchan et al. 2024a). By offering a functional readout of the microbiome's activity, metabolomics serves as a powerful tool for elucidating the intricate interplay between the microbiome and the host, ultimately contributing to the development of novel diagnostic and therapeutic strategies for microbiome-associated diseases.

9.4 Advancements in Analytical Technologies

Microbiome research typically begins with cataloging organisms and genes using DNA or RNA sequencing to analyze microbial community composition (Bauermeister et al. 2021). This approach is complemented by studying how chemical environments influence microbial communities and the functions of microbial-derived molecules (Bauermeister et al. 2021). While sequencing provides insights into microbial presence and metabolic capacity, metabolomics offers a direct readout of system functions by capturing the metabolome, which reflects cellular processes in response to stimuli (Bauermeister et al. 2021). Despite its potential to detect all small molecules under 2000 Da, metabolomics faces practical limitations in extraction, ionization, and detection (Bauermeister et al. 2021). Mass spectrometry (MS) is a key tool in metabolomics due to its sensitivity and capability to analyze complex biological samples (Bauermeister et al. 2021).

9.4.1 Mass Spectrometry (MS) Techniques

Mass spectrometry is another crucial analytical technique in the field of metabolomics. It offers exceptional sensitivity and versatility, making it a powerful tool for examining complex biological samples in microbiome research (El-Aneed et al. 2009). When coupled with separation methods such as gas chromatography and liquid chromatography, mass spectrometry provides a robust analytical platform for the identification and quantification of a vast array of metabolites (Yates 2001; Vandell and Limbach 2016; Brown et al. 2020).

Mass spectrometry in microbiome metabolomics:

- *Principle:* MS measures the mass-to-charge ratio (m/z) of ions. The sample is first ionized, and the resulting ions are then separated based on their mass-to-charge (m/z) ratios in a mass analyzer before being detected (Medhe 2018). The resulting mass spectrum displays the relative abundance of ions at different m/z values, providing a unique profile of the sample's composition (El-Aneed et al. 2009).
- *Advantages in Metabolomics:*
  High Sensitivity: Mass spectrometry demonstrates exceptional sensitivity, capable of detecting metabolites at remarkably low concentrations, making it well-suited for analyzing trace compounds (Bruce et al. 2013). Additionally, MS exhibits broad applicability, enabling the analysis of a diverse range of metabolites, including small molecules, lipids, and peptides (Vandell and Limbach 2016). Coupling MS with separation techniques like gas chromatography and liquid chromatography further enhances its resolving power, empowering the analysis of highly complex mixtures (Brown et al. 2020). Moreover, tandem MS techniques can provide valuable structural information about metabolites, facilitating their identification (Bruce et al. 2013).
- *Applications in Microbiome Research:*
  Metabolite Profiling: Mass spectrometry is employed to profile and quantify metabolites in various biological samples, offering insights into the metabolic activity of the microbiome (Vandell and Limbach 2016).
  Host–Microbe Interactions: Mass spectrometry can be employed to investigate the metabolic interplay between the host organism and its associated microbiome, providing insights into how microbial metabolites modulate host physiological processes.

Disease Biomarker Discovery: Mass spectrometry-based metabolomic studies have the potential to identify candidate biomarkers linked to microbiome-associated diseases, which can then inform diagnostic and prognostic approaches.
Dietary Interventions: Mass spectrometry can be employed to evaluate the influence of dietary modifications on the metabolic activity of the microbiome, informing the development of personalized nutritional strategies.

- *Limitations:*
  Sample Preparation: MS often requires specific sample preparation procedures, which can introduce bias or alter the metabolite composition.
  Data Complexity: Analyzing the large and intricate datasets generated by mass spectrometry requires the use of advanced bioinformatics tools and techniques (Bruce et al. 2013).
  Ionization Techniques: The choice of ionization technique is crucial for the successful analysis of different types of metabolites, as certain methods are more suitable for specific classes of compounds (Medhe 2018).

Mass spectrometry, with its exceptional sensitivity and extensive versatility, is a foundational technique in the field of microbiome metabolomics (Yates 2001; Ebnesajjad 2011). Its unparalleled capability to detect and quantify a vast array of metabolites renders it an essential tool for unveiling the intricate metabolic interplay within the microbiome and between the microbiome and its host.

- *Measurement*: MS measures ions based on mass-to-charge ratio and relative abundance. Untargeted MS helps discover new molecules but presents challenges in data analysis (Bauermeister et al. 2021).
- *High-Resolution MS*: Techniques like TOF, Orbitrap, and FTICR allow detailed elemental composition calculations. Tandem MS (MS/MS) provides fragmentation patterns for deducing chemical structures (Bauermeister et al. 2021).
- *Gas Chromatography-MS (GC-MS)*: Used for detecting volatile molecules; electron impact ionization generates fragmentation patterns. Liquid chromatography (LC) and ion mobility (IMS) are crucial for complex sample separation (Bauermeister et al. 2021).

The human gut microbiome consists of a diverse array of microorganisms, including bacteria, viruses, archaea, and eukaryotes, all of which contribute to various metabolic processes that influence host health (Puig-Castellví et al. 2023). Individual microbiome compositions vary significantly due to factors such as diet, environment, and lifestyle. High-throughput analytical techniques like metagenomics and metabolomics have advanced our understanding of these microbial communities. Metagenomics, utilizing methods like shotgun sequencing, helps characterize microbial genomes, while metabolomics employs techniques such as mass spectrometry (MS) and nuclear magnetic resonance (NMR) to profile metabolites (Puig-Castellví et al. 2023).

Metabolomics Strategies and Tools: Metabolomics employs a variety of methods, including targeted and untargeted approaches, with tandem mass spectrometry (MS/MS) offering detailed structural information. Tools like Data-Dependent Acquisition (DDA) and Selected Reaction Monitoring (SRM) are used for generating spectral data. The Simultaneous Quantification and Discovery (SQUAD) approach integrates annotation and quantification in a single LC-MS run. Standardizing metabolite identifiers and utilizing databases such as HMDB, MassBank, and MiMeDB facilitate data comparison and validation (Puig-Castellví et al. 2023).

9.4.2 Nuclear Magnetic Resonance Spectroscopy

Nuclear magnetic resonance spectroscopy is a versatile and widely employed analytical technique in metabolomics for the study of the microbiome (Dayrit and Dios 2017; Emwas et al. 2019; de Moraes Pontes et al. 2017). This method offers a non-destructive and highly reproducible approach to identify and quantify a wide array of metabolites present in complex biological samples (Benaki and Mikros 2018; Serkova et al. 2019; Zhang et al. 2010). Nuclear magnetic resonance spectroscopy is a principle-based technique that exploits the magnetic properties of atomic nuclei to provide structural and quantitative information about molecules. When a sample is placed in a strong magnetic field, certain atomic nuclei absorb and re-emit electromagnetic radiation at specific frequencies, which are characteristic of their chemical environment (Dayrit and Dios 2017; de Moraes Pontes et al. 2017).

NMR offers several advantages in metabolomics research. It is a non-destructive analysis method that does not require extensive sample preparation or derivatization, preserving the integrity of the sample for further analysis (Dayrit and Dios 2017; Benaki and Mikros 2018). NMR signals are directly proportional to the concentration of metabolites, allowing for accurate quantification (Dayrit and Dios 2017). Additionally, NMR can detect a wide range of metabolites simultaneously without prior knowledge of their presence, minimizing bias (Dayrit and Dios 2017). NMR experiments are also highly reproducible, providing reliable and consistent results (Benaki and Mikros 2018). Furthermore, NMR provides detailed structural information about metabolites, aiding in their identification (de Moraes Pontes et al. 2017).

In the context of microbiome research, NMR-based metabolomics has several applications. It is used to identify and quantify metabolites in various biological samples, such as feces, urine, serum, and tissue extracts, providing insights into the metabolic activity of the microbiome (Kruk et al. 2016; de Moraes Pontes et al. 2017; Benaki and Mikros 2018). NMR can also be employed to study the metabolic exchange between the host and its microbiome, revealing how microbial metabolites influence host physiology (Xie et al. 2013). Additionally, NMR-based metabolomics can identify potential biomarkers associated with microbiome-related diseases, aiding in diagnosis and prognosis (Kruk et al. 2016; Serkova et al. 2019).

Finally, NMR can assess the impact of dietary changes on the microbiome's metabolic activity, guiding personalized nutrition strategies.

Limitations

NMR has lower sensitivity compared to mass spectrometry, making it more challenging to detect metabolites present at low concentrations (Zhang et al. 2010). Overlapping Signals: The complex nature of biological samples can lead to overlapping signals in NMR spectroscopy, which can complicate the resolution and identification of individual metabolites (Emwas et al. 2019).

Although NMR spectroscopy has certain limitations, it continues to be a valuable technique in the field of microbiome metabolomics. This analytical method provides unique advantages for characterizing the metabolic activity of the microbiome and its interactions with the host. Ongoing developments in NMR techniques and technologies are aimed at enhancing its sensitivity and overcoming its inherent limitations, further bolstering its utility in metabolomic research applications (Emwas et al. 2019).

The human gut microbiome, composed of a diverse range of microorganisms such as viruses, bacteria, archaea, and eukaryotes, plays a crucial role in shaping the host's health through its metabolic activities. These microbes produce microbial compounds, bioaccumulate and biotransform drugs, and contribute to a complex microbial ecosystem. Inter-individual variations in microbiome composition are influenced by numerous factors, including diet, health, lifestyle, and environment. The study of the gut microbiome has primarily relied on metagenomics and metabolomics, which analyze the genomes of microorganisms and the metabolites they produce, respectively. Shotgun sequencing, a common metagenomic technique, provides detailed microbial species-level identification, while mass spectrometry (MS) is widely used in metabolomics for its sensitivity, although nuclear magnetic resonance (NMR) spectroscopy is more reproducible. The study of microbiota-derived metabolites has focused on specific chemical classes, such as secondary bile acids, short-chain fatty acids, branched-chain amino acids, trimethylamines, and aromatic amino acid derivatives, due to their relevance in metabolic disorders. Currently, profiling of large numbers of metabolites, typically between 500 and 1500 using LC-MS (liquid chromatography-mass spectrometry) is being carried out. If a metabolite is associated with a disease, various methodologies are applied to determine whether it originates from the microbiome. Databases like KEGG and ChEBI provide information on many metabolites, but some metabolites require experimental validation to confirm their microbial origin. The process for establishing the microbial origin of metabolites involves several steps: biological sample analysis, metabolite signal annotation, molecular networking to improve annotation, and experimental validation of the metabolite–microbiome association (Puig-Castellví et al. 2023).

9.4.3 Metabolomics Experimental Strategies

To achieve comprehensive metabolite characterizations, researchers use a combination of empirical and data analysis strategies. These include various sample preparation methods, instrumental settings, chromatographic columns, and both targeted and untargeted methods. However, these approaches tend to focus on major metabolite classes, such as organic acids and lipids, leaving specialized microbial metabolites underexplored. For example, many microbial enzymes convert L-amino acids to D-configurations, but comprehensive methods to distinguish both enantiomers need further development. In metabolomics, tandem mass spectrometry (MS/MS) is widely used to generate spectral data for structural information. Both targeted and untargeted methods are used, with strategies like iterative data-dependent acquisition (DDA) and selected reaction monitoring (SRM) employed to detect hundreds of metabolites. Semi-targeted methods combine the sensitivity of targeted methods with the broad coverage of untargeted approaches. A recent method, Simultaneous Quantification and Discovery (SQUAD), enables both metabolite annotation and quantification in a single LC-MS injection (Puig-Castellví et al. 2023).

The study employed various technical methods, including metagenomic sequencing using high-throughput technology and metabolite quantification through ultra-performance liquid chromatography coupled with tandem mass spectrometry (UPLC–MS/MS). DNA from stool samples was extracted using a commercial kit, and the sequencing libraries were prepared using a DNA library prep kit. For metabolite analysis, the Q300 Metabolite Array Kit was used, and metabolites were identified with high confidence by comparing them to standard substances (Ning et al. 2023).

Metabolomics involves analyzing the repertoire of small-molecule metabolites in complex biological systems, such as the gut microbiota, using high-throughput analytical methods like mass spectrometry and nuclear magnetic resonance spectroscopy (Ursell et al. 2014). These techniques help researchers identify thousands of metabolites in biological samples such as feces, urine, and tissues. By comparing metabolomic profiles before and after perturbations, researchers can identify which metabolites are significantly affected by experimental variables, helping to understand how treatments influence the host's metabolic status. Platforms like HMDB, METLIN, and MassBank facilitate the identification and annotation of metabolites, accelerating this process. However, assigning spectral features remains challenging, and researchers often rely on multivariate statistics to interpret the results (Ursell et al. 2014).

9.4.4 Metabolomics Techniques: Other Analytical Platforms

Beyond NMR and mass spectrometry, various other analytical platforms contribute valuable insights to microbiome metabolomics research. Although these techniques are often used in conjunction with NMR or mass spectrometry, they offer unique strengths and expand the scope of metabolomic investigations. Chromatographic methods, such as gas chromatography, liquid chromatography, and high-performance liquid chromatography, are frequently coupled with mass spectrometry to separate complex mixtures of metabolites before detection (El-Aneed et al. 2009; Brown et al. 2020). This separation enhances the sensitivity and specificity of mass spectrometric analysis, enabling the identification and quantification of individual metabolites in complex biological samples. Gas chromatography is particularly well suited for volatile and thermally stable compounds, while liquid chromatography and high-performance liquid chromatography are more versatile for analyzing a broader range of metabolites, including polar and nonvolatile compounds (Vandell and Limbach 2016). Capillary electrophoresis separates charged molecules based on their electrophoretic mobility in a capillary filled with an electrolyte solution. This technique offers high resolution and speed, making it suitable for analyzing complex mixtures of metabolites. Capillary electrophoresis can be coupled with mass spectrometry to enhance sensitivity and identification capabilities. Fourier transform infrared spectroscopy measures the absorption of infrared radiation by molecules, providing information about their functional groups and chemical bonds. It is a rapid and nondestructive technique that can be used for metabolic fingerprinting and identifying changes in metabolite profiles associated with microbiome alterations or disease states.

Raman spectroscopy is a complementary vibrational spectroscopic technique to FTIR. It measures the inelastic scattering of light by molecules, providing insights into their molecular structure and composition. This method is particularly useful for analyzing aqueous samples and can be employed for in vivo measurements, offering real-time information about metabolic processes.

Surface plasmon resonance is a label-free approach used to study biomolecular interactions. It can be applied to analyze the binding of metabolites to specific receptors or proteins, offering insights into their biological activity and function. Additionally, SPR can be utilized to investigate the interactions between microbial metabolites and host cells, shedding light on the mechanisms underlying host–microbe interactions. These additional analytical platforms, often combined with NMR or mass spectrometry, provide a comprehensive toolkit for microbiome metabolomics research. Their diverse capabilities enable researchers to explore a broader range of metabolites and investigate the complex metabolic interactions within the microbiome and between the microbiome and its host.

9.5 The Importance of Spatial and Dynamic Microbial Interactions

Most studies have treated body compartments like the skin or gut as homogeneous microbial environments (Dorrestein et al. 2014), but the spatial and dynamic associations between specific microbes, their metabolites, and immune responses are critical for understanding how these sites function. For instance, understanding the microbial colonization of specific areas in the gut, such as crypts, and their impact on stem cell proliferation or conditions like inflammatory bowel disease (IBD) or cancer, provides valuable insights not easily obtainable from stool sample-based microbiome assessments. In IBD, biomarkers identified from mucosal biopsies offer clearer information than those found in stool samples, which are often used due to their non-invasive nature. Furthermore, technologies like imaging mass spectrometry that study how microbes interact to produce unique metabolites could be expanded to understand interkingdom interactions, such as between bacteria and fungi. These spatially defined studies could lead to predictive models for disease diagnosis and patient stratification, providing a powerful mechanism for identifying lead compounds for clinical applications (Dorrestein et al. 2014).

9.5.1 Precision Medicine

Precision medicine aims to provide personalized treatments based on a patient's genetic, clinical, and microbiome data. This individualized approach seeks to enhance treatment efficacy and minimize toxicity. The field, although still in its early stages, has gained momentum with advancements in high-throughput omics technologies (such as genomics, proteomics, and metabolomics) and computational tools that can integrate these diverse datasets for clinical diagnosis and treatment predictions. Notably, the gut microbiome, which contains a vast number of genes far surpassing those of its human host, plays a significant role in health outcomes, influencing up to 46% of the variance in circulating metabolites. Each individual's microbiome is unique, but functional redundancies within microbial species ensure many functions are conserved. However, strain-specific functions in certain individuals may lead to unique variations. Research has linked microbiome metabolites to various diseases, including cardiovascular issues (TMAO), cancer, Type 2 diabetes (branched-chain amino acids), and heart failure (imidazole propionic acid, phenylacetylglutamine). The identification of gene clusters, such as those involved in TMA production, has further highlighted the microbiome's influence on human health. Despite the promising advancements, significant challenges remain in gut microbiome research. The interpretation of complex biological and clinical data is hindered by limited knowledge about the role of each gut microbe, with most research focusing on pathogens and abundant microbes. Low-abundant microbes, however, can also play critical roles in host health, which complicates the understanding of microbiome functionality. In vitro culture and isolation of gut microbes could address this, but many microbes are not culturable, and cultured microbes may not exhibit the same metabolic activity as in vivo. Another major issue is the development of diagnostic tools based on microbiome-related omics data. While gut microbiome analysis shows potential for diagnosing and monitoring diseases and treatment responses, there is a need for further validation and the inclusion of more diverse populations in research cohorts to ensure the wider applicability of these tools (Puig-Castellví et al. 2023).

9.5.2 Multi-Omic Technologies for Understanding Lactation Performance

In order to fully capture the biological mechanisms involved in lactation, a combination of DNA, RNA, and metabolite analyses is used (Zhang et al. 2024a, b). Metabolomics platforms such as ultra-performance liquid chromatography and tandem mass spectrometry (UPLC-MS/MS) allow the identification of a wide variety of metabolites and provide more comprehensive insights into the metabolic changes that occur during lactation, helping to optimize lactation performance (Zhang et al. 2024a; b).

9.5.3 Exploring Microbial Metabolites

Microbial metabolites, such as bile acids, short-chain fatty acids, and aromatic amino acid derivatives, play significant roles in host metabolism and health (Puig-Castellví et al. 2023). Understanding these metabolites requires sophisticated analytical tools:

- *Databases for Metabolite Identification*: Databases like KEGG and ChEBI are essential resources for identifying and cataloging microbial metabolites. They provide structured data that enables researchers to connect metabolites with their potential biological functions (Puig-Castellví et al. 2023).
- *Analytical Tools*: Technologies like LC-MS allow for comprehensive metabolite profiling, while NMR spectroscopy is used for confirming the structure and reproducibility of metabolites (Puig-Castellví et al. 2023).

These tools and databases enable researchers to explore the vast array of microbial metabolites, advancing our understanding of their role in disease and health.

9.5.4 Innovations in Spatial Microbial Studies

Spatial microbiology is an emerging field that focuses on mapping microbial communities within specific anatomical niches, such as the gut crypts. These studies provide valuable insights into microbial interactions and their effects on host physiology. One of the innovative tools in this area is **imaging mass spectrometry (MS)**, which allows researchers to map metabolites and microbial activity within tissues, helping to diagnose diseases like inflammatory bowel disease (IBD) (Dorrestein et al. 2014). This spatial resolution is critical for understanding how microbial communities function in their native environments and how they interact with host tissues.

9.5.5 Metabolomics in Gut Microbiota Studies

Metabolomics is increasingly used to understand the role of gut microbiota in health and disease. By profiling small molecules, researchers can gain insights into how microbial communities affect host metabolism. Tools like MS and NMR spectroscopy are essential in capturing these profiles. In addition to experimental techniques, databases such as the Human Metabolome Database (HMDB) and MassBank provide key resources for metabolite annotation and identification, making it easier to relate specific metabolites to microbial functions (Ning et al. 2023).

9.6 Other Emerging Technologies in Microbiome Research

9.6.1 Spatial and Dynamic Microbial Interactions

Recent studies have begun to explore the dynamic interactions between microbes, metabolites, and the immune system. Spatial studies are instrumental in uncovering relationships that could serve as biomarkers for diseases (Dorrestein et al. 2014). These interactions are especially important for understanding chronic diseases like IBD, where microbial dysbiosis plays a crucial role.

9.6.2 Multi-omics Data Integration

Integrating data from various omics technologies, including metagenomics, metabolomics, and transcriptomics, has provided a more holistic understanding of microbial ecosystems. This multi-omics approach enables researchers to draw more accurate conclusions about microbial behavior, enhancing our ability to predict disease outcomes and develop new therapeutic strategies. Multi-omics datasets and further utilization of omics data with other advanced techniques, such as adaptive immune receptor repertoire sequencing, microbial culturomics, and machine learning, to evaluate important microbiome characteristics in the gut (Whon et al. 2021; Kumar et al. 2024). Systems biology, a computational approach for modeling complex biological systems, is increasingly important for analyzing multi-omics datasets and identifying disease mechanisms. By integrating data from genomics, transcriptomics, proteomics, and metabolomics, systems biology can offer a more comprehensive understanding of disease pathogenesis (Zhang et al. 2024a, b).

9.6.3 Technological Tools

Recent advancements in technologies such as CITE-seq and ATAC-seq have improved the resolution of immune response studies, particularly in microbiome research. These tools provide detailed insights into cellular interactions, offering new perspectives on immune system functioning and microbial dysbiosis (Whon et al. 2021).

9.7 Bioinformatics and Data Analysis

9.7.1 Evolution of Bioinformatics Tools

Bioinformatics tools have evolved significantly to handle the increasing volume and complexity of biological data. One of the most prominent advancements has been in metabolomics, the large-scale study of metabolites within biological samples. Metabolomics data presents unique challenges due to the diversity of metabolites and the complexity of their interactions. Traditional bioinformatics tools were not equipped to process large-scale metabolomics data, but recent advancements have addressed these gaps by integrating multi-omics platforms, including genomics, transcriptomics, and proteomics, for a more comprehensive understanding of the microbiome.

The integration of multi-omics data allows researchers to capture a holistic view of microbial interactions and their influence on host biology. For example, the use of metagenomics and metabolomics together has significantly improved the ability to identify biomarkers linked to diseases like inflammatory bowel disease (IBD). A study utilizing these methods combined fecal metagenomics with metabolomics to analyze gut microbiota alterations in IBD. The study identified 79 shared metabolites across different cohorts, highlighting the potential of multi-omics data in disease diagnosis and biomarker identification (Ning et al. 2023).

Additionally, platforms like GNPS (Global Natural Product Social Molecular Networking) have revolutionized the analysis of microbial metabolites. By utilizing molecular networking, these platforms enable the identification of structurally related metabolites based on their fragmentation patterns, which is a critical step in understanding microbial roles in health and disease. Tools such as SIRIUS and MetFrag facilitate in silico fragmentation to generate reference spectra, further enhancing metabolite annotation (Puig-Castellví et al. 2023). These developments allow bioinformaticians to integrate large datasets from various sources, ensuring a more complete and accurate analysis.

9.7.2 Metabolite Identification and Annotation

Metabolite identification remains one of the major challenges in bioinformatics due to the vast diversity of metabolites produced by both microbial and host-originating processes. The traditional approach to metabolite identification relied on spectral matching, but this method often faces limitations due to the complexity of metabolites and their fragmentation patterns. However, recent advances in machine learning and artificial intelligence (AI) have greatly improved the process of metabolite annotation.

Machine learning algorithms, especially deep learning techniques, have been increasingly employed to identify unknown metabolites from mass spectrometry data. These methods can learn complex patterns in large datasets and make predictions that improve the accuracy of metabolite identification. For instance, AI-based tools are now capable of predicting the most likely structure of metabolites based on their fragmentation patterns and mass-to-charge ratios (m/z). These tools, when combined with databases like the Human Metabolome Database (HMDB), facilitate the annotation of metabolites, making the data interpretation process faster and more reliable (Ning et al. 2023).

Moreover, the integration of AI in metabolomics is not limited to annotation but extends to predicting the functional roles of metabolites in biological systems. By combining metabolomics data with genomic and transcriptomic data, AI models can help elucidate the mechanistic pathways of microbial metabolites and their impact on host health. This approach is crucial for understanding the complex interactions between microbes and their hosts and for identifying potential biomarkers for disease (Dorrestein et al. 2014).

9.7.3 Challenges in Metabolite Identification

Despite these advancements, metabolite identification still faces significant challenges. One of the main issues is the lack of comprehensive spectral libraries for a wide range of metabolites, particularly those that are rare or unique to specific environmental or biological conditions. Moreover, many metabolites exist at low concentrations, making their detection difficult. This limitation is compounded by the phenomenon of ion adducts, where a single compound can generate multiple ion species, further complicating the analysis.

To address these challenges, bioinformatics platforms have integrated molecular networking techniques that help identify relationships between different metabolites based on their spectral features. By connecting related molecules in a network, these platforms can improve the identification process, even for compounds that lack direct reference spectra. In particular, ion identity molecular networking (IIMN) has been introduced to deal with ion adducts, ensuring that the full range of metabolites is considered during analysis (Puig-Castellví et al. 2023).

Furthermore, the field of functional metagenomics has identified the importance of linking genes to their functional roles within microbial communities. Tools like Prodigal and Bakta have been developed to predict protein-coding sequences and annotate genes, but challenges still exist in translating genomic information into functional insights. Misannotations and the lack of activity data for certain genes, especially in environmental or host-associated microbes, often lead to inaccurate predictions. Therefore, integrating functional genomics with metabolomics remains an area of active research, with the aim of improving the prediction of microbial functions in diverse environments (Sardar et al. 2024).

9.7.4 The Role of AI and Machine Learning in Metabolite Annotation

Machine learning (ML) and AI have revolutionized the field of metabolite annotation by enabling the processing of large-scale data and making predictions based on complex patterns. One of the key benefits of AI is its ability to integrate multi-omics data, such as genomics, transcriptomics, and metabolomics, to provide a more complete understanding of microbial functions and their influence on host biology. AI algorithms can learn from existing data and apply this knowledge to newly acquired data, improving the accuracy of metabolite identification and annotation.

For instance, deep learning methods, which use multi-layered neural networks, have shown promise in analyzing large-scale metabolomics data. These methods can recognize subtle patterns in fragmentation spectra, enabling the identification of metabolites that were previously undetectable. This capability is crucial for identifying metabolites that play significant roles in health and disease, such as those produced by the gut microbiome in relation to conditions like inflammatory bowel disease (Whon et al. 2021).

Furthermore, ML algorithms can also predict the functional roles of viral genomes and microbial metabolites by linking them to their genetic and environmental contexts (Kanchan et al. 2024b). This ability to predict how metabolites contribute to disease pathogenesis or health offers new avenues for personalized medicine. For example, ML-based tools have been applied to study the gut microbiome in the context of diseases like IBD, where specific microbial metabolites serve as biomarkers for diagnosis and disease progression (Ning et al. 2023). Table 9.1 presents the analysis pipeline for spectral metabolomic data analysis.

9.8 Challenges and Opportunities

9.8.1 Current Challenges in Metabolomics for Microbiome Analysis

Metabolomics, a comprehensive analysis of metabolites in biological systems, is a pivotal approach for understanding the complex interactions between the microbiome and its host. However, as the field of metabolomics expands, several challenges must be addressed to fully leverage its potential for microbiome analysis. These challenges include issues related to data standardization, reproducibility, and the complexity of metabolome–microbiome interactions. Despite these challenges, there are substantial opportunities for future research, including the integration of advanced technologies, the development of sophisticated bioinformatics tools, and the exploration of novel therapeutic targets.

A significant challenge in metabolomics is the lack of standardization in methodologies. Variations in sample collection methods, population characteristics, and analytical techniques often result in inconsistent findings. This lack of consistency hampers the comparability of results across studies, making it difficult to draw

Table 9.1 The analysis pipeline for spectral metabolomic data. (Aguiar-Pulido et al. 2016)

Step	Description	Databases	Functionality
1. Preprocessing	Denoising and peak-picking to improve data quality before analysis	–	–
2. Statistical analysis	Identifying metabolites and quantifying their concentrations by comparing spectral patterns against databases	–	–
3. Machine learning	Pattern recognition and further data interpretation	–	–
Human Metabolome Database	Cross-referenced database of small metabolites found in the human body	Human Metabolome Database	Focuses on human metabolites
BioMagResBank	Central repository for experimental NMR data, including small metabolites and macromolecules	BioMagResBank	Supports experimental NMR data
Madison-Qingdao Database	An annotated repository of NMR and MS data collected from various databases and literature	Madison-Qingdao Metabolomics Consortium Database	Combines and annotates NMR and MS data
MassBank	Merges spectral data under different conditions to improve precision in compound identification	MassBank	Focuses on precision in compound identification
Golm Metabolome Database	Stores GC-MS spectral data with retention indexes for automated compound identification	Golm Metabolome Database	Enhances automated compound identification for GC-MS
METLIN Database	Contains curated spectral information of biological metabolites without context of the sample's environmental origin	METLIN Metabolite Database	Focuses on curated spectral data of metabolites
Purpose of databases	Maintained by specialized consortia to automate and standardize spectral analysis processes	All databases listed above	Serve as repositories of spectral data and provide links to biological interpretation of metabolomic profiles

reliable conclusions. For instance, in microbiome studies related to diseases like necrotizing enterocolitis (NEC), variations in sample collection protocols can introduce biases, leading to inconsistent findings (Provitera et al. 2024). Similarly, the complexity of metabolomics data, often involving hundreds or thousands of metabolites, necessitates the development of standardized protocols to ensure reproducibility and facilitate data sharing between research groups (Provitera et al. 2024). A more harmonized approach to sample collection, data processing, and analytical techniques will be critical for improving the reliability and comparability of metabolomic studies.

Another major challenge is ensuring the reproducibility of metabolomic analyses. As with many omics technologies, metabolomics suffers from a high level of variability in experimental results. This is particularly true when it comes to complex interactions between microbial metabolites and host processes. For instance, the microbial metabolite p-cresol has been shown to influence the host's ability to metabolize acetaminophen, but the variability in microbial composition between individuals can result in inconsistent findings regarding this interaction (Ursell et al. 2014). Additionally, validating computational models and bioinformatics tools used to analyze metabolomic data remains a difficult task. While methods like feature filtering and co-occurrence analysis can help manage complex datasets, their performance is still under investigation, and they require further validation in real-world microbiome–metabolome interactions (Ursell et al. 2014; Sardar et al. 2024).

9.8.2 Complexity of Metabolome–Microbiome Interactions

Understanding the complexity of metabolome–microbiome interactions remains a critical challenge. The microbiome, with its vast diversity of microorganisms, can significantly affect host metabolism, immune responses, and disease progression. However, deciphering the intricate networks of microbial metabolism, host interactions, and their impact on health is no simple task. For instance, microbial metabolites such as short-chain fatty acids (SCFAs), including butyrate, play a critical role in immune regulation and inflammation, and disruptions in microbial communities can lead to diseases like inflammatory bowel disease (IBD) (Sardar et al. 2024; Mallik et al. 2024). Moreover, microbial molecules such as quorum-sensing molecules and tryptophan derivatives influence systemic immunity and the production of inflammatory mediators, adding another layer of complexity to understanding these interactions (Dorrestein et al. 2014). The bidirectional relationship between the microbiome and the immune system, which influences diseases like colitis, further complicates our ability to interpret these interactions accurately (Dorrestein et al. 2014).

9.9 Opportunities for Future Research

9.9.1 Integration of New Technologies

Despite the challenges, the future of metabolomics in microbiome analysis holds great promise. One of the most significant opportunities lies in the integration of new technologies. Advances in high-throughput sequencing technologies and metagenomic approaches have enabled deeper insights into the interactions between microbiota and host physiology. Multi-omics approaches, combining genomics, proteomics, and metabolomics, are particularly promising for uncovering the complex mechanisms underlying host–microbiome interactions (Zhang et al. 2024a, b). These approaches can provide a more comprehensive understanding of diseases like IBD by profiling microbial communities and their metabolites in relation to host responses. As these technologies continue to be improved, they will offer more accurate and detailed insights into microbial functions, disease mechanisms, and therapeutic targets.

9.9.2 Development of Sophisticated Bioinformatics Tools

The development of more sophisticated bioinformatics tools is another key opportunity. Current computational methods, such as network analysis and machine learning models, show promise in deciphering complex microbiome and metabolome data. However, many of these tools are still in the early stages of development and need refinement. Future research should focus on improving algorithms for the analysis of multi-omics data, particularly for identifying meaningful host–microbiome interactions and predicting therapeutic outcomes. Additionally, the integration of metaproteomics and metatranscriptomics with metabolomics will help researchers gain a more holistic understanding of microbiome functions and their effects on the host (Sardar et al. 2024).

9.9.3 Discovering Novel Therapeutic Targets

The potential for discovering novel therapeutic targets is perhaps the most exciting opportunity for future research in metabolomics and microbiome analysis. As our understanding of microbiome-induced metabolic changes and their effects on health deepens, new therapeutic approaches can be developed. For instance, microbial enzymes have been shown to modulate the efficacy of certain drugs, such as irinotecan, a colorectal cancer drug (Ursell et al. 2014). By identifying specific microbial functions that influence drug metabolism, researchers can develop strategies to enhance drug efficacy and minimize side effects. Furthermore, the discovery of new biomarkers and microbial taxa associated with diseases like IBD or NEC could pave the way for personalized medicine approaches tailored to an individual's microbiome profile (Zhang et al. 2024a, b).

9.9.4 Ethical and Practical Considerations

Ethical considerations in microbiome research are paramount, particularly when it comes to personal privacy, informed consent, and the use of microbiome data in personalized medicine. Since microbiome data may reveal sensitive personal information, privacy concerns are heightened, necessitating robust protections to ensure that individuals' microbiome information is not misused. This is especially critical when microbiome data are integrated with genetic or clinical information, as the combination could expose an individual's health risks or predispositions (Rhodes 2016). Additionally, ethical challenges arise in the context of microbiome interventions, such as fecal microbiota transplantation (FMT), which carries risks and requires donor screening and informed consent (Ejtahed et al. 2023). As microbiome research expands, the potential for interventions like genetically modified probiotics or prebiotics adds another layer of complexity, requiring clear regulations and safety standards (McGuinness et al. 2023). Furthermore, the ethical considerations surrounding biobanks, which store microbiome samples for future research, necessitate transparent consent forms that ensure participants understand the risks, ownership, and future use of their samples (Rhodes 2016). These issues emphasize the need for comprehensive ethical guidelines that balance innovation with the protection of individual rights and the responsible advancement of microbiome science. Ethical frameworks must also address the broader societal implications of microbiome-based treatments and interventions, ensuring they are accessible and beneficial to diverse populations without exacerbating existing health disparities (Rhodes 2016).

9.10 Conclusions

In conclusion, metabolomics approaches have revolutionized microbiome research by enabling comprehensive analysis of microbe–metabolite interactions and their impact on health and disease. From its inception, advancements in sequencing technologies and multi-omics integration have transformed our understanding of microbiome dynamics and their functional roles (Whon et al. 2021). Future research must continue to refine computational methods and foster interdisciplinary collaboration, addressing current challenges such as standardization and ethical considerations (Ejtahed et al. 2023). These efforts will pave the way for groundbreaking discoveries, providing robust frameworks for targeted interventions and personalized medicine (McGuinness et al. 2023). By leveraging these developments, metabolomics holds immense potential to unravel complex biological interactions and translate findings into impactful clinical applications (Ursell et al. 2014).

Short Chapter Summary

The human microbiome, comprising diverse microorganisms, plays a crucial role in health, influencing processes like immunity and diseases. This chapter highlights various omics technologies, such as metagenomics and metabolomics, and their role in understanding microbial communities and their metabolic contributions. Challenges in metabolite identification being addressed using AI and thereby improving disease diagnosis are also discussed. Future research perspectives that hold promise in personalized medicine, microbial biomarkers, and integrating multi-omics data for better health outcomes have also been incorporated.

Author Contributions Susanta Roy and Smrita Singh contributed equally.

References

Aguiar-Pulido V, Huang W, Suarez-Ulloa V, Cickovski T, Mathee K, Narasimhan G (2016) Metagenomics, metatranscriptomics, and metabolomics approaches for microbiome analysis. Evol Bioinforma 121:5–16. https://doi.org/10.4137/ebo.s36436

Bauermeister A, Mannochio-Russo H, Costa-Lotufo LV, Jarmusch AK, Dorrestein PC (2021) Mass spectrometry-based metabolomics in microbiome investigations. Nat Rev Microbiol 20:143–160. https://doi.org/10.1038/s41579-021-00621-9

Benaki D, Mikros E (2018) NMR-based metabolic profiling procedures for biofluids and cell and tissue extracts. In: Methods in molecular biology. Springer Science+Business Media, p 117. https://doi.org/10.1007/978-1-4939-7643-0_8

Brown HM, McDaniel TJ, Fedick PW, Mulligan CC (2020) The current role of mass spectrometry in forensics and future prospects. *Anal Methods* 12(32):3974

Bruce C, Stone KL, Gulcicek EE, Williams KR (2013) Proteomics and the analysis of proteomic data: 2013 overview of current protein-profiling technologies. Curr Protoc Bioinformatics 41(1):13.21.1–13.21.17

Bujak JP, Pereira AL, Azevedo J, Bujak AA, Leshyk V, Pham Gia M, Stadtlander T, Vasconcelos V, Winstead DJ (2024) *Azolla* as a safe food: suppression of cyanotoxin-related genes and cyanotoxin production in its symbiont, *Nostoc azollae*. Plants (Basel) 13(19):2707

Cho I, Blaser MJ (2012) The human microbiome: at the interface of health and disease. Nat Rev Genet 13(4):260

Couvillion S, Agrawal N, Colby S, Brandvold K, Metz T (2020) Who is metabolizing what? Discovering novel biomolecules in the microbiome and the organisms who make them. Front Cell Infect Microbiol 10:388. https://doi.org/10.3389/fcimb.2020.00388

Dayrit FM, Dios AC (2017) ^{1}H and ^{13}C NMR for the profiling of natural product extracts: theory and applications. InTech eBooks. https://doi.org/10.5772/intechopen.71040

de Moraes Pontes JG, Brasil AJM, Cruz G, de Souza RN, Tasić L (2017) NMR-based metabolomics strategies: plants, animals and humans. Anal Methods 9(7):1078. https://doi.org/10.1039/c6ay03102a

Dorrestein PC, Mazmanian SK, Knight R (2014) Finding the missing links among metabolites, microbes, and the host. Immunity 40:824–832. https://doi.org/10.1016/j.immuni.2014.05.015

Ebnesajjad S (2011) Surface and material characterization techniques. Elsevier eBooks, 31. Elsevier BV. https://doi.org/10.1016/b978-1-4377-4461-3.10004-5

Ejtahed HS, Parsa M, Larijani B (2023) Ethical challenges in conducting and the clinical application of human microbiome research. J Med Ethics His Med 16:5. https://doi.org/10.18502/jmehm.v16i5.13313

El-Aneed A, Cohen A, Banoub J (2009) Mass spectrometry, review of the basics: electrospray, MALDI, and commonly used mass analyzers. Appl Spectrosc Rev 44(3):210. https://doi.org/10.1080/05704920902717872

Emwas A, Roy R, McKay RT, Tenori L, Saccenti E, Gowda GAN, Raftery D, Alahmari F, Jaremko L, Jaremko M, Wishart DS (2019) NMR spectroscopy for metabolomics research. *Metabolites* 9(7):123. https://doi.org/10.3390/metabo9070123

Escobar-Zepeda A, De León AV-P, Sanchez-Flores A (2015) The road to metagenomics: from microbiology to DNA sequencing technologies and bioinformatics. Front Genet 6:348. https://doi.org/10.3389/fgene.2015.00348

Ganesan R, Gupta H, Gebru YA, Youn GS, Choi YR, Kim HS, Yoon SJ, Kim DJ, Kim T, Suk KT (2021) Recent advances of microbiome-associated metabolomics profiling in liver disease: principles, mechanisms, and applications. Int J Mol Sci 22(3):1160. https://doi.org/10.3390/ijms22031160

Grice EA, Segre JA (2012) The human microbiome: our second genome. Annu Rev Genom Human Genet 13(1):151. https://doi.org/10.1146/annurev-genom-090711-163814

Hattori M, Taylor TD (2009) The human intestinal microbiome: a new frontier of human biology. DNA Res 16(1):1. https://doi.org/10.1093/dnares/dsn033

Johnson CH, Gonzalez FJ (2011) Challenges and opportunities of metabolomics. J Cell Physiol 227:2975–2981. https://doi.org/10.1002/jcp.24002

Kanchan S, Sinha RP, Chaudière J, Kesheri M (2020) Computational metagenomics: current status and challenges. In: Katara P (ed) Recent trends in 'computational omics': concepts and methodology. Nova Science Publishers, pp 371–395

Kanchan S, Kesheri M, Srivastava U, Katharia H, Ratna-Raj R, Chittoori B, Bogomolnaya L, Sinha RP, Denvir J (2024a) Integrative omics data mining: challenges and opportunities. In: Gupta MK, Katara P, Mondal S, Singh RL (eds) Integrative omics: concepts, methodology, and application. Academic Press (Elsevier), pp 237–255. https://doi.org/10.1016/B978-0-44-16092-9.00015-1

Kanchan S, Ogden E, Kesheri M, Skinner A, Miliken E, Lyman D, Armstrong J, Sciglitano L, Hampikian G (2024b) COVID-19 hospitalizations and deaths predicted by SARS-CoV-2 levels in Boise, Idaho wastewater. Sci Total Environ 907:167742. https://doi.org/10.1016/j.scitotenv.2023.167742

Kashyap P, Chia N, Nelson H, Segal E, Elinav E (2017) Microbiome at the frontier of personalized medicine. Mayo Clin Proc 92(12):1855. https://doi.org/10.1016/j.mayocp.2017.10.004

Kesheri M, Kanchan S, Srivastava S, Ratna-Raj R, Chittoori B, Vaishampayan A, Rastogi RP, Sinha RP, Primerano D (2024a) Ecology and environmental omics. In: Gupta MK, Katara P, Mondal S, Singh RL (eds) Integrative omics: concepts, methodology, and application. Academic Press (Elsevier), pp 315–331. https://doi.org/10.1016/B978-0-443-16092-9.00019-9

Kesheri M, Kanchan S, Kumar A, Srivastava U, Sharda S, Malik B, Mishra T, Kaithal P, Narayan J, Kumar P, Priya P, Sinha RP (2024b) Peculiar endosymbiosis in the cyanobiont *Nostoc azollae* 0708: an in-silico approach. In: Kesheri M, Kanchan S, Salisbury TB, Sinha RP (eds) Microbial omics in environment and health. Springer Nature, Singapore, pp 343–365. https://doi.org/10.1007/978-981-97-1769-9_13

Kim BS, Jeon Y-S, Chun J (2013) Current status and future promise of the human microbiome. Pediatr Gastroenterol Hepatol Nutr 16(2):71. https://doi.org/10.5223/pghn.2013.16.2.71

Kruk J, Doskocz M, Jodłowska E, Zacharzewska A, Łakomiec J, Czaja K, Kujawski J (2016) NMR techniques in metabolomic studies: a quick overview on examples of utilization. Appl Magn Reson 48(1):1. https://doi.org/10.1007/s00723-016-0846-9

Kumar P, Kanchan S, Kesheri M (2024) Multi-omics in human disease biomarker discovery. In: Kesheri M, Kanchan S, Salisbury TB, Sinha RP (eds) Microbial omics in environment and health. Springer, Singapore. https://doi.org/10.1007/978-981-97-1769-9_8

Lange L, Berg G,.Cernava T, Champomier-Vergès M-C, Charles T, Cocolin L, Cotter P, D'Hondt K, Kostic T, Maguin E, Makhalanyane T, Meisner A, Ryan M, Kiran GS, De Souza RS, Sanz Y, Schloter M, Smidt H, Wakelin S, Sessitsch A (2022) Microbiome ethics, guiding principles for microbiome research, use and knowledge management. Environ Microbiome 17:50. https://doi.org/10.1186/s40793-022-00444-y

Lee-Sarwar K, Lasky-Su J, Kelly RS, Litonjua AA, Weiss ST (2020) Metabolome–microbiome crosstalk and human disease. Meta 10(5):181. https://doi.org/10.3390/metabo10050181

Mallik B, Mishra T, Dubey P, Kesheri M, Kanchan S (2024) Exploring the secrets of microbes: unveiling the hidden world through microbial omics in environment and health. In: Kesheri M, Kanchan S, Salisbury TB, Sinha RP (eds) Microbial omics in environment and health. Springer, Singapore. https://doi.org/10.1007/978-981-97-1769-9_10

McGuinness AJ, Stinson LF, Snelson M, Loughman A, Stringer A, Hannan AJ, Cowan CSM, Jama HA, Caparros-Martin JA, West ML, Wardill HR (2023) From hype to hope: considerations in conducting robust microbiome science. Brain Behav Immun 115:120–130. https://doi.org/10.1016/j.bbi.2023.09.022

Medhe S (2018) Ionization techniques in mass spectrometry: a review. Mass Spectrom Purif Tech 4(1). https://doi.org/10.4172/2469-9861.1000126

Mishra T, Mallik B, Kesheri M, Kanchan S (2024) The interplay of gut microbiome in health and diseases. In: Kesheri M, Kanchan S, Salisbury TB, Sinha RP (eds) Microbial omics in environment and health. Springer, Singapore. https://doi.org/10.1007/978-981-97-1769-9_1

Navgire GS, Goel N, Sawhney G, Sharma M, Kaushik P, Mohanta YK, Mohanta TK, Al-Harrasi A (2022) Analysis and interpretation of metagenomics data: an approach. Biol Proced Online 24:18. https://doi.org/10.1186/s12575-022-00179-7

Ning L, Zhou Y-L, Sun H, Zhang Y, Shen C, Wang Z, Xuan B, Zhao Y, Ma Y, Yan Y, Tong T, Huang X, Hu M, Zhu X, Ding J, Zhang Y, Cui Z, Fang J-Y, Chen H, Hong J (2023) Microbiome and metabolome features in inflammatory bowel disease via multi-omics integration analyses across cohorts. Nat Commun 14:7135. https://doi.org/10.1038/s41467-023-42788-0

Peters DL, Wang W, Zhang X, Ning Z, Mayne J, Figeys D (2019) Metaproteomic and metabolomic approaches for characterizing the gut microbiome. Proteomics 19(16):e1800363. https://doi.org/10.1002/pmic.201800363

Petriz B, Franco OL (2017) Metaproteomics as a complementary approach to gut microbiota in health and disease. Front Chem 5:4. https://doi.org/10.3389/fchem.2017.00004

Plotnikoff GA, Riley DR (2014) The human microbiome. Glob Adv Health Med 3(3):4. https://doi.org/10.7453/gahmj.2014.023

Proctor LM, Creasy HH, Fettweis JM, Lloyd-Price J, Mahurkar A, Zhou W, Buck GA, Snyder M, Strauss JF, Weinstock GM, White O, Huttenhower C (2019) The integrative human microbiome project. Nature 569(7758):641. https://doi.org/10.1038/s41586-019-1238-8

Provitera L, Tomaselli A, Algieri F, Tripodi M, Raffaeli G, Amodeo I, Raymo L, Bronzoni CV, Fumagalli M, Garrido F, Cavallaro G (2024) Gut microbiota-derived metabolites and their role in the pathogenesis of necrotizing enterocolitis in preterm infants: a narrative review. Meta 14:570. https://doi.org/10.3390/metabo14110570

Puig-Castellví F, Pacheco-Tapia R, Deslande M, Jia M, Andrikopoulos P, Chechi K, Bonnefond A, Froguel P, Dumas M-E (2023) Advances in the integration of metabolomics and metagenomics for human gut microbiome and their clinical applications. Trends Anal Chem 167:117248. https://doi.org/10.1016/j.trac.2023.117248

Rhodes R (2016) Ethical issues in microbiome research and medicine. BMC Med 14:156. https://doi.org/10.1186/s12916-016-0702-7

Sahu N, Mishra S, Kesheri M, Kanchan S, Sinha RP (2023) Identification of cyanobacteria- based natural inhibitors against SARS-CoV-2 druggable target ACE2 using molecular docking study, ADME and toxicity analysis. Indian J Clin Biochem 8(3):361–373. https://doi.org/10.1007/s12291-022-01056-65.8

Sardar P, Almeida A, Pedicord VA (2024) Integrating functional metagenomics to decipher microbiome–immune interactions. Immunol Cell Biol 102:680–691. https://doi.org/10.1111/imcb.12798

Serkova NJ, Davis D, Steiner J, Agarwal R (2019) Quantitative NMR-based metabolomics on tissue biomarkers and its translation into in vivo magnetic resonance spectroscopy. Methods Mol Biol 1978:369–387. https://doi.org/10.1007/978-1-4939-9236-2_23

Shukla N, Bhatia A, Kesheri M, Kanchan S, Narayan J (2024) Planet in peril: unveiling the microbial consequences of environmental degradation on human health through omics approaches. In: Kesheri M, Kanchan S, Salisbury TB, Sinha RP (eds) Microbial omics in environment and health. Springer, Singapore. https://doi.org/10.1007/978-981-97-1769-9_2

Srivastava U, Kesheri M, Kanchan S, Singh S (2024) Computational omics protocol for the comparative study of microbiome analysis. In: Kesheri M, Kanchan S, Salisbury TB, Sinha RP (eds) Microbial omics in environment and health. Springer, Singapore. https://doi.org/10.1007/978-981-97-1769-9_5

Ursell LK, Haiser HJ, Van Treuren W, Garg N, Reddivari L, Vanamala J, Dorrestein PC, Turnbaugh PJ, Knight R (2014) The intestinal metabolome: an intersection between microbiota and host. Gastroenterology 146:1470–1476. https://doi.org/10.1053/j.gastro.2014.03.00

Vandell VE, Limbach PA (2016) Overview of biochemical applications of mass spectrometry. Elsevier BV, p 514

Whon TW, Shin N-R, Kim JY, Roh SW (2021) Omics in gut microbiome analysis. J Microbiol 59:292–297. https://doi.org/10.1007/s12275-021-1004-0
Xie G, Zhang S, Zheng X, Wang J (2013) Metabolomics approaches for characterizing metabolic interactions between host and its commensal microbes. Electrophoresis 34(19):2787. https://doi.org/10.1002/elps.201300017
Yates JR (2001) Mass spectrometry in biology. Encyclopedia of Life Sciences. https://doi.org/10.1038/npg.els.0000999
Zhang S, Gowda GAN, Tao Y, Raftery D (2010) Advances in NMR-based biofluid analysis and metabolite profiling. Analyst 135(7):1490. https://doi.org/10.1039/c000091d
Zhang C, Liu H, Jiang X, Zhang Z, Hou X, Wang Y, Wang D, Li Z, Cao Y, Wu S, Huws SA, Yao J (2024a) An integrated microbiome- and metabolome-genome-wide association study reveals the role of heritable ruminal microbial carbohydrate metabolism in lactation performance in Holstein dairy cows. Microbiome 12:232. https://doi.org/10.1186/s40168-024-01937-3
Zhang Y, Thomas JP, Korcsmaros T, Gul L (2024b) Integrating multi-omics to unravel host-microbiome interactions in inflammatory bowel disease. Cell Rep Med 5:101738. https://doi.org/10.1016/j.xcrm.2024.101738

Brain Microbiome in Health and Disease: A Multi-omics Perspective

10

Bhagaban Mallik, Poornima Dubey, Sajad Bhat, Tarun Mishra, Ashok Kumar Sharma, Swarna Kanchan, and Minu Kesheri

Abstract

The concept of a brain microbiome has emerged as a dynamic area of research, challenging the long-held view of the brain as a sterile organ. This paradigm shift, enabled by sequencing and imaging technologies advancements, has revealed microbial DNA, RNA, and intact microorganisms within brain tissue. Early skepticism about these findings, rooted in the protective nature of the blood-brain barrier and the brain's immune privilege, has given way to evidence suggesting a localized microbiome with potential roles in health and disease. Initial studies focused on identifying these microbes and investigating their origins, proposing mechanisms such as hematogenous dissemination or transloca-

B. Mallik (✉)
Department of Biological Sciences, IISER, Bhopal, Madhya Pradesh, India

Department of Anatomy and Cell Biology, University of Iowa, Iowa City, IA, USA

P. Dubey
Department of Nanotechnology, Indian Institute of Technology, Uttarakhand, India, Roorkee

Department of Biological Engineering, Massachusetts Institute of Technology, Cambridge, MA, USA

S. Bhat
Department of Adult Neurosurgery, Stanford School of Medicine, Stanford, CA, USA

T. Mishra
Department of Biological Sciences, IISER, Bhopal, Madhya Pradesh, India

Department of Microbiology and Immunology, University of Iowa, Iowa City, IA, USA

A. K. Sharma
Research Bioinformatician II, Cedars Sinai Medical Center, Los Angeles, CA, USA

S. Kanchan · M. Kesheri
Department of Biomedical Sciences, Marshall University, Huntington, WV, USA

M. Kesheri et al. (eds.), *Multi-Omics in Biomedical Sciences and Environmental Sustainability*, https://doi.org/10.1007/978-981-96-7067-3_10

tion from the gut via the vagus nerve. Research has uncovered associations between microbial signatures and neurodegenerative diseases, including Alzheimer's, Parkinson's, and multiple sclerosis. As of 2025, the field is evolving to explore the functional implications of brain-resident microorganisms. Emerging studies highlight their potential to influence central nervous system processes, such as immune modulation, neurotransmitter synthesis, and neurogenesis. Evidence also suggests their involvement in neurodevelopment and psychiatric conditions, including depression and schizophrenia. These findings underscore the importance of the brain microbiome in shaping behavior and cognitive functions. Future research aims to unravel these microorganism's origins, mechanisms, and therapeutic potential, promising novel interventions in neurology and psychiatry and redefining our understanding of brain health through a microbial lens.

Keywords

Alzheimer's disease · Brain · Microbiome · Microorganisms · Vagus nerve

Objectives

- Elucidate the existence and composition of the brain microbiome.
- Investigate the functional roles of the brain microbiome.
- Examine the gut-brain axis as a gateway to understanding brain microbial interactions.
- Novel methodologies and future directions in studying the brain microbiome.

10.1 Introduction

The brain microbiome is the concept that microorganisms, including bacteria, fungi, viruses, and other microbial entities, may exist within the brain and influence its function (Mallik et al. 2024a; Mishra et al. 2024). Traditionally, the brain was considered an immune-privileged organ, protected by the blood-brain barrier (BBB) from potential microbial invaders. However, emerging research suggests that the brain might not be completely sterile, and microorganisms or their components could play a role in various neurological and cognitive processes. The significance of the brain microbiome lies in its potential impact on brain health and disease. If microorganisms can influence brain function, they may contribute to neuroinflammation, neurodegeneration, and cognitive disorders like Alzheimer's disease, Parkinson's disease, depression, and anxiety (Galland 2014; Osadchiy et al. 2019; Wang et al. 2023). Understanding this connection could offer novel insights into how the brain communicates with the body and introduce potential therapeutic targets for brain-related conditions. While research in this area is in its early stages, the implications are vast. Just as the gut microbiome has been shown to affect metabolism, immune responses, and overall health, the brain microbiome could directly or indirectly affect neural development, plasticity, and mental health (Lynch and Pedersen 2016; Berding et al. 2021; Zhang et al. 2022). Studying these relationships holds promise for revolutionizing our understanding of brain physiology and pathology.

Moreover, studying microorganisms has long focused on areas like the gut, skin, and oral cavity, where diverse microbial communities thrive (Mallik et al. 2024a; Mishra et al. 2024). The idea of a microbiome existing within the brain was not seriously considered for much of scientific history, mainly because the brain was believed to be isolated from microbes by the blood-brain barrier. This barrier was thought to strictly prevent the entry of pathogens and other foreign substances into the brain (Alahmari 2021). The first key shift in thinking occurred with the discovery of the gut-brain axis. The gut-brain axis (GBA) revealed that gut microbes could communicate with the brain, challenging the idea of complete separation between the microbiota and the central nervous system (CNS) (Lee et al. 2023). However, it wasn't until more recent studies using advanced techniques, such as next-generation sequencing and metagenomics, that the concept of a brain microbiome gained traction. Throughout the twentieth century, advances in microbiology, neuroscience, and immunology set the stage for this new paradigm (Mallik et al. 2022; Mallik et al. 2024b). It became increasingly clear that many neurological conditions involved inflammation or immune responses, suggesting potential interactions between microbial components and the brain. Detecting bacterial DNA in postmortem brain samples in the early twenty-first century ignited the debate on whether these microorganisms represented normal brain flora or merely contaminants. Contemporary studies investigate how microbes might reach the brain through a disrupted blood-brain barrier, immune cells, or olfactory or cranial nerves (Kim 2008). As more evidence accumulates, it is becoming clear that microbes in the brain are not a mere anomaly but may play a functional role in health and disease (Osadchiy et al. 2019). However, definitive conclusions remain elusive, and much research is required to fully understand this brain-associated microbiome's implications.

The gut-brain axis (GBA) is a bidirectional communication system linking the gastrointestinal (GI) tract and the central nervous system. It has been the subject of extensive research, especially concerning how gut microbes affect mental and cognitive health. The GBA integrates neural, hormonal, and immune pathways, enabling the gut microbiota to influence brain function (Osadchiy et al. 2019). Several vital discoveries have established the GBA as a foundational concept for understanding the brain microbiome, such as:

10.2 Neurotransmitter Production by Gut Microbes

Gut bacteria produce neurotransmitters, for example, serotonin, gamma-aminobutyric acid (GABA), and dopamine, all of which regulate mood, behavior, and cognition. Since the brain relies on these chemicals for proper functioning, disruptions in the gut microbiome can lead to depression or anxiety (Chen and Xu 2021; Mallik and Frank 2023).

10.2.1 Microbial Metabolites and Brain Function

Short-chain fatty acids (SCFAs), by-products of gut microbial fermentation, can influence brain function by modulating inflammation and the permeability of the blood-brain barrier. Thus, producing these metabolites provides another link between gut health and cognitive function (Knox et al. 2022).

10.2.2 Immune Modulation and Neuroinflammation

The gut microbiome profoundly impacts the immune system, which, in turn, can influence brain health. Inflammatory cytokines produced in response to microbial activity can cross the BBB or signal the brain through vagal nerve pathways, potentially contributing to neuroinflammation and neurological conditions like multiple sclerosis and Alzheimer's (Rutsch et al. 2020).

10.2.3 The Vagus Nerve

The vagus nerve is a crucial pathway for gut-brain communication. It connects the gut and the brainstem, and gut-derived signals can influence mood, stress responses, and cognition. Animal studies have shown that vagus nerve stimulation can affect behavior, supporting the hypothesis that gut microbes can modulate brain function via neural pathways. With these mechanisms in mind, it is evident that microbes residing outside the brain can affect neurological health. This laid the foundation for the hypothesis that microbes or microbial products within the brain could directly affect brain cells. The gut-brain axis provided a critical precedent: if gut microbes can influence the brain without being physically present, it is plausible that brain-associated microbes might exert even more immediate effects (Bonaz et al. 2018).

Ongoing studies explore whether the brain microbiome communicates with gut microbes and how these relationships affect brain health. The "microbiome-gut-brain axis" suggests a complex interplay between microbial communities in different body parts, with profound implications for neuroscience, mental health, and neurology. Understanding the brain microbiome requires a holistic view of microbial ecosystems throughout the body and their integrated roles in health and disease (Longo et al. 2023).

10.3 Human Microbiome

The human microbiome can be found throughout the body, including the skin, respiratory tract, oral cavity, and urogenital tract. However, the gut microbiome is the most extensively studied and complex microbial environment. The gut, particularly the large intestine, is home to a vast ecosystem of microbes that perform functions ranging from breaking down indigestible fibers to producing vitamins and other essential nutrients.

The human microbiome begins to form at birth, with significant variation in microbial composition depending on delivery mode (vaginal vs. cesarean) and the infant's initial diet (breastfeeding vs. formula). As individuals grow, their microbiomes become more stable but remain dynamic throughout life, influenced by diet, environment, antibiotics, and lifestyle choices. The gut microbiome is the focal point of most research due to its profound influence on many bodily systems, including digestion, metabolism, immune function, and mental health (Davenport et al. 2017; Requena and Velasco 2021). It is now understood that disruptions in the balance of gut microbiota, a condition known as dysbiosis, can lead to various health issues such as gastrointestinal disorders, irritable bowel syndrome (IBS), and chronic diseases like obesity and diabetes (Iebba et al. 2016).

10.4 The Gut-Brain Axis

The gut-brain axis (GBA) represents a bidirectional communication network connecting the gastrointestinal and central nervous systems (CNS). This complex relationship involves neural, hormonal, immune, and microbial signaling pathways that work together to regulate gut and brain functions. The GBA underscores the role of gut health in influencing mental health and neurological processes (Fig. 10.1) (Mayer et al. 2022).

Historically, the gut-brain connection was understood primarily through anatomical pathways like the vagus nerve. However, recent research has uncovered a more intricate system where gut microbiota is essential in regulating brain activity, mood, cognition, and behavior (Goralczyk-Binkowska et al. 2022).

10.5 Regulators of the Gut-Brain Axis: Influences and Modulators

The gut-brain axis (GBA) is a bidirectional communication system between the gastrointestinal tract and the central nervous system, significantly influenced by biological and environmental factors (Fig. 10.2). These regulators shape the axis's functionality, impacting neurodevelopment, mental health, and disease susceptibility (Liu et al. 2022).

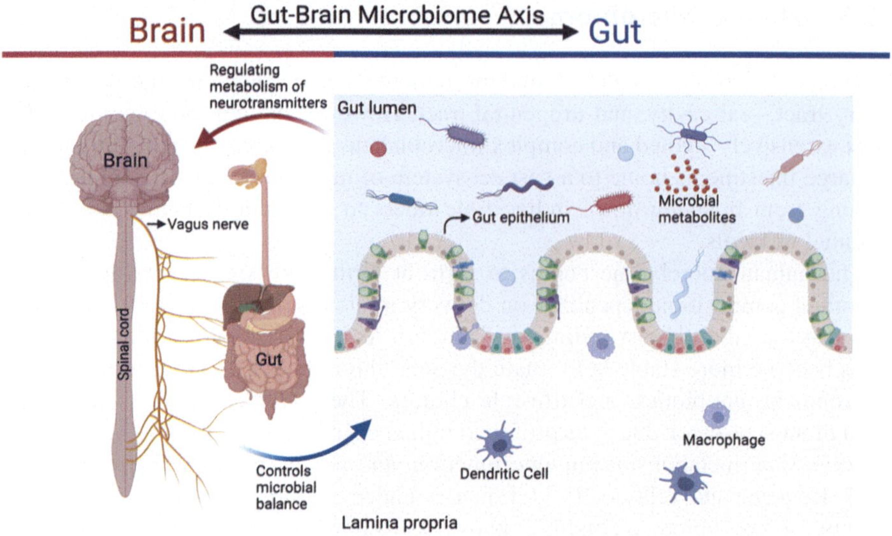

Fig. 10.1 Brain regulation of microbial balance through the gut-brain axis: The central nervous system regulates gut microbial balance via neural, hormonal, and immune pathways. The brain communicates with the gut through the vagus nerve, modulating intestinal motility, secretion, and permeability, directly influencing the gut microbiota composition. Furthermore, the gut influences the production and metabolism of neurotransmitters that support the growth of gut microbiota. Neurotransmitters such as norepinephrine and serotonin modulate microbial growth and activity, maintaining or disrupting microbial homeostasis depending on the signaling context. Note: The figure was made with the help of the BioRender tool

10.5.1 Sex Differences

Hormonal influences cause sex-specific variations in gut microbiota composition and GBA signaling (Caldarelli et al. 2024). Estrogens, for instance, modulate gut microbial diversity and metabolism, affecting neurotransmitter production and inflammatory pathways. These differences contribute to the sex-biased prevalence of disorders like anxiety, depression, and irritable bowel syndrome.

10.5.2 Diet

Nutritional intake profoundly shapes the gut microbiome, a central player in the GBA. Diets rich in fiber promote short-chain fatty acid (SCFA) production, which enhances gut barrier integrity and reduces inflammation. Conversely, high-fat or high-sugar diets can induce dysbiosis, altering gut permeability and influencing mood disorders via increased pro-inflammatory cytokines and altered vagal signaling (van de Wouw et al. 2018).

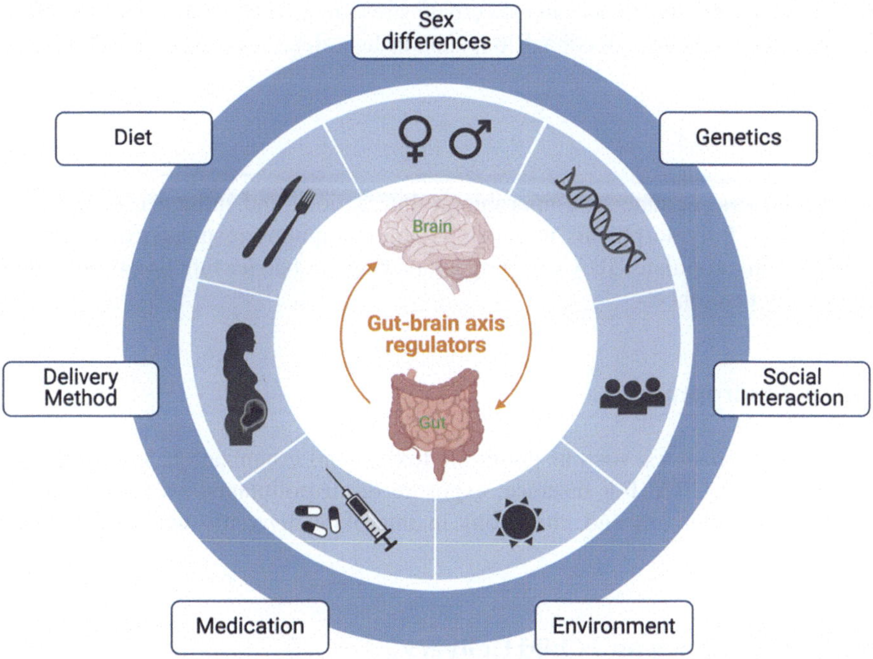

Fig. 10.2 Multifactorial regulation of the gut-brain axis: The gut-brain axis is influenced by a complex interplay of genetic, environmental, and lifestyle factors. Genetic variations affect host-microbiome interactions, modulating gut microbial diversity and function. Sex differences, driven by hormonal fluctuations, impact microbiota composition and neurotransmitter signaling, leading to divergent responses to stress and disease between males and females. Social interactions and chronic stress alter the hypothalamic-pituitary-adrenal (HPA) axis, influencing microbial balance and gut permeability. Environmental exposures, such as pollutants and toxins, disrupt microbial homeostasis, contributing to systemic inflammation and neurobehavioral changes (Cryan et al. 2019). Note: The figure was made with the help of the BioRender tool

10.5.3 Medications

Antibiotics, proton pump inhibitors, and psychotropic drugs significantly alter gut microbial communities, sometimes leading to long-term dysbiosis. Antibiotics, for example, can disrupt microbial diversity, impairing the microbiome's ability to produce neurotransmitters such as serotonin. Psychotropic medications, like selective serotonin reuptake inhibitors (SSRIs), may have direct and indirect effects on the microbiota, influencing GBA communication.

10.5.4 Social Interaction

Social environments influence GBA via stress pathways. Chronic social stress alters the hypothalamic-pituitary-adrenal (HPA) axis, disrupting gut microbiota composition. This dysbiosis can exacerbate stress-related behaviors and inflammatory responses, creating a feedback loop that further affects mental health (Zhu 2023).

10.5.5 Genetics

Genetic factors regulate host-microbiome interactions and influence microbiome composition. Polymorphisms in genes related to immune responses or metabolic pathways can modulate GBA signaling by altering gut barrier function or microbial community resilience.

10.5.6 Environment

Environmental factors, such as pollution, toxins, and urbanization, impact the gut microbiota and GBA. For instance, exposure to air pollutants has been linked to systemic inflammation and alterations in microbial diversity, which can impair brain function.

10.5.7 Pregnancy and Child Delivery

Maternal microbiota significantly influences the neonatal gut microbiome, particularly during vaginal delivery, where the infant is exposed to maternal vaginal and gut flora. Cesarean sections can delay gut colonization, which may alter GBA development and increase risks for immune and neurodevelopmental disorders. Moreover, pregnancy stress influences maternal and neonatal gut microbiota, shaping the child's neuroimmune development.

These regulators collectively underscore the dynamic and integrative nature of the gut-brain axis, emphasizing the need for a multifactorial approach to understanding and managing its influence on human health.

10.6 Overview of the Bidirectional Communication Between the Gut and the Brain

The gut and the brain communicate via several pathways:

10.6.1 Neural Pathways

The vagus nerve is the main highway for communication between the gut and the brain. It allows gut signals to be transmitted to the brainstem and other regions of the CNS, impacting mood, stress responses, and cognitive function (Bonaz et al. 2018).

10.6.2 Endocrine Pathways

The hypothalamic-pituitary-adrenal (HPA) axis is a major endocrine pathway connecting the gut and brain. The HPA axis responds to stress, and dysregulation of this axis can alter gut motility, permeability, and immune response. Stress-induced changes in gut physiology can also affect the microbiome and brain function.

10.6.3 Immune Pathways

The immune system plays a pivotal role in gut-brain communication. Inflammatory cytokines produced in response to dysbiosis or infection can cross the blood-brain barrier, leading to neuroinflammation. Gut microbes also regulate systemic immune responses, influencing brain health by controlling inflammation levels.

10.6.4 Microbial Metabolites

Metabolites such as SCFAs produced by gut bacteria directly affect the brain, which can cross the blood-brain barrier and influence neuroinflammation, mood, and cognitive function (van de Wouw et al. 2018). Other microbial products, such as lipopolysaccharides (LPS), can trigger immune responses that impact the CNS (Fig. 10.3; Table 10.1).

10.7 Role of the Vagus Nerve, Immune System, and Endocrine Pathways

10.7.1 Vagus Nerve

The vagus nerve is the most direct link between the gut and the brain. This nerve relays signals from the gut to the brainstem, influencing emotional responses, cognitive processes, and even pain perception. For instance, animal studies have demonstrated that altering the gut microbiota can modulate vagus nerve activity and affect mood and behavior. Vagal nerve stimulation is being explored as a therapeutic option for conditions such as depression, epilepsy, and inflammatory bowel disease (Van Hinsbergh et al. 1986; Gonzalez et al. 2019; Claudino Dos Santos et al. 2023).

10.7.2 Immune System

Gut microbiota regulates the immune system by promoting the production of regulatory T-cells and anti-inflammatory cytokines and maintaining the integrity of the gut barrier. A dysregulated immune response, often resulting from dysbiosis, can lead to increased intestinal permeability (frequently called "leaky gut"), allowing

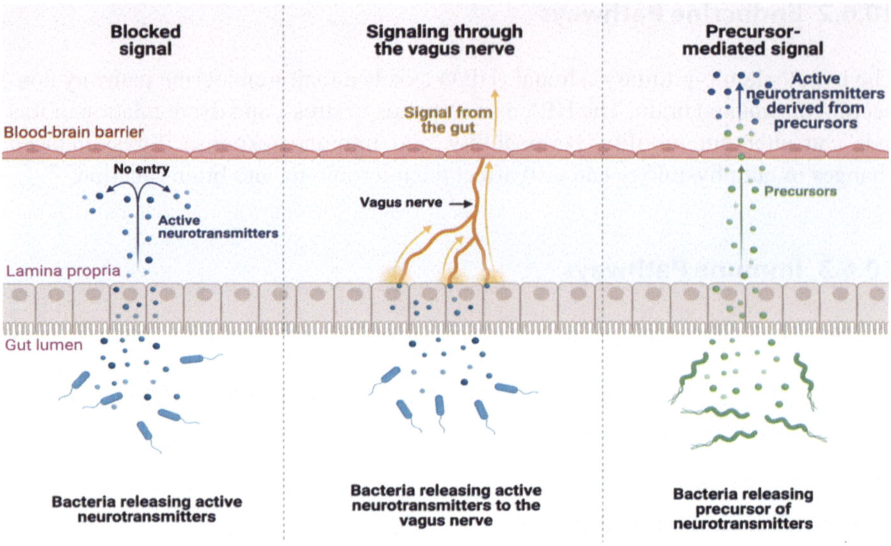

Fig. 10.3 Microbial production of neurotransmitters in the gut lumen and their influence on the central nervous system: Neurotransmitters produced by gut microorganisms, such as serotonin, dopamine, and gamma-aminobutyric acid (GABA), represent a potential mechanism for influencing brain function and behavior. Although most neurotransmitters cannot directly cross the blood-brain barrier, alternative pathways facilitate their impact. For instance, neurotransmitters can signal through the vagus nerve's afferent neurons, linking gut-derived signals to central nervous system activity. Additionally, microbial production of neurotransmitter precursors, such as tryptophan, enables their transport across the blood-brain barrier, where they are converted into active neurotransmitters. The availability of tryptophan is particularly significant, as blood plasma levels correlate with serotonergic signaling in the brain. This highlights the gut microbiota's role in modulating brain chemistry and behavior through direct and indirect mechanisms (Johnson and Foster 2018). The figure was made with the help of the BioRender tool

bacterial products like LPS to enter the bloodstream and cause systemic inflammation (Giri et al. 2022; Mrazek et al. 2024). Chronic inflammation is associated with neurodegenerative diseases and mental health disorders, highlighting the importance of gut health in maintaining brain health.

10.7.3 Endocrine Pathways

The HPA axis is activated in response to stress, and gut microbes can influence this pathway by regulating cortisol levels. Dysbiosis or stress-induced alterations in gut bacteria can lead to increased production of stress hormones, affecting mood, anxiety, and cognitive function. Conversely, gut microbiota can modulate the HPA axis by producing metabolites that reduce stress and improve mental health.

Table 10.1 Role of microbiota in nervous system function

Gut Bacteria	Microbial-derived neuroactive metabolites	Types of nerve involved	Effects on the central nervous system and behavior
Lactobacillus reuteri	Increased production of oxytocin	Vagus nerve	Increases neuronal plasticity, social behavior and oxytocin levels
Lactobacillus rhamnosus	Increased production of GABA	Vagus nerve	Decreases anxiety and depressive-like behavior, and stress responsiveness
Bifidobacterium longum NCC3001	Increased production of BDNF	Vagus nerve	Decreases anxiety-like and depressive behavior and excitability ENS neurons
Bacteroides fragilis	Decreased production of 4-EPS	Not known	Decreases anxiety-like behavior, and repetitive behavior and enhances communication
SCFA-producing bacteria	Increased production of SCFA	Not known	Decreases stress responsiveness, anxiety, and depressive-like behavior

10.8 Evidence of Gut Microbiota Influencing Brain Function and Behavior

A growing body of research supports the idea that the gut microbiome has profound effects on brain function and behavior:

10.8.1 Mood Disorders

Earlier studies have shown that patients with depression or anxiety often exhibit altered gut microbiota compositions compared to healthy individuals. Animal studies using germ-free mice (lacking any microbiota) have revealed that these mice show increased anxiety-like behaviors, which can be normalized by introducing a healthy microbiome (Dinan and Cryan 2017).

10.8.2 Cognitive Function

There is emerging evidence that the gut microbiota can influence cognitive processes such as learning, memory, and executive function. For example, probiotic interventions in animals and humans have improved cognitive performance, suggesting a role for gut microbes in brain plasticity (Kolobaric et al. 2024).

10.8.3 Autism Spectrum Disorders (ASD)

Research has found a link between gut dysbiosis and the severity of symptoms in children with ASD. Studies suggest that interventions aimed at restoring gut microbiota balance could mitigate behavioral and cognitive impairments. The gut-brain axis exemplifies the complexity of microbiome interactions, illustrating how a healthy microbiome is crucial for maintaining gut health and mental and neurological well-being (Alharthi et al. 2022).

Table 10.1. Role of microbiota and microbial metabolites in shaping host behavior and nervous system dynamics

Microorganisms and their metabolites play critical roles in gut-brain signaling, influencing host behavior and nervous system function through multiple pathways. Microbial-derived neuroactive compounds, such as γ-aminobutyric acid (GABA) and short-chain fatty acids (SCFAs), signal to the brain via the vagus nerve or by modulating the immune system. Examples include *Lactobacillus reuteri*, which enhances oxytocin levels and social behavior in mice, and *Lactobacillus rhamnosus*, which produces GABA and regulates brain GABA receptors to reduce anxiety-like behavior in mice. *Bifidobacterium longum NCC3001* upregulates brain-derived neurotrophic factor (BDNF), improving mood and neuronal plasticity in the enteric nervous system (ENS) and the brain. In contrast, *Bacteroides fragilis* alleviates anxiety-like behaviors by reducing 4-ethylphenylsulfate (4-EPS). Additionally, SCFAs regulate microglia maturation and morphology, illustrating the diverse mechanisms by which microbiota influence the gut-brain axis (Morais et al. 2021).

10.9 The Concept of the Brain Microbiome

10.9.1 Is there a Brain Microbiome?

The brain microbiome includes bacteria, viruses, fungi, and other microbes that may reside in the brain and influence its function. This idea challenges the long-standing belief that the brain is a sterile, immune-privileged organ, isolated from the rest of the body by the blood-brain barrier (BBB). For decades, the prevailing view has been that microbial communities were confined to specific body compartments, such as the gut, skin, and respiratory tract. At the same time, the central nervous system (CNS) remained protected from such exposure. However, recent advancements in microbial detection techniques have sparked interest in the possibility of a brain-associated microbiome.

This emerging field has gained attention because of the potential implications for brain health and disease. If microorganisms can colonize the brain or their metabolic by-products reach the brain, they could contribute to neuroinflammation, neurodegeneration, and various psychiatric and neurological conditions (Link 2021; Arabi et al. 2023). However, the existence of a brain microbiome remains highly controversial, and the scientific community is divided on whether the microorganisms found in brain tissues are functionally significant residents or merely transient contaminants.

10.9.2 Debate and Skepticism: Does the Brain Have Its Microbiome?

The debate over whether the brain harbors a microbiome centers around several key issues. The brain is protected from microbial invasion by the blood-brain barrier. The BBB is a highly selective membrane that separates circulating blood from the brain's extracellular fluid, preventing harmful pathogens and toxins from entering the CNS. This structure led many scientists to believe the brain should remain sterile without microbial inhabitants. However, some researchers now argue that this view might be outdated.

Discovering microbial DNA and other microbial signatures in brain tissues has fueled the debate. Detractors of the brain microbiome concept suggest that detecting microbial components in brain tissue samples could result from contamination during sample collection or processing rather than evidence of a resident brain microbiome. The highly sensitive nature of modern molecular techniques, such as next-generation sequencing, means that even trace amounts of contaminating DNA can be detected. Critics also point out that there is currently limited evidence of live, actively replicating microorganisms within brain tissues.

Moreover, the immune system's role within the CNS presents another layer of complexity. While the brain is considered an immune-privileged organ, there is growing recognition that immune cells, particularly microglia, play a crucial role in regulating brain homeostasis and responding to infection or injury. Some researchers speculate that microorganisms in the brain would likely trigger an immune response, resulting in inflammation or tissue damage. The absence of widespread neuroinflammation in healthy brains is often cited as evidence against the existence of a significant brain microbiome.

Despite these concerns, the growing body of evidence suggesting microbial presence in the brain has prompted further investigation. Proponents of the brain microbiome hypothesis argue that low levels of microbial colonization, potentially with beneficial functions, may exist without triggering a robust immune response (Link 2021). Alternatively, microbial components may reach the brain without the microorganisms colonizing the CNS.

10.9.3 Recent Discoveries: Detection of Microbial DNA and Other Evidence in Brain Tissues

Recent studies have used advanced molecular techniques, such as metagenomics, to investigate the possible presence of microbes in the brain. A study led by Dr. Rosalinda Roberts (2018) detected bacterial DNA in postmortem brain samples of both healthy individuals and patients with schizophrenia. The bacteria were found primarily in perivascular spaces and along astrocytic foot processes, suggesting a potential interaction with the blood-brain barrier. The researchers hypothesized that these bacteria might not be random contaminants but could be functional in brain physiology or pathology.

Another study published in Frontiers in Aging Neuroscience in 2018 reported finding bacterial DNA in the brains of patients with Alzheimer's disease, suggesting a possible link between microbial presence and neurodegenerative processes (Alonso et al. 2018). These findings have raised questions about whether microbial colonization could contribute to or exacerbate neuroinflammation, amyloid plaque formation, or cognitive decline.

While the presence of microbial DNA is intriguing, it remains unclear whether these microbial components represent live microorganisms, degraded cellular debris, or transient products transported to the brain from other body sites, such as the gut. The lack of culturable bacteria from brain tissue samples further complicates the picture, raising questions about whether these microbes are alive and active in the brain environment.

One theory posits that bacterial products, such as cell wall components or metabolites, could travel to the brain through compromised barriers, such as the BBB, without active colonization. These microbial by-products could then interact with the brain's immune cells or neurons, potentially influencing brain function or triggering pathological processes. This idea aligns with findings that gut-derived metabolites, such as short-chain fatty acids (SCFAs), can affect brain function through the gut-brain axis, suggesting a more indirect role for microbes in the CNS (van de Wouw et al. 2018).

10.10 Potential Origins of Microbes in the Brain

10.10.1 Blood-Brain Barrier Breaches

The BBB is the primary defense mechanism that protects the brain from harmful pathogens in the bloodstream. However, under certain conditions, the integrity of the BBB can be compromised, allowing microbes or microbial products to enter the brain. Infections, inflammation, or diseases like multiple sclerosis can disrupt the tight junctions of the BBB, providing a pathway for microorganisms to cross into the CNS (Kermode et al. 1990; Kanda 2003). Once inside, these microbes could interact with neurons, glial cells, or immune cells, potentially contributing to neuroinflammation or neurodegeneration.

10.10.2 Neuroinvasion via the Olfactory Nerve

The olfactory nerve, which connects the nasal cavity to the brain, is another potential route for microbial entry. Several pathogens, including viruses and bacteria, travel along the olfactory nerve, bypassing the BBB and gaining direct access to the brain (Kristensson et al. 2013). This pathway is exciting because it potentially links respiratory infections, nasal colonization, and brain health. For example, the herpes simplex virus (HSV) has been shown to invade the CNS via the olfactory nerve, and some researchers hypothesize that other microbes could also use this route (Mori et al. 2006).

10.10.3 Translocation from the Gut

The gut-brain axis provides another possible route for microbial components to reach the brain. The gut and brain are closely connected through neural, endocrine, and immune pathways and gut health disruptions can significantly affect brain function. Dysbiosis, or imbalances in gut microbial communities, can lead to increased intestinal permeability (often referred to as "leaky gut"), allowing microbial products such as lipopolysaccharides (LPS) to enter the bloodstream. Once in circulation, these products could cross a weakened BBB and influence brain function (Hu et al. 2022). There is also evidence that gut-derived microbial metabolites, such as SCFAs, can cross the BBB and modulate brain processes. However, this does not necessarily require the physical presence of live microbes in the brain.

10.10.4 Peripheral Infection

Peripheral infections, such as those in the respiratory or urinary tracts, can trigger systemic inflammation that affects the brain. This is sometimes referred to as the "systemic inflammation hypothesis." In this scenario, cytokines and other immune signals produced in response to peripheral infections can travel to the brain, which may induce neuroinflammatory responses (Westhoff et al. 2019). Additionally, microbial products from peripheral infections can potentially cross the BBB if it is compromised by inflammation or disease.

10.11 Brain Microbiome and Neurological Health

Several studies have indicated that the brain microbiome could potentially affect neurological health in various ways, such as:

10.11.1 Microbiome's Role in Neurodevelopment

The microbiome plays a significant role in neurodevelopment, particularly in early life when the brain is most plastic. During this critical period, the gut microbiome and its metabolites influence the maturation of neural circuits, synaptic pruning, and neurogenesis—processes essential for normal brain function (Borre et al. 2014).

10.11.2 Early Life Exposure and Brain Development

The microbial environment that infants are exposed to from birth is crucial for proper brain development. Studies on germ-free animals (those raised without exposure to microorganisms) have shown significant deficits in brain structure and function, including altered synaptic density and reduced expression of neurotrophic factors like brain-derived neurotrophic factor (BDNF). These animals also exhibit behavioral changes such as heightened anxiety and impaired social interactions, highlighting the role of the microbiome in modulating stress responses and social behavior. Early-life disruptions in the microbiome, such as cesarean delivery, antibiotic use, or formula feeding, can lead to long-lasting alterations in the gut-brain axis. These changes are associated with cognitive deficits, increased susceptibility to neuro-developmental disorders, and altered stress responses in adulthood. The microbiome's influence on the hypothalamic-pituitary-adrenal (HPA) axis is significant during early development, as it shapes the brain's stress reactivity and emotional regulation (Ratsika et al. 2021).

10.11.3 Influence on Neural Circuitry and Neurogenesis

Gut bacteria produce metabolites, such as short-chain fatty acids (SCFAs), with neuroactive properties. SCFAs can cross the blood-brain barrier and regulate neuroinflammatory pathways, impacting neurogenesis and synaptic plasticity. Microglia are also influenced by gut-derived signals and play a key role in reorganizing neural circuits during development (Sarubbo et al. 2022). Dysregulation of microglial activity due to altered gut microbiota can impair synaptic pruning, leading to cognitive and behavioral dysfunction. The microbiome also influences neurotransmitter systems. For instance, gut microbes can produce gamma-aminobutyric acid (GABA) and serotonin, which are critical for mood regulation, cognitive function, and neural plasticity (Mallik et al. 2017; Guzzetta et al. 2022; Mallik and Frank 2022; Mallik et al. 2023). These findings suggest that the gut microbiome contributes to establishing neural circuitry that governs mood, cognition, and social behavior.

10.11.4 Implications for Neurodegenerative Diseases

Mounting evidence suggests that the gut microbiome is linked to the development of neurodegenerative diseases, such as Alzheimer's disease, Parkinson's disease, and multiple sclerosis (MS). The gut-brain axis is pivotal in modulating neuroinflammatory processes and other pathological mechanisms in these conditions.

10.11.5 Alzheimer's Disease

Alzheimer's disease (AD) is characterized by the accumulation of amyloid-beta plaques and tau tangles in the brain, along with chronic neuroinflammation. Research indicates that gut microbiota may contribute to these pathological processes. Studies have shown that alterations in the gut microbiome—such as an increased abundance of pro-inflammatory bacteria—are associated with higher levels of circulating lipopolysaccharides (LPS) and other microbial components. These molecules can cross the blood-brain barrier and activate microglia, leading to a neuroinflammatory response that accelerates amyloid-beta aggregation and tau phosphorylation. SCFAs produced by beneficial gut bacteria have been shown to have neuroprotective effects by reducing inflammation and enhancing microglial clearance of amyloid-beta. Conversely, dysbiosis may impair these protective pathways, contributing to AD progression. Gut-targeted therapies, including probiotics and fecal microbiota transplantation (FMT), are being explored as potential interventions to modulate neuroinflammation and amyloid pathology in AD (Jiang et al. 2017).

10.11.6 Parkinson's Disease

Parkinson's disease (PD) involves the progressive loss of dopaminergic neurons in the substantia nigra, leading to motor dysfunction. Emerging evidence suggests that the gut microbiome may play a role in the pathogenesis of PD by producing neurotoxins or by triggering systemic inflammation. The presence of alpha-synuclein aggregates, a hallmark of PD, has been detected in patients' enteric nervous system (ENS), raising the possibility that PD may originate in the gut and spread to the brain via the vagus nerve (Salim et al. 2023).

Dysbiosis in PD is associated with increased intestinal permeability ("leaky gut") and elevated levels of pro-inflammatory cytokines, which may contribute to neurodegeneration. SCFAs have also been implicated in PD pathogenesis; while these molecules are generally neuroprotective, a decrease in SCFA-producing bacteria has been observed in PD patients. This disruption may impair the gut-brain communication necessary for maintaining neuronal integrity (Tan et al. 2014).

10.11.7 Multiple Sclerosis

Multiple sclerosis (MS) is an autoimmune disease characterized by demyelination and chronic neuroinflammation in the CNS. Recent studies suggest that the gut microbiome may influence the onset and progression of MS by modulating the immune response. Dysbiosis in MS patients is associated with an imbalance in regulatory and pro-inflammatory immune cells, including T helper 17 (Th17) cells, which are known to promote Gut bacteria influence the differentiation of these immune cells, and altered gut microbiota can exacerbate immune-mediated damage in the CNS. Restoring gut microbial balance through diet, probiotics, or FMT may represent a novel therapeutic strategy for modulating immune responses and slowing disease progression in MS (Ordonez-Rodriguez et al. 2023).

10.11.8 Impact on Mental Health

The gut microbiome has also been implicated in mental health disorders, including depression, anxiety, and schizophrenia. Increasing evidence links gut dysbiosis to neuropsychiatric conditions, with microbial imbalance contributing to neuroinflammation, altered neurotransmitter systems, and HPA axis dysregulation (McGinnis 1988).

10.11.9 Depression and Anxiety

Dysbiosis is commonly observed in patients with major depressive disorder (MDD) and anxiety. Gut microbiota influences the production of key neurotransmitters, including serotonin and GABA, which are critical for mood regulation. Approximately 90% of the body's serotonin is produced in the gut, and gut microbes play a direct role in this process. Imbalances in microbial populations can lead to altered levels of these neurotransmitters, contributing to mood dysregulation and anxiety. Neuroinflammation is another critical mechanism linking the gut microbiome to depression and anxiety (Peirce and Alvina 2019). Increased intestinal permeability, often seen in dysbiosis, allows microbial products like LPS to enter the bloodstream, triggering systemic inflammation. Pro-inflammatory cytokines, such as IL-6 and TNF-α, can cross the BBB and activate microglia, leading to neuroinflammation and disruption of neuronal circuits involved in mood regulation. This inflammatory response is thought to contribute to the pathophysiology of depression and anxiety.

10.12 Mechanisms of Interaction Between the Brain and Microbiome

The gut-brain axis represents a complex communication network between the central nervous system (CNS) and the gut microbiota. Multiple mechanisms mediate this interaction, including immune system modulation and neurotransmitter production. Understanding these mechanisms provides insight into how the gut microbiome influences brain function and contributes to neurological and psychiatric disorders.

10.12.1 Immune System Interactions

The immune system is a central player in the gut-brain axis, with the gut microbiota significantly influencing immune responses that affect brain health. These immune interactions primarily involve microglial activation and the production of cytokines and chemokines, which serve as molecular messengers between the gut and brain.

10.12.2 Microglia Activation and Neuron-Inflammation

Microglia, the resident immune cells of the CNS, are critically involved in maintaining neural homeostasis and responding to injury or infection. Under normal conditions, microglia function in synaptic pruning, neurogenesis, and repair processes. However, signals from the gut microbiota can influence microglial activation, particularly in the context of dysbiosis, an imbalance in the gut microbial composition. For example, bacterial components like lipopolysaccharides (LPS) from Gram-negative bacteria can activate microglia through toll-like receptor 4 (TLR4) signaling, triggering a pro-inflammatory response in the brain (Yu et al. 2023).

Chronic activation of microglia, often seen in the presence of dysbiosis, contributes to neuroinflammation, a hallmark of many neurodegenerative diseases, including Alzheimer's disease and Parkinson's disease. Neuroinflammation, driven by activated microglia, can exacerbate neuronal damage and lead to the progression of these diseases. The gut microbiota, therefore, plays a pivotal role in promoting neuroprotection or contributing to neurodegeneration, depending on its composition and pro-inflammatory microbial components (Kim et al. 2023).

10.12.3 Cytokines and Chemokines as Mediators

Cytokines and chemokines are critical signaling molecules mediating communication between the immune system and the CNS. The gut microbiota influences the systemic production of these immune mediators, which can cross the blood-brain barrier (BBB) or affect brain endothelial cells to modulate neuroinflammatory processes. In a healthy gut, beneficial microbes stimulate the production of anti-inflammatory cytokines such as IL-10 and TGF-β, which help to maintain microglial quiescence and prevent excessive inflammation. However, dysbiosis increases the production of pro-inflammatory cytokines such as IL-6, IL-1β, and TNF-α. These cytokines can enter the CNS through the BBB or interact with peripheral nerves, initiating neuroinflammatory events (Schirmer et al. 2016). This chronic, low-grade inflammation in the brain, termed neuroinflammation, is a contributing factor to various CNS disorders, including neurodegeneration and psychiatric diseases. By modulating the production of cytokines and chemokines, the gut microbiota significantly influences brain health and disease progression (Sun et al. 2022).

10.12.4 Neurotransmitter Production

The gut microbiota directly produces and regulates neurotransmitters essential for brain function. These neurotransmitters, produced by gut bacteria and the host, include serotonin, gamma-aminobutyric acid (GABA), and dopamine which play crucial roles in mood, cognition, and behavior (O'Mahony et al. 2015; Mittal et al. 2017).

10.12.5 Microbiota-Produced Neurotransmitters

The gut microbiota can synthesize various neurotransmitters, including serotonin (5-HT), GABA, and dopamine. Although primarily produced in the gut, these neurotransmitters can influence brain function through multiple pathways, including the enteric nervous system (ENS), vagus nerve, and circulation (Chen and Xu 2021).

10.12.6 Serotonin (5-HT)

Nearly 90% of the body's serotonin is produced in the gut, specifically by enterochromaffin cells regulated by gut bacteria. Certain microbial species, such as *Escherichia coli* and *Lactobacillus rhamnosus*, have been shown to influence serotonin synthesis. Serotonin plays a crucial role in mood regulation, anxiety, and depression, and alterations in gut microbiota composition can impact the levels of serotonin available for CNS function (O'Mahony et al. 2015).

10.12.7 GABA

GABA is the primary inhibitory neurotransmitter in the brain, and gut microbes also influence its production. *Bifidobacterium* and *Lactobacillus* produce GABA, which can affect brain function by modulating neural circuits involved in anxiety and mood regulation. Dysbiosis can lead to imbalances in GABA levels, potentially contributing to mood disorders such as anxiety and depression (Bravo et al. 2011).

10.12.8 Dopamine

Dopamine is another neurotransmitter whose production can be influenced by gut microbiota. Although dopamine produced in the gut does not cross the BBB, gut bacteria can modulate its precursors and metabolites, indirectly affecting brain dopamine levels. Altered dopamine signaling is implicated in several psychiatric disorders, including schizophrenia and depression. The gut microbiota directly impacts brain function and behavior by producing these neurotransmitters (Hamamah et al. 2022). Disruptions in microbial populations can lead to imbalances in neurotransmitter production, contributing to the pathophysiology of mental health disorders.

10.13 Role of Mitochondria and Energy Sources in the Brain Microbiome

Mitochondria, often called the cell's powerhouses, are essential for maintaining energy homeostasis and supporting neuronal function in the brain (Cheng et al. 2022). Their role in the brain extends beyond energy production, integrating metabolic and signaling pathways critical for cellular communication and survival. In the context of the brain microbiome, mitochondria and microbial metabolites converge to influence brain metabolism, neuronal health, and overall cognitive function.

Neurons are highly energy-dependent, relying predominantly on oxidative phosphorylation (OXPHOS) within mitochondria to meet their energy demands (Brand et al. 2013). Disruptions in mitochondrial function can compromise ATP production, impairing neuronal signaling, synaptic plasticity, neuronal homeostasis, and axonal transport (Mallik et al. 2023). In parallel, the brain microbiome contributes to energy balance by producing metabolites such as short-chain fatty acids (SCFAs) and lactate, which can serve as alternative energy sources for neurons. SCFAs, including butyrate, acetate, and propionate, are derived from the fermentation of dietary fibers by gut microbiota and have been shown to enhance mitochondrial efficiency and regulate neuroinflammatory processes.

Mitochondrial dysfunction is also intricately linked to neurodegenerative diseases, where disruptions in energy metabolism exacerbate pathological processes such as amyloid-beta aggregation in Alzheimer's disease or α-synuclein accumulation in Parkinson's disease (Klemmensen et al. 2024). Emerging evidence suggests that gut and brain microbiota dysbiosis may exacerbate these mitochondrial impairments by promoting systemic inflammation, oxidative stress, and altered nutrient availability. For instance, increased gut lining permeability in dysbiosis allows inflammatory cytokines and microbial toxins, such as lipopolysaccharides (LPS), to reach the brain, disrupting mitochondrial dynamics and exacerbating neuronal damage.

Furthermore, mitochondria are central to regulating cellular redox states, critical for maintaining the delicate balance of excitatory and inhibitory neurotransmitter signaling. Microbiota-derived metabolites influence this balance by modulating mitochondrial reactive oxygen species (ROS) production. Excessive ROS can lead to oxidative stress, damaging mitochondrial DNA and proteins, while controlled ROS signaling is essential for normal synaptic transmission and neurogenesis (Mallik and Frank 2024). The interplay between mitochondria and the brain microbiome also extends to neuroimmune interactions. Microbial metabolites regulate microglial activation and maturation, processes heavily dependent on mitochondrial function. Dysregulated microglial activity resulting from mitochondrial or microbial dysfunction can drive neuroinflammation, a hallmark of numerous neuropsychiatric and neurodegenerative conditions.

In summary, mitochondria are a critical nexus linking energy metabolism, microbial activity, and brain function (Qiao et al. 2024).Their interactions with the brain microbiome underscore the importance of maintaining metabolic and microbial homeostasis to support neuronal health and prevent disease. Understanding these dynamics offers potential therapeutic avenues for targeting mitochondrial dysfunction and microbiota imbalances in brain-related disorders.

10.14 Techniques in Studying the Brain Microbiome

Although relatively new, the emerging field of the brain microbiome has rapidly gained traction due to advancements in sequencing technologies and experimental models. Below is an overview of the current research methodologies used in studying the brain microbiome, including sequencing technologies, imaging techniques, and animal models.

10.14.1 Advances in Microbiome Sequencing and Analysis

One of the most significant breakthroughs in microbiome research has been the development of high-throughput sequencing technologies, particularly next-generation sequencing (NGS). By identifying bacterial DNA or RNA sequences, these technologies have enabled researchers to profile complex microbial communities in various tissues, including the gut and potentially the brain (Salzberg et al. 2016; Suchodolski 2022).

10.14.2 16S rRNA Gene Sequencing

This is the most commonly used technique for identifying bacteria based on the highly conserved 16S ribosomal RNA gene (Kurilshikov et al. 2021). Specific variable regions of the 16S rRNA gene are amplified and sequenced to classify bacteria to the genus or species level. This technique has been instrumental in identifying bacterial communities in low-biomass samples, including brain tissues. However, it is limited by its inability to detect nonbacterial microbes such as viruses and fungi.

10.14.3 Metagenomics

Metagenomic sequencing allows for a more comprehensive analysis of all genetic material present in a sample. This approach identifies bacteria and detects viruses, fungi, and archaea. By sequencing the entire microbial genome, metagenomics provides insights into the functional potential of microbial communities. This technique has been increasingly used in brain microbiome research to investigate whether microbial DNA exists in healthy or diseased brain tissues (Liu et al. 2024).

10.14.4 Metatranscriptomics and Metaproteomics

While metagenomics provides a snapshot of the microbial genetic landscape, metatranscriptomics (sequencing microbial RNA) and metaproteomics (analyzing microbial proteins) offer insights into the activity and functional roles of the microbiota. These advanced techniques can reveal whether microbes in the brain are metabolically active, producing specific proteins or signaling molecules that could influence brain function (Ramirez-Acosta et al. 2021). These methods, however, require empathetic tools, given the low biomass of microbial presence in the brain, and are still in the early stages of being applied to brain microbiome studies.

10.14.5 Imaging Techniques for Observing Microbial Activity in the Brain

In addition to sequencing technologies, imaging techniques have been developed to visualize microbial presence and activity in brain tissues. These methods provide direct evidence of microbial existence in the brain and allow researchers to study potential microbial interactions with brain cells (Sekhar et al. 2019).

10.14.6 Fluorescence In Situ Hybridization (FISH)

FISH is a technique that uses fluorescent probes targeting specific microbial DNA or RNA sequences to visualize microorganisms within tissues. In the context of brain microbiome research, FISH has been used to detect bacterial cells in brain tissue sections, helping to confirm the presence of microbes in the CNS. Although FISH can provide spatial resolution and identify specific microbes, it requires particular probes and optimized protocols for low-biomass environments like the brain (Mirza et al. 2020).

10.14.7 Electron Microscopy

Electron microscopy offers high-resolution imaging that can detect individual microbial cells. This method has been used to detect potential bacterial or viral particles in brain tissues. However, electron microscopy alone does not provide specific identification of the organisms and must be used alongside molecular techniques to confirm microbial identity (Wang et al. 2020).

10.14.8 Optical Imaging and MRI

While not typically used to visualize individual microbes, advanced imaging modalities such as magnetic resonance imaging (MRI) and optical imaging techniques can track inflammation and immune responses in the brain, which could be influenced by microbial presence. Researchers are investigating how these imaging techniques can be integrated with molecular methods to observe how microbial activity might affect neural function (Campisciano and Biffi 2022).

10.15 Future Research Directions

Future research directions in brain microbiome studies include creating a comprehensive map of the brain microbiome and developing microbiome-based biomarkers for diagnosing and monitoring neurological diseases.

10.15.1 Mapping the Brain Microbiome

A key objective for future research is to establish a detailed map of microbial populations in the brain, including their composition, localization, and function. Advances in sequencing and imaging technologies will be crucial for identifying the specific microbes present in different brain regions and understanding their roles in health and disease. Such mapping efforts could reveal whether certain microbial communities are associated with specific neurological conditions and how they interact with brain cells, including neurons, astrocytes, and microglia (Hulme et al. 2022).

10.15.2 Microbiome-Based Biomarkers

Another promising area of research is the development of microbiome-based biomarkers for brain diseases (Mishra et al. 2018; Ananthakrishnan 2020; Chen et al. 2023; Gopal et al. 2024). These biomarkers could be used to diagnose neurological conditions at earlier stages, monitor disease progression, or predict responses to treatment. For example, changes in gut microbiota composition or microbial metabolites in the blood or cerebrospinal fluid (CSF) could serve as non-invasive indicators of neuroinflammation or neurodegeneration. By identifying specific microbial signatures associated with brain disorders, researchers hope to develop diagnostic tools that are more accurate and less invasive than current methods.

Brain microbiome research holds significant therapeutic potential, with approaches like probiotics, prebiotics, and FMT showing promise for treating neurological disorders. However, challenges remain in translating these therapies to clinical practice, particularly regarding safety, complexity, and ethical considerations. Future research aims to address open questions about the brain microbiome's role in disease, establish comprehensive maps of microbial populations in the brain, and develop biomarkers that could revolutionize the diagnosis and treatment of neurological diseases. With continued advancements in research methodologies, the brain microbiome may play an increasingly important role in our understanding and treatment of brain health in the years to come.

10.16 Conclusion

The brain microbiome represents a groundbreaking area of neuroscience, challenging the long-held belief in the brain's sterility. Over the past decade, research has revealed profound connections between microbial communities, particularly in the gut, and brain function. While the existence of a distinct brain microbiome remains under investigation, evidence suggests that microbes may influence brain health via immune modulation, neurotransmitter production, and metabolic processes.

The gut-brain axis exemplifies this interplay, with the gut microbiome regulating neurotransmitters like serotonin and GABA and modulating immune signaling that crosses the blood-brain barrier. These discoveries have spurred interest in microbiome-based therapies for conditions like Alzheimer's, Parkinson's, depression, and anxiety.

Despite these advancements, distinguishing causality from correlation remains a challenge. Sophisticated research tools are essential to unraveling the complex, dynamic interactions between microbes and the brain, which offer transformative potential for neuroscience.

Short Chapter Summary

The brain microbiome represents a transformative frontier in neuroscience, challenging the notion of the brain as sterile. Recent studies reveal that microbial communities, especially in the gut, influence brain function through immune modulation, neurotransmitter production, and metabolism, impacting neurodevelopment, mental health, and neurodegenerative diseases. The gut-brain axis highlights these connections, though causality remains unclear. Advanced techniques are vital to unravel these dynamics, offering groundbreaking insights and novel therapies for brain disorders.

Acknowledgments We are deeply grateful to all those who contributed to the completion of this book chapter. First and foremost, we thank Dr. Swarna Kanchan and Dr. Minu Kesheri for their invaluable insights, guidance, and encouragement throughout this process. We also acknowledge the collaborative spirit of our colleagues, whose constructive feedback and discussions have enriched the content and perspective presented here. Thanks to our families and friends for their unwavering patience and support during the countless hours dedicated to this endeavor. Finally, we thank the editors and publishers for their meticulous review and for providing the platform to share this work with the broader academic community. This chapter reflects collective efforts, and we remain indebted to everyone who contributed, directly or indirectly, to its realization.

Authors Contributions Chapter concept and design, Bhagaban Mallik, Poornima Dubey, and Sajad Bhat; chapter writing, Bhagaban Mallik, Poornima Dubey, Sajad Bhat, and Tarun Mishra; chapter editing and review, Bhagaban Mallik, Poornima Dubey, Sajad Bhat, Ashok Kumar Sharma, Swarna Kanchan, and Minu Kesheri.

References

Alahmari A (2021) Blood-brain barrier overview: structural and functional correlation. Neural Plast 2021:6564585. https://doi.org/10.1155/2021/6564585

Alharthi A, Alhazmi S, Alburae N, Bahieldin A (2022) The human gut microbiome as a potential factor in autism spectrum disorder. Int J Mol Sci 23(3):1363. https://doi.org/10.3390/ijms23031363

Alonso R, Pisa D, Fernandez-Fernandez AM, Carrasco L (2018) Infection of fungi and bacteria in brain tissue from elderly persons and patients with alzheimer's disease. Front Aging Neurosci 10:159. https://doi.org/10.3389/fnagi.2018.00159

Ananthakrishnan AN (2020) Microbiome-based biomarkers for IBD. Inflamm Bowel Dis 26(10):1463–1469. https://doi.org/10.1093/ibd/izaa071

Arabi TZ, Alabdulqader AA, Sabbah BN, Ouban A (2023) Brain-inhabiting bacteria and neurodegenerative diseases: the "brain microbiome" theory. Front Aging Neurosci 15:1240945. https://doi.org/10.3389/fnagi.2023.1240945

Berding K, Vlckova K, Marx W, Schellekens H, Stanton C, Clarke G, Jacka F, Dinan TG, Cryan JF (2021) Diet and the microbiota-gut-brain axis: sowing the seeds of good mental health. Adv Nutr 12(4):1239–1285. https://doi.org/10.1093/advances/nmaa181

Bonaz B, Bazin T, Pellissier S (2018) The vagus nerve at the interface of the microbiota-gut-brain axis. Front Neurosci 12:49. https://doi.org/10.3389/fnins.2018.00049

Borre YE, O'Keeffe GW, Clarke G, Stanton C, Dinan TG, Cryan JF (2014) Microbiota and neurodevelopmental windows: implications for brain disorders. Trends Mol Med 20(9):509–518. https://doi.org/10.1016/j.molmed.2014.05.002

Brand MD, Orr AL, Perevoshchikova IV, Quinlan CL (2013) The role of mitochondrial function and cellular bioenergetics in ageing and disease. Br J Dermatol 169(Suppl 2):1–8. https://doi.org/10.1111/bjd.12208

Bravo JA, Forsythe P, Chew MV, Escaravage E, Savignac HM, Dinan TG, Bienenstock J, Cryan JF (2011) Ingestion of *Lactobacillus* strain regulates emotional behavior and central GABA receptor expression in a mouse via the vagus nerve. Proc Natl Acad Sci USA 108(38):16050–16055. https://doi.org/10.1073/pnas.1102999108

Caldarelli M, Rio P, Marrone A, Ocarino F, Chiantore M, Candelli M, Gasbarrini A, Gambassi G, Cianci R (2024) Gut-brain axis: focus on sex differences in neuroinflammation. Int J Mol Sci 25(10):5377. https://doi.org/10.3390/ijms25105377

Campisciano G, Biffi S (2022) Microbiota in vivo imaging approaches to study host-microbe interactions in preclinical and clinical setting. Heliyon 8(12):e12511. https://doi.org/10.1016/j.heliyon.2022.e12511

Chen Y, Xu J (2021) Regulation of neurotransmitters by the gut microbiota and effects on cognition in neurological disorders. Nutrients 13(6):2099. https://doi.org/10.3390/nu13062099

Chen J, Siliceo SL, Ni Y, Nielsen HB, Xu A, Panagiotou G (2023) Identification of robust and generalizable biomarkers for microbiome-based stratification in lifestyle interventions. Microbiome 11(1):178. https://doi.org/10.1186/s40168-023-01604-z

Cheng XT, Huang N, Sheng ZH (2022) Programming axonal mitochondrial maintenance and bioenergetics in neurodegeneration and regeneration. Neuron 110(12):1899–1923. https://doi.org/10.1016/j.neuron.2022.03.015

Claudino Dos Santos JC, Oliveira LF, Noleto FM, Gusmao CTP, Brito GAC, Viana GSB (2023) Gut-microbiome-brain axis: the crosstalk between the vagus nerve, alpha-synuclein and the brain in Parkinson's disease. Neural Regen Res 18(12):2611–2614. https://doi.org/10.4103/1673-5374.373673

Cryan JF, O'Riordan KJ, Cowan CSM, Sandhu KV, Bastiaanssen TFS, Boehme M, Codagnone MG, Cussotto S, Fulling C, Golubeva AV, Guzzetta KE, Jaggar M, Long-Smith CM, Lyte JM, Martin JA, Molinero-Perez A, Moloney G, Morelli E, Morillas E, O'Connor R, Cruz-Pereira JS, Peterson VL, Rea K, Ritz NL, Sherwin E, Spichak S, Teichman EM, van de Wouw M,

Ventura-Silva AP, Wallace-Fitzsimons SE, Hyland N, Clarke G, Dinan TG (2019) The microbiota-gut-brain axis. Physiol Rev 99(4):1877–2013. https://doi.org/10.1152/physrev.00018.2018
Davenport ER, Sanders JG, Song SJ, Amato KR, Clark AG, Knight R (2017) The human microbiome in evolution. BMC Biol 15(1):127. https://doi.org/10.1186/s12915-017-0454-7
Dinan TG, Cryan JF (2017) The microbiome-gut-brain axis in health and disease. Gastroenterol Clin N Am 46(1):77–89. https://doi.org/10.1016/j.gtc.2016.09.007
Galland L (2014) The gut microbiome and the brain. J Med Food 17(12):1261–1272. https://doi.org/10.1089/jmf.2014.7000
Giri J, Basu M, Roy S, Mishra T, Jana K, Chande A, Ukil A (2022) Translationally controlled tumor protein-mediated stabilization of host antiapoptotic protein MCL-1 is critical for establishment of infection by intramacrophage parasite *Leishmania donovani*. J Immunol 208(11):2540–2548. https://doi.org/10.4049/jimmunol.2100748
Gonzalez HFJ, Yengo-Kahn A, Englot DJ (2019) Vagus nerve stimulation for the treatment of epilepsy. Neurosurg Clin N Am 30(2):219–230. https://doi.org/10.1016/j.nec.2018.12.005
Gopal RK, Sankar Ganesh P, Pathoor NN, Viswanathan A (2024) Enhancing microbiome-based biomarkers: challenges and opportunities. Lancet Microbe 6:101032. https://doi.org/10.1016/j.lanmic.2024.101032
Goralczyk-Binkowska A, Szmajda-Krygier D, Kozlowska E (2022) The microbiota-gut-brain axis in psychiatric disorders. Int J Mol Sci 23(19):11245. https://doi.org/10.3390/ijms231911245
Guzzetta KE, Cryan JF, O'Leary OF (2022) Microbiota-gut-brain axis regulation of adult hippocampal neurogenesis. Brain Plast 8(1):97–119. https://doi.org/10.3233/BPL-220141
Hamamah S, Aghazarian A, Nazaryan A, Hajnal A, Covasa M (2022) Role of microbiota-gut-brain axis in regulating dopaminergic signaling. Biomedicines 10(2):436. https://doi.org/10.3390/biomedicines10020436
Hu W, Kong X, Wang H, Li Y, Luo Y (2022) Ischemic stroke and intestinal flora: an insight into brain-gut axis. Eur J Med Res 27(1):73. https://doi.org/10.1186/s40001-022-00691-2
Hulme H, Meikle LM, Strittmatter N, Swales J, Hamm G, Brown SL, Milling S, MacDonald AS, Goodwin RJA, Burchmore R, Wall DM (2022) Mapping the influence of the gut microbiota on small molecules across the microbiome gut brain Axis. J Am Soc Mass Spectrom 33(4):649–659. https://doi.org/10.1021/jasms.1c00298
Iebba V, Totino V, Gagliardi A, Santangelo F, Cacciotti F, Trancassini M, Mancini C, Cicerone C, Corazziari E, Pantanella F, Schippa S (2016) Eubiosis and dysbiosis: the two sides of the microbiota. New Microbiol 39(1):1–12
Jiang C, Li G, Huang P, Liu Z, Zhao B (2017) The gut microbiota and alzheimer's disease. J Alzheimers Dis 58(1):1–15. https://doi.org/10.3233/JAD-161141
Johnson KV, Foster KR (2018) Why does the microbiome affect behaviour? Nat Rev Microbiol 16(10):647–655. https://doi.org/10.1038/s41579-018-0014-3
Kanda T (2003) Blood-brain barrier in multiple sclerosis: mechanisms of its breakdown and repair. Nihon Rinsho 61(8):1402–1408
Kermode AG, Tofts PS, Thompson AJ, MacManus DG, Rudge P, Kendall BE, Kingsley DP, Moseley IF, du Boulay EP, McDonald WI (1990) Heterogeneity of blood-brain barrier changes in multiple sclerosis: an MRI study with gadolinium-DTPA enhancement. Neurology 40(2):229–235. https://doi.org/10.1212/wnl.40.2.229
Kim KS (2008) Mechanisms of microbial traversal of the blood-brain barrier. Nat Rev Microbiol 6(8):625–634. https://doi.org/10.1038/nrmicro1952
Kim CS, Jung S, Hwang GS, Shin DM (2023) Gut microbiota indole-3-propionic acid mediates neuroprotective effect of probiotic consumption in healthy elderly: a randomized, double-blind, placebo-controlled, multicenter trial and in vitro study. Clin Nutr 42(6):1025–1033. https://doi.org/10.1016/j.clnu.2023.04.001
Klemmensen MM, Borrowman SH, Pearce C, Pyles B, Chandra B (2024) Mitochondrial dysfunction in neurodegenerative disorders. Neurotherapeutics 21(1):e00292. https://doi.org/10.1016/j.neurot.2023.10.002

Knox EG, Aburto MR, Tessier C, Nagpal J, Clarke G, O'Driscoll CM, Cryan JF (2022) Microbial-derived metabolites induce actin cytoskeletal rearrangement and protect blood-brain barrier function. iScience 25(12):105648. https://doi.org/10.1016/j.isci.2022.105648

Kolobaric A, Andreescu C, Jasarevic E, Hong CH, Roh HW, Cheong JY, Kim YK, Shin TS, Kang CS, Kwon CO, Yoon SY, Hong SW, Aizenstein HJ, Karim HT, Son SJ (2024) Gut microbiome predicts cognitive function and depressive symptoms in late life. Mol Psychiatry 29(10):3064–3075. https://doi.org/10.1038/s41380-024-02551-3

Kristensson K, Masocha W, Bentivoglio M (2013) Mechanisms of CNS invasion and damage by parasites. Handb Clin Neurol 114:11–22. https://doi.org/10.1016/B978-0-444-53490-3.00002-9

Kurilshikov A, Medina-Gomez C, Bacigalupe R, Radjabzadeh D, Wang J, Demirkan A, Le Roy CI, Raygoza Garay JA, Finnicum CT, Liu X, Zhernakova DV, Bonder MJ, Hansen TH, Frost F, Ruhlemann MC, Turpin W, Moon JY, Kim HN, Lull K, Barkan E, Shah SA, Fornage M, Szopinska-Tokov J, Wallen ZD, Borisevich D, Agreus L, Andreasson A, Bang C, Bedrani L, Bell JT, Bisgaard H, Boehnke M, Boomsma DI, Burk RD, Claringbould A, Croitoru K, Davies GE, van Duijn CM, Duijts L, Falony G, Fu J, van der Graaf A, Hansen T, Homuth G, Hughes DA, Ijzerman RG, Jackson MA, Jaddoe VWV, Joossens M, Jorgensen T, Keszthelyi D, Knight R, Laakso M, Laudes M, Launer LJ, Lieb W, Lusis AJ, Masclee AAM, Moll HA, Mujagic Z, Qibin Q, Rothschild D, Shin H, Sorensen SJ, Steves CJ, Thorsen J, Timpson NJ, Tito RY, Vieira-Silva S, Volker U, Volzke H, Vosa U, Wade KH, Walter S, Watanabe K, Weiss S, Weiss FU, Weissbrod O, Westra HJ, Willemsen G, Payami H, Jonkers D, Arias Vasquez A, de Geus EJC, Meyer KA, Stokholm J, Segal E, Org E, Wijmenga C, Kim HL, Kaplan RC, Spector TD, Uitterlinden AG, Rivadeneira F, Franke A, Lerch MM, Franke L, Sanna S, D'Amato M, Pedersen O, Paterson AD, Kraaij R, Raes J, Zhernakova A (2021) Large-scale association analyses identify host factors influencing human gut microbiome composition. Nat Genet 53(2):156–165. https://doi.org/10.1038/s41588-020-00763-1

Lee A, Lee JY, Jung SW, Shin SY, Ryu HS, Jang SH, Kwon JG, Kim YS (2023) Brain-gut-microbiota axis. Korean J Gastroenterol 81(4):145–153. https://doi.org/10.4166/kjg.2023.028

Link CD (2021) Is there a brain microbiome? Neurosci Insights 16:26331055211018709. https://doi.org/10.1177/26331055211018709

Liu L, Huh JR, Shah K (2022) Microbiota and the gut-brain-axis: implications for new therapeutic design in the CNS. EBioMedicine 77:103908. https://doi.org/10.1016/j.ebiom.2022.103908

Liu X, Liu Y, Liu J, Zhang H, Shan C, Guo Y, Gong X, Cui M, Li X, Tang M (2024) Correlation between the gut microbiome and neurodegenerative diseases: a review of metagenomics evidence. Neural Regen Res 19(4):833–845. https://doi.org/10.4103/1673-5374.382223

Longo S, Rizza S, Federici M (2023) Microbiota-gut-brain axis: relationships among the vagus nerve, gut microbiota, obesity, and diabetes. Acta Diabetol 60(8):1007–1017. https://doi.org/10.1007/s00592-023-02088-x

Lynch SV, Pedersen O (2016) The human intestinal microbiome in health and disease. N Engl J Med 375(24):2369–2379. https://doi.org/10.1056/NEJMra1600266

Mallik B, Frank CA (2022) Roles for mitochondrial complex I subunits in regulating synaptic transmission and growth. Front Neurosci 16:846425. https://doi.org/10.3389/fnins.2022.846425

Mallik B, Frank CA (2023) Is creatine a CNS neurotransmitter? elife 12:e91824. https://doi.org/10.7554/eLife.91824

Mallik B, Frank CA (2024) Mitochondrial complex I and ROS control synapse function through opposing pre- and postsynaptic mechanisms. bioRxiv:2024.2012.2030.630694. https://doi.org/10.1101/2024.12.30.630694

Mallik B, Dwivedi MK, Mushtaq Z, Kumari M, Verma PK, Kumar V (2017) Regulation of neuromuscular junction organization by Rab2 and its effector ICA69 in *Drosophila*. Development 144(11):2032–2044. https://doi.org/10.1242/dev.145920

Mallik B, Bhat S, Kumar V (2022) Role of bin-amphiphysin-rvs (BAR) domain proteins in mediating neuronal signaling and disease. Synapse 76(11–12):e22248. https://doi.org/10.1002/syn.22248

Mallik B, Brusich DJ, Heyrman G, Frank CA (2023) Precise mapping of one classic and three novel GluRIIA mutants in *Drosophila melanogaster*. MicroPubl Biol 2023. https://doi.org/10.17912/micropub.biology.000784

Mallik B, Mishra T, Dubey P, Kesheri M, Kanchan S (2024a) Exploring the secrets of microbes: unveiling the hidden world through microbial omics in environment and health. In: Kesheri M, Kanchan S, Salisbury TB, Sinha RP (eds) Microbial omics in environment and health. Springer Nature, Singapore, pp 269–294. https://doi.org/10.1007/978-981-97-1769-9_10

Mallik B, Pippadpally S, Bisht A, Bhat S, Mukherjee S, Kumar V (2024b) Distinct bin/Amphiphysin/Rvs (BAR) family proteins may assemble on the same tubule to regulate membrane organization in vivo. Heliyon 10(13):e33672. https://doi.org/10.1016/j.heliyon.2024.e33672

Mayer EA, Nance K, Chen S (2022) The gut-brain axis. Annu Rev Med 73:439–453. https://doi.org/10.1146/annurev-med-042320-014032

McGinnis JM (1988) Public health objectives for the year 2000. Am Pharm NS28(3):46–51. https://doi.org/10.1016/s0160-3450(15)32002-x

Mirza A, Forbes JD, Zhu F, Bernstein CN, Van Domselaar G, Graham M, Waubant E, Tremlett H (2020) The multiple sclerosis gut microbiota: a systematic review. Mult Scler Relat Disord 37:101427. https://doi.org/10.1016/j.msard.2019.101427

Mishra R, Mushtaq Z, Regar R, Mallik B, Kumar V, Sankar J (2018) Selective imaging of lipids in adipocytes by using an imidazolyl derivative of perylene bisimide. Chembiochem 19(13):1386–1390. https://doi.org/10.1002/cbic.201800134

Mishra T, Mallik B, Kesheri M, Kanchan S (2024) The interplay of gut microbiome in health and diseases. In: Kesheri M, Kanchan S, Salisbury TB, Sinha RP (eds) Microbial omics in environment and health. Springer Nature, Singapore, pp 1–34. https://doi.org/10.1007/978-981-97-1769-9_1

Mittal R, Debs LH, Patel AP, Nguyen D, Patel K, O'Connor G, Grati M, Mittal J, Yan D, Eshraghi AA, Deo SK, Daunert S, Liu XZ (2017) Neurotransmitters: the critical modulators regulating gut-brain axis. J Cell Physiol 232(9):2359–2372. https://doi.org/10.1002/jcp.25518

Morais LH, Schreiber HL, Mazmanian SK (2021) The gut microbiota-brain axis in behaviour and brain disorders. Nat Rev Microbiol 19(4):241–255. https://doi.org/10.1038/s41579-020-00460-0

Mori I, Goshima F, Watanabe D, Ito H, Koide N, Yoshida T, Liu B, Kimura Y, Yokochi T, Nishiyama Y (2006) Herpes simplex virus US3 protein kinase regulates virus-induced apoptosis in olfactory and vomeronasal chemosensory neurons in vivo. Microbes Infect 8(7):1806–1812. https://doi.org/10.1016/j.micinf.2006.02.018

Mrazek J, Mrazkova L, Mekadim C, Jarosikova T, Krayem I, Sohrabi Y, Demant P, Lipoldova M (2024) Effects of *Leishmania* major infection on the gut microbiome of resistant and susceptible mice. Appl Microbiol Biotechnol 108(1):145. https://doi.org/10.1007/s00253-024-13002-y

O'Mahony SM, Clarke G, Borre YE, Dinan TG, Cryan JF (2015) Serotonin, tryptophan metabolism and the brain-gut-microbiome axis. Behav Brain Res 277:32–48. https://doi.org/10.1016/j.bbr.2014.07.027

Ordonez-Rodriguez A, Roman P, Rueda-Ruzafa L, Campos-Rios A, Cardona D (2023) Changes in gut microbiota and multiple sclerosis: a systematic review. Int J Environ Res Public Health 20(5):4624. https://doi.org/10.3390/ijerph20054624

Osadchiy V, Martin CR, Mayer EA (2019) The gut-brain axis and the microbiome: mechanisms and clinical implications. Clin Gastroenterol Hepatol 17(2):322–332. https://doi.org/10.1016/j.cgh.2018.10.002

Peirce JM, Alvina K (2019) The role of inflammation and the gut microbiome in depression and anxiety. J Neurosci Res 97(10):1223–1241. https://doi.org/10.1002/jnr.24476

Qiao L, Yang G, Wang P, Xu C (2024) The potential role of mitochondria in the microbiota-gut-brain axis: implications for brain health. Pharmacol Res 209:107434. https://doi.org/10.1016/j.phrs.2024.107434

Ramirez-Acosta S, Arias-Borrego A, Navarro-Roldan F, Selma-Royo M, Calatayud M, Collado MC, Huertas-Abril PV, Abril N, Barrera TG (2021) Omic methodologies for assessing metal(-loid)s-host-microbiota interplay: a review. Anal Chim Acta 1176:338620. https://doi.org/10.1016/j.aca.2021.338620

Ratsika A, Codagnone MC, O'Mahony S, Stanton C, Cryan JF (2021) Priming for life: early life nutrition and the microbiota-gut-brain axis. Nutrients 13(2):423. https://doi.org/10.3390/nu13020423

Requena T, Velasco M (2021) The human microbiome in sickness and in health. Rev Clin Esp (Barc) 221(4):233–240. https://doi.org/10.1016/j.rceng.2019.07.018

Rutsch A, Kantsjo JB, Ronchi F (2020) The gut-brain axis: how microbiota and host inflammasome influence brain physiology and pathology. Front Immunol 11:604179. https://doi.org/10.3389/fimmu.2020.604179

Salim S, Ahmad F, Banu A, Mohammad F (2023) Gut microbiome and Parkinson's disease: perspective on pathogenesis and treatment. J Adv Res 50:83–105. https://doi.org/10.1016/j.jare.2022.10.013

Salzberg SL, Breitwieser FP, Kumar A, Hao H, Burger P, Rodriguez FJ, Lim M, Quinones-Hinojosa A, Gallia GL, Tornheim JA, Melia MT, Sears CL, Pardo CA (2016) Next-generation sequencing in neuropathologic diagnosis of infections of the nervous system. Neurol Neuroimmunol Neuroinflamm 3(4):e251. https://doi.org/10.1212/NXI.0000000000000251

Sarubbo F, Cavallucci V, Pani G (2022) The influence of gut microbiota on neurogenesis: evidence and hopes. Cells 11(3):382. https://doi.org/10.3390/cells11030382

Schirmer M, Smeekens SP, Vlamakis H, Jaeger M, Oosting M, Franzosa EA, Ter Horst R, Jansen T, Jacobs L, Bonder MJ, Kurilshikov A, Fu J, Joosten LAB, Zhernakova A, Huttenhower C, Wijmenga C, Netea MG, Xavier RJ (2016) Linking the human gut microbiome to inflammatory cytokine production capacity. Cell 167(4):1125–1136. e1128. https://doi.org/10.1016/j.cell.2016.10.020

Sekhar AR, Mallik B, Kumar V, Sankar J (2019) A cell-permeant small molecule for the super-resolution imaging of the endoplasmic reticulum in live cells. Org Biomol Chem 17(15):3732–3736. https://doi.org/10.1039/c9ob00408d

Suchodolski JS (2022) Analysis of the gut microbiome in dogs and cats. Vet Clin Pathol 50(Suppl 1):6–17. https://doi.org/10.1111/vcp.13031

Sun J, Zhang Y, Kong Y, Ye T, Yu Q, Kumaran Satyanarayanan S, Su KP, Liu J (2022) Microbiota-derived metabolite indoles induced aryl hydrocarbon receptor activation and inhibited neuroinflammation in APP/PS1 mice. Brain Behav Immun 106:76–88. https://doi.org/10.1016/j.bbi.2022.08.003

Tan J, McKenzie C, Potamitis M, Thorburn AN, Mackay CR, Macia L (2014) The role of short-chain fatty acids in health and disease. Adv Immunol 121:91–119. https://doi.org/10.1016/B978-0-12-800100-4.00003-9

van de Wouw M, Boehme M, Lyte JM, Wiley N, Strain C, O'Sullivan O, Clarke G, Stanton C, Dinan TG, Cryan JF (2018) Short-chain fatty acids: microbial metabolites that alleviate stress-induced brain-gut axis alterations. J Physiol 596(20):4923–4944. https://doi.org/10.1113/JP276431

van Hinsbergh VW, Mommaas-Kienhuis AM, Weinstein R, Maciag T (1986) Propagation and morphologic phenotypes of human umbilical cord artery endothelial cells. Eur J Cell Biol 42(1):101–110

Wang Y, An Y, Ma W, Yu H, Lu Y, Zhang X, Liu W, Wang T, Xiao R (2020) 27-hydroxycholesterol contributes to cognitive deficits in APP/PS1 transgenic mice through microbiota dysbiosis and intestinal barrier dysfunction. J Neuroinflammation 17(1):199. https://doi.org/10.1186/s12974-020-01873-7

Wang Q, Yang Q, Liu X (2023) The microbiota-gut-brain axis and neurodevelopmental disorders. Protein Cell 14(10):762–775. https://doi.org/10.1093/procel/pwad026

Westhoff D, Engelen-Lee JY, Hoogland ICM, Aronica EMA, van Westerloo DJ, van de Beek D, van Gool WA (2019) Systemic infection and microglia activation: a prospective postmortem study in sepsis patients. Immun Ageing 16:18. https://doi.org/10.1186/s12979-019-0158-7

Yu Z, Chen W, Zhang L, Chen Y, Meng S, Lu L, Han Y, Shi J (2023) Gut-derived bacterial LPS attenuates incubation of methamphetamine craving via modulating microglia. Brain Behav Immun 111:101–115. https://doi.org/10.1016/j.bbi.2023.03.027

Zhang F, Aschenbrenner D, Yoo JY, Zuo T (2022) The gut mycobiome in health, disease, and clinical applications in association with the gut bacterial microbiome assembly. Lancet Microbe 3(12):e969–e983. https://doi.org/10.1016/S2666-5247(22)00203-8

Zhu L (2023) Editorial: animal social behaviour and gut microbiome. Front Microbiol 14:1210717. https://doi.org/10.3389/fmicb.2023.1210717

Microbiome Dynamics in Colorectal Cancer: Implications for Personalized Treatment and Multi-Omics Analysis

11

Chandandeep Kaur, Minu Kesheri, and Swarna Kanchan

Abstract

Colorectal cancer (CRC) is a complex disease influenced by various genetic, epigenetic, and environmental factors. The integration of multi-omics approaches, which include genomics, transcriptomics, proteomics, and metabolomics, has significantly advanced our understanding of CRC development, progression, and the molecular landscape of CRC tumors, leading to the identification of potential biomarkers and therapeutic targets. One of the key areas of interest in CRC research is the role of the gut microbiome in modulating tumor development, maintaining gut homeostasis, immune function, and response to treatment. An imbalance in the gut microbiome has been linked to various diseases, including CRC, and modulating the gut microbiome could be used as a novel therapeutic approach for CRC. By targeting specific microbes or microbial pathways, researchers aim to enhance the antitumor immune response and inhibit tumor growth. By integrating data, researchers can identify microbial signatures associated with CRC development and progression. These signatures may serve as biomarkers for early detection or prognosis, as well as potential targets for therapy. This chapter illuminates the basic concepts pertaining to CRC with a focus on elaborating various multi-omics approaches that are instrumental in elucidating the complex interactions between the gut microbiome and CRC.

C. Kaur (✉)
Happiest Health, Bangalore, India

M. Kesheri · S. Kanchan
Boise State University, Boise, ID, USA

Joan C. Edwards School of Medicine, Marshall University, Huntington, WV, USA

M. Kesheri et al. (eds.), *Multi-Omics in Biomedical Sciences and Environmental Sustainability*, https://doi.org/10.1007/978-981-96-7067-3_11

Keywords

Cancer treatment response · Chemotherapy · Colorectal Cancer · Intratumor microbiome · Multi-omics

Objectives

- To understand the symptoms, early detection, genetic factors, and tumor development of colorectal cancer
- To explore the diagnostics and treatment challenges of colorectal cancer
- To understand the role of intratumor microbiomes in cancer development and therapeutics
- To explore the computational framework for multi-omics studies and implications in cancer multi-omics studies

11.1 Introduction

Despite decades of research on cancer treatments, the disease remains a formidable public health challenge, standing as the second leading cause of mortality. It is influenced by a multitude of internal and external factors, leading to individualized genetic alterations in DNA, RNA, proteins, metabolites, and molecular networks. These alterations disrupt cellular regulation, promoting cancer cell proliferation and contributing to genetic variability within the cell population.

Typically, cancer treatment takes place in primary care settings, where decisions are largely based on clinicopathological factors and histopathological analyses. Molecular assessments, although limited, are sometimes incorporated into the process. Current cancer treatment protocols mainly depend on basic factors like tumor stage, grade, and specific biomarker expression. These factors help in evaluating tumor characteristics such as size, spread, and biological behavior. However, personalized treatment approaches for individual patients are still not fully developed. Therefore, there is a need for significant efforts in cancer research and practice to better understand the specific cancer characteristics or phenotypes that drive tumor growth and behavior.

Translating what is known about cancer biology into clinical use is tough. So, there has been a move in how to approach therapy and diagnosis. There is a shift from looking at just one thing at a time to considering many factors together. This approach, called personalized medicine, is becoming more important in cancer research and treatment. It is expected that by focusing on predictive, preventive, and personalized medicine (PPPM), better predictions can be made of how patients will respond to treatment, sorting patients into groups based on their needs, and customizing medical care accordingly.

The advancement of omics methodologies (like genomics, transcriptomics, proteomics, and metabolomics), along with progress in data technologies, has greatly improved the understanding of tumor biology and how to manage cancer clinically (Kaithal et al. 2024; Srivastava et al. 2024). These omics techniques have revealed important mutations and molecular pathways and have identified and measured different molecules that are expressed differently, regardless of what caused those differences. As a result, many potential targets for treatment have been found. By comprehensively analyzing multidimensional omics data in individuals using computational tools and bioinformatics, drug targets and optimized treatments can be pinpointed. Colorectal cancer (CRC) serves as a prime example for exploring these personalized approaches due to its prevalence and the complexity of its genetic and environmental factors.

11.2 Colorectal Cancer: An Overview

Colorectal cancer, also known as colon or bowel cancer, is a prevalent and serious malignancy affecting the colon (large intestine) or rectum (the last part of the large intestine). Uncontrolled growth of cells in the lining of the colon/rectum leads to the development of colorectal cancer. The disease is categorized as colon cancer when it originates in the colon or rectal cancer if it begins in the rectum (Henson et al. 2004).

Globally, colorectal cancer ranks as the third most common cancer and the second leading cause of cancer-related deaths. In 2018, the projected global incidence of colorectal cancer (CRC) was 1.85 million new cases, carrying a substantial economic impact and contributing to approximately 10% of total cancer-related costs in the European Union. The prevalence of colorectal cancer exhibits global variations, with developed countries experiencing higher rates attributed to lifestyle factors.

Diverse risk factors contribute to differing incidence rates, including age, family history, genetic syndromes, and inflammatory bowel disease. The risk of colorectal cancer increases with advancing age, particularly in individuals over 50. A family history of colorectal cancer or specific genetic syndromes like Lynch syndrome and familial adenomatous polyposis (FAP) significantly raises an individual's risk. Conditions such as Crohn's disease or ulcerative colitis, collectively known as inflammatory bowel disease (IBD), are associated with an increased risk.

Lifestyle choices play a crucial role in CRC development, with a higher prevalence observed in regions characterized by Westernized lifestyles. The World Cancer Research Fund/American Institute for Cancer Research (WCRF/AICR) panel has identified specific lifestyle factors associated with CRC risk, including physical activity, diet, body weight, smoking, and alcohol consumption. Notably, men exhibit stronger associations compared to women, particularly concerning rectal cancer. According to one study, adherence to all five healthy lifestyle factors could potentially prevent 16% of new CRC cases, with variations by sex and cancer site.

11.3 Symptoms and Early Detection

Colorectal cancer manifests itself through a variety of symptoms that can indicate its presence. Patients may notice persistent changes in their bowel habits, such as diarrhea, constipation, or a change in stool consistency, that last for more than a few days. Rectal bleeding or the presence of blood in the stool, often resulting in stools that appear dark or tar-like, is another common symptom. Abdominal discomfort or pain, including cramping, bloating, or a feeling of fullness, may also occur, particularly in the lower abdomen.

Unexplained weight loss, not attributable to changes in diet or physical activity, is another potential sign of colorectal cancer. This weight loss can be significant and may occur rapidly. Fatigue, or a feeling of extreme tiredness or weakness, is also frequently reported by individuals with colorectal cancer. These symptoms can be subtle and are not always indicative of cancer; however, they warrant attention, especially when they persist or worsen over time.

Early detection of colorectal cancer is critical for improving prognosis and treatment outcomes. Screening tests, such as colonoscopies, fecal occult blood tests, and sigmoidoscopies, can detect cancer at an early stage when treatment is most effective. Regular screening, as recommended by healthcare providers based on individual risk factors, can help detect colorectal cancer in its early stages or even prevent it by identifying and removing precancerous polyps. O'Connell et al. (2004) emphasized the importance of early detection and regular screening in their research, highlighting the impact these practices can have on improving patient outcomes and reducing mortality rates associated with colorectal cancer.

11.4 Genetic Factors and Tumor Development

Colorectal cancer typically evolves through a multistep process that involves the development of polyps in the colon. Polyps are small, noncancerous growths that can form on the inner lining of the colon. Over time, some polyps can undergo malignant transformation into cancerous tumors due to genetic mutations and alterations in specific oncogenes and tumor suppressor genes (Fearon 2011).

The well-established adenoma-carcinoma sequence describes the stepwise progression of colorectal cancer, suggesting that the majority of cases arise from preexisting adenomatous polyps. These polyps, classified into tubular adenomas, villous adenomas, and tubulovillous adenomas based on microscopic appearance, are considered precancerous. The risk of malignant transformation increases with the size and histological characteristics of these polyps (Giardiello et al. 2002).

The Adenomatous Polyposis Coli (APC) gene, a tumor suppressor gene, is mutated in the majority of colorectal cancers. Inheriting a defective APC gene leads to familial adenomatous polyposis (FAP), a rare hereditary condition characterized by the early development of numerous polyps in the colon and rectum, significantly elevating the risk of colorectal cancer (Burt 2000). Mutations in oncogenes, such as Kirsten Rat Sarcoma Viral Oncogene Homolog (KRAS) and Neuroblastoma RAS

Viral Oncogene Homolog (NRAS), found in a substantial proportion of colorectal cancers, contribute to uncontrolled cell growth (Andreyev et al. 2001). The TP53 (Tumor Protein 53) gene, often termed the "guardian of the genome," is another tumor suppressor gene, and its mutations are associated with more advanced and aggressive colorectal cancers (Iacopetta 2003). DNA mismatch repair (MMR) genes, including MLH1 (MutL Homolog 1), MSH2 (MutS Homolog 2), MSH6 (MutS Homolog 6), and PMS2 (Postmeiotic Segregation 2), play a crucial role in repairing errors during DNA replication (Kanchan et al. 2014). Mutations in these genes lead to Lynch syndrome, a hereditary condition associated with an increased risk of colorectal cancer and other cancers (Lynch and de la Chapelle 1993; Nilbert et al. 2000).

11.5 Diagnostic and Treatment Challenges

Expanding our understanding beyond genetic predisposition, recent research has shed light on the significant role of environmental factors, particularly the gut microbiome (Kesheri et al. 2024), in colorectal cancer development. Lee et al. (2023) identified specific gut bacteria associated with the growth of precancerous colon polyps, highlighting the intricate interplay between microbial dysbiosis and tumorigenesis.

Furthermore, chronic inflammation, as observed in conditions like inflammatory bowel disease, significantly heightens the risk of colorectal cancer (Terzić et al. 2010). These environmental influences not only modulate genetic pathways but also present diagnostic challenges by influencing disease presentation and response to treatment. In tandem with genetic and environmental influences, epigenetic changes, such as DNA methylation and histone modifications, emerge as critical contributors to colorectal cancer pathogenesis. Eads et al. (2001) demonstrated how these epigenetic alterations, alongside genetic mutations, influence polyp formation by modulating gene expression patterns. Given the potential for polyps to progress to colorectal cancer, regular screening is essential for early detection and prevention. Below are commonly used methods:

11.5.1 Colonoscopies, Fecal Occult Blood Tests, and Other Screening Methods

These screening methods are crucial for identifying and removing precancerous polyps before they develop into colorectal cancer (Knudsen et al. 2016). Colonoscopies involve inserting a flexible tube with a camera, known as a colonoscope, into the colon to examine its inner lining thoroughly. This procedure allows for the detection and removal of polyps, which are abnormal growths that may potentially become cancerous over time. Additionally, fecal occult blood tests are noninvasive tests used to detect small amounts of blood in the stool, which can be an early indicator of colorectal cancer or polyps.

11.5.2 Staging and TNM Classification System

Staging plays a crucial role in determining the extent of colorectal cancer and guiding treatment decisions (DeSantis et al. 2014). The Tumor, Node, Metastasis (TNM) classification system is a standardized method used for staging colorectal cancer. It evaluates three key factors: tumor size (T), lymph Node involvement (N), and the presence of distant Metastases (M). By assessing these parameters, clinicians can determine the stage of the cancer, which is essential for selecting appropriate treatment options and predicting patient prognosis.

11.5.3 Treatment Guidelines and Multidisciplinary Approach

Treatment guidelines provided by organizations such as the National Comprehensive Cancer Network (NCCN) and the European Society for Medical Oncology (ESMO) play a crucial role in informing treatment decisions for colorectal cancer. These guidelines emphasize a multidisciplinary approach due to the complexity of patient cases. Patient risk of recurrence is closely tied to the pathological stage, with variations in survival rates across stages I, II, and III.

11.5.4 Adjuvant Chemotherapy and Research Efforts

Adjuvant chemotherapy, which is chemotherapy given after surgery, is commonly used in the treatment of colorectal cancer, particularly for stage II patients. However, various benefits associated with adjuvant chemotherapy are not fully understood. Ongoing research explores genetic testing and emerging prognostic tools like ploidy status and immune system parameters to assess chemotherapy-related adverse events and optimize treatment outcomes.

11.5.5 Challenges in Diagnosing Symptomatic Patients and Biomarkers

Despite the success of screening programs, diagnosing symptomatic patients remains challenging. The prioritization of colonoscopy based on risk scores is hindered by increased demand and limited resources. Ongoing research explores noninvasive tools like the fecal immunochemical test as biomarkers for diagnosing symptomatic patients. The molecular heterogeneity of colorectal tumors poses challenges in identifying universal biomarkers applicable across different genetic and environmental contexts.

11.5.6 Mucosal Biopsy Specimens and Transitional Mucosa

Mucosal biopsy specimens involve obtaining tissue samples from the inner lining of the colon for examination under a microscope. These samples can help identify specific markers or abnormalities associated with colorectal cancer or precancerous changes. The debate around the significance of transitional mucosa and the distinct features of adenomas in various genetic conditions highlights the complexity of this scientific endeavor. Ongoing research aims to address these challenges and enhance the ability to detect and manage high-risk conditions for colorectal cancer.

11.5.7 Signaling Pathways and Drug Development

Understanding and intervening in the signaling pathways involved in colorectal cancer progression offers potential avenues for drug development targeting the uncontrolled cell division characteristic of colon cancer. These pathways represent promising targets for the development of novel therapeutic interventions aimed at improving patient outcomes and reducing treatment-related toxicities.

With this understanding, various routes through which bacteria enter tumors are explored, which shed light on mechanisms of bacterial involvement in cancer development and progression.

11.6 Routes of Bacterial Entry into Tumors

The routes through which bacteria enter tumors from common parasitic sites are yet to be clearly defined.

11.6.1 Vascular Leakage as a Route of Entry

This theory posits that bacteria can access tumor sites through vascular leakage. When the mucosa is damaged or intestinal permeability is increased, bacteria have the opportunity to penetrate the blood or lymphatic vessels. Various cell types, including dendritic cells, intestinal microfold (M) cells, or goblet cells, may facilitate this process. Importantly, this can occur even in the absence of significant intestinal damage. For example, Cummins and Tangney (2013) highlighted the role of dendritic cells in capturing and transporting bacteria, suggesting that these cells could play a crucial role in bacterial dissemination to tumor sites. Castillo et al. (2019) also suggested that intestinal microfold cells may be involved in transporting bacteria to lymphoid tissues, further supporting this pathway as a means for bacterial entry into tumors.

11.6.2 Dendritic Cell (DC)-Mediated Pathway

Dendritic cells are known for their ability to capture and present antigens to T cells, playing a critical role in immune responses. In the context of bacterial entry into tumors, dendritic cells can capture bacteria expelled by M cells into lymph nodes. Additionally, they can capture bacteria that penetrate the intestinal epithelial monolayer through tight junction proteins. This pathway is significant as it provides a potential mechanism for microorganisms associated with other body sites, such as the skin, mouth, and intestine, to access tumor sites. Research by Yang et al. (2021) and Fernández et al. (2020) supports the idea of dendritic cell-mediated transport of bacteria to tumor sites, suggesting its relevance in the context of cancer development and progression.

11.6.3 Hematogenous Pathway

The hematogenous pathway involves bacterial migration through the bloodstream to reach tumor sites. Studies involving the injection of *Fusobacterium nucleatum* into a mouse model's tail vein have shown a higher bacterial content in tumor tissue compared to adjacent normal tissues. This finding indicates that hematogenous spread may be a primary route for certain bacteria, like *F. nucleatum*, to reach tumors. Abed et al. (2016) demonstrated this phenomenon and suggested that understanding the mechanisms of hematogenous spread could provide insights into the role of bacteria in cancer development.

11.6.4 Alternative Hypothesis

Host Cell-Mediated Transport: An alternative hypothesis proposes that host cells play a role in transporting bacteria to tumor tissues. This mechanism, if proven, could have significant implications for our understanding of bacterial colonization of tumors and its impact on cancer progression and treatment. Further research is needed to explore the potential mechanisms by which host cells may facilitate bacterial transport to tumor sites and to determine the significance of this pathway in the context of cancer biology.

11.7 The Role of Intratumor Microbiomes in Cancer Development

The earliest historical records linked cancer and microbes. Approximately 10^{12} known microbial species on Earth exist (Sepich-Poore et al. 2021; Locey and Lennon 2016). However, only 11 have been classified as "oncomicrobes" by the International Agency for Research on Cancer (IARC). These oncomicrobes are responsible for approximately 2.2 million cancer cases annually, representing around 13% of cancer cases globally.

Interestingly, live microbes have been discovered in various types of tumors, including breast, colorectal, hepatocellular carcinoma, pancreatic, skin, renal cell carcinoma, gastric, lung, prostate, and nasopharyngeal carcinomas, among others. Additionally, intratumorally, microbes exhibit significant similarities within the same tumor type.

Though research into the gut microbiome's influence on cancer has been extensive, investigations into the intratumor microbiome's role remain limited. Nonetheless, grasping their presence and dysbiosis holds enormous potential. These microbiomes affect various mechanisms, including the induction of DNA damage and mutations, activation of oncogenic pathways, catabolism of antitumor drugs, and modulation of the immune system which can lead to carcinogenesis (Nejman et al. 2020; Erdman et al. 2010; Zeller et al. 2014; Rao et al. 2006; Zackular et al. 2013) (Fig. 11.1). Additionally, they modulate host pathways such as the Wingless-Int (Wnt)/β-catenin pathway, well-known for its involvement in cancer stemness and growth regulation (Pai et al. 2017).

Discovered in 1982 during studies on mouse mammary tumor virus (MMTV), the Wnt/β-catenin pathway revealed its crucial role in cancer development, especially in colorectal cancer (CRC). Activation of this pathway involves Wnt proteins binding to Frizzled receptors and low-density lipoprotein receptor-related protein co-receptors (LRP), which inhibits glycogen synthase kinase-3β (GSK-3β), resulting in β-catenin accumulation and subsequent nuclear translocation. This initiates T cell factor/lymphoid enhancer factor-1 (TCF/Lef1)-mediated gene transcription, fostering tumorigenesis. In CRC, mutations in the adenomatous polyposis coli (APC) gene commonly disrupt this pathway, leading to aberrant β-catenin accumulation and nuclear localization. This dysregulation drives early carcinogenesis and aggressive cancer biology, often associated with poor prognosis.

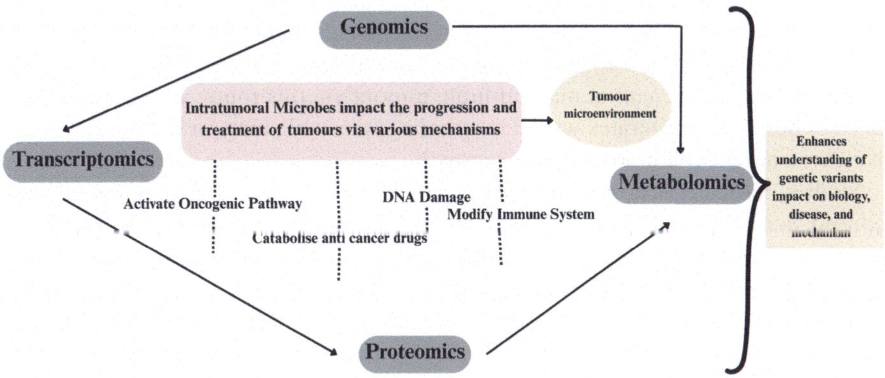

Fig. 11.1 Multi-omics approaches in studying the tumor microenvironment

Furthermore, interaction with the Hippo/YAP (Yes-associated protein pathway) further amplifies CRC progression through various means. The canonical regulator for YAP is the Hippo pathway, acting as a brake by phosphorylating and sequestering YAP in the cytoplasm, limiting its oncogenic potential. Mutations or deletions in the Hippo pathway components can disrupt YAP regulation, promoting oncogenesis, including CRC progression. YAP can also stimulate the transcription of target genes involved in its own activation, creating a feed-forward loop that enhances YAP activity and CRC progression. Tyrosine phosphorylation also regulates YAP subcellular localization and activity, with Steroid Receptor Coactivator (Src) family kinases (SFKs) identified as important regulators. Dysregulation of SFK-mediated YAP activation can contribute to CRC progression independently of the Hippo pathway. Various experimental strategies have been explored to inhibit YAP activity and suppress CRC progression, including interrupting feed-forward signaling cascades and targeting novel regulatory mechanisms such as noncoding RNAs.

Elevated Wnt pathway activity near stromal myofibroblasts in tumors maintain cancer stem cell (CSC) properties, fuelling tumor growth and therapy resistance. Understanding the significance of the Wnt/β-catenin pathway in CRC illuminates potential therapeutic targets for inhibiting tumor growth and overcoming treatment resistance. Additionally, intratumor microbiomes promote chronic inflammation, characterized by prolonged immune response activation, inhibiting antitumor immune responses, and stimulating pro-tumor immune responses, significantly contributing to oncogenesis. Intratumor microbiomes impact carcinogenesis by inducing DNA damage and mutations, modulating host pathways like the Wnt/β-catenin pathway, and promoting chronic inflammation. These processes inhibit antitumor immune responses while stimulating pro-tumor immune responses, thereby significantly contributing to oncogenesis.

11.7.1 Specific Microbial Players

Within the complex landscape of cancer biology, *F. nucleatum* emerges as a significant factor in colorectal cancer (CRC) (Sun et al. 2018), particularly through its adept manipulation of host cells such as macrophages and epithelial cells via macropinocytosis, allowing it to evade immune surveillance. Originating commonly in the oral cavity, *F. nucleatum* may infiltrate tumors via this route, where it not only survives but also proliferates within macrophages, establishing an immunosuppressive environment conducive to tumor growth.

Studies indicate a correlation between elevated levels of *F. nucleatum* and reduced densities of CD3+ T cells within CRC tissue, suggesting an inhibitory impact on the immune system. Furthermore, *F. nucleatum* directly undermines immune cell function, notably natural killer (NK) cells, employing mechanisms like the Fap2 protein thus aiding tumor cells in evading immune detection. Additionally, *F. nucleatum* influences the behavior of tumor-associated macrophages (TAMs), skewing the microenvironment toward tumor progression by suppressing antitumor M1 TAMs.

In terms of implications for CRC, research suggests that bacteria invading tumor cells can disseminate to other tissues. Within CRC, *F. nucleatum* instigates a pro-inflammatory milieu, perpetuating continuous cell proliferation and oxidative stress, thereby facilitating tumor development. This inflammatory environment not only sustains the survival and growth of cancer cells but also fosters the acquisition of aggressive phenotypes. Moreover, *F. nucleatum* has been implicated in promoting resistance to chemotherapy, posing challenges for treatment strategies.

11.7.2 Expanding Perspectives

Several bacterial species such as *Escherichia coli, Fusobacterium nucleatum, Bacteroides fragilis, Enterococcus faecalis*, and *Salmonella* sp. were observed to be present in cancer microenvironments and contribute into cancer microbiome. These bacteria can contribute to the initiation and development of colorectal cancer through various mechanisms.

Strains of *E. coli* containing the polyketide synthase (pks) island have been associated with colorectal cancer (CRC). These strains produce a harmful compound named colibactin, which causes double-strand breaks (DSBs) in the host's DNA thus playing a role in carcinogenesis. Colibactin-producing *E. coli* can induce DNA damage in colon epithelial cells, correlating with an elevated CRC risk. The genotoxic nature of colibactin likely contributes significantly to the initiation and progression of CRC. Moreover, *E. coli* can indirectly impact CRC by triggering signaling pathways such as the Wnt/β-catenin pathway, which is linked to cell proliferation and tumor development.

Certain strains of *Bacteroides fragilis*, particularly those producing *B. fragilis* toxin (BFT), have been implicated in CRC development. BFT disrupts the intestinal epithelium's integrity and activates pro-inflammatory pathways, fostering chronic inflammation and tumorigenesis. BFT also induces DNA damage and triggers signaling pathways like NF-κB, promoting cell proliferation and survival. Colonization by *B. fragilis* has been linked to a heightened CRC risk, particularly in individuals with defective mismatch repair mechanisms.

E. faecalis generates reactive oxygen species (ROS) and other damaging compounds that harm host DNA, thus facilitating CRC initiation and progression. Additionally, *E. faecalis* can activate inflammatory and cell proliferation pathways, contributing to tumor growth. The presence of *E. faecalis* has been associated with an increased CRC risk, and inhibiting its activity has shown promise in reducing intestinal polyps in animal models.

Various *Salmonella* strains, including *Salmonella enteritidis* and *F. nucleatum* have been linked to a heightened CRC risk, especially in younger individuals. *Salmonella* secretes genotoxic substances such as typhoid toxin (TT), which causes DNA damage in host cells, leading to genomic instability and CRC development. *Salmonella* can also activate signaling pathways like PI3K, fostering increased cell proliferation and tumor growth. As the inquiry deepens, attention turns to the colonization process once bacteria reach the tumor site. Fap2 lectin recognition and binding to tumor-expressed D-galactose-β (1−3)-N-acetyl-D-galactosamine (Gal-GalNAc) implantation in adenomas with high Gal-GalNAc expression, such as CRC and breast cancer (Abed et al. 2016; Parhi et al. 2020; Ali et al. 2024).

11.7.3 Integration with Immunotherapy

The classification of CRC based on immune-cell infiltration (ICI) subtypes offers valuable predictive information regarding responses to immunotherapy (Wu et al. 2023). Patients exhibiting specific immune-cell infiltration patterns, such as the ICI-2 subtype, demonstrate a higher response rate to immune checkpoint blockade (ICB) therapy compared to those classified as ICI-3 (Petitprez et al. 2020). The disparity in response to immune checkpoint blockade (ICB) therapy between patients categorized as Immune Checkpoint Inhibitor-2 (ICI-2) and Immune Checkpoint Inhibitor-3 (ICI-3) subtypes can be ascribed to various factors:

11.7.4 Immune Cell Composition

The makeup of immune cells within the tumor microenvironment (TME) is pivotal in dictating the response to ICB therapy. Patients labelled as ICI-2 might harbor a more favorable assortment of immune cells, including higher proportions of cytotoxic T cells like CD8+ T cells and activated dendritic cells. These cells are known for their antitumor activity and are linked to a better response to ICB therapy. Conversely, patients classified as ICI-3 may possess a less favorable immune cell composition, marked by elevated levels of immunosuppressive cell types such as regulatory T cells and M2 macrophages, which can suppress the antitumor immune response and diminish the effectiveness of ICB therapy.

11.7.5 Tumor Microenvironment Characteristics

The TME is a complex milieu consisting of diverse cell types, signaling molecules, and extracellular matrix components. Disparities in the TME between ICI-2 and ICI-3 subtypes may impact the response to ICB therapy. Tumors with an inflamed TME, characterized by heightened immune cell infiltration and pro-inflammatory cytokine release, tend to respond better to ICB therapy. Conversely, tumors with an immunosuppressive TME, characterized by the presence of inhibitory immune checkpoints such as Programmed Death-Ligand 1 (PD-L1) expression and immunosuppressive cell populations, typically exhibit a reduced response to ICB therapy.

11.7.6 Genomic and Molecular Features

Variations in the genomic and molecular characteristics of tumors classified as ICI-2 versus ICI-3 may also contribute to differing responses to ICB therapy. For instance, tumors with high mutational burden, microsatellite instability (MSI-H), or defects in DNA mismatch repair (dMMR) are more likely to respond to ICB therapy due to increased neoantigen formation and subsequent immune recognition. Conversely, tumors with low mutational burden or genomic instability may display reduced responsiveness to ICB therapy.

11.7.7 Biomarker Expression

The expression levels of specific biomarkers within the TME, such as PD-L1 or immune-related gene signatures, may vary between ICI-2 and ICI-3 subtypes, impacting the response to ICB therapy. Higher expression of PD-L1 or enrichment of immune-related gene signatures within the TME is associated with heightened response rates to ICB therapy.

The differential responses observed underscore the importance of considering intratumor microbial composition in tailoring immunotherapeutic strategies for individual patients. The intricate interplay between intratumor microbiomes and cancer development underscores the importance of considering microbial influences in understanding tumor biology and designing effective therapeutic interventions. By elucidating these relationships, a way can be paved for more precise and personalized approaches to cancer management. Additionally, research extending beyond bacteria to include fungi and viruses offers further insights into the complexity of cancer biology and potential avenues for therapeutic intervention.

Studies have shown that mycoplasma infections are prevalent in various cancer types, including colon carcinoma, gastric, esophageal, lung, breast, prostate, ovarian, cervical, kidney, pancreatic cancers, and glioma. Significant accumulation of mycoplasma has been observed in cancer tissues compared to healthy controls. In vitro studies have demonstrated that mycoplasma infections induce transformation and tumorigenicity in human cell lines from different organs, contributing to cancer progression. However, the exact pathological role of mycoplasmas in tumorigenesis remains debated. Additionally, mycoplasma infections have been linked to drug resistance, further complicating cancer treatment.

The fungal presence has been associated with cancer risk, and their abundance, particularly the *Malassezia* spp., is significantly higher in pancreatic ductal adenocarcinoma (PDA) compared to normal pancreatic tissues. Studies in mouse models have shown that Malassezia accelerates oncogenesis in PDA. Other fungi, such as Candida, have been linked to cancer risk, and proposed mechanisms include altering cell proliferation and secreting cytokines that modulate the immune response and promote tumor-cell adhesion. While fungi have been detected in various cancer types, the underlying mechanisms require further investigation to better understand their contribution to tumorigenesis (Vallianou et al. 2021).

The association between the virome and cancer has been established in solid tumors such as colon, hepatocellular, oral, breast, cervical, esophageal, gastric, and lung cancers (Schiller and Lowy 2021). Several viruses have been identified as cancer-related, including Epstein-Barr virus, Kaposi sarcoma herpesvirus, Human Papillomavirus (HPV), human T cell lymphotropic virus, hepatitis B virus, hepatitis C virus, and Merkel cell polyomavirus. HPV, in particular, has been confirmed to have a causal effect on cervical cancer oncogenesis and correlates with the progression of other cancer types. Bacteriophages might also play a role in cancer development, with specific phages detected in colorectal cancer patients compared to controls.

11.8 Microbiome-Based Therapeutic Interventions

Modulating the tumor microbiome to treat colorectal cancer (CRC) is a new, promising avenue of research. The gut microbiome, consisting of trillions of microorganisms in the gastrointestinal tract, is closely linked to CRC development and progression. Here is how adjusting the tumor microbiome might help treat CRC:

11.8.1 Immune System Regulation

The gut microbiome plays a crucial role in immune system regulation (Mallik et al. 2024). By altering its composition, it may be possible to enhance the antitumor immune response in CRC. For example, certain bacteria like *Bacteroides fragilis* can promote the differentiation of regulatory T cells (Tregs) and suppress pro-inflammatory Th17 cells, leading to an overall anti-inflammatory effect. This immune modulation within the tumor microenvironment could help control CRC growth (Round and Mazmanian 2009; Mishra et al. 2024).

11.8.2 Metabolite Production

The gut microbiome produces metabolites, including short-chain fatty acids (SCFAs), through dietary fiber fermentation. SCFAs have anti-inflammatory and antitumor effects (Son and Cho 2023). Adjusting the gut microbiome to increase SCFA production might create an environment less favorable for CRC development and progression.

Also, certain strains of *F. nucleatum* can adhere to and invade tumor cells, potentially influencing their growth and progression. Modulating the abundance of these bacteria could impact CRC development. Adjusting the tumor microbiome is a novel approach to CRC treatment. Understanding the complex interactions between the gut microbiome and CRC may lead to targeted therapies that improve patient outcomes and reduce the disease burden.

11.8.3 Probiotics

Probiotics is defined as live microorganisms offering health benefits when consumed adequately, and have gained attention for their potential impact on gut microbiota and cancer. Certain strains like *Lactobacillus* and *Bifidobacterium* demonstrate antitumor and antimutagenic activities, potentially enhancing cancer therapy efficacy and reducing its side effects thus protecting the intestinal mucosal barrier.

In a study, increased levels of tumor suppressor miRNAs (miR-26b, miR-18a, APC, PTEN) were observed as well as decreased oncogene expression (miR135b, KRAS) after administering probiotics (*Lactobacillus acidophilus, Bifidobacterium bifidum*) in an animal colon cancer model. Lactic acid bacteria (LAB) like *L. acidophilus* and *Lactobacillus casei* increased the apoptotic activity of 5-fluorouracil (5-FU) in CRC cell lines by converting lactate and acetate to butyrate, an anticarcinogenic agent.

Studies have shown that LAB secretory metabolites inhibit the invasiveness of human colorectal cell lines, with lactobacilli supernatants reducing cancer cell proliferation and inducing apoptosis. Conversely, lactobacilli extract induced necrosis but increased cancer cell migration and invasion. In another study, *L. plantarum* cell-free supernatants enhanced chemosensitivity in 5-FU-resistant colon cancer cell lines.

Sivamaruthi et al. (2020) demonstrated that *Bifidobacterium infantis* has a chemoprotective effect and anti-colorectal cancer (CRC) properties in a mouse model. They found that it protected against chemotherapy-induced health damages through changes in T cell immunity profiles. Despite their potential benefits, caution is advised with probiotic use, particularly in patients with compromised intestinal barriers or immunity, to avoid bacterial translocation and systemic infections.

Next-generation probiotics (NGPs), identified through comparative microbiota analysis, offer a new frontier in CRC treatment. Strains like *Bifidobacterium* spp., *B. fragilis, Akkermansia municiphila*, and *Faecalibacterium prausnitzii* show promise. NGPs, including *B. fragilis,* may enhance immune checkpoint inhibitor efficacy, but caution is warranted due to the potential negative effects of enterotoxin-containing strains (Kaźmierczak-Siedlecka et al. 2022). Probiotics can mitigate chemotherapy-induced toxicities such as diarrhea and intestinal mucositis by maintaining a low pH in the intestinal lumen. They may also reduce opportunistic infections in neutropenic patients receiving chemotherapy.

Perioperative probiotic use in CRC patients undergoing surgery has shown benefits, including reduced postoperative complications and infections, faster recovery of normal gut function, and reduced diarrhea. Overall, probiotics offer a range of potential benefits in CRC treatment, but further research is needed to understand their mechanisms and optimal use.

11.8.4 Prebiotics

Prebiotics are nondigestible oligosaccharides that selectively ferment in the gut, altering the composition and activity of gut microbiota to provide health benefits. These carbohydrates include fructooligosaccharides (FOS), xylooligosaccharides, galactooligosaccharides (GOS), inulin, and fructans, with FOS and GOS being the most studied (Aachary et al. 2011). They stimulate beneficial gut bacteria, leading to short-chain fatty acid (SCFA) production, regulate immune response, control gene expression in bacterial cells, improve micronutrient absorption in the colon, and modulate xenobiotic-metabolizing enzymes.

Combining prebiotic inulin enriched with oligofructose with probiotics like LGG and *Bifidobacterium lactis* has shown antitumorigenic activity in rat models of colon carcinogenesis. Clinical studies have also demonstrated that preoperative use of prebiotics in CRC patients can improve commensal microbiota abundance and serum immunologic indicators. Despite these findings, there is limited clinical research on the direct effectiveness of prebiotics in treating colon cancer.

Further research, particularly on bacterial species like *Bacteroides*, is needed to understand the mechanisms of action of prebiotics. Prebiotics produced by trans glycosylation of β-galactosidases or β-glucosidases, such as β (1–4) GOS, lactulose, and FOS, are expected to play a role in CRC prevention.

11.8.5 Synbiotics

Synbiotics, which combine probiotics and prebiotics, are food ingredients or dietary supplements that exhibit synergistic effects. In colorectal cancer (CRC), synbiotics are thought to be beneficial due to their immunomodulatory properties and their ability to reduce postoperative infection rates. For example, in rats, a cocktail containing oligofructose-maltodextrin (prebiotics) enriched with *Lactobacillus acidophilus*, *Bifidobacterium bifidum,* and *Bifidobacterium infantum* reduced cancer growth, increased mucin secretion, preserved tight junctions, and inhibited inflammation. This cocktail also modulated gut microbiota compositions (Raman et al. 2013).

In a recent study, using a colon-specific cancer mouse model, synbiotic treatment suppressed dextran sodium sulfate-induced colitis in mice, reducing mortality and inhibiting tumorigenesis. Additionally, in a clinical study involving 37 CRC patients, administering a synbiotic formula (containing LGG, *Bifidobacterium lactis* Bb12, and inulin) decreased tumor proliferation and led to positive changes in the composition of intestinal microbiota. Synbiotics, by combining probiotics and prebiotics, can enhance their beneficial effects on the gut microbiome, potentially influencing the tumor microbiome in CRC.

These interventions aim to promote a balanced and healthy gut microbiome, which may help prevent or reduce the risk of CRC and improve outcomes in CRC patients undergoing treatment. However, further research is needed to fully understand the mechanisms underlying the effects of probiotics, prebiotics, and synbiotics on the tumor microbiome in CRC.

Ongoing clinical trials explore innovative strategies to manipulate the microbiota as a means of augmenting antitumor immune responses and improving cancer therapy outcomes. Interventions such as fecal microbial transplantation (FMT) or administration of defined bacterial isolates aim at modulating the microbiota in a manner that enhances the efficacy of immunotherapies while minimizing the impact of immune-regulatory bacteria. Additionally, the use of antibiotics and bacteriophages offers targeted approaches to selectively eliminate cancer-associated microbes, further enhancing treatment responses.

11.8.6 Integration of Microbial Data into Clinical Practice

The integration of intratumor microbial data into clinical practice holds promise for personalized cancer treatments tailored to individual microbial profiles. By leveraging microbial information, oncologists can optimize treatment strategies to enhance efficacy while minimizing adverse effects. The exploration of combination therapies, which combine microbiome modulation with conventional cancer treatments, represents a step toward achieving more durable and robust treatment responses in cancer patients.

11.9 Advancements in Multi-Omics Studies

Genomics, which studies the structure and function of genomes through DNA sequencing and genetic variation analysis, is particularly important (Gasperskaja and Kučinskas 2017). Using next-generation sequencing (NGS) technologies in a systematic genomic approach helps us understand how genotype relates to phenotype compared to just looking at one type of data. The completion of the Human Genome Project has deepened our understanding of cancer, especially regarding how DNA mutations contribute to different types of cancer, helping us develop and anticipate the effectiveness of new treatments. Transcriptomics looks at the transcriptome—the full set of RNA transcripts produced by a genome, which is a precursor to the proteome, the entire set of proteins expressed by a genome. By using techniques like RNA interference (RNAi) screening and RNA sequencing (RNA-seq or WTSS), along with NGS technologies, it is possible to identify biomarkers that can predict how cancer might respond to treatment (Wang et al. 2009).

Genomics serves as a fundamental and relatively rapid starting point for other clinical-omics strategies. While sequencing provides valuable information for predictive, preventive, and personalized medicine (PPPM) (Krzyszczyk et al. 2018), it lacks functional insights, such as posttranslational modifications (PTMs). Therefore, research and advancement in proteome-based biomarker discovery are imperative (Kumar et al. 2024). Proteins orchestrate most biological processes, and the dynamic variations in protein expression and PTMs reflect disease activity and status, systematically analyzed in individual cancer patients using bioinformatics tools. Although knowledge of protein pathways and network information flow within cells is beneficial, the considerable and dynamic proteomic variations involved in regulating biological functions present challenges. The metabolome, as the final downstream product and the most predictive marker of phenotypes in PPPM, is another invaluable resource for identifying unknown compounds and alterations in cancer patients. Apart from genomics, proteomics, and metabolomics, other regulatory information like the transcriptome and epigenome also hold significance in cancer PPPM. No method has been devised to simultaneously assay the entire genome, transcriptome, proteome, and metabolome; hence, systematic multi-omics strategies with integrated data systems are indispensable for cancer PPPM (Hu et al. 2013; Dai and Shen 2022). Complementing the insights gained from genomics and other-omics strategies, the inclusion of the tumor microbiome enhances the understanding of cancer biology and improves early diagnosis and treatment strategies.

Rather than being temporary residents, the tumor microbiome is recognized as a lasting influence, impacting cancer development, metastasis, and responses to therapy. Given its potential for early cancer diagnosis and treatment, recent research into the tumor microbiota has drawn extensive attention across scientific disciplines, leading to notable progress facilitated by interdisciplinary technologies. However, significant challenges persist, including the low levels of bacteria within tumors and the inability to culture certain microbial species. To address these challenges, new methods with enhanced spatial and temporal resolution are urgently needed, particularly due to the complexity of tumor microbiome research, such as the heterogeneity of the tumor microenvironment (TME). Among emerging techniques, multi-omics technologies (integrating genomics, transcriptomics, proteomics, and metabolomics) stand out as powerful tools, offering insights into the tumor microbiome at various levels of the central dogma (Zhang et al. 2023). Consequently, multi-omics, especially single-cell omics, is expected to have profound impacts on future studies exploring the interactions between microbes and the tumor microenvironment.

These methodologies enable researchers to meticulously scrutinize the genomes and transcriptomes of diverse organisms. By harnessing advanced sequencing technologies such as Illumina and PacBio, scientists delve into the genetic composition of tumor microbes, unravelling their intricate interactions with cancer cells and the immune system.

DNA sequencing techniques enable the examination of complex samples like the human microbiome, revealing microbial functions and metabolic pathways. 16S rRNA sequencing, on the other hand, targets bacterial and archaeal components, aiding in the identification of distinct species within tumor communities. RNA sequencing (RNA-seq) provides insights into active genes and their expressions, offering profound insights into protein production and RNA structures.

Epigenetics explores alterations in gene expression without modifying the DNA sequence. Unlike genetics, epigenetic changes, including DNA/RNA modifications and histone modifications, are reversible and do not impact the genetic code. Chromatin, composed of repeating units called nucleosomes, undergoes covalent modifications on DNA, RNA, and histones, orchestrating gene activation and suppression. These epigenetic mechanisms are pivotal in maintaining chromatin structure during cellular processes like replication, transcription, and DNA repair, intimately associated with diseases like cancer.

DNA methylation, a prevalent epigenetic change, influences gene expression without altering the DNA sequence, potentially triggering tumorigenesis by silencing tumor suppressor genes. Aberrant DNA methylation serves as a diagnostic biomarker across various cancer types, with techniques like whole-genome bisulfite sequencing facilitating the exploration of DNA methylation patterns (Paska and Hudler 2015).

Histone posttranslational modifications, encompassing methylation, phosphorylation, and acetylation, play crucial roles in cancer development. For instance, microorganism-derived enzymes can modulate tumor progression by altering histones (Liang et al. 2023). Chromatin immunoprecipitation sequencing (ChIP-seq) emerges as a potent method to probe the effects of histone modifications on gene transcription. It identifies DNA regions interacting with specific histones, fostering a deeper understanding of epigenetic changes in cancer cells and the tumor microenvironment.

Three-dimensional (3D) genomics delves into the spatial organization of genomes and their regulatory mechanisms, including those within bacterial chromosomes in tumor environments. Gene editing, synonymous with genetic modification or engineering, entails altering genomic sequences using biochemical methods. This encompasses inserting or deleting specific gene fragments to modify the genetic makeup of a cell. Various tools, often termed "genetic scalpels," have been developed for this purpose, including artificial restriction enzymes, zinc finger nucleases (ZFNs), transcription activator-like effector nucleases (TALENs), and the CRISPR-Cas9 system (Zhang et al. 2023). CRISPR-Cas9 has become a very popular gene editing technology nowadays due to its adaptability and precision. Additionally, Cas9 variants like dead Cas9 (dCas9) serve as gene location guides across various biological systems, including cancer research (Fonfara et al. 2014).

In tumor microbiome studies, gene editing is deployed on both host cells and microbes to grasp genomic variations. For instance, the composition of intestinal microorganisms exhibits wide variability, and scrutinizing these variances using techniques like CRISPR-Cas9 can yield valuable insights into microbial functions and their involvement in cancer development. Traditional genome and transcriptome methodologies offer expansive coverage but may overlook low-abundance sequences in intricate samples. Single-cell sequencing addresses this gap by analyzing individual cells and unveiling gene structures and expressions crucial for comprehending cellular heterogeneity (Wang et al. 2021).

For tumor microbiome research, four primary single-cell techniques are employed: agar plate dilution, fine capillary pipetting, droplet microfluidics, and enhanced fluorescence in situ hybridization (FISH). Agar plate dilution isolates single cells from mixed suspensions, fostering pure cultures. Fine capillary pipetting visually segregates cells under a microscope, facilitating genetic analysis. Droplet microfluidics encapsulates single cells in micro-droplets, enabling high-throughput sequencing. Improved FISH labels target RNA in single cells, enhancing resolution (Hou et al. 2012; Xu et al. 2012; Lan et al. 2017: Shi et al. 2020).

On the other hand, proteomics scrutinizes protein expression, fundamental for comprehending cellular function. Techniques like liquid chromatography-mass spectrometry (LC-MS) aid in protein analysis. Data-dependent acquisition (DDA) and data-independent acquisition (DIA) strategies provide differing insights, with DIA preferred for unbiased proteome analysis (Chapman et al. 2014). Targeted proteomics quantifies specific proteins using methods like multiple reaction monitoring (MRM) or parallel reaction monitoring (PRM), aiding in biomarker discovery (Chambers et al. 2014; Zhu et al. 2020).

Additionally, metabolomics identifies metabolites impacting tumor development. Techniques like nuclear magnetic resonance (NMR) and liquid chromatography-mass spectrometry (LC-MS) reveal metabolite profiles. Targeted and nontargeted approaches analyze specific or all metabolites in a sample, aiding in biomarker identification and pathway elucidation. Metabolic quenching and extraction ensure accurate analysis, essential for understanding microbe-host interactions.

Single-cell analysis of proteins and metabolites is indispensable for comprehending cellular functions at a granular level. Mass spectrometry instruments like Orbitrap-based MS offer heightened sensitivity and accuracy. Matrix-assisted laser desorption/ionization (MALDI) mass spectrometry imaging maps metabolites in tissues with high resolution, aiding in spatial analysis (Sun et al. 2018). Overall, integrating single-cell genomics, proteomics, and metabolomics furnishes comprehensive insights into cellular functions and microbe-host interactions in tumor microbiome research.

One type of omics is not enough to understand complex processes like the tumor microbiome. Integrated multi-omics offers a new direction to explore these topics comprehensively. For instance, Microbial metabolites such as methylglyoxal (MGO) play a significant role in tumorigenesis by altering proteins, genes, and histones, leading to DNA damage and disruptions in gene expression. Chemical proteomics and glycation studies shed light on MGO's impact on chromatin structure and cellular functions, highlighting the importance of interdisciplinary approaches in understanding its role in cancer development (Galligan et al. 2018; Sanghvi et al. 2019).

Looking ahead, understanding the interplay between microbes, tumor cells, and immune cells is crucial for cancer treatment. Multi-omics techniques aid in identifying cancer biomarkers and developing diagnostic and treatment strategies. However, challenges like immature research techniques and lack of standard processes remain. Single-cell multi-omics will be pivotal in advancing tumor microbiome research, though method development is ongoing.

11.10 Computational Frameworks for Multi-Omics Studies

It is clear that advancements in high-throughput measurements have revolutionized the profiling of various molecular patterns, including somatic mutations, copy number variations (CNVs), and the expression of mRNA, non-coding RNA, and proteins. This has paved the way for the development of computational algorithms for multi-view clustering, aimed at identifying coherent features from diverse inputs. In the field of biomedicine, these technologies have significantly contributed to defining clinical subtypes in complex disorders like cancer. By identifying co-expressed gene modules and patient subgroups within specific diseases, clustering methods have played a crucial role (Langfelder and Horvath 2008). Integrating multi-omics datasets from the same set of samples has been instrumental in revealing nuanced structures that would have been overlooked by analyzing single data types alone.

For example, by combining gene expression and mutation profiles from the same patients, it is possible to classify cancer subtypes based on multi-omics datasets. This integration of multi-omics data can mitigate potential bias or noise inherent in a single omics dataset, offering a more comprehensive representation of cellular aspects ranging from the genomic to the epigenomic level (Nguyen and Wang 2020). Various tools, which are listed in Table 11.1, have been developed for multi-omics datasets with several objectives, including identifying disease subtypes or classifying subgroups, identifying potential biomarkers for diagnostics and driver genes for diseases, and gaining insights into disease biology. These frameworks primarily rely on Bayesian statistics, similarity networks, joint nonnegative matrix factorization, and sparse canonical correlation analysis. While many of these tools are widely used and show superior performance for subtype prediction and survival analysis, their efficacy can vary significantly based on the biological characteristics of the study objects, even when using the same data. This presents a computational and biological challenge, requiring researchers to carefully select the appropriate multi-omics tools for their analyses.

11.10.1 iCluster

iCluster is an early multi-omics integration method that integrates multiple inputs and identifies multi-omics clusters by using clustering and expectation–maximization algorithms. The updated version of iCluster is called iCluster+, which includes more features and provides full Bayesian regularization for clustering. iCluster+ identified colorectal cancer subtypes with different cancer progression pathways, highlighting pathways that may not require aggressive drug treatment in addition to surgery.

11.10.2 iOmicsPASS

iOmicsPASS is an innovative algorithm that integrates genome-based networks with multi-omics datasets (Koh et al. 2019). It calculates scores for biological interactions by transforming omics datasets, which are then used to construct networks. iOmicsPASS has proven effective in enhancing the identification of subtypes of breast invasive ductal carcinoma (IDC) by integrating mRNA expression and protein abundance data. This integrated analysis revealed a new transcriptional

Table 11.1 Multi-omics analysis tools: a summary of findings and methodologies

Study	Findings	Dataset	Principles
iCluster	Novel subgroups from 2000 breast tumors	mRNA expression, CNV	Joint latent variable model-based clustering method
iOmicsPASS	Novel transcriptional regulatory network	mRNA expression, CNV, Protein expression	Network construction using a modified nearest shrunken centroid algorithm
SALMON	Improved survival analysis	Mutation, mRNA/miRNA expression, CNV	Deep learning based on co-expression modules
SNF	Subtype classification of clinical relevance	mRNA/miRNA expression, DNA methylation	Patient similarity networks using an iterative procedure based on message passing
NEMO	Novel subtypes from even partial AML datasets	mRNA/miRNA expression, DNA methylation	Sample clustering from partial datasets using an adjusted Rand index
MONET	Module detection of patient subtypes	mRNA/miRNA expression, DNA methylation	Detect similar modules commonly present across multi-omics datasets
PARADIGM	Detection of pathways affected by cancer	mRNA expression, CNV	Pathway recognition algorithm applied to multi-omics datasets
LRAcluster	Subtype detection in both pan-cancer analysis	Mutation, mRNA expression, CNV, DNA methylation	Performance of low-rank approximation from probabilistic models
BCC	Detection of patient subtypes in response to survival rates	mRNA/miRNA expression, DNA methylation, Protein expression	Bayesian framework for estimation of an integrative clustering model

regulatory network specific to a certain breast cancer subtype, a discovery that could not have been made through single-omics analysis alone.

11.10.3 SALMON (Survival Analysis Learning with Multi-Omics Neural Networks)

SALMON is a deep learning-based method (Huang et al. 2019) that utilizes multi-omics datasets from cancer patients to compute eigengenes from co-expression modules. This approach addresses the issue of overfitting that often arises in multi-omics studies with many features but few samples. For example, by analyzing mRNA and miRNA datasets from 583 female breast invasive carcinoma patients, SALMON accurately predicted survival outcomes.

11.10.4 SNF (Similarity Network Fusion)

SNF is a novel algorithm for generating patient similarity networks using an iterative procedure based on message passing (Wang et al. 2014). It calculates similarity networks for individual patients and then merges them to identify disease subtypes and predict phenotypes. Unlike early integration methods, SNF leverages individual omics datasets to construct independent single-omics networks. This approach helps find coherent modules sourced from similar biological features across patients with similar clinical features. SNF has demonstrated high efficiency in identifying clinical subtypes of cancers and other disorders such as autism.

11.10.5 NEMO (Neighborhood Based Multi-Omics Clustering)

NEMO is a multi-omics clustering method that can be used for partial datasets without the need for data imputation (Rappoport and Shamir 2019). It first calculates an inter-patient similarity matrix for each omics dataset and then combines the matrices of different omics datasets into a single matrix. Clusters are identified using an adjusted Rand index to compute the similarity between patients by distance. NEMO has shown superior performance compared to other multi-omics clustering algorithms, particularly in enhancing cluster detection from partial datasets.

11.10.6 MONET (Multi-Omic Clustering by Non-exhaustive Types)

MONET detects similar modules commonly present across multi-omics datasets (Rappoport et al. 2020). By utilizing three omics datasets (mRNA expression, DNA methylation, and miRNA expression), MONET computes an edge-weighted graph per omics dataset, where nodes represent samples and edges represent the similarity between samples. It then identifies a disjoint set of modules for patients from multiple omics graphs. MONET was able to reveal four sample modules representing venous invasion status and survival rates in 287 patients with ovarian serous cystadenocarcinoma.

11.10.7 PARADIGM (PAthway Recognition Algorithm Using Data Integration on Genomic Models)

This method uses multi omics dataset for identifying biological pathways (Vaske et al. 2010). It combines multi-omics-scale values derived from an individual sample with gene activities, products, and pathway interactions included in the National Cancer Institute (NCI) database. PARADIGM utilizes factor graphs to create probabilistic graphical models, inferring significant and nonsignificant interactions between pathways involving different entities. This tool has been efficient in revealing four subtypes of glioblastoma, leading to significantly different survival outcomes based on perturbated pathways.

11.10.8 LRAcluster (Low Rank Approximation-Based Multi-Omics Data Clustering)

LRAcluster is a multi-omics approach that integrates data on somatic mutations, CNVs, DNA methylation, and gene expression (Wu et al. 2015). It performs low-rank approximation from the probabilistic models of various molecular features, transforming all molecular features into variables and arranging them in a parameter matrix. Dimension reduction is then conducted to reveal clusters associated with distinct clinical subtypes. LRAcluster has outperformed other existing methods in terms of both time and classification accuracy when tested on multi-omics datasets of breast invasive carcinoma, colon adenocarcinoma, and lung adenocarcinoma (LUAD).

11.10.9 BCC (Bayesian Consensus Clustering)

BCC is a data-driven approach that performs consensus clustering across multi-omics datasets (Lock and Dunson 2013). Based on the finite Dirichlet mixture model, BCC explains overall consensus clustering as well as important features inherent to an individual omics dataset. By seeking an integrative point for adherence to an overall cluster, BCC effectively reveals cancer subtypes associated with specific clinical features in breast cancer patients.

11.11 Latest Findings and Implications in Cancer Multi-Omics Studies

Advances in omics technologies from genomics to transcriptomics have enabled cancer researchers to leverage multiple omics datasets from the same patients. Multi-omics approaches provide a unique opportunity to identify molecular and clinical features of cancer patients. Recent advances in proteomics have introduced proteogenomic approaches, integrating genomic data with proteomics and information on posttranslational modifications. Large-scale proteogenomic research has been conducted to unravel new biological mechanisms in cancers and provide fundamental information on multi-omics approaches for the development of integration strategies or computational algorithms.

Multi-omics clustering has refined the association between molecular profiles and clinical features among cancer patients. The identification of coherent subtypes across multiple dataset layers could have major implications for predicting clinical relevance or therapeutic response regardless of the overall tumor mutational load. Integration of proteomics datasets enables the identification of a direct connection between mutations and phenotypes, thereby increasing the resolution of clustering patterns across samples.

11.12 Conclusion

In conclusion, the exploration of genomics, transcriptomics, proteomics, metagenomics, and radiomics has fundamentally transformed our comprehension of CRC. Genomic studies have pinpointed critical genes like APC, TP53, KRAS, and BRAF, unravelling their roles in cancer development and progression, providing potential biomarkers for targeted therapies. Transcriptomic analyses have categorized CRC into distinct molecular subtypes (CMSs), crucial for identifying specific treatment targets. Proteomic studies have identified biomarkers for diagnosis, prognosis, and treatment response, albeit requiring further validation. Metagenomic investigations have underscored the impact of the gut microbiota on CRC, suggesting novel therapeutic strategies. Additionally, radiomics, leveraging advanced imaging technologies, has facilitated the development of models predicting genomic mutations, aiding in treatment decisions. These holistic approaches offer a comprehensive understanding of CRC, fostering personalized treatment strategies and enhancing patient outcomes.

Short Chapter Summary
This chapter reviews the symptoms, early detection, genetic factors, tumor development, diagnostics being used, and various treatment-based challenges of CRC. This chapter also emphasizes the role of intratumor microbiomes in cancer development, various therapeutics being used, computational framework for cancer-based multi-omics studies as well as various implications of multi-omics studies in cancer.

Author Contributions Chapter concept and design, Chandandeep Kaur; chapter writing, Chandandeep Kaur; chapter editing and review, Swarna Kanchan and Minu Kesheri.

References

Aachary AA, Gobinath D, Prapulla SG (2011) Short chain xylooligosaccharides: a potential prebiotic used to improve batter fermentation and its effect on the quality attributes of idli, a cereal–legume-based Indian traditional food. Int J Food Sci Technol 46(7):1346–1355. https://doi.org/10.1111/j.1365-2621.2011.02591.x

Abed J, Emgård JE, Zamir G, Faroja M, Almogy G, Grenov A, Sol A, Naor R, Pikarsky E, Atlan KA, Mellul A (2016) Fap2 mediates *Fusobacterium nucleatum* colorectal adenocarcinoma enrichment by binding to tumor-expressed Gal-GalNAc. Cell Host Microbe 20(2):215–225. https://doi.org/10.1016/j.chom.2016.07.006

Ali N, Wolf C, Kanchan S, Veerabhadraiah SR, Bond L, Turner MW, Jorcyk CL, Hampikian G (2024) 9S1R nullomer peptide induces mitochondrial pathology, metabolic suppression, and enhanced immune cell infiltration, in triple-negative breast cancer mouse model. Biomed Pharmacother 170:115997. https://doi.org/10.1016/j.biopha.2023.115997

Andreyev HJ, Norman AR, Cunningham D, Oates JR, Dix BR, Iacopetta BJ, Young J, Walsh T, Ward R, Hawkins N, Beranek M (2001) Kirsten ras mutations in patients with colorectal cancer: the 'RASCAL II' study. Br J Cancer 85(5):692–696

Ayyappan AA, Prapulla SG (2011) Xylooligosaccharides (XOS) as an emerging prebiotic: microbial synthesis, utilization, structural characterization, bioactive properties, and applications. Compr Rev Food Sci Food Saf 10(1):2–16. https://doi.org/10.1111/j.1541-4337.2010.00135.x

Burt RW (2000) Colon cancer screening. Gastroenterology 119(3):837–853. https://doi.org/10.1053/gast.2000.16508
Castillo NA, Perdigón G, de Moreno de LeBlanc A, Zurita-Turk M (2019) The mucosal immune system and its regulation by autophagy. Appl Microbiol Biotechnol 103(11):4371–4388. https://doi.org/10.1007/s00253-019-09759-7
Chambers AG, Percy AJ, Simon R, Borchers CH (2014) MRM for the verification of cancer biomarker proteins: recent applications to human plasma and serum. Expert Rev Proteomics 11(2):137–148. https://doi.org/10.1586/14789450.2014.877346
Chapman JD, Goodlett DR, Masselon CD (2014) Multiplexed and data-independent tandem mass spectrometry for global proteome profiling. Mass Spectrom Rev 33(6):452–470. https://doi.org/10.1002/mas.21400
Cummins J, Tangney M (2013) Bacteria and tumors: causative agents or opportunistic inhabitants? Infect Agent Cancer 8(1):11. https://doi.org/10.1186/1750-9378-8-11
Dai X, Shen L (2022) Advances and trends in omics technology development. Front Med (Lausanne) 9:911861. https://doi.org/10.3389/fmed.2022.911861
DeSantis CE, Lin CC, Mariotto AB, Siegel RL, Stein KD, Kramer JL, Alteri R, Robbins AS, Jemal A (2014) Cancer treatment and survivorship statistics. CA Cancer J Clin 64(4):252–271. https://doi.org/10.3322/caac.21235
Eads CA, Lord RV, Wickramasinghe K, Long TI, Kurumboor SK, Bernstein L, Peters JH, DeMeester SR, DeMeester TR, Skinner KA, Laird PW (2001) Epigenetic patterns in the progression of esophageal adenocarcinoma. Cancer Res 61(8):3410–3418
Erdman SE, Rao VP, Olipitz W, Taylor CL, Jackson EA, Levkovich T, Lee CW, Horwitz BH, Fox JG, Ge Z, Poutahidis T (2010) Unifying roles for regulatory T cells and inflammation in cancer. Int J Cancer 126(7):1651–1665. https://doi.org/10.1002/ijc.24923
Fearon ER (2011) Molecular genetics of colorectal cancer. Annu Rev Pathol 6:479–507. https://doi.org/10.1146/annurev-pathol-011110-130235
Fernández MI, Pedron T, Tournebize R, Olivo-Marin JC, Sansonetti PJ, Phalipon A (2020) A bile-induced 'supershedding' phenotype identified in a *Salmonella enterica* serovar Typhi culture. Microb Pathog 147:104337. https://doi.org/10.1016/j.micpath.2020.104337
Fonfara I, Le Rhun A, Chylinski K, Makarova KS, Lecrivain AL, Bzdrenga J, Koonin EV, Charpentier E (2014) Phylogeny of Cas9 determines functional exchangeability of dual-RNA and Cas9 among orthologous type II CRISPR-Cas systems. Nucleic Acids Res 42(4):2577–2590. https://doi.org/10.1093/nar/gkt1074
Galligan JJ, Wepy JA, Streeter MD, Kingsley PJ, Mitchener MM, Wauchope OR, Beavers WN, Rose KL, Wang T, Spiegel DA, Marnett LJ (2018) Methylglyoxal-derived posttranslational arginine modifications are abundant histone marks. Proc Natl Acad Sci USA 115(37):9228–9233. https://doi.org/10.1073/pnas.1802901115
Gasperskaja E, Kučinskas V (2017) The most frequent sites of cancer-associated somatic mutations in human cancer genomes. Genes 8(12):372
Giardiello FM, Yang VW, Hylind LM, Krush AJ, Petersen GM, Trimbath JD, Piantadosi S, Garrett E, Geiman DE, Hubbard W, Offerhaus GJA (2002) Primary chemoprevention of familial adenomatous polyposis with sulindac. N Engl J Med 346(14):1054–1059. https://doi.org/10.1056/NEJMoa012015
Henson DE, Dittus C, Younes M, Nguyen H, Albores-Saavedra J (2004) Differential trends in the intestinal and diffuse types of gastric carcinoma in the United States, 1973–2000: increase in the signet ring cell type. Arch Pathol Lab Med 128(7):765–770. https://doi.org/10.5858/2004-128-765-DTITIA
Hou Y, Song L, Zhu P, Zhang B, Tao Y, Xu X, Li F, Wu K, Liang J, Shao D, Wu H (2012) Single-cell exome sequencing and monoclonal evolution of a JAK2-negative myeloproliferative neoplasm. Cell 148(5):873–885. https://doi.org/10.1016/j.cell.2012.02.028
Hu R, Wang X, Zhan X (2013) Multi-parameter systematic strategies for predictive, preventive and personalised medicine in cancer. EPMA J 4(1):2. https://doi.org/10.1186/1878-5085-4-2

Huang Z, Zhan X, Xiang S, Johnso TS, Helm B, Yu CY, Zhang J, Salama P, Rizkalla M, Han Z, Huang K (2019) SALMON: survival analysis learning with multi-omics neural networks on breast cancer. Front Genet 10:166. https://doi.org/10.3389/fgene.2019.00166

Iacopetta B (2003) TP53 mutation in colorectal cancer. Hum Mutat 21(3):271–276. https://doi.org/10.1002/humu.10175

Kaithal P, Kanchan S, Kesheri M (2024) Recent advances in biological omics databases and tools in human health. In: Kesheri M, Kanchan S, Salisbury TB, Sinha RP (eds) Microbial omics in environment and health. Springer, Singapore. https://doi.org/10.1007/978-981-97-1769-9_12

Kanchan S, Mehrotra R, Chowdhury S (2014) Evolutionary pattern of four representative DNA repair proteins across six model organisms: an in silico analysis. Netw Model Anal Health Inform Bioinforma 3:70. https://doi.org/10.1007/s13721-014-0070-1

Kaźmierczak-Siedlecka K, Marano L, Merola E, Roviello F, Połom K (2022) Sodium butyrate in both prevention and supportive treatment of colorectal cancer. Front Cell Infect Microbiol 12:1023806. https://doi.org/10.3389/fcimb.2022.1023806

Kesheri M, Kanchan S, Salisbury TB, Sinha RP (2024) Microbial omics in environment and health. Springer Nature Singapore. https://doi.org/10.1007/978-981-97-1769-9

Knudsen AB, Zauber AG, Rutter CM et al (2016) Estimation of benefits, burden, and harms of colorectal cancer screening strategies: modeling study for the US preventive services task force. JAMA 315(23):2595–2609. https://doi.org/10.1001/jama.2016.6828

Koh HWL, Fermin D, Vogel C, Choi KP, Ewing RM, Choi H (2019) iOmicsPASS: network-based integration of multiomics data for predictive subnetwork discovery. NPJ Syst Biol Appl 5:22. https://doi.org/10.1038/s41540-019-0099-y

Krzyszczyk P, Acevedo A, Davidoff EJ, Timmins LM, Marrero-Berrios I, Patel M, White C, Lowe C, Sherba JJ, Hartmanshenn C, O'Neill KM (2018) The growing role of precision and personalized medicine for cancer treatment. Technology (Singap World Sci) 6(3–4):79–100. https://doi.org/10.1142/S2339547818300020

Kumar P, Kanchan S, Kesheri M (2024) Multi-omics in human disease biomarker discovery. In: Kesheri M, Kanchan S, Salisbury TB, Sinha RP (eds) Microbial omics in environment and health. Springer, Singapore. https://doi.org/10.1007/978-981-97-1769-9_8

Lan F, Demaree B, Ahmed N, Abate AR (2017) Single-cell genome sequencing at ultra-high-throughput with microfluidic droplet barcoding. Nat Biotechnol 35(7):640–646. https://doi.org/10.1038/nbt.3880

Langfelder P, Horvath S (2008) WGCNA: an R package for weighted correlation network analysis. BMC Bioinformatics 9:559. https://doi.org/10.1186/1471-2105-9-559

Lee YK, Kang BS, Yoon JH (2023) Association between gut microbiome composition and the development of colon polyps: a case-control study. PLoS One 18(1):e0262234

Liang J, Wang X, Yang J, Sun P, Sun J, Cheng S, Liu J, Ren Z, Ren M (2023) Identification of disulfidptosis-related subtypes, characterization of tumor microenvironment infiltration, and development of a prognosis model in breast cancer. Front Immunol 14:1198826. https://doi.org/10.3389/fimmu.2023.1198826

Locey KJ, Lennon JT (2016) Scaling laws predict global microbial diversity. Proc Natl Acad Sci USA 113(21):5970–5975. https://doi.org/10.1073/pnas.1521291113

Lock EF, Dunson DB (2013) Bayesian consensus clustering. Bioinformatics 29(20):2610–2616. https://doi.org/10.1093/bioinformatics/btt425

Lynch HT, de la Chapelle A (1993) Hereditary colorectal cancer. N Engl J Med 329(10):681–692

Mallik B, Mishra T, Dubey P, Kesheri M, Kanchan S (2024) Exploring the secrets of microbes: unveiling the hidden world through microbial omics in environment and health. In: Kesheri M, Kanchan S, Salisbury TB, Sinha RP (eds) Microbial omics in environment and health. Springer, Singapore. https://doi.org/10.1007/978-981-97-1769-9_10

Mishra T, Mallik B, Kesheri M, Kanchan S (2024) The interplay of gut microbiome in health and diseases. In: Kesheri M, Kanchan S, Salisbury TB, Sinha RP (eds) Microbial omics in environment and health. Springer, Singapore. https://doi.org/10.1007/978-981-97-1769-9_1

Nejman D, Livyatan I, Fuks G, Gavert N, Zwang Y, Geller LT, Rotter-Maskowitz A, Weiser R, Mallel G, Gigi E, Meltser A (2020) The human tumor microbiome is composed of tumor

type-specific intracellular bacteria. Science 368(6494):973–980. https://doi.org/10.1126/science.aay9189
Nguyen ND, Wang D (2020) Multiview learning for understanding functional multiomics. PLoS Comput Biol 16(4):e1007677. https://doi.org/10.1371/journal.pcbi.1007677
Nilbert M, Planck M, Fernebro E (2000) A mutation analysis of the APC gene in colorectal cancer pedigrees from northern Sweden. N Engl J Med 82(3):523–527
O'Connell JB, Maggard MA, Ko CY (2004) Colon cancer survival rates with the new American Joint Committee on Cancer sixth edition staging. J Natl Cancer Inst 96(19):1420–1425. https://doi.org/10.1093/jnci/djh275
Pai SG, Carneiro BA, Mota JM, Costa R, Leite CA, Barroso-Sousa R, Kaplan JB, Chae YK, Giles FJ (2017) Wnt/beta-catenin pathway: modulating anticancer immune response. J Hematol Oncol 10(1):101. https://doi.org/10.1186/s13045-017-0471-6
Parhi L, Alon-Maimon T, Sol A, Nejman D, Shhadeh A, Fainsod-Levi T, Yajuk O, Isaacson B, Abed J, Maalouf N, Nissan A (2020) Breast cancer colonization by *Fusobacterium nucleatum* accelerates tumor growth and metastatic progression. Nat Commun 11(1):3259. https://doi.org/10.1038/s41467-020-16967-2
Paska AV, Hudler P (2015) Aberrant methylation patterns in cancer: a clinical view. Biochem Med (Zagreb) 25(2):161–176. https://doi.org/10.11613/BM.2015.017
Petitprez F, Meylan M, de Reyniès A, Sautès-Fridman C, Fridman WH (2020) The tumor microenvironment in the response to immune checkpoint blockade therapies. Front Immunol 11:784. https://doi.org/10.3389/fimmu.2020.00784
Raman M, Ambalam P, Kondepudi KK, Pithva S, Kothari C, Patel AT, Purama RK, Dave JM, Vyas BRM (2013) Potential of probiotics, prebiotics and synbiotics for management of colorectal cancer. Gut Microbes 4(3):181–192. https://doi.org/10.4161/gmic.23919
Rao VP, Poutahidis T, Ge Z (2006) Proinflammatory CD4+ CD45RBhi lymphocytes promote mammary and intestinal carcinogenesis in ApcMin/+ mice. Cancer Res 66(1):57–61. https://doi.org/10.1158/0008-5472.CAN-05-3035
Rappoport N, Shamir R (2019) NEMO: cancer subtyping by integration of partial multi-omics data. Bioinformatics 35(18):3348–3356. https://doi.org/10.1093/bioinformatics/btz058
Rappoport N, Safra R, Shamir R (2020) MONET: multi-omic module discovery by omic selection. PLoS Comput Biol 16(9):e1008182. https://doi.org/10.1371/journal.pcbi.1008182
Round JL, Mazmanian SK (2009) The gut microbiota shapes intestinal immune responses during health and disease. Nat Rev Immunol 9(5):313–323. https://doi.org/10.1038/nri2515
Sanghvi VR, Leibold J, Mina M, Mohan P, Berishaj M, Li Z, Miele MM, Lailler N, Zhao C, de Stanchina E, Viale A (2019) The oncogenic action of NRF2 depends on de-glycation by fructosamine-3-kinase. Cell 178(4):807–819.e21. https://doi.org/10.1016/j.cell.2019.07.031
Schiller JT, Lowy DR (2021) An introduction to virus infections and human cancer. Recent Results Cancer Res 217:1–11. https://doi.org/10.1007/978-3-030-57362-1_1
Sepich-Poore GD, Zitvogel L, Straussman R, Hasty J, Wargo JA (2021) A microbial perspective of human developmental biology. Nature 593(7857):211–220. https://doi.org/10.1038/s41586-021-03364-1
Shi H, Shi Q, Grodner B, Lenz JS, Zipfel WR, Brito IL, De Vlaminck I (2020) Highly multiplexed spatial mapping of microbial communities. Nature 588:676–681
Sivamaruthi BS, Kesika P, Chaiyasut C (2020) The role of probiotics in colorectal cancer management. Evid Based Complement Alternat Med 2020:3535982. https://doi.org/10.1155/2020/3535982
Son MY, Cho HS (2023) Anticancer effects of gut microbiota-derived short-chain fatty acids in cancers. J Microbiol Biotechnol 33(7):849–856. https://doi.org/10.4014/jmb.2301.01031
Srivastava U, Kanchan S, Kesheri M, Gupta MK, Singh S (2024) Types of omics data: genomics, metagenomics, epigenomics, transcriptomics, proteomics, metabolomics, and phenomics. In: Gupta MK, Katara P, Mondal S, Singh RL (eds) Integrative omics: concepts, methodology, and application. Chapter 2. Academic Press, United States, pp 13–34. https://doi.org/10.1016/B978-0-443-16092-9.00002-3

Sun S, Zhu L, Hu Y, Liu Y (2018) Studies on the metabolism of paeoniflorin in human intestinal microflora by high performance liquid chromatography/electrospray ionization/Fourier transform ion cyclotron resonance mass spectrometry and quadrupole time-of-flight mass spectrometry. J Chromatogr B Analyt Technol Biomed Life Sci 1085:63–71. https://doi.org/10.1016/j.jchromb.2018.03.042

Terzić J, Grivennikov S, Karin E, Karin M (2010) Inflammation and colon cancer. Gastroenterology 138(6):2101–2114.e5. https://doi.org/10.1053/j.gastro.2010.01.058

Vallianou N, Christodoulatos GS, Karampela I, Tsilingiris D, Magkos F, Stratigou T, Kounatidis D, Dalamaga M (2021) Understanding the role of the gut microbiome and microbial metabolites in non-alcoholic fatty liver disease: current evidence and perspectives. Biomol Ther 12(1):56. https://doi.org/10.3390/biom12010056

Vaske CJ, Benz SC, Sanborn JZ, Earl D, Szeto C, Zhu J, Haussler D, Stuart JM (2010) Inference of patient-specific pathway activities from multi-dimensional cancer genomics data using PARADIGM. Bioinformatics 26(12):i237–i245. https://doi.org/10.1093/bioinformatics/btq182

Wang Z, Gerstein M, Snyder M (2009) RNA-Seq: a revolutionary tool for transcriptomics. Nat Rev Genet 10(1):57–63. https://doi.org/10.1038/nrg2484

Wang B, Mezlini AM, Demir F, Fiume M, Tu Z, Brudno M, Haibe-Kains B, Goldenberg A (2014) Similarity network fusion for aggregating data types on a genomic scale. Nat Methods 11(3):333–337. https://doi.org/10.1038/nmeth.2810

Wang Z, Chai C, Wang R, Feng Y, Huang L, Zhang XX, Yang S, Zhang Y, Zhang X (2021) Single-cell transcriptome atlas of human mesenchymal stem cells exploring cellular heterogeneity. Clin Transl Med 11(12):e650. https://doi.org/10.1002/ctm2.650

Wu D, Wang D, Zhang MQ, Gu J (2015) Fast dimension reduction and integrative clustering of multi-omics data using low-rank approximation: application to cancer molecular classification. BMC Genomics 16:1022. https://doi.org/10.1186/s12864-015-2223-8

Wu X, Li J, Zhang Y, Cheng Y, Wu Z, Zhan W, Deng Y (2023) Identification of immune cell infiltration landscape for predicting prognosis of colorectal cancer. Gastroenterol Rep (Oxf) 11:goad014. https://doi.org/10.1093/gastro/goad014

Xu X, Hou Y, Yin X, Bao L, Tang A, Song L, Li F, Tsang S, Wu K, Wu H, He W (2012) Single-cell exome sequencing reveals single-nucleotide mutation characteristics of a kidney tumor. Cell 148(5):886–895. https://doi.org/10.1016/j.cell.2012.02.025

Yang J, Liu KQ, Zhou M, Wu Y, Tian Y, Wu J (2021) Dendritic cells in tumour microenvironment and progression of cancer: a critical review. Clin Dev Immunol. https://doi.org/10.1155/2021/4480891

Zackular JP, Baxter NT, Iverson KD, Sadler WD, Petrosino JF, Chen GY, Schloss PD (2013) The gut microbiome modulates colon tumorigenesis. mBio 4(6):e00692–e00613. https://doi.org/10.1128/mBio.00692-13

Zeller G, Tap J, Voigt AY, Sunagawa S, Kultima JR, Costea PI, Amiot A, Böhm J, Brunetti F, Habermann N, Hercog R (2014) Potential of fecal microbiota for early-stage detection of colorectal cancer. Mol Syst Biol 10(11):766. https://doi.org/10.15252/msb.20145645

Zhang N, Kandalai S, Zhou X, Hossain F, Zheng Q (2023) Applying multi-omics toward tumor microbiome research. Imeta 2(1):e73. https://doi.org/10.1002/imt2.73

Zhu T, Zhu Y, Xuan Y, Gao H, Cai X, Piersma SR, Pham TV, Schelfhorst T, Haas RR, Bijnsdorp IV, Sun R (2020) DPHL: a DIA pan-human protein mass spectrometry library for robust biomarker discovery. Genom Proteom Bioinform 18(2):104–119. https://doi.org/10.1016/j.gpb.2019.11.008

Human–Environment Interactions: A Multi-Omics and Interactome Perspective

12

Swarna Kanchan, Minu Kesheri, Poonam Kaithal, Upasna Srivastava, Harleen Kaur, Jainendra Pathak, and Rajeshwar P. Sinha

Abstract

To understand the effects of the environment as a whole (envirome), it is important to explore various specific domains to assess how, individually and collectively, these domains affect human health. Natural environment features such as sunlight, altitude, diurnal rhythms, vegetation, and biodiversity affect human health. However, the effects of the natural environment are moderated by the social environment, which comprises built environments, agricultural and industrial activities, pollutants and contaminants, culture, economic activities, and social networks that affect health by influencing access to health care. An understanding of the interactions between different domains of the envirome and their

S. Kanchan · M. Kesheri (✉)
Boise State University, Boise, ID, USA

Joan C. Edwards School of Medicine, Marshall University, Huntington, WV, USA
e-mail: minukesheri@boiestate.edu

P. Kaithal
Department of Molecular and Cellular Engineering, Jacob Institute of Biotechnology and Bioengineering SHUATS, Prayagraj, India

U. Srivastava
Yale School of Medicine, New Haven, CT, USA

H. Kaur
CSIR-National Physical Laboratory, New Delhi, India

J. Pathak
Department of Botany, Pt. Jawaharlal Nehru College (Affiliated to Bundelkhand University, Jhansi), Banda, India

R. P. Sinha
Laboratory of Photobiology and Molecular Microbiology, Department of Botany, Institute of Science, Banaras Hindu University, Varanasi, India

M. Kesheri et al. (eds.), *Multi-Omics in Biomedical Sciences and Environmental Sustainability*, https://doi.org/10.1007/978-981-96-7067-3_12

integrated effects on human health using various muti-omics and their integration studies could help in exploring the human–environment interactions. This chapter also discusses the concept of interactome and its role in unrevealing human diseases. This chapter illuminates the basic concepts of human health, focusing on elaborating various multi-omics approaches that are instrumental in elucidating the complex interactions between humans and the environment.

Keywords

Diseases · Environment · Human health · Interactome · Multi-Omics

Objectives

- To understand the complex human–environment interaction considering the biological, social, and cultural aspect
- To explore the concept of exposome, and the effect of different environmental factors, such as pollution, on human health
- To use multi-omics approaches to understand the effect of environmental factors on human health
- To explore the concept of interactome and its applications in human health

12.1 Introduction

Humans live in complex environments that are unique combinations of culture and history. During evolution, humans adapted to such cultures and environments. Understanding and quantifying environmental determinants of health for humans becomes difficult due to the lack of a universal conceptual framework for def ining the human environment. Human–environment interaction is multidimensional and is controlled by various environmental factors. Understanding such interaction is important because that affects human health and disease susceptibility. The psychological state of every individual is internal, but it is also affected by the environment.

In biology, the environment is defined as the overall surroundings, which is due to a complex interplay of physical, chemical, and biotic factors. Such an environment determines the overall health and survival of an organism or a community.

12.2 Environment

12.2.1 The Natural Environment

The natural environment includes all living and nonliving things that are natural (i.e., not man-made). Our environment is composed of five main systems or spheres on earth. The first system or sphere of the environment is called the geosphere. The geosphere includes the solid portions of the earth, such as the interior and surface of the earth, both of which are made up of rocks. The geosphere is also defined in terms of longitude, latitude, and altitude. Geosphere differ in their climate, sunlight, temperature, rainfall, and seasons. An individual's geosphere is also characterized by the diurnal cycles (night and day) of an individual. Diurnal cycles are well synchronized to the circadian rhythms, which refer to the physical, mental, and behavioral changes within an individual over a 24-hour cycle. A mismatch between circadian rhythms and the diurnal cycle increases the risk of several diseases, such as cardiovascular diseases (Gangwisch et al. 2006), diabetes (Knutson et al. 2006), and obesity (Kohatsu et al. 2006).

The second system or sphere of the environment is called as biosphere, which is composed of all plants, animals, and microbes. Due to the presence of environmental microbial diversity and their interaction with humans, the biosphere influences our health. Various environmental microorganisms interact with the human gut microbiome and play important beneficial immunomodulatory roles (von Hertzen et al. 2011). Therefore, a biosphere that is rich in microbial biodiversity protects against allergic and autoimmune diseases (Rook 2009).

12.2.2 The Social Environment

Within the natural environment, humans create diverse social environments. The social environment refers to the various activities/relationships humans establish with nature. Interactions of humans with nature result in activities, such as agriculture, and various social environments have also played important roles in human health and helped in their survival.

Current approaches for studying environmental effects on human health are mostly based on social determinants of health, such as socioeconomic, demographic, psychological, and geographical. The socioeconomic status (SES) of an individual within a society has an important impact on health. Individuals with lower SES were generally observed to have poor health. Individuals who face discrimination, and disparities such as African Americans in the United States have a high risk of cardiovascular health (Lukachko et al. 2014; Beatty Moody et al. 2016). To sustain the social environment, humans create complex and artificially built environments, such as the construction of houses, to ensure comfort and safety. Focusing more on complex and artificially built environments has led to the progressive separation of humans from their natural ecosystems. Nowadays, more than 50% of the world's population lives in artificially built environments in cities/towns. Therefore, artificially built environments determine human health, safety, and well-being. Artificially built environments that lack clean water, energy supplies, drainage systems, hygiene, and waste disposal could lead not only to the transmission of infectious diseases but improper management of artificially built environments is also responsible for enhancing the risk of various diseases.

12.2.3 The Personal Environment

Personal environments are built by individuals, which include occupation, diet, nutrition, physical activity, and chemical use choices. The personal environment refers to microenvironments and is limited to an individual and differs from the natural and social environment, which includes larger domains of the environment (the ecosystem or the society). The personal environment includes various choices such as nutrition, physical activity, the use of alcohol, drugs, and tobacco, which affect human health and are responsible for the high risk of various diseases. The personal environment has been shown as an important determinant of cardiovascular health due to lifestyle choices related to nutrition.

12.3 Exposome

The exposome (Wild 2005) refers to environmental exposures (including lifestyle factors) in totality (exposure from conception to death). Various types of environmental exposures, whether external or internal factors, contribute to human health and disease. The environmental exposures are categorized into three categories, which are called internal, specific external, and general external (Rappaport and Smith 2010; Prüss-Ustün et al. 2019). The internal environment refers to the body's internal chemical environment, composed of biologically active chemicals produced by metabolism, inflammation, oxidative stress, etc. Specific external exposures consist of environmental pollutants, tobacco smoke, infectious agents, alcohol, and diet. Lastly, general external exposures in the exposome include climate conditions, socioeconomic status, urban/rural environment, noise levels, and broader social determinants of health that affect a population at large without considering the individual lifestyle.

12.4 Pollution

Pollution is defined as unwanted materials that could be dangerous and introduced into the natural environment due to various human activities that threaten human health. Pollution harms the natural environment, and it is considered the largest role player in human disease and premature mortality. Pollution is responsible for several diseases and premature mortality. Both urban and rural areas face pollution from several resources, including land, water, and air.

12.4.1 Air Pollution

Air pollution contributes to greenhouse gases (carbon dioxide (CO_2), methane (CH_4), and nitrous oxide (N_2O), in the environment which contribute to global warming, along with affecting human health. The respiratory tract was observed as the main organ affected by air pollution, and it is studied the most among all organs. In a recent study, exposure to air pollution illustrated the association of air pollution with lung cancer regardless of smoking habits, which are considered the main cause of lung cancer. Air pollution contributed to 500,000 lung cancer deaths and 1.6 million chronic obstructive pulmonary disease (COPD) deaths in the United States in 2018 (Schraufnagel et al. 2019). Continuous exposure to fine particulate matter (PM <2.5 μm) was found to be linked to adverse cardiovascular diseases such as myocardial infarction, stroke, and sudden cardiac death. Overall, air pollution accounts for 19% of all cardiovascular deaths and 21% of all stroke deaths. The long-term effects of exposure to air pollution on the risk of cardiovascular diseases are also studied (Newby et al. 2015). Environmental exposures may increase the risk of autoimmune diseases (Farhat et al. 2011). Air pollution also plays a role in bone density and mineralization. The National Health Database of Taiwan and the Taiwan Environmental Protection Agency found an association between exposure to carbon monoxide (CO) and NO_2 and osteoporosis (Chang et al. 2015). Air pollution has adverse effect on human brain which affects cognitive function in older individuals and increases the risk of dementia such as Alzheimer's and Parkinson's diseases. Air pollution can damage the developing brain, which is especially concerning because this damage can impair cognitive function across the lifespan (Schraufnagel et al. 2019). Older adults more heavily exposed to air pollution perform poorly on their cognitive testing, and air pollution also increases the risk of dementia compared with less exposed adults (Wellenius et al. 2012).

The coronavirus pandemic caused by SARS-CoV-2 illustrated the impact of air pollution on human health by decreasing the number of premature deaths in Europe (Dąbrowiecki et al. 2022). During the lockdown, PM2.5 concentrations decreased on average by 17.1% in Europe compared with pre-pandemic daily PM2.5 concentrations in 2016–2019, which reduced around 2190 premature deaths in Europe within 3 months in 2020 (Dąbrowiecki et al. 2022).

12.4.2 Soil Pollution, Water Pollution, and Health Effects

Soil pollution refers to the presence of more pollutants than normal concentrations due to various human activities that affect human and ecosystem health. The main reasons for soil pollution are due to macro- and microplastic, overfertilization, deforestation, heavy metals, pesticides, etc. Toxic substances such as heavy (transition) metals and pesticides pollute soil and initiate oxidative stress in living beings. Heavy metals such as cadmium, lead, and arsenic further induce cardiovascular diseases. Chemical soil pollutants such as polychlorinated biphenyls, benzene, and bisphenol A (BPA) originate from industries and have adverse effects on human health. Adverse health effects that are mediated by various mechanisms such as induction of oxidative stress, inflammation, or epigenetic dysregulations (e.g., via microRNAs) could increase the risk for cancer, atherosclerosis, apoptosis, obesity, neurodegenerative disorders, and other cardiometabolic diseases in men and women (Drechsel and Patel 2008; Miguel et al. 2018). Water pollution affects the health of both women and men, which induces hormonal changes, accelerates aging, and induces various age-related health problems (Bochynska et al. 2024). Moreover, certain water pollutants induce neurological and psychiatric disorders, such as mood swings, depression, and anxiety, impacting the health of both women and men. The oxidative stress induced by pollutants inhibits spermatogenesis, and therefore reduces the sperm count in men (Bochynska et al. 2024).

12.5 Role of Multi-Omics in Studying Effects of the Environment on Human Health

Omics encompasses various high-throughput technologies such as genomics, epigenomics, transcriptomics, proteomics, metabolomics, and metagenomics. (Srivastava et al. 2024). To fully understand the complex human–environment interactions that could explain the etiology of various diseases, information across various levels must be integrated (Noble et al. 2022). Individual profiles from the genome, epigenome, transcriptome, proteome, metabolome, and microbiome—collectively known as multi-omics should be integrated (Kanchan et al. 2019). Integration of molecular profiles from various omics technologies should also be integrated with exposome to understand the effect of the environment on human health and diseases (Wu et al. 2023). Multi-omics technologies in the environment and human health and their interaction research have resulted in a better understanding of environmental toxicity-based health issues and diseases (Kesheri et al. 2024b) by discovering novel biomarkers as shown in Fig. 12.1. Moreover, multi-omics technologies helped us better understand the development of diseases due to environmental chemical (individual/mixture) exposure. The environment causes gene mutations; these mutated genes synthesize mutated proteins, which in turn synthesize lipids, carbohydrates, and metabolites that respond to environmental changes.

Genomics provides us with a better understanding of the genetic makeup of various organisms present in the environment. Genomics includes genome assembly, gene prediction, and functional annotation using various soft computing tools as well as machine learning methods (Kumar et al. 2024; Kanchan et al. 2019). Genomics has been enhancing our understanding of various diseases by investigating genetic variants such as SNPs, insertions-deletions, and structural variants. Genomics is also applied to various toxicity studies due to chemicals/pesticides and helps us in investigating novel biomarkers and disease pathogenesis (Kesheri et al. 2024a; Kumar et al. 2024).

Transcriptome refers to all RNA transcripts, including coding and noncoding types, which provide comprehensive insights about mRNA expression, alternative splicing, etc. Transcriptomics studies identify differentially expressed genes in response to individual chemicals or a mixture of chemicals present in the environment. Integrated transcriptomics and genomics are used to examine human aortic endothelial cells' response to oxidized phospholipids that are responsible for atherosclerosis (Romanoski et al. 2010). In this integrated transcriptomics and genomics study, gene expression responses to environmental stimuli at a few genetic loci

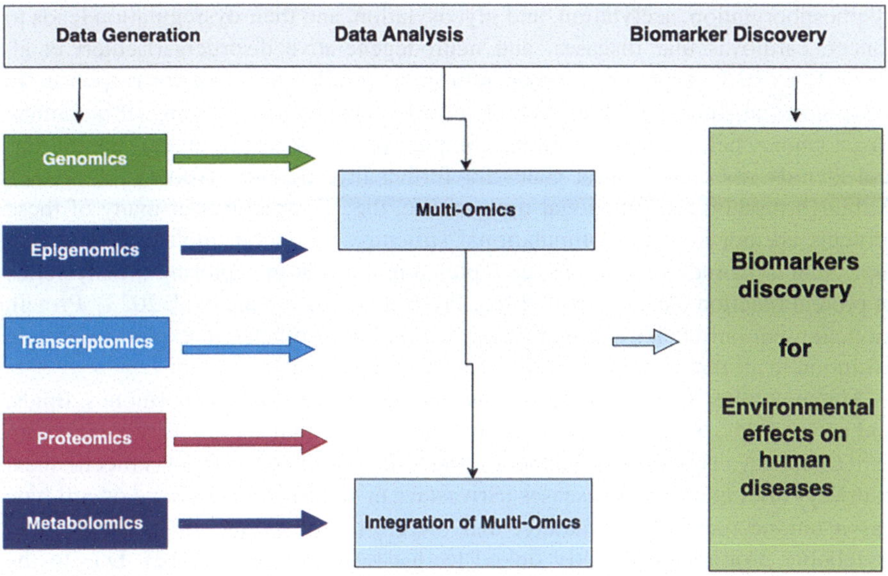

Fig. 12.1 Multi-omics approaches and its integration to explore the novel biomarkers to understand the effects of environment on human health and diseases

were studied. Various studies also illustrate how integrating genomic and transcriptomic data could help us in exploring complex gene–environment interactions in cardiovascular diseases and other diseases such as Alzheimer's diseases (Orozco et al. 2012; Dermitzakis 2015; Eid et al. 2019).

Epigenetic modifications regulate gene expression without altering the DNA sequence and are involved in various cellular events and tumorigenesis (Stricker et al. 2016; Mehrmohamadi et al. 2021). Epigenomics studies various epigenetic modifications, such as DNA methylation and histone modification, and plays a crucial role in understanding how environmental factors and genetic makeup interact to influence the development of various diseases as well as increase the risk of cancer (Shiek et al. 2021; Stepanyan et al. 2023). Similarly, prenatal dietary formulations such as folate intake could alter the epigenome, affect fetal development, and increase the risk for various diseases in life (Kim et al. 2009; Choi and Friso 2010; Lee 2015). Epigenomics could provide a molecular framework to understand how the human gene interacts and responds to environmental factors and participates in disease etiology. Moreover, by studying various epigenetic modifications, researchers could discover novel genes and pathways influenced by genetic factors and environmental exposures.

Proteomics explores the entire array of proteins produced by an organism, offering insights into disease development and progression. Unlike the genome, the proteome is highly dynamic, exhibiting a significant amount of variability due to processes like alternative splicing and protein modifications. Proteins are modified by phosphorylation, acetylation, and glycosylation, and their dysregulation leads to cancer, cardiovascular diseases, and neurodegenerative disorders (Leutert et al. 2021; Lee 2015). Proteomics-based molecular profiles also capture responses to environmental stimuli, such as diet, chemical exposure, and smoking. Proteomics-based studies help in understanding complex proteome–environment interactions and identify associated target genes for further investigation (Noble et al. 2022). Although proteins are functional units of life, the 3D structures of many of these proteins are not known. Computational structure prediction, molecular docking, molecular dynamics simulations, etc., play a major role in exploring the mysteries of protein function (Kesheri et al. 2015; Priya et al. 2017; Sahu et al. 2023). Protein modeling not only helps in function prediction but also helps in exploring protein evolution in all three domains of life (Kanchan et al. 2019).

Metabolomics focuses on small-molecule metabolites such as hormones, lipids, and amino acids, which are produced as intermediates and end products of metabolism (Srivastava et al. 2023). Metabolites are directly linked to various biochemical pathways driving complex diseases (Srivastava et al. 2024). Each tissue or cell type has a unique metabolic signature, and therefore any changes in tissue-specific metabolite profiles are directly linked to disease (Patti et al. 2012). Due to the dynamic nature of the metabolites, these metabolites respond quickly to factors, such as diet and chemical exposure, making metabolomics-based studies significant for studying gene–environment interactions in various diseases (Gieger et al. 2008). Integrating metabolomics data with genomics and proteomics enhances our ability to map metabolic pathways efficiently. Metabolomics enables the integrated study of both endogenous metabolites with environmental chemicals and pollutants thus helping us to understand how external exposures impact our internal biological systems (Rattray et al. 2018). Analyzing these external compounds along with changes in the endogenous metabolome reveals the mysteries of the biological effects of environmental exposures due to harmful chemicals/pesticides. A key challenge in metabolomics is the identification and quantification of metabolites, but recent advancements in various metabolomics experimental techniques have significantly removed these limitations (Patti et al. 2012; Rattray et al. 2018).

12.6 The Interactome

In any organism, proteins are large macromolecules responsible for a variety of functions that govern various biological/molecular processes, which result in a signaling cascade through protein–protein interactions. Understanding such protein–protein interactions is the first step to understanding the concept of interactome. Interactome refers to various protein–protein interactions that form protein–protein interaction networks (PPINs). Studying interactome provides a powerful tool for understanding the complex relationships of protein–protein interactions governing cellular processes and other events inside the cells (Alemu et al. 2025).

The human genome includes approximately 20,000 protein-coding genes, which provide 200 million unique protein interactions without including protein isoforms. Various experimental techniques were discovered to identify protein–protein interactions. Low throughput techniques such as co-immunoprecipitation (co-IP) have failed because it is resource-intensive and yield high false-positive results. On the other hand, various high–throughput techniques such as yeast two-hybrid (Y2H) and tandem affinity purification mass-spectrometry (TAP-MS) performed better in identifying the protein–protein interactions and producing a low number of false-positive results.

Computational approaches have also been discovered that utilize amino acid-based protein sequences. Sequence-based predictors rely on features computed from amino acid sequences such as physicochemical properties. Non-sequence-based predictors perform better by utilizing functional, protein subcellular location, structural, and other biological annotations to predict the protein–protein interactions (Dunham and Ganapathiraju 2022). Several recently published predictors claimed their accuracy, as high as 98% (Dunham and Ganapathiraju 2022).

Protein–protein interaction networks help discover functional enrichment, and identify drug targets, etiology, and disease progression to various complex diseases such as cancer, autoimmune disorders, and diabetes (Safari-Alighiarloo et al. 2014).

Currently, various human PPINs databases are available such as HINT (Das and Yu 2012), IntAct (Orchard et al. 2014), Reactome (Wu et al. 2010), and STRING (Szklarczyk et al. 2023). HINT is a database of high-quality protein–protein interactomes for humans, yeast, and *Oryza sativa*. String database includes known and predicted protein–protein interactions, including direct (physical) and indirect (functional) associations, which include more than six million interactions. IntAct database includes the second-highest number of interactions. Database of interacting proteins (DIPs: http://dip.doe-mbi.ucla.edu/dip/Main.cgi) includes protein–protein interaction pairs that combine information from a variety of sources. HitPredict (http://hintdb.hgc.jp/htp/) includes experimentally determined protein–protein interactions from *Homo sapiens* and many other model organisms. Interactome3D (http://interactome3d.irbbarcelona.org/) database provides the structural annotation of protein–protein interaction networks.

12.7 Conclusion

This chapter includes various spheres of the environment such as the Biosphere and its interaction with human. This chapter discusses the complex human–environment interaction considering the biological, social, and cultural aspects. In this chapter, the concept of exposome is discussed, other different environmental factors such as pollution, and their effect on human health is well explained. Various multi-omics approaches are discussed to understand the effect of environmental factors on human health and diseases. The concept of interactome, various databases related to protein–protein applications, and use of interaction in understanding various diseases are well explained.

Short Chapter Summary

This chapter aims to understand the complex human–environment interaction considering the biological, social, and cultural aspect. Concept of exposome, and the effect of different environmental factors, such as pollution, on human health and diseases are discussed further in this chapter. Multi-omics approaches are discussed further to understand the effect of environmental factors on human health. Moreover, the interactome and its applications in human health are explained to understand human diseases.

Author Contributions Concept, design, and writing, Swarna Kanchan and Minu Kesheri; Review and editing, Rajeshwar P. Sinha, Poonam Kaithal, Upasna Srivastava, Harleen Kaur, Jainendra Pathak, Swarna Kanchan, and Minu Kesheri and contributed equally as first author.

References

Alemu R, Sharew NT, Arsano YY, Ahmed M, Tekola-Ayele F, Mersha TB, Amare AT (2025) Multi-omics approaches for understanding gene-environment interactions in noncommunicable diseases: techniques, translation, and equity issues. Hum Genomics 19(1):8. https://doi.org/10.1186/s40246-025-00718-9

Beatty Moody DL, Waldstein SR, Tobin JN, Cassells A, Schwartz JC, Brondolo E (2016) Lifetime racial/ethnic discrimination and ambulatory blood pressure: the moderating effect of age. Health Psychol 35:333–342

Bochynska S, Duszewska A, Maciejewska-Jeske M, Wrona M, Szeliga A, Budzik M, Szczesnowicz A, Bala G, Trzcinski M, Meczekalski B, Smolarczyk R (2024) The impact of water pollution on the health of older people. Maturitas 185:107981. https://doi.org/10.1016/j.maturitas.2024.107981

Chang KH, Chang MY, Muo CH, Wu TN, Hwang BF, Chen CY, Lin TH, Kao CH (2015) Exposure to air pollution increases the risk of osteoporosis: a nationwide longitudinal study. Medicine (Baltimore) 94(17):e733. https://doi.org/10.1097/MD.0000000000000733

Choi SW, Friso S (2010) Epigenetics: a new bridge between nutrition and health. Adv Nutr 1:8–16

Dąbrowiecki P, Badyda A, Chciałowski A, Czechowski PO, Wrotek A (2022) Influence of selected air pollutants on mortality and pneumonia burden in three polish cities over the years 2011–2018. J Clin Med 11:3084

Das J, Yu H (2012) HINT: high-quality protein interactomes and their applications in understanding human disease. BMC Syst Biol 6:92. https://doi.org/10.1186/1752-0509-6-92

Dermitzakis ET (2015) Gene-gene and gene-environment interactions detected by transcriptome sequence analysis in twins. Nat Genet 47:88–91

Drechsel DA, Patel M (2008) Role of reactive oxygen species in the neurotoxicity of environmental agents implicated in Parkinson's disease. Free Radic Biol Med 44:1873–1886

Dunham B, Ganapathiraju MK (2022) Benchmark evaluation of protein-protein interaction prediction algorithms. Molecules 27(1):41. https://doi.org/10.3390/molecules27010041

Eid A, Mhatre I, Richardson JR (2019) Gene-environment interactions in Alzheimer's disease: a potential path to precision medicine. Pharmacol Therap 199:173–187

Farhat SC, Silva CA, Orione MA, Campos LM, Sallum AM, Braga AL (2011) Air pollution in autoimmune rheumatic diseases: a review. Autoimmun Rev 11(1):14–21. https://doi.org/10.1016/j.autrev.2011.06.008

Gangwisch JE, Heymsfield SB, Boden-Albala B, Buijs RM, Kreier F, Pickering TG, Rundle AG, Zammit GK, Malaspina D (2006) Short sleep duration as a risk factor for hypertension: analyses of the first National Health and nutrition examination survey. Hypertension 47:833–839

Gieger C, Geistlinger L, Altmaier E, Hrabé de Angelis M, Kronenberg F, Meitinger T, Mewes HW, Wichmann HE, Weinberger KM, Adamski J, Illig T (2008) Genetics meets metabolomics: a genome-wide association study of metabolite profiles in human serum. PLoS Genet 4(11):e1000282

Here's the reference formatted according to the style of the "Annals of Forest Science" journal:

Kanchan S, Sharma P, Chowdhury S (2019) Evolution of endonuclease IV protein family: an in-silico analysis. 3 Biotech 9:168. https://doi.org/10.1007/s13205-019-1696-6

Kesheri M, Kanchan S, Chowdhury S, Sinha RP (2015) Secondary and tertiary structure prediction of proteins: a bioinformatic approach. In: Zhu Q, Azar AT (eds) Complex system modelling and control through intelligent soft computations. Springer, Germany, pp 541–569. https://doi.org/10.1007/978-3-319-12883-2_19

Kesheri M, Kanchan S, Salisbury TB, Sinha RP (2024a) Microbial omics in environment and health. Springer Nature Singapore. https://doi.org/10.1007/978-981-97-1769-9

Kesheri M, Kanchan S, Srivastava S, Ratna-Raj R, Chittoori B, Vaishampayan A, Rastogi RP, Sinha RP, Primerano D (2024b) Ecology and environmental omics. In: Gupta MK, Katara P, Mondal S, Singh RL (eds) Integrative omics: concepts, methodology, and application. Academic Press, pp 315–331. https://doi.org/10.1016/B978-0-443-16092-9.00019-9

Kim K, Friso S, Choi SW (2009) DNA methylation, an epigenetic mechanism connecting folate to healthy embryonic development and aging. J Nutr Biochem 20:917–926

Knutson KL, Ryden AM, Mander BA, Van Cauter E (2006) Role of sleep duration and quality in the risk and severity of type 2 diabetes mellitus. Arch Intern Med 166:1768–1774

Kohatsu ND, Tsai R, Young T, VanGilder R, Burmeister LF, Stromquist AM, Merchant JA (2006) Sleep duration and body mass index in a rural population. Arch Intern Med 166:1701–1705

Kumar P, Kanchan S, Kesheri M (2024) Multi-omics in human disease biomarker discovery. In: Kesheri M, Kanchan S, Salisbury TB, Sinha RP (eds) Microbial omics in environment and health. Springer, Singapore. https://doi.org/10.1007/978-981-97-1769-9_8

Lee HS (2015) Impact of maternal diet on the epigenome during in utero life and the developmental programming of diseases in childhood and adulthood. Nutrients 7:9492–9507

Leutert M, Entwisle SW, Villén J (2021) Decoding Post-Translational Modification Crosstalk With Proteomics. Mol Cell Proteomics (20)100129. doi: https://doi.org/10.1016/j.mcpro.2021.100129

Lukachko A, Hatzenbuehler ML, Keyes KM (2014) Structural racism and myocardial infarction in the United States. Soc Sci Med 103:42–50

Mehrmohamadi M, Sepehri MH, Nazer N, Norouzi MR (2021) A comparative overview of epigenomic profiling methods. Front Cell Dev Biol 9:26

Miguel V, Cui JY, Daimiel L, Espinosa-Diez C, Fernandez-Hernando C, Kavanagh TJ, Lamas S (2018) The role of microRNAs in environmental risk factors, noise-induced hearing loss, and mental stress. Antioxid Redox Signal 28:773–796

Newby DE, Mannucci PM, Tell GS (2015) Expert position paper on air pollution and cardiovascular disease. Eur Heart J 36(2):83b–93b. https://doi.org/10.1093/eurheartj/ehu458

Noble AJ, Purcell RV, Adams AT, Lam YK, Ring PM, Anderson JR, Osborne AJ (2022) A final frontier in environment-genome interactions? Integrated, multi-omic approaches to predictions of non-communicable disease risk. Front Genet 13:214

Orchard S, Ammari M, Aranda B, Breuza L, Briganti L, Broackes-Carter F, Campbell NH, Chavali G, Chen C, Del-Toro N, Duesbury M (2014) The MIntAct project–IntAct as a common curation platform for 11 molecular interaction databases. Nucleic Acids Res 42(Database issue):D358–D363. https://doi.org/10.1093/nar/gkt1115

Orozco LD, Bennett BJ, Farber CR, Ghazalpour A, Pan C, Che N, Wen P, Qi HX, Mutukulu A, Siemers N, Neuhaus I (2012) Unraveling inflammatory responses using systems genetics and gene-environment interactions in macrophages. Cell 151:658–670

Patti GJ, Yanes O, Siuzdak G (2012) Innovation: metabolomics: the apogee of the omics trilogy. Nat Rev Mol Cell Biol 13:263–269

Priya P, Kesheri M, Sinha RP, Kanchan S (2017) Molecular dynamics simulations for biological systems. In: Karâa WBA, Dey N (eds) Biomedical image analysis and mining techniques for improved health outcomes. IGI Global, pp 286–313. https://doi.org/10.4018/978-1-4666-8811-7.ch014

Prüss-Ustün A, van Deventer E, Mudu P, Campbell-Lendrum D, Vickers C, Ivanov I, Forastiere F, Gumy S, Dora C, Adair-Rohani H, Neira M (2019) Environmental risks and non-communicable diseases. BMJ 364:l265. https://doi.org/10.1136/bmj.l265

Rappaport SM, Smith MT (2010) Environment and disease risks. Science 330:460–461

Rattray NJW, Deziel NC, Wallach JD, Khan SA, Vasiliou V, Ioannidis JP, Johnson CH (2018) Beyond genomics: understanding exposotypes through metabolomics. Hum Genomics 12:215

Romanoski CE, Lee S, Kim MJ, Ingram-Drake L, Plaisier CL, Yordanova R, Tilford C, Guan B, He A, Gargalovic PS, Kirchgessner TG (2010) Systems genetics analysis of gene-by-environment interactions in human cells. Am J Hum Genet 86:399–410

Rook GA (2009) Review series on helminths, immune modulation and the hygiene hypothesis: the broader implications of the hygiene hypothesis. Immunology 126:3–11

Safari-Alighiarloo N, Taghizadeh M, Rezaei-Tavirani M, Goliaei B, Peyvandi AA (2014) Protein-protein interaction networks (PPI) and complex diseases. Gastroenterol Hepatol Bed Bench 7(1):17

Sahu N, Mishra S, Minu K, Kanchan S, Sinha RP (2023) Identification of cyanobacteria-based natural inhibitors against SARS-CoV-2 druggable target ACE2 using molecular docking study, ADME and toxicity analysis. Indian J Clin Biochem 38:361–373. https://doi.org/10.1007/s12291-022-01056-6

Schraufnagel DE, Balmes JR, Cowl CT et al (2019) Air pollution and noncommunicable diseases: a review by the forum of international respiratory societies' environmental committee, part 2: air pollution and organ systems. Chest 155(2):417–426. https://doi.org/10.1016/j.chest.2018.10.041

Shiek SS, Mani MS, Kabekkodu SP, Dsouza HS (2021) Health repercussions of environmental exposure to lead: methylation perspective. Toxicology 461:96

Srivastava U, Kanchan S, Kesheri M, Singh S (2023) Nutrimetabolomics: metabolomics in nutrition research. In: Soni V, Hartman TE (eds) Metabolomics. Springer, Cham, pp 241–268. https://doi.org/10.1007/978-3-031-39094-4_8

Srivastava U, Kanchan S, Kesheri M, Gupta MK, Singh S (2024) Types of omics data: genomics, metagenomics, epigenomics, transcriptomics, proteomics, metabolomics, and phenomics. In: Gupta MK, Katara P, Mondal S, Singh RL (eds) Integrative omics: concepts, methodology, and application. Academic Press, pp 13–34. https://doi.org/10.1016/B978-0-443-16092-9.00002-3

Stepanyan A, Petrackova A, Hakobyan S, Savara J, Davitavyan S, Kriegova E, Arakelyan A (2023) Long-term environmental metal exposure is associated with hypomethylation of CpG sites in NFKB1 and other genes related to oncogenesis. Clin Epigenetics 15:52

Stricker SH, Köferle A, Beck S (2016) From profiles to function in epigenomics. Nat Rev Genet 18:51–66

Szklarczyk D, Kirsch R, Koutrouli M, Nastou K, Mehryary F, Hachilif R, Gable AL, Fang T, Doncheva NT, Pyysalo S, Bork P (2023) The STRING database in 2023: protein-protein association networks and functional enrichment analyses for any sequenced genome of interest. Nucleic Acids Res 51(D1):D638–D646. https://doi.org/10.1093/nar/gkac1000

von Hertzen L, Hanski I, Haahtela T (2011) Natural immunity. Biodiversity loss and inflammatory diseases are two global megatrends that might be related. EMBO Rep 12:1089–1093

Wellenius GA, Boyle LD, Coull BA (2012) Residential proximity to nearest major roadway and cognitive function in community-dwelling seniors: results from the MOBILIZE Boston Study. J Am Geriatr Soc 60(11):2075–2080. https://doi.org/10.1111/j.1532-5415.2012.04195.x

Wild CP (2005) Complementing the genome with an "exposome": the outstanding challenge of environmental exposure measurement in molecular epidemiology. Environ Health Perspect 113:A166–A167

Wu G, Feng X, Stein L (2010) A human functional protein interaction network and its application to cancer data analysis. Genome Biol 11(5):R53. https://doi.org/10.1186/gb-2010-11-5-r53
Wu H, Eckhardt CM, Baccarelli AA (2023) Molecular mechanisms of environmental exposures and human disease. Nat Rev Genet 24:332–344

Bioremediation: A Boon for a Sustainable Environment

13

Pallavi Singhal

Abstract

Humans are advancing rapidly in science and technology. But this advancement has its price, which is being paid by the environment. Our desire to have a comfortable life can be challenging for the environment. Removing hazardous pollutants we release seems to be a strenuous task. The strategy for utilizing resources in such a manner that they are available for both present and future generations can be understood by the concept of sustainability. The need of the hour is to have a sustainable environment for our present and future generations. It is of utmost urgency to encourage activities to achieve it. Bioremediation is a microbial approach to remove pollutants generated by industrialization, transportation, and overpopulation. The efficiency of bioremediation can be enhanced by administering advanced omics technologies such as proteomics, genomics, and metabolomics. This chapter focuses on these omics tools that provide cost-effective means to facilitate the removal of pollutants and to achieve environmental sustainability.

Keywords

Bioremediation · Microbial omics · Phytoremediation · Proteomics · Sustainable environment

P. Singhal (✉)
Meerut Institute of Technology, Meerut, Uttar Pradesh, India

M. Kesheri et al. (eds.), *Multi-Omics in Biomedical Sciences and Environmental Sustainability*, https://doi.org/10.1007/978-981-96-7067-3_13

Objectives

- To highlight the concept of bioremediation, its types, and its role in dealing with environmental pollution
- To address the use of microbial, plant, and other nontoxic methods for the remediation of contamination from the environment
- To explore various objectives and applications of bioremediation for sustainable development of the environment in a green and cost-effective manner
- To highlight the use of omics technologies and nanotechnology for the enhancement of bioremediation efficiency

13.1 Introduction

Environment in brief may be defined as "our natural surroundings." These surroundings comprise biotic and abiotic factors that interact together to form a suitable environment for our sustenance. Ironically, humans are the culprits to ruin it. Overpopulation and human desires for affluence need rapid industrialization, urbanization, better transportation, etc., which can have their demerits regarding environmental sustainability if not properly utilized. Also, huge populations require land to live on and need to grow more food. The pressure to increase food productivity causes excessive use of chemicals in the crop field. Chemicals-enriched pesticides and fertilizers not only degrade the soil quality but also affect the whole ecosystem indirectly. Various pollutants released from different industries are highly toxic because they contain heavy metals and man-made chemicals in high concentrations. Being dumped without treatment, these pollutants can cause soil and water pollution and may further create serious health problems for humans. Therefore, it is necessary to process waste before dumping it. Humans have developed various chemical and physicochemical processes for decontamination, but they may have their limitations (Abdulsalam et al. 2012). They can be costly, less efficient, and can produce toxic by-products. These by-products still can be harmful to nature (Khan et al. 2019). Consequently, the shortcomings of these treatment processes need to be resolved. A different technology that removes toxic pollutants with the help of living organisms known as bioremediation can overcome the shortcomings of physicochemical processes. This method is economical and eco-friendly which proves its sustainability to the environment (Bhat et al. 2018; Olawale et al. 2020). Several environmental remediation strategies are being explored. The implementation of these strategies is limited due to inadequate information about desirable microbes' growth factors and metabolism. We are indeed developing, and this development is necessary. But in need of immediate benefits, we are neglecting its long-term impact on the environment and ourselves. Ultimately, humans' negligence toward the environment can cause extreme harm to the environment, which cannot be recovered by any means. Sustainability is the key to our issues. We need to create a sustainable environment for future generations. The concept of bioremediation for achieving a sustainable environment is reviewed in this chapter.

13.2 Bioremediation

Bioremediation is a biological process for detoxifying the environment from pollutants released by different industries, such as chemicals (petrochemicals, textiles, tanneries, and pharmaceuticals) and agro-based industries (Singh et al. 2020). Toxic chemicals such as polycyclic aromatic hydrocarbons (PAHs) and polychlorinated biphenyls (PCBs) are frequently released into the environment due to anthropogenic activities. If these xenobiotic chemicals excreted by these industries are not properly treated and dumped, they can contaminate our environment. Moreover, their unprocessed dumping can induce perilous health hazards for humans. The soil and underground water leached with heavy metals can have a detrimental effect on human health and the ecosystem. There are different approaches developed so far to mitigate these pollutants from the environment. Among them, biological treatment or bioremediation proves to be more efficient. The biosphere is rich with microorganisms that have enzymes to degrade complex forms into simpler forms (Nojiri and Tsuda 2005; Thakur et al. 2019). Microorganisms are adaptive and nutritionally flexible, which makes them a significant tool in bioremediation techniques. Microorganisms such as yeast, algae, and bacteria are used to clean up contaminated water and soil by bioremediation. In this process, microorganisms reduce the contaminants and convert them from lethal to less harmful forms. The conventional treatment methods are somehow lacking to provide a sustainable environment whereas bioremediation has proven effective and reliable. Nanoparticles from various biological sources have emerged as potential detoxifiers especially in the case of wastewater (Solomon et al. 2024). Bioremediation is an economical, eco-friendly strategy to detoxify pollutants (Lovely 1991). Thus, bioremediation technologies are extensively used and explored. Microbial bioremediation utilizes microorganisms for the detoxification of pollutants at contaminated sites. These microorganisms can be either present there indigenously or can be added at the site for remediation. Bioremediation is a restoration process that reduces the expansion of pollution in a natural ecosystem (Yerushalmi et al. 2003; Kumar et al. 2018). Microorganisms can decompose organic waste and prevent its accumulation. These microorganisms are known as natural cleaners and decompose organic matter and mix into the soil. Decomposers help in nutrient recycling and maintain soil fertility. The technologies that are used for bioremediation include mechanisms like biosequestration, biodegradation, phytohydraulics, biological extraction, and volatilization. Figure 13.1 lists various purposes, objectives, and applications of bioremediation to keep our environment sustainable and safe. Based on the application site of this strategy, bioremediation is of three types.

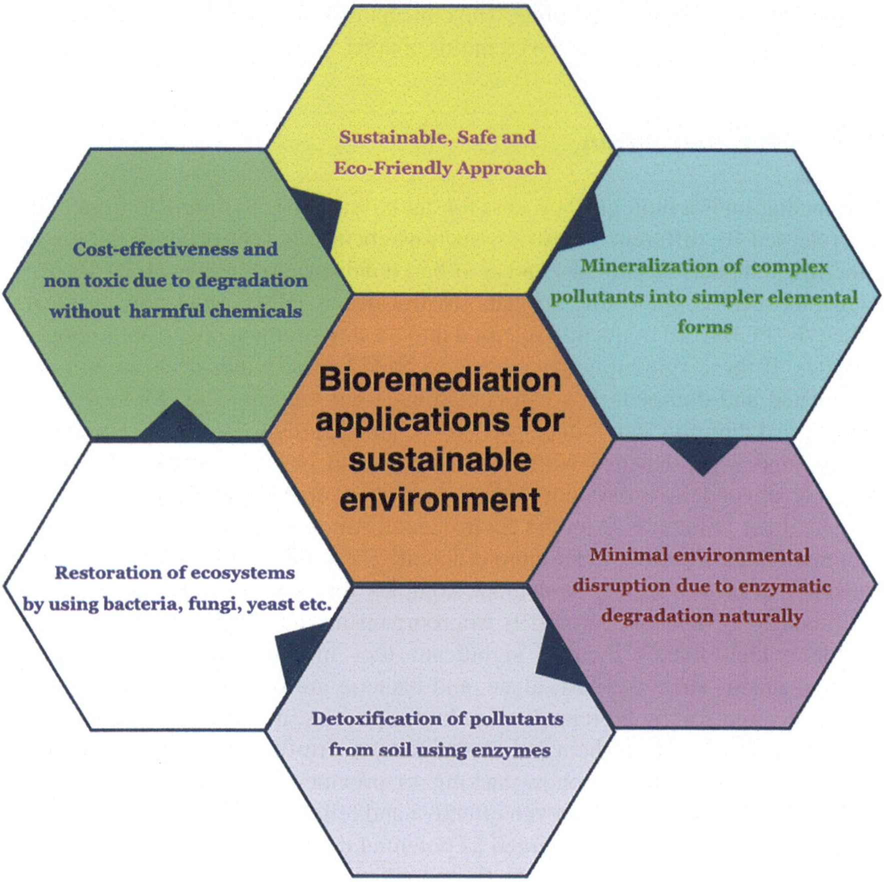

Fig. 13.1 Various applications of bioremediation to keep our environment sustainable

13.2.1 In Situ Bioremediation

In situ remediation is a bioremediation strategy where pollutants are treated at the site/point of contamination itself. It is divided into intrinsic in situ bioremediation and enhanced/engineered in situ bioremediation (Azubuike et al. 2016). In situ bioremediation is economically feasible as it does not require the cost of extraction and transportation. This method is effectively used in the treatment of areas contaminated by dyes, heavy metals, chlorinated solvents, and hydrocarbons (Kim et al. 2014). The limitation of in situ bioremediation is that it is less effective than ex situ bioremediation.

- *Intrinsic In Situ Bioremediation*
  The intrinsic in situ bioremediation process is the method where pollutants are mitigated naturally without any human intervention. Thus, this method is also

known as natural reduction. This technique exploits the capability of microbial organisms to function with or without oxygen for the biodegradation of contaminants (Rittmann and McCarty 2001).

- *Enhanced/Engineered In Situ Bioremediation*
  This type of in situ bioremediation involves humans for the inoculation of certain microorganisms into the polluted site. The improvement of polluted sites can be done using genetically engineered microorganisms, which enhance the physico-chemical conditions by the nutrients and air supplementation to enhance microbial growth for the biodegradation process. Enhanced in situ bioremediation process can be further divided into techniques such as bioaugmentation (Ali et al. 2020; Awasthi et al. 2020) biostimulation, bio slurping, biosparging, and bioventing (Tiwari and Singh 2014).
- *Bioaugmentation*
  Augmentation is a process to enhance the ability. Similarly, bioaugmentation is an augmentation process done with the help of living organisms. Bioaugmentation is a technique in which microorganisms are added either indigenously or exogenously to amplify the degradation of the contaminant at the pollution site. Genetically engineered microorganisms can also be used for bioaugmentation. The municipal wastewater can be treated and organic contaminants such as benzene and toluene can be removed using this technique (Major et al. 2002). Bioaugmentation is indeed a good process, but is limited by certain factors. Bioaugmentation falls short in its application because it is not suitable for all soil compositions and abiotic environmental factors (pH, temperature, moisture, etc.). The metabolic activity and diversity of the microbial community is reduced by excessive supplementation of nutrients (Wang et al. 2012). The bioaugmentation process may be inhibited by competitive and predation interactions between indigenous and exogenous microbial species. Therefore, proper management is required for the treatment (Sharma 2012).
- *Biostimulation*
  The biostimulation process is the addition of nutrients to stimulate naturally occurring microbial populations at the contamination site. The stimulation is done by optimizing physical parameters, which further favors the growth of microorganisms (Margesin et al. 2000). The stimulated microorganisms enhance the degradation of lower hydrocarbons and metals present in contaminated soil and water (Kao et al. 2008). Biostimulation is helpful in the degradation of contaminants without disturbing the environment, and thus it is an eco-friendly and cost-effective method (Mohee and Mudhoo 2012). The biostimulation technique utilizes preexisting native microorganisms that are already adapted to the environment (Adams et al. 2015). It requires long-term monitoring of contaminated sites. Biosparging and bioventing are improvised biostimulation techniques.
- *Bioventing*
  Bioventing is one of the most explored in situ remediation strategies (Höhener and Ponsin 2014). The oxygen and nutrients are introduced at low rates to stimulate the efficiency of microorganisms for the transformation of toxic pollutants (Atlas and Phillips 2005). Bioventing is commercially utilized for the treatment

of aerobically biodegradable substances such as simple petroleum hydrocarbons. Bioventing is inefficient when the water table is in proximity. It increases moisture in the soil and decreases its permeability. It may result in vapor formation which hinders the degradation process. Moreover, regular monitoring of vapors will be necessary.

- *Biosparging*
  Bioremediation techniques, biosparging and bioventing, are similar in principle. In both techniques, the air is injected at a low rate to enhance the microbial activity. The air injected in the saturated zone helps in the upward movement of volatile organic compounds to the unsaturated zone. This stimulates the biodegradation process. The limitation of bioventing can be overcome by the biosparging technique. Here, air is introduced below the water table. It increases the oxygen concentration of groundwater and enhances the aerobic degradation of contaminants. The credibility of biosparging depends on soil permeability and pollutant degradability (Philp and Atlas 2005). The biosparging process has been broadly used for the treatment of aquifers polluted with simple hydrocarbons (benzene, xylene, toluene, and ethylbenzene) and mineral oils (Kao et al. 2008).
- *Bioslurping*
  The technique of bioslurping combines characteristics of vacuum-enhanced pumping, soil vapor extraction, and bioventing. It is used to achieve soil and groundwater remediation by providing oxygen indirectly and stimulating biodegradation of contaminants (Gidarakos and Aivalioti 2007). Bioslurping is different from bioventing and biosparging. Here, oxygen is added indirectly, while it is added directly in bioventing and biosparging. Bioslurping is used to remove volatile and semi-volatile compounds from the soil.

13.2.2 Phytoremediation

Phytoremediation can be used to remediate heavy metal contamination in soil and water. This contamination may be a result of natural and anthropogenic actions. A few examples of such heavy metals are cadmium, lead, arsenic, and mercury. They are highly toxic and may be leached into the soil by chemical fertilizers or through weathering of rocks. Heavy metal contamination poses a serious threat to agricultural lands. It disturbs the physiological and biochemical processes of crops and thus reduces crop productivity (Clemens 2006). Further, these toxic elements can either leach to underground water and contaminate it or they can enter the food chain and accumulate in the feeder's body. The accumulation of these heavy metals in the body of living organisms through diet is known as biomagnification. The high concentration of heavy metals in humans can cause lethal diseases such as cancer and renal dysfunction (Sarwar et al. 2010; Rehman et al. 2017). Some of the examples of diseases caused in humans by heavy metal accumulation are Minamata disease (mercury accumulation), Itai-Itai disease (cadmium accumulation), and Plumbism (lead accumulation). Therefore, it is necessary to remove these pollutants from soil and water.

Phytoremediation is a promising bioremediation technique that explores the use of plants and their subparts for the treatment of contaminated soil and water. It is being reported that the plants absorb pollutants from the soil through a passive process. They are translocated from roots with water through xylem tissue. The contaminants are then accumulated in the shoot (Miguel et al. 2013). Recently, phytoremediation has been extensively applied for the decontamination of soil and water. It is an eco-friendly and economical technique that cleans the environment naturally (Chandra Shekhar et al. 2003).

Plants behave as biofilters. They metabolize pollutants by degrading, accumulating, immobilizing, and transforming them into less harmful forms (Meagher 2000). The advanced root system of plants can even absorb low concentrations of ionic compounds present in the soil. A rhizosphere ecosystem is established by the extended root system of plants. It accumulates heavy metals and metabolizes them. Thus, the polluted soil is reclaimed, and soil fertility is stabilized (Ali et al. 2013; Jacob et al. 2018; DalCorso et al. 2019).

The phytoremediation process incorporates different techniques to mitigate a variety of pollutants. Elemental pollutants can be stripped from soil using plants (Khan et al. 2000; Clemens et al. 2002; Cobbett and Goldsbrough 2002). Elemental pollutants can be hyper-accumulated or adsorbed by plant species such as sunflower, tobacco, mustard, maize, and sand rocket (Meagher 2000).

Phytoremediation is a natural, solar energy-dependent, and cost-effective remediation technique suitable for low levels of contamination. Phytoremediation is more convenient than conventional methods for the mitigation of elemental and organic pollutants (Ahalya and Ramachandra 2006). Phytoremediation can be conveniently used at very large field sites where other methods of remediation are not suitable or practicable. Phytoremediation does have some limitations. It can be a time-consuming method, and it may cause potential contamination of the food chain and vegetation (Sutar and Das 2012). Therefore, it is necessary to consider these shortcomings before considering phytoremediation for site remediation.

Different mechanisms are involved during phytoremediation such as Phyto stabilization, phytoextraction, phytovolatilization, and phytofiltration (Alkorta et al. 2004). These mechanisms differ from each other based on plants' capacity to remove, immobilize, or degrade pollutants.

- *Phytoextraction/Phytoaccumulation*
  It is the process used by plants to absorb contaminants through harvestable tissues such as roots and translocate them into aboveground shoots or leaves. Phytoaccumulation saves tremendous remediation costs as low levels of contaminants are released during this process. Plant biomass and contaminants (usually metals) produced during this process are transported further for disposal or recycling.
- *Phytodegradation/Rhizodegradation*
  Phytodegradation is the metabolism and degradation of contaminants with the help of photo-enzymes. This breakdown of contaminants takes place within the plant tissues. Rhizodegradation is the degradation of contaminants in the rhizo-

sphere. The plants and microbes have a symbiotic relationship. This interaction helps in the breakdown of contaminants in the rhizosphere. The proteins and enzymes produced by plants provide nutrition to soil microorganisms, while microbes provide nutrient-enriched soil to the plants. These techniques are commonly used for the remediation of soil and water.

- *Rhizofiltration*
  It is a type of phytoremediation method generally used for water remediation. This method is used to clean different kinds of liquid waste released due to human activities. In this method, plant roots are used as a filter to adsorb pollutants from the contaminated liquids. The advantage of hemofiltration is that it is an economical and eco-friendly technique.
- *Phytovolatilization*
  The contaminants are absorbed from the soil by the plant's roots. They are metabolized and released in the atmosphere. This method of removing contaminants from the soil is known as phytovolatilization (Moreno et al. 2005). Mainly, plants evapotranspirate heavy metals such as selenium, mercury, and volatile hydrocarbons. Selenium (Se) is a vital micronutrient for plants. But at higher concentrations, it becomes highly poisonous. Se is structurally like sulfur, and it may lead to selenium uptake and assimilation by plants through sulfate transporters. It causes increased concentration of selenium in plants and further leads to toxicity to a great extent. The toxicity of selenium can be remediated by using the approach of greener biotechnology methods, such as phytoextraction and phytovolatilization. Selenium is absorbed by the plants from the soil and is converted into volatile compounds such as dimethyl selenide. These compounds are less toxic than their previous form and less harmful to humans.
- *Phytosequestration/Phytostabilization*
  The phytostabilization mechanism involves the binding of a plant substrate on the surface of polluted sites. This enables the stabilization of inorganic contaminants like heavy metalloids in the soil. It further reduces their movement in soil and water. This method requires regular screening of the site so that stabilizing conditions are maintained. It is an environmentally friendly method. It does not generate any waste residue and increases soil fertility too.

13.2.3 Ex Situ Bioremediation

When in situ techniques fail, ex situ bioremediation techniques are introduced into the environment. These techniques function in a controlled environment. The soil is excavated from the polluted site and transported to sites used for the treatment of pollutants. Certain parameters are considered before using the ex situ bioremediation process. These parameters are geographical location of the polluted site, extent of pollution in the zone, nature of pollutants present, degree of contamination, and total expenses for the treatment (Atlas and Philip 2005). A few examples of conventional and innovative methods used for ex situ bioremediation are landfarming, composting, biopiling, and the use of bioreactors and biofilters. The residue left after the remediation technique can be used for landscaping purposes.

- *Landfarming*
  The landfarming technique is used to remediate pollutants by excavation, spreading, and tilling or plowing of contaminated soil. The excavated polluted soils are carefully applied on a fixed layer of support above the ground surface. This allows aerobic biodegradation of pollutants by indigenous microorganisms (Philp and Atlas 2005). Plowing provides optimum growth factors to stimulate microbial activity during landfarming (Volpe et al. 2012). This technique is gaining popularity as an alternative dumping method because it requires less infrastructure, financial support, and maintenance for its operation (Williams 2006). The depth of pollutants in the soil is used to determine whether land farming will be carried out in an ex situ way or in situ way. If a pollutant lies less than 1 m below the ground surface, an in situ bioremediation technique might be used. Similarly, if a pollutant lies more than 1.7 m below the ground surface, it needs to be extracted and then transported to bioremediation treatment plants for proper mitigation (Nikolopoulou et al. 2013). Land farming is usually used for the remediation of noncyclic and polycyclic aromatic hydrocarbons in PCB-polluted sites (Silva-Castro et al. 2012).
- *Biopiling*
  Biopiling is an amalgamation of land farming and composting methods. It is also named as bioheaps, biocells, or biomounds technology (Das and Dash 2014). The low molecular weight petroleum pollutants present in soil and sediments, such as BTEX, PAHs, and phenols, are enormously treated using this technique. It is advantageous over other techniques because it reduces the utility cost of water and electricity. Moreover, its ability to control conditions makes it easier to maintain and monitor the process (Whelan et al. 2015). The process of biopiling involves the assembling of pollutant-rich soils into piles. The soil is then aerated to optimize pH, moisture level, and replenishment of nitrogen and phosphorus. It stimulates the biodegradation activity of microbial populations (Jørgensen et al. 2000; Li et al. 2004). As an outcome, the microbial activity is increased, and degradation of the volatile organic pollutants and fuel hydrocarbons is done (Gomez and Sartaj 2013). This technique may be constructively used in the remediation of cold-region contamination (Dias et al. 2015).
- *Biofilters*
  Biofilters are one of the oldest and most common bioremediation techniques. The organic contaminants are separated from air and water using this method (Srivastava et al. 2021). The gaseous pollutants are eliminated using microorganisms (Boopathy 2000). A biofilter is a layer of medium on which microorganisms grow to form biofilms. In this technique, pollutants are passed through columns embedded with microorganisms. These microorganisms are capable of degradation of pollutants. Air pollution and water pollution are controlled using biofilters. The use of the trickling filter method enables the treatment of secondary waste in water.
- *Bioreactors*
  The bioreactor is a vessel in which pollutants are degraded under controlled conditions. Bioreactors are used for the remediation of pollutants present in soil or

water. Volatile organic pollutants such as benzene, toluene, ethylbenzene, and xylene undergo aerobic microbial degradation easily. Slurry bioreactors are extensively used for the remediation process. Bioremediation done by using these bioreactors is more advantageous than other techniques. The reaction takes place in a closed environment where environmental factors such as temperature, pH, aeration, substrate, and nutrient concentration can be controlled, managed, and manipulated (Azubuike et al. 2016). The bioreactors with their accommodating designs allow maximum biological degradation and produce maximum yield (Mohan et al. 2004).

13.3 Factors Affecting Bioremediation

Various chemicals from the environment are detoxified with the help of bioremediation. It allows the possibility of degrading, removing, transforming, and immobilizing pollutants through the action of bacteria (Verma et al. 2006; Sung et al. 2016) (Boruvka and Vacha 2006), plants, and fungi (Kvesitadze et al. 2006). Microorganisms play a vital role in the degradation or transformation of hazardous contaminants into less harmful forms. Bioremediation is an application of the metabolic activity of microbes. Microbial degradation is facilitated by enzymes. The enzymes released by bacteria, fungi, and plants act as biocatalysts and, through their enzymatic pathways, facilitate the process of biodegradation reactions. Bioremediation processes are only applicable in living environments because microbes need a variety of compounds for their nutrition and energy.

The efficiency of bioremediation is determined by several factors. These factors include the chemical nature and concentration of pollutants, the physicochemical characteristics of the environment, and their availability to microorganisms (El Fantroussi and Agathos 2005). Moreover, the bioremediation process is affected by factors like suitable microbial population, availability of contaminants to the microbial population, temperature, and edaphic factors.

The degradation or remediation of environmental hazards is affected by the adaptability of microbes and other biological systems. These biological organisms can be divided into the following subgroups:

- *Aerobic Microorganisms*: These microbes function in the presence of oxygen. They can degrade pesticides and hydrocarbons (both alkanes and polyaromatic compounds). A few examples of aerobic microorganisms are *Pseudomonas*, *Alcaligenes*, *Sphingomonas*, *Rhodococcus*, and *Mycobacterium*. The contaminants are used as the sole source of carbon and energy by these bacteria.
- *Anaerobic Microorganisms*: Anaerobic bacteria can function in the absence of oxygen. They metabolize organic contaminants into methane, carbon dioxide, and hydrogen gas. They are used for bioremediation of polychlorinated biphenyls (PCBs) in river sediments, dechlorination of the solvent trichloroethylene (TCE), and chloroform in halogenated contaminants (Vidali 2001). Metal contaminants in estuaries and seas or oceans are degraded by *Paraclostridium bifer-*

mentans strain B1 (Binish et al. 2022). *Escherichia coli, Staphylococcus* genus, and *Clostridium* genus are some examples of anaerobic microorganisms.
- *Ligninolytic fungi*: Ligninolytic fungi can be used to treat contaminants in water and soil. Fungi such as the white rot fungus *Phanaerochaete chrysosporium* may be able to degrade a wide range of pesticides, synthetic dyes, PCBs, PAHs, etc., from the environment (Adenipekun and Fasidi 2005).
- *Methylotrophs*: Methylotrophs are aerobic bacteria that grow by utilizing methane for carbon and energy. Methane monooxygenase is the initial enzyme in the pathway for aerobic degradation (Vidali 2001). It has a broad substrate range and is active against a wide range of compounds, including the chlorinated aliphatics trichloroethylene and 1,2-dichloroethane. Some examples of methylotrophs are *Methylobacter*, *Methylococcus*, and *Methylocella*

The turbidity and chemicals from the domestic wastewater can be removed by the action of microorganisms. Aerobic bacteria like *Pseudomonas, Corynebacteria, Flavobacteria, Chlorobacteria, Acinetobacter, and Streptomyces and anaerobic bacteria like Bacilli, Aeromonas, Macrobenthos,* and other aquatic plants such as *E. crassipes* and *L. hoffmeisteri* removes waste from water (Mangunwardoyo et al. 2013). Essential nutrients produce enzymes to break down pollutants in a polluted environment. The inorganic compounds such as nitrogen, phosphate, sulfur, iron, and potassium stimulate cellular metabolism and microbial growth (Atagan et al. 2003).

Bioremediation processes can be affected by the physical state of pollutants. The properties of the pollutants, such as their toxicity and chemical nature of compounds, affect the efficiency of bioremediation. They can be organic or inorganic pollutants, heavy metals, polycyclic aromatic hydrocarbons, pesticides, or chlorinated solvents. Similarly, the nature of the polluted area is crucial for bioremediation because it may affect the quality of the remediation process.

Physicochemical attributes of pollutants, such as temperature and pH, greatly influence the rate and extent of the biodegradation process. The optimum range of these parameters is suitable for the growth of microorganisms and the removal of contaminants (Vidali 2001). The ideal pH for biodegradation in most aquatic and terrestrial systems is 6.5–8.5, but it can occur under a wide range of pH. The efficiency of bioremediation is also dependent on moisture content. The rate of contaminant metabolism is influenced by moisture content. The moisture level is determined by the nature and concentration of soluble materials that are available as well as the osmotic pressure and pH of terrestrial and aquatic systems (Cases and De Lorenzo 2005).

Biological degradation can be carried out both in aerobic and anaerobic conditions. In most cases, aerobic conditions enhance hydrocarbon metabolism (Macaulay 2014), whereas anaerobic conditions support degradation of chlorate compounds. The amount of oxygen can be increased in soil by tilling or sparging air. In some cases, hydrogen peroxide or magnesium peroxide can be introduced into the environment to increase oxygen levels.

13.4 Merits of Bioremediation

- Bioremediation does not involve human intervention naturally. Microbes degrade the contaminant, and the biodegradative populations become reduced. Water, carbon dioxide, and cell biomass are by-products of the process. These residues can be used as manure or to produce biofuel.
- The process of bioremediation requires very little effort. It can be easily carried out on site itself, without disturbing normal microbial activities. This reduces the transportation cost and the possible threats to human health and the environment during transportation.
- The bioremediation process is free of harmful chemicals. Nutrients, especially fertilizers, are only added to promote active and fast microbial growth. Due to this, less amount of toxic chemicals and compounds are released during bioremediation process.
- It does not require intensive labor and therefore is cost effective.
- It does not disturb the area of contamination.
- Bioremediation helps in ecological restoration.
- Bioremediation acts as an eco-friendly, sustainable opportunity to remediate pollutants at a large-scale level.

13.5 Demerits of Bioremediation

- Bioremediation is restricted to biodegradable compounds.
- During the remediation process, many new products may be produced, which can be more toxic and accumulate in the environment.
- This process is highly specific. Bioremediation requires the presence of metabolically active microbial populations, suitable environmental growth conditions, and availability of nutrients and contaminants for working.
- Bioremediation can be a time-consuming method.
- Limitations like the release of toxic products during the process can be overcome with the help of advanced research.

13.6 Omics Tools in Bioremediation

Several research studies have been conducted to explore the various approaches/methods for environmental remediation. However, these approaches could not be executed due to insufficient information about beneficial microorganisms participating in bioremediation, their functional profile, and their metabolism. Recently, advancements in omics technologies have enhanced our understanding of the genetic diversity as well as the genes participating in the functional profile of microorganisms from the natural environment participating in bioremediation (Handelsman 2004; Meena et al. 2019; Gilbert and Dupont 2011). Various Omics technologies, namely, genomics, epigenomics, transcriptomics, metagenomics, proteomics, and fluxomics, have created novel opportunities for environmental management of pollution through bioremediation (Rodríguez et al. 2020; Plewniak et al. 2018; Kesheri et al. 2024a). These omics techniques, especially metagenomics, have also accelerated the study of microbial communities, which was previously dependent on bacterial culture (Iwamoto and Nasu 2001; Gutleben et al. 2018). Nowadays, multi-omics strategies are frequently being utilized to investigate various microbial communities because a single omics analysis is not enough to explore the functional activity of the microbial community in a particular habitat (Shah et al. 2013; Meena et al. 2018). Various "omics" technologies are applied to understand the mechanisms behind the survival of the microorganisms in their native environments, such as polluted water, gut microbiomes (Gill et al. 2006; Mishra et al. 2024; Satya et al. 2024), heavy metal contaminated soil, deep-sea sediments (Hu et al. 2010), groundwater and marine systems (DeLong 2005; Benndorf et al. 2007; Hemme et al. 2010), and extreme environments (Baker and Banfield 2003).

13.6.1 Metagenomics

Metagenomics is a culture-independent omics technology to quantify and analyze the total genomic content of microorganisms in a particular habitat (Daniel 2005; Kanchan et al. 2020; Srivastava et al. 2024b). Metagenomics has many applications in both medical and environmental sciences; wastewater surveillance could be considered the best example of the same. In environmental microbiology, metagenomics studies the diversity of microbial communities by analyzing rRNA genes. Metagenomics studies also study how bacterial communities respond to changes in the physical/chemical properties of these environments (Delmont et al. 2011; Mallik et al. 2024). Metagenomics is beneficial, especially in exploring nonculturable microbes growing in different environments (Oulas et al. 2015; Bilal et al. 2018). Metagenomics helps in identifying DNA sequences of various novel uncultured microbes growing in various environments. The metagenomic information helps researchers in combining genomics, transcriptomics, and proteomics of bacterial communities (Hodkinson and Grice 2015). Metagenomics also helps in discovering unique novel genes and diverse pathways associated with novel enzymes containing highly specific catalytic properties (Scholz et al. 2012; Yergeau et al. 2017; Awasthi et al. 2020).

Metagenomics studies include selecting diverse ecological environments of interest that are considered the habitat of the diverse microbial communities that might have potential ecological and biotechnological applications. Metagenomic researchers are mostly interested in extreme conditions or unique environmental conditions such as extreme acidic or alkaline pH, high pressures, radiation, salinity, temperature, or high metal concentrations (Mirete et al. 2016). Metagenomics studies are also involved in isolating and identifying specific microbial genes involved in various bioremediation processes for eliminating environmental pollutants and contaminants (Jaiswal and Shukla 2020; Phulpoto et al. 2020; Kumar and Chandra 2020).

Metagenomics has a huge potential for bioremediation, which could be helpful in several interconnected ways (Satyanarayana et al. 2012; Tripathi et al. 2018).

13.6.2 Metatranscriptomics

The genetic makeup of the microorganisms inhabiting any environment can be identified using metagenomics, but it cannot analyze their gene expression and activity (Gosalbes et al. 2011; Bashiardes et al. 2016). Since metagenomics provides restricted clarification about gene activity, there is a need for a more advanced approach. Transcriptome is an RNA-based approach that acts as a connecting link between genomics and proteomics and aids in the regulation of gene expression (Milward et al. 2016). Metatranscriptomics strategy complements metagenomics (Franzosa et al. 2015) to predict and explore the functional aspects of microorganisms (Poretsky and Ann Moran 2011; Shukla et al. 2024). Metatranscriptomics is of enormous interest concerning bioremediation processes. Recent improvements in the metatranscriptomics approach have been applied to an herbicide degrading the tdfA gene. It has made the identification and quantification of actively expressing functional genes easier and traceable. It was reported that this gene codes for dioxygenase and helps in the degradation of herbicides (Nicolaisen et al. 2008).

13.6.3 Proteomics

Proteomics is an interdisciplinary field of science related to proteomes. A set of proteins formed inside a cell, tissue, organ, or organism is known as a proteome (Srivastava et al. 2024a; Kesheri et al. 2024b). Proteomics explores the structure and functioning of proteins present in an organism at the cellular level. Accelerated and significant protein identification through proteomics ensures microbial remediation (Basu and Stolz 2014). The analysis of a complete set of proteins in microbial communities from a specific environment is known as metaproteomics or community proteomics (Hart et al. 2018). The proteomic analysis contributes to the assessment of the molecular functioning of a microorganism by quantitative and qualitative protein characterization. Metaproteomic analyses are more effective than conventional genomics. Metaproteomics relies on effective protein extraction, protein fractionation from an intricate mixture, explicit identification of fractionated peptides/proteins, database search, and data interpretation (Verberkmoes et al. 2009).

Metaproteomics has enabled the identification and evaluation of the global expression of proteins in microorganisms residing in contaminated areas (Kim et al. 2004). Two-dimensional (2D) gel electrophoresis and mass spectrometry are used to evaluate microbial proteins present in cadmium-contaminated water (Lacerda et al. 2007). Novel genes are identified during the anaerobic degradation of toluene and ethylbenzene with the help of proteomic studies (Kühner et al. 2005). The identification of *Geobacter* sp. was reported by Konopka and Wilkins using metaproteomics in the stimulation of carbon (Konopka and Wilkins 2012). Peroxidase enzyme was identified from the PYR-1 *Mycobacterium* sp. strain when exposed to pyrene aromatic hydrocarbon (Wang et al. 2020). NG80-2 is a thermophilic strain isolated from an oil reservoir in Northern China that has alkane degradation potential (Feng et al. 2007). It is beneficial for the removal of contaminants present in the environment.

13.7 Sustainable Environment

The World Commission on Environment and Development 1987 coined the term sustainability in a report, "Our Common Future." This report was published by the UN Brundtland Commission. Sustainability means the ability to maintain. In this report, the term sustainability was used in the context of environmental resources. Due to the ongoing need of mankind to utilize preexisting resources, enormous pressure has been placed on us and our future generations. Sustainable development is a strategy to reduce this pressure by utilizing the resources in such a manner that these resources can meet today's world's needs without compromising the needs of future generations.

Lately, we are fruiting the stability of the sustainable climate of our ancestors (Revell 2020). But this stability has been disturbed by the negligent behavior of our present generation. Due to this, we are facing the problem of climate change. This drastic climate change is seen with the rise of temperature globally, which further results in the melting of polar ice (Hansen et al. 2015), and, thus, loss of biodiversity (Handa et al. 2014; Underwood et al. 2009; Delgado-Baquerizo et al. 2020). All this leads to ruining our environment, and we need to recover from this soon else we will not be able to survive (Revell 2020). Corals are invertebrates that inhabit marine water. (Wagner et al. 2020). Corals are basic structures of reefs, where they have symbiotic relationships with photosynthetic unicellular dinoflagellates. They are dying due to high ocean pH or ocean acidification and coral bleaching (Hoegh-Guldberg et al. 2017). Plastic waste dumped in the ocean and leakage of crude oils from cargo are the major factors responsible for marine water pollution (Price et al. 2003), and this affects the coral population (Monteiro et al. 2018; Forrest et al. 2019). Oil spills in oceans and on land can cause huge amounts of destruction and endanger life and the quality of natural ecosystems. Oil spills cover the surface of oceans and restrict the penetration of oxygen and sunlight into the water body; dissolved oxygen is reduced, and the death of flora and fauna is caused. Increased greenhouse gas emission from fossil fuel combustion is impacting us through global warming. The sudden rise in temperature is further causing the melting of polar ice, which leads to floods and the rise of sea water levels. It further disturbs the rainfall pattern and causes droughts and desertification. Excessive floods increase soil salinity and disturb vegetation growth.

All these human actions have harshly injured the environment, in such a manner that our future generations will not be able to utilize the natural resources which we are utilizing. To keep our environment healthy for our future generations, we need to rectify all these issues. We need to create a sustainable environment in which our future generations can thrive. This can be achieved by adopting measures that can not only reduce the source of pollution but can also provide proper treatment to the wastes released into the environment due to anthropogenic needs. Bioremediation is a suitable technique to protect the environment from pollutants. Moreover, phytomanagement can prove to be a better strategy to protect the environment. It is more beneficial economically, environmentally, and socially. (Burges et al. 2018).

13.8 Future Perspectives

Recently, our ecosystems have been seriously threatened by burgeoning organic and inorganic contaminants. These contaminants are released because of our detrimental activities. To safeguard our ecosystems, we human beings are sorely in need of evaluating our actions and innovating present waste removal methods with technologies that can handle this pressure. The existing technologies can remove these pollutants but somehow, they are not enough. They have their shortcomings. In this hour, we need such technologies which can be both sustainable and effective. Lately, the co-integration of bioremediation and nanotechnology has been established as a significant, novel, and fruitful method. The use of nanomaterials in bioremediation techniques may prove to be a pivotal tool in the future. Nanomaterials are characterized by improved catalysis, high reactivity, and adsorption qualities that promise a sustainable future (Wu et al. 2019). Nano-bioremediation is a futuristic technique to deal with the restoration of damaged ecosystems (Ramezani et al. 2021). It utilizes microorganisms' culture components and extracts as catalysts for the sustainable production of nanomaterials. The immobilization or synthesis of enzymes aids in bioremediation of toxic compounds.

The bioremediation efficiency can be improved by utilizing nanomaterials and nanotechnological methods (Vázquez-Núñez et al. 2020). The contamination in water can be treated with this approach (Yogalakshmi et al. 2020). Soil contaminated with heavy metals and hydrocarbon-contaminated sediments can be refined using nanotechnology (De Gisi et al. 2017). The organic or inorganic compounds in the soil can be refined through this technology (Bharagava et al. 2020). The use of nanomaterials reduces the production of intermediates during the bioremediation process. This lowers the associated cost of mitigating their adverse impacts on the environment.

13.9 Conclusion

This chapter describes the concept of bioremediation, its applications, and the use of various bioremediation techniques to address industrial effluents, contaminated soils, and groundwater. In this chapter, microbial, plant, and other nontoxic approaches were discussed for the eradication of pollutants from the environment. This chapter also explains the use of various omics technologies like proteomics, metagenomics, and metatranscriptomics for increasing bioremediation efficiency. Moreover, this chapter rationalizes the concept of sustainable development in an environmentally friendly and cost-effective way.

Short Chapter Summary

This chapter discusses the mitigation of pollutants from the environment in two different ways. Firstly, the use of bioremediation techniques in a cost-effective and eco-friendly manner for detoxifying the environment effectively is discussed. This chapter also highlights the future improvements in the bioremediation technique to strengthen its efficiency and relevance in treating anthropogenic waste. Various types of bioremediations, their objectives, and applications for a sustainable environment were also elaborated. Secondly, the utilization of omics technologies like proteomics, metagenomics, and metatranscriptomics for the enhancement of bioremediation efficiency. It is also estimated that the co-integration of nanotechnology and bioremediation can offer more practical methods for enhancing the capability of bioremediation.

References

Abdulsalam S, Adefila SS, Bugaje 1M, Ibrhaim S (2012) Bioremediation of soil contaminated with used motor oil in a closed system. J Bioremediation Biodegrad 03(12)

Adams GO, Fufeyin PT, Okoro SE, Ehinomen I (2015) Bioremediation, biostimulation and bioaugmentation: a review. Int J Environ Bioremediat Biodegrad 3(1):28–39

Adenipekun C, Fasidi I (2005) Bioremediation of oil-polluted soil by *Lentinus subnudus*, a Nigerian white-rot fungus. Afr J Biotechnol 4(8):796–798

Ahalya N, Ramachandra TV (2006) Phytoremediation: processes and mechanisms. J Ecobiol 18(1):33–38

Ali H, Khan E, Sajad MA (2013) Phytoremediation of heavy metals-concepts and applications. Chemosphere 91:869–881

Ali M, Ishqi HM, Husain Q (2020) Enzyme engineering: reshaping the biocatalytic functions. Biotechnol Bioeng 117:1877–1894

Alkorta I, Hernandez-Allica J, Becerril JM, Amezaga I, Albizu I, Garbisu C (2004) Recent findings on the phytoremediation of soils contaminated with environmentally toxic heavy metals and metalloids such as zinc, cadmium, lead, and arsenic. Rev Environ Sci Biotechnol 3:71–90

Atagan HI, Hayne R, Wallis F (2003) Optimization of soil physical and chemical conditions for the bioremediation of creosote-contaminated soil. Biodegradation 14(4):297–307

Atlas RM and Philip J 2005 Bioremediation: Applied Microbial Solutions for Real world Environmental Cleanup. American society for Microbiology (ASM) Press, Washington, DC, pp.78–105

Awasthi MK, Ravindran B, Sarsaiya S, Chen H, Wainaina S, Singh E (2020) Metagenomics for taxonomy profiling: tools and approaches. Bioengineered 11:356–374

Azubuike CC, Chikere CB, Okpokwasili GC (2016) Bioremediation techniques–classification based on site of application: principles, advantages, limitations and prospects. World J Microbiol Biotechnol 32(11):1–18

Baker BJ, Banfield JF (2003) Microbial communities in acid mine drainage. FEMS Microbiol Ecol 44:139–152

Bashiardes S, Zilberman Schapira G, Elinav E (2016) Use of metatranscriptomics in microbiome research. Bioinform Biol Insights 10:BBI.S34610

Basu P, Stolz JF (2014) Application of proteomics in bioremediation. In: Chovenec P (ed) Microbial metal and metalloid metabolism. ASM Press, United States, pp 247–272

Benndorf D, Balcke GU, Harms H, von Bergen M (2007) Functional metaproteome analysis of protein extracts from contaminated soil and groundwater. ISME J 1:224–234

Bharagava RN, Saxena G, Mulla SI (2020) Bioremediation of Industrial Waste for Environmental Safety. Springer; Singapore: Introduction to industrial wastes containing organic and inorganic pollutants and bioremediation approaches for environmental management; pp. 1–18

Bhat SA, Singh S, Singh J, Kumar S, Vig AP (2018) Bioremediation and detoxification of industrial wastes by earthworms: vermicompost as a powerful crop nutrient in sustainable agriculture. Bioresour Technol 252:172–179

Bilal T, Malik B, Hakeem KR (2018) Metagenomic analysis of uncultured microorganisms and their enzymatic attributes. J Microbiol Methods 155:65–69

Binish MB, Pinu B, Gopikrishna VG, Mohan M (2022) Potential of anaerobic bacteria in bioremediation of metal contaminated marine and estuarine environment. In: Microbial biodegradation and bioremediation, vol 2, pp 305–326. https://doi.org/10.1016/B978-0-323-85455-9.00015-1

Boopathy R (2000) Factors limiting bioremediation technologies. Bioresour Technol 74(1):63–67

Boruvka L, Vacha R (2006) Litavka river alluvium as a model area heavily polluted with potentially risk elements. In: Morel JL, Echevarria G, Goncharova N (eds) Phytoremediation of metal-contaminated soils. Springer, Germany

Burges A, Alkorta I, Epelde L, Garbisu C (2018) From phytoremediation of soil contaminants to phytomanagement of ecosystem services in metal contaminated sites. Int J Phytoremediation 20:384–397. https://doi.org/10.1080/15226514.2017.1365340

Cases I, De Lorenzo V (2005) Genetically modified organisms for the environment: stories of success and failure and what we have learned from them. Int Microbiol 8:213–222

Chandra Shekhar K, Kamala CT, Chary NS, Anjaneyulu Y (2003) Removal of heavy metals using a plant biomass with reference to environmental control. Int J Miner Process 68(1–4):37–45

Clemens S, Palmgren MG, Krämer U (2002) A long way ahead: understanding and engineering plant metal accumulation. Trends Plant Sci 7: 309–315

Clemens S (2006) Toxic metal accumulation, responses to exposure and mechanisms of tolerance in plants. Biochimie 88:1707–1719

Cobbett C, Goldsbrough P (2002) Phytochelatins and metallothioneins: roles in heavy metal detoxification and homeostasis. Annu Rev Plant Biol 53: 159–182

DalCorso G, Fasani E, Manara A, Visioli G, Furini A (2019) Heavy metal pollutions: state of the art and innovation in phytoremediation. Int J Mol Sci 20:3412

Daniel R (2005) The metagenomics of soil. Nat Rev Microbiol 3(6):470–478

Das S, Dash HR (2014) Microbial biodegradation and bioremediation: a potential tool for restoration of contaminated areas. Elsevier Science Publishing Co Inc, London

De Gisi S, Casella P, Cellamare CM, Ferraris M, Petta L, Notarnicola M (2017) Wastewater reuse. In: Abraham MA (ed) Encyclopedia of sustainable technologies. Elsevier, Oxford, pp 53–68

Delgado-Baquerizo M, Reich PB, Trivedi C, Eldridge DJ, Abades S, Alfaro FD (2020) Multiple elements of soil biodiversity drive ecosystem functions across biomes. Nat Ecol Evol 4:210–220

Delmont TO, Robe P, Cecillon S, Clark IM, Constancias F, Simonet P, Hirsch PR, Vogel TM (2011) Accessing the soil metagenome for studies of microbial diversity. Appl Environ Microbiol 77(4):1315–1324

DeLong EF (2005) Microbial community genomics in the ocean. Nat Rev Microbiol 3:459–469

Dias RL, Ruberto L, Calabró A, Balbo AL, Del Panno MT, Mac Cormack WP (2015) Hydrocarbon removal and bacterial community structure in on-site biostimulated biopile systems designed for bioremediation of diesel contaminated Antarctic soil. Polar Biol 38:677–687

El Fantroussi S, Agathos SN (2005) Is bioaugmentation a feasible strategy for pollutant removal and site remediation? Curr Opin Microbiol 8:268–275

Feng et al .(2007) Feng L, Wang W, Cheng J, Ren Y(2007) Genome and proteome of long- chain alkane degrading Geobacillus thermodenitrificans NG80-2 isolated from a deep subsurface oil reservoir. Proccedings of National Academy of Sciences 104(13):5602–7

Forrest A, Giacovazzi, Dunlop S, Reisser J, Tickler D, Jamieson A (2019) Eliminating plastic pollution: how a voluntary contribution from industry will drive the circular plastics economy. Front Mar Sci 6:627

Franzosa EA, Hsu T, Sirota-Madi A, Shafquat A, Abu-Ali G, Morgan XC, Huttenhower C (2015) Sequencing and beyond: integrating molecular "omics" for microbial community profiling. Nat Rev Microbiol 13(6):360–372

Gidarakos E, Aivalioti M (2007) Large scale and long-term application of bioslurping: the case of a Greek petroleum refinery site. J Hazard Mater 149:574–581

Gilbert JA, Dupont CL (2011) Microbial metagenomics: beyond the genome. Annu Rev Mar Sci 3:347–371

Gill S, Pop MT, Deboy R, Eckburg P, Turnbaugh P, Samuel B (2006) Metagenomic analysis of the human distal gut microbiome. Science 312:1355–1359

Gomez F, Sartaj M (2013) Field scale ex-situ bioremediation of petroleum contaminated soil under cold climate conditions. Int Biodeterior Biodegradation 85:375–382

Gosalbes MJ, Durbán A, Pignatelli M, Abellan JJ, Jiménez-Hernández N, Pérez-Cobas AE, Latorre A, Moya A (2011) Metatranscriptomic approach to analyze the functional human gut microbiota. PLoS One 6(3):e17447

Gutleben J, Chaib De Mares M, van Elsas JD, Smidt H, Overmann J, Sipkema D (2018) The multi-omics promise in context: from sequence to microbial isolate. Crit Rev Microbiol 44:212–229

Handa IT, Aerts R, Berendse F, Berg MP, Bruder A, Butenschoen O (2014) Consequences of biodiversity loss for litter decomposition across biomes. Nature 509:218–221

Handelsman J (2004) Metagenomics: application of genomics to uncultured microorganisms. Microbiol Mol Biol Rev 68:669–685

Hansen J, Sato M, Hearty P, Ruedy R, Kelley M, Masson-Delmotte V et al (2015) Ice melt, sea level rise and superstorms: evidence from paleoclimate data, climate modeling, and modern observations that 2° C global warming is highly dangerous. Atmos Chem Phys Discus 15:3761–3812

Hart EH, Creevey CJ, Hitch T, Kingston-Smith AH (2018) Metaproteomics of rumen microbiota indicates niche compartmentalisation and functional dominance in a limited number of metabolic pathways between abundant bacteria. Sci Rep 8:10504

Hemme CL, Deng Y, Gentry TJ, Fields MW, Wu L, Barua S et al (2010) Metagenomic insights into evolution of a heavy metal contaminated groundwater microbial community. ISME J 4:660–667

Hodkinson BP, Grice EA (2015) Next-generation sequencing: a review of technologies and tools for wound microbiome research. Adv Wound Care 4:50–58

Hoegh-Guldberg O, Poloczanska ES, Skirving W, Dove S (2017) Coral reef ecosystems under climate change and ocean acidification. Front Mar Sci 4:158

Höhener P, Ponsin V (2014) In situ vadose zone bioremediation. Curr Opin Biotechnol 27:1–7

Hu Y, Fu C, Yin Y, Cheng G, Lei F, Yang X et al (2010) Construction and preliminary analysis of a deep-sea sediment metagenomic fosmid library from Qiongdongnan Basin, South China Sea. Mar Biotechnol 12:719–727

Iwamoto T, Nasu M (2001) Current bioremediation practice and perspective. J Biosci Bioeng 92:1–8

Jacob JM, Karthik C, Saratale RG, Kumar SS, Prabakar D, Kadirvelu K et al (2018) Biological approaches to tackle heavy metal pollution: a survey of literature. J Environ Manag 217:56–70

Jaiswal S, Shukla P (2020) Alternative strategies for microbial remediation of pollutants via synthetic biology. Front Microbiol 11:808

Jørgensen KS, Puustinen J, Suortti AM (2000) Bioremediation of petroleum hydrocarbon-contaminated soil by composting in biopiles. Environ Pollut 107:245–254

Kanchan S, Sinha RP, Chaudière J, Kesheri M (2020) Computational metagenomics: current status and challenges. In: Katara P (ed) Recent trends in 'computational omics: concepts and methodology'. Nova Science Publishers, United States, pp 371–395

Kao CM, Chen CY, Chen SC, Chien HY, Chen YL (2008) Application of in-situ biosparging to remediate a petroleum-hydrocarbon spill site: field and microbial evaluation. Chemosphere 70(8):1492–1499

Kesheri M, Kanchan S, Srivastava S, Ratna-Raj R, Chittoori B, Vaishampayan A, Rastogi R, Sinha RP, Primerano D (2024a) Ecology and environmental omics. In: Gupta MK, Katara P, Mondal S, Singh R (eds) Integrative omics: concepts, methodology, and application. Academic Press, United States, pp 315–331. https://doi.org/10.1016/B978-0-443-16092-9.00019-9

Kesheri M, Kanchan S, Salisbury TB, Sinha RP (2024b) Microbial omics in environment and health. Springer Nature, Singapore. https://doi.org/10.1007/978-981-97-1769-9

Khan AG, Kuek C, Chaudhry TM, Khoo CS, Hayes WJ (2000) Role of plants, mycorrhizae and phytochelators in heavy metal contaminated land remediation. Chemosphere, 41(1-2):197–207. doi: https://doi.org/10.1016/S0045-6535(99)00412-9

Khan I, Aftab M, Shakir SU, Ali M, Qayyum S, Rehman MU, Haleem KS, Touseef I (2019) Mycoremediation of heavy metal (Cd and Cr)–polluted soil through indigenous metallotolerant fungal isolates. Environ Monit Assess 191(9):1–11

Kim HJ, Ishidou E, Kitagawa E, Momose Y, Iwahashi H (2004) A yeast DNA microarray for the evaluation of toxicity in environmental water containing burned ash. Environ Monit Assess 92:253–272

Kim S, Krajmalnik-Brown R, Kim J-O, Chung J (2014) Remediation of petroleum hydrocarbon-contaminated sites by DNA diagnosis-based bioslurping technology. Sci Total Environ. 497:250–259. doi:https://doi.org/10.1016/j.scitotenv.2014.08.002

Konopka A, Wilkins MJ (2012) Application of meta-transcriptomics and proteomics to analysis of in situ physiological state. Front Microbiol 3:184

Kühner S, Wöhlbrand L, Fritz I, Wruck W, Hultschig C, Hufnagel P, Kube M, Reinhardt R, Rabus R (2005) Substrate-dependent regulation of anaerobic degradation pathways for toluene and ethylbenzene in a denitrifying bacterium, strain EbN1. J Bacteriol 187:1493–1503

Kumar V, Chandra R (2020) Bioremediation of melanoidins containing distillery waste for environmental safety. In: Bharagava R, Saxena G (eds) Bioremediation of industrial waste for environmental safety. Springer, Singapore

Kumar V, Shahi SK, Singh S (2018) Bioremediation: an eco-sustainable approach for restoration of contaminated sites. In: Singh J, Sharma D, Kumar G, Sharma N (eds) Microbial bioprospecting for sustainable development. Springer, Singapore

Kvesitadze G, Khatisashvili G, Sadunishvili T, Ramsden JJ (2006) Biochemical mechanisms of detoxification in higher plants. Springer, Singapore, p 256

Lacerda CMR, Choe LH, and Reardon KF (2007) Metaproteomic analysis of a bacterial community response to cadmium exposure. J. Proteome Res. 6, 1145–1152. doi: https://doi.org/10.1021/pr060477v

Li L, Cunningham CJ, Pas V, Philp JC, Barry DA, and Anderson P (2004) Field trial of a new aeration system for enhancing biodegradation in a biopile. Waste Manag. 24, 127–137. doi: https://doi.org/10.1016/j.wasman.2003.06.001

Lovely DR, Phillips EJP, Gorby YA, and Landa ER (1991) Microbial reduction of uranium. Nature 350, 413–416. doi: 10.1038/350413a0

Macaulay BM (2014) Understanding the behavior of oil-degrading microorganisms to enhance the microbial remediation of spilled petroleum. Appl Ecol Environ Res 13:247–262

Major DW, McMaster ML, Cox EE, Edwards EA, Dworatzek SM, Hendrickson E, Starr MG, Payne JA, Buonamici LW (2002) Field demonstration of successful bioaugmentation to achieve dechlorination of tetrachloroethene to ethene. Environ Sci Technol 36:5106–5116

Mallik B, Mishra T, Dubey P, Kesheri M, Kanchan S (2024) Exploring the secrets of microbes: unveiling the hidden world through microbial omics in environment and health. In: Kesheri M, Kanchan S, Salisbury TB, Sinha RP (eds) Microbial omics in environment and health. Springer, Singapore, pp 269–294. https://doi.org/10.1007/978-981-97-1769-9_10

Mangunwardoyo W, Sudjarwo T, Patria MP (2013) Bioremediation of effluent wastewater treatment plant Bojongsoang Bandung Indonesia using consortium aquatic plants and animals. Int J Res Rev Appl Sci 14(1):150–160

Margesin R, Zimmerbauer A, Schinner F (2000) Monitoring of bioremediation by soil biological activities Chemosphere, 40, pp. 339-346

Meagher RB (2000) Phytoremediation of toxic elemental organic pollutants. Curr Opin Plant Biol 3:153–162

Meena M, Aamir M, Vikas K, Swapnil P, Upadhyay RS (2018) Evaluation of morpho-physiological growth parameters of tomato in response to Cd induced toxicity and characterization of metal sensitive NRAMP3 transporter protein. Environ Exp Bot 148:144–167

Meena M, Divyanshu K, Kumar S, Swapnil P, Zehra A, Shukla V, et al (2019) Regulation of L-proline biosynthesis, signal transduction, transport, accumulation and its vital role in plants during variable environmental conditions. Heliyon 5:e02951. doi: https://doi.org/10.1016/j.heliyon.2019.e02952

Miguel AS, Ravanel P, Raveton M (2013) A comparative study on the uptake and translocation of organochlorines by *Phragmites australis*. J Hazard Mater 244:60–69

Milward EA, Shahandeh A, Heidari M, Johnstone DM, Daneshi N, Hondermarck H (2016) Transcriptomics. Encycl Cell Biol 4:160–165

Mirete S, Morgante V, González-Pastor JE (2016) Functional metagenomics of extreme environments. Curr Opin Biotechnol 38:143–149

Mishra T, Mallik B, Kesheri M, Kanchan S (2024) The interplay of gut microbiome in health and diseases. In: Kesheri M, Kanchan S, Salisbury TB, Sinha RP (eds) Microbial omics in environment and health. Springer, Singapore, pp 1–34. https://doi.org/10.1007/978-981-97-1769-9_1

Mohan SV, Sirisha K, Rao NC, Sarma PN, Reddy SJ (2004) Degradation of chlorpyrifos contaminated soil by bioslurry reactor operated in sequencing batch mode: bioprocess monitoring. J Hazard Mater 116:39–48

Mohee R, Mudhoo A (2012) Bioremediation and sustainability: research and applications. John Wiley & Sons, Hoboken

Monteiro RC, do Sul JAI, Costa MF (2018) Plastic pollution in islands of the Atlantic Ocean. Environ Pollut 238:103–110

Moreno FN, Anderson CWN, Stewart RB, and Robinson BH (2005) Mercury volatilisation and phytoextraction from base-metal mine tailings. Environ. Pollut. 136, 341–352

Nicolaisen MH, Bælum J, Jacobsen CS, Sørensen J (2008) Transcription dynamics of the functional tfdA gene during MCPA herbicide degradation by *Cupriavidus necator* AEO106 (pRO101) in agricultural soil. Environ Microbiol 10(3):571–579

Nikolopoulou M, Pasadakis N, Norf H, Kalogerakis N (2013) Enhanced ex situ bioremediation of crude oil contaminated beach sand by supplementation with nutrients and rhamnolipids. Mar Pollut Bull 77:37–44

Nojiri H, and Tsuda M (2005) Functional evolution of bacteria in degradation of environmental pollutants. Tanpakushitsu Kakusan Koso 50, 1505–1509

Olawale O, Obayomi KS, Dahunsi SO, Folarin O (2020) Bioremediation of artificially contaminated soil with petroleum using animal waste: cow and poultry dung. Cogent Eng 7(1):1721409

Oulas A, Pavloudi C, Polymenakou P, Pavlopoulos GA, Papanikolaou N, Kotoulas G (2015) Metagenomics: tools and insights for analyzing next generation sequencing data derived from biodiversity studies. Bioinform Biol Insights 9:75–88

Philp JC, Atlas RM (2005) Bioremediation of contaminated soils and aquifers. In: Atlas RM, Philp JC (eds) Bioremediation: applied microbial solutions for real-world environmental cleanup. American Society for Microbiology (ASM), Washington, pp 139–236

Phulpoto AH, Maitlo MA, Kanhar NA (2020) Culture-dependent to culture independent approaches for the bioremediation of paints: a review. Int J Environ Sci Technol 18(3):241–262

Plewniak F, Crognale S, Rossetti S, Bertin PN (2018) A genomic outlook on bioremediation: the case of arsenic removal. Front Microbiol 9:820

Poretsky R, Ann Moran M (2011) Comparative day/night metatranscriptomic analysis of microbial communities in the north pacific subtropical gyre. Environ Microbiol 11:1358–1375

Price JM, Johnson WR, Marshall CF, Ji ZG, Rainey GB (2003) Overview of the oil spill risk analysis (OSRA) model for environmental impact assessment. Spill Sci Technol Bull 8:529–533

Ramezani M, Enayati M, Ramezani M, Ghorbani A (2021) A study of different strategical views into heavy metal(oid) removal in the environment. Arab J Geosci 14:2225

Rehman MZU, Rizwan M, Ali S, Ok YS, Ishaque W, Saifullah (2017) Remediation of heavy metal contaminated soils by using *Solanum nigrum*: a review. Ecotoxicol Environ Saf 143:236–248

Revell T (2020) Attenborough's rallying cry. New Sci 245:30

Rittmann BE, McCarty PL (2001) Environmental biotechnology: principles and applications. McGraw-Hill, New York

Rodríguez A, Castrejón-Godínez ML, Salazar-Bustamante E, GamaMartínez Y, Sánchez-Salinas E, Mussali-Galante P (2020) Omics approaches to pesticide biodegradation. Curr Microbiol 77:545–563

Sarwar N, Malhi SS, Zia MH, Naeem A, Bibi S, Farid G (2010) Role of mineral nutrition in minimizing cadmium accumulation by plants. J Sci Food Agric 90(6):925–937

Satya S, Sharma S, Choudhary G, Kaushik G (2024) Advances in environmental microbiology: a multi-omic perspective. In: Kesheri M, Kanchan S, Salisbury TB, Sinha RP (eds) Microbial omics in environment and health. Springer, Singapore, pp 175–204. https://doi.org/10.1007/978-981-97-1769-9_7

Satyanarayana T, Prakash A, Johri BN (2012) Microorganisms in environmental management: microbes and environment. Springer, Netherlands

Scholz MB, Lo CC, Chain PS (2012) Next generation sequencing and bioinformatic bottlenecks: the current state of metagenomic data analysis. Curr Opin Biotechnol 23:9–15

Shah V, Zakrzewski M, Wibberg D, Eikmeyer F, Schlüter A, Madamwar D (2013) Taxonomic profiling and metagenome analysis of a microbial community from a habitat contaminated with industrial discharges. Microb Ecol 66:533–550

Sharma S (2012) Bioremediation: features, strategies and applications. Asian J Pharm Life Sci 2231:4423

Shukla N, Bhatia A, Kesheri M, Kanchan S, Narayan J (2024) Planet in peril: unveiling the microbial consequences of environmental degradation on human health through omics approaches. In: Kesheri M, Kanchan S, Salisbury TB, Sinha RP (eds) Microbial omics in environment and health. Springer, Singapore, pp 35–56. https://doi.org/10.1007/978-981-97-1769-9_2

Silva-Castro GA, Uad I, Gonzalez-Lopez J, Fandino CG, Toledo FL, Calvo C (2012) Application of selected microbial consortia combined with inorganic and oleophilic fertilizers to recuperate oil-polluted soil using land farming technology. Clean Technol Environ Policy 14:719–726

Singh P, Singh VK, Singh R, Borthakur A, Madhav S, Ahamad A, Kumar A , Pal DB, Tiwary D, Mishra PK (2020) Bioremediation: a sustainable approach for management of environmental contaminants Abat. Environ. Pollut: Trends Strateg, pp. 1–23

Solomon NO, Kanchan S, Kesheri M (2024) Nanoparticles as detoxifiers for industrial wastewater. Water Air Soil Pollut 235:214. https://doi.org/10.1007/s11270-024-07016-5

Srivastava AK, Singh RK, Singh D (2021) Woodhead Publishing Series Food Science, Technology and Nutrition, pp 241–253

Srivastava U, Kanchan S, Kesheri M, Gupta MK, Singh S (2024a) Types of omics data: genomics, metagenomics, epigenomics, transcriptomics, proteomics, metabolomics, and phenomics. In: Gupta MK, Katara P, Mondal S, Singh RL (eds) Integrative omics: concepts, methodology, and application. Academic Press, United States, pp 13–34. https://doi.org/10.1016/B978-0-443-16092-9.00002-3

Srivastava U, Kesheri M, Kanchan S, Singh S (2024b) Computational omics protocol for the comparative study of microbiome analysis. In: Kesheri M, Kanchan S, Salisbury TB, Sinha RP (eds) Microbial omics in environment and health. Springer, Singapore, pp 109–133. https://doi.org/10.1007/978-981-97-1769-9_5

Sung CY, Yang CY, Ye JA, Cheng CM (2016) Integrated circuit-based biofabrication with common biomaterials for probing cellular biomechanics. Trends Biotechnol 34(2):171–186

Sutar H, Das CK (2012) A review on bioremediation. Int J Res Chem Environ 2(1):13–21

Thakur M, Medintz IL, and Walper SA (2019) Enzymatic bioremediation of organophosphate compounds—progress and remaining challenges. Front. Bioeng. Biotechnol. 7, 289. doi: https://doi.org/10.3389/fbioe.2019.00289

Tiwari G, Singh SP (2014) Application of bioremediation on solid waste management: a review. J Bioremed Biodeg 5(6):248–1172

Tripathi M, Singh D, Vikram S, Singh V, Kumar S (2018) Metagenomic approach towards bioprospection of novel biomolecule(s) and environmental bioremediation. Annu Res Rev Biol 22:1–12

Underwood EC, Viers JH, Klausmeyer KR, Cox RL, Shaw MR (2009) Threats and biodiversity in the mediterranean biome. Divers Distrib 15:188–197

Vázquez-Núñez E, Molina-Guerrero CE, Peña-Castro JM, Fernández-Luqueño F, de la Rosa-Álvarez M (2020) Use of nanotechnology for the bioremediation of contaminants: a review. Processes 8(7):826

Verberkmoes NC, Denef VJ, Hettich RL, Banfield JF (2009) Systems biology: functional analysis of natural microbial consortia using community proteomics. Nat Rev Microbiol 7(3):196–205

Verma P, George K, Singh S, Juwarkar A, Singh RN (2006) Modeling rhizofiltration: heavy metal uptake by plant roots. Environ Model Assess 11:387–394
Vidali M (2001) Bioremediation: an overview. Pure Appl Chem 73(7):1163–1172
Volpe A, Moro GD, Rossetti S, Tandoi V (2012) Fingerprinting Hydrocarbons in a Contaminated Soil from an Italian Natural Reserve and Assessment of the Performance of a Low-Impact Bioremediation Approach. Springer Nature Link 223: 1773-1782
Wagner D, Friedlander AM, Pyle RL, Brooks CM, Gjerde KM, Wilhelm TA (2020) Coral reefs of the high seas: hidden biodiversity hotspots in need of protection. Front Mar Sci 7:776
Wang RF, Wennerstrom D, Cao WW, Khan AA, Cerniglia CE (2020) Cloning, expression, and characterization of the katG gene, encoding catalase-peroxidase, from the polycyclic aromatic hydrocarbon-degrading bacterium Mycobacterium sp. strain PYR-1. Appl Environ Microbiol 66(10):4300–4304
Whelan MJ, Coulon F, Hince G, Rayner J, McWatters R, Spedding T (2015) Fate and transport of petroleum hydrocarbons in engineered biopiles in polar regions. Chemosphere 131:232–240
Williams J (2006) Bioremediation of contaminated soils: a comparison of in situ and ex situ techniques. Semantic Scholar, pp 421–521
Wu J, Wang X, Wang Q, Lou Z, Li S, Zhu Y, Qin L, Wei H (2019) Nanomaterials with enzyme-like characteristics (nanozymes): next-generation artificial enzymes (II). Chem Soc Rev 48(4):1004–1076
Yergeau E, Michel C, Tremblay J, Niemi A, King TL, Wyglinski J (2017) Metagenomic survey of the taxonomic and functional microbial communities of seawater and sea ice from the Canadian Arctic. Sci Rep 7:42242
Yerushalmi L, Rocheleau S, Cimpoia R, Sarrazin M, Sunahara G, Peisajovich A, Leclair G, Guiot SR (2003) Enhanced biodegradation of petroleum hydrocarbons in contaminated soil. Biorem J 7(1):37–51
Yogalakshmi KN, Das A, Rani G, Jaswal V, Randhawa JS (2020) Nano-bioremediation: a new age technology for the treatment of dyes in textile effluents, bioremediation of industrial waste for environmental safety. Springer, Singapore, pp 313–347

Multi-Omics Applications in Environmental Sustainability

14

Harleen Kaur, Minu Kesheri, and Swarna Kanchan

Abstract

Omics alludes to various high-throughput technologies that can analyze the complete set of biological molecules in a given organism or environment, including genes, transcripts, proteins, and metabolites. These technologies have emerged as powerful tools for understanding the complex interactions between organisms and their environments and have the potential to contribute significantly to environmental sustainability efforts. One of the most important applications of omics in environmental sustainability is the study of microbial communities. This chapter discusses the crucial role of microbes in maintaining the health and productivity of various ecosystems through the window of omics technologies to provide insights into their diversity, metabolic capabilities, and interactions with other organisms. Moreover, several applications of omics technologies in unravelling the cords of environmental sustainability, ranging from the study of bioremediation to the monitoring of pollutants and the assessment of the impact of environmental changes on ecosystems are also highlighted. This chapter serves as a conduit of the state of the art of dynamism that these omics technologies are bestowed with to serve as a beacon to protect and preserve our planet.

Keywords

Climate change · Ecosystem · Environmental sustainability · High-throughput technologies · Metagenomics

---

H. Kaur
CSIR National Physical Laboratory, New Delhi, India

M. Kesheri · S. Kanchan (✉)
Boise State University, Boise, ID, USA

Joan C. Edwards School of Medicine, Marshall University, Huntington, WV, USA
e-mail: kanchan@marshall.edu

M. Kesheri et al. (eds.), *Multi-Omics in Biomedical Sciences and Environmental Sustainability*, https://doi.org/10.1007/978-981-96-7067-3_14

Objectives

- To understand and explore various types of potential health hazards caused by the environment that affects human health
- To explore various types of bioremediations, their advantages, objectives, and role in environmental sustainability, and their limitations
- To understand the role of genomics, transcriptomics, proteomics, and metabolomics in bioremediation
- To explore the advantages, applications, and limitations of various omics technologies in bioremediation

14.1 Introduction

The increasing population, rapid industrialization, and changes in human lifestyles have led to the accumulation of various contaminants in the environment. The increase in human activity worldwide has resulted in the release of harmful substances into the environment, leading to pollution and a rise in health-related issues. Harmful pollutants such as heavy metals, industrial effluents, hydrocarbons, polychlorinated biphenyls, and xenobiotics pose significant risks to human health, plant life, and overall ecosystem well-being. The economic and technological advancements in industries have led to adverse effects on human cultures and the environment. Contamination of the environment (soil and water) is an ongoing problem causing significant challenges to human well-being. Therefore, addressing these environmental risks is crucial on a global scale to ensure the protection of human health and the preservation of ecosystems. The conventional techniques of environmental sustainability deploy the use of microorganisms to degrade and remove pollutants from the environment known as bioremediation.

Microbes play crucial roles in regulating biogeochemical cycles, promoting plant growth, suppressing plant diseases, and cleansing the environment of pollutants. However, despite the promise of bioremediation, there are still challenges to be addressed. Bioremediation methods that work in one location may not be effective in another, and laboratory success does not always translate to field applications. Other methods include chemical treatments such as chemical oxidation, which involves the use of oxidizing agents to convert pollutants into less harmful or nontoxic substances. Common oxidizing agents used in environmental remediation include hydrogen peroxide, ozone, and potassium permanganate. These agents react with pollutants, such as organic contaminants or heavy metals, to break them down into simpler, less harmful compounds. Coagulation and flocculation are processes used for water and wastewater treatment. Chemical coagulants, such as alum (aluminum sulfate) or ferric chloride, are added to water contaminated with suspended particles or pollutants. These chemicals destabilize the particles, causing them to clump together (coagulation) and form larger particles called flocs (flocculation). The flocs can then be easily removed through sedimentation or filtration processes.

Adsorption is a process where pollutants are removed from a liquid or gas phase by binding to a solid material called an adsorbent. Activated carbon is a commonly used adsorbent for various pollutants, including organic compounds, dyes, and heavy metals. The pollutants adhere to the surface of the activated carbon, effectively removing them from the environment. Precipitation is used to remove dissolved pollutants, particularly heavy metals, from wastewater or contaminated water. Chemicals, such as lime (calcium hydroxide) or sodium hydroxide, are added to the water to raise the pH and form insoluble metal hydroxide precipitates. These precipitates can then be separated and removed from the water. Lastly, neutralization to adjust the pH of acidic or alkaline pollutants to a neutral level is done. This process is commonly used for the treatment of acid mine drainage or industrial effluents with extreme pH levels. Acidic solutions can be neutralized using bases, such as lime or sodium hydroxide, while alkaline solutions can be neutralized using acids. However, the major drawback of chemical treatment is that these chemicals contain hydrocarbons, which are a significant contributor to pollution. Improper handling, storage, or disposal of these chemicals can lead to their release into the environment, contaminating soil and water resources. Hydrocarbon spills and leaks from industrial activities can have severe ecological consequences, harming aquatic life, contaminating groundwater, and affecting the overall balance of ecosystems. In this regard, microbial-based bioremediation approaches, including the use of nanoparticles (Solomon et al. 2024) have garnered potential solutions for removing toxins from the environment and restoring damaged ecosystems. To gain a deeper understanding of the underlying mechanisms involved in bioremediation and to develop innovative approaches, researchers are increasingly employing various omics techniques, including metabolomics, genomics, transcriptomics, proteomics, and fluxomics. These techniques provide valuable insights into metabolic activities, genetic information, gene expression patterns, protein profiles, and fluxes of metabolites within microbial communities involved in bioremediation processes. These approaches have revolutionized the field of environmental microbiology by enabling comprehensive studies of microbial communities and their responses to contaminants. Next-generation sequencing (NGS) technologies have revolutionized the field of microbial ecology and have significantly contributed to our understanding of microbial communities in polluted environments. NGS enables the rapid and cost-effective sequencing of DNA or RNA molecules, allowing researchers to analyze the genetic information of entire microbial communities present in a given sample. Omics techniques have revolutionized the discovery of novel bacteria with the capability to degrade a wide range of environmental pollutants. Traditional culture-based methods rely on isolating and culturing microorganisms in the laboratory, which often restricts the types of bacteria that can be identified and studied. Many microorganisms, especially those with specialized or unique metabolic capabilities, are difficult to culture using traditional methods.

Omics techniques, such as metagenomics and metatranscriptomics, allow researchers to study the genetic material and gene expression of entire microbial communities directly from environmental samples (Kanchan et al. 2024a).

This approach provides a more comprehensive view of the microbial diversity and functional potential within a given ecosystem. By sequencing and analyzing the DNA or RNA extracted from these samples, researchers can identify the genetic sequences and gene products associated with pollutant degradation.

In recent years, there has been a significant focus on studying microbial consortia in various habitats using high-throughput omics approaches for bioremediation research. These omics techniques, including genomics, transcriptomics, proteomics, metabolomics, and fluxomics, have provided valuable insights into the dynamics, growth, metabolism, structure, and activities of microorganisms involved in sophisticated bioremediation systems (Sharma et al. 2022). The integration of fluxomics, genomics, metabolomics, meta-proteomics, meta-transcriptomics, and in silico analyses allows for comprehensive and cost-effective studies of metagenomic samples containing diverse microbial communities. The studies that utilize omics techniques in bioremediation research provide valuable insights into the range of microbial species present in contaminated sites and their functional capabilities. By analyzing the metagenomic, metatranscriptomic, proteomic, and metabolomic data, researchers can identify the microbial taxa and genes involved in pollutant degradation pathways. The information obtained from omics techniques allows a deeper understanding of the metabolic activities of microbial communities during bioremediation processes. For example, by identifying specific genes or enzymes that are responsible for the breakdown of a particular pollutant, researchers can focus on engineering or selecting microbial strains that possess those genes or enzymes. This targeted approach can enhance the efficiency and effectiveness of bioremediation by ensuring that the microbial community has the necessary metabolic machinery to efficiently degrade the pollutants of interest. Furthermore, omics data can help in understanding the regulatory elements and mechanisms that control the metabolism of pollutants in microbial communities. This information can be used to design strategies to manipulate and optimize microbial activities during bioremediation processes.

Overall, the integration of omics techniques in bioremediation research allows for a more comprehensive understanding of the metabolic activities of microbial communities in contaminated environments (Kesheri et al. 2024a). This knowledge can be harnessed to develop novel approaches, improve bioremediation strategies, and optimize the selection or engineering of microorganisms for enhanced pollutant degradation.

14.2 Potential Health Hazards Caused by the Environment

The environment plays a significant role in human health, and exposure to certain environmental factors can pose potential health hazards. Here are some examples of potential health hazards caused by the environment:

14.2.1 Air Pollution

Air pollution caused by industries is a significant concern due to its adverse effects on human health. Industries release a variety of pollutants into the air, including particulate matter, nitrogen oxides, sulfur oxides, volatile organic compounds, heavy metals, and other toxic substances. These pollutants can have detrimental effects on the respiratory and cardiovascular systems, leading to a range of health issues (Shruti and Kesheri 2016). Exposure to industrial air pollution can result in respiratory problems such as asthma, bronchitis, and chronic obstructive pulmonary disease (COPD). The inhalation of pollutants irritates the airways, causing inflammation and constriction, and making it difficult to breathe. Prolonged exposure to industrial air pollution has also been linked to an increased risk of lung cancer, particularly due to the presence of carcinogenic substances in emissions. Furthermore, industrial air pollution can have cardiovascular effects. Fine particulate matter (PM2.5) and other pollutants can enter the bloodstream, leading to systemic inflammation, oxidative stress, and the development or exacerbation of cardiovascular diseases. Studies have shown associations between industrial air pollution and conditions such as heart disease, stroke, high blood pressure, and impaired heart function. Children are particularly vulnerable to the health effects of industrial air pollution. Their developing respiratory systems are more sensitive to pollutants, and exposure during critical stages of lung development can lead to long-term respiratory issues and impaired lung function. Addressing industrial air pollution requires effective regulations, emission control technologies, and cleaner production practices. Implementing stringent air quality standards, promoting the use of cleaner energy sources, improving industrial processes, and adopting pollution control measures can significantly reduce the release of harmful pollutants (Lee et al. 2020; Xue et al. 2023).

14.2.2 Water Pollution and Associated Contamination

Water pollution caused by industries is a major environmental concern with severe implications for ecosystems and human health. Industrial activities generate various pollutants that find their way into water bodies, including rivers, lakes, and groundwater sources. These pollutants can have detrimental effects on aquatic life, such as causing eutrophication, leading to luxuriant growth of harmful cyanobacterial blooms, for example, *Microcystis aeruginosa* (Kesheri et al. 2022) and *Oscillatoria* sp. (Kesheri et al. 2021), and pose significant risks to human populations that rely on contaminated water sources. Industrial activities contribute to water pollution through the discharge of untreated or inadequately treated wastewater containing a range of harmful substances. These may include heavy metals, toxic chemicals, organic pollutants, pesticides, and industrial by-products. The pollutants can disrupt the natural balance of aquatic ecosystems, leading to the degradation of water quality and the loss of biodiversity. The impact of industrial water pollution on human health is significant. Consuming or coming into contact with contaminated water can lead to various waterborne diseases, such as gastrointestinal infections, cholera, typhoid, and hepatitis. Exposure to toxic substances present in industrial wastewater can have long-term health effects, including organ damage, developmental issues, hormonal imbalances, and an increased risk of cancer. Furthermore, industrial water pollution can contaminate agricultural lands and disrupt food chains, affecting the safety and quality of the food supply. The accumulation of pollutants in fish and seafood from polluted water bodies can pose health risks for those consuming contaminated aquatic organisms. Addressing industrial water pollution requires strict regulations, effective monitoring, and the implementation of appropriate treatment technologies. Industries must adopt responsible wastewater management practices, including proper treatment before discharge, recycling and reusing water, and implementing pollution prevention measures to minimize the release of hazardous substances. Government bodies, environmental agencies, and industry stakeholders need to collaborate to enforce regulations, encourage sustainable industrial practices, and promote the use of clean technologies. Furthermore, public awareness and education campaigns play a crucial role in fostering responsible industrial behavior and encouraging individuals to conserve water resources and advocate for clean water initiatives. Contamination of water sources with pollutants, including heavy metals, pesticides, industrial chemicals, and microbial pathogens, can lead to waterborne diseases such as diarrhea, cholera, hepatitis, and other gastrointestinal infections. Prolonged exposure to certain water contaminants may also contribute to the development of chronic health conditions, including organ damage, neurological disorders, and certain types of cancer (Choi et al. 2017; Koop and van Leeuwen 2017; Saravanan et al. 2021).

14.2.3 Chemical Exposure

Exposure to hazardous chemicals, either through occupational settings or through environmental contamination, impacts human health. Certain chemicals, such as lead, asbestos, benzene, and pesticides, have been linked to various health problems, including neurological disorders, cancer, reproductive abnormalities, and respiratory illnesses. Workers in industrial settings are at higher risk of chemical exposure due to direct contact with hazardous substances. These chemicals can enter the body through various routes. Inhalation is a common route when chemicals are in the form of vapors, gases, or airborne particles (Saeed et al. 2018). Skin contact can occur through direct handling or contact with contaminated surfaces. Ingestion may occur if workers inadvertently consume chemicals through contaminated hands or food. Industries such as manufacturing, construction, agriculture, and chemical production carry an increased risk of chemical exposure for workers. The health effects of industrial chemical exposure can vary depending on the specific chemicals involved, the concentration and duration of exposure, and individual susceptibility. Short-term exposure to high levels of certain chemicals can cause acute effects such as irritation of the skin, eyes, or respiratory system, dizziness, nausea, or chemical burns. Long-term exposure to lower levels of certain chemicals may lead to chronic health conditions, including respiratory disorders, neurological issues, reproductive problems, organ damage, and an increased risk of cancer. Preventing industrial chemical exposure requires implementing proper safety measures and protocols. This includes using engineering controls, such as ventilation systems and enclosed workspaces, to minimize airborne contaminants. Personal protective equipment (PPE) like gloves, goggles, respirators, and protective clothing should be provided and used appropriately. Proper training, education, and monitoring programs are essential to ensure worker safety and minimize the risk of chemical exposure (Hopf et al. 2021; Johnston and Cushing 2020; Leso et al. 2017).

14.2.4 Radiation Exposure

Industrial radiation exposure refers to the exposure of workers to ionizing radiation in various industrial settings. While radiation has beneficial uses in industries such as medicine and energy production, it also poses potential health risks. Exposure to ionizing radiation, including X-rays, gamma rays, and alpha and beta particles, whether from natural sources (e.g., radon) or human-made sources (e.g., nuclear accidents, medical imaging), can have acute and long-term effects on human health (Laskowski et al. 2022). High levels of radiation can cause acute effects such as radiation sickness, burns, and an increased risk of cancer. Chronic exposure to lower levels of radiation may contribute to the development of cancer over time. Further, radiation exposure can also cause genetic mutations that may be passed on to future generations, increasing the risk of hereditary disorders (Rushton and Hutchings 2017).

14.2.5 Climate Change

Climate change, driven by human activities and greenhouse gas emissions, poses a wide range of health hazards. Rising temperatures can lead to heat-related illnesses and heatwaves, particularly affecting vulnerable populations such as the elderly and those with preexisting health conditions. Changes in precipitation patterns can impact water availability and quality, leading to waterborne diseases and food insecurity. Climate change can also exacerbate the spread of vector-borne diseases, allergies, and respiratory disorders. Study elucidates that climate change affects stress responses in cyanobacteria, evident by the presence of antioxidants such as superoxide dismutase, which is the first line of defense (Kesheri et al. 2014).

14.2.6 Exposure to Hazardous Waste

Improper disposal and mishandling of hazardous waste, including industrial waste, electronic waste, and chemical waste, can contaminate soil, water, and air, posing serious health risks to nearby communities. Exposure to hazardous waste can lead to a range of health problems, including respiratory issues, neurological disorders, birth defects, and an increased risk of cancer (Barton and Ainerua 2020). Certain hazardous substances, such as asbestos, lead, mercury, and PCBs, are known to have serious long-term health impacts (Sarigiannis 2017). Communities living near industrial facilities that generate or handle hazardous waste may also be at risk of exposure. Contaminated air, water, and soil can affect nearby residents through inhalation, consumption of contaminated food or water, or direct contact with contaminated surfaces. Disproportionate exposure risks often affect disadvantaged and marginalized communities, leading to environmental justice concerns. Addressing exposure to hazardous waste requires a comprehensive approach involving effective waste management practices, stringent regulatory enforcement, worker protection measures, community engagement, and environmental monitoring. By prioritizing the reduction, recycling, and safe disposal of hazardous waste, monitoring the entry of run-off containing hazardous waste to agricultural lands (Kumari et al. 2018), we can minimize the potential risks associated with exposure and protect both human health and the environment (Hassan and Saleh 2022; Heacock et al. 2018; Kesheri et al. 2024b).

Therefore, it is vital to mitigate these environmental hazards through effective environmental policies, regulations, and public awareness. Implementing pollution control measures, promoting sustainable practices, and adopting clean energy sources can help minimize the potential health risks associated with environmental factors. Table 14.1 outlines the health hazards caused by the environment.

14.3 Bioremediation and Its Types

Bioremediation is a process that uses living organisms, such as microorganisms or plants, to degrade or remove contaminants from the environment. It is a sustainable and environmentally friendly approach to remediate polluted sites and restore them to their natural state. Bioremediation can be applied to various types of contaminants, including organic pollutants, heavy metals, and petroleum hydrocarbons (Azubuike et al. 2016; Hazen 2018). Three main types of bioremediation are described and listed below in Fig. 14.1.

A. *Phytoremediation*: Phytoremediation involves the use of plants and their associated microorganisms to remediate contaminated sites. Plants have the ability to absorb, stabilize, and metabolize contaminants through processes like phytoextraction, rhizofiltration, phytostabilization, and phytodegradation. Plant roots release exudates that promote microbial activity, further enhancing pollutant degradation. Phytoremediation is particularly useful for pollutants in soil, groundwater, and sediments.
B. *Microbial Bioremediation*: This approach utilizes microorganisms, such as bacteria, fungi, and archaea, to degrade or transform contaminants into less harmful substances. Microbes possess enzymes and metabolic pathways that can break down complex pollutants, converting them into simpler compounds. The microbial community can naturally exist in the contaminated environment or can be

Table 14.1 Potential environmental health hazards affecting human health

Environmental hazard	Potential health effects
Air pollution	Respiratory diseases (asthma, bronchitis, lung cancer)
	Cardiovascular diseases
	Impaired lung development in children
Water contamination	Waterborne diseases (diarrhea, cholera, hepatitis)
	Chronic organ damage
	Neurological disorders
Chemical exposure	Neurological disorders
	Cancer
	Reproductive abnormalities
	Respiratory illnesses
Radiation exposure	Radiation sickness
	Burns
	Increased risk of cancer
Climate change	Heat-related illnesses and heatwaves
	Waterborne diseases and food insecurity
	Spread of vector-borne diseases
	Allergies and respiratory disorders
Exposure to hazardous waste	Respiratory issues
	Neurological disorders

introduced or stimulated to enhance biodegradation. Factors such as temperature, pH, oxygen availability, nutrient levels, and the presence of co-substrates can influence microbial activity and the effectiveness of bioremediation.

C. *Hybrid or Integrated approaches*: It refers to combined bioremediation techniques utilizing phytoremediation and microbial bioremediation, to achieve a more comprehensive and effective cleanup of polluted area. It is beneficial to address mixed types of pollutants present in a contaminated site. It is faster and cost-effective, site-specific contaminant removal compared to using a single method. Hybrid approaches are sometimes good because they have synergistic effects, where one process enhances the effectiveness of another.

14.3.1 Objectives, Advantages, and Applications of Bioremediation

Bioremediation is a process of providing a sustainable and eco-friendly remedy using a living organism to create a healthy environment suitable for all biotic communities. Bioremediation reduces the contaminants to nontoxic forms, keeps the pollutants within tolerable levels and regulatory limits (Kuppan et al. 2024). Bioremediation is also responsible for the mineralization of complex organic pollutants into their elemental form to keep our environment safe by making it sustainable

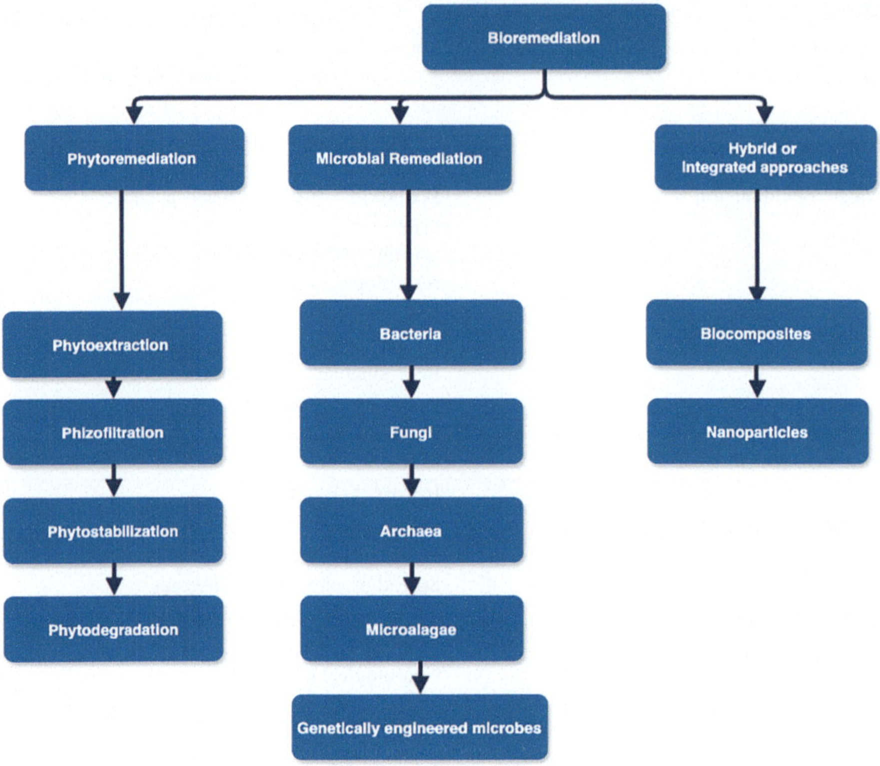

Fig. 14.1 Different types of bioremediations

(Kuppan et al. 2024). Figure 14.2 lists various purposes and objectives of bioremediation to keep our environment sustainable and safe.

Benefits of bioremediation include its cost-effectiveness, minimal environmental disruption, and the potential to treat contaminants in situ (on-site) without the need for excavation or transportation. It can be applied to various environments, including soil, water, and air. Bioremediation is also compatible with other remediation techniques, such as physical and chemical methods, as part of a comprehensive approach to address complex contamination scenarios. However, the success of bioremediation depends on several factors, including the type and concentration of contaminants, environmental conditions, the availability of suitable microorganisms or plants, and the site-specific characteristics. Monitoring and optimization of the bioremediation process are essential to ensure its effectiveness (Abatenh et al. 2017).

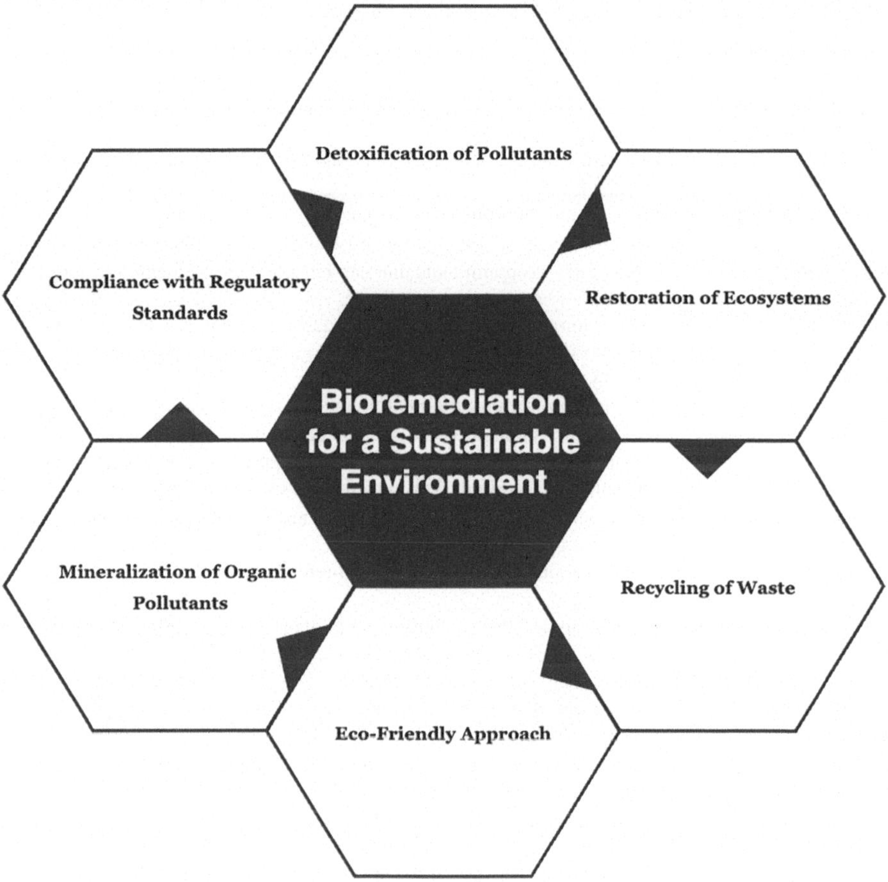

Fig. 14.2 Objectives/purposes of bioremediation for a sustainable environment

14.3.2 Limitations of Bioremediation

While bioremediation is a valuable and sustainable approach to environmental cleanup, it also has certain limitations (Azubuike et al. 2016). Some of the key limitations of bioremediation (Table 14.2) include the following:

A. *Contaminant Compatibility*: Bioremediation may not be suitable for all types of contaminants. Some highly toxic or persistent pollutants may be resistant to degradation by natural microbial processes, limiting the effectiveness of bioremediation. Additionally, certain contaminants, such as heavy metals, may require specific microbial strains or plants with metal-tolerance mechanisms to be effectively remediated.
B. *Site Conditions*: The success of bioremediation is highly dependent on site-specific conditions. Factors such as temperature, pH, moisture content, and

Table 14.2 Limitations of various bioremediation techniques

Bioremediation technique	Limitations
Microbial bioremediation	Limited effectiveness for certain contaminants that are resistant to microbial degradation Requires suitable environmental conditions (e.g., temperature, pH, oxygen levels) Slow degradation rates for complex or persistent pollutants Difficulty in accessing deep soils or inaccessible contaminated areas
Phytoremediation	Restricted to contaminants that can be taken up and metabolized by plants May require a long time to achieve desired cleanup goals Site-specific factors (e.g., soil conditions, plant species) can affect efficacy Limited application for contaminants in water or air
Bioaugmentation	Risk of introducing non-native or genetically modified microorganisms Competition between introduced and indigenous microorganisms Difficulty in sustaining introduced microorganisms in the environment
Biostimulation	Dependence on environmental factors and availability of nutrients for microbial growth and activity Limited control over microbial community dynamics and metabolic pathways
Bioventing	Limited application to contaminants in unsaturated soils Requires careful monitoring to prevent the release of volatile contaminants into the atmosphere
Landfarming	Space limitations and land availability for large-scale treatment Potential for leaching of contaminants into groundwater Limited effectiveness for certain pollutants that are resistant to microbial degradation
In situ chemical oxidation	Potential for the generation of harmful by-products during oxidation reactions Difficulty in achieving uniform distribution of oxidants throughout the contaminated area

nutrient availability can significantly influence the activity and growth of microbial populations or plant performance. Unsuitable environmental conditions can limit the effectiveness of bioremediation or slow down the degradation process.

C. *Time Considerations*: Bioremediation is often a slower process compared to other remediation techniques. Microbial degradation rates can vary depending on the type and concentration of contaminants, as well as the metabolic capabilities of the microorganisms involved. In some cases, bioremediation may require longer timeframes to achieve desired cleanup goals, which may not be feasible in time-sensitive situations.
D. *Site Accessibility*: The accessibility of contaminated sites can pose challenges for implementing bioremediation. If contaminants are located in deep soils, sediments, or inaccessible areas, it may be difficult to introduce microbial populations or plants to the contaminated area effectively. Excavation or other physical methods may be required to provide better access for bioremediation.

E. *Monitoring and Control*: Effective monitoring and control of bioremediation processes are essential to ensure their success. Monitoring microbial activity, contaminant levels, and environmental conditions requires regular sampling and analysis. Implementing appropriate monitoring strategies and control measures can be resource-intensive and may require technical expertise.
F. *Regulatory Considerations*: Bioremediation projects may need to comply with regulatory requirements and obtain permits for implementation. The regulatory process can introduce additional complexities and delays, impacting the feasibility and timeline of bioremediation projects.

14.4 Omics in Bioremediation

Omics studies encompass a wide range of high-throughput technologies and analytical approaches that aim to comprehensively analyze and understand the various molecules within a biological system. The main principle behind omics studies is to view biological systems as a whole, considering the interactions and relationships among genes, proteins, metabolites, and other molecules. The advancement of omics, such as genomics, transcriptomics, proteomics, and other related omics techniques, has further led to new approaches that help in building new strategies for environmental management and promoting sustainability.

Through omics approaches, researchers can collect large amounts of data representing the entirety of molecules within a biological sample. This data can then be analyzed using bioinformatics and statistical tools to extract meaningful insights. Omics studies enable a non-biased and nontargeted investigation of the entire molecular profile of a biological system, allowing for a comprehensive understanding of its complexity.

By adopting a holistic approach, omics studies contribute to the field of systems biology, which aims to understand biological systems as integrated networks of molecules and their interactions (Pandey et al. 2019). This approach goes beyond studying individual components in isolation and seeks to uncover the intricate relationships and dynamics within the system as a whole. These studies provide a wealth of information and insights into biological processes, disease mechanisms, and the effects of various interventions or treatments. Overall, omics studies play a crucial role in advancing our understanding of complex biological systems, uncovering biomarkers, identifying therapeutic targets, and guiding personalized medicine approaches. These technologies continue to evolve and contribute significantly to various fields, including genomics, transcriptomics, proteomics, metabolomics, and more (Chandran et al. 2020; Sharma et al. 2022).

14.5 Genomics in Bioremediation

Genomics is a field of biological sciences that focuses on the study of genomes, which refers to the complete set of genes or genetic material present in an organism. It involves the analysis and interpretation of DNA sequences to understand the structure, function, and evolution of genes and genomes. Genomics plays a significant role in advancing the field of bioremediation, which involves the use of microorganisms to remove or degrade pollutants from the environment (Plewniak et al. 2018). A list of various ways by which genomics could contribute to bioremediation is listed below:

A. *Microbial Community Analysis*: Genomics allows for the comprehensive analysis of microbial communities present in contaminated environments. Through metagenomic sequencing and its bioinformatic data analysis, the collective genetic material of the microbial community can be studied (Srivastava et al. 2024), providing insights into the diversity, composition, and functional potential of the microbes involved in bioremediation processes. This information helps identify key microbial players and their metabolic capabilities.
B. *Functional Gene Identification*: Genomic analysis enables the identification of specific genes and gene clusters responsible for pollutant degradation or transformation. By comparing the genomes of known pollutant-degrading microorganisms with environmental samples, researchers can identify novel genes and enzymes involved in bioremediation (Shukla et al. 2024). This information helps in the design of tailored approaches for targeted pollutant removal.
C. *Gene Expression Profiling*: Genomics techniques, such as transcriptomics, allow for the study of gene expression patterns in response to pollutant exposure. By comparing the gene expression profiles of microorganisms under different conditions, researchers can understand the molecular mechanisms involved in pollutant degradation and the factors that influence microbial activity in contaminated environments (Mallik et al. 2024).
D. *Genetic Engineering*: Genomics provides the foundation for genetic engineering approaches in bioremediation. By identifying and manipulating specific genes and genetic pathways involved in pollutant degradation, researchers can enhance the capabilities of microorganisms or introduce novel functions to improve bioremediation efficiency. This includes genetic modifications to enhance enzyme production, increase pollutant uptake, or optimize metabolic pathways.
E. *Predictive Modeling and Simulation*: Genomic data, combined with computational modelling and simulation (Kesheri et al. 2015), allows for the prediction of microbial community dynamics and pollutant degradation rates in different environmental conditions. This helps in designing effective bioremediation strategies and optimizing treatment protocols.
F. *Bioinformatics and Data Integration*: Genomics generates large volumes of data that require sophisticated bioinformatics tools for data analysis, interpretation, and integration. Bioinformatics enables the identification of functional genes,

metabolic pathways, and regulatory networks involved in bioremediation processes (Kumari et al. 2018). It also facilitates the integration of genomics data with other omics approaches, such as metatranscriptomics, metaproteomics, and metabolomics, for a comprehensive understanding of microbial activities in contaminated sites.

By leveraging genomics technologies and approaches, researchers can gain insights into the genetic potential of microbial communities in polluted environments. This knowledge aids in the development of tailored bioremediation strategies, including the selection of appropriate microorganisms, optimization of environmental conditions, and genetic engineering of microbial strains to enhance pollutant degradation efficiency. Overall, genomics has revolutionized the field of bioremediation and holds immense potential for addressing environmental contamination challenges. Genomic techniques such as DNA hybridization, metabolic engineering, and molecular connectivity are deployed. Further, polymerase chain reaction (PCR)-based techniques, such as single strand conformation polymorphism (SSCP), terminal-restriction fragment length polymorphism (T-RFLP), and amplified fragment length polymorphisms (AFLP), are available. A PCR-based quantitative analysis is used as an indicator for the detection and identifying the presence of microbial taxa or functional genes in the soil. This technique allows for the direct analysis of molecular biomarker genes to assess the composition and diversity of soil microbial communities (Singh et al. 2017; Chakdar et al. 2022).

14.5.1 Advantages of Genomics in Bioremediation

Genomics plays a crucial role in bioremediation and offers several advantages and applications (Table 14.3) that contribute to the effectiveness and efficiency of the bioremediation process:

A. *Identification of Biodegradation Genes*: Genomics allows for the identification of genes and enzymes involved in the biodegradation of specific pollutants. This knowledge helps in selecting appropriate microorganisms with the genetic potential to degrade the target contaminants.
B. *Metabolic Pathway Analysis*: Genomics provides insights into the metabolic pathways used by microorganisms to break down pollutants. Understanding these pathways helps in designing targeted bioremediation strategies and optimizing the conditions for pollutant degradation.
C. *Predicting Biodegradation Efficiency*: By analyzing the genetic makeup of microbial communities, genomics can help predict the efficiency of bioremediation processes for specific pollutants. This information aids in selecting the most suitable microorganisms for effective treatment.
D. *Characterization of Microbial Communities*: Genomics allows for the comprehensive characterization of microbial communities present in contaminated

Table 14.3 Applications of genomics in bioremediation

Pollutant	Microorganism	Technique used	Reference
Crude oil	*Achromobacter* sp. HZ01	Whole Genome Sequencing (WGS)	Hong et al. (2017)
Heavy metals (Cr, Cu, Ni, Pb)	Metal-resistant bacteria, fungi, and algae (e.g., *Bacillus cereus*, *Aspregillus*, *Oedogonium*, cyanobacteria)	Comparative genomics	Chakdar et al. (2022)
Sodium benzoate, selenite reduction, chloroalkane	*Haloferax* sp.	Shotgun sequencing, 16 s rRNA	Naik et al. (2021)
Methanol, methane	Methanotrophs	DGGE, DNA-SIP	Hwangbo et al. (2023)

environments. This understanding helps in assessing the diversity and functional potential of these communities for bioremediation purposes.

E. *Development of Engineered Microbes*: Genomics facilitates the genetic engineering of microorganisms to enhance their bioremediation capabilities. Researchers can introduce specific genes or modify existing ones to improve pollutant degradation efficiency.
F. *Monitoring Bioremediation Progress*: Genomic tools enable real-time monitoring of microbial communities during the bioremediation process. Changes in gene expression patterns can indicate the effectiveness of treatment and guide any necessary adjustments.
G. *Environmental Genomics*: Metagenomics, a branch of genomics, allows for the direct sequencing and analysis of DNA from environmental samples. This technique helps in studying microbial communities without the need for culturing, providing a more comprehensive understanding of the potential biodegradation capabilities (Kanchan et al. 2020).
H. *Cost and Time Efficiency*: Genomic analysis can provide rapid and cost-effective assessments of microbial communities, allowing for quicker decision-making in bioremediation strategies.
I. *Integration with Other Omics Techniques*: Genomics can be integrated with other omics approaches, such as transcriptomics, proteomics, and metabolomics, providing a more holistic view of microbial responses during bioremediation.
J. *Development of Targeted Bioremediation Strategies*: By understanding the genetic basis of pollutant degradation, genomics enables the development of targeted bioremediation strategies that focus on specific contaminants and microbial communities.

14.6 Transcriptomics in Bioremediation

Transcriptomics is a powerful genomics technique used in bioremediation to study the expression pattern of genes in microbial communities during the process of pollutant degradation. It provides valuable insights into the response of microorganisms to environmental pollutants and helps identify the genes and pathways involved in biodegradation. For performing transcription analysis, it is necessary to isolate mRNA to synthesize cDNA following sequencing. In the context of bioremediation, transcriptomics involves the following steps (Chandran et al. 2020; Malik et al. 2021; Sharma et al. 2022):

A. *Sampling*: Environmental samples, such as soil, water, or sediment, are collected from contaminated sites where bioremediation is being conducted or intended to be implemented.
B. *RNA Extraction*: Total RNA is extracted from the collected samples. RNA represents the active genes in the microbial community at the time of sampling.
C. *cDNA Synthesis*: The extracted RNA is converted into complementary DNA (cDNA) using reverse transcription. cDNA serves as a representation of the gene expression levels in the microbial community. For example, wastewater surveillance pertaining to the recent COVID-19 pandemic used RTqpcr for quantifying COVID-19 viral load in wastewater and used machine learning (Kesheri et al. 2016) for future prediction of the pandemic (Kanchan et al. 2024b; Meadows et al. 2025).
D. *Sequencing*: The cDNA is then subjected to next-generation sequencing (NGS) or other high-throughput techniques to determine the sequence and abundance of different transcripts.
E. *Data Analysis*: Bioinformatic tools and software are used to analyze the sequencing data and identify differentially expressed genes. These are genes that are upregulated or downregulated in response to pollutant exposure.

14.6.1 Advantages of Transcriptomics in Bioremediation

Transcriptomics in bioremediation offers several benefits and applications (Table 14.4):

A. *Gene Expression Profiling*: It allows researchers to profile the gene expression patterns of microbial communities during biodegradation. This helps in understanding which genes are activated or suppressed in response to pollutants.
B. *Identifying Key Players*: Transcriptomics helps in identifying key microbial species and functional genes involved in the breakdown of specific contaminants. This information can be used to optimize bioremediation strategies.
C. *Predicting Biodegradation Pathways*: By analyzing the expression patterns of different genes, researchers can predict the metabolic pathways used by microorganisms to degrade pollutants.

Table 14.4 Application of transcriptomics in bioremediation

Pollutant	Microorganism	Technique used	Reference
Crude oil	Hydrocarbon-degrading bacteria (e.g., *Alcanivorax, Bacillus, Georgenia, Micrococcus, Stappia)*	RNA-sequencing, microarrays	Das et al. (2023)
Heavy metals (Cr, Cu, Ni, Pb)	Metal-resistant bacteria, fungi, and algae (e.g., *Bacillus cereus, Aspregillus, Oedogonium)*	RNA-sequencing, qRT-PCR	Peng et al. (2021)
Toluene	*Cladophialophora immunda*	RT-PCR, sequencing	Blasi et al. (2017)
Polycyclic Aromatic Hydrocarbons (PAHs)	PAH-degrading bacteria (e.g., *Sphingobium, Mycobacterium, Rhodococcus*)	RNA-sequencing, microarrays, qRT-PCR	Volpicella et al. (2017)
Perchlorate	Perchlorate-reducing bacteria (e.g., *Dechloromonas, Zoogloea*)	RNA-sequencing, microarrays, qRT-PCR	Nitz et al. (2020)
Methyl parathion	*Burkholderia cenocepacia*	RNA-sequencing, RNA mapping	Ortiz-Hernández et al. (2021)

D. *Monitoring Bioremediation Progress*: Transcriptomics can be used as a monitoring tool during bioremediation to assess the effectiveness of the process and to make necessary adjustments if required.

14.7 Proteomics in Bioremediation

Proteomics is a powerful omics technique used in bioremediation to study the entire set of proteins expressed by microorganisms during the process of pollutant degradation (Pandey et al. 2019; Chakdar et al. 2022). It provides valuable information about the proteins involved in biodegradation pathways and their regulation, complementing other omics approaches such as genomics and transcriptomics (Sharma et al. 2022).

In the context of bioremediation, proteomics involves the following steps:

A. *Sampling*: Environmental samples, such as soil, water, or sediment, are collected from contaminated sites where bioremediation is being conducted or intended to be implemented.
B. *Protein Extraction*: Total proteins are extracted from the collected samples. Proteins represent the functional molecules in the microbial community and can provide insights into the active metabolic pathways.
C. *Protein Separation*: Proteins are separated and fractionated using techniques such as gel electrophoresis or liquid chromatography. This separation allows for the identification of individual proteins.

D. *Mass Spectrometry*: The separated proteins are subjected to mass spectrometry, which identifies and quantifies the proteins present in the samples. Mass spectrometry helps in identifying the specific proteins involved in pollutant degradation.
E. *Data Analysis*: Bioinformatic tools and software are used to analyze the mass spectrometry data and identify the proteins and their functions. This analysis helps in understanding the response of microbial communities to pollutant exposure.

14.7.1 Advantages of Proteomics in Bioremediation

Proteomics offers several advantages and applications (Table 14.5) in the field of bioremediation, enhancing our understanding of microbial responses during pollutant degradation and contributing to the optimization of bioremediation strategies. Some of the key advantages of proteomics in bioremediation include the following:

A. *Comprehensive Analysis*: Proteomics allows for the study of the entire set of proteins expressed by microorganisms during bioremediation. This comprehensive analysis provides a holistic view of the functional proteins involved in pollutant degradation.
B. *Protein Identification*: Proteomics helps in identifying the specific proteins responsible for pollutant degradation. This information is crucial for understanding the molecular mechanisms and metabolic pathways involved in the biodegradation process.
C. *Functional Characterization*: Proteomics provides insights into the functions of microbial communities during bioremediation. It helps in understanding how different proteins work together to efficiently degrade pollutants.
D. *Biomarker Discovery*: Proteomics can lead to the discovery of protein biomarkers that serve as indicators of bioremediation efficiency and microbial activity.

Table 14.5 Applications of proteomics in bioremediation

Pollutant	Microorganism	Technique used	Reference
Crude oil	*Pseudomonas putida* KT2440	Two-dimensional gel electrophoresis (2D-PAGE)	Bhaganna et al. (2016); Możejko-Ciesielska and Mostek (2019)
Heavy metals (Cr, Cu, Ni, Pb)	*Cupriavidus metallidurans*	Liquid chromatography-mass spectrometry (LC-MS)	Tong et al. (2023)
Chlorinated compounds	*Dehalococcoides ethenogenes*	Isobaric Tags for Relative and Absolute Quantitation (iTRAQ)	Chang et al. (2021)
Radionuclides	*Geobacter sulfurreducens*	Stable isotope labeling by amino acids in cell culture (SILAC)	Sun et al. (2018)

These biomarkers can be used to monitor the progress of bioremediation in real time.

E. *Optimization of Bioremediation Strategies*: Proteomics data can be used to optimize bioremediation strategies by identifying key proteins and metabolic pathways involved in pollutant degradation. This information helps in designing targeted and effective bioremediation approaches.
F. *Real-Time Monitoring*: Proteomics allows for real-time monitoring of microbial responses during bioremediation. Changes in protein expression patterns can be tracked to assess the effectiveness of treatment and make necessary adjustments.
G. *Identification of Novel Enzymes*: Proteomics can reveal novel enzymes involved in pollutant degradation. Discovering new enzymes expands our understanding of microbial capabilities and may lead to the development of more efficient bioremediation processes.
H. *Integration with other Omics Techniques*: Proteomics can be integrated with other omics approaches, such as genomics and transcriptomics, providing a more comprehensive understanding of microbial responses and interactions during bioremediation.
I. *Enhanced Bioremediation Strategies*: Proteomics data can be used to engineer microorganisms for enhanced bioremediation capabilities. Modifying or introducing specific proteins can improve pollutant degradation efficiency.
J. *Reduced Environmental Impact*: By understanding the protein expression patterns of microbial communities during bioremediation, strategies can be developed to minimize any potential negative impacts on the environment (Kesheri et al. 2015).

14.8 Metabolomics in Bioremediation

Metabolomics is a powerful tool in bioremediation that focuses on studying the small-molecule metabolites produced by microorganisms during the degradation of pollutants (Jeevanandam and Osborne 2022; Patankar et al. 2022). It provides valuable insights into the metabolic changes and pathways involved in pollutant transformation and bioremediation processes. Some of the key applications of metabolomics in bioremediation include the following:

A. *Monitoring Metabolic Changes*: Metabolomics allows for the real-time monitoring of metabolic changes in microbial communities during bioremediation. It provides a comprehensive view of the metabolic response to pollutant exposure.
B. *Identification of Metabolites*: Metabolomics helps in identifying specific metabolites produced during the biodegradation of pollutants. This information aids in understanding the metabolic pathways involved in pollutant transformation.
C. *Biomarker Discovery*: Metabolomics can lead to the discovery of specific metabolite biomarkers that serve as indicators of bioremediation efficiency and microbial activity. These biomarkers can be used to assess the progress of bioremediation.

D. *Assessment of Environmental Stress*: Metabolomics can be used to assess the impact of pollutants on microbial metabolism and detect stress responses in microbial communities.
E. *Optimization of Bioremediation Strategies*: Metabolomics data can be used to optimize the bioremediation strategies by identifying key metabolic pathways and metabolites involved in pollutant degradation. This information helps in designing targeted and efficient bioremediation approaches.
F. *Identification of Novel Pathways*: Metabolomics can reveal novel metabolic pathways involved in pollutant degradation. This expands our understanding of microbial capabilities and may lead to the development of more effective bioremediation processes.
G. *Integrative Omics Studies*: Metabolomics can be integrated with other omics techniques, such as genomics, proteomics, and transcriptomics, to provide a more holistic view of microbial responses during bioremediation.
H. *Impact on the Environment*: Metabolomics data can help assess the impact of bioremediation on the overall environmental metabolome, providing insights into any potential unintended consequences.
I. *Targeted Engineering of Microbes*: By understanding the metabolic pathways involved in pollutant degradation, metabolomics facilitates the targeted engineering of microorganisms to enhance bioremediation capabilities.

Overall, metabolomics offers a comprehensive approach to studying the metabolic responses of microorganisms during bioremediation, leading to a deeper understanding of the processes involved and the development of more effective and sustainable bioremediation strategies.

14.8.1 Advantages of Metabolomics in Bioremediation

Metabolomics offers several advantages and applications (Table 14.6) in the field of bioremediation, providing valuable insights into the metabolic responses of

Table 14.6 Applications of metabolomics in bioremediation

Pollutant	Microorganism	Technique used	Reference
Crude oil	*Pseudomonas putida*	Nuclear Magnetic Resonance (NMR) Spectroscopy	Kumar (2016); Schwechheimer et al. (2018)
Phenanthrene	*Sinorhizobium* sp. C4	Gas Chromatography-Mass Spectrometry (GC-MS)	Sakshi (2020)
Chlorinated compounds	*Dehalococcoides ethenogenes*	Liquid Chromatography-Mass Spectrometry (LC-MS)	Adrian and Marco-Urrea (2016); Kleindienst et al. (2019)
Radionuclides	*Geobacter sulfurreducens*	Capillary Electrophoresis-Mass Spectrometry (CE-MS)	Howley et al. (2022)

microorganisms during pollutant degradation. Some of the key advantages of metabolomics in bioremediation include the following:

A. *Comprehensive View of Metabolites*: Metabolomics provides a comprehensive view of the small-molecule metabolites produced by microorganisms. This enables the simultaneous analysis of multiple metabolites, giving a holistic understanding of the metabolic changes during bioremediation (Rawat and Rangarajan 2019).
B. *Real-Time Monitoring*: Metabolomics allows for real-time monitoring of microbial metabolic changes during bioremediation. This provides immediate feedback on the efficiency of the bioremediation process and allows for timely adjustments if needed.
C. *Identification of Key Metabolic Pathways*: Metabolomics helps in identifying the key metabolic pathways involved in pollutant degradation. Understanding these pathways is crucial for optimizing bioremediation strategies and enhancing pollutant removal efficiency.
D. *Biomarker Discovery*: Metabolomics can lead to the discovery of specific metabolite biomarkers that indicate the presence of pollutants and the progress of bioremediation. These biomarkers serve as indicators of bioremediation effectiveness.
E. *Assessment of Microbial Health*: Metabolomics data can be used to assess the overall health and status of microbial communities during bioremediation. Changes in metabolite profiles can indicate stress or adaptation to pollutant exposure.
F. *Integration with other Omics Techniques*: Metabolomics can be integrated with other omics techniques, such as genomics, proteomics, and transcriptomics, providing a more comprehensive understanding of microbial responses during bioremediation.
G. *Impact on Environmental Metabolome*: Metabolomics helps in assessing the impact of bioremediation on the overall environmental metabolome. This information is essential for understanding the wider consequences of bioremediation on the ecosystem.
H. *Targeted Engineering of Microbes*: By understanding the specific metabolites involved in pollutant degradation, metabolomics facilitates the targeted engineering of microorganisms to enhance their bioremediation capabilities.
I. *Nondestructive Analysis*: Metabolomics allows for nondestructive analysis of samples, enabling repeated measurements from the same microbial community over time, which is beneficial for time-course studies in bioremediation processes.
J. *Validation of Predicted Pathways*: Metabolomics data can validate the predicted metabolic pathways based on genomics and transcriptomics information. This validation enhances our confidence in the understanding of microbial metabolism during bioremediation.

14.9 Conclusion

Mitigation of various types of environmental hazards, such as air pollution, water pollution, chemical pollution, and radiation pollution, and their detrimental effects by converting complex pollutants into simpler elemental forms with the aid of microbes and biological processes is referred to as bioremediation. Such bioremediation practices are cost-effective and eco-friendly, and ensure sustainable environments. Moreover, various types of bioremediation strategies utilize omics technologies, such as genomics, transcriptomics, proteomics, and metabolomics, leveraging platforms for environmental research and pave the way toward a sustainable environment.

Chapter Summary

This chapter describes various types of bioremediation approaches, their roles, advantages, and applications in environmental sustainability. Contribution of various omics technologies such as genomics, transcriptomics, proteomics, metabolomics, and bioinformatics in shaping bioremediation research, its broad outreach, limitations, and its contribution in achieving the goal of a sustainable environment have also been incorporated. Efforts have been made to elucidate strategies to cope with various types of environmental hazards affecting human health. In a nutshell, this chapter is a conduit of the state of the art of bioremediation approaches fueled by omics technologies.

References

Abatenh E, Gizaw B, Tsegaye Z, Wassie M (2017) The role of microorganisms in bioremediation-a review. Open J Environ Biol 2:038–046

Adrian L, Marco-Urrea E (2016) Isotopes in geobiochemistry: tracing metabolic pathways in microorganisms of environmental relevance with stable isotopes. Curr Opin Biotechnol 41:19–25

Azubuike CC, Chikere CB, Okpokwasili GC (2016) Bioremediation techniques–classification based on site of application: principles, advantages, limitations and prospects. World J Microbiol Biotechnol 32:1–18

Barton CC, Ainerua MO (2020) Environmental toxicology: hazardous waste. In: Information resources in toxicology. Academic Press, pp 321–329

Bhaganna P, Bielecka A, Molinari G, Hallsworth JE (2016) Protective role of glycerol against benzene stress: insights from the *Pseudomonas putida* proteome. Curr Genet 62:419–429

Blasi B, Tafer H, Kustor C, Poyntner C, Lopandic K, Sterflinger K (2017) Genomic and transcriptomic analysis of the toluene degrading black yeast *Cladophialophora immunda*. Sci Rep 7:11436

Chakdar H, Thapa S, Srivastava A, Shukla P (2022) Genomic and proteomic insights into the heavy metal bioremediation by cyanobacteria. J Hazard Mater 424:127609

Chandran H, Meena M, Sharma K (2020) Microbial biodiversity and bioremediation assessment through omics approaches. Front Environ Chem 1:570326

Chang SC, Wu MH, Chen TW (2021) Emulsion-enhanced remediation of lindane and DDT in soils. J Soils Sediments 21:469–486

Choi YY, Baek SR, Kim JI, Choi JW, Hur J, Lee TU, Park CJ, Lee BJ (2017) Characteristics and biodegradability of wastewater organic matter in municipal wastewater treatment plants collecting domestic wastewater and industrial discharge. Water 9(6):409

Das N, Das A, Das S, Bhatawadekar V, Pandey P, Choure K, Damare S, Pandey P (2023) Petroleum hydrocarbon catabolic pathways as targets for metabolic engineering strategies for enhanced bioremediation of crude-oil-contaminated environments. Fermentation 9:196

Hassan AI, Saleh HM (2022) Toxicity and hazardous waste regulations. In: Hazardous waste management. Elsevier, pp 165–182

Hazen TC (2018) Microbiology of the terrestrial deep subsurface. In: Bioremediation. CRC Press, pp 247–266

Heacock M, Trottier B, Adhikary S, Asante KA, Basu N, Brune M-N, Caravanos J, Carpenter D, Cazabon D, Chakraborty P (2018) Prevention-intervention strategies to reduce exposure to e-waste. Rev Environ Health 33:219–228

Hong YH, Ye CC, Zhou QZ, Wu XY, Yuan JP, Peng J, Deng H, Wang JH (2017) Genome sequencing reveals the potential of *Achromobacter* sp. HZ01 for bioremediation. Front Microbiol 8:1507

Hopf NB, Fustinoni S, Harris R (2021) Biological monitoring of exposure to industrial chemicals. Patty's industrial hygiene Evaluation and Control USA: Wiley:3

Howley E, Ki D, Krajmalnik-Brown R, Torres CI (2022) *Geobacter sulfurreducens*' unique metabolism results in cells with a high iron and lipid content. Microbiol Spectr 10:e02593–e02522

Hwangbo M, Shao Y, Hatzinger PB, Chu KH (2023) Acidophilic methanotrophs: occurrence, diversity, and possible bioremediation applications. Environ Microbiol Rep 15(4):265–281

Jeevanandam V, Osborne J (2022) Understanding the fundamentals of microbial remediation with emphasize on metabolomics. Prep Biochem Biotechnol 52:351–363

Johnston J, Cushing L (2020) Chemical exposures, health, and environmental justice in communities living on the fenceline of industry. Curr Environ Health Rep 7:48–57

Kanchan S, Sinha RP, Chaudière J, Kesheri M (2020) Computational metagenomics: current status and challenges. In: Katara P (ed) Recent trends in 'computational omics': concepts and methodology. Nova Science Publishers, pp 371–395

Kanchan S, Kesheri M, Srivastava U, Katharia H, Ratna-Raj R, Chittoori B, Bogomolnaya L, Sinha RP, Denvir J (2024a) Integrative omics data mining: challenges and opportunities. In: Gupta MK, Katara P, Mondal S, Singh RL (eds) Integrative omics: concepts, methodology, and application. Academic Press (Elsevier), pp 237–255

Kanchan S, Ogden E, Kesheri M, Skinner A, Miliken E, Lyman D, Armstrong J, Sciglitano L, Hampikian G (2024b) COVID-19 hospitalizations and deaths predicted by SARS-CoV-2 levels in Boise, Idaho wastewater. Sci Total Environ 907:167742

Kesheri M, Kanchan S, Richa, Sinha RP (2014) Isolation and in-silico analysis of Fe-superoxide dismutase in *Nostoc commune*. Gene 553(2):117–125

Kesheri M, Kanchan S, Chowdhury S, Sinha RP (2015) Secondary and tertiary structure prediction of proteins: a bioinformatic approach. In: Zhu O, Azar AT (eds) Complex system modelling and control through intelligent soft computations, vol 319. Springer, pp 541–569

Kesheri M, Sinha RP, Kanchan S (2016) Advances in soft computing approaches for gene prediction: a bioinformatics approach. In: Dey N, Bhateja V, Hassanien AE (eds) Advancements in bio-medical sensing, imaging, measurements and instrumentation, vol 651. Springer, pp 383–405

Kesheri M, Kanchan S, Sinha RP (2021) Isolation and in-silico analysis of antioxidants in response to temporal variations in the cyanobacterium *Oscillatoria* sp. Gene Rep 23:101023

Kesheri M, Kanchan S, Sinha RP (2022) Responses of antioxidants for resilience to temporal variations in the cyanobacterium *Microcystis aeruginosa*. S Afr J Bot 148:190–199

Kesheri M, Kanchan S, Srivastava S, Ratna-Raj R, Chittoori B, Vaishampayan A, Rastogi RP, Sinha RP, Primerano D (2024a) Ecology and environmental omics. In: Gupta MK, Katara P, Mondal S, Singh RL (eds) Integrative omics: concepts, methodology, and application. Academic Press (Elsevier), pp 315–331

Kesheri M, Kanchan S, Salisbury TB, Sinha RP (2024b) Microbial omics in environment and health. Springer Nature, Singapore. (ISBN: 978-981-97-1768-2)

Kleindienst S, Chourey K, Chen G, Murdoch RW, Higgins SA, Iyer R, Campagna SR, Mack EE, Seger ES, Hettich RL (2019) Proteogenomics reveals novel reductive dehalogenases and methyltransferases expressed during anaerobic dichloromethane metabolism. Appl Environ Microbiol 85:e02768–e02718

Koop SH, van Leeuwen CJ (2017) The challenges of water, waste and climate change in cities. Environ Dev Sustain 19:385–418

Kumar D (2016) Nuclear magnetic resonance (NMR) spectroscopy for metabolic profiling of medicinal plants and their products. Crit Rev Analy Chem 46:400–412

Kumari A, Kesheri M, Sinha RP, Kanchan S (2018) Integration of soft computing approach in plant biology and its applications in agriculture. In: Purohit HJ, Raje DV, More RP, Kalia VC (eds) Soft computing in biological sciences. Springer Nature Singapore, pp 265–228

Kuppan N, Padman M, Mahadeva M, Srinivasan S, Devarajan R (2024) A comprehensive review of sustainable bioremediation techniques: eco-friendly solutions for waste and pollution management. Waste Manage Bull 2(3):154–171

Laskowski L, Williams D, Seymour C, Mothersill C (2022) Environmental and industrial developments in radiation cataractogenesis. Int J Radiat Biol 98:1074–1082

Lee KK, Bing R, Kiang J, Bashir S, Spath N, Stelzle D, Mortimer K, Bularga A, Doudesis D, Joshi SS (2020) Adverse health effects associated with household air pollution: a systematic review, meta-analysis, and burden estimation study. Lancet Glob Health 8:e1427–e1434

Leso V, Capitanelli I, Lops EA, Ricciardi W, Iavicoli I (2017) Occupational chemical exposure and diabetes mellitus risk. Toxicol Ind Health 33:222–249

Malik G, Arora R, Chaturvedi R, Paul MS (2021) Implementation of genetic engineering and novel omics approaches to enhance bioremediation: a focused review. Bull Environ Contam Toxicol 108(3):443–450

Mallik B, Mishra T, Dubey P, Kesheri M, Kanchan S (2024) Exploring the secrets of microbes: unveiling the hidden world through microbial omics in environment and health. In: Kesheri M, Kanchan S, Salisbury TB, Sinha RP (eds) Microbial omics in environment and health. Springer, Singapore

Meadows T, Coats ER, Narum S, Top EM, Ridenhour BJ, Stalder T (2025) Epidemiological model can forecast COVID-19 outbreaks from wastewater-based surveillance in rural communities. Water Res 268:122671

Możejko-Ciesielska J, Mostek A (2019) Time-course proteomic analysis of *Pseudomonas putida* KT2440 during mcl-polyhydroxyalkanoate synthesis under nitrogen deficiency. Polymers 11:748

Naik MM, Imran M, Vaigankar DC, Mujawar SY, Malik AD, Gaonkar SK (2021) Genome guided bioprospecting of extremely halophilic *Haloferax* sp. AS1 for CAZymes, bioremediation and study metabolic versatility. Proc Natl Acad Sci India Sect B Biol Sci 91:297–305

Nitz H, Duarte M, Jauregui R, Pieper DH, Müller JA, Kästner M (2020) Identification of benzene-degrading Proteobacteria in a constructed wetland by employing in situ microcosms and RNA-stable isotope probing. Appl Microbiol Biotechnol 104:1809–1820

Ortiz-Hernández ML, Gama-Martínez Y, Fernández-López M, Castrejón-Godínez ML, Encarnación S, Tovar-Sánchez E, Salazar E, Rodríguez A, Mussali-Galante P (2021) Transcriptomic analysis of *Burkholderia cenocepacia* CEIB S5-2 during methyl parathion degradation. Environ Sci Pollut Res 28:42414–42431

Pandey A, Tripathi PH, Tripathi AH, Pandey SC, Gangola S (2019) Omics technology to study bioremediation and respective enzymes. In: Smart bioremediation technologies. Academic Press, pp 23–43

Patankar RS, Reshi N, Kutty R (2022) Bioremediation: gaining insights through metabolomics. In: Kumar V, Thakur IS (eds) Omics insights in environmental bioremediation. Springer Nature Singapore, Singapore, pp 37–46

Peng H, Gao J, Song X (2021) Identification of heavy metal-responsive genes in radish (*Raphanus sativus* L.) through RNA-Seq meta-analysis. Sci Hortic 288:110402

Plewniak F, Crognale S, Rossetti S, Bertin PN (2018) A genomic outlook on bioremediation: the case of arsenic removal. Front Microbiol 9:820
Rawat M, Rangarajan S (2019) Omics approaches for elucidating molecular mechanisms of microbial bioremediation. In: Smart bioremediation technologies. Academic Press, pp 191–203
Rushton L, Hutchings J (2017) The burden of occupationally-related cutaneous malignant melanoma in Britain due to solar radiation. Br J Cancer 116(4):536–539
Saeed O, Boyer NL, Pamplin JC, Driscoll IR, DellaVolpe J, Cannon J, Cancio LC (2018) Inhalation injury and toxic industrial chemical exposure. Mil Med 183:130–132
Sakshi HA (2020) A comprehensive review of metabolic and genomic aspects of PAH-degradation. Arch Microbiol 202:2033–2058
Saravanan A, Kumar PS, Jeevanantham S, Karishma S, Tajsabreen B, Yaashikaa P, Reshma B (2021) Effective water/wastewater treatment methodologies for toxic pollutants removal: processes and applications towards sustainable development. Chemosphere 280:130595
Sarigiannis DA (2017) Assessing the impact of hazardous waste on children's health: the exposome paradigm. Environ Res 158:531–541
Schwechheimer SK, Becker J, Peyriga L, Portais JC, Sauer D, Müller R, Hoff B, Haefner S, Schröder H, Zelder O (2018) Improved riboflavin production with *Ashbya gossypii* from vegetable oil based on ^{13}C metabolic network analysis with combined labeling analysis by GC/MS, LC/MS, 1D, and 2D NMR. Metab Eng 47:357–373
Sharma P, Singh SP, Iqbal HM, Tong YW (2022) Omics approaches in bioremediation of environmental contaminants: an integrated approach for environmental safety and sustainability. Environ Res 211:113102
Shruti MNK, Kesheri M (2016) Forecast analysis of the potential and availability of renewable energy in India: a review. Int J Industr Electron Electr Eng 4(10):17–22
Shukla N, Bhatia A, Kesheri M, Kanchan S, Narayan J (2024) Planet in peril: unveiling the microbial consequences of environmental degradation on human health through omics approaches. In: Kesheri M, Kanchan S, Salisbury TB, Sinha RP (eds) Microbial omics in environment and health. Springer, Singapore, pp 35–56
Singh RP, Manchanda G, Li ZF, Rai AR (2017) Insight of proteomics and genomics in environmental bioremediation. In: Handbook of research on inventive bioremediation techniques. IGI Global, pp 46–69
Solomon NO, Kanchan S, Kesheri M (2024) Nanoparticles as detoxifiers for industrial wastewater. Water Air Soil Pollut 235(3):214
Srivastava U, Kesheri M, Kanchan S, Singh S (2024) Computational omics protocol for the comparative study of microbiome analysis. In: Kesheri M, Kanchan S, Salisbury TB, Sinha RP (eds) Microbial omics in environment and health. Springer, Singapore, pp 109–133
Sun W, Krumins V, Dong Y, Gao P, Ma C, Hu M, Li B, Xia B, He Z, Xiong S (2018) A combination of stable isotope probing, Illumina sequencing, and co-occurrence network to investigate thermophilic acetate-and lactate-utilizing bacteria. Microb Ecol 75:113–122
Tong Y, Wu Y, Bai H, Li S, Jiang L, Zhou Q, Chen C (2023) Highly efficient and simultaneous magnetic solid phase extraction of heavy metal ions from water samples with l-cysteine modified magnetic polyamidoamine dendrimers prior to high performance liquid chromatography. Chemosphere 313:137340
Volpicella M, Leoni C, Manzari C, Chiara M, Picardi E, Piancone E, Italiano F, D'Erchia A, Trotta M, Horner D (2017) Transcriptomic analysis of nickel exposure in *Sphingobium* sp. ba1 cells using RNA-seq. Sci Rep 7:8262
Xue W, Lei Y, Liu X, Shi X, Liu Z, Xu Y, Chen X, Song X, Zheng Y, Zhang Y (2023) Synergistic assessment of air pollution and carbon emissions from the economic perspective in China. Sci Total Environ 858:159736

A Review on Root Phenomics: Understanding Plant Root Systems

15

Ratnaprabha Ratna-Raj, Minu Kesheri, Swarna Kanchan, and Bhaskar Chandra Srinivas Chittoori

Abstract

Root phenomics refers to comprehensive study of plant root characteristics. Root phenomics integrated with genomics help identify molecular pathways of specific root traits and discover potential genes and genetic markers. This information can be used in plant breeding programs in developing genetically engineered crop varieties that possess desirable root traits. Traditional destructive analysis of root characteristics is very challenging as it is very capital-, labor-, and time-intensive. Two-dimensional (2D) technologies like root scanners and rhizotrons, 3D approaches like magnetic resonance imaging (MRI), X-ray computed tomography (CT), and 4D modeling help visualize and analyze root structures, in situ and nondestructively. High-throughput automated systems and robotics data pertaining to root traits help in collecting data quickly and efficiently, allowing for the analysis of large numbers of plants and root systems simultaneously, without any human error. This chapter provides an overview of various approaches used in root phenomics and their applications for improving plant growth using multi-omics technologies.

Keywords

2D Root technology · 3D Roots approaches · 4D Root modelling · Multi-omics · Root phenomics · Root phenotype · Root system architecture

R. Ratna-Raj (✉) · B. C. S. Chittoori
Department of Civil Engineering, Boise State University, Boise, ID, USA
e-mail: ratnaraj@boisestate.edu

M. Kesheri · S. Kanchan
Biology Department, Boise State University, Boise, ID, USA

Department of Biomedical Sciences, Marshall University, Huntington, WV, USA

M. Kesheri et al. (eds.), *Multi-Omics in Biomedical Sciences and Environmental Sustainability*, https://doi.org/10.1007/978-981-96-7067-3_15

Objectives

- Root phenomics: concept of plant root phenotypes/traits such as root length, diameter, root surface area, root hair development, root tip density, root branching patterns, and root architecture.
- Root growth, development, function, and strategies to cope with stress in the environment.
- Recent methods for efficiently growing plants using root phenomics and quantification of root traits using traditional, 2D, 3D, and 4D levels.
- Applications of root phenomics in selecting and breeding new and improved crop varieties with desirable root traits.

15.1 Introduction

Roots anchor plants to the soil and serve as the major conduit of water and nutrient uptake thus influencing crop yield (Taiz et al. 2018). They determine plants' ability to sense and cope with a wide range of abiotic and biotic stressors in the environment (Hodge et al. 2009). Abiotic stresses like salinity, drought, and extreme temperatures (Kesheri et al. 2014), eutrophication (Kesheri et al. 2022), and harmful UV-radiations (Kesheri et al. 2021) have been known to significantly affect photosynthetic organisms, including resilient cyanobacteria to higher plants, and have shown to reduce the crop yield by up to 70% (Rahman et al. 2015). Plants with effective root systems, proposed as "second green revolution," could significantly increase crop production (Lynch 2007). According to Food & Agricultural Organization (FAO), by 2050, crop production has to increase by 70% to feed a population of 9.1 billion (Food and Agricultural Organization (FAO) 2009). Advanced research in omics technologies have been crucial for plant growth and development; therefore, studying root phenotyping/root phenomics is critical for all research that aims at improving crop yield (Priya et al. 2017; Kesheri et al. 2024).

Root phenomics involves comprehensive analysis of root morphological traits such as root length, root radius, root surface area, presence of root hairs, number of root tips, root branching, and root architecture. Studying root traits helps in understanding root growth, development, and its functions and response to environmental factors like drought, soil salinity, and extreme temperatures (Kesheri et al. 2011). The knowledge gained could be used to develop plants with effective root systems with better coping mechanisms to environmental conditions thus producing improved yield. High-throughput genotyping has become standard in plant breeding, but collecting phenotypic data, especially in root traits, has been challenging. This poses a major bottleneck in plant breeding creating a wide gap between root phenomics and genomics. High-throughput phenotypic data of about 100–500 individual plants are necessary to narrow this gap either through quantitative trait loci (QTL) or by association mapping (Rafalski 2010). Recent breakthrough in technologies have made it possible to collect accurate, fast, and high-throughput root data, thus reducing the root phenotype to genotype gap. The data obtained, especially of root system architecture, finds direct application in quantitative trait locus (QTL) approaches for identifying molecular basis of the root architecture and crop performances in varying environmental conditions. The information gained will significantly advance plant breeding, crop improvement, and sustainable agriculture. This chapter attempts to provide an overview of various methods of root phenotyping, which could be used singly or in combination for genetic gain.

15.2 Traditional Root Phenotyping (Ex Situ)

Traditional methods for root phenotyping in the field are very time-consuming, labor-intensive, and subjected to root loss and human error. These approaches are usually destructive in nature as they involve the removal of some roots or the entire roots from the soil. The traditional methods include *soil trench*, *soil cores*, *core-break*, *mesh bags*, *monoliths*, and *shovelomics*.

Soil trench method is the oldest method of root phenotyping (Weaver et al. 1922). It involves digging a 5- to 9-feet-deep trench by the side of the plants to be examined, thus exposing half of the root system. For each case, several plants at the same stage are examined with the possibility of securing some root systems in their entirety. Root systems were discussed in detail and drawn on papers, in the field. This method is tedious and requires more patience and experience with soil texture. *Soil core* is the common method for root phenotyping. Soil core or auger is inserted next to the root system to extract a cylindrical sample (Kücke et al. 1995). The soil core is carefully washed and dissected, and the roots within are measured. One main drawback of this method is the heavy loss of roots during washing. As an extension to soil-core method, *core-break* method was developed (Bland 1989). Here, a vertical core of soil from a specific depth is taken and broken to reveal a cross-section, and the number of roots visible at that depth is counted. This method provides a rapid assessment of root length density. In *Mesh bag* approach, bags typically made of nylon or other porous material are filled with soil (Steen 1991). Holes are dug next to where the seeds are sown, and the bags are placed inside, allowing the roots to grow into them. These bags allow water, nutrients, and air to penetrate while retaining the roots inside. The bags are carefully extracted, and the roots are measured. One of the setbacks of this method is that the bags can be placed only next to the shallow roots on top. *Monolith* involves the extraction of an entire soil block or monolith, either shaped as a square or a cylinder, containing the entire root system. Hence, it is considered nondestructive when compared to all the other traditional methods (Böhm 1979). Depending on the hardness of the soil and the size of the root system, this approach may be power-driven, thus providing a rapid method for root phenotyping (Teramoto et al. 2019). *In Shovelomics*, a shovel or a spade is used to carefully dig up and excavate a soil cylinder, 40 cm in diameter and 25 cm in depth, using the plant base as the horizontal center of the soil cylinder (Trachsel et al. 2011). The adhering soil mass is removed by soaking the soil cylinder in mild detergent, and the roots are later washed and examined (Colombi et al. 2015). It is simple and high throughput compared to previous methods.

15.3 2D Root Phenotyping (Ex Situ and In Situ)

Two-dimensional root approaches involve studying plant root traits in 2D space. Here, the plants' roots are either (1) grown in soil medium and then are excavated, washed, and placed on a 2D flat scanning surface or (2) grown in soil-less medium along transparent walls. Either way, the roots are photographed, and images are analyzed for their root traits using software. The software may be manual, semiautomated, or fully automated and uses pixels in the image which are transformed to approximate the root traits. These approaches are quick, nondestructive, and high throughput, enabling batch analysis of several samples at a time.

15.3.1 Soil-Less Media for 2D Root Phenotyping

Soil-less medium is widely used for root phenotyping of smaller plants or seedling plants as it provides a good estimate of the root architectural traits for comparison. Also, it gives a clear and visible root system for imaging with a good contrast between the roots and the rooting medium. The advantage of a soil-less system is that it is noninvasive, and the roots can be imaged at a much younger stage. The main disadvantage is that the roots do not represent the actual roots of mature plants.

Agar is one of the widely used soil-less mediums as roots are clearly visible for imaging (Bengough et al. 2004). Chambers made of plastic plates are usually used with one transparent side for viewing roots. The chambers separated by spacers are filled with transparent gel and pre-germinated seeds are sown in the media. The holes on the sides provide aeration and allow shoot growth. A variation of agar plates is the Petri plate method (Nagel et al. 2009). Sterile agar medium enriched with nutrient solution is filled in Petri dishes (Nagel et al. 2009). Surface-sterilized seeds are inserted halfway into the agar medium through the holes on the sides of the Petri dishes. These holes are later closed with tape to prevent external contamination. The Petri dishes are provided for a 12-hour photoperiod. After the seeds germinate, the tape is removed, and the seeds are moved to the holes on the sides of the Petri dishes. The Petri dishes are mounted vertically allowing shoots to grow upward through the hole and roots to grow in the agar. The roots are imaged using a desktop flat-bed scanner (Bengough et al. 2004) or a digital camera (Nagel et al. 2009), and the images are analyzed for root traits. One main advantage of agar media is that it provides impedance to roots similar to those grown in soil. Also, agar keeps roots in place preventing overlap and minimizing errors during analyses. The main drawback of this medium is the limitation in space. In addition, keeping the media free of any pathogen is a laborious process. Hence, studies conducted are mostly of short duration. Furthermore, agar medium can produce ethylene, resulting in an uncommon plant structure (Eliasson and Bollmark 1988).

Hydroponics media is widely used in experiments studying nutrient effects on roots/shoots because there is full control over the nutrient media. With roots growing in a controlled environment, the results are highly reproducible. Also, it minimizes the loss of roots while washing. Further, the entire root system is transparent and can be optically visualized and accessed for sampling, imaging, or measuring the root traits. One main disadvantage is that the hydroponics media does not provide mechanical impedance like in soil, thus resulting in root growth and branching different from the soil. Besides, hydroponics cannot be recommended for studying root hairs as it is not well formed in this media. Inexpensive and high-throughput "*rhizoscope*" hydroponic systems addresses the absence of impedance (Audebert et al. 2010; Courtois et al. 2013). Here, plants are grown in "nail board rhizoboxes" filled with glass beads to mimic field mechanical impedance and the impedance can be manipulated by changing the bead size. The "rhizoboxes" are filled with aerated nutrient solution while growing. The roots are photographed after removing beads, and images are analyzed for root traits using computer. This system is at par with comparative studies using soil-filled tubes for rice and is generic and may be used to obtain relevant observations in any other plant species. A hydroponics rhizotron, *Rhizoponics* studies plants throughout their life cycle instead of a few early growth stages (Mathieu et al. 2015). It consists of nylon mesh supported by an aluminum frame. This frame is slid through the slot on the tank lid and immersed in the tank filled with nutrient solution. The frame is taken out and placed horizontally in a dark room with indirect backlighting, and the roots are imaged using a digital camera and analyzed for root characteristics. Rhizoponics allows for time-lapse studies of not just the roots, but also the shoots. In addition, the support keeps the roots tangle-free, thus allowing global analyses like total root size and depth and local analyses like lateral root length and lateral root density.

When most soil-less cultures study shallow roots, *a semi-hydroponic* bin system studies the deep root system to up to 1.2 m (Chen et al. 2011). This system is constructed using readily available and cost-effective materials like plastic mobile trash bins and clear acrylic panels covered with black calico cloth. The black cloth retains moisture from a controlled watering system and serves as growth pouches for seedlings to grow. The roots are imaged after removing the cloth and the traits are analyzed. The main attribute of this methodology is that it efficiently screens 40 plant genotypes in a compact space of about 0.5 m^2. *RhizoTubes*, another semi-hydroponic system, is high throughput capable of imaging up to 1100 plants per day (Jeudy et al. 2016). It consists of two concentric tubes sandwiched by a substrate that soaks up water and nutrients. The seeds are sown on the top of the substrate and roots grow in between the outer pipe and the substrate. The tube can grow one to six plants for up to 8 weeks with a maximum rooting depth of 1.1 m. The RhizoTube mounted on a conveyor is imaged using a camera inside an imaging cabin called RhizoCab. Semi-hydroponic bin system and RhizoTubes lack mechanical impedance just like hydroponics system (Ma et al. 2019). Hence, it does not provide the exact root architecture and physiology like the roots grown in the soil. Inexpensive and mechanically stable *hydrogel* or "*transparent soil*" addresses the impedance issue. The roots grown in hydrogel are more like those grown in the field than those grown in hydroponics because of similar pore size and easy scalability. Also, the roots in hydrogel can be imaged in vivo using both photography and microscopy, unlike the other two.

Aeroponics is widely used as a rapid method of propagation for potato seed production as it conserves water and energy (Singh et al. 2010; Buckseth et al. 2016; Tiwari et al. 2020). It is now used in studying the root structure of Cassava tubers (Selvaraj et al. 2019). The aeroponic chamber consists of a low-cost, recyclable tank painted black to block light and wrapped in metallic polyester to minimize heat absorption. The perforated foam lid supports the plants, and a pump connected to a sprinkler system delivers nutrients as a mist to the roots. During root phenotyping, the plants are taken out of the tank to the image setup where it is hung freely from a foam on a metallic support. Roots are imaged with a black curtain in the backdrop and images analyzed using software for their root traits (Abramoff et al. 2003). The main advantage of aeroponics makes it possible to study a large number of plants over a long period of time, extending to the maturity phase. Hence, it allows for time-series analysis needed for studying gene-expression analyses and novel gene identification. Also, just like hydrogel, the roots growing in aeroponics are very similar to soil-grown roots than hydroponically grown roots (Block et al. 2017). Moreover, sampling roots in this method is the simplest and practical when compared to all the previous methods.

Semi-aeroponics system is an aeroponics system modified with tanks half-filled with nutrient solution to mimic the semi-aeration condition (Selvaraj et al. 2019). The lower roots bath in the nutrient solution throughout the growing period, while the upper roots remain exposed to air. Root phenotyping and analysis are the same as in Aeroponics. Semi-aeroponic system can be exclusively used for studying storage root research in cassava, as roots exposed to air can trigger the growth of starch root (Selvaraj et al. 2019).

Dripponics, yet another variation to the aeroponics system, has nutrients delivered to the roots using a drip system instead of mist (Selvaraj et al. 2019). The drip system had a hole big enough to wet the roots and tubers at regular intervals. The plants were set up and grown, and roots were imaged and analyzed as in aeroponics (Selvaraj et al. 2019). The main disadvantage of this system is that there is less uptake of nutrients and oxygen by the cassava roots because water flow has very little contact when compared to the large surface area of the storage root.

15.4 Various Setup for 2D Root Phenotyping

A variety of setups are developed with the goal of using low-cost ingredients to grow large number of plants in a small space, and thus, image higher number of plant roots. *Pouch-and-wick system/growth patches* use germinated seeds moistened with blotting paper and placed in pouches made of strong opaque black polyethylene foil, supported by paper clips (Hund et al. 2009; Falk et al. 2020). The pouches are covered with aluminum foil to prevent overheating and a hole in the upper side of the pouch allows the emergence of the coleoptile. The pouches are hung in plastic containers containing nutrient solution with blotting paper touching the solution at lower bottom. The roots are imaged using a scanner, photoshopped, and analyzed using software (Hund et al. 2009; Falk et al. 2020). This is a high throughput allowing hundreds of plants to grow individually in pouches. However, root length can be overestimated by over 100% creating errors, if the difference in germination is not considered. *Root box/Rhizotron* is a specialized frame or a structure with transparent walls made of acrylic or plexiglass 20 × 20 × 0.5 cm (James et al. 1985; Neufeld et al. 1989). Plants are grown inside the structure, and the roots are sandwiched in between inside a nylon mesh, which is permeable to the nutrients or water but not to the roots. The roots are monitored and studied through the glass without disturbing the plant or soil. The entire root system is drawn onto clear acetate sheets using different colored permanent markers. The root image is digitized, but root traits are manually analyzed. The major setback of this approach is that some of the roots are hidden in the soil, and so it is severely restricted in detecting roots (Neufeld et al. 1989). *GROWSCREEN-Rhizo/GROWSCREEN-ROOT* is an upgraded version of rootbox/rhizotron. Plants can be grown in this to a maximum root depth of 90 cm, unlike 20 cm in a rootbox (Nagel et al. 2012). The roots are imaged, and root traits are analyzed using GROWSCREEN-ROOT software (Mühlich et al. 2008; Nagel et al. 2009). It is high throughput with a maximum capacity of 5472 rice seedlings simultaneously and automatically imaging up to 60 roots per hour in soil-filled rhizotrons (Nagel et al. 2012). However, it has the same drawback as rhizotron with part of the roots hidden and undetectable in the soil. In GROWSCREEN-Rhizo, 72 individual plants can be grown at once, with a phenotyping rate of 60 plants per hour (Nagel et al. 2012). *Rhizoslides* uses two rigid PVC bars with watering channels are spaced out using washer to create a slit for growing plants (Le Marié et al. 2014). Two plexiglass sheets covered with germination papers are attached to the support which serves as a substrate for plant roots to grow. These plexiglasses are covered again with a foil allowing gaseous exchange and for a controlled environment. Germinated seeds are placed in the central slot on the upper edge of the sheets with the help of paper clips. Root images are taken using digital camera and the preprocessed images on MatLab is analyzed using WinRhizo software (Regents Instruments, Inc.). Unlike previous methods, the "rhizoslide sandwich" of germination papers visually identifies the growth of embryonic roots from postembryonic roots in different layers. Hence, it facilitates the study of a particular root type to changes in space and time of the root environment. The studies on postembryonic roots are very important as they give rise to the mature root system. Rhizoslide approach facilitates the study of root plasticity and root dynamics too in a varying environment. The upgraded version of rhizoslides is *RADIX* which is high throughput (Le Marié et al. 2016). The platform consists of an aluminum rack with 100 compartments to house 100 rhizoslides. The rhizoslides are enclosed in black foil to provide a dark environment for the roots to grow and are roofed with a cover plate with holes in it for shoots to grow. The plants are irrigated through a micro-drip system, which is automated for the delivery of water and nutrients. The roots and shoots are imaged at the same time using a digital camera, and images are analyzed using MatLab and SmartRoot software (Lobet et al. 2011). RADIX studies both root and shoot traits and their dynamics to split-root application of varying environmental conditions, unlike Rhizoslide and previous methods. This method is a pioneer in evaluating the differential response using a larger number of plants. Unlike all previous platforms where cultivation of plants and imaging of roots happen in different places, *RhizoChamber-Monitor* combines a cultivation system and the imaging system into one single unit (Wu et al. 2018). It consists of a rhizobox made of two transparent panels separated by spacers, for easy viewing of roots. The spacers grip an ethylene propylene diene monomer sheet covered by a red piece of cloth. The sheet also runs an irrigation tube at the top for delivering water and nutrients. The slot in between the spacers allows plants to grow, and the sheet presses against their roots keeping it in place. Several rhizoboxes are held on a sliding rhizobox frame housed inside a cuboid main frame. The frame consists of horizontal and vertical beams with a gap in between to accommodate the rhizoboxes. The roots are imaged using a digital camera and the root traits are analyzed using a customized software, extracted in MatLab. Unlike other platforms, the frequency and duration of the irrigation can be adjusted, which allows the investigation of root responses to water stress. *WinRoots* system consists of multiple independent circulatory systems and several plant culture units (Zhang et al. 2022). Each plant culture unit accommodates multiple culture cases. The transparent bottom of each culture case is filled with a soil matrix and covered using a cover plate. The culture box, which houses multiple culture cases, functions as a container for water or culture solution. The plants growing at the center of these tops reveal the root system through the transparent side. The roots are manually photographed, and images are analyzed using ImageJ software (Kimura and Yamasaki 2003). It is high throughput accommodating 2160 soybean seedlings 12 m^2 chamber and imaging 800 seedlings per hour using an RGB camera. WinRoots provides a consistent microenvironment unlike most root phenotyping methods in pots (a plant/pot) and germination paper (several plants/paper) which provides great root visibility unlike other phenotyping methods in soil. It has precise control of soil salt stress conditions providing highly homogenous conditions for collecting accurate quantification of phenotypes. *PhenoRoots* is similar to WinRoots allowing easy monitoring and observation of root growth (Martins et al. 2019). However, it uses cost-effective material, Aluminum Composite Material (ACM) plate, and a glass plate separated with the help of aluminum spacers. The plates and aluminum spacers are held together using binder clips. The roots visible through glass are photographed using a digital camera and analyzed using WinRHIZO software and ImageJ software (Abramoff et al. 2003). The main highlight of this method is that the image analyses show high correlation with those generated with laborious and time-consuming destructive methods. Also, this can be applied to a wide variety of plants including cotton, soybean, bean, corn, rice, sorghum, and pearl millet. *Rhizopot* exclusively studies the finer roots (less than 2 mm in diameter) along with root hairs using a high-resolution scanning system (Xiao et al. 2020). This setup consists of transparent acrylic plates and an oblique plane serving as an observing root growth. For a dark root growth environment, black shaded cardboard is used to shield the trapezoidal acrylic board area formed by the two scanners. A scanner (Epson Perfection V39, Suwa, Japan) aligned with an observation window is used for root imaging, and the images are analyzed using WinRHIZOTron MF 2012a (Regent Instruments Inc.) and Adobe Photoshop CC 2019 (Adobe Systems Incorporated) software. This system is ideal for drought treatments and allows the observation and imaging of finer roots and root hairs without interference.

15.5 2D Softwares for Root Phenotyping

WinRHIZO™ (Regent Instruments Inc.) is most widely used manual root image and analysis software (Fig. 15.1). It uses an optical scanner to capture high-quality images of plant roots. It offers various lighting options to suit specific research needs and is user friendly, which does not require frequent recalibration when changing optical settings or resolution. The system detects and compensates for overlapping root segments and thus ensures accurate measurements even when roots intersect. The main advantage is that it provides batch processing without additional user interference and however, for in-depth topology analysis, it requires intense user interaction, which is time-consuming. *ROOTEDGE* measures the length of digitized binary images of roots with the help of a desktop scanner (Kaspar and Ewing 1997).

It employs the edge chord algorithm and is designed to be insensitive to the orientation of objects in the images. This software reduces the need for manual exposure threshold adjustments and provides image processing techniques to improve accuracy. However, its main limitation is in capturing very fine root structures. *Digital Imaging of Root Traits (DIRT)* combines a field-imaging protocol with an algorithmic approach to analyze mature root systems in their natural field soil, unlike other methods, which analyze roots of younger seedlings of plants grown in specialized containers in controlled conditions (Bucksch et al. 2014). As an extension to the traditional shovelomics approach, roots are excavated, washed, and

Fig. 15.1 A 2D scanned root image using WinRhizo software (Chittoori 2012)

imaged on an imaging board using a digital camera on a tripod. The images are converted to binary images to deduce various root traits. The main advantage is that it produces quick, low-cost with reproducible trait estimation in field conditions. *Root Estimator for Shovelomics Traits (REST)* is an improvement over the shovelomics DIRT approach with enhanced digital imaging and software (Colombi et al. 2015). Rootstocks are collected with the novel approach of splitting the root crown lengthwise to remove the soil. This splitting facilitates the discrimination of the root and background during imaging as it coincides with the focal plane of the camera. The rootstocks are photographed under controlled illumination inside an imaging tent. The images are analyzed using fully automated, high-throughput REST software. The image processing serves as a powerful phenotypic approach for mature maize root crowns. However, this software provides only an estimate and not the actual measurement. *Fluorescence spectroscopy is* the upgraded version of the core-break method, where soil cores are broken at the desired depth and the roots revealed at the broken cross-section are manually counted (Wasson et al. 2016). In the upgraded version, a "BlueBox" is introduced as a portable fluorescence imaging system that illuminates the roots and automates counting of roots revealed in the soil cores. The system includes specialized lighting to produce high-contrast root images, providing a rapid assessment. Compared to manual conventional coring, this approach is 16% more labor efficient, 19% more accurate, and 12% cheaper. While Fluorescence spectroscopy illuminates roots in the field, *GLO-Roots (The Growth and Luminescence Observatory for Roots)* illuminates roots grown in the lab (Rellán-Álvarez et al. 2015). The roots are bioluminescent from a constitutively expressed luciferase transgene, and these "luminescent" markers help visualize the roots. Plants are grown in soil media/agar gel in rhizotrons and GLO1, a semiautomated luminescence-imaging system, illuminates the roots and images them. The roots were imaged at various stages and analyzed using GLO-RIA (GLO-Root Image Analyzer) for their root position, length, direction, shape, and depth. The main advantage of this software is that the image analysis algorithms allow the spatial integration of root architecture, soil characteristics, and gene expression thus aiding studies with a multidimensional approach. *GLO-Bot* is the upgraded version of GLO-Roots where the robotics-assisted phenotyping platform and the image analysis pipeline are both automated (LaRue et al. 2022). The image analysis pipeline uses a modified version of GLO-RIA (Growth and Luminescence Observatory Root Image Analysis) considering image alignment, signal optimization, and background noise removal which enables measurement of robust traits. *GLO-Bot* images 96 rhizotrons per day at maximum capacity. Because of the automation, there is increased throughput of the time-lapse imaging of roots which allows the study of dynamic root growth. *ChronoRoot* is an innovative high-throughput software to automatically identify and extract root structures from 2D images using 3D-printed devices and advanced machine learning (ML) techniques (Gaggion et al. 2021). The approach used plants grown in agar media. The software is open source and is designed to work with low-cost cameras. It considers changes in root structure and growth over time, making it suitable to analyze dynamic processes. It is an ideal platform for root phenotyping for genetics, natural variation studies, and for screening of clock-related mutants. *PlaRom (Plant Root monitoring)* includes (a) Phyto chamber, (b) a root imaging platform, and (c) software to analyze the root traits (Yazdanbakhsh and Fisahn 2009). Phyto chamber contains plants growing on agar medium inside Petri dishes. The root imaging platform is automated and high throughput enabling high-resolution monitoring of 50 plants. The software allows imaging the plants over time, and the software analyzes the images for root growth dynamics including lateral root formation, root architecture, and root hair development. One of the main advantages of this system is that PlaRom software controls the light and temperature growth conditions of the phyto chamber as well thus revealing the root extension pattern in constant growth conditions of various genotypes. *DART (Data Analysis of Root Tracings)* provides manual 2D analysis of root systems using manual software written in JAVA based on human vision to identify roots. It displays similarities with software developed for 3D scanning of plant shoots (Bot et al. 2010). Unlike other softwares, it can analyze the lifespan of individual roots by measuring the entire root system in detail by keeping track of them using different colors. *SmartRoot* is a semiautomated root image analyzer based on ImageJ, an open-source image processing program (Lobet et al. 2011). Unlike all other softwares, it uses vectorial representation for root objects instead of pixel-based data allowing for more accurate and flexible analysis. The software translates pixel coordinates and gray levels from images into meaningful biological attributes, such as segment diameter, orientation, distance between segments, and topological position, making it easier to interpret and analyze the data in biological terms. It includes a powerful tracing algorithm that can handle a wide range of image sources and quality. It permits both tagging and tracing of individual roots. The software also processes time-lapse images allowing for the study of root dynamics and hence is suited for studies monitoring the changes in the growth rates of individual roots. *RootNav* is also a semiautomated root image analysis like SmartRoot, but with two contributions (Pound et al. 2013). First, it employs a top-down approach using the expectation maximization (EM) classification algorithm to analyze regions within input images. This algorithm calculates the likelihood that certain pixels in the image correspond to roots. Once the likelihood is determined, it applies an optimization approach to detect and quantify roots within the image. Second, it provides user-friendly interface where the user marks the end points of root branches through automatic tip detection guides, and a rich set of architectural traits is provided in a separate viewer tool. SmartRoot also has user interaction. However, it is more intuitive in RootNav than SmartRoot. *General Image Analysis of Roots (GIA Roots)* is a semiautomated, high-throughput tool with a user interface like SmartRoot and RootNav (Galkovskyi et al. 2012). However, unlike the other two, users interact with the software in the initial steps of image analysis, segment the images, and distinguish the roots from the background. Afterward, users select traits of interest and instruct the software to automatically measure these selected traits. This linear design helps users to keep track of the progress and process the images automatically. Also, users can add novel traits to the software using the application programming interface and extend the root traits of interest. GiARoots just like WinRhizo offers the advantage of simple batch processing without additional user interference. But it requires images with a good contrast to prevent false tracing. EZ-Rhizo provides about 20 primary and derived root traits in a rapid, straightforward, and accurate manner (Armengaud 2009). The main highlight of this software is that it can access the growth rates of specific roots. This allows for the study of roots in various nutrient and environmental conditions. One demerit is that it cannot correct for root overlap. To address this, ML algorithms that will remember the temporal and spatial "history" of roots will be needed for future versions of this software. *RootTrace* software exclusively measures the lengths of primary roots over time-lapse image sequences allowing the study of the root dynamics (French et al. 2009). The major setback of this version was that it does not account for the curved or agravitropic roots. In the advanced version, a particle-based filter is used to measure roots from the base to the apex of plants grown on agar plates. The software tracks, identifies, and counts the number of curved roots and the lateral roots growing off straight main roots (Naeem et al. 2011). RootTrace finds application in root screening studies of germplasms that require counting the lateral and highly curved roots. *Rhizovision Explorer* is semiautomated software with two modes of measurements (Seethepalli et al. 2021). The default broken root method uses plant roots grown in soil media which is washed and imaged on a desktop scanner and analyzed for their root length, volume, and diameter. The second mode is the "whole root" mode for analyzing the entire root system analyzing the root crowns for the convex hull, root angle, and depth. The software has a simple user interface allowing users to define regions of interest for in-depth analysis. Also, it has batch analysis for rapid image processing of multiple sets. Further, it provides grouped measurements as well, based on diameter, which is helpful for categorizing roots based on their thickness. However, complex images from rhizoboxes, blue background screens, or minirhizotrons may not work with this software. While all softwares rely on a standard resolution of 32 pixels/mm to accurately quantify the root traits, *DigiRoot* accurately determines root traits images with a low resolution of 12 pixels/mm or 300 dots/inch (Stefanelli et al. 2009). The accuracy is adequate even at 1 pixel/mm beyond which it declines. One main advantage of this software is that it can analyze any type of image, photographed or scanned. Also, it requires minimal user interaction for identifying roots. *RootReader2D is a* free and open software that can analyze both individual roots and the entire root system (Clark et al. 2013). The platform is versatile, and it can be used on plants grown in a wide range of media like soil, gel, paper, pouches, and hydroponics. The software has both automated and minimal user-interaction settings for efficient root phenotyping. It finds application on a variety of monocots like rice, sorghum, maize, and dicots like Arabidopsis and tomato. The imaging station is novel with a digital camera mounted on a copy stand, facing down a light box and roots are placed in a tray, filled with a layer of nutrient solution, and imaged. *Phytomorph* is a fully automated high-throughput system imaging upto 1700 roots growing in Petri dishes on agar medium (Subramanian et al. 2013). It consists of a grid which can load up to 36 Petri dishes containing 2–4 seedlings per Petri dish. The grid is illuminated by a backlight and the robot images the roots. Phytomorph requires very little space and is adaptable to various plants and environmental conditions. It has very high resolution and can maintain great experimental consistency. The main drawback of this system is that at a constant working distance, some samples go out of focus depending on the thickness of the agar. *GrowScreen-PaGe (GrowScreen-Page Germination)* is a high-throughput phenotyping system using germination paper to study both root traits and root dynamics in various plant species in monocot and dicot (Gioia et al. 2016). This system consists of polymethyl methacrylate (PMMA) plates enclosed on either side with wet germination paper soaked in water, nutrients, and antifungal solution. Germinated seedlings are placed on top, fixed in place by wet filter paper strips. Twenty-five to a maximum of 200 plates per square meter, with seedlings, are immersed in opaque polypropylene containers with nutrient solution. Containers with growing plants are phenotyped inside a climate chamber. It can assess four plants per minute allowing for the acquisition of time series of up to 180 plants a day. This system may be used exclusively to study the correlation among root traits and root plasticity by quantifying root adaptation to varying nutrient levels.

15.6 3D Root Phenotyping (Ex Situ and In Situ)

Root system is three-dimensional and 3D features like network perimeter (Iyer-Pascuzzi et al. 2010), solidity, convex area, network area, bushiness index, angle, and convex hull are not conveyed in 2D images. Thus, 3D root phenotyping provides a better understanding of root development, growth, and their interaction with soil, much better than the 2D images. 3D approaches are noninvasive and high throughput. *X-ray Computed Tomography (CT)/Computed Axial Tomography (CAT)/CT scanning* is a versatile medical imaging technique that uses X-rays to create detailed cross-sectional images of the internal structures of the human body (Hounsfield 1973). It was first used in medical fields but now it finds application in several fields including root phenotyping (Heeraman et al. 1997; Chapman et al. 2012; Bao et al. 2014; Rogers et al. 2016; Keyes et al. 2017). It helps in nondestructive imaging of roots while plants are in the field, in the soil medium. Thus, it surpasses the opacity of soil which is a major limitation in traditional and 2D approaches. *Magnetic Resonance Imaging (MRI)* is also a medical imaging technique similar to CT (Lauterbur 1973). Unlike CT, which uses X-rays, MRI uses magnetic fields and radio waves to generate detailed images of the internal structures (Lauterbur 1973). MRI is now used in imaging roots of sugar beet, radish, maize, and barley in soil (Jahnke et al. 2009; van Dusschoten et al. 2016; Kumari et al. 2016, 2018). The NMRooting software developed for analyzing MRI data has adequacy like gravimetric and WinRHIZO measurements (Jahnke et al. 2009; van Dusschoten et al. 2016). One major drawback of this approach is the manual thresholding of MRI images with low contrast-to-noise ratios (CNR), which results in gaps in the root structure and loss of roots (Selzner et al. 2023). Advanced U-Net segmentation improves root recovery rate to up to 14% enhancing the root visibility. However, it fails in identifying thinner roots of higher orders. *Ground Penetrating Radar (GPR)*, also known as Ground Probing Radar or georadar, is a geophysical tool that uses high-frequency radio waves to detect the shapes of the subsurface (Baker et al. 2007). It consists of a transmitter and antenna that emits electromagnetic energy into the ground. When the energy encounters a buried object, it creates a variation in the return signal, which is recorded by the antenna (Baker et al. 2007). It has numerous applications in military, archaeology, civil engineering, and geosciences (Annan 2002; Solla et al. 2012, 2013; Rasol et al. 2020; LaRue et al. 2022). Now, GPR finds application in plant roots (Zhu et al. 2014; Zhang et al. 2019; Agbona et al. 2021). One great advantage of GPR is that it distinguishes live roots from dead roots, which is important in studies looking into the effects of pathogens (Zhang et al. 2019). However, it distinguishes roots only if their distance is more than 10 cm horizontally and 5 cm vertically. Also, the moisture content of the media influences the quality of the root images (Pflugfelder et al. 2017). *Electrical capacitance* refers to the capability of an object to collect and store energy in the form of an electrical charge (Cseresnyés et al. 2019). When electricity is passed through a plant tissue, it acts as a capacitor, and its charge can be quantified as electric potential measured between the tissue and ground. In wheat (Streda et al. 2020), mustard, carrot, maize, sunflower, sorghum, oats, onion, and rape seed (Cseresnyés et al. 2019) showed a relation between electrical capacitance being an adequate method for examining root growth. Like GPR, the main drawback of this method is the water content in the cells, which influences the permittivity and thus the electrical capacitance. *Electrical Resistivity Tomography (ERT)/Electrical resistivity imaging (ERI)* is a geophysical tool like GPR which images subsurface structures (Zhou 2019). Unlike GPR, which uses radio waves, ERT injects an electric current that is passed to the ground using current electrodes. The resulting difference in potential or electrical resistivity is measured using potential electrodes (Zhou 2019). The data is processed, and images are created. ERT is now used to study root phenotyping of trees with larger roots, just like GPR (Cao et al. 2010; Rossi et al. 2011). However, the major limitation is in the small number of probes disturbing the soil and electrical cabling, making it low throughput. ERT is an inexpensive, low-intrusion, and nonionizing method providing data both spatially and temporally. *Electrical Impedance Tomography (EIT)* is very similar to ERT in terms of concept. However, EIT uses alternating current to image subsurface structures while ERT uses direct current (Weigand and Kemna 2017). EIT is used in quantifying root surface area, imaging the root system, and examining changes associated with root physiological processes. The main limitation of EIT is that the spatial resolution is much lower than X-ray and nuclear magnetic resonance-based methods. *Electromagnetic Inductance (EMI)* refers to the ability of a material to use a magnetic field to cause a change in electric current (Mouritsen and Ritz 2005). Inductive coupling measures electrical conductivity at different soil depths resulting from roots drying the soil (Mester et al. 2011; Shanahan et al. 2015). Changes in conductivity show consistency with measured changes in soil water status, and thus, EMI helps study root activity (Shanahan et al. 2015). EMI is high throughput than Electrical Resistivity Tomography (ERT) as it does not use probes, and it does not require disturbing the soil. Hence, it is quick and repeatable and can be used for studying soil–root interactions.

15.7 3D Softwares for Root Phenotyping (Ex Situ and In Situ)

Three-dimensional root imaging softwares uses 2D images of roots taken at different angles to get a 3D representation of the root system (Fig. 15.2) (Iyer-Pascuzzi et al. 2010; Chittoori 2012). Plants grown in gel media inside a cylindrical glass container are taken to a dark chamber where it is mounted on a turntable and illuminated using a light source. As the turntable rotates, the roots are imaged from different angles using a digital camera. The 2D images obtained are processed using software to yield various root traits. This approach is ideal for screening germplasms at young stages for Quantitative Trait Loci (QTL) studies targeted at identifying genes in the root system architecture. *RootReader3D* uses plants grown in a cylindrical glass container in gellan gum, which is mounted on a turntable and illuminated using a lightbox (Clark et al. 2011). Using a digital camera, the roots are imaged every 9° for a total of 40 images for each complete 360° rotation. The images are analyzed in *RootReader3D* software. This software facilitates both automated and semiautomated analysis of roots. Also, it offers high spatial and temporal resolution for measuring root traits.

Moreover, both static and dynamic 3D root system architectural traits can be studied as well. However, significant differences in the root traits of plants grown in gellan medium, sand, and hydroponics demand more in-depth evaluation of the system. *DynamicRoots* uses a similar approach for growing and imaging the roots, just like RootReader (Symonova et al. 2015). However, it uses just 20 out of the 40 images to reconstruct the root system represented as a collection of voxels in 3D space (Zheng et al. 2011). Plants are grown in gel, but it can be applied to other growing media and other setups for imaging roots. The software is automatic, and it quantifies root branches over time allowing to study root dynamics. It yields a complete branching hierarchy along with 3D root characteristics for each hierarchy. But

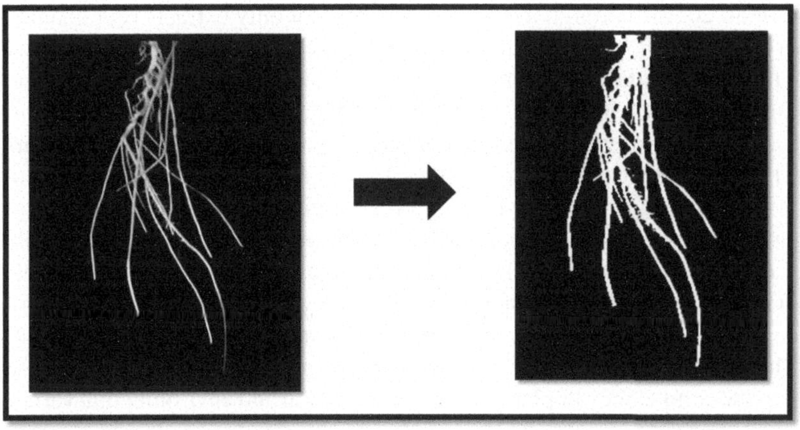

Fig. 15.2 3D root imaging. Original 2D root images in color are converted to binary images using adaptive thresholding in MATLAB software (Iyer-Pascuzzi et al. 2010; Chittoori 2012)

it does not account for the touching root branches resulting in an overlap. Unlike previous approaches done on younger roots in the lab, *3D/DIRT* can be done on field plants (Liu et al. 2021). The roots are excavated using shovelomics method (Trachsel et al. 2011). They are then placed on a 3D root scanner equipped with ten synchronized cameras, which takes 2000 images of the root. The images are computed in the software to yield a 3D root system, and the software analysis calculates the root traits. The main advantage of this method is that it measures highly obstructed and mature traits (nodal roots and whorl distances) in dense root crowns in maize. However, the main limitation is in achieving the resolution for detecting nodes within the root crowns. *3D/DIRT* is an invasive approach to field plants, but RootTrak is noninvasive (Mairhofer et al. 2012). X-ray Microcomputed Tomography is used to obtain the scans/images of the roots growing in the soil, and the images are analyzed in RootTrak software. The software segments the images and uses a visual tracking algorithm to identify and track roots in the images thereby mapping the entire root system in 3D. This method is fast and uses a minimal user interface. It is less susceptible to errors arising from root overlaps as in DynamicRoots. *TopoRoot* gives a full set of root hierarchy similar to RootTrak and, in addition, produces fine-grained characteristics of the mature root of maize using a single 3D image (Zeng et al. 2021). It is a fully automated and high-throughput software. Unlike DynamicRoots, TopoRoot deals with topological errors without the need for time series. Also, it is more accurate in counting the number of nodal roots than DynamicRoots and more accurate in coarse-grained and fine-grained traits than DynamicRoots and GiaRoots.

15.8 4D Softwares for Root Phenotyping

4DRoot phenotyping software uses 3D root scans obtained from X-ray Computed Tomography along with space scan or time scan as the fourth dimension (Herrero-Huerta et al. 2022). Through a cylinder fitting, it not only extracts root characteristics but also gives the topological and hierarchical root traits. It is automatic and has a high throughput. 4DRoot examines the time series of 3D Computed Tomography scans and estimates both spatial and temporal dynamics of roots. This software has direct application in crop improvement targeted at mapping genes and markers. Table 15.1 enumerates details of various root phenotyping methods. Recently, a Python-based package developed pipelines leveraging root trait phenotyping by utilizing a pose estimation-based approach for root system architecture (RSA) phenotyping (Berrigan et al. 2024). The spatial configuration or distribution of plants' roots within the root system or RSA signifies the volume of soil that the roots can explore and, in turn, depicts the availability of water and minerals to the root system. Therefore, an overview of RSA efficiently indicates the resilience strategies and adaptive competency of plants to various temporal and environmental variations.

Table 15.1 Overview of various root phenotyping methods

Root phenotyping	Approach		Reference
	Type	Name	
Traditional	Process	Soil trench	Weaver et al. (1922)
		Core-break	Bland (1989)
		Monolith	Böhm (1979)
		Backhoe-assisted monolith	Teramoto et al. (2019)
		Showelomics	Colombi et al. (2015)
2D	Set up	Rhizotron	James et al. (1985)
		Rootbox	Neufeld et al. (1989)
		GROWSCREEN-Rhizo	Nagel et al. (2012)
		Rhizoslides	Le Marié et al. (2014)
		RADIX	Le Marié et al. (2016)
		RhizoChamber-Monitor	Wu et al. (2018)
		WinRoots	Zhang et al. (2022)
		PhenoRoots	Martins et al. (2019)
		Rhizopot	Xiao et al. (2020)
	Software	ROOTEDGE	Kaspar and Ewing (1997)
		DIRT	Bucksch et al. (2014)
		REST	Colombi et al. (2015)
		Fluorescence Spectroscopy	Wasson et al. (2016)
		GLO-Roots	Rellán-Álvarez et al. (2015)
		GLO-Bot	LaRue et al. (2022)
		ChronoRoot	Gaggion et al. (2021)
		PlaRom	Yazdanbakhsh and Fisahn (2009)
		DART	Bot et al. (2010)
		SmartRoot	Lobet et al. (2011)
		RootNav	Pound et al. (2013)
		GIA-Roots	Galkovskyi et al. (2012)
		Ez-Rhizo	Armengaud (2009)
		RootTrace	French et al. (2009)
		Rhizovision Explorer	Seethepalli et al. (2021)
		DigiRoot	Stefanelli et al. (2009)
		RootReader2D	Clark et al. (2013)
		Phytomorph	Subramanian et al. (2013)
		GrowScreen-PaGe	Gioia et al. (2016)

(continued)

Table 15.1 (continued)

Root phenotyping	Approach		Reference
	Type	Name	
3D	Technology	X-ray CT	Hounsfield (1973)
		MRI	Lauterbur (1973)
		Ground Penetrating Radar (GPR)	Baker et al. (2007)
		Electrical Capacitance (EC)	Cseresnyés et al. (2019)
		Electrical Resistivity Tomography (ERT)	Zhou (2019)
		Electrical Impedance Tomography (EIT)	Weigand and Kemna (2017)
		Electro Magnetic Inductance (EMI)	Shanahan et al. (2015)
	Software	RootReader3D	Clark et al. (2011)
		Dynamic Roots	Zheng et al. (2011)
		3D/DIRT	Liu et al. (2021)
		Rootrak	Mairhofer et al. (2012)
		TopoRoot	Zeng et al. (2021)
4D	Software	4DRoot	Herrero-Huerta et al. (2022)

15.9 Conclusion

Root phenomics is an important field of research that investigates the complex world of plant roots. It is targeted at unlocking their potential for improving crop performance, food security, and environmental sustainability. Root phenomics is considered key to narrowing the knowledge gap in plant breeding between genotype-to-phenotype. Root phenomics has witnessed breakthrough technologies leading to the fast, accurate, automatic, and high-throughput 4D technology. These advances help us in the deeper understanding of roots and their integration into genetic and molecular biology. These developments will drive us positively in addressing global challenges in agriculture and environmental sustainability.

Short Chapter Summary

Roots are crucial to plant growth and thereby serve as key players in agricultural development and environmental sustainability. They play a pivotal role in plants' stress resilience as well as adaptation to changing environmental conditions and temporal variations as recently elucidated by pose estimation-based approach for RSA phenotyping. Root phenomics has encompassed breakthrough technologies for root imaging paving way from labor-, time- and capital-intensive traditional methods to the fast, accurate, automatic, and high-throughput 4D technology thereby narrowing the knowledge gap in plant breeding between genotype-to-phenotype. Overall, this chapter summarizes different root-growing media and the most recent techniques and softwares for analyzing root characteristics along with various root phenotyping methods.

Acknowledgments The corresponding author is grateful to the editors for this opportunity and also thanks the coauthors for their timely and valuable contributions.

Author Contributions Chapter concept and design, Minu Kesheri, Swarna Kanchan, Bhaskar C.S. Chitoori, Ratnaprabha Ratna-Raj; chapter writing, Ratnaprabha Ratna-Raj.

References

Abramoff M, Magalhães P, Ram SJ (2003) Image processing with imageJ. Biophotonics Int 11:36–42

Agbona A, Teare B, Ruiz Guzman H, Dobreva I, Everett M, Adams T, Montesinos-López O, Kulakow P, Hays D (2021) Prediction of root biomass in cassava based on ground penetrating radar phenomics. Remote Sens 13:4908

Annan AP (2002) GPR-history, trends, and future developments. Subsurf Sens Technol Appl 3:253–270

Armengaud P (2009) EZ-Rhizo software. Plant Signal Behav 4:139–141

Audebert AY, Ghneim T, Roques S, Thaunay P, Flueriot JP (2010) Development of a high-throughput system for phenotyping rice roots traits. In: The 28th international rice research conference, climate change and rice agriculture, November 8–12, 2010, Hanoi, Vietnam. IRRI

Baker GS, Jordan TE, Pardy J (2007) An introduction to ground penetrating radar (GPR). In: Baker GS, Jol HM (eds) Stratigraphic analyses using GPR, vol 432. Geological Society of America, pp 1–18

Bao Y, Aggarwal P, Robbins NE, Sturrock CJ, Thompson MC, Tan HQ, Tham C, Duan L, Rodriguez PL, Vernoux T, Mooney SJ (2014) Plant roots use a patterning mechanism to position lateral root branches toward available water. Proc Natl Acad Sci USA 111(25):9319–9324

Bengough AG, Gordon DC, Al-Menaie H, Ellis RP, Allan D, Keith R, Thomas WTB, Forster BP (2004) Gel observation chamber for rapid screening of root traits in cereal seedlings. Plant Soil 262:63–70

Berrigan EM, Wang L, Carrillo H, Echegoyen K, Kappes M, Torres J, Ai-Perreira A, McCoy E, Shane E, Copeland CD, Ragel L (2024) Fast and efficient root phenotyping via pose estimation. Plant Phenomics 6:0175

Bland W (1989) Estimating root length density by the core-break method. Soil Sci Soc Am J 53:1595–1597

Block A, Christensen S, Hunter C, Alborn H (2017) Herbivore-derived fatty-acid amides elicit reactive oxygen species burst in plants. J Exp Bot 69:1235–1245

Böhm W (1979) Monolith methods. In: Methods of studying root systems. Springer, Berlin, Heidelberg, pp 20–38

Bot J, Serra V, Fabre J, Draye X, Adamowicz S, Pagès L (2010) DART: a software to analyse root system architecture and development from captured images. Plant Soil 326:261–273

Bucksch A, Burridge J, York LM, Das A, Nord E, Weitz JS, Lynch JP (2014) Image-based high-throughput field phenotyping of crop roots. Plant Physiol 166:470–486

Buckseth T, Sharma A, Pandey KK, Singh BP, Muthuraj R (2016) Methods of pre-basic seed potato production with special reference to aeroponics – a review. Sci Hortic 204:79–87

Cao Y, Repo T, Silvennoinen R, Lehto T, Pelkonen P (2010) An appraisal of the electrical resistance method for assessing root surface area. J Exp Bot 61:2491–2497

Chapman N, Miller A, Lindsey K, Whalley R (2012) Roots, water, and nutrient acquisition: let's get physical. Trends Plant Sci 17:701–710

Chen Y, Dunbabin V, Diggle A, Siddique K, Rengel Z (2011) Development of a novel semi-hydroponic phenotyping system for studying root architecture. Funct Plant Biol 38:355–363

Chittoori R (2012) Root morphological and physiological bases to understand genotypic control of mineral acquisition in rice grains. Doctoral dissertation, Texas A&M University

Clark RT, MacCurdy RB, Jung JK, Shaff JE, McCouch SR, Aneshansley DJ, Kochian LV (2011) Three-dimensional root phenotyping with a novel imaging and software platform. Plant Physiol 156:455–465

Clark RT, Famoso AN, Zhao K, Shaff JE, Craft EJ, Bustamante CD, McCouch SR, Aneshansley DJ, Kochian LV (2013) High-throughput two-dimensional root system phenotyping platform facilitates genetic analysis of root growth and development. Plant Cell Environ 36:454–466

Colombi T, Kirchgessner N, Le Marié CA, York LM, Lynch JP, Hund A (2015) Next generation shovelomics: set up a tent and REST. Plant Soil 388:1–20

Courtois B, Audebert A, Dardou A, Roques S, Ghneim T, Droc G, Frouin J, Rouan L, Goze E, Kilian A, Ahmadi N (2013) Genome-wide association mapping of root traits in a japonica rice panel. PLoS One 8(11):e78037

Cseresnyés I, Takács T, Sepovics B, Kovács R, Füzy A, Paradi I, Rajkai K (2019) Electrical characterization of the root system: a noninvasive approach to study plant stress responses. Acta Physiol Plant 41:169

Eliasson L, Bollmark M (1988) Ethylene as a possible mediator of light-induced inhibition of root growth. Physiol Plant 72:605–609

Falk KG, Jubery TZ, Mirnezami SV, Parmley KA, Sarkar S, Singh A, Ganapathysubramanian B, Singh A (2020) Computer vision and machine learning enabled soybean root phenotyping pipeline. Plant Methods 16:1–19

Food and Agricultural Organization (FAO) (2009) How to feed the world 2050. FAO

French A, Ubeda-Tomás S, Holman TJ, Bennett MJ, Pridmore T (2009) High-throughput quantification of root growth using a novel image-analysis tool. Plant Physiol 150:1784–1795

Gaggion N, Ariel F, Daric V, Lambert É, Legendre S, Roulé T, Camoirano A, Milone DH, Crespi M, Blein T et al (2021) ChronoRoot: high-throughput phenotyping by deep segmentation networks reveals novel temporal parameters of plant root system architecture. GigaScience 10:giab052

Galkovskyi T, Mileyko Y, Bucksch A, Moore B, Symonova O, Price CA, Topp CN, Iyer-Pascuzzi AS, Zurek PR, Fang S et al (2012) GiA Roots: software for the high throughput analysis of plant root system architecture. BMC Plant Biol 12:116

Gioia T, Galinski A, Lenz H, Müller C, Lentz J, Heinz K, Briese C, Putz A, Fiorani F, Watt M, Schurr U (2016) GrowScreen-PaGe, a non-invasive, high-throughput phenotyping system based on germination paper to quantify crop phenotypic diversity and plasticity of root traits under varying nutrient supply. Funct Plant Biol 44:76–93

Heeraman DA, Hopmans JW, Clausnitzer V (1997) Three-dimensional imaging of plant roots in situ with X-ray Computed Tomography. Plant Soil 189:167–179

Herrero-Huerta M, Raumonen P, González-Aguilera D (2022) 4DRoot: root phenotyping software for temporal 3D scans by X-ray computed tomography. Front Plant Sci 13:986856

Hodge A, Berta G, Doussan C, Merchan F, Crespi M (2009) Plant root growth, architecture and function. Plant Soil 321:153–187

Hounsfield GN (1973) Computerized transverse axial scanning (tomography): Part I. Description of system. Br J Radiol 46(552):1016–1022

Hund A, Trachsel S, Stamp P (2009) Growth of axile and lateral roots of maize: I development of a phenotyping platform. Plant Soil 325:335–349

Iyer-Pascuzzi A, Symonova O, Mileyko Y, Hao Y, Belcher H, Harer J, Weitz J, Benfey P (2010) Imaging and analysis platform for automatic phenotyping and trait ranking of plant root systems. Plant Physiol 152:1148–1157

Jahnke S, Menzel MI, Dusschoten D, Roeb GW, Bühler J, Minwuyelet S, Blümler P, Temperton VM, Hombach T, Streun M, Beer S (2009) Combined MRI-PET dissects dynamic changes in plant structures and functions. Plant J Cell Mol Biol 59(4):634–644

James BR, Bartlett RJ, Amadon JF (1985) A root observation and sampling chamber (rhizotron) for pot studies. Plant Soil 85:291–293

Jeudy C, Adrian M, Baussard C, Bernard C, Bernaud E, Bourion V, Busset H, Cabrera-Bosquet L, Cointault F, Han S, Lamboeuf M (2016) RhizoTubes as a new tool for high throughput imaging

of plant root development and architecture: test, comparison with pot grown plants and validation. Plant Methods 12:1–8
Kaspar TC, Ewing RP (1997) ROOTEDGE: software for measuring root length from desktop scanner images. Agron J 86(6):932–940
Kesheri M, Richa, Sinha RP (2011) Antioxidants as natural arsenal against multiple stresses in cyanobacteria. Int J Pharma Bio Sci 2(2):B168–B187
Kesheri M, Kanchan S, Richa, Sinha RP (2014) Isolation and in-silico analysis of Fe-superoxide dismutase in the cyanobacterium *Nostoc commune*. Gene 553(2):117–125
Kesheri M, Kanchan S, Sinha RP (2021) Isolation and in-silico analysis of antioxidants in response to temporal variations in the cyanobacterium *Oscillatoria* sp. Gene Rep 23:101023
Kesheri M, Kanchan S, Sinha RP (2022) Responses of antioxidants for resilience to temporal variations in the cyanobacterium *Microcystis aeruginosa*. S Afr J Bot 148:190–199
Kesheri M, Kanchan S, Salisbury TB, Sinha RP (2024) Microbial omics in environment and health. Springer Nature, Singapore. ISBN: 978-981-97-1768-2
Keyes S, Zygalakis K, Roose T (2017) An explicit structural model of root hair and soil interactions parameterised by synchrotron X-ray computed tomography. Bull Math Biol 79:2785–2813
Kimura K, Yamasaki S (2003) Accurate root length and diameter measurement using NIH Image: use of Pythagorean distance for diameter estimation. Plant Soil 254:305–315
Kücke M, Schmid H, Spiess A (1995) A comparison of four methods for measuring roots of field crops in three contrasting soils. Plant Soil 172:63–71
Kumari A, Kanchan S, Kesheri M (2016) Applications of bio-molecular databases in bioinformatics. In: Dey N, Bhateja V, Hassanien AE (eds) Advancements in bio-medical sensing, imaging, measurements and instrumentation, vol 651. Springer-Verlag, Berlin, Heidelberg, pp 329–351
Kumari A, Kesheri M, Sinha RP, Kanchan S (2018) Integration of soft computing approach in plant biology and its applications in agriculture. In: Purohit HJ, Raje DV, More RP, Kalia VC (eds) Soft computing in biological sciences. Springer Nature, Singapore, pp 265–281
LaRue T, Lindner H, Srinivas A, Exposito-Alonso M, Lobet G, Dinneny JR (2022) Uncovering natural variation in root system architecture and growth dynamics using a robotics-assisted phenomics platform. elife 11:e76968
Lauterbur PC (1973) Image formation by induced local interactions: examples employing nuclear magnetic resonance. Nature 242:190–191
Le Marié C, Kirchgessner N, Marschall D, Walter A, Hund A (2014) Rhizoslides: paper-based growth system for non-destructive, high throughput phenotyping of root development by means of image analysis. Plant Methods 10:13
Le Marié C, Kirchgessner N, Flütsch P, Pfeifer J, Walter A, Hund A (2016) RADIX: rhizoslide platform allowing high throughput digital image analysis of root system expansion. Plant Methods 12:40
Liu S, Barrow CS, Hanlon M, Lynch JP, Bucksch A (2021) DIRT/3D: 3D root phenotyping for field-grown maize (*Zea mays*). Plant Physiol 187:739–757
Lobet G, Pagès L, Draye X (2011) A novel image analysis toolbox enabling quantitative analysis of root system architecture. Plant Physiol 157:29–39
Lynch JP (2007) Roots of the second green revolution. Aust J Bot 55:493–512
Ma L, Shi Y, Siemianowski O, Yuan B, Egner T, Mirnezami SV, Lind K, Ganapathysubramanian B, Venditti V, Cademartiri L (2019) Hydrogel-based transparent soils for root phenotyping in vivo. Proc Natl Acad Sci 116:201820334
Mairhofer S, Zappala S, Tracy SR, Sturrock C, Bennett M, Mooney SJ, Pridmore T (2012) RooTrak: automated recovery of three-dimensional plant root architecture in soil from X-ray microcomputed tomography images using visual tracking. Plant Physiol 158:561–569
Martins S, Brito GGD, Gonçalves WDC, Tripode BMD, Lartaud M, Duarte JB, Morello CDL, Giband M (2019) PhenoRoots: an inexpensive non-invasive phenotyping system to assess the variability of the root system architecture. Sci Agric 77:20180420
Mathieu L, Lobet G, Tocquin P, Périlleux C (2015) "Rhizoponics": a novel hydroponic rhizotron for root system analyses on mature *Arabidopsis thaliana* plants. Plant Methods 11:3

Mester A, Kruk J, Zimmermann E, Vereecken H (2011) Quantitative two-layer conductivity inversion of multi-configuration electromagnetic induction measurements. Vadose Zone J 10:1319
Mouritsen H, Ritz T (2005) Magnetoreception and its use in bird navigation. Curr Opin Neurobiol 15:406–414
Mühlich M, Truhn D, Nagel K, Walter A, Scharr H, Aach T (2008) Measuring plant root growth. In: Pattern recognition: 30th DAGM symposium Munich. Springer, Berlin, Heidelberg, pp 497–506
Naeem A, French AP, Wells DM, Pridmore TP (2011) High-throughput feature counting and measurement of roots. Bioinformatics 27:1337–1338
Nagel KA, Kastenholz B, Jahnke S, van Dusschoten D, Aach T, Mühlich M, Truhn D, Scharr H, Terjung S, Walter A, Schurr U (2009) Temperature responses of roots: impact on growth, root system architecture and implications for phenotyping. Funct Plant Biol 36(11):947–959
Nagel K, Putz A, Gilmer F, Kathrin H, Fischbach A, Pfeifer J, Marc F, Blossfeld S, Michaela E, Dimaki C et al (2012) GROWSCREEN-Rhizo is a novel phenotyping robot enabling simultaneous measurements of root and shoot growth for plants grown in soil-filled rhizotrons. Funct Plant Biol 39:891
Neufeld H, Durall D, Rich P, Tingey D (1989) A rootbox for quantitative observations on intact entire root systems. Plant Soil 117:295–298
Pflugfelder D, Metzner R, van Dusschoten D, Reichel R, Jahnke S, Koller R (2017) Non-invasive imaging of plant roots in different soils using magnetic resonance imaging (MRI). Plant Methods 13:102
Pound MP, French AP, Atkinson JA, Wells DM, Bennett MJ, Pridmore T (2013) RootNav: navigating images of complex root architectures. Plant Physiol 162:1802–1814
Priya P, Kesheri M, Sinha RP, Kanchan S (2017) Molecular dynamics simulations for biological systems. In: Karâa WBA, Dey N (eds) Biomedical image analysis and mining techniques for improved health outcomes. IGI Global, Hershey, pp 286–313
Rafalski JA (2010) Association genetics in crop improvement. Curr Opin Plant Biol 13:174–180
Rahman H, Ramanathan V, Jagadeeshselvam N, Ramasamy S, Rajendran S, Ramachandran M, Sudheer PD, Chauhan S, Natesan S, Muthurajan R (2015) Phenomics: technologies and applications in plant and agriculture. In: Barh D, Khan MS, Davies E (eds) PlantOmics: the omics of plant science. Springer, New Delhi, pp 385–411
Rasol MA, Pérez-Gracia V, Fernandes FM, Pais JC, Santos-Assunçao S, Santos C, Sossa V (2020) GPR laboratory tests and numerical models to characterize cracks in cement concrete specimens, exemplifying damage in rigid pavement. Measurement 158:107662
Rellán-Álvarez R, Lobet G, Lindner H, Pradier PL, Sebastian J, Yee MC, Geng Y, Trontin C, LaRue T, Schrager-Lavelle A, Haney CH (2015) GLO-Roots: an imaging platform enabling multidimensional characterization of soil-grown root systems. elife 4:e07597
Rogers ED, Monaenkova D, Mijar M, Nori A, Goldman DI, Benfey PN (2016) X-ray computed tomography reveals the response of root system architecture to soil texture. Plant Physiol 171:2028–2040
Rossi R, Amato M, Bitella G, Bochicchio R, Ferreira Gomes JJ, Lovelli S, Martorella E, Favale P (2011) Electrical resistivity tomography as a non-destructive method for mapping root biomass in an orchard. Eur J Soil Sci 62:206–215
Seethepalli A, Dhakal K, Griffiths M, Guo H, Freschet G, York L (2021) RhizoVision Explorer: open-source software for root image analysis and measurement standardization. AoB Plants 13(6):plab056
Selvaraj MG, Montoya M, Atanbori J, French AP, Pridmore T (2019) A low-cost aeroponic phenotyping system for storage root development: unravelling the below-ground secrets of cassava (*Manihot esculenta*). Plant Methods 15:1–13
Selzner T, Horn J, Landl M, Pohlmeier A, Helmrich D, Huber K, Vanderborght J, Vereecken H, Behnke S, Schnepf A (2023) 3D U-Net segmentation improves root system reconstruction from 3D MRI images in automated and manual virtual reality work flows. Plant Phenomics 5:0076

Shanahan P, Binley A, Whalley R, Watts CW (2015) The use of electromagnetic induction to monitor changes in soil moisture profiles beneath different wheat genotypes. Soil Sci Soc Am J 79(2):459–466

Singh S, Singh V, Singh BP, Pandey SK (2010) Aeroponic for potato seed production. ICAR News 16:1–2

Solla M, Lorenzo H, Rial FI, Novo A (2012) Ground-penetrating radar for the structural evaluation of masonry bridges: results and interpretational tools. Constr Build Mater 29:458–465

Solla M, Lagüela S, Riveiro B, Lorenzo H (2013) Non-destructive testing for the analysis of moisture in the masonry arch bridge of Lubians (Spain). Struct Control Health Monit 20(11):1366–1376

Steen E (1991) Usefulness of the mesh bag method in quantitative root studies. Cambridge University Press, Cambridge

Stefanelli D, Fridman Y, Perry RL (2009) DigiRoot (TM): new software for root studies. Eur J Hortic Sci 74:169–174

Streda T, Haberle J, Klimesova J, Klimek-Kopyra A, Stredova H, Bodner G, Chloupek O (2020) Field phenotyping of plant roots by electrical capacitance-a standardized methodological protocol for application in plant breeding: a review. Int Agrophys 34:173–184

Subramanian R, Spalding EP, Ferrier NJ (2013) A high throughput robot system for machine vision-based plant phenotype studies. Mach Vis Appl 24:619–636

Symonova O, Topp CN, Edelsbrunner H (2015) DynamicRoots: a software platform for the reconstruction and analysis of growing plant roots. PLoS One 10:e0127657

Taiz L, Zeiger E, Miller M, Muphy A (2018) Fundamentals of plant physiology. Sinauer Associates Inc., New York

Teramoto S, Kitomi Y, Nishijima R, Takayasu S, Maruyama N, Uga Y (2019) Backhoe-assisted monolith method for plant root phenotyping under upland conditions. Breed Sci 69:508–513

Tiwari J, Devi S, Buckseth T, Ali N, Singh R, Zinta R, Dua V, Chakrabarti S (2020) Precision phenotyping of contrasting potato (*Solanum tuberosum* L.) varieties in a novel aeroponics system for improving nitrogen use efficiency: in search of key traits and genes. J Integr Agric 19:51–61

Trachsel S, Kaeppler SM, Brown KM, Lynch JP (2011) Shovelomics: high throughput phenotyping of maize (*Zea mays* L.) root architecture in the field. Plant Soil 341:75–87

van Dusschoten D, Metzner R, Kochs J, Postma JA, Pflugfelder D, Bühler J, Schurr U, Jahnke S (2016) Quantitative 3D analysis of plant roots growing in soil using magnetic resonance imaging. Plant Physiol 170:1176–1188

Wasson A, Bischof L, Zwart A, Watt M (2016) A portable fluorescence spectroscopy imaging system for automated root phenotyping in soil cores in the field. J Exp Bot 67:1033–1043

Weaver JE, Weaver JE, Jean FC, Crist JW (1922) Development and activities of roots of crop plants; a study in crop ecology. Carnegie Institution of Washington, Washington, DC

Weigand M, Kemna A (2017) Multi-frequency electrical impedance tomography as a non-invasive tool to characterize and monitor crop root systems. Biogeosciences 14:921–939

Wu J, Wu Q, Pagès L, Yuan Y, Zhang X, Du M, Tian X, Li Z (2018) RhizoChamber-Monitor: a robotic platform and software enabling characterization of root growth. Plant Methods 14:1–15

Xiao S, Liu L, Zhang Y, Sun H, Zhang K, Bai Z, Dong H, Li C (2020) Fine root and root hair morphology of cotton under drought stress revealed with RhizoPot. J Agron Crop Sci 206:679–693

Yazdanbakhsh N, Fisahn AJ (2009) High throughput phenotyping of root growth dynamics, lateral root formation, root architecture and root hair development enabled by PlaRoM. Funct Plant Biol 36(11):938–946

Zeng D, Li M, Jiang N, Ju Y, Schreiber H, Chambers E, Letscher D, Ju T, Topp C (2021) TopoRoot: a method for computing hierarchy and fine-grained traits of maize roots from 3D imaging. Plant Methods 17:1–17

Zhang X, Derival M, Albrecht U, Ampatzidis Y (2019) Evaluation of a ground penetrating radar to map the root architecture of HLB-infected citrus trees. Agronomy 9(7):354

Zhang Y, Zhang W, Cao Q, Zheng X, Yang J, Xue T, Sun W, Du X, Wang L, Wang J, Zhao F (2022) WinRoots: a high-throughput cultivation and phenotyping system for plant phenomics studies under soil stress. Front Plant Sci 12:794020

Zheng Y, Gu S, Edelsbrunner H, Tomasi C, Benfey P (2011) Detailed reconstruction of 3D plant root shape. In: 2011 International conference on computer vision. IEEE, pp 2026–2033
Zhou B (2019) Electrical resistivity tomography: a subsurface-imaging technique. In: Kanlı AI (ed) Applied geophysics with case studies on environmental, exploration and engineering geophysics. IntechOpen
Zhu S, Huang C, Su Y, Sato M (2014) 3D ground penetrating radar to detect tree roots and estimate root biomass in the field. Remote Sens 6:5754–5773

Envisioning Production of Alginates Through the Lens of Multi-Omics

16

Minu Kesheri, Swarna Kanchan, Bhagaban Mallik, Tarun Mishra, Ratnaprabha Ratna-Raj, Bhaskar Chandra Srinivas Chittoori, and Rajeshwar P. Sinha

Abstract

Microbial communities secrete exopolysaccharides such as alginate forming biofilms at various surfaces/environments, which could be the lungs of people with cystic fibrosis, or wounds, that support bacterial growth and survival. The discovery of alginate, various sources of alginate, alginate biosynthesis, structure, and properties are explained in detail. Modifications in bacterial alginates could be facilitated during acetylation and epimerization for desired properties. Various omics approaches leverage tailor-made alginate production having desired physicochemical properties. Machine learning (ML) methods with examples to modify alginates of industrial importance are explained in this chapter. Various omics

M. Kesheri · S. Kanchan (✉)
Boise State University, Boise, ID, USA

Joan C. Edwards School of Medicine, Marshall University, Huntington, WV, USA
e-mail: kanchan@marshall.edu

B. Mallik
Department of Biological Sciences, IISER, Bhopal, Madhya Pradesh, India

Department of Anatomy and Cell Biology, University of Iowa, Iowa City, IA, USA

T. Mishra
Department of Biological Sciences, IISER, Bhopal, Madhya Pradesh, India

Department of Microbiology and Immunology, University of Iowa, Iowa City, IA, USA

R. Ratna-Raj · B. C. S. Chittoori
Boise State University, Boise, ID, USA

R. P. Sinha
Laboratory of Photobiology and Molecular Microbiology, Department of Botany, Institute of Science, Banaras Hindu University, Varanasi, India

M. Kesheri et al. (eds.), *Multi-Omics in Biomedical Sciences and Environmental Sustainability*, https://doi.org/10.1007/978-981-96-7067-3_16

studies are discussed, which contribute to exploring biofilm development, host biofilm interaction, antibiotic resistance, etc. In this chapter, we provide insights into the state-of-the-art elucidating the production of alginates and highlight their roles in biomedical and other industrial applications.

Keywords

Alginate · Biomedical applications · Machine learning · Omics · Tailor-made alginates

Objectives

- To explore the various sources, biosynthesis, structure, and properties of alginates
- To understand the methods to design tailor-made alginates by modifying acetylation and epimerization processes for desired physicochemical properties
- To explore ML methods to modify alginates of industrial importance
- Use of omics techniques to explore biofilm development, host biofilm interaction, antibiotic resistance, etc.
- To explore various industrial applications of alginates in biomedical, pharmaceutical, food sciences, etc.

16.1 Introduction

Alginates were discovered by an English chemist, E.C.C. Stanford, in 1883 (Stanford ECC 1883). In 1929, Kelco Company from San Diego started producing pure alginates a large and commercial scale (Clare 1993). Later, Merck and Co. Inc., USA, became the largest alginate producer worldwide after acquiring a few industries in this sector (Clare 1993). Linker and Jones reported the alginate production by bacteria for the first time in 1964. They observed alginates on plates having large mucoid colonies collected from cystic fibrosis patients (Linker and Jones 1964). Later, Gorin and coworkers reported that *Azotobacter vinelandii, Pseudomonas putida, P. fluorescens*, and *P. mendocina* bacterial species were responsible for the production of alginic acid (Gorin and Spencer 1966). Alginates production was also reported from the cell wall of brown seaweed such as *Laminaria digitata, Ascophyllum nodosum, Microcystis pyrifera, Laminaria hyperborea*, and *Laminaria japonica* (Hay et al. 2013; Lee and Mooney 2012). Other than brown algae/seaweeds, bacterial organisms such as *Acetobacter* and *Pseudomonas* sp. were illustrated to be directly responsible for the secretion of alginates.

Alginate is a polymer containing two different kinds of monosaccharide units, namely L-guluronic (G) and D-mannuronic (M) acids, joined by 1,4-glycosidic linkages. Alginates are usually present as heteropolymers, including M and G sugar residues. At least 200 distinct alginates are produced by the various combinations of M and G blocks and their lengths. However, the design of alginates also depends on their source of origin (Zhang et al. 2021). Algal alginates are usually rich in G blocks, whereas alginates produced from bacterial species called *Pseudomonas aeruginosa* are devoid of G blocks (Urtuvia et al. 2017). Different physicochemical properties are associated with differences in the composition of alginate molecules. Guluronic acid residues (G Units) interact with divalent ions and form crosslinks making alginate stiff and hard. Therefore, alginates without G-blocks form softer gels (Fu et al. 2011). Different combinations of M/G blocks in the alginate molecule reduce the viscoelasticity of alginates in the order MG block > MM block > GG block. Alginate from bacterial species such as *Pseudomonas* sp. is acetylated compared to algal alginates (Linker and Jones 1966). Acetylation of alginate molecules is frequently observed with bacterial alginates. Since acetylated alginate interacts better with water molecules, it absorbs more water and becomes more soluble. (Skjåk-Bræk et al. 1989; Moradali et al. 2015). Alginates also bind to various divalent cations and affinity increases in the order $Mg^{2+} < Mn^{2+} < Ca^{2+} < Sr^{2+} < Ba^{2+} < Cu^{2+} < Pb^{2+}$. (Haug and Smidsrød 1967; Haug and Smidsrod 1970).

This chapter encapsulates various sources, biosynthesis, structure, and properties of alginates. Various methods to design tailor-made alginates by modifying acetylation and epimerization processes for desired physicochemical properties are discussed. Emphasis has been made to elucidate machine learning-based methods to modify alginates of industrial importance. The use of omics techniques to explore biofilm development, host biofilm interaction, antibiotic resistance, etc., is also highlighted. Various applications of alginates in biomedical, pharmaceutical, food sciences, etc., have also been discussed.

16.2 Alginate Biosynthesis, Polymerization, Acetylation, Epimerization, and Degradation

Various genes responsible for alginate production are called algK, algE (algJ), algD, alg8, alg44, algI, algJ (algV), algG, algX, algL, algF, algA, which are clustered together to form an operon (Chitnis and Ohman 1993; Hay et al. 2014) except algC in various bacterial genomes. The promoter present upstream to algD transcribes various genes for alginate production. Since alginate is a polymer, therefore, Alg8 (polymerase) and Alg44 (co-polymerase) are responsible for the polymerization of the alginate molecules. Various modification events, such as acetylation and epimerization, are performed after polymerization and translocation through periplasmic space. After polymerization, acetylation as well as epimerization are performed, which are responsible for the material properties of alginates.

Acetylation of alginates has been observed only in various bacterial species such as *Pseudomonas aeruginosa*. AlgX and other enzymes such as AlgI, AlgJ, and AlgF play an essential role in acetylation (Baker et al. 2014; Franklin and Ohman 2002).

AlgG-type epimerizes M blocks, and therefore such alginates are devoid of G blocks (Gimmestad et al. 2003; Jain et al. 2003). Ca2 + -dependent AlgE-type enzymes catalyze the conversion of M to G residues, which increases the number of G blocks in the alginate blocks.

Alginate lyases are also called alginate-degrading enzymes because they break down the alginate molecules. Based on their structural diversity, alginate lyases are categorized as either mannuronate or guluronate lyases (Wong et al. 2000; Ertesvåg 2015). Alginate lyases have been isolated from various organisms such as algae (except brown algae), terrestrial microorganisms, and marine invertebrates. Alginate lyases digest alginates and utilize them as carbon sources (Wong et al. 2000; Ertesvåg 2015).

16.3 Tailor-Made Alginate Production

Alginates from brown algae are introduced with desired physicochemical properties, mainly chemical modifications and treatments that are uncontrolled. Hence, alginate modifications by their associated proteins can build up the foundation to produce tailor-made alginates. Alg8 and Alg44 are mainly responsible for polymerization, whereas AlgX (acetyltransferase) and AlgG (epimerase) are responsible for various kinds of modifications in the alginate molecules. The three-dimensional (3D) structure of AlgG (*Pseudomonas syringae* pv. tomato str. DC3000) (Wolfram et al. 2014) is available at the protein databank (Pdb ID: 4NK6) as shown in Fig. 16.1. Similarly, Whitney et al. 2015 elucidated the 3D structure of Alg44 (*Pseudomonas aeruginosa* PAO1) which is also available at protein databank (Pdb ID: 4RT0) as depicted in Fig. 16.2. Moreover, 3D structure for Alg8 is not available in protein databank, which could be predicted using threading/homology modeling

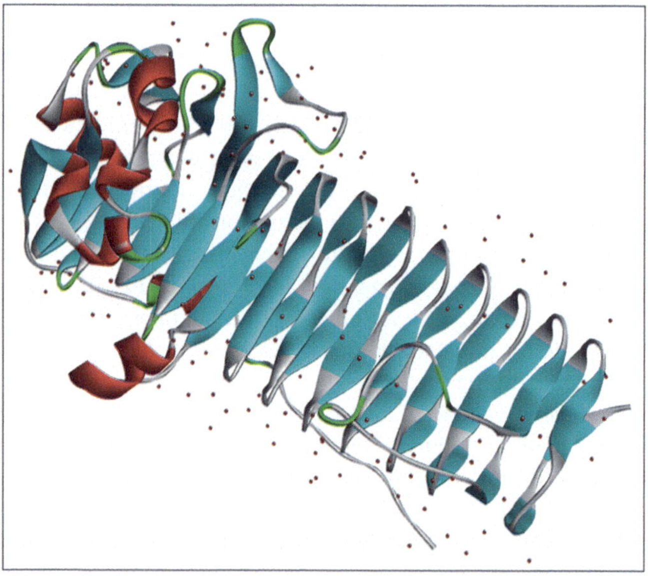

Fig. 16.1 Crystal structure of the periplasmic alginate epimerase AlgG (PDB ID: 4NK6)

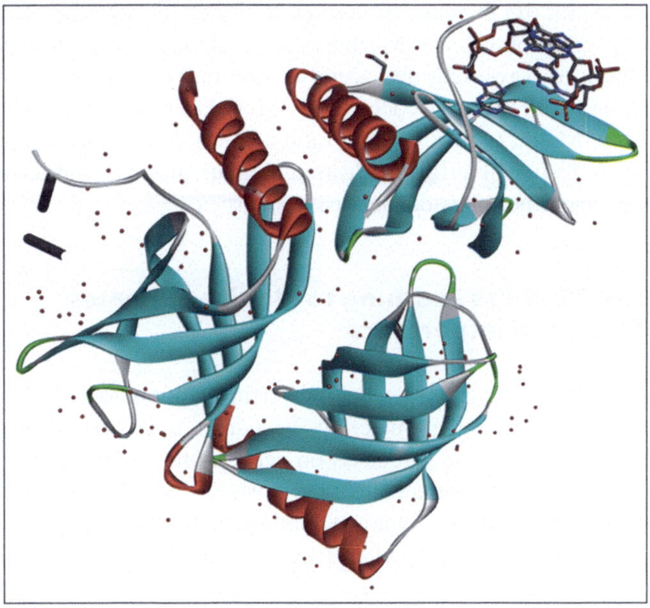

Fig. 16.2 Structure of the alg44 pilz domain from *Pseudomonas aeruginosa* PAO1 in complex with c-di-gmp (pdb id: 4rt0)

and the quality of the 3D model could be further improved by molecular dynamics simulation methods (Kesheri et al. 2015; Priya et al. 2017).

Studies illustrated that alginate polymerization is usually intersected by epimerization and produces alginates with lower molecular mass (2755 kDa) (Moradali et al. 2015). By eliminating epimerization due to the introduction of a catalytically inactive variant of AlgG, alginate polymerization leads to the production of a very high molecular mass of alginate (4653 kDa) (Moradali et al. 2015).

Moreover, studies illustrated that after eliminating epimerization, alginate (yield) production was about threefold higher compared to normal epimerization.

Additionally, it was demonstrated that the introduction of a catalytically inactive form of AlgX (S269A) eliminated acetylation and produced alginate with a lower molecular mass (2086 kDa).

Active acetylation is positively correlated with alginate polymerization, producing a higher molecular mass (2460 kDa) of alginate (Moradali et al. 2015). Additionally, the ratio of M blocks increased because of the overproduction of Alg8 and Alg44, which affected polymerization but increased the acetylation events (Moradali et al. 2015; Moradali et al. 2017).

Generally, the alginates without G residues and a high number of M blocks produce alginates of higher molecular mass, having the highest viscoelasticity. Moreover, increasing the epimerization event and the frequency of G residues in *P. aeruginosa* led to the production of alginate with the lowest viscoelastic behavior when a high copy of alginate acetyl transferase AlgX was introduced (Moradali et al. 2015). Overall, the increase in viscoelasticity of the alginate molecules was observed by the addition of each M unit into the alginate molecule. On the other hand, the addition of G residues and acetyl groups in the alginate chain was found to be responsible for lowered viscoelasticity (Moradali et al. 2015). Therefore, the molar fraction of M and G residues, the ratio of M and G blocks, acetylation, and epimerization are responsible for the desired alginate properties in terms of viscoelasticity, mass, and other physicochemical properties.

16.4 Use of Machine Learning to Modify Alginates of Industrial Importance

Machine learning (ML) is considered a powerful computational tool for various complex tasks such as protein engineering (Feehan et al. 2021). ML has become a popular tool of choice because it focuses on patterns that could be applied to alginate engineering. ML uses biological data to retrieve multiple patterns and utilizes various information such as the organism name, protein structure, protein sequence of the protein responsible for alginate production, acetylation, and epimerization events, amino acid residues, atoms, etc. Such information could be used in predicting desired properties, optimal microenvironments for various novel properties in alginates, etc.

ML is categorized into three main types: supervised learning, which utilizes labeled data for predictions as output, whereas unsupervised learning utilizes unlabeled data to identify patterns as output. Reinforcement learning is based on predicting the correct output by maximizing its reward. Some common examples of supervised learning-based ML algorithms include random forests, decision trees, support vector machines (SVM) (Boser et al. 1992), and linear regression. Unsupervised learning is used for hierarchical and K-means clustering, principal component analysis (PCA), and association (i.e., using hidden Markov models (HMMs) (Baum and Eagon 1967). An artificial neural network includes 1 or 2 hidden layers and is used for both supervised and unsupervised learning, which is inspired by the human brain and various connections in the brain in the form of neurons (McCulloch and Pitts 1943). Artificial neural networks with multiple hidden layers of interconnected neurons are used to solve more complex and large problems, which is know as deep learning (Kanchan et al. 2024). However, sometimes deep learning/artificial neural networks require large amounts of training data, which could be used to explore the sequence-function landscape of alginates with desired properties.

A recent study by Xu et al. (2024) investigated applications in flexible electronics by optimizing a dual-network hydrogel employing a blend of acrylamide and alginate using ML methods as shown in Fig. 16.3. A series of experiments was performed and validated experimentally, after several optimization and validation experiments, a hydrogel with superior properties was retrieved. This hydrogel was observed to have elevated strain sensitivity and flexibility. Additionally, a regression model was developed to demonstrate its utility in predicting the relationship. The

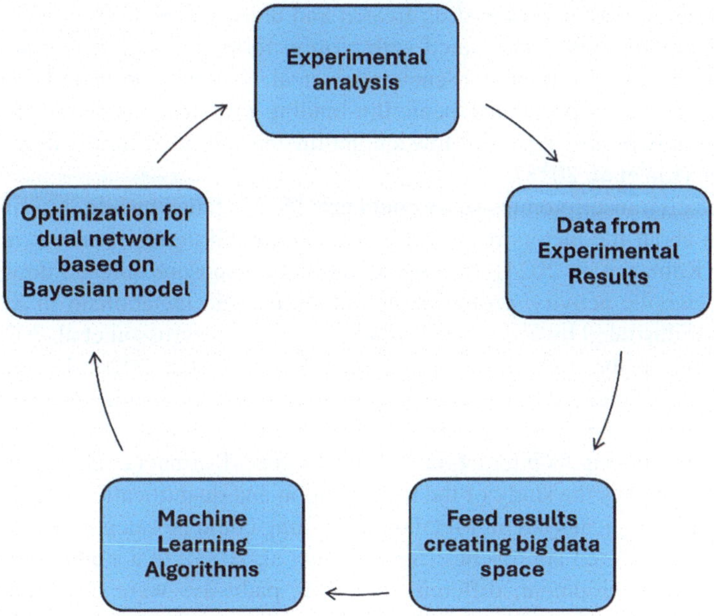

Fig. 16.3 ML-based optimization for production of hydrogels from alginates

regression model was developed utilizing various experimental parameters and hydrogel properties, which was finally validated through experiments. ML leveraged production of tailored hydrogel has a crucial application in bioprinting (Ege and Boccaccini 2024). Application of nanoparticles for wastewater treatment has been a novel approach (Solomon et al. 2024) by utilizing microbial products by employing ML for optimization of a sodium alginate-based hydrogel facilitating adsorption of methylene blue dye present in the form of water pollutant (Umar et al. 2024). ML approaches could also identify the transcriptome signatures for distinguishing between human infection and in vitro cultures of *P. aeruginosa* (Seneviratne et al. 2020).

16.5 Omics Applications to Alginate

Omics technologies (Srivastava et al. 2024a) have helped us better understand microbial biology, particularly in the last 20 years (Kesheri et al. 2024). Alginate biofilms are bacterial communities that use alginate to form a protective barrier. Such biofilm provides several advantages to microorganisms and helps microbial communities survive. Alginate biofilms protect microbial communities under various stress conditions such as nutrient deprivation and antibiotic treatment (Seneviratne et al. 2020).

Transcriptomics refers to the study of transcriptome, which was utilized to study bacterial biofilm and host interactions and drug resistance mechanisms. In a transcriptomics study between drug-treated and untreated biofilms, various differentially expressed genes associated with biofilm formation and drug resistance in *E. faecalis* were investigated (Seneviratne et al. 2017). Significantly increased expression of genes producing penicillin-binding proteins was found in another RNA-seq study on the impact of low ampicillin concentration on S. aureus biofilm formation. (Liu et al. 2018).

Single-cell transcriptomics study could provide a detailed insight into single-cell resolution about the factors responsible for antibiotic/drug resistance in microbial biofilms (Kumar et al. 2024). Prolonged antibiotic exposure induced downregulation of metabolic activity, proliferation and amino acid metabolism in *S. aureus*, which was illustrated by a single cell RNA-seq study. (Peyrusson et al. 2020).

Proteomics is the study of the "proteome" (Kanchan et al. 2020). Utilizing a co-culture model of multilayered gingival epithelium interacting with a 10-species subgingival biofilm, a proteomics study showed that the gingival epithelium released fewer proteins during its interaction with the biofilm (Bostanci et al. 2015).

Metabolomics is the study of the identification and quantification of metabolites present in a biological system. Metabolite profiling could be categorized into either untargeted or targeted approaches (Srivastava et al. 2023). In a study of *P. aeruginosa* biofilm development, different metabolic pathways were discovered (Wan et al. 2018).

Metagenomics is the study of an overall microbial community, which includes extracting the DNA of all the microbes, followed by sequencing of marker genes/whole genomes (Kanchan et al. 2020; Shukla et al. 2024). Metagenomics identifies all the microbes and functions of the genes that participate in and contribute to disease progression (Mallik et al. 2024; Srivastava et al. 2024b). A metagenomics analysis of subgingival biofilm samples revealed unique molecular signatures associated with the progression of periodontitis (Yost et al. 2015). A bioinformatics tool, such as STRING, unravels the protein–protein interaction of algG. During the production of the linear polysaccharide alginate, Pseudomonas aeruginosa alginate-c5-mannuronan-epimerase AlgG is engaged in the epimerization of beta-D-mannuronate to alpha-L-guluronate as illustrated in Fig. 16.4.

16.6 Broad Applications of Alginates

Alginate is an FDA-approved biodegradable, renewable, nonimmunogenic, easy gelation capacity, and biocompatible natural product. Alginate has industrial applications in various fields listed in Table 16.1.

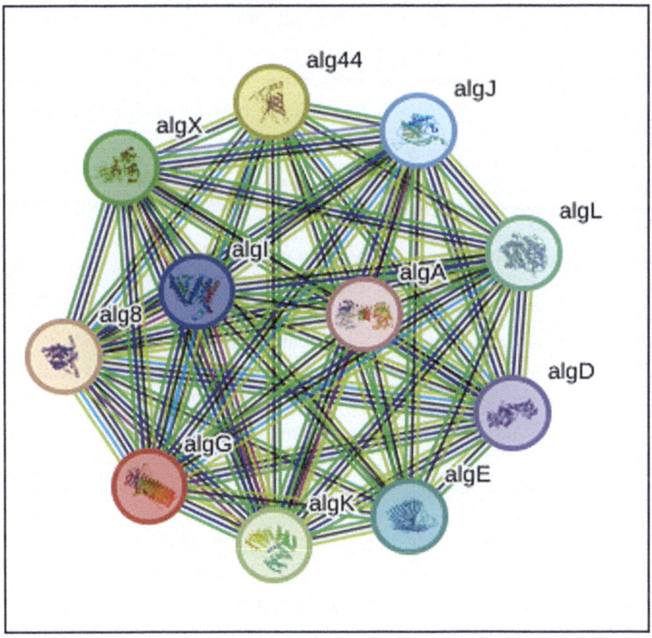

Fig. 16.4 Protein–protein interaction of algG from *Pseudomonas aeruginosa* involved in catalysis of epimerization of beta-D-mannuronate to alpha-L-guluronate during alginate synthesis

16.7 Alginate in Biological Soft Tissue Engineering and Regeneration

Various organs of the human body are composed of soft tissues that can regenerate, and organ transplants are not required in such cases. However, the regeneration of soft tissues is difficult due to the complicated tissue structure and anatomy. Nowadays, hydrogel-based biomaterials are used for the manufacturing of tissue-engineered functional organs. Three-dimensional bioprinting technology uses such biomaterials to manufacture tissue-engineered functional organs. Hydrogels have become popular to be used as soft tissues because of their material, controlled biodegradation, and mechanical properties. The 3D design of a hydrogel as a network is suitable for hydrogel-based tissue-engineered functional organs due to various properties such as crosslinking density and degradation rate that could be measured and controlled for such hydrogel-based functional organs (Hu et al. 2019). Sodium alginate (SA) has become a popular compound to manufacture such hydrogels, and it has become a promising material for use in soft tissue engineering (Pahlevanzadeh et al. 2020). Alginate can be used as a biodegradable material in soft tissue scaffolds. Alginate enhances regeneration in damaged tissues and contributes to various dermatological applications (Amini et al. 2022).

Table 16.1 List of various specific applications of alginate

S. no.	Alginate-specific applications	References
1.	3D bioprinting	Gonzalez-Fernandez et al. (2021)
2.	Adhesives, binders, and building insulation	Lacoste et al. (2018)
3.	Beer foam stabilizer	Jackson (1980)
4.	Biodegradable packaging	Amariei et al. (2022)
5.	Cancer therapy	Mirrahimi et al. (2020)
6.	Controlled drug release	Silva et al. (2016)
7.	Cosmetics	Song and Chen (2018); Langley et al. (2019)
8.	Detergents	Khodagholi et al. (2008)
9.	Food packing materials	Kontominas (2020)
10.	Food supplements	Bi et al. (2022)
11.	Immobilization of enzymes and cells	Mirrahimi et al. (2020)
12.	Scaffolding in tissue engineering	Kostenko et al. (2022)
13.	Thickener	Gaggero et al. (2021)
14.	Biosensors	Tavakoli and Tang (2017)
15.	Dressings for wounds/burns	Ramdhan et al. (2020); Aguero et al. (2021)
16.	Ice cream production	Goraya et al. (2021)

However, Sodium alginate has a few disadvantages. Firstly, Sodium alginate does not offer adequate mechanical support in many demanding requirements and secondly, Sodium alginate deteriorates gradually, which is unstoppable (Hurtado et al. 2022).

16.8 Alginate in Wound Healing and Dressing

Alginates are converted into gels by adding divalent/trivalent cation (calcium, barium, and iron ions) solution; these gels are used for wound healing and dressing (Ramdhan et al. 2020). Among various gel types, Calcium alginate is used as the most popular wound dressing material currently (Aguero et al. 2021). Alginates without G-blocks form softer gels; therefore, such soft gels are popular in the market. Alginate gel-based wound dressing materials are not only available as soft gel but are also available in the form of films, foams, etc. Alginate-based wound dressing materials are considered better compared to other wound dressing materials because they absorb large amounts of wound fluids and minimize bacterial infections, which could also stay as a gel for a longer duration at the wound site.

16.9 Alginates in Food Industry

Alginates are used as a packaging material and for the preservation of various food commodities as reported in many literature. Alginates are natural polymers, biodegradable, safe, and environmentally friendly, and therefore, alginates could also be used as food packing material. Alginates are used to increase the shelf life of freshly cut potato strips and apples (Amaral et al. 2017); pineapple, peaches (Maftoonazad et al. 2008), vegetables and melons (Rojas-Grau et al. 2009); and Mozzarella cheese (Mastromatteo et al. 2014). Alginates are also used in freshly cut mangoes for storage (Chiumarelli et al. 2011) and for the quality retention of sweet cherries (Chiabrando and Giacalone 2015).

A few drawbacks are associated with alginates being used as food packaging materials, which must be overcome to use alginates as a popular food packaging material. To improve the high-water vapor transmission rate, cellulose nanoparticles or montmorillonite clay are added to the alginate. The other improvement was made by the addition of $CaCl_2$ to the alginate matrix for the improvement of various properties such as mechanical strength, water vapor transmission rate, and water resistance. Glycerol and sorbitol were also added to the alginate matrix to improve the flexibility of the material. To improve the shelf life, silver-montmorillonite nanoparticles were used in the alginate-based material due to being antimicrobial. Polylactic acid was added to the alginate matrix to increase oxygen barrier properties. Lipids such as corn/olive oil are added to alginate to minimize water loss. Organic acids are added to the alginate matrix to protect foods by inhibiting the growth of food pathogens such as *Listeria monocytogenes*, *Escherichia coli*, and *Salmonella typhimurium*.

16.10 Alginate in Drug Delivery and Cancer Treatment

Due to having various properties such as biocompatibility, chelating ability, nonimmunogenicity, being cheaper in terms of cost, and water solubility, alginates have become popular materials to be used in various industries for different purposes. Alginates as hydrogels are hydrophilic polymers that crosslink together to form 3D networks that absorb water or other fluids to maintain their 3D network structure. The resulting 3D cross-linked hydrogel was useful in the controlled release of the bioactive molecules (Blandino et al. 2000; Lamas et al. 2001). In addition, due to the biocompatibility of alginate with human tissue and long-term and better-controlled drug delivery, alginate-based wound dressings could be used as a drug carrier to treat mucosal tissue injuries that require (Severino et al. 2019).

Alignate with properties associated with drug delivery could be added by blending with certain materials making alginates an excellent material for drug delivery. Due to the high water absorption capability of alginate, it allows sustained and gradual release of the oral drug by forming a hydrogel via diffusion upon contact with gastrointestinal fluids. Alginate could be used to deliver the drug (active ingredients) into the gastrointestinal tract without releasing it into the stomach.

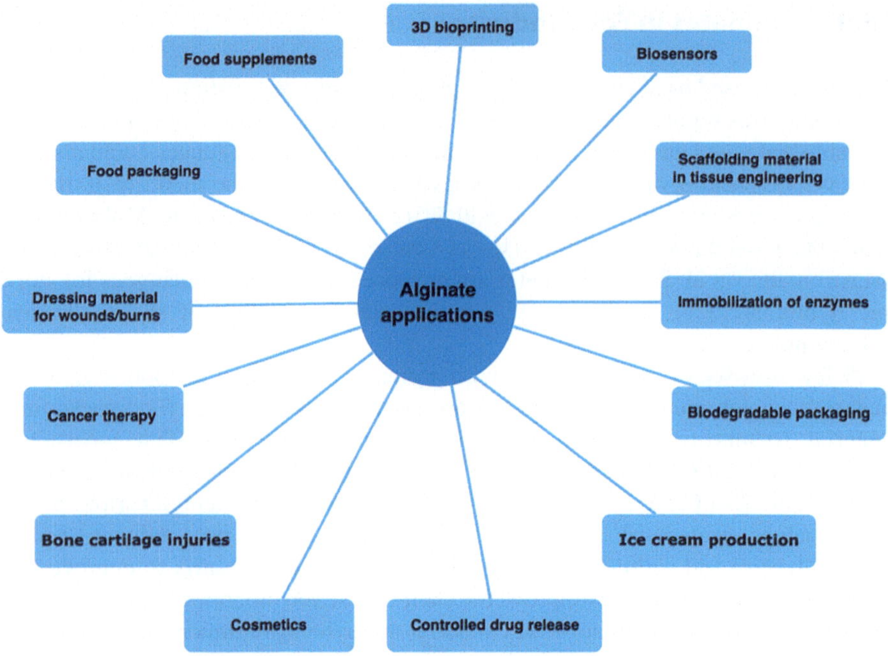

Fig. 16.5 Applications of alginates in biomedical, pharmaceutical, food, environmental sciences, etc.

Alginate-based hydrogels are also used to deliver growth factors for better angiogenesis in engineered constructs after implantation.

Cancer is a major public health concern worldwide. Chemotherapy, radiation therapy, or both are used primarily in cancer treatment. The drugs used in chemotherapy are nonselective and due to the poor bioavailability of such drugs to the tumor site, it is not 100% effective. Various cancer drugs cause severe toxicity to normal cells becomes the main reason for the high mortality rate of cancer patients. Hydrogels increase bioavailability, improve treatment efficacy, and avoid toxicity to normal cells via enhanced permeability and retention. Alginate hydrogels include various outstanding properties, such as high-water retention, nontoxicity, biodegradability, and biocompatibility, making them good drug carriers. Such hydrogels deliver low molecular weight drugs and macromolecules such as peptides to the target site.

Depending on the pH, alginate hydrogels act differently. At low pH (as in the stomach), hydrogel shrinks and produces a viscous acidic gel (which does not release its encapsulated drugs). On the other hand, at higher pH, for example, in the gastrointestinal tract, the alginate is converted to a soluble viscous gel, and drug dissolution and release to the target site take place. This helps in controlling systemic toxicity due to the release of the drug over a sustained period. In 2015, alginate-PAMAM dendrimer (AG-G5) hybrid nanogel was investigated as a suitable material for anticancer drug delivery. Furthermore, the in vitro cytotoxicity study of the AG-G5 nanogels illustrated the induced cell death through apoptosis (Matai and Gopinath 2016). Bevacizumab-loaded calcium alginate hydrogel was observed to be effective for anti-VEGF cancer therapy. This hydrogel was considered to improve antiangiogenic and antineoplastic activity and was observed as pH-independent present in many tumors. A triple combination of heat, drug, and radiation uses alginate hydrogel co-loaded with gold nanoparticles, which is used for cancer therapy (Mirrahimi et al. 2020). A list of usage of alginates in biomedical, pharmaceutical, food, environment, etc., is shown in Fig. 16.5.

16.11 Conclusion

Alginates are polysaccharides secreted by bacterial, algal, and brown algal species. Secretion of alginates is responsible for biofilm development around lung diseases which protects the microbial community from antibiotics, etc. Polymerization, acetylation, and epimerization events are responsible for the production of different varieties of alginates having different biochemical properties. ML-based investigations leverage the production of different varieties of alginates. Various omics studies could also be used to explore biofilm development, host biofilm interaction, antibiotic resistance, etc. Moreover, alginates have broad biomedical, pharmaceutical, food, and environmental applications in various industries.

Chapter Summary

This chapter explains the discovery, various sources, biosynthesis, structure, and properties of alginates. This chapter also explains various ways to design tailor-made alginates by modifying the acetylation and epimerization processes or by including different numbers of M and G units for desired physicochemical properties. ML applications are also explained which could be used to modify alginates of industrial importance to be used as wound and dressing material, soft tissue engineering material, food packaging material, drug delivery material useful for cancer, etc.

Author Contributions Concept, design, and writing, Minu Kesheri and Swarna Kanchan; Literature review, Bhagaban Mallik and Tarun Mishra. Review and editing, Rajeshwar P. Sinha, Ratnaprabha Ratna-Raj, and Bhaskar Chittoori. Minu Kesheri and Swarna Kanchan contributed equally as the first author.

References

Aguero L, Alpdagtas S, Ilhan E, Zaldivar-Silva D, Gunduz O (2021) Functional role of cross-linking in alginate scaffold for drug delivery and tissue engineering: a review. Eur Polym J 160:110807

Amaral RDA, Achaerio I, Benedetti BC, Pujola M (2017) The influence of edible coatings, blanching and ultrasound treatments on quality attributes shelf life of vacuum packaged potato strips. LWT Food Sci Technol 85:449–455

Amariei S, Ursachi F, Petraru A (2022) Development of new biodegradable agar-alginate membranes for food packaging. Membranes 12(6):576

Amini N, Milan PB, Sarmadi VH, Derakhshanmehr B, Hivechi A, Khodaei F, Hamidi M, Ashraf S, Larijani G, Rezapour A (2022) Microorganism-derived biological macromolecules for tissue engineering. Front Med 16(3):358–377

Baker P, Ricer T, Moynihan PJ, Kitova EN, Walvoort MT, Little DJ, Whitney JC, Dawson K, Weadge JT, Robinson H, Ohman DE (2014) *P. aeruginosa* SGNH hydrolase-like proteins AlgJ and AlgX have similar topology but separate and distinct roles in alginate acetylation. PLoS Pathog 10(8):e1004334

Baum LE, Eagon JA (1967) An inequality with applications to statistical estimation for probabilistic functions of Markov processes and to a model for ecology. Bull Am Math Soc 73:360–363

Bi D, Yang X, Yao L, et al (2022) Potential food and nutraceutical applications of alginate: A Review. Mar Drugs 20(9):564. Published 2022 Sep 4. https://doi.org/10.3390/md20090564

Blandino A, Macias M, Cantero D (2000) Glucose oxidase release from calcium alginate gel capsules. Enzym Microb Technol 27(3–5):319–324

Boser BE, Guyon IM, Vapnik VNA (1992) Proceedings of the fifth annual workshop on computational learning theory, pp 144–152

Bostanci N, Bao K, Wahlander A, Grossmann J, Thurnheer T, Belibasakis GN (2015) Secretome of gingival epithelium in response to subgingival biofilms. Mol Oral Microbiol 30(4):323–335

Chiabrando V, Giacalone G (2015) Effects of alginate edible coating on quality and antioxidant properties in sweet cherry during postharvest storage. Ital J Food Sci 27:173–180

Chitnis CE, Ohman DE (1993) Genetic analysis of the alginate biosynthetic gene cluster of Pseudomonas aeruginosa shows evidence of an operonic structure. Mol Microbiol 8(3):583–590

Chiumarelli M, Ferrari CC, Sarantopoulos CI, Hubinger MD (2011) Fresh cut "Tommy Atkins" mango pre-treated with citric acid coated with cassava (*Manihot esculenta* Crantz) starch or sodium alginate. Innov Food Sci Emerg Technol 12:381–387

Clare K (1993) Algin. In: Industrial gums. Academic Press, pp 105–143

Ege D, Boccaccini AR (2024) Investigating the effect of processing and material parameters of alginate dialdehyde-gelatin (ADA-GEL)-based hydrogels on stiffness by XGB machine learning model. Bioengineering 11(5):415

Ertesvåg H (2015) Alginate-modifying enzymes: biological roles and biotechnological uses. Front Microbiol 6:523

Feehan R, Montezano D, Slusky JSG (2021) ML for enzyme engineering, selection and design. Protein Eng Des Sel 34:gzab019

Franklin MJ, Ohman DE (2002) Mutant analysis and cellular localization of the AlgI, AlgJ, and AlgF proteins required for O-acetylation of alginate in *Pseudomonas aeruginosa*. J Bacteriol 184(11):3000–3007

Fu S, Thacker A, Sperger DM, Boni RL, Buckner IS, Velankar S, Munson EJ, Block LH (2011) Relevance of rheological properties of sodium alginate in solution to calcium alginate gel properties. Aaps Pharm Sci Tech 12:453–460

Gaggero G, Delucchi M, Allegretta G, Vicini S, Botter R (2021) Interaction of sodium alginate thickener with components of architectural water-based coatings. Prog Org Coat 151:106016

Gimmestad M, Sletta H, Ertesvåg H, Bakkevig K, Jain S, Suh S-j, Skjåk-Bræk G, Ellingsen TE, Ohman DE, Valla S (2003) The *Pseudomonas fluorescens* AlgG protein, but not its mannuronan C-5-epimerase activity, is needed for alginate polymer formation. J Bacteriol 185(12):3515–3523

Gonzalez-Fernandez T, Tenorio AJ, Campbell KT, Silva EA, Leach JK (2021) Alginate-based bioinks for 3D bioprinting and fabrication of anatomically accurate bone grafts. Tissue Eng Part A 27(17–18):1168–1181

Goraya RK, Singla M, Bajwa U, Kaur A, Pathania S (2021) Impact of sodium alginate gelling and ingredient amalgamating order on ingredient interactions and structural stability of ice cream. LWT 147:111558

Gorin P, Spencer J (1966) Exocellular alginic acid from *Azotobacter vinelandii*. Can J Chem 44(9):993–998

Haug A, Smidsrød O (1967) Strontium-calcium selectivity of alginates. Nature 215(5102):757–757
Haug A, Smidsrod O (1970) Selectivity of some anionic polymers for divalent metal ions. Acta Chem Scand 24(3):843–854
Hay ID, Rehman ZU, Moradali MF, Wang Y, Rehm BHA (2013) Microbial alginate production, modification and its applications. Microb Biotechnol 6:637–650
Hay ID, Wang Y, Moradali MF, Rehman ZU, Rehm BH (2014) Genetics and regulation of bacterial alginate production. Environ Microbiol 16(10):2997–3011
Hu W, Wang Z, Xiao Y, Zhang S, Wang J (2019) Advances in crosslinking strategies of biomedical hydrogels. Biomater Sci 7:843–855
Hurtado A, Aljabali AA, Mishra V, Tambuwala MM, Serrano-Aroca Á (2022) Alginate: enhancement strategies for advanced applications. Int J Mol Sci 23:4486
Jain S, Franklin MJ, Ertesvåg H, Valla S, Ohman DE (2003) The dual roles of AlgG in C-5-epimerization and secretion of alginate polymers. Mol Microbiol 47(4):1123–1133
Jackson G, Roberts RT, Wainwright T (1980) Mechanism of beer foam stabilization by propylene glycol alginate. J Inst Brew, 86(1): 34–37. Accessed May 20, 2025. https://doi.org/10.1002/j.2050-0416.1980.tb03953.x
Kanchan S, Sinha RP, Chaudière J, Kesheri M (2020) Computational metagenomics: current status and challenges. In: Katara P (ed) Recent trends in 'computational omics: concepts and methodology. Nova Science Publishers, pp 371–395
Kanchan S, Ogden E, Kesheri M, Skinner A, Miliken E, Lyman D, Armstrong J, Sciglitano L, Hampikian G (2024) COVID-19 hospitalizations and deaths predicted by SARS-CoV-2 levels in Boise, Idaho wastewater. Sci Total Environ 907:167742
Kesheri M, Kanchan S, Chowdhury S, Sinha RP (2015) Secondary and tertiary structure prediction of proteins: a bioinformatic approach. In: Zhu Q, Azar AT (eds) Complex system modelling and control through intelligent soft computations, vol 319. Springer, pp 541–569
Kesheri M, Kanchan S, Salisbury TB, Sinha RP (eds) (2024) Microbial omics in environment and health. Springer Nature Singapore
Khodagholi F, Eftekharzadeh B, Yazdanparast R (2008) A new artificial chaperone for protein refolding: sequential use of detergent and alginate. Protein J 27(2):123-129. https://doi.org/10.1007/s10930-007-9115-y
Kontominas MG (2020) Use of alginates as food packaging materials (9)1440. https://doi.org/10.3390/foods9101440
Kostenko A, Swioklo S, Connon CJ (2022) Alginate in corneal tissue engineering. Biomed Mater.17(2). https://doi.org/10.1088/1748-605X/ac4d7b
Kumar P, Kanchan S, Kesheri M (2024) Multi-omics in human disease biomarker discovery. In: Kesheri M, Kanchan S, Salisbury TB, Sinha RP (eds) Microbial omics in environment and health. Springer Nature Singapore, pp 205–239
Lacoste C, El Hage R, Bergeret A, Corn S, Lacroix P (2018) Sodium alginate adhesives as binders in wood fibers/textile waste fibers biocomposites for building insulation. Carbohydr Polym 184:1–8
Lamas MC, Bregni C, Daquino M, Degrossi J, Firenstein R (2001) Calcium alginate microspheres of Bacillus subtilis. Drug Dev Ind Pharm 27(8):825–829
Langley N, Michniak-Kohn B, Osborne DW (eds) (2019) The role of microstructure in topical drug product development. Springer International Publishing. ISBN 978-3-030-17354-8
Lee KY, Mooney DJ (2012) Alginate: properties and biomedical applications. Prog Polym Sci 37:106–126
Linker A, Jones RSA (1964) Polysaccharide resembling alginic acid from a *Pseudomonas* microorganism. Nature 204:187–188
Linker A, Jones RSA (1966) New polysaccharide resembling alginic acid isolated from *Pseudomonads*. J Biol Chem 241:3845–3851
Liu J, Yang L, Hou Y, Soteyome T, Zeng B, Su J, Li L, Li B, Chen D, Li Y, Wu A (2018) Transcriptomics study on *Staphylococcus aureus* biofilm under low concentrations of ampicillin. Front Microbiol 9:2413

Maftoonazad N, Ramaswamy HS, Marcotte M (2008) Shelf-life extension of peaches through sodium alginate and methyl cellulose edible coatings. Int J Food Sci Technol 43:951–957

Mallik B, Mishra T, Dubey P, Kesheri M, Kanchan S (2024) Exploring the secrets of microbes: unveiling the hidden world through microbial omics in environment and health. In: Kesheri M, Kanchan S, Salisbury TB, Sinha RP (eds) Microbial omics in environment and health. Springer, Singapore, pp 269–294

Mastromatteo M, Conte A, Faccia M, del Nobile MA, Zambrini AV (2014) Combined effect of active coating and modified atmosphere packaging on prolonging the shelf life of low-moisture Mozzarella cheese. J Dairy Sci 97:36–45

Matai I, Gopinath P (2016) Chemically cross-linked hybrid nanogels of alginate and PAMAM dendrimers as efficient anticancer drug delivery vehicles. ACS Biomater Sci Eng 2(2):213–223

McCulloch WS, Pitts W (1943) A logical calculus of the ideas immanent in nervous activity. Bull Math Biophys 5:115–133

Mirrahimi M, Beik J, Mirrahimi M, Alamzadeh Z, Teymouri S, Mahabadi VP, Eslahi N, Tazehmahalleh FE, Ghaznavi H, Shakeri-Zadeh A, Moustakis C (2020) Triple combination of heat, drug and radiation using alginate hydrogel co-loaded with gold nanoparticles and cisplatin for locally synergistic cancer therapy. Int J Biol Macromol 158:617–626

Moradali MF, Donati I, Sims IM, Ghods S, Rehm BH (2015) Alginate polymerization and modification are linked in *Pseudomonas aeruginosa*. MBio 6(3):10–1128

Moradali MF, Ghods S, Rehm BH (2017) Activation mechanism and cellular localization of membrane-anchored alginate polymerase in *Pseudomonas aeruginosa*. Appl Environ Microbiol 83:03499–03416

Pahlevanzadeh F, Mokhtari H, Bakhsheshi-Rad HR, Emadi R, Kharaziha M, Valiani A, Poursamar SA, Ismail AF, Rama Krishna S, Berto F (2020) Recent trends in three-dimensional bioinks based on alginate for biomedical applications. Materials 13:3980

Peyrusson F, Varet H, Nguyen TK, Legendre R, Sismeiro O, Coppée JY, Wolz C, Tenson T, Van Bambeke F (2020) Intracellular *Staphylococcus aureus* persisters upon antibiotic exposure. Nat Commun 11(1):2200

Priya P, Kesheri M, Sinha RP, Kanchan S (2017) Molecular dynamics simulations for biological systems. In: Karâa WBA, Dey N (eds) . Pharmaceutical sciences: Breakthroughs in research and practice, IGI Global, pp 1044–1071

Ramdhan T, Ching SH, Prakash S, Bhandari B (2020) Physical and mechanical properties of alginate-based composite gels. Trends Food Sci Technol 106:150–159

Rojas-Grau M, Oms-Oliu G, Soliva-Fortuny R, Martin-Belloso O (2009) The use of packaging techniques to maintain freshness in fresh-cut fruits vegetables: a review. Int J Food Sci Technol 4:875–889

Seneviratne CJ, Suriyanarayanan T, Swarup S, Chia KHB, Nagarajan N, Zhang C (2017) Transcriptomics analysis reveals putative genes involved in biofilm formation and biofilm-associated drug resistance of *Enterococcus faecalis*. J Endodont 43(6):949–955

Seneviratne CJ, Suriyanarayanan T, Widyarman AS, Lee LS, Lau M, Ching J, Delaney C, Ramage G (2020) Multi-omics tools for studying microbial biofilms: current perspectives and future directions. Crit Rev Microbiol 46(6):759–778

Severino P, da Silva CF, Andrade LN, de Lima Oliveira D, Campos J, Souto EB (2019) Alginate nanoparticles for drug delivery and targeting. Curr Pharm Des 25:1312–1334

Shukla N, Bhatia A, Kesheri M, Kanchan S, Narayan J (2024) Planet in peril: unveiling the microbial consequences of environmental degradation on human health through omics approaches. In: Kesheri M, Kanchan S, Salisbury TB, Sinha RP (eds) Microbial omics in environment and health. Springer, Singapore, pp 35–56

Silva D, Pinto LFV, Bozukova D, Santos LF, Serro AP, Saramago B (2016) Chitosan/alginate based multilayers to control drug release from ophthalmic lens. Colloids Surf B Biointerfaces 147:81–89

Skjåk-Bræk G, Paoletti S, Gianferrara T (1989) Selective acetylation of mannuronic acid residues in calcium alginate gels. Carbohydr Res 185(1):119–129

Solomon NO, Kanchan S, Kesheri M (2024) Nanoparticles as detoxifiers for industrial wastewater. Water Air Soil Pollut 235(3):214

Song J, Chen H (2018) Preparation of aroma microcapsules with sodium alginate and tetradecylallyldimethylammonium bromide (TADAB) and its potential applications in cosmetics. Flavour Fragr J 33:160–165

Srivastava U, Kanchan S, Kesheri M, Singh S (2023) Nutrimetabolomics: metabolomics in nutrition research. In: Soni V, Hartman TE (eds) Metabolomics. Springer, Cham, pp 241–268

Srivastava U, Kanchan S, Kesheri M, Gupta MK, Singh S (2024a) Types of omics data: genomics, metagenomics, epigenomics, transcriptomics, proteomics, metabolomics, and phenomics. In: Gupta MK, Katara P, Mondal S, Singh RL (eds) Integrative omics: concepts, methodology, and application. Academic Press (Elsevier), pp 13–34

Srivastava U, Kesheri M, Kanchan S, Singh S (2024b) Computational omics protocol for the comparative study of microbiome analysis. In: Kesheri M, Kanchan S, Salisbury TB, Sinha RP (eds) Microbial omics in environment and health. Springer Nature Singapore, pp 109–133

Stanford ECC (1883) On align: a new substance obtained from some of the commoner species of marine algae. Chem News 47:254–257

Tavakoli J, Tang Y (2017) Hydrogel based sensors for biomedical applications: an updated review. Polymers (Basel) 9(8):364

Umar M, Khan H, Hussain S, Arshad M, Choi H, Lima EC (2024) Integrating DFT and ML for the design and optimization of sodium alginate-based hydrogel adsorbents: efficient removal of pollutants from wastewater. Environ Res 247:118219

Urtuvia V, Maturana N, Acevedo F, Peña C, Díaz-Barrera AB (2017) Bacterial alginate production: an overview of its biosynthesis and potential industrial production. World J Microbiol Biotechnol 33:1–10

Wan N, Wang H, Ng CK, Mukherjee M, Ren D, Cao B, Tang YJ (2018) Bacterial metabolism during biofilm growth investigated by 13C tracing. Front Microbiol 9:2657

Whitney JC, Whitfield GB, Marmont LS, Yip P, Neculai AM, Lobsanov YD, Robinson H, Ohman DE, Howell PL (2015) Dimeric c-di-GMP is required for post-translational regulation of alginate production in *Pseudomonas aeruginosa*. J Biol Chem 290(20):12451–12462

Wolfram F, Kitova EN, Robinson H, Walvoort MT, Codée JD, Klassen JS, Howell PL (2014) Catalytic mechanism and mode of action of the periplasmic alginate epimerase AlgG. J Biol Chem 289(9):6006–6019

Wong TY, Preston LA, Schiller NL (2000) Alginate lyase: review of major sources and enzyme characteristics, structure-function analysis, biological roles, and applications. Ann Rev Microbiol 54(1):289–340

Xu S, Chen X, Wang S, Chen Z, Pan P, Huang Q (2024) Integrating ML for the optimization of polyacrylamide/alginate hydrogel. Regen Biomater 11:rbae109

Yost S, Duran-Pinedo AE, Teles R, Krishnan K, Frias-Lopez J (2015) Functional signatures of oral dysbiosis during periodontitis progression revealed by microbial metatranscriptome analysis. Genome Med 7(1):1–19

Zhang H, Cheng J, Ao Q (2021) Preparation of alginate-based biomaterials and their applications in biomedicine. Mar Drugs 19:264

Proteomics and Bioinformatics Approaches for Exploring Resilience Strategies in Cyanobacteria

17

Minu Kesheri, Swarna Kanchan, Donat-P. Häder, and Rajeshwar P. Sinha

Abstract

This chapter envisages the responses of cyanobacteria to different abiotic stresses with special reference to the cyanobacterium *Scytonema tolypothrichoides* VB-61278. We describe the in silico approach to compare Fe and Mn Super Oxide Dismutase (SOD) for deciphering their role in promoting resilience in *Scytonema tolypothrichoides* VB-61278. Various physicochemical parameters such as molecular weight, instability index, isoelectric point (pI), aliphatic index, and GRand Average of Hydropathy (GRAVY) were predicted and explained in an illustrative manner. The secondary structure of the protein was predicted followed by the prediction of three-dimensional (3D) structural models using Modeller and trailblazing artificial intelligence (AI)-powered AlphaFold2 tools have been illustrated. Exhaustive explanations of validation methods for the models using Prosa-web, Verify-3D, Ramachandran plot, and LDDT score have been highlighted. Studies emphasizing the conservation and prediction of metal binding positions, domains, and motifs, along with residues of structural and functional importance, as well as the generation of phylogenetic trees, have been well explained. This chapter serves as a state-of-the-art about methodologies,

M. Kesheri · S. Kanchan
Biology Department, Boise State University, Boise, ID, USA

Joan C. Edwards School of Medicine, Marshall University, Huntington, WV, USA

D.-P. Häder
Emeritus from Friedrich-Alexander University, Department of Biology, Möhrendorf, Germany

R. P. Sinha (✉)
Laboratory of Photobiology and Molecular Microbiology, Department of Botany, Institute of Science, Banaras Hindu University, Varanasi, India

M. Kesheri et al. (eds.), *Multi-Omics in Biomedical Sciences and Environmental Sustainability*, https://doi.org/10.1007/978-981-96-7067-3_17

ensuing results, and inferences drawn from tools used in bioinformatics to decipher protein modeling and relevant biotechnological applications in a lucid and illustrative way.

Keywords

AlphaFold · Antioxidants · Cyanobacteria · Homology modeling · Superoxide dismutase

Objectives

- Aims to summarize the predicted physicochemical properties, secondary structure, tertiary structure, motifs, and evolutionarily conserved residues in superoxide dismutase protein.
- Bioinformatics helps us identify drug targets and predict drug–target interactions.
- Aims to highlight the computational strategies, methods, and techniques for bioinformatics analysis of protein structures.

17.1 Introduction

Cyanobacteria possess the potential to survive the harmful effects of climate change and other biotic and abiotic stress responses owing to an array of defense mechanisms. Being ubiquitous, cyanobacteria, along with other microorganisms, have evolved mechanisms to eradicate the harmful effects of climate change and play an important role in various aquatic, terrestrial, and agricultural ecosystems (Bastiaansen et al. 2020). Moreover, continuous cyanobacterial blooms throughout the world contribute to eutrophication and global climate change (Lürling et al. 2017; Huisman et al. 2018; Burberg et al. 2020; Häder and Gao 2023). Deciphering the temporal dynamics of cyanobacterial communities might help us comprehend our readiness toward the potential incidence of toxic blooms (Kesheri et al. 2021; Carratala et al. 2023). Cyanobacteria in their natural habitats are frequently challenged with numerous eco-physiological stresses such as UV-radiations, drought, salinity, extremes of temperatures, and heavy metal toxicity at their natural habitat (Kesheri and Richa 2011; Rajneesh et al. 2019). Various abiotic stresses, as well as certain developmental and physiological processes, stimulate the generation of various reactive oxygen species (ROS) such as superoxide ion, hydroxyl radical, hydrogen peroxide (H_2O_2), and singlet oxygen that, in turn, are responsible for causing oxidative stress (Kesheri et al. 2021). ROS are highly reactive free radicals and are produced as by-products of general cellular metabolism such as respiration, photosynthesis, growth, and various developmental processes (Oliveira et al. 2019). Moreover, several abiotic factors also lead to the generation of ROS leading to oxidative stress in the cell (Kesheri et al. 2014). Inequity between the generation of ROS and their hunting by antioxidants is responsible for oxidative stress and disruption in redox homeostasis. Accumulation of ROS leads to various modifications in biomolecules such as DNA, proteins, metabolites, and lipids (Foyer and Noctor 2005; Sabadin et al. 2019). Excessive irradiation exceeding the photosynthesis capacity culminates in photooxidative damage (Latifi et al. 2009). However, cyanobacteria possess a strong resilience power comprising several layers of defense mechanisms (Kesheri and Richa 2011).

The abiotic and biotic stresses arising because of human activities as well as global climate change serve as potential threats to agricultural production and lead to global food insecurity (Sharma et al. 2022). Cyanobacteria may serve as a crucial candidate for the development of adaptive microbial communities (SynComs), an evolving scientific approach to designing inoculants that would enrich sustainable agricultural practices by incorporating concepts from microbiology and genetics (de Souza et al. 2020). *Nostoc calcicola* HN9 extract, containing enzymatic as well as nonenzymatic antioxidants, enhanced the adaptive perspective to protect soybeans from oxidative stress and thereby elevated the resistance of *G. max* cv. "Nam Dan" to *A. craccivora* (Mai et al. 2017).

Trailblazing developments in cyanobacterial research have been accomplished by setting foot in the "omics era." Advances in genomics provided significant insights into cyanobacterial genome plasticity (Chen et al. 2021). Transcriptomics not only opened doors to cyanobacterial resilience strategies to an array of biotic and abiotic stresses but also aided in targeting regulatory tools in the form of transcription factors for regulating gene expression in cyanobacteria (Gordon and Pfleger 2018; Farias et al. 2024). Research envisaging transcription factors has enhanced the industrial production of cyanobacterial compounds crucial to producing biodiesel, antioxidative, feed, and food supplements (Gupta et al. 2024). Harnessing metabolomics and meta-transcriptomics analysis paved the way for novel druggable targets (Lin et al. 2014; Sahu et al. 2023), bioactive compounds, and biomarkers pivotal to biomedical and environmental sustainability (Le et al. 2024). Coupling various omics approaches in synergy with bioinformatics incorporating data mining, next-generation sequencing (NGS), a quality check of the sequence reads, sequence assembly, metagenomics (Kanchan et al. 2020), protein modeling (Kesheri et al. 2024) and docking, gene prediction, and functional assignment, taxonomic binning, genetic diversity, metabolic pathway reconstruction, and pipeline development to analyze micro genomic data shall leverage vast avenues for identification of beneficial microbial traits in improvising sustainable agricultural practices (Kumari et al. 2016; Kanchan et al. 2020; Cai et al. 2023), improved biomedical research and well informed as well as robust health care system. In this novel era of artificial intelligence (AI), cyanobacterial research has been immensely impacted by various machine learning (ML) tools (Ning et al. 2022) such as artificial neural networks, deep learning, random forest, convolutional neural networks, support vector machines, and recurrent neural networks that have been used for the predictive modeling of transcription factor binding sites elucidating molecular mechanisms unveiling interactions between transcription factors and genetic material such as DNA (Farias et al. 2024).

This chapter discusses various ecophysiological stresses faced by cyanobacteria in their natural habitat and the defense layers destined to protect these organisms from the vagaries of nature and their physiological stresses due to growth and development. Moreover, recent advances in methodologies, ensuing results, and inferences drawn from the state-of-the-art tools used in bioinformatics to decipher protein modeling and relevant information crucial to biotechnological applications have been highly emphasized in an illustrative and lucid style.

17.2 Defense Barriers in Cyanobacteria

Four layers of defense shields enable cyanobacteria to endure ecophysiological stresses. Avoidance or escapist strategy forms the first and foremost rescue comprising mat formation, migration, morphological changes, amorphous silica matrices for enclosing cyanobacterial filaments, and production of extracellular polysaccharides. *Scytonema tolypothrichoides* is a mesophilic cyanobacterium, and as an adaptation to desiccation owing to extremely high temperature and enhanced irradiation (Dabravolski and Isayenkov 2022), it produces and accumulates extracellular polysaccharides (EPS) as a first layer of defense strategy (Kvíderová et al. 2018). These EPS molecules, bestowed with the property to bind with water, play an important role in desiccation tolerance as well as carbon sink during unbalanced growth (Otero and Vincenzini 2004). EPS also assists sunscreen pigments such as scytonemin, which safeguards the cyanobacteria from harmful UV radiation (Häder 2001).

Antioxidant scavengers comprising enzymatic and nonenzymatic antioxidants confer a second layer of shielding against oxidative stress. The significant role of enzymatic and nonenzymatic antioxidants as an effective weapon against several stresses has been reported in various cyanobacterial species in their natural habitat. Antioxidative enzymes comprise catalase, superoxide dismutase, ascorbic peroxidase, and glutathione reductase. Nonenzymatic antioxidants comprise phycobiliproteins, carotenoids, and certain sunscreen pigments such as scytonemin. In *Nostoc commune* enzymatic antioxidants such as catalase and superoxide dismutase have been shown to confer shielding against the extreme cold temperature in their natural habitat in Antarctica (Kesheri et al. 2014). *Oscillatoria* sp. inhabiting sewage water also possesses catalase and superoxide dismutase as an arsenal to combat stresses arising due to temporal variations faced by them in their natural habitat (Kesheri et al. 2021). Extensive screening by photoprotective compounds (Priyadarshani and Rath 2012) and those such as biopterin glucoside confer a third layer of shielding against stresses such as those caused by harmful UV-B radiations. Repair and resynthesis form the last and final rescue shield of the defense mechanisms (Irankhahi et al. 2024).

17.3 In Silico Physicochemical Characterization

The complete protein sequences of Fe-SOD as well as Mn-SOD in *Scytonema tolypothrichoides* VB-61278 were collected from the protein databank of the National Centre for Biotechnology Information (NCBI) protein database. Physicochemical properties such as number of amino acids, molecular weight, theoretical isoelectric point (pI), aliphatic index, grand average of hydropathicity (GRAVY), and instability index were predicted for Fe-SOD and Mn-SOD from cyanobacteria *Scytonema tolypothrichoides* VB-61278 protein sequences using ExPASy-ProtParam tool (Gasteiger et al. 2005). Fe-SOD protein (Genbank accession no. KAB8334667.1) and Mn-SOD protein (Genbank accession no. *KAB8335432.1*) from *Scytonema tolypothrichoides* VB-61278 were 200 and 247 amino acids long, respectively. Physicochemical parameters for molecular weight, pI, instability index, aliphatic index, and GRAVY were predicted as 22290.84 Da, 5.06, 24.33, 79.65, and −0.421, respectively, for Fe-SOD whereas 27463.00 Da, 9.47, 35.01, 72.02 and −0.536, respectively, were predicted for Mn-SOD in *Scytonema tolypothrichoides* VB-61278. Secondary structure elements and their arrangements in Fe-SOD and Mn-SoD in *Scytonema tolypothrichoides* VB-61278 were predicted in terms of α-helices, β-sheets, and β turns using SOPMA (Self-Optimized Prediction Method with Alignment) available at https://npsa-prabi.ibcp.fr/cgi-bin/npsa_automat.pl?page=/NPSA/npsa_sopma.html (Geourjon and Deleage 1995).

17.4 Homology Modeling and Model Structure Validation

Fe-SOD and Mn-SOD protein sequences (*Scytonema tolypothrichoides* VB-61278) may be used as a query for retrieving homologous sequences from the Protein Data Bank (PDB) database using the blastp program (https://blast.ncbi.nlm.nih.gov/Blast.cgi) by keeping the rest of the parameters as default (Altschul et al. 1997). Template protein was selected based on the highest percent identity and coverage with the query protein in blastp search. Based on the sequence alignment, the three-dimensional (3D) structural protein models of Fe-SOD and Mn-SOD proteins in *Scytonema tolypothrichoides* VB-61278 may be constructed using homology modeling by MODELLER 9v23 program (Sali and Blundell 1993). The coordinates of the 3D structure of the template protein were collected from the protein data bank available at http://www.rcsb.org/pdb/. The 3D protein structural models of FESOD and MNSOD in *Scytonema tolypothrichoides* VB-61278 were visualized by UCSF Chimera version 1.14 (Pettersen et al. 2004). The structural qualities of 3D structural models of Fe-SOD and MN-SOD in *Scytonema tolypothrichoides* VB-61278 may be validated utilizing various computational programs. Protein 3D structural superposition and root mean square deviation (RMSD) between target and template 3D proteins may be performed using UCSF Chimera (Pettersen et al. 2004). The stereochemical quality of 3D models of Fe-SOD and Mn-SOD proteins in *Scytonema tolypothrichoides* VB-61278 may be validated through the generation of Ramachandran plot by PROCHECK (Laskowski et al. 1993) and Verify 3D (Bowie et al. 1991; Lüthy et al. 1992) programs. The structural quality of the model may be further evaluated concerning stereochemical geometry and energy by the ProSA-web (Wiederstein and Sippl 2007).

The homology model for Fe-SOD in *Scytonema tolypothrichoides* was built for almost the entire protein length spanning from residue no. 2 to 198 amino acid positions. Primary pairwise sequence alignment by blastp between target protein Fe-SOD (*Scytonema tolypothrichoides)* indicated 65.3% identity with template Fe-SOD (*Thermosynechococcus elongatus* having PDB ID: 1MY6, chain A). Similarly, the homology model for Mn-SOD in *Scytonema tolypothrichoides* was built for the catalytic domain region spanning from residue no. 25 to 233 amino acid positions. Primary pairwise sequence alignment between Mn-SOD observed 64% identity with template Mn-SOD (*Nostoc* sp. PCC 7120, PDB ID: 1GV3, chain A). A single 3D structural protein model for each Fe-SOD (Fig. 17.1a), as well as Mn-SOD (Fig. 17.1b) in *Scytonema tolypothrichoides* VB-61278, was constructed. The RMSD value generated for Fe-SOD and Mn-SOD was found to be 0.235 Å and 0.589 Å, respectively. 190 and 196 Cα atoms were involved in RMSD calculation of Fe-SOD and Mn-SOD in *Scytonema tolypothrichoides,* respectively.

Ramachandran plot generated observed 93.6% residues (most favored regions), 4.7% residues (additional allowed regions), and 1.1% residues (generously allowed regions), while 0.6% residues (disallowed regions) in the plot for Fe-SOD in *Scytonema tolypothrichoides* (Fig. 17.2a). However, in the case of Mn-SOD in *Scytonema tolypothrichoides,* Ramachandran plot observed 94.8% residues (most favored regions), 3.5% residues (additional allowed regions), 1.2% residues (generously allowed regions), while 0.6% residues (disallowed regions) of the plot (Fig. 17.2b).

Evaluation of the built 3D models by Verify 3D observed an averaged 3D-1D score equal to/higher than 0.2 for 100.0% and 99.01% of residues for Fe-SOD and Mn-SOD in *Scytonema tolypothrichoides,* respectively. ProSA-web generated Z scores of −8.67 and − 9.27, respectively, for 3D protein models of Fe-SOD and Mn-SOD in *Scytonema tolypothrichoides,* which were illustrated in the plot with a dark black point (Fig. 17.3a, b).

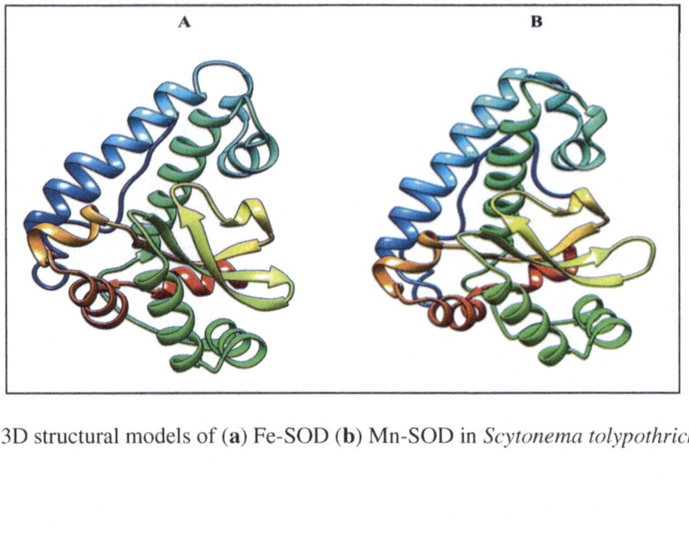

Fig. 17.1 3D structural models of (**a**) Fe-SOD (**b**) Mn-SOD in *Scytonema tolypothrichoides*

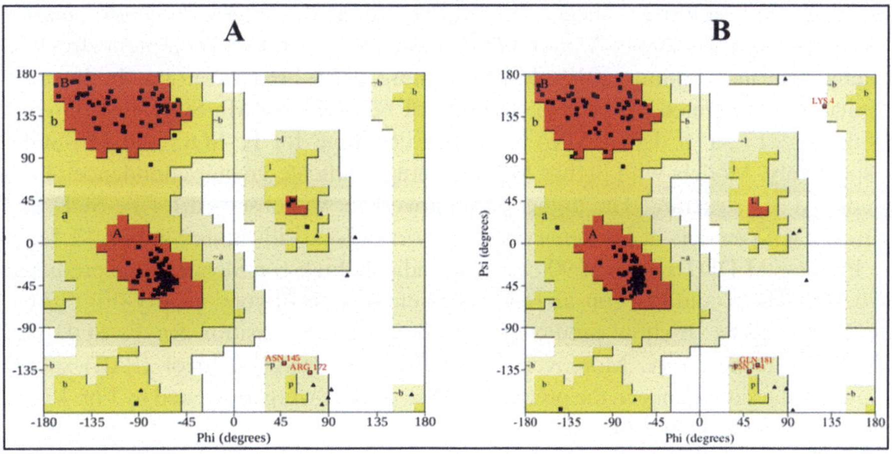

Fig. 17.2 Ramachandran plot of 3D structures models of (**a**) Fe-SOD (**b**) Mn-SOD in *Scytonema tolypothrichoides* VB-61278

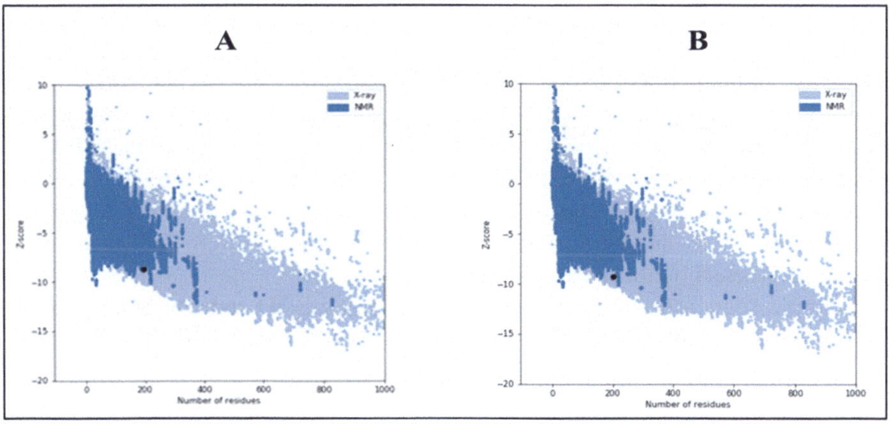

Fig. 17.3 The Dark black point represents Z scores of the modeled protein (**a**) Fe-Sod (**b**) Mn-SOD in *Scytonema tolypothrichoides*

17.5 Protein Modeling and Model Structure Validation Using AlphaFold2

Among the ML-based modeling strategies that use known protein structures for predicting the structures of target proteins, Deep learning (DL) has been the most developed because of the rapidly increasing volume of data, escalating computing power, and ongoing optimization of DL algorithms (Kanchan et al. 2024). AlphaFold (AF) is a DL-based modeling method used for protein structure prediction (Senior et al. 2020; Jumper et al. 2021). AF2 is the most advanced protein structure prediction that uses a deep neural network to train the model utilizing protein homologous structures and multiple sequence alignments. Taking *Scytonema tolypothrichoides* VB-61278 as an example, Alphafold2 was used to generate a 3D protein structure (Steinegger and Söding 2017; Mirdita et al. 2022). Figure 17.4a, b illustrate Alphafold2 generated 3D protein structure of Fe-SOD and Mn-SOD, respectively, while Fig. 17.5a, b depict MSA sequence coverage for Fe-SOD and Mn-SOD, respectively. Models with higher MSA coverage indicate greater confidence in the predicted 3D structure. The higher MSA coverage was observed for Fe-SOD and Mn-SOD 3D models of *Scytonema tolypothrichoides as* illustrated in Fig. 17.5a, b.

Predicted LDDT scores of 90 or higher indicate high confidence in the predicted structure which could be considered 3D protein models of good quality (Fig. 17.6a, b). Figure 17.6a, b show predicted LDDT scores per position for Fe-SOD and Mn-SOD, respectively. Moreover, lower Predicted Aligned Error (PAE) scores show greater confidence in the predicted 3D structures were illustrated by Fig. 17.7a, b for Fe-SOD and Mn-SOD, respectively.

17.6 Functional Analysis

Domains are called functional units of proteins. Domains in Fe-SOD as well as Mn-SOD proteins in *Scytonema tolypothrichoides* VB-61278 may be predicted by conserved domain (CD) search tool (Marchler-Bauer and Bryant 2004) using Conserved domain databases CDDv3.18 (Marchler-Bauer et al. 2015). Functional motifs present in the retrieved proteins may be predicted using the Multiple Extraction-Maximization for Motif Elicitation (MEME) tool (Bailey et al. 2009). Homologous protein sequences of Fe-SOD and Mn-SOD in *Scytonema tolypothrichoides* VB-61278 among various cyanobacterial species were collected using default parameters by the blastp program.

Multiple sequence alignments of homologous Fe-SOD and Mn-SOD proteins were generated using ClustalW available with MegaX version 10.0.5 (Thompson

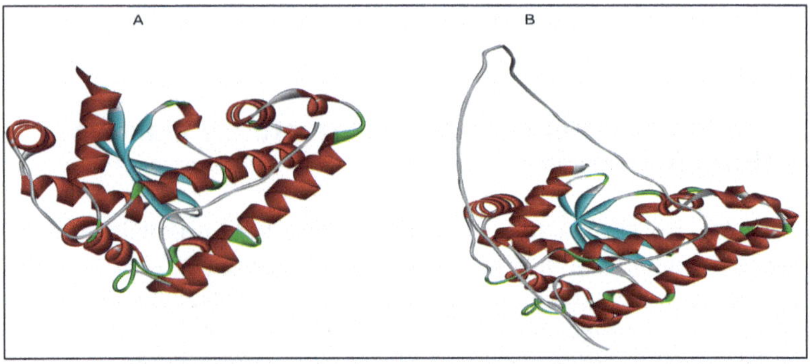

Fig. 17.4 3D structural models by AlphaFold2 of (**a**) Fe-SOD and (**b**) Mn-SOD in *Scytonema tolypothrichoides*

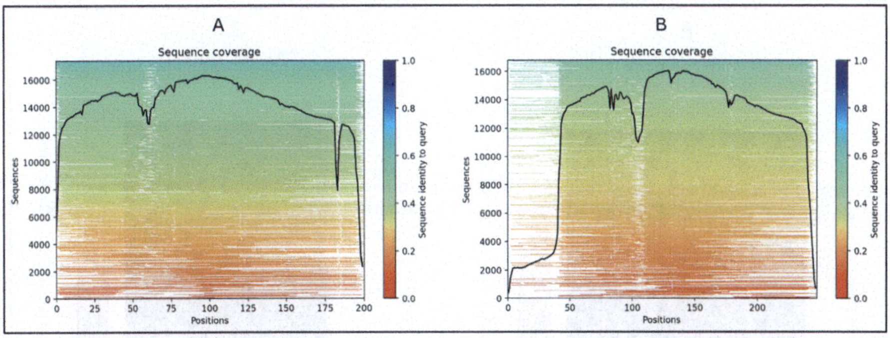

Fig. 17.5 MSA Sequence Coverage of (**a**) Fe-SOD and (**b**) Mn-SOD in *Scytonema tolypothrichoides*

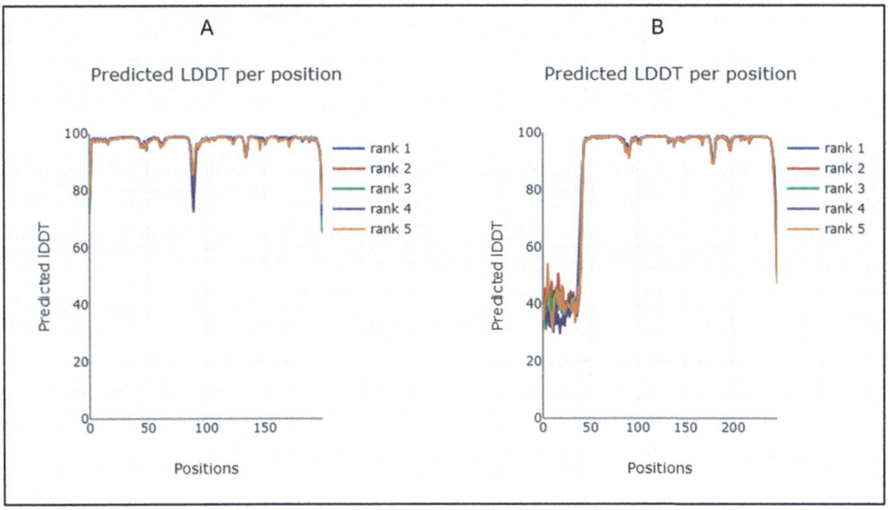

Fig. 17.6 Predicted LDDT score per position of (**a**) Fe-SOD and (**b**) Mn-SOD in *Scytonema tolypothrichoides*

et al. 1994; Kumar et al. 2018). Colorful multiple-sequence alignment was further generated using MView (Madeira et al. 2019).

Evolutionary conservation of various amino acid positions was further investigated, and functional and structural residues along with their environment of Fe-SOD as well as Mn-SOD proteins in *Scytonema tolypothrichoides* VB-61278 may be investigated by Consurf, available at http://consurf.tau.ac.il.Consurf. Consurf uses multiple sequence alignment of homologous cyanobacterial protein sequences and phylogenetic tree (if available) to generate the conservation score for each amino acid position. Every rapidly evolving amino acid position is associated with a variable conservation score while slowly evolving amino acid positions are considered as conserved and highly conserved. A conservational score on a scale of

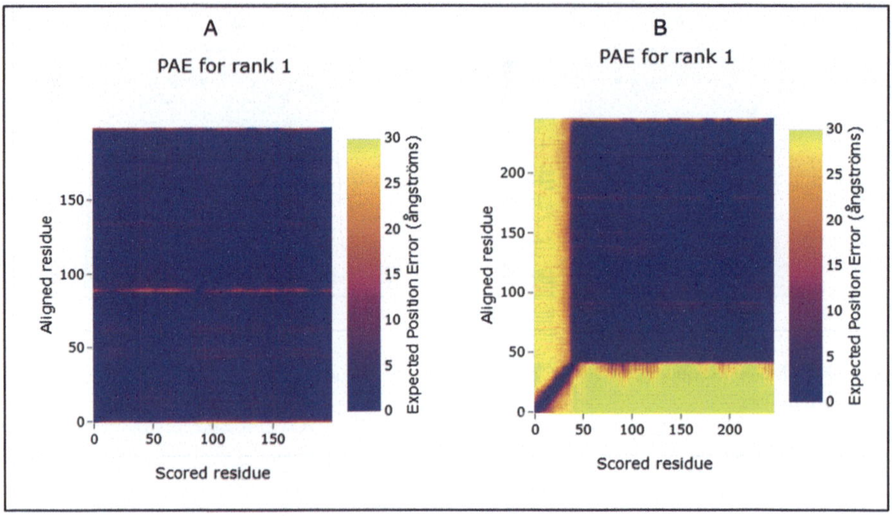

Fig. 17.7 Predicted Aligned Error of (**a**) Fe-SOD and (**b**) Mn-SOD in *Scytonema tolypothrichoides*

1–9 is associated with the multiple sequence alignment as generated by Consurf, where 1–3 (most variable positions), 4–6 (intermediately conserved positions), and 7–9 (highly conserved positions) signify different degree of conservation of amino acid positions (Ashkenazy et al. 2010; Ashkenazy et al. 2016).

MEME predicted the three motifs (identical in both Fe-SOD and Mn-SOD in *Scytonema tolypothrichoides)*. In Fe-SOD in *Scytonema tolypothrichoides*, motif 1 (residue numbers 1–36), motif 2 (residue numbers 111–131), and motif 3 (residue numbers 160–177) were investigated (Fig. 17.8). In Mn-SOD in *Scytonema tolypothrichoides*, motif 1 (residue number 42–77), motif 2 (residue number 156–176), and motif 3 (residue number 206–223) are illustrated in Fig. 17.8.

Evolutionary conservation using Consurf for Fe-SOD in *Scytonema tolypothrichoides* predicted 13 amino acid residues as functional and exposed residues (highly conserved with conservation score: 9) (Fig. 17.9a). Consurf also predicted 20 amino residues as structural and buried which were found highly conserved (highly conserved with conservation score: 9) in Fe-SOD present in *Scytonema tolypothrichoides*. Similarly, evolutionary conservation analysis using Consurf for Mn-SOD in *Scytonema tolypothrichoides* predicted 18 amino acid residues as functional and exposed residues (highly conserved with conservation score: 9) (Fig. 17.9b). Consurf also predicted 20 amino residues as structural and buried residues (highly conserved with conservation score: 9) in Mn-SOD present in *Scytonema tolypothrichoides*.

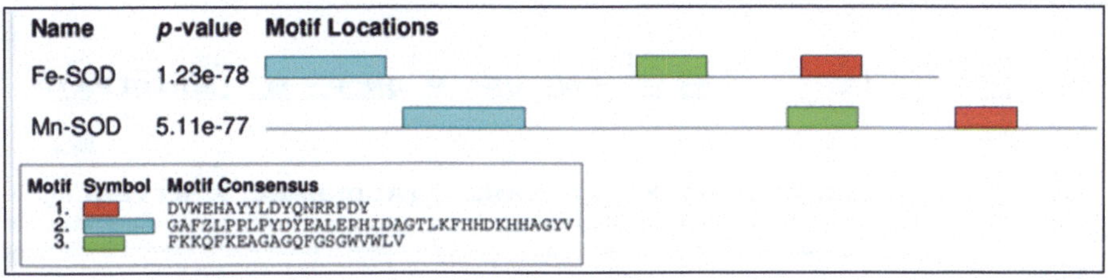

Fig. 17.8 Three identical motifs identified in Fe-SOD and Mn-SOD in *Scytonema tolypothrichoides,* Motif 1 (red) Motif 2 (cyan), and Motif 3 (green), were shown in Fe-SOD and Mn-SOD in *Scytonema tolypothrichoides*

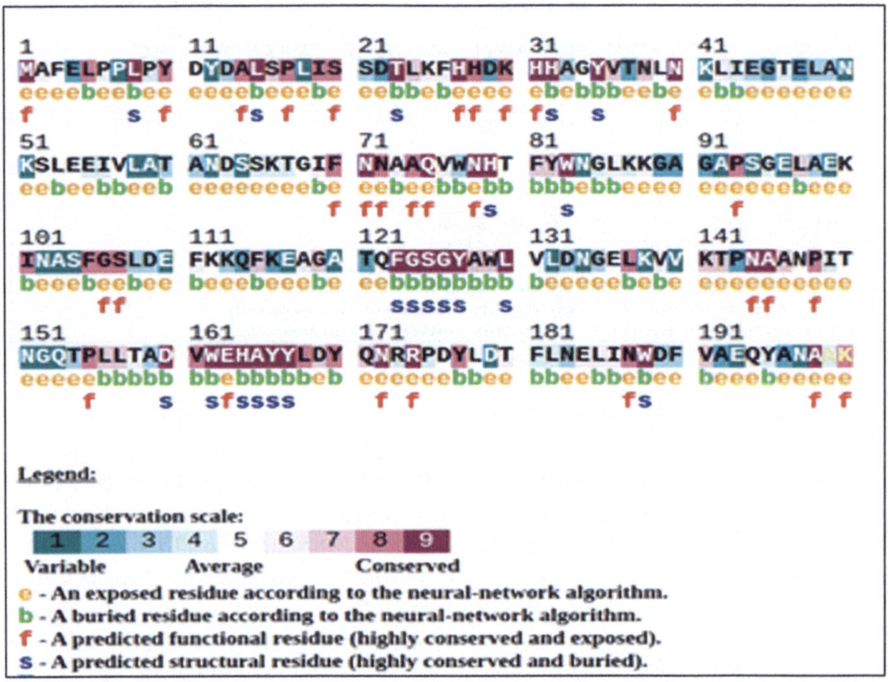

Fig. 17.9 (**a**) Evolutionary conservation of amino acids in the Fe-SOD in *Scytonema tolypothrichoides* determined by the ConSurf server. (**b**) Evolutionary conservation of amino acids in the Mn-SOD in *Scytonema tolypothrichoides* determined by the ConSurf server

17.7 Phylogenetic Analysis

A list of homologous Fe-SOD protein sequences (including Fe-SOD in *Scytonema tolypothrichoides*) in 16 cyanobacterial species with their accession no. and sequence length was retrieved from blast search tool using the nr database to generate the phylogenetic tree for Fe-SOD in *Scytonema tolypothrichoides* as shown in Table 17.1.

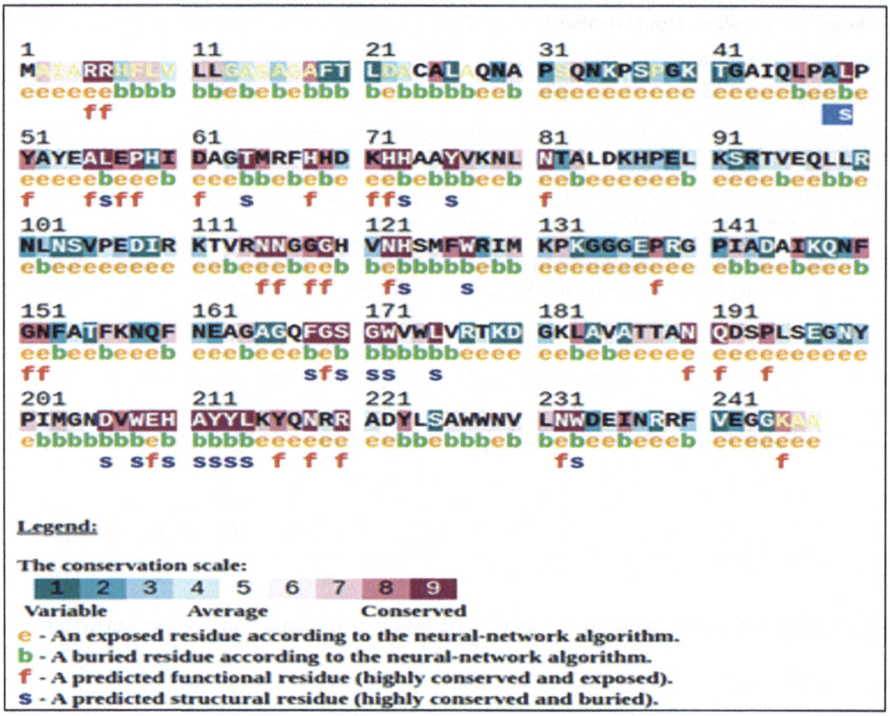

Fig. 17.9 (continued)

The phylogenetic tree of Fe-SOD and Mn-SOD homologous proteins may be constructed using the neighbor-joining (NJ) method (Saitou and Nei 1987) using MegaX version 10.0.5 (Kumar et al. 2018). NJ phylogenetic trees were subjected to a bootstrap test with 1000 replicates (Felsenstein 1985) and the JTT matrix-based method (Jones et al. 1992) may be used for calculating evolutionary distances.

The phylogenetic tree for Fe-SOD illustrates *Brasilonema sennae* as the closest evolutionary homolog of *Scytonema tolypothrichoides* (Fig. 17.10a). A list of homologous Mn-SOD protein sequences (including Mn-SOD in *Scytonema tolypothrichoides*) in 16 cyanobacterial species with their accession no. and sequence length was retrieved from blast search tool using the nr database to generate the phylogenetic tree for Mn-SOD in *Scytonema tolypothrichoides* as shown in Table 17.2.

The phylogenetic tree for Mn-SOD depicts *Tolypothrix campylonemoides* as the closest evolutionary homolog of *Scytonema tolypothrichoides* as shown in Fig. 17.10b.

Table 17.1 List of homologous Fe-SOD protein sequences (including Fe-SOD in *Scytonema tolypothrichoides*) in 16 cyanobacterial species with their NCBI accession no. and protein sequence lengths

S. no.	Accession number	Cyanobacterial sp.	Protein length
1.	*WP_048870051.1*	*Scytonema tolypothrichoides*	200
2.	WP_171976354.1	*Brasilonema sennae*	200
3.	WP_041036004.1	*Tolypothrix campylonemoides*	199
4.	MBW4634974.1	*Iphinoe* sp. HA4291-MV1	200
5.	WP_017314806.1	*Mastigocladopsis repens*	199
6.	ARV62440.1	*Nostocales cyanobacterium* HT-58-2	200
7.	WP_190796444.1	*Leptolyngbya* sp. FACHB-541	200
8.	WP_190722069.1	*Oscillatoria* sp. FACHB-1406	200
9.	WP_073599074.1	*Hydrococcus rivularis*	200
10.	WP_015956466.1	*Gloeothece citriformis*	199
11.	RFP63026.1	*Limnothrix* sp. CACIAM 69d	200
12.	WP_207090370.1	*Phormidium pseudopriestleyi*	200
13.	WP_168571301.1	*Oxynema aestuarii*	200
14.	WP_124975449.1	*Aphanothece sacrum*	200
15.	OON12525.1	*Synechococcus lacustris* str. *Tous*	199
16.	*WP_096384730.1*	*Stanieria* sp. NIES-3757	199

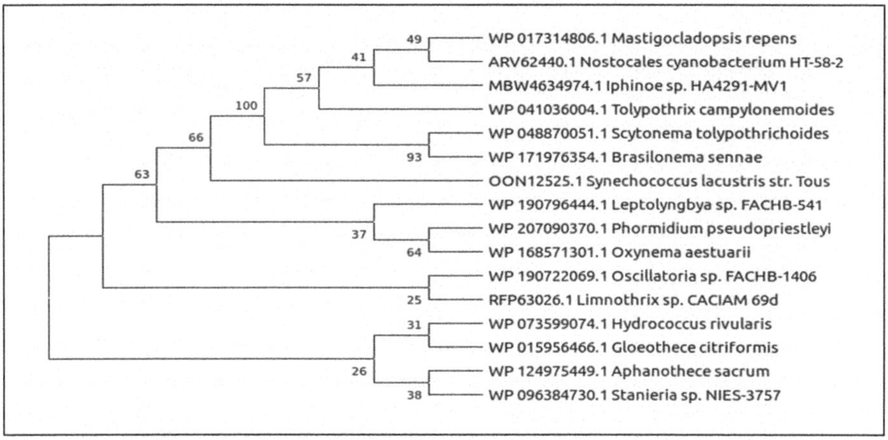

Fig. 17.10 (**a**) Phylogenetic tree of Fe-SOD homologs including that of *Scytonema tolypothrichoides*. Bootstrap support values are shown close to branches. Each taxon starts with the genebank accession numbers followed by the name of the species. (**b**) Phylogenetic tree of Mn-SOD homologs including that of *Scytonema tolypothrichoides*. Bootstrap support values are shown close to branches. Each taxon starts with the genebank accession numbers followed by the name of the species

17.8 Conclusion

Cutting-edge omics technologies leveraged path-breaking research encompassing a wide range including the identification of novel taxonomic diversity, CO_2-concentrating mechanism (Kupriyanova et al. 2023), physiological and

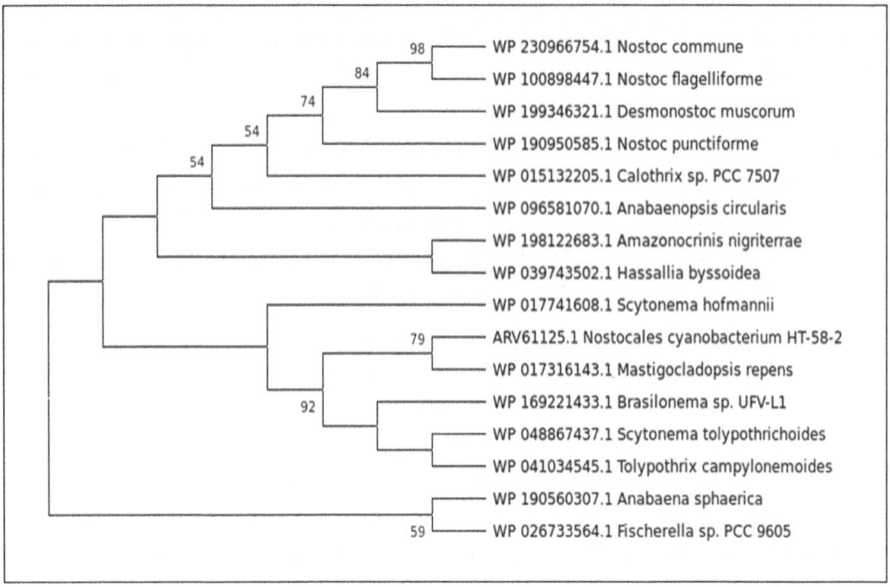

Fig. 17.10 (continued)

Table 17.2 List of homologous Mn-SOD protein sequences (including Mn-SOD in *Scytonema tolypothrichoides*) in 16 cyanobacterial species with their accession no. and sequence length

S.No.	Accession number	Organism name	Protein length
1	WP_048867437.1	*Scytonema tolypothrichoides.*	247
2	WP_041034545.1	*Tolypothrix campylonemoides*	247
3	ARV61125.1	*Nostocales cyanobacterium* HT-58-2	247
4	WP_017316143.1	*Mastigocladopsis repens*	248
5	WP_169221433.1	*Brasilonema* sp. UFV-L1	253
6	WP_015132205.1	*Calothrix* sp. PCC 7507	252
7	WP_190950585.1	*Nostoc punctiforme*	253
8	WP_230966754.1	*Nostoc commune*	247
9	WP_026733564.1	*Fischerella* sp. PCC 9605	255
10	WP_017741608.1	*Scytonema hofmannii*	246
11	WP_096581070.1	*Anabaenopsis circularis*	246
12	WP_198122683.1	*Amazonocrinis nigriterrae*	249
13	WP_199346321.1	*Desmonostoc muscorum*	259
14	WP_190560307.1	*Anabaena sphaerica*	240
15	WP_100898447.1	*Nostoc flagelliforme*	255
16	WP_039743502.1	*Hassallia byssoidea*	241

biotechnological engineering for engendering resources for biofuel (Nagi et al. 2021; Maneechote et al. 2021; Karkala et al. 2021), and biofertilizer and several druggable bioactive compounds (Lin et al. 2014; Dolganyuk et al. 2020; Adamczyk-Grochala et al. 2020; Favas et al. 2021; Martínez-Ruiz et al. 2022;

Santiesteban-Romero et al. 2022) to enhance environmental sustainability. Bioinformatics powered by AI has further stood up as a beacon in unraveling crucial information regarding protein modeling, and protein–protein interactions (Kesheri et al. 2022). Computational prediction of physicochemical properties of Fe-SOD and Mn-SOD in *Scytonema tolypothrichoides* may further unravel the nature and behavior of the protein. The instability index having a value lower than 40 is considered as more stable compared to a value higher than 40. In this context, Fe-SOD (Instability index: 24.33) is more stable than Mn-SOD (Instability index: 35.01). Negative GRAVY for Fe-SOD and Mn-SOD illustrated these proteins as hydrophilic. The higher aliphatic index of Fe-SOD and Mn-SOD in *Scytonema tolypothrichoides* illustrates the thermostable nature of these proteins. Fe-SOD and Mn-SOD include α-helices in the majority; therefore, these proteins could have many protein–protein interactions (Azzarito et al. 2013). Fe-SOD and Mn-SOD of *Nostoc azollae* also revealed the thermostable nature of the protein due to the presence of α-helices in the majority and the structurally stable nature of these proteins (Petukhov et al. 1997; Escobedo et al. 2019; Kesheri et al. 2024).

Superimposition between target and template structures illustrated conserved protein folds in Fe-Sod and Mn-Sod in *Scytonema tolypothrichoides*. For a model structure to be considered "good," more than 90% of its residues should be in the most preferred regions of the Ramachandran plot. (Sali and Blundell 1993; Kesheri et al. 2015). Thus, the Ramachandran plot results from this work indicated that the Fe-SOD and Mn-SOD in *Scytonema tolypothrichoides* were stereochemical stable, with over 93% of the amino acids being located in the most preferred regions of both Fe-SOD and Mn-SOD. Verify 3D predicted the compatibility of a structural class based on its location and environment (alpha, beta, loop, polar, nonpolar, etc.), with good compatibility scores above zero for the majority of residues in the modeled protein. An acceptable model should have a Verify 3D score of at least 0.2 for 80% of amino acid residues, which indicates excellent resolution (about 2.5–3.0 Å). Our Verify 3D results for FeS-OD and Mn-SOD were consistent (>99% in both) which suggested the decent quality of these 3D protein structural models. ProSA-web in terms of Z-score validates the 3D structure by checking the local compatibility of the model related to good 3D protein structure in pdb. The Z-score of Fe-SOD and Mn-SOD in *Scytonema tolypothrichoides* predicted in the acceptable range of −10 to 10 and a negative which suggests a good quality of these 3D protein structure models. The predicted 3D models for Fe-SOD and Mn-SOD in *Scytonema tolypothrichoides* were found to be in the range of good quality, according to the strict confirmation tests.

Three identical motifs were investigated in both Fe-SOD and Mn-SOD in *Scytonema tolypothrichoides* suggesting functional conservation of these proteins. Consurf's evolutionary conservation analysis identified several highly conserved structural (hidden) and functional (exposed) residues in *Scytonema tolypothrichoides*, which helped to clarify the proteins' functions. In silico structural and functional analysis of Fe-SOD and Mn-SOD in *Scytonema tolypothrichoides* would further establish a correlation between the antioxidant Fe-SOD enzyme and oxidative stress.

Conclusively, the role of bioinformatics powered by AI has further stood up as a beacon in unraveling crucial information regarding protein modeling, protein–protein interactions, and various other information facilitating developments in cyanobacterial research.

17.9 Future Prospects

Cyanobacteria may serve as a crucial candidate for the development of adaptive microbial communities (SynComs), an emerging scientific approach to designing inoculants that would enrich sustainable agricultural practices by incorporating concepts from microbiology and genetics (de Souza et al. 2020). Production of nanoparticles utilizing environmental microbes is a prospective technology for achieving crop improvement and thereby stress tolerance (Abd-Elsalam 2024; Solomon et al. 2024; Srivastava et al. 2025). Bioinformatics encompassing data mining, NGS of sequence reads and their quality check, sequence assembly, metagenomics, protein modeling and docking, gene prediction and functional assignment, taxonomic binning, metabolic pathway reconstruction, and pipeline development to analyze microgenomic data shall leverage vast avenues for identification of beneficial microbial traits in improvising sustainable agricultural practices. In this era of AI, ML tools based on deep learning have fueled AlphaFold2 leveraging protein modeling of almost all proteins leveraging crucial information about structure, design, function, and targets significant for drug discovery and development (Yang et al. 2023). Owing to the cyanobacterial potential for eco-friendly sustainable agriculture, they may serve as crucial candidates for the mass production of cyanobacterial biofertilizers (Chittora et al. 2020). The antioxidative properties of cyanobacterial phycobiliproteins, mycosporine-like amino acids, and UV-screening pigment scytonemin have been shown to possess numerous biotechnological properties (Kim et al. 2018; Zaytseva et al. 2021; Mahata et al. 2022; Kumari et al. 2023). In silico study of Fe-SOD as well as Mn-SOD in *Scytonema tolypothrichoides* could be used as a model for stress biology studies and survival of life in its natural habitat. Thus, cyanobacterial resilience strategies may open the door to a conduit of sustainable resources for several industrial processes exploiting their biomedical, and biotechnological applications in the future by deploying multi-omics and ML approaches.

Short Chapter Summary

Cyanobacterial research envisioned through the lens of omics has the potential to leverage trailblazing developments crucial to biomedical sciences and environmental sustainability. From time immemorial, cyanobacteria have been bestowed with the power of resilience and adaptivity that has been fueled by enzymatic as well as nonenzymatic antioxidative and bioactive compounds. This chapter delineates various stresses faced and resilience strategies of cyanobacteria to cope with biotic and abiotic stresses. In silico approach to navigate various physicochemical parameters such as molecular weight, GRAVY, instability index, pI, and aliphatic index is explained in an illustrative manner. Owing to the current demand for bioinformatics analysis which is the key to the treasure of various omics data, investigation of the secondary structure of proteins and state-of-the-art in the prediction of 3D structural models using modeler and trailblazing AI-powered AlphaFold tools have been illustrated using cyanobacterium *Scytonema tolypothrichoides* VB-61278 as a case study. Exhaustive explanations of validation methods for the models using Prosa-web, Verify-3D, Ramachandran plot, and LDDT score have been highlighted. Studies emphasizing the conservation and prediction of metal binding positions, domains, and motifs along with residues of structural and functional importance, as well as the generation of phylogenetic trees, have been well explained.

Author Contributions Concept, design, and writing, Minu Kesheri and Swarna Kanchan; Review and editing, Donat-P. Häder and Rajeshwar P. Sinha. Minu Kesheri and Swarna Kanchan contributed equally as first author.

References

Abd-Elsalam KA (2024) Special issue: microbial nanotechnology. Microorganisms 12(2):352

Adamczyk-Grochala J, Wnuk M, Duda M, Zuczek J, Lewinska A (2020) Treatment with modified extracts of the microalga *Planktochlorella nurekis* attenuates the development of stress-induced senescence in human skin cells. Nutrients 12(4):1005

Altschul SF, Madden TL, Schäffer AA, Zhang J, Zhang Z, Miller W, Lipman DJ (1997) Gapped BLAST and PSI-BLAST: a new generation of protein database search programs. Nucleic Acids Res 25:3389–3402

Ashkenazy H, Abadi S, Martz E, Chay O, Mayrose I, Pupko T, Ben-Tal N (2016) ConSurf 2016: an improved methodology to estimate and visualize evolutionary conservation in macromolecules. Nucleic Acids Res 44(W1):W344–W350

Ashkenazy H, Erez E, Martz E, Pupko T, Ben-Tal N (2010) ConSurf 2010: calculating evolutionary conservation in sequence and structure of proteins and nucleic acids. Nucleic Acids Res 38(Web Server issue):W529–W533

Azzarito V, Long K, Murphy NS, Wilson AJ (2013) Inhibition of alpha-helix-mediated protein-protein interactions using designed molecules. Nat Chem 5:161–173

Bailey TL, Boden M, Buske FA, Frith M, Grant CE, Clementi L, Ren J, Li WW, Noble WS (2009) MEME SUITE: tools for motif discovery and searching. Nucleic Acids Res 37:W202–W208

Bastiaansen R, Doelman A, Eppinga MB, Rietkerk M (2020) The effect of climate change on the resilience of ecosystems with adaptive spatial pattern formation. Ecol Lett 23:414–429

Bowie JU, Lüthy R, Eisenberg D (1991) A method to identify protein sequences that fold into a known three-dimensional structure. Science 253:164–170

Burberg C, Petzoldt T, Elert EV (2020) Phosphate limitation increases content of protease inhibitors in the cyanobacterium *Microcystis aeruginosa*. Toxins 12:33

Cai H, McLimans CJ, Beyer JE, Krumholz LR, Hambright KD (2023) *Microcystis pangenome* reveals cryptic diversity within and across morphospecies. Sci Adv 9:eadd3783

Carratala A, Chappelier C, Selmoni O, Guillaume AS, Chmiel HE, Pasche N, Weil C, Kohn T, Joost S (2023) Vertical distribution and seasonal dynamics of planktonic cyanobacteria communities in a water column of deep mesotrophic Lake Geneva. Front Microbiol 14:1295193

Chen MY, Teng WK, Zhao L, Hu CX, Zhou YK, Han BP, Song LR, Shu WS (2021) Comparative genomics reveals insights into cyanobacterial evolution and habitat adaptation. ISME J 15:211–227

Chittora D, Meena M, Barupal T, Swapnil P, Sharma K (2020) Cyanobacteria as a source of biofertilizers for sustainable agriculture. Biochem Biophys Rep 22:100737

Dabravolski SA, Isayenkov SV (2022) Metabolites facilitating adaptation of desert cyanobacteria to extremely arid environments. Plan Theory 11:3225

de Souza RSC, Armanhi JSL, Arruda P (2020) From microbiome to traits: designing synthetic microbial communities for improved crop resiliency. Front Plant Sci 11:1179

Dolganyuk V, Belova D, Babich O, Prosekov A, Ivanova S, Katserov D, Patyukov N, Sukhikh S (2020) Microalgae: a promising source of valuable bioproducts. Biomol Ther 10(8):1153

Escobedo A, Topal B, Kunze M, Aranda J, Chiesa G, Mungianu D, Bernardo-Seisdedos G, Eftekharzadeh B, Gairí M, Pierattelli R, Felli IC, Diercks T, Millet O, García J, Orozco M, Crehuet R, Lindorff-Larsen K, Salvatella X (2019) Side chain to main chain hydrogen bonds stabilize a polyglutamine helix in a transcription factor. Nat Commun 10:2034

Farias AB, Martinez GS, Galán-Vásquez E, Nicolás MF, Pérez-Rueda E (2024) Predicting bacterial transcription factor binding sites through machine learning and structural characterization based on DNA duplex stability. Brief Bioinform 25(6):bbae581

Favas R, Morone J, Martins R, Vasconcelos V, Lopes G (2021) Cyanobacteria and microalgae bioactive compounds in skin-ageing: potential to restore extracellular matrix filling and overcome hyperpigmentation. J Enzyme Inhib Med Chem 36(1):1829–1838

Felsenstein J (1985) Confidence limits on phylogenies: an approach using the bootstrap. Evolution 39:783–791

Foyer CH, Noctor G (2005) Redox homeostasis and antioxidant signaling: a metabolic interface between stress perception and physiological responses. Plant Cell. 17(7):1866–1875. https://doi.org/10.1105/tpc.105.033589

Gasteiger E, Hoogland C, Gattiker A, Duvaud S, Wilkins MR, Appel RD, Bairoch A (2005) Protein identification and analysis tools on the ExPASy server. In: Walker JM (ed) The proteomics protocols handbook. Humana Press, pp 571–607

Geourjon C, Deleage G (1995) SOPMA: significant improvements in protein secondary structure prediction by consensus prediction from multiple alignments. Comput Appl Biosci 11(6):681–684

Gordon GC, Pfleger BF (2018) Regulatory tools for controlling gene expression in cyanobacteria. Adv Exp Med Biol 1080:281–315

Gupta A, Kang K, Pathania R, Saxton L, Saucedo B, Malik A, Torres-Tiji Y, Diaz CJ, Dutra Molino JV, Mayfield SP (2024) Harnessing genetic engineering to drive economic bioproduct production in algae. Front Bioeng Biotechnol 12:1350722

Häder D-P (2001) Adaptation to UV stress in algae. In: Rai LC, Gaur JP (eds) Algal adaptation to environmental stresses, physiological, biochemical and molecular mechanisms. Springer, New York, pp 173–202

Häder D-P, Gao K (2023) Aquatic productivity under multiple stressors. Water 15(4):817

Huisman J, Codd GA, Paerl GA, Ibelings BW, Verspagen JMH, Visser PM (2018) Cyanobacterial blooms. Nat Rev Microbiol 16:471

Irankhahi P, Riahi H, Hassani SB, Eskafi M, Azimzadeh Irani M, Shariatmadari Z (2024) The role of the protective shield against UV-C radiation and its molecular interactions in *Nostoc* species (Cyanobacteria). Sci Rep 14(1):19258

Jones DT, Taylor WR, Thornton JM (1992) The rapid generation of mutation data matrices from protein sequences. Comput Appl Biosci 8:275–282

Jumper J, Evans R, Pritzel A, Green T, Figurnov M, Ronneberger O, Tunyasuvunakool K, Bates R, Žídek A, Potapenko A, Bridgland A, Meyer C, Kohl SAA, Ballard AJ, Cowie A, Romera-Paredes B, Nikolov S, Jain R, Adler J, Back T, Petersen S, Reiman D, Clancy E, Zielinski M, Steinegger M, Pacholska M, Berghammer T, Bodenstein S, Silver D, Vinyals O, Senior AW, Kavukcuoglu K, Kohli P, Hassabis D (2021) Highly accurate protein structure prediction with AlphaFold. Nature 596:583–589

Kanchan S, Kesheri M, Chaudière J, Sinha RP (2020) Computational metagenomics: current status and challenges. In: Katara P (ed) Recent trends in computational omics: concepts and methodology. Nova Science Publishers, pp 371–395

Kanchan S, Ogden E, Kesheri M, Skinner A, Miliken E, Lyman D, Armstrong J, Sciglitano L, Hampikian G (2024) COVID-19 hospitalizations and deaths predicted by SARS-CoV-2 levels in Boise, Idaho wastewater. Sci Total Environ 907:167742

Karkala S, D'Souza L, Nivas S (2021) Bioprospecting microalgae harnessed from the coastal belt of Mangalore, India as prospective nutraceutical and biofuel candidates. Appl Phycol 2:60–73

Kesheri M, Kanchan S, Chowdhury S, Sinha RP (2015) Secondary and tertiary structure prediction of proteins: a bioinformatic approach. In: Zhu Q, Azar AT (eds) Complex system modelling and control through intelligent soft computations. Springer, pp 541–569

Kesheri M, Kanchan S, Kumar A, Srivastava U, Sharda S, Malik B, Mishra T, Kaithal P, Narayan J, Kumar P, Priya P, Sinha RP (2024) Peculiar endosymbiosis in the cyanobiont *Nostoc azollae* 0708: an *in-silico* approach. In: Kesheri M, Kanchan S, Salisbury TB, Sinha RP (eds) Microbial omics in environment and health. Springer Nature Singapore, Singapore, pp 343–365

Kesheri M, Kanchan S, Richa SRP (2014) Isolation and *in-silico* analysis of Fe-superoxide dismutase in *Nostoc commune*. Gene 553(2):117–125

Kesheri M, Kanchan S, Sinha RP (2021) Isolation and *in-silico* analysis of antioxidants in response to temporal variations in the cyanobacterium *Oscillatoria* sp. Gene Rep 23:101023

Kesheri M, Kanchan S, Sinha RP (2022) Responses of antioxidants for resilience to temporal variations in the cyanobacterium *Microcystis aeruginosa*. S Afr J Bot 148:190–199

Kesheri M, Richa SRP (2011) Antioxidants as natural arsenal against multiple stresses in cyanobacteria. Int J Pharma Biosci 2(2):B168–B187

Kim EY, Choi YH, Nam TJ (2018) Identification and antioxidant activity of synthetic peptides from phycobiliproteins of *Pyropia yezoensis*. Int J Mol Med 42:789–798

Kumar S, Stecher G, Li M, Knyaz C, Tamura K (2018) MEGA X: molecular evolutionary genetics analysis across computing platforms. Mol Biol Evol 35:1547–1549

Kumari A, Kesheri M, Sinha RP, Kanchan S (2016) Integration of soft computing approach in plant biology and its applications in agriculture. In: Purohit HJ, Raje DV, More RP, Kalia VC (eds) Soft computing in biological sciences. Springer, Singapore, pp 265–278

Kumari N, Pandey A, Gupta A, Mishra S, Sinha RP (2023) Characterization of UV-screening pigment scytonemin from cyanobacteria inhabiting diverse habitats of Varanasi, India. Biologia 78:319–330

Kupriyanova EV, Pronina NA, Los DA (2023) Adapting from low to high: an update to CO_2-concentrating mechanisms of cyanobacteria and microalgae. Plan Theory 12(7):1569

Kvíderová J, Kumar D, Lukavský J, Kaštánek P, Adhikary SP (2018) Estimation of growth and exopolysaccharide production by two soil cyanobacteria, *Scytonema tolypothrichoides* and *Tolypothrix bouteillei* as determined by cultivation in irradiance and temperature crossed gradients. Eng Life Sci 19(3):184–195

Laskowski RA, MacArthur MW, Moss DS, Thornton JM (1993) PROCHECK: a program to check the stereochemical quality of protein structure. J Appl Crystallogr 26:283–291

Latifi A, Ruiz M, Zhang CC (2009) Oxidative stress in cyanobacteria. FEMS Microbiol Rev 33:258–278

Le VV, Tran QG, Ko SR, Oh HM, Ahn CY (2024) Insights into cyanobacterial blooms through the lens of omics. Sci Total Environ 934:173028

Lin LT, Wen CH, Lin CC (2014) Antiviral natural products and herbal medicines. J Tradit Complement Med 4(1):24–35

Lürling M, Van Oosterhout F, Faassen EJ (2017) Eutrophication and warming boost cyanobacterial biomass and microcystins. Toxins 9:64

Lüthy R, Bowie JU, Eisenberg D (1992) Assessment of protein models with three-dimensional profiles. Nature 356:83–85

Madeira F, Park YM, Lee J, Buso N, Gur T, Madhusoodanan N, Basutkar P, Tivey AR, Potter SC, Finn RD, Lopez R (2019) The EMBL-EBI search and sequence analysis tools APIs in 2019. Nucleic Acids Res 47(W1):W636–W641

Mahata C, Das P, Khan S, Thaher MIA, Abdul Quadir M, Annamalai SN, Al Jabri H (2022) The potential of marine microalgae for the production of food, feed, and fuel (3F). Fermentation 8(7):316

Mai VC, Nguyen BH, Nguyen DD, Nguyen LA (2017) *Nostoc calcicola* extract improved the antioxidative response of soybean to cowpea aphid. Bot Stud 58:55

Maneechote W, Cheirsilp B, Srinuanpan S, Pathom-aree W (2021) Optimizing physicochemical factors for two-stage cultivation of newly isolated oleaginous microalgae from local lake as promising sources of pigments, PUFAs and biodiesel feedstocks. Bioresour Technol Rep 15:100738

Marchler-Bauer A, Bryant SH (2004) CD-search: protein domain annotations on the fly. Nucleic Acids Res 32:W327–W331

Marchler-Bauer A, Derbyshire MK, Gonzales NR, Lu S, Chitsaz F, Geer LY, Geer RC, He J, Gwadz M, Hurwitz DI, Lanczycki CJ, Lu F, Marchler GH, Song JS, Thanki N, Wang Z, Yamashita RA, Zhang D, Zheng C, Bryant SH (2015) CDD:NCBI's conserved domain database. Nucleic Acids Res 43(Database issue):D222–D226

Martínez-Ruiz M, Martínez-González CA, Kim DH, Santiesteban-Romero B, Reyes-Pardo H, Villaseñor-Zepeda KR, Meléndez-Sánchez ER, Ramírez-Gamboa D, Díaz-Zamorano AL, Sosa-Hernández JE, Coronado-Apodaca KG, Gámez-Méndez AM, Iqbal HMN, Parra-Saldivar R (2022) Microalgae bioactive compounds to topical applications products-a review. Molecules 27(11):3512

Mirdita M, Schütze K, Moriwaki Y, Heo L, Ovchinnikov S, Steinegger M (2022) ColabFold: making protein folding accessible to all. Nat Methods 19:679–682

Nagi GK, Minhas AK, Gaur S, Jain P, Mandal S (2021) Integration of algal biofuels with bioremediation coupled industrial commodities towards cost-effectiveness. Front Energy Res 9:489

Ning H, Li R, Zhou T (2022) Machine learning for microalgae detection and utilization. Front Mar Sci 9:947394

Oliveira R, de Andrade AS, Imparato DO, de Lima J, de Almeida R, Lima J, Pasquali M, Dalmolin R (2019) Analysis of *Arabidopsis thaliana* Redox Gene network indicates evolutionary expansion of class iii peroxidase in plants. Sci Rep 9:15741

Otero A, Vincenzini M (2004) *Nostoc* (Cyanophyceae) goes nude: extracellular polysaccharides serve as a sink for reducing power under unbalanced C/N metabolism. J Phycol 40:74–81

Pettersen EF, Goddard TD, Huang CC, Couch GS, Greenblatt DM, Meng EC, Ferrin TE (2004) UCSF Chimera- a visualization system for exploratory research and analysis. J Comput Chem 25:1605–1612

Petukhov M, Kil Y, Kuramitsu S, Lanzov V (1997) Insights into thermal resistance of proteins from the intrinsic stability of their α-helices. Proteins 29:309–320

Priyadarshani I, Rath B (2012) Commercial and industrial applications of microalgae-a review. J Algal Biomass Util 3:89–100

Rajneesh, Pathak J, Richa, Häder D-P, Sinha RP (2019) Impacts of ultraviolet radiation on certain physiological and biochemical processes in cyanobacteria inhabiting diverse habitats. Environ Exp Bot 161:375–387

Sabadin GA, Xavier GA, Vaz Júnior IS (2019) Control of redox homeostasis in tick blood feeding. Acta Sci Vet 47:1678

Sahu, N., Mishra, S., Kesheri, M. et al. Identification of Cyanobacteria-Based Natural Inhibitors Against SARS-CoV-2 Druggable Target ACE2 Using Molecular Docking Study, ADME and Toxicity Analysis. Ind J Clin Biochem 38:361–373 (2023). https://doi.org/10.1007/s12291-022-01056-6

Saitou N, Nei M (1987) The neighbor-joining method: a new method for reconstructing phylogenetic trees. Mol Biol Evol 4:406–425

Sali A, Blundell TL (1993) Comparative protein modeling by satisfaction of spatial restraints. J Mol Biol 234:779–815

Santiesteban-Romero B, Martínez-Ruiz M, Sosa-Hernández JE, Parra-Saldívar R, Iqbal HM (2022) Microalgae photo-protectants and related bio-carriers loaded with bioactive entities for skin applications-an insight of microalgae biotechnology. Mar Drugs 20:487

Senior AW, Evans R, Jumper J, Kirkpatrick J, Sifre L, Green T, Qin C, Žídek A, Nelson AWR, Bridgland A, Penedones H, Petersen S, Simonyan K, Crossan S, Kohli P, Jones DT, Silver D, Kavukcuoglu K, Hassabis D (2020) Improved protein structure prediction using potentials from deep learning. Nature 577:706–710

Sharma B, Singh BN, Dwivedi P, Rajawat MVS (2022) Interference of climate change on plant-microbe interaction: present and future prospects. Front Agron 3:725804

Solomon NO, Kanchan S, Kesheri M (2024) Nanoparticles as detoxifiers for industrial wastewater. Water Air Soil Pollut 235:214

Srivastava U, Kanchan S, Kesheri M, Sharma A (2025) Single-cell technology for improvement in significant crops. In: Chen JT (ed) Guide to plant single-cell technology. Academic Press, pp 299–320

Steinegger M, Söding J (2017) MMseqs2 enables sensitive protein sequence searching for the analysis of massive data sets. Nat Biotechnol 35:1026–1028

Thompson JD, Higgins DG, Gibson TJ (1994) CLUSTAL W: improving the sensitivity of progressive multiple sequence alignment through sequence weighting, position-specific gap penalties and weight matrix choice. Nucleic Acids Res 22(22):4673–4680

Wiederstein M, Sippl MJ (2007) ProSA-web: interactive web service for the recognition of errors in three dimensional structures of proteins. Nucleic Acids Res 35:W407–W410

Yang Z, Zeng X, Zhao Y, Chen R (2023) AlphaFold2 and its applications in the fields of biology and medicine. Signal Transduct Target Ther 8:115

Zaytseva A, Chekanov K, Zaytsev P, Bakhareva D, Gorelova O, Kochkin D, Lobakova E (2021) Sunscreen effect exerted by secondary carotenoids and mycosporine-like amino acids in the aeroterrestrial chlorophyte *Coelastrella rubescens* under high light and UV-A irradiation. Plan Theory 10:2601

Advancements and Trends in Omics of Cyanobacterial Ultraviolet-Screening Compounds, Scytonemin and Mycosporine-Like Amino Acids

18

Haseen Ahmed, Minu Kesheri, Swarna Kanchan, Rajeshwar P. Sinha, and Jainendra Pathak

Abstract

Cyanobacteria, on exposure to ultraviolet radiations (UVR), synthesize UV-absorbing compounds such as scytonemin and mycosporine-like amino acids (MAAs) to mitigate their harmful effects. Only a few cyanobacteria are capable of synthesizing scytonemin, an extracellular polysaccharide sheath pigment. It is a lipophilic pigment that consists of phenolic and indolic monomer components connected by an olefinic carbon atom. In MAAs, the nitrogen moiety of an amino acid or its amino alcohol is coupled with an aminocyclohexinimine or aminocyclohexenone chrompohore. These are small, colorless, hydrophilic molecules. These compounds have strong antioxidative potential, are photostable, and are effective photoprotectants. The complex stress response pathway and environmental signals that contribute to the synthesis and induction of these secondary metabolites are poorly understood, despite the fact that a great deal of work has been invested in the past ten years to bio-prospect and describe them. Cyanobacterial molecular biology has greatly benefited from the applica-

H. Ahmed
Department of Botany, Rajkiya Mahila Snatkottar Mahavidyalay (Affiliated to Veer Bahadur Singh Purvanchal University, Jaunpur, U.P.), Ghazipur, UP, India

M. Kesheri · S. Kanchan
Department of Biological Sciences, Boise State University, Boise, ID, USA

Department of Biomedical Sciences, Joan C. Edwards School of Medicine, Marshall University, Huntington, WV, USA

R. P. Sinha
Laboratory of Photobiology and Molecular Microbiology, Centre of Advanced Study in Botany, Institute of Science, Banaras Hindu University, Varanasi, India

J. Pathak (✉)
Department of Botany, Pt. Jawaharlal Nehru College (Affiliated to Bundelkhand University, Jhansi), Banda, India

M. Kesheri et al. (eds.), *Multi-Omics in Biomedical Sciences and Environmental Sustainability*, https://doi.org/10.1007/978-981-96-7067-3_18

tion of "omics" approaches. Although targeted metabolome approaches only measure small groups of metabolites, metabolomics research provides an investigative tool for the quick exploration, discovery, and characterization of secondary metabolites produced by an organism. Metabolomics has emerged as a key technique in synthetic biology for identifying new molecules from microorganisms, such as cyanobacteria. Additionally, it makes it possible to manufacture whole sets of metabolites or their precursors with the appropriate titers. This chapter discusses developments made in the field of metabolomics of cyanobacterial ultraviolet-screening compounds, scytonemin, and MAAs in detail.

Keywords

Cyanobacteria · Metabolomics · Mycosporine-like amino acids · Scytonemin · Sunscreen

Objectives

- Omics technologies can help better understand the genetics and biosynthetic pathways of MAAs and scytonemin.
- Omics approach provides the effective yield enhancement strategies by metabolic pathway engineering and provides advanced techniques for the characterization of cyanobacterial bioactive compounds.
- Metabolomics can be useful for the quick detection, identification, and characterization of secondary metabolites produced by an organism.
- Aims to enlist the different techniques used in metabolomic analysis and the advancements in the field of cyanobacterial metabolomics.

18.1 Introduction

Like eukaryotic algae and higher plants, cyanobacteria are the unique assemblage of prokaryotes that have chlorophyll (Chl) and can carry out oxidative phosphorylation using water as a source of protons. These creatures were responsible for the emergence of oxygenic environmental conditions on Earth (Ahmed et al. 2021; Pathak et al. 2022; Singh et al. 2022). Over the past few decades, the depletion of the ozone layer caused by rising atmospheric pollution has intensified the harmful impacts of prolonged exposure to solar ultraviolet radiation (UVR), making skin cancer a significant public health issue (Milito et al. 2021). Strong links exist between climate change and stratospheric ozone depletion, which is mostly brought about by rising greenhouse gas concentrations. This leads to higher UVR, which has a deleterious effect on aquatic ecosystems (Häder et al. 2015; Pathak et al. 2019). There are several ways that UVR, particularly UV-B (280–315 nm), harms cyanobacteria. UV-B causes lipid peroxidation, DNA strand breaking, thymine dimer production, and decreased orientation and motility in the cyanobacterial cells (Sancar 2016). The primary target of UVR in photosynthetic species is the most critical element of the photosynthetic machinery, which is PSII (Barbato et al. 2000; Schneider et al. 2022). Through photosensitizing interactions with chloride and phycobilins, UV-A radiation affects the cell directly. Because MAAs and scytonemin proved to be extremely photostable and have strong antioxidant and photoprotective properties, the cosmetics sector can use them biotechnologically (Krishnan et al. 2021).

The use of "omics" techniques greatly expanded our understanding of cyanobacterial molecular biology. Because of their unusual structures and distinct biological actions, photoprotective compounds like MAAs and scytonemin remain an unmatched source of pharmacologically active principal compounds (Pathak et al. 2017). Omics-based techniques, such as transcriptomics, proteomics, metabolomics, genomics, as well as bioinformatics, have been acknowledged as useful instruments required to develop novel approaches for identifying photoprotective chemicals from a wide range of animals. Despite significant advancements, the successful identification of natural product targets is still a difficult and time-consuming process that increasingly depends on the efficient integration of technologies based on multi-omics to develop new panomics-based approaches (Zhang et al. 2021). The goal of omics analysis is to thoroughly describe and measure a system's molecular pools. Omics techniques offer a thorough understanding of the molecules in a sample, as opposed to traditional technologies that examine one or a small number of molecules in a single examination. Novel bioproduct discovery has increased exponentially as a result of recent technological advancements, such as the use of omics technology. The idea and methods of omics are not restricted to biological or clinical materials, despite their extensive use in biological systems. Comprehensive analyses of sizable pools of organic chemicals have been accomplished, and there has been an increase in the use of omics techniques for environmental sample analysis (Tang et al. 2020).

The metabolites found in biological fluids were primarily examined using nuclear magnetic resonance spectroscopy (NMR) (Nicholson et al. 1983) and mass spectrometry (MS) (Greef et al. 1983) before the development of "omics" approaches. Multivariate statistics were used to interpret these intricate data sets (Gartland et al. 1990, 1991). Pre-separation of the metabolic components is necessary for MS, and this can be accomplished using analytical methods like gas chromatography (GC), liquid chromatography (LC), or its more sophisticated variant, ultrahigh pressure LC (UPLC). Although specialized methods such as arrayed electrochemical detection and Fourier transformed infrared spectroscopy (FTIR) have been employed in certain instances, their application is restricted by their lack of precise molecule identification. These days, metabolite identification from cell extracts or samples is typically accomplished with the use of high-performance liquid chromatography (HPLC) and mass spectrometry (MS). For both pure compounds as well as complex mixtures, NMR spectroscopy provides comprehensive information on molecular structure by analyzing the change in the spin of H or C atoms (Burja et al. 2003). All of the bioactive compounds can be identified and quantified without bias or targeting using metabolomics, while a limited number of already explored metabolites that are typically associated with a particular metabolic pathway can be identified and quantified using metabolic profiling (Ellis and Goodacre 2012). The targeted metabolome technique, which may be utilized to offer thorough and systematic profiling of metabolite levels in biofluids and tissues, is a method for quantifying limited collections of metabolites (Nicholson and Wilson 1989). Natural bioactive substances like scytonemin and MAAs may be used as clinical medications to improve human well-being and as a strong photoprotectant in the cosmetics industry (Rastogi et al. 2015; Geraldes and Pinto 2021). Therefore, the primary focus of future studies is the characterization, biosynthesis, bioanalysis, and profiling of these naturally occurring UV-absorbing compounds. This chapter on cyanobacterial UV protective chemicals, scytonemin, and MAAs covers the structure, synthesis, role in photoprotection, and several techniques for characterizing these special molecules. A thorough discussion of recent advancements in the field of characterizing these compounds is given, with particular attention paid to the methods used to characterize these multifunctional compounds.

18.2 Omics Technologies in Characterization of Cyanobacterial UV Protective Compounds

Our knowledge of cyanobacterial UV-protective secondary metabolites has been completely transformed by omics technologies, such as transcriptomics, proteomics, and metabolomics. These substances proved to be essential for cyanobacteria's ability to survive in high UVR conditions, and omics techniques have shed light on their synthesis, regulation, and role. Cyanobacteria have evolved a variety of defense mechanisms against UVR damage, ranging from molecular to behavioral strategies (Singh et al. 2010a). One of the many adaptive methods to combat the toxicity of UVR is the biosynthesis of secondary metabolites that absorb UV radiation, such as scytonemin and MAAs. Numerous cyanobacteria species produce these photoprotective metabolites, which are important for UVR screening (Pathak et al. 2017). These substances are regarded as natural photoprotectants and are "multipurpose" secondary metabolites in the truest sense. They make great candidates for usage in the cosmetics sector and for more environmentally friendly skin protection. Consequently, the pharmaceutical and cosmetics industries may use them biotechnologically. The creation of new drugs could greatly benefit from a better understanding of cyanobacterial metabolites and how to use them. Regretfully, the rate of advancements in microorganism sequencing and functional characterization of microbial metabolism has not kept pace. Sequencing has advanced much beyond the characterization of microbial metabolites (Galperin and Koonin 2010). In addition to aiding in the functional assignment of genes, metabolite profiling frequently results in the identification of new metabolic capacities. Techniques like HPLC, electrospray ionization mass spectrometry (ESI-MS), FTIR, and NMR are used to detect novel metabolites or metabolites frequently, which manifests uncharacterized biosynthetic capabilities (Baran et al. 2010; Bowen and Northen 2010; Kind and Fiehn 2010). Over the past ten years, the procedures of discovering and characterizing natural products have benefited greatly from the use of genomic, transcriptomic, proteomic, metatranscriptomic, and metabolomic methods (Atanasov et al. 2021; Rosic et al. 2023).

The molecular structures of extracted substances and, in turn, their biological activity can be influenced by the extraction technique employed. Additionally, biological extracts frequently contain a number of different compounds. In order to enhance the production, it is essential to perform studies in a thorough and precise manner while reducing the possibility of contamination and unstable molecular disintegration. The sciences of genomics as well as transcriptomics, which study genes and the accompanying RNA (or transcripts), have begun adopting next-generation sequencing (NGS) after decades of using Sanger sequencing methodology (Ambrosino et al. 2019). NGS makes it possible to investigate a number of DNA fragments at once quickly and effectively. It has been demonstrated that using transcriptomics and proteomics to apply bioinformatics pipelines enables speedier exploration and functional characterization of novel MNP, such as marine toxins (Xie et al. 2017). Metabolomics combines LC-MS-based metabolomics with other platforms to conduct a thorough investigation of metabolites in biological sample studies (Kuehnbaum and Britz-McKibbin 2013; Clish 2015). In this developing technology, molecules are measured according to their physical characteristics, and metabolites are sorted according to their chemical structure, polarity, or functional groups (Kuehnbaum and Britz-McKibbin 2013). Several gene transcripts generated by interactions between microbes and their hosts were evaluated in metatranscriptomics. This strategy is beneficial since the majority of MNPs are discovered to be the consequence of symbiotic relationships between microbes and other marine species, and alterations in gene expression can be used to identify altered pathways crucial to the creation of bioactive compounds (Jiménez 2018). The variance in ocean microbial communities was also effectively estimated using the metatranscriptome technique (Salazar et al. 2019). On utilizing the already available tools of proteomics and metabolomics pipelines, this method can enhance the extraction of novel bioactive compounds (Fig. 18.1). Secondary metabolite discoveries have increased dramatically as a result of improvements in analytical techniques. Particularly, MS-based metabolomics has emerged as a potent technique for both identifying new compounds and detecting metabolites in intricate biological systems. Recently, metabolomics was used to model cyanobacteria in a systematic manner. To describe alterations in its main metabolism in various environmental settings or in specific mutations, the cyanobacterial metabolome was examined (Schwarz et al. 2013).

18.3 Cyanobacterial UV Protective Compounds MAAs and Scytonemin

As one of their many defense mechanisms against UVR, cyanobacteria can protect themselves from UVR by biosynthesizing UV-absorbing secondary compounds, including scytonemin and MAAs.

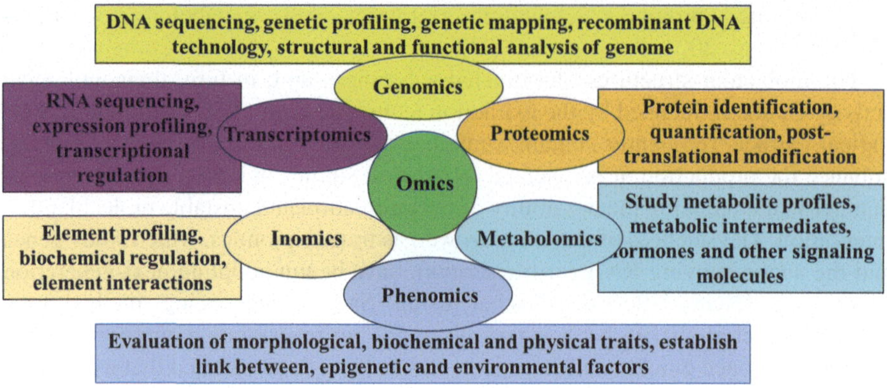

Fig. 18.1 Major branches of omics with main components employed in various integrated techniques

18.3.1 Mycosporine-Like Amino Acids (MAAs)

MAAs are colorless, low-molecular-weight, polar, typically intracellular secondary metabolites. MAAs and a related class of chemicals known as mycosporines are found throughout the biological universe (Rosic and Dove 2011; Wada et al. 2015; Weiss et al. 2022; Bonin et al. 2024; Lalegerie et al. 2024). In terms of chemistry, MAAs are alicyclic molecules that share a methoxy group at C-2 and a core 5-dihydroxy, 5-hydroxymethyl, cyclohex-1, 2-ene ring. Oxo-mycosporines are created by substituting an amino molecule (often an amino acid or amino alcohol) for this core structure in C-3. The imino-mycosporine group is defined by the substitution of a second amino molecule for a ketone.

The variety of chemical structures within both groups is due to the N-substitution on C-3 with various amino acids or amino alcohols (Sinha et al. 2007). UV-absorbing MAAs have been particularly credited with photoprotective functions because they appear to have no role in photosynthesis (Moisan and Mitchell 2001). MAAs exhibit absorption peaks at about 310 nm for oxo-mycosporines and approximately 330 nm for simple imino-mycosporines. MAAs exhibit typical UV absorption spectra (Table 18.1). They have distinct spectral characteristics because of their special molecular bonding and structure. MAAs are great options for enhanced skin protection in an environmentally responsible manner as well as for application in the cosmetics industry (Vega et al. 2021; Raj et al. 2021). MAAs are extremely varied, water-soluble substances with good heat tolerance and photostability. Environmental factors, such as variations in UVR levels, can alter the production of MAAs (Rosic 2021; Nishida et al. 2021). Due to their photoprotective action, MAAs are regarded as natural sunscreens and antioxidants, making them primarily utilized in anti-UV skincare products in the cosmetics industry (Ghedifa et al. 2021).

Table 18.1 Molecular structure of MAAs and their corresponding absorption maxima (modified from Sinha et al. 2007)

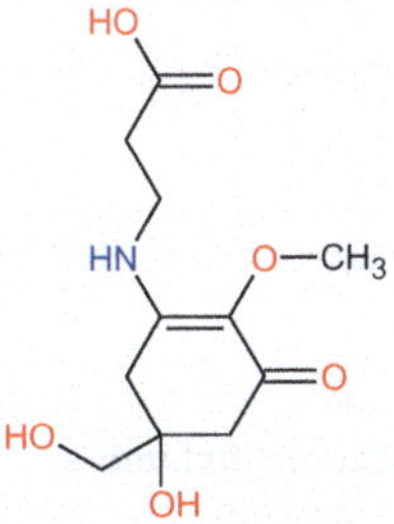 Mycosporine-glycine λ_{max} (310 nm)	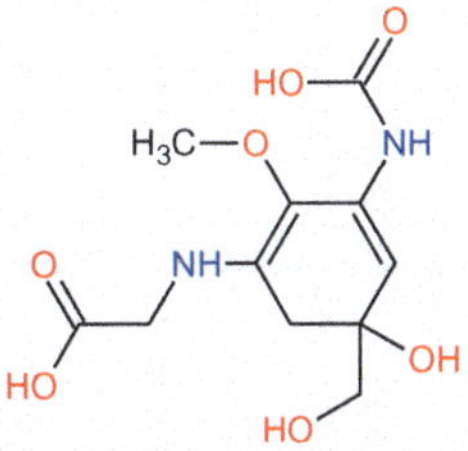 Mycosporine-2-glycine λ_{max} (334 nm)	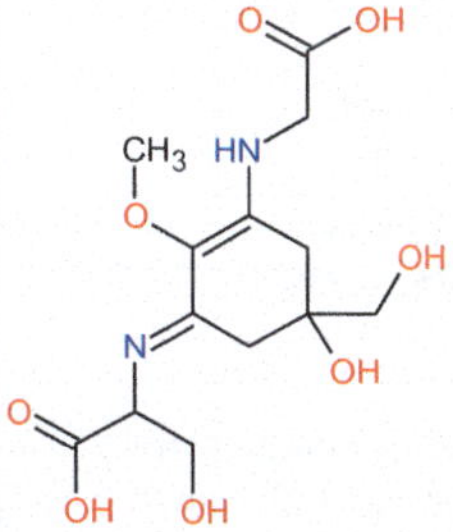 Shinorine λ_{max} (334 nm)
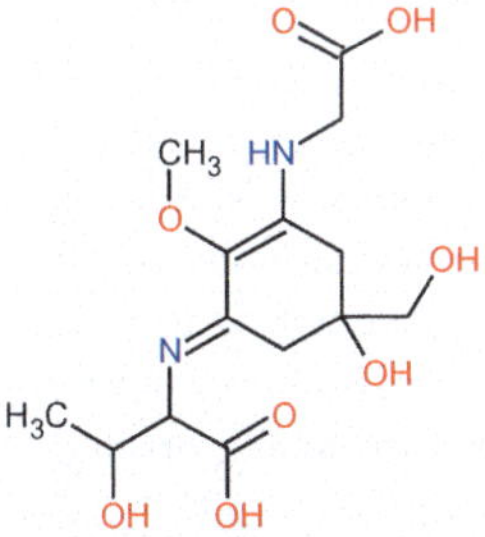 Porphyra-334 λ_{max} (334 nm)	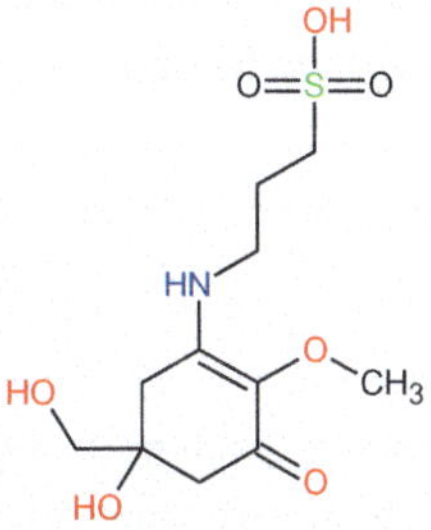 **Mycosporine-taurine** λ_{max} **(309 nm)**	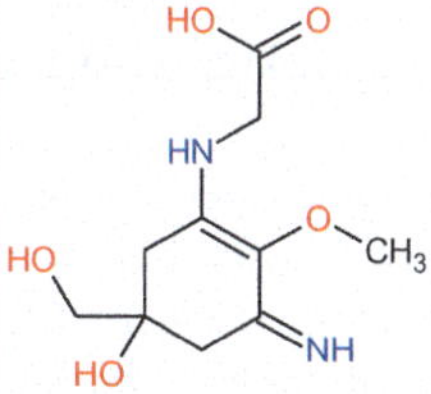 **Palythine** λ_{max} **(320 nm)**
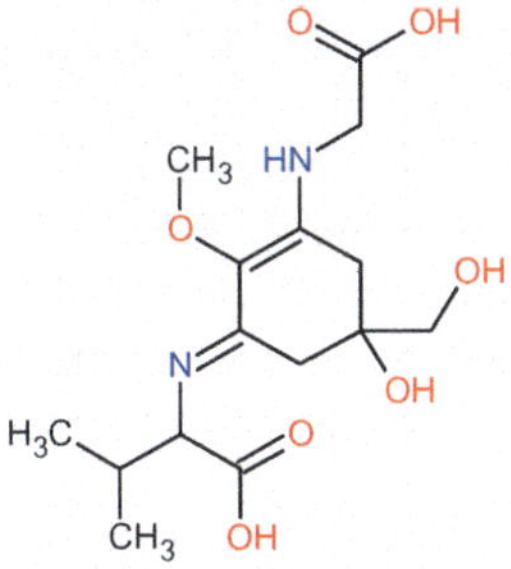 Mycosporine-glycine-valine λ_{max} (335 nm)	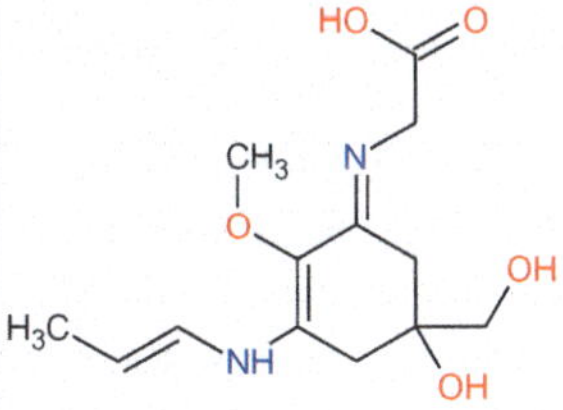 **Palythene** λ_{max} **(360 nm)**	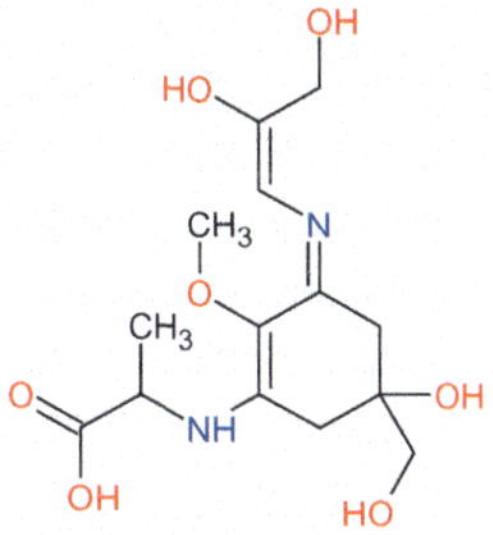 **Euhalothece-362** λ_{max} **(360 nm)**
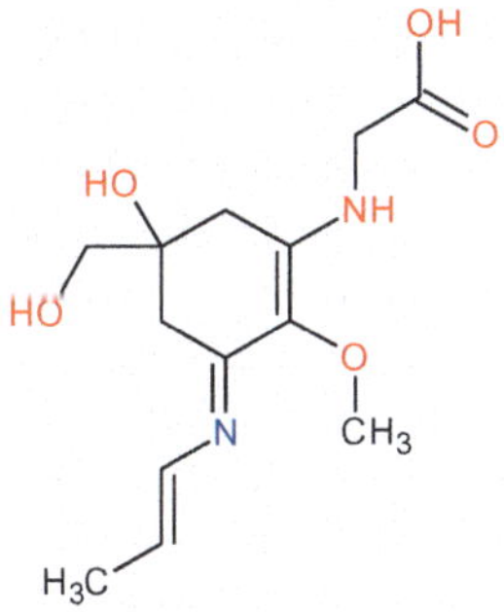 Usujirene λ_{max} (357 nm)	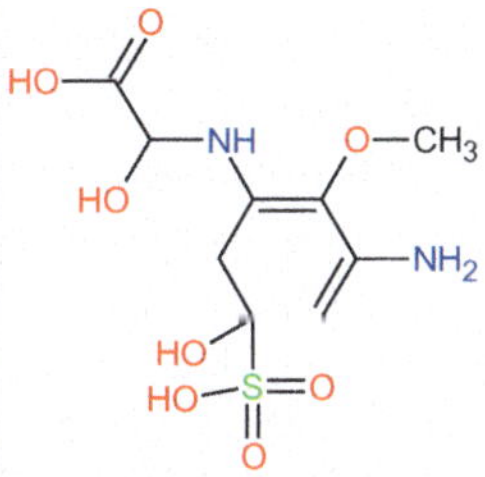 **Palythine-serine-sulfate** λ_{max} **(320 nm)**	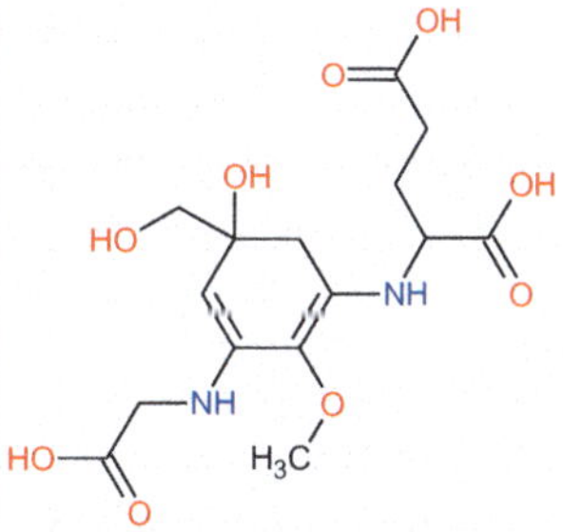 **Mycosporine-glutamic acid-glycine** λ_{max} **(330 nm)**

(continued)

Table 18.1 (continued)

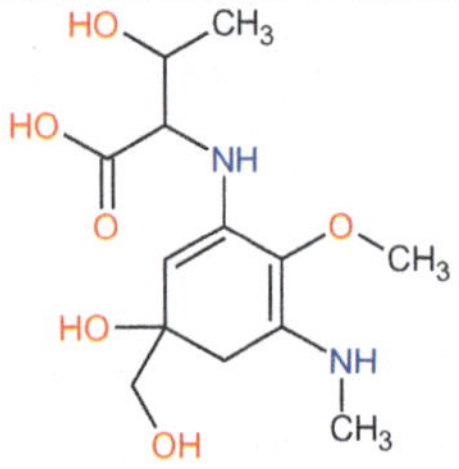 Mycosporine-methylamine-threonine λ_{max} (327 nm)	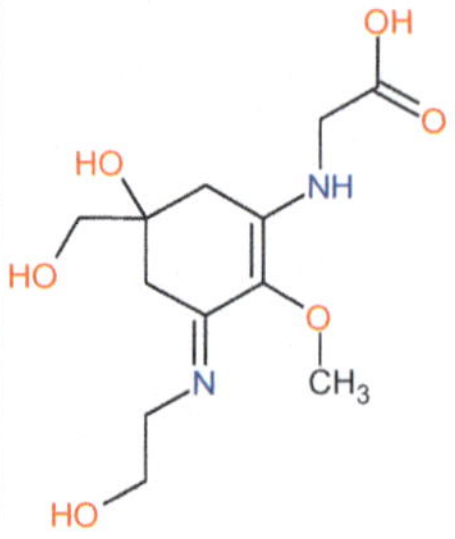 **Asterina-330 λ_{max} (330 nm)**	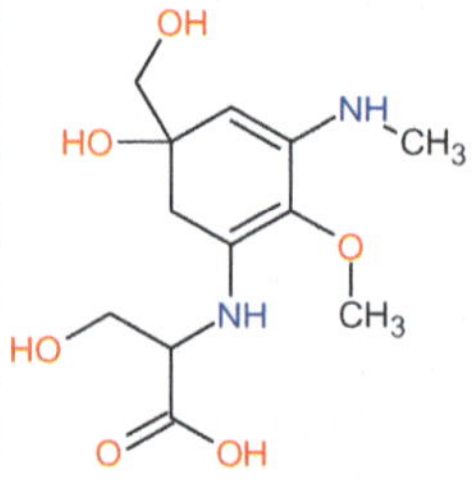 **Mycosporine-methylamine-serine λ_{max} (327 nm)**

18.3.2 Genetics and Biosynthesis of MAAs

Studies have shown that the enzymatic mechanism for the synthesis of MAA originated from a cyanobacterial ancestor. Multiple lines of evidence suggest that the biosynthesis process for MAAs is the shikimic acid pathway. According to Favre-Bonvin et al. (1987), the plant fungal pathogen *Trichothecium roseum* particularly absorbed the shikimate intermediate [U-^{14}C]3-dehydroquinic acid, which produced the identical structure related to fungal mycosporines. Similarly, radiolabeled MAAs with high specificity were formed in *Chlorogloeopsis* sp. PCC 6912, when [^{14}C] pyruvate was examined for MAAs biosynthesis via phosphoenol pyruvate, an obligatory precursor of the shikimic acid pathway. In contrast, [^{14}C] acetate (polyketide pathway) did not show the same observations (Portwich and Garcia-Pichel 2003). Furthermore, the cyanobacterium's production of MAAs was suppressed by an external supply of tyrosine, a known inhibitor of the shikimic acid pathway. When *Nostoc commune* was treated with glyphosate (an inhibitor of the shikimic acid pathway), Sinha et al. (2003) saw the same outcome. Based on these findings, it was determined that the shikimic acid pathway's branch-point intermediate is where MAA production starts (Fig. 18.2).

Balskus and Walsh (2010) described the first steps in the biosynthesis of MAAs in *Anabaena variabilis*. The cyanobacterium *Anabaena variabilis* ATCC 29413 was reported to produce shinorine (ava_3855 to ava_3858), through a gene cluster consisting of four genes (Balskus and Walsh 2010). These genes produce metabolites that change the pentose phosphate pathway intermediate sedoheptulose 7-phosphate (SH7-P) into shinorine. The precursor is transformed to 4-deoxygadusol by 2-epi-5-epi-valiolone synthase (a homologue of dehydroquinate synthase, or DHQS) and an O-methyltransferase (O-MT). Mycosporine-glycine and serine are next converted to serine by an ATP-grasp homologue. The complete in-frame deletion of the O-MT gene from *A. variabilis* ATCC 29413 revealed the intimate relationship between MAA synthesis and the pentose phosphate and shikimate pathways. Previous research using cyanobacteria and fungi showed this, and it was shown that the shikimate route was the main mechanism for MAA production in response to UV radiation (Pope et al. 2015). *Anabaena doliolum* uses mycosporine-glycine as a

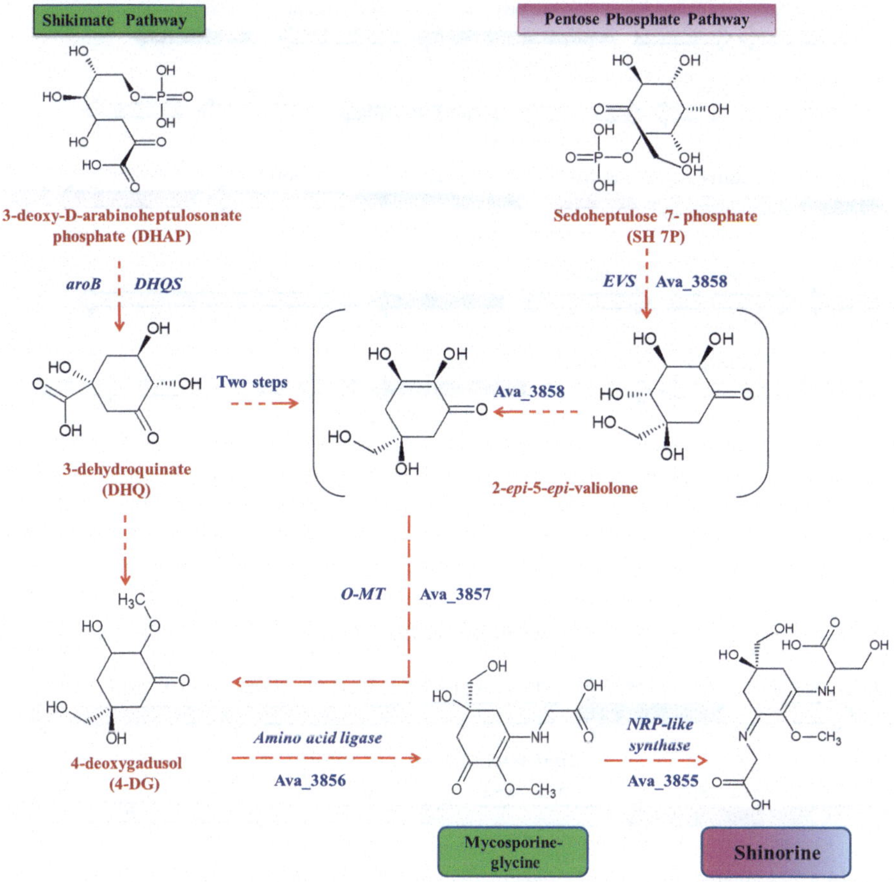

Fig. 18.2 Diagrammatic illustration of the processes by which cyanobacteria produce MAAs. It displays synthesis from the pentose phosphate intermediate and the shikimic acid pathway intermediate. On both sides of the arrows are references to the genes and enzymes that are part of the bioprocess. DAHP stands for 3-deoxy-D-arabino-heptulosonate; S7P for sedoheptulose-7-phosphate; 3-DHQ for 3-dehydroquinate; EV for 2-epi-5-epi-valiolone; PEP for phosphoenolpyruvate; and E4P for erythrose 4-phosphate (modified from Pope et al. 2015)

precursor for the UV-B-induced synthesis of porphyra-334 and shinorine (Singh et al. 2010b).

An attempt was made to identify potential genes involved in MAA production by using bioinformatics methods to analyze the genomes of cyanobacteria that synthesize MAAs and those that do not (Singh et al. 2010b). The capacity of four cyanobacteria to manufacture MAA was examined: *Anabaena* sp. PCC 7120, *Anabaena variabilis* PCC 7937, *Synechococcus* sp. PCC 6301, and *Synechocystis* sp. PCC 6803. *A. variabilis* PCC 7937 was the only cyanobacteria that could manufacture MAA out of all those that were investigated. Two sets of the DHQS genes YP-324358 and YP-324879 were present in *A. variabilis* PCC 7937

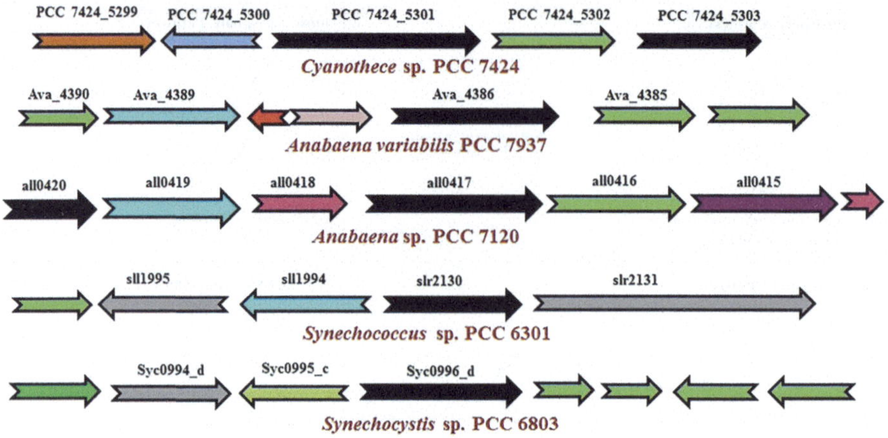

Fig. 18.3 The cyanobacterium *Cyanothece* sp. PCC 7424's genomic region is compared to that of *Anabaena variabilis* PCC 7937, *Anabaena* sp. PCC 7120, *Synechococcus* sp. PCC 6301, and *Synechocystis* PCC 6803 for the gene PCC 7424_5301 (WP_015956914) DHQ synthase, which is not involved in the manufacture of MAAs. (Modified from Singh et al. 2010b)

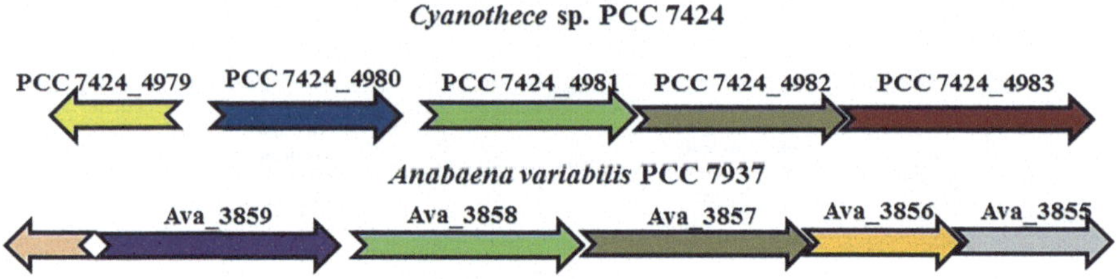

Fig. 18.4 *Anabaena variabilis* PCC 7937 and *Cyanothece* sp. PCC 7424's genomic regions are compared for the gene PCC 7424_4981 (WP_015956914) DHQ synthase, which is involved in the production of MAAs. (Modified from Ahmed et al. 2017)

(Fig. 18.3). Gene YP-324358 has an O-MT gene YP-324357 downstream to it, although cyanobacteria did not have this combination (Fig. 18.4).

The results of Balskus and Walsh (2010) contradict the long-standing assumption that MAA biosynthesis involved a shikimate pathway intermediate. These authors failed to observe any 6-deoxygadusol production when the putative substrate 3-dehydroquinate was incubated with the DHQS and O-MT homologues (NpR 5600 and NpR 5599respectively) from *Nostoc punctiforme* (ATCC 29133) and the typical cofactors S-adenosylmethionine (SAM), nicotinamide adenine dinucleotide (NAD^+), and Co^{2+}. In addition, treatment of the intermediate of the pentose phosphate pathway: SH7-P with NpR 5600 and NpR 5599 in the presence of SAM, NAD^+, and Co^{2+}, produces a single product, 6-deoxygadusol. The results of Balskus and Walsh (Balskus and Walsh 2010) confirm that the expression of the gene Ava_3856 from *A. variabilis* can convert 6-deoxygadusol and glycine into mycosporine-glycine in the presence of Adenosine Triphosphate (ATP) and Mg^{2+}

cofactors. This study also demonstrated that the enzyme Ava_3855, which functions similarly to non-ribosomal peptide synthetase (NRPS), is involved in the biosynthesis of shinorine through the ATP-mediated interconversion of serine as well as mycosporine-glycine. However, little is known about how mycosporine-taurine, one of the many other well-known oxo-carbonyl MAAs, condenses amino acids.

The pathway, including the homologous genes (NpR5598–5600) involved in MAA production in *N. punctiforme* ATCC 29133, is comparable to that of *A. variabilis* (Katoch et al. 2016). Although NpR5598 functions as a homologue of Ava_3856 in *N. punctiforme*, its exact action is unknown. The final step, the condensation of serine to mycosporine-glycine, is an exception, though, as it is regulated by a protein that resembles D-ala-D-ala ligase (NpF5597) instead of a non-ribosomal peptide synthetase found in *A. variabilis* (Balskus and Walsh 2010; Singh et al. 2010b) has been identified as a major contradiction in the biosynthesis of imino-mycosporines (Gao and Garcia-Pichel 2011). We conclude that shinorine production in *N. punctiforme* is carried out by a different enzyme or pathway. The enzymes 2-epi 5-epivaliolone synthase (EVS) and O-MT convert the sedoheptulose-7-phosphate (an intermediary of the pentose phosphate cycle) to 4-deoxygadusol, the parent core structure of MAAs in cyanobacteria. Constitutive production of MAAs was only detected in strains where genes formed a colinear cluster (Dextro et al. 2023). Genome mining has identified homologues of the EVS gene, which are responsible for the production of MAAs in cyanobacteria and are typically lacking in nonproducers (Singh et al. 2010b; Rosic 2012; Rosic 2022). Despite lacking an EVS gene, *Synechocystis* sp. PCC6803's classification is debatable because it produced three unique MAAs after being exposed to UVR for a few hours (Zhang et al. 2007). The *A. variabilis* ATCC 29413 was used to clarify the genetics of MAAs and the primary processes engaged in their manufacture (Spence et al. 2012). Based on a knockout experiment in *A. variabilis* ATCC 29413, Spence et al. (2012) postulated the presence of functionally redundant but separate routes in the production of MAAs. Both the pentose phosphate and, most likely, the shikimic acid pathways are involved in MAA biosynthesis, which produces shinorine through the convergent intermediate 4-deoxygadusol. In their study, Mogany et al. (2022) conducted a bioinformatics analysis of the genome of *Euhalothece* sp. revealing a unique mys gene cluster consisting of six genes (mysA to mysE), in contrast to the four genes typically found in other MAA-producing cyanobacteria. Notably, the presence of both EEVS (Erythrose-4-phosphate/Erythrose-4-phosphate synthase) and DHQ (3-Dehydroquinate) synthase suggests that this organism has the capability to produce mycosporine-like amino acids (MAAs) through both the shikimate pathway and the sedoheptulose 7-phosphate pathway.

18.3.3 Scytonemin

Nägeli was the first to report this pigment in certain terrestrial cyanobacteria (Nägeli 1849). About 300 cyanobacteria have been found to have this yellow-brown hydrophobic pigment in their extracellular polysaccharide sheaths (Kang et al. 2020).

Scytonemin is lipid-soluble and extremely durable. When the stratospheric ozone layer was absent during the Precambrian period, scytonemin helped cyanobacteria thrive and shielded them from the deadly effects of UV radiation (Irankhahi et al. 2024). In addition to photoprotection, scytonemin has been shown to have an antiproliferative effect on various types of malignant cells (Evans et al. 2021). Scytonemin is a stable UV-protective pigment with potential applications in medicine and as a natural, eco-friendly sunscreen. However, its industrial application has been limited due to low productivity (Madrahi and Naeimpoor 2023). In contrast to pure scytonemin, which exhibits considerable absorption at 252, 278, and 300 nm and a maximum absorption at 386 nm, the highest absorption of scytonemin in vivo occurs at 370 nm (Fig. 18.5). A number of environmental factors play role in the biosynthesis of scytonemin and to upscale its production for its biotechnological and commercial application is an important area of research (Gao et al. 2021; Bennett and Soule 2022; Casero et al. 2024). In their study, Ručová et al. (2023) demonstrated that scytonemin from *Nostoc commune* is a promising natural compound for use in sunscreen products due to its high SPF value of 33.34 ± 0.02 and potent antioxidant properties. Its ability to protect against UV radiation and reduce oxidative stress makes it an attractive ingredient for both cosmetic and therapeutic applications (Sen and Mallick 2022). Various computational techniques were employed to predict the binding affinity of Scytonemin and its derivatives to the active site of filarial thioredoxin reductase (TrxR). Scytonemin and its derivatives could be developed into novel treatments for lymphatic filariasis and other diseases caused by filarial parasites (Sen et al. 2023).

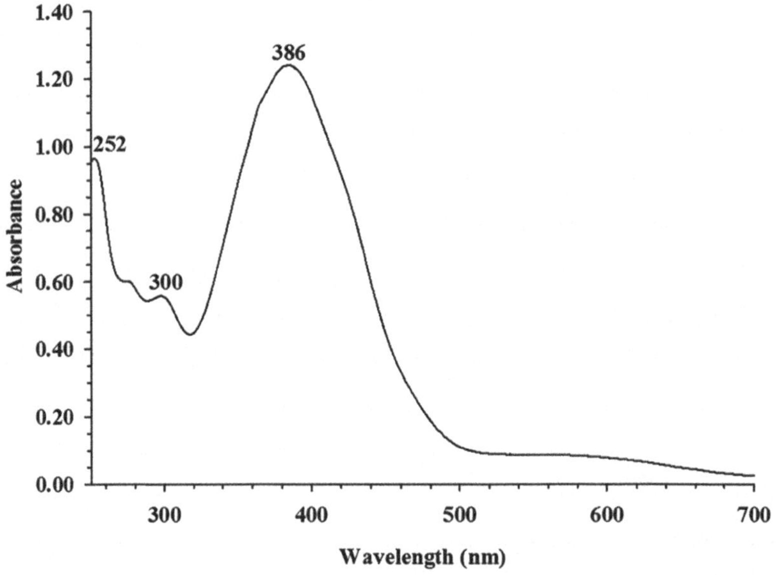

Fig. 18.5 Absorption spectrum of scytonemin having peaks at 252, 300, and 386 nm

It has a molecular mass of 544 Da and is a dimer made up of phenolic and indole monomers. Unlike other natural compounds, scytonemin has an olefinic carbon atom connecting both monomers. This connection, known as "the scytoneman skeleton," gives scytonemin a novel ring structure (Proteau et al. 1993). Fuscochlorin and fuscorhodin are the names given to the two forms of scytonemin, which are oxidized (green) and reduced (red) (Kylim 1927, 1937; Garcia-Pichel and Castenholz 1991). Throughout the extraction process, redox and acid-base conditions dictate the presence of both forms. Based on ^{1}H and ^{13}C NMR and MS investigations, three novel compounds—dimethoxyscytonemin, tetramethoxyscytonemin, and scytonin—were discovered in the chemical extracts of *Scytonema* sp. (Bultel-Poncé et al. 2004). The "scytoneman skeleton" of the scytonemin is the source of all these forms. Scytonemin-3a-imine (Table 18.2), a novel member of the "scytoneman skeleton," was recently discovered in the cyanobacterium *Scytonema hoffmani*, which grows in extremely bright light (Grant and Louda 2013).

18.3.4 Genetics and Biosynthesis of Scytonemin

Using a random transposon mutagenesis technique, the genetics of biosynthesis of scytonemin in the cyanobacterium *N. punctiforme* was published (Soule et al. 2007). The 18-gene cluster's genomic area (Npun_R1276 to Npun_R1259) in *N. punctiforme* has been discovered for the manufacturing of scytonemin (Soule et al. 2007). Soule et al. have examined the gene cluster linked to *Nostoc punctiforme* scytonemin production and the related two-component regulatory system (Soule et al. 2009a, b). Tryptophan and tyrosine were proposed as precursors of the biosynthesis of scytonemin based on the chemical structure of scytonemin, previous feedback inhibition, and radiotracer investigations (Proteau et al. 1993). Along with genes producing proteins involved in the shikimic acid pathway (AroB and AroG), the identified gene cluster also contains proteins implicated in the production of tryptophan (TrpE, TrpC, TrpA, TrpB, and TrpD) and tyrosine (TyrP) (Soule et al. 2007; Soule et al. 2009a). Genomic analysis identified additional copies of genes encoding the proteins AroB, AroG, TrpE, TrpC, TrpA, TrpB, TrpD, and TyrP elsewhere in the genome of *N. punctiforme*. These genes are exclusively involved in the biosynthesis of scytonemins because they provide the precursor amino acids (Soule et al. 2007; Soule et al. 2009a).

For the production of adequate raw materials required for the synthesis of scytonemin as well as necessary amino acids, the genome of *N. punctiforme* contains some replica genes of the shikimic acid pathway. Comparative genomic analysis, biochemical, genetic, and sequence similarity investigations demonstrated the presence of genes linked to scytonemin biosynthesis in many cyanobacteria, including *Anabaena* sp. PCC 7120, *Lyngbya* sp. PCC 8106, *Chlorogloeopsis* sp. CCMEE 5094, and *Nodularia* sp. CCY 9414 (Soule et al. 2007; Soule et al. 2009a). The assembly of scytonemin from core metabolites was also suggested to be facilitated by six putative genes of the scytonmein gene cluster (Npun_R1276 to Npun_R1271 in *N. punctiforme*), designated as scyA, scyB, scyC, scyD, scyE, and scyF

Table 18.2 The molecular makeup and absorption maxima of the many types of scytonemin that are frequently found in cyanobacteria

UV-absorbing pigment (Scytonemins)	Structure	λ_{max} (nm)	ε ($M^{-1}\ cm^{-1}$)	Reference
Reduced scytonemin	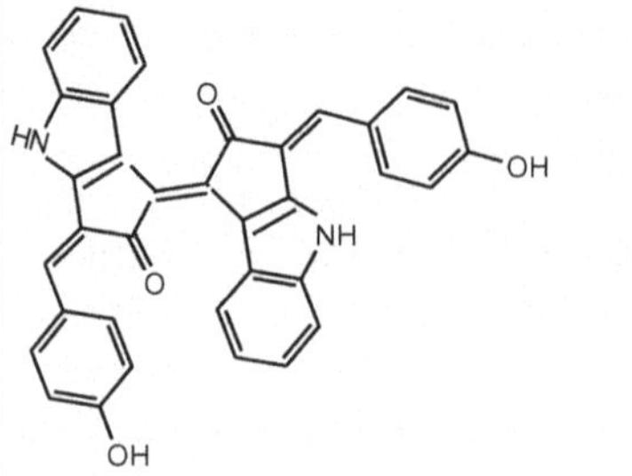	246 276 314 378 474 572	30,000 14,000 15,000 22,000 14,000 7600	Varnali and Edwards (2014)
Dimethoxyscytonemin	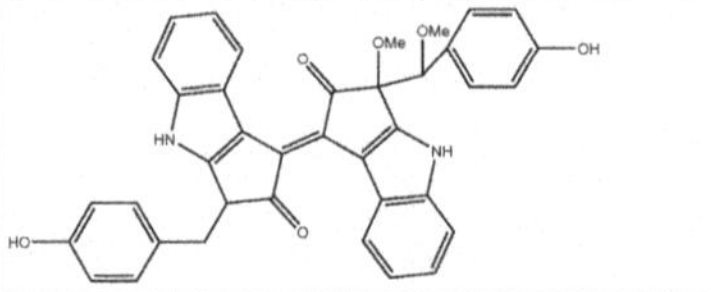	316 422	18,143 23,015	Varnali and Edwards (2014)
Tetramethoxyscytonemin	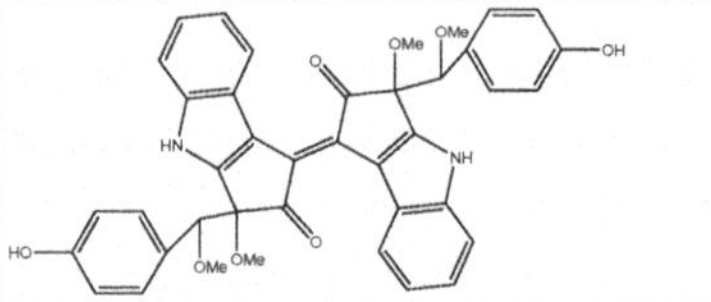	562	5944	Varnali and Edwards (2014)
Scytonemin-3a-imine	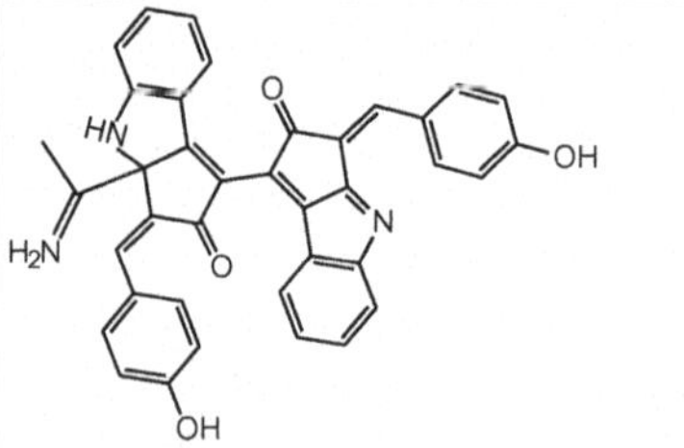	237 366 437 564		Grant and Louda (2013)
Scytonin	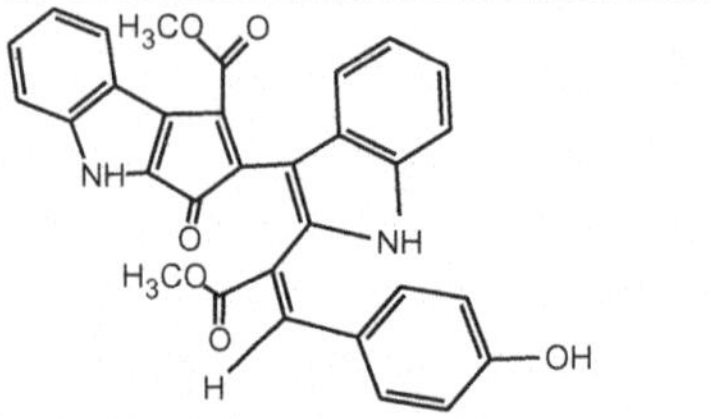	225 270	37,054 22,484	Varnali and Edwards (2014)
Oxidized scytonemin	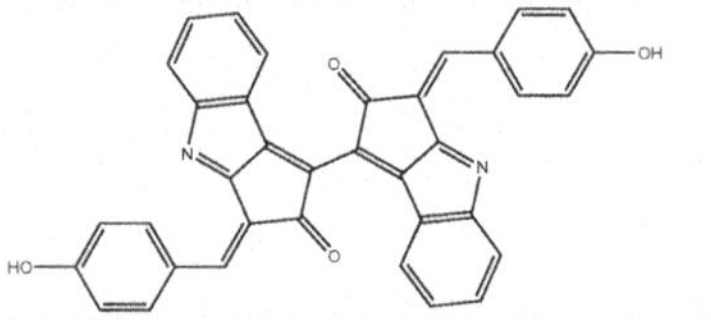			Varnali and Edwards (2014)

(continued)

Table 18.2 (continued)

UV-absorbing pigment (Scytonemins)	Structure	λ_{max} (nm)	ε ($M^{-1}\ cm^{-1}$)	Reference
Scytonemin A	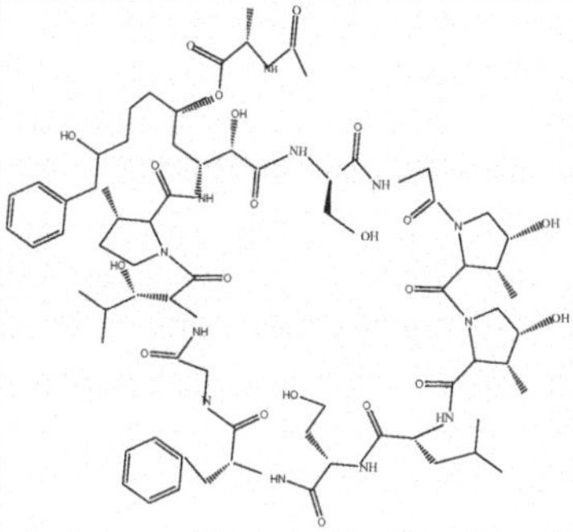			Helms et al. (1988)

Modified from Simeonov and Michaelian (2017)

(Soule et al. 2009a). ScyD, ScyE, and ScyF are among the recently discovered proteins that have export signal domains, while ScyA, TyrP, and NPR1259 have transmembrane domains. These results suggest that the periplasm and the cytoplasm both contribute to the production of scytonemin.

A two-component regulatory mechanism (Npun_F1277/Npun_F1278) has been identified directly upstream of the genes involved in the biosynthesis of scytonemin in the genome of *N. punctiforme* (Soule et al. 2009a). Comparative genomic research demonstrated their function in the control of scytonemin production. The analysis of gene expression showed that exposure to UV-A elevated the genes that were included in the identified 18-gene cluster, confirming that UV-A radiation has a role in inducing scytonemin biosynthesis. Through the overexpression of respective genes in *E. coli*, protein purification, and subsequent spectrophotometric and HPLC analysis, the ScyA, ScyB, and ScyC proteins were characterized in vitro and demonstrated their importance in the first stage of assembly of scytonemin (Balskus and Walsh 2008, 2009). ScyB (Npun_R1275), a homologue of leucine dehydrogenase, catalyzes the first biosynthetic step by oxidatively deaminating L-tryptophan to produce indole-3-pyruvic acid. The labile β-keto acid precursor is then produced by the acyloin coupling of p-hydroxyphenyl pyruvic acid and indole-3-pyruvic acid, which is mediated by a thiamin-dependent enzyme called ScyA (Npun_R1276) (Balskus and Walsh 2008). The β-ketoacid precursor is then cyclized and decarboxylated by ScyC (Npun_R1274) to produce acetone. ScyC produces a molecule that is oxidized to produce a precursor of scytonemin monomeric (Balskus and Walsh 2009). The ScyC phylogeny provides the most precise evaluation of organisms potentially capable of producing scytonemin, as it includes the smallest group of taxa. All taxa with a ScyC homolog are also found in the ScyA and ScyB phylogenies. However, some taxa present in the ScyA and ScyB trees lack a ScyC homolog, suggesting they cannot produce scytonemin. Additionally, each taxon possesses only a single copy of ScyC (Tamre and Fournier 2022). The *N. punctiforme* genome and the many biochemical processes engaged in the production of scytonemin have been thoroughly investigated (Soule et al. 2007; Balskus and Walsh 2008). The three *N. punctiforme* genes, namely scyABC, have recently been heterologously expressed in *E. coli*, leading to the formation of the monomer group of scytonemin with the help of *E. coli* when tryptophan and tyrosine are supplemented (Malla and Sommer 2014). This discovery implies that the ScyA, ScyB, and ScyC proteins are necessary for the heterologous synthesis of scytonemin monomer in *E. coli* and further supports the role of the scyABC gene product in scytonemin biosynthesis. The indigenous enzymes of *E. coli* can perform the extra reaction needed for monomer production (Malla and Sommer 2014). It was discovered that the ΔscyE strain did not produce scytonemin nor did it accumulate any of the anticipated stages of the scytonemin biosynthesis pathway. Scytonemin production in the ΔscyD and ΔscyF strains was found to be comparable to that of the WT; however, it was somewhat inhibited in the ΔscyF strain. Together, these results imply that ScyD and ScyF are not necessary for *N. punctiforme* latter stages of scytonemin production. The idea that Npun_F1277 and Npun_F1278 products control the production of scytonemin was further supported by this discovery. Naurin et al. (2016) investigated the function of the response regulator (RR) protein, encoded by Npun_F1278, which was previously discovered using comparative genomic analysis and was thought to control scytonemin production. Even after being stimulated by UV-A, the mutant strain of Npun_F1278 of *N. punctiforme* was unable to synthesize any scytonemin at all, indicating that RR Npun_F1278 is necessary for the production of scytonemin. Based on comparative genomics research, a highly conserved cluster of five genes known as the "ebo cluster" was recently discovered in the scytonemin operon (Klicki et al. 2018). Although this "ebo cluster" is common and well-preserved throughout a number of phyla, its exact role is not yet understood. Klicki et al. (2018) created in-frame deletion mutants in *Nostoc punctiforme*, one for each ebo gene and one for the complete cluster. Their investigation clarified the ebo cluster's role. All ebo genes are involved in the production of scytonemin, as evidenced by the fact that all ebo mutants acquired scytonemin monomer while being scytoneminless upon UV-A induction. Gao et al. (2023) investigated the effects of periodic short-term solar irradiation with varying intervals and durations on the cell culture of *Nostoc flagelliforme*. The study revealed that a daily exposure of 30 minutes to solar irradiation was the most effective strategy for enhancing scytonemin production while minimizing growth inhibition. Additionally, the technological potential could be improved by incorporating $NaHCO_3$ or tryptophan into the culture medium. This research demonstrates a practical approach to leveraging environmental solar radiation for the efficient production of UV-induced biochemicals in cyanobacteria.

18.4 Techniques Used for Metabolic Profiling of UV-Screening Compounds

For the identification of metabolites, new advances in analytical technology for secondary metabolite characterization can yield additional information from the experimental data produced. These techniques are fast and high-resolution separation systems like ultra-performance liquid chromatography (UPLC), higher sensitivity NMR systems, and wide dynamic range, high mass accuracy MS equipments (Moco et al. 2007). Compared to chromatographic procedures, which require longer analysis times, techniques like NMR, FTIR, and direct injection mass spectrometry (DIMS) are better suited for larger replication numbers since they use high-throughput, short analysis time tactics (Dunn et al. 2005). The desired metabolites can be extracted using either an aqueous solvent (for water-soluble compounds like MAAs) or an organic solvent (for lipid-soluble molecules like scytonemin). The chemical characteristics of solvents determine the kind of metabolites that are extracted (Moco et al. 2007). The secondary metabolites, including MAAs, scytonemin, chloride, and carotenoids, are frequently analyzed using UV-visible spectrophotometry. It calculates how much UVR, or visible light, is absorbed by a material in solution. If any recorded data is available, a spectrophotometer can be used to identify chemical molecules in qualitative analysis. Quantitative spectrophotometric analysis is done to quantify the number of molecular entities that have captured the radiation (Behera et al. 2012). The molecular environment of a molecule's chromophores determines its absorbance strength and wavelength of absorption, in addition to its chemical structure (Schmid 2001). By interpreting and comparing mass spectra, mass spectrometry offers highly sensitive metabolite detection and identification. Accurate mass measurements are also used to determine chemical formulas (Dunn et al. 2005). MS analyzes a certain phase called the gaseous ionic state by moving the analytes from the condensed phase to the gas phase and then ionizing them (Banerjee and Mazumdar 2012). The benefits of MS-based metabolic profiling techniques include high throughput, depth of coverage, and excellent sensitivity and selectivity. Prior to the invention of ESI-MS, a number of ionization techniques (such as chemical ionization and electron ionization) were employed, but none of them could overcome the tendency of analyte fragmentation. In 1989, Fenn created electrospray ionization, which uses repeated charges to ionize entire chemical entities (Fenn et al. 1989). Chromatographic separation is not necessary because electrospray and atmospheric pressure chemical ionization liquid chromatography (APCI) procedures are methods that generate molecular ions with little fragmentation. There is no analyte fragmentation during soft ionization, and the analyte retains very little of its residual energy. Additionally, the gas phase preserves very weak noncovalent interactions (Przybylski and Glocker 1996). The sequential charge causes the resulting ions' m/z values to decrease, and they fall inside the mass ranges of all traditional mass analyzers. ESI proved to be beneficial in the generation of gas-phase ions from biological materials. Additionally, it aids in quick identification of metabolites based on molecular mass and the following MS analysis of metabolites for structural characterization (Przybylski and Glocker 1996). Nevertheless, it is also damaging, in contrast to FTIR and NMR. Solvents made of nonvolatile buffers may cause contamination issues in ESI-MS. The only technique that can currently reach the desired mass resolutions (greater than 100,000 FWHM) is Fourier transform ion cyclotron resonance mass spectrometry (FTICR-MS). Fourier transform (FT) mass spectrometry (MS) provides an alternative technique with a mass accuracy of better than 1 ppm and exceptionally high resolution (Dunn et al. 2005). However, the instrument's application has been restricted thus far due to its high cost. Prior to mass detection, the majority of MS applications in metabolomics include a separation technique, usually capillary electrophoresis (CE), GC, or LC. The exploration of volatile and thermally stable polar as well as nonpolar compounds using GC-MS-based metabolomic approaches has rapidly grown. High-resolution chromatographic separations using capillary gas chromatography columns at temperatures as high as 350 °C advance the analyses. Chemical ionization mass spectrometry or electron impact spectrometry is used for the sensitive subsequent identification of components. Metabolite identification is done by comparing retention time or index with pure chemicals as well as interpreting or comparing mass spectra utilizing retention index/mass spectral library databases (Zhang et al. 2012). Both GC-MS and LC-MS are common methods that are capable of detecting a large number of different chemicals. However, because of the different ionization processes employed, the MS instrument configurations for these two approaches are different. Large spectrum libraries, improved retention of tiny molecules that get separated fast with the solvent front in reverse-phase LC-MS processes, and greater chromatographic resolution are just a few of the benefits that GC-MS offers over LC-MS-based methods (Zhang et al. 2012). GC-MS has a bias for low molecular weight (350 Da) metabolites that are volatile and thermally stable. GC-MS cannot readily analyze the metabolites because most of them are polar and nonvolatile.

For metabolomic applications, GC-MS is a relatively low-throughput technology due to derivatization, lengthy chromatographic method (10–60 min per run), and substantial sample preparation (Zhang et al. 2012). The use of FTIR in biological systems has grown, especially for evaluating the functional groups found in metabolites. Because FTIR is a "fingerprint technique," no two samples have the same number of peaks or greatest peak intensities in their FTIR spectra. Water can exhibit significant absorptions in FTIR; hence, attenuated total reflection (ATR) or drying samples to exclude water are used. Under some circumstances, NMR-based metabolomics can offer a "holistic view" of the metabolites, making it a suitable and beneficial method for metabolomic research (Zhang et al. 2012). NMR is said to be one of the most selective techniques for identifying substances because each molecule has a unique NMR spectrum (Moco et al. 2007). It exploits the atomic nucleus' spin characteristics. Metabolite-profiling techniques like NMR or MS can help identify metabolites; however, combining metabolite data from several sources is frequently necessary to provide a complete chemical description of a molecule (Moco et al. 2007). Each unique resonance in an NMR spectrum is exclusive to a specific molecule. Since the analysis is nondestructive, there is no need to preselect the criteria. It has made it possible to examine silent phenotypes by providing more thorough coverage of the metabolic profile. Sample pretreatment for NMR is minimal, and analyte chemical derivatization is not necessarily required beforehand. UV protection metabolites from cyanobacteria can also be characterized using Raman spectroscopy. In their study Jehlička et al. (2023) utilized Raman spectroscopy to

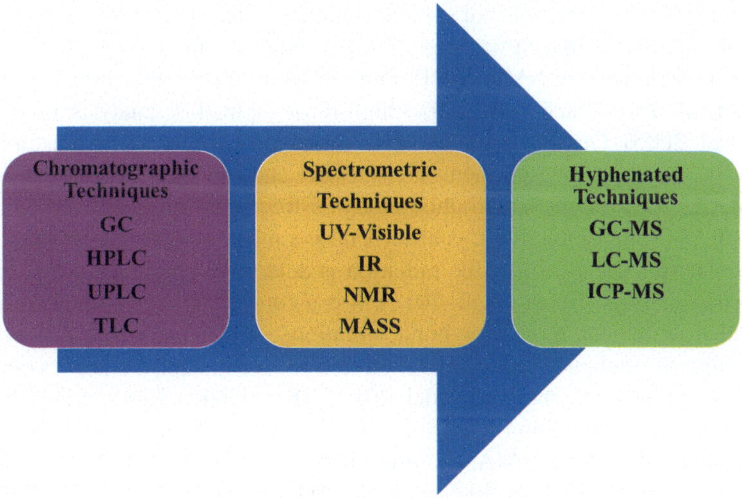

Fig. 18.6 Different analytical techniques employed for simultaneous estimation of samples

identify and map the distribution of scytonemin and carotenoids in epilithic and endolithic cyanobacterial communities inhabiting marble, serpentinite, and gypsum substrates, demonstrated the effectiveness of Raman spectroscopy in detecting these pigments within various lithic environments, highlighting its potential for noninvasive analysis of microbial colonization in different rock types. The application of excitation-laser-line-scanning spontaneous Raman scattering spectral microscopy enabled the detailed visualization of scytonemin's redox status. This technique allows for the detailed analysis of the chemical state of scytonemin within the cells. The redox changes of scytonemin were confirmed by comparing the Raman spectra of cyanobacterial cells with solid-state scytonemin. This comparison provided insights into how the redox state affects the spectral characteristics of scytonemin (Tamamizu et al. 2023). It is a completely noninvasive, label-free method that stimulates molecular bond vibrations. It has also made it possible to perform high-throughput, reagent-free, nondestructive, and quick analyses on a variety of sample types (Downes and Elfick 2010). Therefore, UV-VIS spectroscopy, followed by HPLC, ESIMS/MS, FTIR, and NMR, represent the key techniques for characterizing UV-protecting metabolites including MAAs and scytonemin (Fig. 18.6).

18.5 Techniques Used for Characterization of MAAs

In order to identify metabolites, new advances in analytical technology for secondary metabolite characterization can yield additional information from the experimental data produced. Higher sensitivity NMR systems, rapid and high-resolution separation systems like UPLC, and large dynamic-range, high-mass accuracy MS instruments are some examples of these techniques (Table 18.3) (Moco et al. 2007). Compared to methods like chromatographic procedures, which require longer analysis times, techniques like NMR, FTIR, and DIMS are better appropriate for bigger replication numbers since they utilize high-throughput, fast analysis time tactics (Dunn et al. 2005). Only minimal amounts of samples are needed for examination because spectroscopic measurements are sensitive and nondestructive. For the analysis of MAAs, UV-visible spectrophotometry is frequently employed. It calculates how much UVR, or visible light, is absorbed by a material in solution. The number of molecular species absorbing the radiation is determined by quantitative spectrophotometric analysis (Behera et al. 2012). It is a completely label-free, noninvasive method that stimulates molecular bond vibrations. It has also made it possible to perform high-throughput, reagent-free, nondestructive, and quick analyses on a variety of sample types (Downes and Elfick 2010). The retention duration in HPLC and the distinctive UV absorption spectra acquired using diode array detection (DAD) are typically used to identify MAAs. Furthermore, a prevalent challenge has been the absence of commercially available standards, whose creation is expensive and time-consuming (Carreto and Carignan 2011). Care should be taken when characterizing these substances (Bandaranayake 1998; Whitehead and Hedges 2002). Nonetheless, the identification process is still complicated by the comparable absorption maxima and retention times of several closely related unknown substances. Therefore, to investigate the structural variety of MAAs, NMR and LC combined with ESI-MS (LC-ESI-MS) are frequently used (Pathak et al. 2017). In liquid extraction, metabolites of interest are extracted using either an organic solvent (for lipid-soluble compounds like scytonemin), an aqueous solvent (for water-soluble compounds such MAAs), or a mixture of solvents. The solvent's chemical characteristics determine the kind of metabolites that are extracted (Moco et al. 2007). Therefore, UV-VIS spectroscopy, followed by HPLC, ESI-MS/MS, FTIR, and NMR, are important analytical methods for characterizing MAAs. However, X-ray investigation was only able to definitively define the crystal and molecular structure of palythine (Furusaki et al. 1980) and palythene (Uemura et al. 1980), including its absolute configuration. The structures of both compounds were identified as inner salts, and the characteristics of charge delocalization were analyzed. For instance, in palythene, the positive and negative charges within the inner salt tend to be positioned close to each other, suggesting that this compound can be represented as a resonance hybrid. Additionally, the X-ray research showed that the twelve various kinds of hydrogen bonds that hold the water and palythine molecules together create a three-dimensional hydrogen-bonded structure (Furusaki et al. 1980).

The structure was defined by the molecular mass of the isolated chemical and the amino acids produced by its alkaline hydrolysis because the MAAs were frequently isolated in very small amounts. As a result of laser-flash photolysis studies, the excited triplet states of the MAAs shinorine, porphyra-334, and palythine have been described (Conde et al. 2000, 2004, 2007). This finding explains the low likelihood of triplets producing reactive intermediates that could initiate additional chemical reactions, consistent with the absence of radical detection noted by Shick et al. (2000). By analyzing the rate constants for the energy transfer processes with donors

Table 18.3 Characterization of the most common MAAs using different techniques

MAAs	Chemical assay	Optical activity	UV-Vis spectroscopy	IR spectroscopy	X Ray diffraction	MS spectroscopy (ESI-MS/ MS^n)	NMR spectroscopy (^1H-NMR, ^{13}C-NMR)	References
Mycosporine-glycine	Amino acid composition after neutral or alkaline hydrolysis Methylation	+	+	–	–	ESI-MS (high resolution)	^1H-NMR ^{13}C-NMR	Ito and Hirata (1977); Bandaranayake et al. (1996); Balskus and Walsh (2010)
Palythine	Amino acid composition after alkaline hydrolysis Interchange H/D	+	+	+	+	ESI-MS (high resolution) ESI-MS^n	^1H-NMR ^{13}C-NMR	Takano et al. (1978); Tsujino et al. (1978); Furusaki et al. (1980); Whitehead et al. (2001); Cardozo et al. (2006)
Mycosporine-taurine	Amino acid composition after hydrolysis (not detailed)	–	+	–	–	ESI-MS (high resolution)	–	Stochaj et al. (1994)
Palythine-serine	No	–	+	–	–	ESI-MS^n	^1H-NMR	Teai et al. (1997)
Palythine-serine sulfate	No	–	–	–	–	ESI-MS (high resolution)	^1H-NMR ^{13}C-NMR	Wu et al. (1997)

(continued)

Table 18.3 (continued)

MAAs	Chemical assay	Optical activity	UV-Vis spectroscopy	IR spectroscopy	X Ray diffraction	MS spectroscopy (ESI-MS/MS^n)	NMR spectroscopy (^{1}H-NMR, ^{13}C-NMR)	References
Mycosporine-methylamine-serine	No	–	+	–	–	ESI-MS^n	^{1}H-NMR	Teai et al. (1997)
Mycosporine-methylamine-threonine	Amino acid composition after hydrolysis (not detailed)	–	+	–	–	ESI-MS (high resolution)	^{1}H-NMR ^{13}C-NMR	Wu et al. (1997)
Asterina-330	Methylation H/D interchange	–	+	–	–	EI-MS (high resolution) ESI-MS (high resolution) ESI-MS^n	^{1}H-NMR	Nakamura et al. (1981); Bandaranayake (1998); Whitehead et al. (2001); Cardozo et al. (2006)
Palythinol	Amino acid composition after acid hydrolysis Methylation Interchange H/D	+	+	+	–	ESI-FTICR-MS^n ESI-MS^n	^{1}H-NMR ^{13}C-NMR	Takano et al. (1978); Whitehead et al. (2001); Cardozo et al. (2006); Cardozo et al. (2008)
Mycosporine-glutamic acid-glycine	Amino acid composition after alkaline hydrolysis	–	+	–	–	SI-MS	^{1}H-NMR ^{13}C-NMR	Bandaranayake et al. (1996)
Shinorine	Amino acid composition after acid hydrolysis Interchange H/D	+	+	+	–	ESI-MS ESI-MS^n FD-MS	^{1}H-NMR ^{13}C-NMR	Tsujino et al. (1978); Whitehead et al. (2001); Cardozo et al. (2006)

Mycosporine-2-glycine	Amino acid composition after acid hydrolysis	–	+	–	–	ESI-MSn (high resolution)	^{1}H-NMR ^{13}C-NMR	Stochaj et al. (1994); Liu et al. (2004); Volkmann et al. (2006)
Mycosporine-glycine-valine	Amino acid composition after alkaline hydrolysis	–	–	–	–	–	–	Karentz et al. (1991)
Porphyra-334	Amino acid composition after acid hydrolysis Methylation	+	+	+	–	ESI-MSn	^{1}H-NMR ^{13}C-NMR	Takano et al. (1979); Whitehead et al. (2001); Klisch et al. (2007)
Z-palythenic acid	Amino acid composition after hydrolysis (not detailed)	+	+	–	–	FD-MS ESI-MS^{n}	+	Kobayashi et al. (1981); Whitehead et al. (2001)
Usujirene	Amino acid composition after alkaline hydrolysis MAA residue after acid hydrolysis Catalytic hydrogenation	+	+	+	–	FD-MS (high resolution) FAB-MS (high resolution)	^{1}H-NMR ^{13}C-NMR	Sekikawa et al. (1986); Nakayama et al. (1999)

(continued)

Table 18.3 (continued)

MAAs	Chemical assay	Optical activity	UV-Vis spectroscopy	IR spectroscopy	X Ray diffraction	MS spectroscopy (ESI-MS/ MS^n)	NMR spectroscopy (^1H-NMR, ^{13}C-NMR)	References
Palythene	MAA residue after acid hydrolysis Catalytic hydrogenation	+	+	+	+	ESI-MSn	^1H-NMR ^{13}C-NMR	Takano et al. (1978); Whitehead et al. (2001)
Euhalothece-362	Amino acid composition after acid hydrolysis	–	+	–	–	ESI-MSn (high resolution)	+	Volkmann et al. (2006)
Dehydroxyl-usujirene	No	–	+	–	–	ESI-MS/MS	–	Zhang et al. (2007)
Palythine-threonine	Amino acid composition after alkaline hydrolysis Interchange H/D	–	+	–	–	ESI-MS (high resolution) ESI-MSn	–	Carignan et al. (2009)
Palythine-threonine sulphate	No	–	–	–	–	ESI-MS (high resolution)	^1H-NMR ^{13}C-NMR	Wu et al. (1997)

Adapted from Pathak et al. (2017)

of various triplet energies, the energy levels of the triplet states (ET) were estimated. Accordingly, ET for palythine provides around 330 kJ•mol^{-1} in aqueous solution, while ET for porphyra-334 is limited by a value of less than 250 kJ•mol^{-1} (Conde et al. 2000, 2007). As a result, porphyra-334 may effectively quench the excited state of thymine in DNA, which has a triplet energy of 270 kJ•mol^{-1}, whereas palythine cannot (Bosca et al. 2006). Apart from the obvious filtering effect, Misonou et al. (2003) proved that the quenching reaction of MAAs acts as a protective mechanism against UV-induced damage to *Porphyra yezoensis*, a red alga. By using modern omics approaches, we could discover more about the diversity and functional relevance of MAAs.

The primary basis for identifying the various MAAs that have been discovered thus far is their absorbance maxima. Table 18.2 lists the molecular structures of some MAAs along with the associated absorbance maxima. The existence of MAAs and their induction were examined in *Scytonema* sp. HKAR-3, which was isolated from an Indian hot spring (Silverstein et al. 2005). With retention times of 4.1 and 8.7 minutes, respectively, HPLC showed the presence of two MAAs: mycosporine-glycine (λmax = 310 nm) and an unidentified MAA-334 (λmax = 334 nm). A common peak at 334 nm was found by HPLC analysis, confirming the presence of MAAs in *Anabaena* species. The chromatograms of the two species showed a single peak that corresponded to normal shinorine. An Amazon SL mass spectrometer was used to record the mass spectra. It was discovered that a cone voltage of 30 V may create $(M + H)^{1+}$ with a mass range of 100–1000 m/z. The program DataAnalysis 4.0 was used to examine the data. In order to perform FTIR, lyophilized MAAs were fused in a 1:100 ratio with oven-dried potassium bromide (which was kept in a desiccator). Shinorine was identified by ESI-MS analysis of the peak when a value of m/z 3321^{+} was obtained from both *Anabaena* species. After ESI-MS analysis of the other two peaks from *A. doliolum*, m/z values of 2471^{+} and 3481^{+} were obtained. FTIR investigation of the strong common peak from both *Anabaena* species, which had a retention time of 2.1–2.8 min and a m/z of 3321^{+}, revealed six distinct bands. NMR studies showed that the purified chemical was shinorine based on the locations of hydrogen (^{1}H) and carbon (^{13}C). The compound's proton NMR (^{1}H) investigation revealed that neither aromatic nor heteroaromatic protons were present.

The compound's identity as shinorine was further confirmed by FTIR analysis of the functional group of the purified MAA (m/z as 332^{1+}), which revealed spectral bands resembling those reported in published reports (Takano et al. 1978; Chioccara et al. 1979; Tsujino et al. 1978; Grant et al. 1985; Bandaranayake et al. 1996; Kedar et al. 2002; Whitehead and Hedges 2002; Oyamada et al. 2008). *Gloeocapsa* sp. CU-2556 is a cyanobacterium that can produce natural sunscreening compounds (Rastogi and Incharoensadki 2013). With RT of 5.9 and 6.4 minutes, respectively, HPLC with PDA detection demonstrated the production of two MAAs: shinorine ($UV\lambda_{max}$ 333 nm) and an unidentified MAA known as M-307 ($UV\lambda_{max}$ 307 nm). Richa and Sinha (2015) made an effort to compare the MAA profiles of two *Nostoc* sp. strains that live in various environments. Two MAAs, shinorine and porphyra-334, were detected in the HPLC study of both *Nostoc* sp. One MAA (RT 3.1–3.8 min, λ_{max} 334 nm) from each of the two *Nostoc* species was identified and characterized using methods including absorption spectroscopy, HPLC, ESI-MS, FTIR, and NMR. The purified fraction of common MAA of both *Nostoc* sp. has an m/z value of 333.1^+ according to ESI-MS analysis. The porphyra-334 was frequently and consistently found to have minor mass losses of 44 Da (CO_2) and 18 (H_2O) (Cardozo et al. 2006; Carignan et al. 2009). For porphyra-334, the NMR spectra were found to match chemical shift data that had already been published (Oyamada et al. 2008; Yoshiki et al. 2009). *Fischerella muscicola* TISTR8215, a filamentous cyanobacterium, was examined by Rastogi and Incharoensadki (2013) for the presence of MAAs and how UVR induces them. An autoinjector was used to inject the 50 μL of samples. The detection wavelength was 330 nm, with a mobile phase consisting of 0.02% (v/v) acetic acid and a flow rate of 1.0 mL min^{-1}. The wavelength range for the PDA scan was 200–400 nm. Every second, absorption spectra were captured directly on the peaks that had been separated by HPLC. Photodiode-array detection tests in conjunction with reverse-phase HPLC demonstrated the presence of an MAA with an absorption maximum at 332 nm and an RT of around 16.1 min. The compound's absorption maximum led to its designation as M-332. Rastogi et al. (2013) carried out a similar investigation on *Rivularia* sp. HKAR-4 for dose-dependent production of MAAs under PAR and PAB conditions. Under both PAR and UVR, mycosporine-glycine (MG; λ_{max} 310 nm) and an unidentified UV-absorbing substance (λ_{max} 335 nm) were detected. HPLC-DAD analysis and absorption spectroscopy were used to identify the chemicals. Using LC-MS and HPLC-PDA analysis, MAAs was identified in the cyanobacterium *Nostoc* sp. R76DM (Rastogi et al. 2016). Every second, the absorption spectra of MAAs were captured directly on the peaks that were separated by HPLC. Software called CLASS-VP Chromatography was used to interpret the results. A fraction collector was used to gather and lyophilize the fraction that contained the pure MAA. LC-ESI-MS was used to further characterize the pure lyophilized MAA. Using a Shimadzu LC–MS-2020, LC-MS analysis was carried out. Under UV-B radiation, it was quite evident that four distinct MAAs—asterina (λ_{max} 331 nm, m/z 289), palythine (λ_{max} 320 nm, m/z 245), palythene (λ_{max} 359 nm, m/z 285), and porphyra (λ_{max} 333 nm, m/z 347)—occurred. Their retention times were 4.71, 2.9, 7.05, and 6.29 minutes, respectively. The presence of a single MAA, shinorine (λ_{max} 332.3 ± 2 nm, m/z 333.1), with retention time 1.157, that concurrently carries out UV-screening and quenching actions was discovered by Ahmed et al. (2021) using HPLC-PDA detection, ESI-MS, FTIR, and NMR analysis (Fig. 18.7).

By using cyclic voltammetry, shinorine's electrochemical characterization was established. In 0.1 M, pH 7.0 phosphate buffer, the shinorine molecule exhibits diffusion-controlled activity and electrochemical activity. *Fischerella* sp. HKAR-5 and *Anabaena* sp. HKAR-7 were found to contain MAAs, according to Singh et al. (2022). In contrast to *Anabaena* sp., which had a sharp peak at retention time 3.21 min (porphyra 334; P-334) with UV λ_{max} 334 nm, *Fischerella* sp. exhibited a variety of MAA forms at retention times 1.16 (shinorine), 2.18 (mycosporine glycine-310), and 3.14 min (palythinol) with UV λ_{max} 310, 332, and 334 nm, respectively. HPLC studies and UV-VIS spectrophotometry demonstrated the existence of

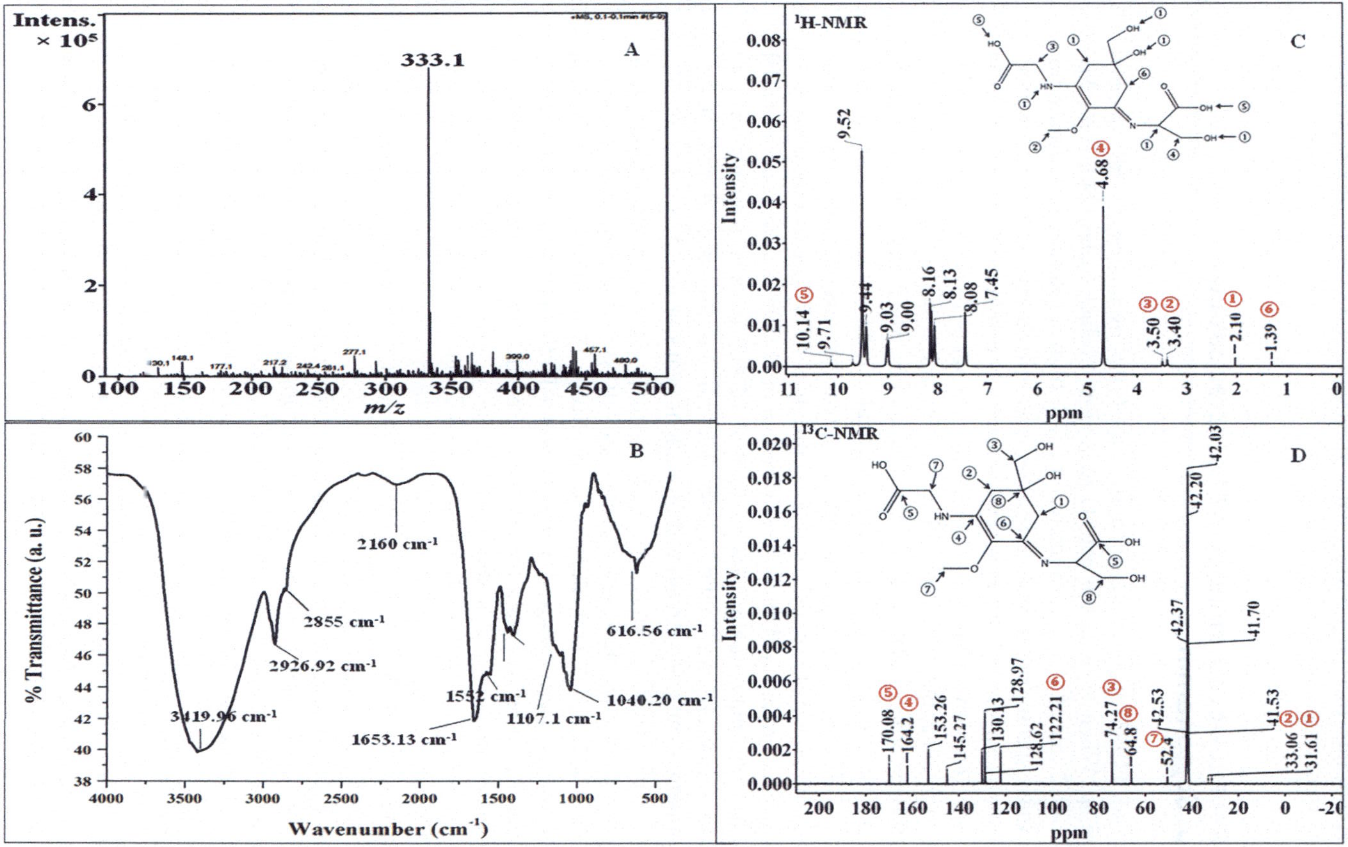

Fig. 18.7 The collected pure fraction was identified as shinorine by Electro spray ionization-mass spectrometry (ESI-MS) of HPLC purified fractions, which produced a peak with a m/z value of 333.1 + 1 (**a**), Fourier Transform Infrared (FTIR) radiation spectroscopy showed the presence of functional groups in the purified portion of MAA (**b**), The ^{1}H (**c**), and ^{13}C (**d**). ^{1}D nuclear magnetic resonance (NMR) spectra of the purified fraction of shinorine were recorded in D_2O solution, and the NMR spectra showed a correlation with the chemical shift data of shinorine. (Adapted from Ahmed et al. 2021)

an MAA with a retention duration of 1.5 minutes and an absorption peak of λ_{max} 334 nm. According to FTIR, NMR, and ESI-MS (m/z: 333 + 1), the MAA was determined to be shinorine.

18.6 Recent Developments in Characterization of Scytonemin

Proteau et al. (1993) isolated scytonemin from *Lyngbya* sp., *Stigonema* sp., and *Scytonema* sp. When scytonemin and its reduction product $C_{36}H_{22}N_2O_4$ (obs. M^+ 546.1578, −0.2 mamu dev.) were analyzed using high resolution fast atom bombardment mass spectrometry (HR FAB MS), it was discovered to have 27 degrees of unsaturation (Nägeli 1849). Since scytonemin only displayed 16 carbon lines in ^{13}C NMR spectroscopy, it has several symmetry elements. According to ^{1}H and ^{13}C NMR, the pigment contained a para-substituted phenol, which explained the degeneracy of the two carbon and two proton absorption lines. The pigment was a homodimer of $C_{18}H_{10}NO_2$ units, as indicated by empirical carbon and hydrogen molecular formula of $C_{18}H_{10}$ that was inferred from the preceding results. All of the carbon atoms in scytonemin were sp_2 hybridized, according to ^{13}C NMR spectroscopy, suggesting that the pigment had eight rings (Proteau et al. 1993). According to NMR and thin-layer chromatography, Stevenson et al. (2002) isolated scytonemin from *Stigonema* sp. and refined it to 95% purity. According to Squier et al. (2004), oxidized scytonemin (m/z 545.2) showed prominent peaks at m/z 528, 517, and 489 (Fig. 18.8). The chemical extracts of *Scytonema* sp. were found to include three novel pigments, tetramethoxyscytonemin, dimethoxyscytonemin, and scytonin, with m/z $[M + H]^+$ 671.2, 609, and 519, respectively (Bultel-Poncé et al. 2004). In MeOH, IR spectra were captured using a Nicolet FTIR. The Bruker Avance 400 spectrometer was used to obtain the ^{1}H and ^{13}C NMR spectra using conventional pulse sequences that operated at 400 and 100 MHz, respectively. The coupling constants are reported in Hz, and the chemical shift values are expressed as δ (ppm units). A JEOL 700 spectrometer was used to measure HRMS (positive mode). Kieselgel 60 (230–400 mesh, E. Merck) was used for Si gel CC, while Polygoprep 60–50 (Macherey-Nagel) was used for RP-18 gel CC.

A second compound (5.6 mg) eluted at a retention time of 21 minutes was separated and purified by subjecting the pigment-containing fraction to semipreparative reversed-phase HPLC and open column reversed-phase RP-18. The primary methods used to identify and characterize these novel scytonemin forms were ^{1}H and ^{13}C NMR and MS.

Both tyrosine and tryptophan were reported to be incorporated into scytonemin by Jones et al. (2011). Additionally, in vivo data were presented that confirmed the tryptophan origin of the ketone carbon implicated in the condensation of the two precursors. During the production of scytonemin, the inclusion of isotopically labeled tyrosine was tracked using the MALDI-TOF mass spectrometry approach. A Bruker program called Clinprotools was used to examine the MALDITOF data. The following isotopically enriched substrates were acquired from Cambridge

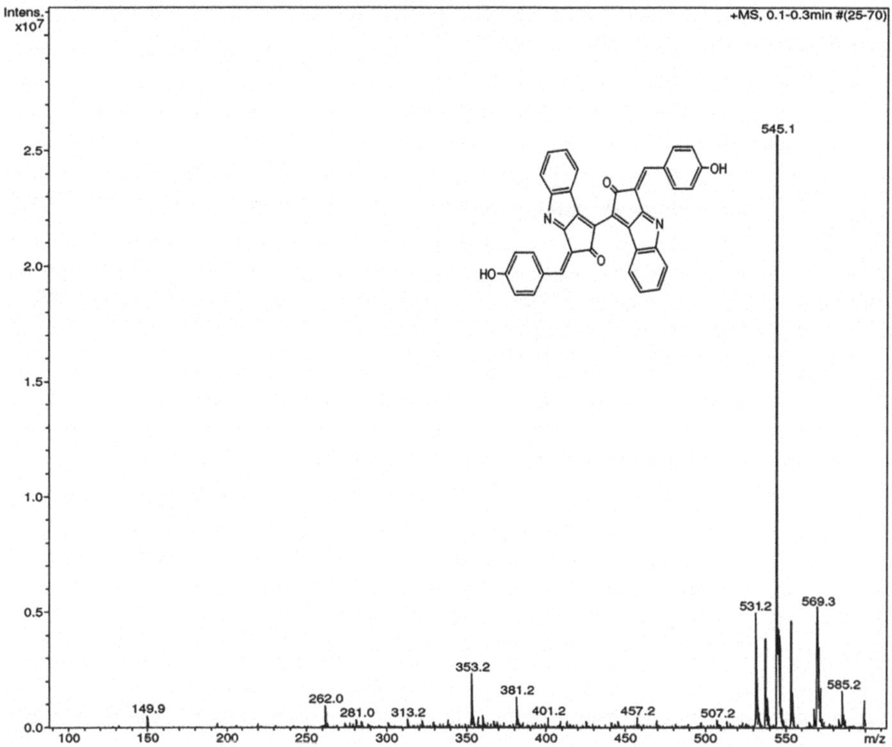

Fig. 18.8 The partially purified scytonemin's mass spectrum displays a noticeable peak at m/z 545.1 along with the associated structure. (Adapted from Pandey et al. 2020)

Isotope Laboratories: L-tryptophan (indole-D_5, 98%); L-tyrosine (ring 3, 5-d_2, 98%); and L-tyrosine (U-$^{13}C_9$, 97–99%, ^{15}N, 97–99%) (Jones et al. 2011).

According to Matsui et al. (2012), scytonemin was isolated from the terrestrial cyanobacterium *Nostoc commune*. At a flow rate of 3 mL per minute, the mobile phase consisted of acetonitrile/acetic acid (99.8: 0.2, v/v). A Hitachi L-4200 UV-VIS monitor was used to measure the A_{380}. The scytonemin-containing fraction was extracted and kept overnight at −30 °C to cause the scytonemin to precipitate. Following 5 min at 4 °C and 21,500 x g centrifugation, the pellet was dissolved in DMSO. Hundred percent methanol was used as the mobile phase in an HPLC using a reverse-phase column to evaluate the final scytonemin product at a flow rate of 1 mL min^{-1}. A UV-VIS detector was used to find the A_{380}. L-ascorbate was added in an equivalent molar amount to minimize scytonemin. Using a Hitachi U-2800 spectrophotometer, UV-VIS spectra were captured. A JOEL ECA600 spectrometer was used to record the scytonemin spectra in $(CD_3)_2SO$. An internal NMR standard, 3-(trimethylsilyl) propanoic acid (TMP), was employed. With an absorption peak at 380 nm, the purified scytonemin displayed a distinctive UV absorption spectrum. Its reduced version also displayed a distinctive UV-vis absorption spectrum. Purified scytonemin's HPLC analysis revealed a single elution peak, signifying its high purity. The purified scytonemin's NMR spectra matched the previously published ones exactly. HPLC-ESI-MS was used to examine the cyanobacterium *Nostoc commune*'s synthesis of oxidized scytonemin (Chen et al. 2013). Finnigan Surveyor and TSQ Quantum Access system with ESI interfaced triple quadrupole mass spectrometer were used for the HPLC-MS analysis. The positive ionization mode was used for the LC-MS, and oxidized scytonemin transitions were observed at m/z 545.1–517.2 and 488.8. A popular method for both qualitative and quantitative ingredient evaluations is HPLC-MS, which combines the potent structural characterization of MS with the effective separation potential of HPLC. Furthermore, compared to UV and evaporative light-scattering detection, the application of MS in the SRM mode yields great selectivity and enhanced sensitivity (Chen et al. 2013).

Rastogi and Incharoensakdi identified ultraviolet screening chemicals from the cyanobacterium *Lyngbya* sp. CU2555 (Rastogi and Incharoensadki 2013). They used ion trap LC/MS analysis and HPLC in conjunction with a diode array detector to partially describe these chemicals. Spectroscopic analysis was carried out with a UV-VIS spectrophotometer in the 200–800 nm range. An auto-injector was used to inject 50 μL of MAAs and scytonemin samples into the HPLC column. The temperature of the column oven (CTO-10A VP) was set at 40 °C. The mobile phase for MAAs was 10% aqueous methanol plus 0.1% HPLC-grade acetic acid, and it was run isocratically at a flow rate of 1.0 mL min^{-1}. The PDA scan wavelength was between 200 and 400 nm, and the MAA detection wavelength was at 330 nm. The PDA scan wavelength ranged from 200 to 800 nm, while the detection wavelength was at 380 nm. A set of trichromatic equations, as outlined by Garcia-Pichel and Castenholz (1991), was used to quantify the concentration of scytonemin. The HPLC system's degasser (Shimadzu, DGU-14A) was used to degas the solvent. The HPLC-separated peaks were used to immediately measure the absorption spectra of UV-absorbing compounds every second. The fraction that contained either scytonemin or pure MAAs was gathered and lyophilized. LC-ESI-MS was used to further characterize the scytonemin and pure lyophilized MAAs. Rastogi et al. (2013) used a similar method to partially characterize scytonemin from the cyanobacterium *Rivularia* sp. HKAR-4. The presence of a UV-absorbing chemical was discovered in HPLC using photodiode-array detection tests, with the absorption peak occurring at 386 nm. The compound's identity as scytonemin was verified by its ion trap LC/MS analysis and absorption spectrum. A noticeable ion peak of the protonated molecule ($[M + H]^+$) was detected at m/z 545 by ESI-LC-MS analysis. The online LC/MS spectra of a cyanobacterial mat extract displaying scytonemin at m/z 545.2 were found to be consistent with the presence of a significant ion (m/z 545.1) in the mass spectrum of *Rivularia* sp. The reduced scytonemin (R-scy), which has been demonstrated to inhibit the proliferation of human T-lymphoid Jurkat cells, was isolated from *Nostoc commune* Vaucher by Itoh et al. (2013). Scy and R-scy were obtained by separating the ethanol extract from the dried algal powder of *N. commune* using column chromatography (CC) on silica gel, ODS, or Sephadex LH-20. These chemicals were primarily characterized using FTIR, MS, and NMR spectra. Scy and R-scy's FABMS spectra showed molecular ion peaks at m/z 544 $[M]^+$ and 546 $[M]^+$, respectively. Scytonemin-3a-imine, a novel type of scytonemin, was identified from the cyanobacterium *Scytonema hoffmani*, which was cultivated in high to strong light conditions, according to Grant and Louda (2013). UV/vis spectrophotometry, FTIR, MALDI, ESI-MS (MS and MSn), and both 1- and 2-D ^{1}H NMR were used for characterization. To validate structural features, derivatization ($NaBH_4$ reduction, acetylation, and deuterium exchange) was used. Absorption maxima were observed at 237, 366, 437, and 564 nm in the UV/vis spectrum. Strong visible absorption bands (437 and 564 nm) indicated a different or additional physiological role for the pigment. During HPLC, the pigment's entire UV/vis spectrum showed maxima in Me_2CO and the eluting solvent at 237, 366, 437, and 562–4 nm. According to the ratios in parentheses, the peaks in EtOH were at 248 (0.14), 305 (1.00), 364 (0.24), 440 (0.26), and 553 (0.15) nm. Purified scytonemin-3a-imine's fluorescence excitation/emission spectra revealed one emission band peaking at 724 nm and excitation maxima at 293 and 345 nm. The phenol functionality was indicated by a large absorption at 3411.8 cm^{-1} in the FTIR spectrum of scytonmein-3a-imine. Generally, signals for amines and imines' N-H functional groups were seen between 3300 and 3500 cm^{-1}. The absorption at 2962 cm^{-1} and 2922 cm^{-1} was characteristic of alkanes' C-H stretching vibrations. At 1665–1710 cm^{-1}, 1640–1680 cm^{-1}, 1500–1700 cm^{-1}, 1620–1690 cm^{-1}, and 1580–1650 cm^{-1}, respectively, C=O of ketones, C=C of aromatics and aliphatics, C=N of imines, and N-H of primary amines were detected (Silverstein et al. 2005) (Fig. 18.9).

MS and NMR studies verified the existence of the functional groups in the suggested structure. The bulk extracts were purified using semi-prep LC, and many MS studies were performed. According to the nitrogen rule, MALDI-TOF consistently produced $[M + H]^+$ at m/z 602.2 and $[M + Na]^+$ at m/z 624, indicating a molecule

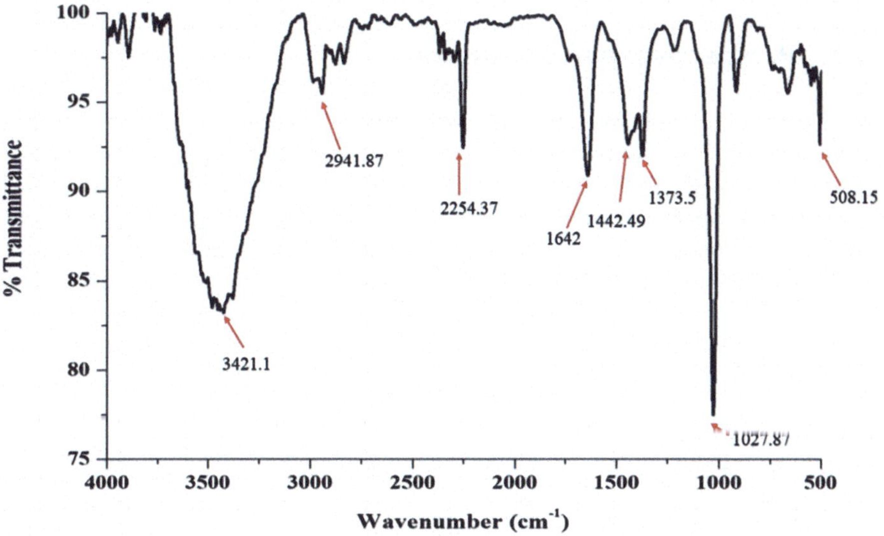

Fig. 18.9 Fourier Transform Infrared (FTIR) radiation spectrum (spectral bands shown by arrow) of purified scytonemin

with an odd number of N atoms (McLafferty 1993). Only two N are present in all known pigments with the "scytoneman" nucleus. The mass of 601.2074 Da, which corresponds to $C_{39}H_{27}N_3O_4$, was determined by high-resolution ESI-TOF-MS, indicating the existence of a third N functionality, possibly an imine. Via loss of 56 Da, a conspicuous m/z 546 was created by positive ESI-MS_2 of $[M + H]^+$ at m/z 602. At m/z 546, $[M\text{-}56 + H]^+$ generated m/z 528, 517/518, and 489/490. The pigment's ^1H NMR spectra ($C_{39}H_{27}N_3O_4$: ESI-TOF-MS) showed two geminal protons as doublets at δ 3.20 (17a-H) and 3.67 (17-H) and a methyl singlet at δ 1.91. At δ 175.83 (C-18), a main ketimine signal was detected, and at δ 54.76 (C-17, 17a), a methylene signal was observed. Vibrational spectroscopy was used by Edward et al. to characterize scytonemin (Edwards et al. 2000). A white light source was used for calibration, and wavenumbers are accurate to 91 cm^{-1}. They investigated scytonemin using Raman spectroscopy and suggested names for the main vibrational bands. Pandey et al. (2020), also used HPLC and ESI-MS to partially describe it (retention time: 1.7 min; UV λ_{max}: 386 nm; $[M + H]^+$ m/z: 545.1). Using HPLC, FTIR, and HRMS, Kumari et al. (2023) recently described scytonemin from biological crusts that live in various natural settings. All samples included scytonemin, and HPLC chromatograms revealed absorption peaks at 386 nm with retention times varying between 5.3 and 5.9 min. Significant absorption also occurred at 252, 278, and 300 nm. $[M + H]^+$ m/z: 545.1 revealed scytonemin, according to HRMS. Parailloux et al. (2020) present an untargeted screening method for the analysis of mycosporines and MAAs using hydrophilic interaction liquid chromatography (HILIC) combined with multistage electrospray mass spectrometry (MS2/MS3). The extraction of MAAs was performed using a 50% methanol solution and sonication, optimizing the conditions for effective separation and detection.

18.7 Perspectives and Conclusion

The production of photoprotective substances like MAAs and scytonemin mitigates the negative effects of short-wavelength solar radiation, allowing cyanobacteria and algae to grow and thrive in harsh environments like rocks and building walls exposed to continuous, intense sunlight. Strong UV-shielding and antioxidant properties, MAAs and scytonemin, are important components in many pharmaceutical companies for the creation of new medications and natural sun protection products. The cost-effective and sustainable industrial production of these cyanobacterial secondary metabolites depends heavily on the results of metabolomics investigations. As a result, there are many opportunities and potential to pursue in this area of study.

Short Chapter Summary
MAAs and scytonemin are two of the several photoprotective substances found in cyanobacteria. A variety of cyanobacterial species have yielded a variety of MAAs and scytonemins that have been identified and described using various methods combined with the omics approach. The goal of advances in the omics of cyanobacterial UV screening chemicals is to employ transcriptomics and genomics to

discover the genes and processes involved in their manufacture. Therefore, the optimum instrument for developing the characterization, metabolic profile, and application aspects of cyanobacterial photoprotective chemicals is omics technology.

Acknowledgements Dr. Rajeshwar P. Sinha, Senior Professor, Center of Advanced Study in Botany, Institute of Science, Banaras Hindu University, Varanasi, is greatly acknowledged for providing necessary suggestions and guidance for the chapter. Incentive grant received from IoE (Scheme no. 6031), Banaras Hindu University, Varanasi, India, to Rajeshwar P. Sinha is highly acknowledged.

Author Contributions Study concept and design, Haseen Ahmed and Jainendra Pathak; chapter writing, Haseen Ahmed Minu Kesheri, and Swarna Kanchan; chapter editing, Haseen Ahmed, Rajeshwar P. Sinha, and Jainendra Pathak; chapter review Haseen Ahmed, Rajeshwar P. Sinha, and Jainendra Pathak.

References

Ahmed H, Padey A, Pathak J, Sinha RP (2017) Comparative in silico analysis of different dehydroquinate synthase proteins in the cyanobacterium *Cyanothece* sp. PCC 7424. J Bioinform Genomics Proteomics 2(2):1018

Ahmed H, Pathak J, Rajneesh SPK, Ganesan V, Häder D-P, Sinha RP (2021) Responses of a hot spring cyanobacterium under ultraviolet and photosynthetically active radiation: photosynthetic performance, antioxidative enzymes, mycosporine-like amino acid profiling and its antioxidative potentials. 3 Biotech 11(1):10

Ambrosino L, Tangherlini M, Colantuono C, Esposito A, Sangiovanni M, Miralto M, Sansone C, Chiusano ML (2019) Bioinformatics for marine products: an overview of resources, bottlenecks, and perspectives. Mar Drugs 17(10):576

Atanasov AG, Zotchev SB, Dirsch VM (2021) Natural products in drug discovery: advances and opportunities. Nat Rev Drug Discov 20:200–216

Balskus EP, Walsh CT (2008) Investigating the initial steps in the biosynthesis of cyanobacterial sunscreen scytonemin. J Am Chem Soc 130(46):15260–15261

Balskus EP, Walsh CT (2009) An enzymatic cyclopentyl [b] indole formation involved in scytonemin biosynthesis. J Am Chem Soc 131(41):14648–14649

Balskus EP, Walsh CT (2010) The genetic and molecular basis for sunscreen biosynthesis in cyanobacteria. Science 329(5999):1653–1656

Bandaranayake W (1998) Mycosporines: are they nature's sunscreens? Nat Prod Rep 15(2):159–172

Bandaranayake WM, Bemis JE, Bourne DJ (1996) Ultraviolet absorbing pigments from the marine sponge *Dysidea herbacea:* isolation and structure of a new mycosporine. Comp Biochem Physiol C Toxicol Pharmacol 115(3):281–286

Banerjee S, Mazumdar S (2012) Electrospray ionization mass spectrometry: a technique to access the information beyond the molecular weight of the analyte. Int J Anal Chem 2012(1):282574

Baran R, Bowen BP, Bouskill NJ, Brodie EL, Yannone SM, Northen TR (2010) Metabolite identification in *Synechococcus* sp. PCC 7002 using untargeted stable isotope assisted metabolite profiling. Anal Chem 82(21):9034–9042

Barbato R, Bergo E, Szabo I, della Vecchia F, Giacometti GM (2000) Ultraviolet B exposure of whole leaves of barley affects structure and functional organization of photosystem II. J Biol Chem 275(15):10976–10982

Behera S, Ghanty S, Ahmad F, Santra S, Banerjee S (2012) UV-visible spectrophotometric method development and validation of assay of paracetamol tablet formulation. J Anal Bioanal Tech 3(6):151–157

Bennett J, Soule T (2022) Expression of scytonemin biosynthesis genes under alternative stress conditions in the cyanobacterium *Nostoc punctiforme*. Microorganisms 10:427

Bonin J, Hammerle FJ, Ganzera M, Krumme U Karsten U (2024) UV-absorbing mycosporine-like amino acids in the eyes of temperate marine and freshwater fish species. Front Mar Sci 11:1426861

Bosca F, Lhiaubet-Vallet V, Cuquerella MC, Castell JV, Miranda MA (2006) The triplet energy of thymine in DNA. J Am Chem Soc 128(19):6318–6319

Bowen BP, Northen TR (2010) Dealing with the unknown: metabolomics and metabolite atlases. J Am Soc Mass Spectrom 21:1471–1476

Bultel-Poncé V, Felix-Theodore F, Sarthon C, Ponge JF, Bodo B (2004) New pigments from the terrestrial cyanobacterium *Scytonema* sp. collected on the Mitaraka Inselberg, French Guyana. J Nat Prod 67(4):678–681

Burja AM, Dhamwichukorn S, Wright PC (2003) Cyanobacterial postgenomic research and systems biology. Trends Biotechnol 21(11):504–511

Cardozo KH, Carvalho VM, Pinto E, Colepicolo P (2006) Fragmentation of mycosporine-like amino acids by hydrogen/deuterium exchange and electrospray ionisation tandem mass spectrometry. Rapid Commun Mass Spectrom 20(2):253–258

Cardozo KH, Vessecchi R, Carvalho VM, Pinto E, Gates PJ, Colepicolo P, Galembeck SE, Lopes NP (2008) A theoretical and MAAs spectrometry study of the fragmentation of mycosporine-like amino acids. Int J Mass Spectrom 273:11–19

Carignan MO, Cardozo KH, Oliveira-Silva D, Colepicolo P, Carreto JI (2009) Palythine-threonine, a major novel mycosporine-like amino acid (MAA) isolated from the hermatypic coral *Pocillopora capitata*. J Photochem Photobiol B Biol 94(3):191–200

Carreto JI, Carignan MO (2011) Mycosporine-like amino acids: relevant secondary metabolites. Chemical and ecological aspects. Mar Drugs 9(3):387–446

Casero MC, Herrero MÁ, De la Roche JP, Quesada A, Velázquez D, Cirés S (2024) Effect of salinity on scytonemin yield in endolithic cyanobacteria from the Atacama Desert. Sci Rep 14:9731

Chen J, Zhao L, Xu J, Yang R, He S, Yan X (2013) Determination of oxidized scytonemin in *Nostoc commune* Vauch cultured on different conditions by high performance liquid chromatography coupled with triple quadrupole mass spectrometry. J Appl Phycol 25:1001–1007

Chioccara F, Miscuraca G, Novellino E, Prota G (1979) Occurrence of two new mycosporine-like amino acids, mytilins a and b in the edible mussel, *Mytilus galloprovincialis*. Tetrahedron Lett 34:3181–3182

Clish CB (2015) Metabolomics: an emerging but powerful tool for precision medicine. Cold Spring Harb Mol Case Stud 1(1):a000588–a000588

Conde FR, Churio MS, Previtali CM (2000) The photoreceptor mechanism of mycosporine-like amino acids. Excited-state properties and photostability of porphyra-334 in aqueous solution. Photochem Photobiol B Biol 56:139–144

Conde FR, Churio MS, Previtali CM (2004) The deactivation pathways of the excited-states of the mycosporine-like amino acids shinorine and porphyra-334 in aqueous solution. Photochem Photobiol Sci 3:960–967

Conde FR, Churio MS, Previtali CM (2007) Experimental study of excited-state properties and photostability of the mycosporine-like amino acid palythine in water solution. Photochem Photobiol Sci 6(6):669–674

Dextro RB, Delbaje E, Geraldes V, Pinto E, Long PF, Fiore MF (2023) Exploring the relationship between biosynthetic gene clusters and constitutive production of mycosporine-like amino acids in Brazilian cyanobacteria. Mol 28(3):1420

Downes A, Elfick A (2010) Raman spectroscopy and related techniques in biomedicine. Sensors 10(3):1871–1889

Dunn WB, Bailey NJC, Johnson HE (2005) Measuring the metabolome: current analytical technologies. Analyst 130(5):606–625

Edwards HGM, Garcia-Pichel F, Newton EM, Wynn-Williams DD (2000) Vibrational raman spectroscopic study of scytonemin, the UV-protective cyanobacterial pigment. Spectrochim Acta - A: Mol Biomol Spectrosc 56(1):193–200

Ellis DI, Goodacre R (2012) Metabolomics-assisted synthetic biology. Curr Opin Biotechnol 23(1):22–28

Evans J, Jones A, Blumenthal E, Soule T (2021) Anti-proliferation of melanoma cells and immune stimulation by the cyanobacterial indole-alkaloid scytonemin. Fine Focus 7:54–63

Favre-Bonvin J, Favre-Bonvin J, Bernillon J, Salin N, Arpin N (1987) Biosynthesis of mycosporines: mycosporine glutaminol in *Trichothecium roseum*. Phytochemistry 26(9):2509–2514

Fenn JB, Mann M, Meng CK, Wong SF, Whitehouse CM (1989) Electrospray ionization for mass spectrometry of large biomolecules. Science 246(4926):64–71

Furusaki A, Matsumoto T, Tsujino I, Sekikawa I (1980) The crystal and molecular structure of palythine, trihydrate. Bull Chem Soc Jpn 53(2):319–323

Galperin MY, Koonin EV (2010) From complete genome sequence to complete understanding? Trends Biotechnol 28(8):398–406

Gao Q, Garcia-Pichel F (2011) An ATP-grasp ligase involved in the last biosynthetic step of the imino-mycosporine shinorine in *Nostoc punctiforme* ATCC 29133. J Bacteriol 193(21):5923–5928

Gao X, Jing X, Liu X, Lindblad P (2021) Biotechnological production of the sunscreen pigment scytonemin in cyanobacteria: progress and strategy. Mar Drugs 19(3):129

Gao X, Yuan X, Zheng T, Ji B (2023) Promoting efficient production of scytonemin in cell culture of *Nostoc flagelliforme* by periodic short-term solar irradiation. Bioresour Technol Rep 21:101352

Garcia-Pichel F, Castenholz RW (1991) Characterization and biological implication of scytonemin, a cyanobacterial sheath pigment. J Phycol 27(3):395–409

Gartland KPR, Sanins SM, Nicholson JK, Sweatman BC, Beddell CR, Lindon JC (1990) Pattern recognition analysis of high resolution ^{1}H NMR spectra of urine. A nonlinear mapping approach to the classification of toxicological data. NMR Biomed 3(4):166–172

Gartland KPR, Beddell CR, Lindon JC, Nicholson JK (1991) Application of pattern recognition methods to the analysis and classification of toxicological data derived from proton nuclear magnetic resonance spectroscopy of urine. Mol Pharmacol 39(5):629–642

Geraldes V, Pinto E (2021) Mycosporine-like amino acids (MAAs): biology, chemistry and identification features. Pharmaceuticals 14(1):63

Ghedifa AB, Vega J, Korbee N, Mensi F, Figueroa FL, Sadok S (2021) Effects of light quality on the photosynthetic activity and biochemical composition of *Gracilaria gracilis* (Rhodophyta). J Appl Phycol 33:3413–3425

Grant CS, Louda JW (2013) Scytonemin-imine, a mahogany-colored UV/Vis sunscreen of cyanobacteria exposed to intense solar radiation. Org Geochem 65:29–36

Grant PT, Middleton C, Plack PA, Thomson RH (1985) The isolation of four aminocyclohexenimines (mycosporines) and a structurally related derivative of cyclohexane-1:3-dione (gadusol) from the brine shrimp, *Artemsia*. Comp Biochem Physiol B Biochem Mol Biol 80(4):755–759

Greef JVD, Tas AC, Bouwman J, Noever T, de Brauw MC, Schreurs WHP (1983) Evaluation of field-desorption and fast atom-bombardment mass spectrometric profiles by pattern recognition techniques. Anal Chim Acta 150(1983):45–52

Häder D-P, Williamson CE, Wängberg S, Rautio M, Rose KC, Gao K, Helbling EW, Sinha RP, Worrest R (2015) Effects of UV radiation on aquatic ecosystems and interactions with other environmental factors. Photochem Photobiol Sci 14(1):108–126

Helms GL, Moore RE, Niemczura WP, Patterson GML, Tomer KB, Gross ML (1988) Scytonemin A, a novel calcium antagonist from a blue-green alga. J Org Chem 53(6):1298–1307

Irankhahi P, Riahi H, Hassani SB, Eskafi M, Irani MA, Shariatmadari Z (2024) The role of the protective shield against UV-C radiation and its molecular interactions in *Nostoc* species (Cyanobacteria). Sci Rep 14:19258

Ito S, Hirata Y (1977) Isolation and structure of a mycosporine from the zoanthidian *Palythoa tuberculosa*. Tetrahedron Lett 28:2429–2430

Itoh T, Tsuzuki R, Tanaka T, Ninomiya M, Yamaguchi Y, Takenaka H, Ando M, Tsukamasa Y, Koketsu M (2013) Reduced scytonemin isolated from *Nostoc commune* induces autophagic cell death in human T-lymphoid cell line jurkat cells. Food Chem Toxicol 60:76–82

Jehlička J, Culka A, Němečková K, Mareš J (2023) Using raman spectroscopy to detect scytonemin of epiliths and endoliths from marble, serpentinite and gypsum. J Raman Spectrosc 54:1280–1296

Jiménez C (2018) Marine natural products in medicinal chemistry. ACS Med Chem Lett 9(10):959–961

Jones CS, Esquenazi E, Dorrestein PC, Gerwick WH (2011) Probing the in vivo biosynthesis of scytonemin, a cyanobacterial ultraviolet radiation sunscreen, through small scale stable isotope incubation studies and MALDI-TOF mass spectrometry. Bioorg Med Chem 19(22):6620–6627

Kang MR, Jo SA, Lee H, Yoon YD, Kwon JH, Yang JW, Choi BJ, Park KH, Lee MY, Lee CW, Lee KR, Kang JS (2020) Inhibition of skin inflammation by scytonemin, an ultraviolet sunscreen pigment. Mar Drugs 18(6):300

Karentz D, Mc Euen FS, Land MV, Dunlap WC (1991) Survey of mycosporine-like amino acid compounds in Antarctic marine organisms: potential protection from ultraviolet exposure. Mar Biol 108:157–166

Katoch M, Mazmouz R, Chau R, Pearson LA (2016) Heterologous production of cyanobacterial mycosporine-like amino acids mycosporine- ornithine and mycosporine-lysine in *Escherichia coli*. Appl Environ Microbiol 82(20):6167–6173

Kedar L, Kashman Y, Oren A (2002) Mycosporine-2-glycine is the major mycosporine-like amino acid in a unicellular cyanobacterium (*Euhalothece* sp.) isolated from a gypsum crust in a hypersaline saltern pond. FEMS Microbiol Lett 208(2):233–237

Kind T, Fiehn O (2010) Advances in structure elucidation of small molecules using mass spectrometry. Bioanal Rev 2:23–60

Klicki K, Ferreira D, Hamill D, Dirks B, Mitchell N, Garcia-Pichel F (2018) The widely conserved ebo cluster is involved in precursor transport to the periplasm during scytonemin synthesis in *Nostoc punctiforme*. MBio 9(6):10–1128

Klisch M, Richter P, Puchta R, Häder D-P, Bauer W (2007) The stereostructure of porphyra-334: an experimental and calculational NMR investigation. Evidence for an efficient-proton sponge. Helv Chim Acta 90(3):488–511

Kobayashi JI, Nakamura H, Hirata Y (1981) Isolation and structure of a UV-absorbing substance 337 from the ascidian *Halocynthia roretzi*. Tetrahedron Lett 22(31):3001–3002

Krishnan S, Salian A, Dutta S, Mandal S (2021) A roadmap to UV-protective natural resources: classification, characteristics, and applications. Mater Chem Front 5(21):7696–7723

Kuehnbaum NL, Britz-McKibbin P (2013) New advances in separation science for metabolomics: resolving chemical diversity in a postgenomic era. Chem Rev 113(4):2437–2468

Kumari N, Pandey A, Gupta A, Mishra S, Sinha RP (2023) Characterization of UV-screening pigment scytonemin from cyanobacteria inhabiting diverse habitats of Varanasi, India. Biologia 78(2):319–330

Kylim H (1927) Über die karotinoide Farbstoffe der Algen. Hoppe-Scyler's Zeitschr. Physiol Chem 166:33–77

Kylim H (1937) Über die Farbstoffe und die Farbe der Cyanophyceen. Fysiogr Sällsk Förhandl 7:131–158

Lalegerie F, Stiger-Pouvreau V, Connan S (2024) Mycosporine-like amino acids in *Palmaria palmata* (Rhodophyta): specific implication of usujirene in photoprotection. Mar Drugs 22(3):121

Liu Z, Häder D-P, Sommaruga R (2004) Occurrence of mycosporine-like amino acids (MAAs) in the bloom-forming cyanobacterium *Microcystis aeruginosa*. J Plankton Res 26(8):963–966

Madrahi GS, Naeimpoor F (2023) Overproduction of cyano-sunscreen scytonemin by *Leptolyngbya mycodia* in two stage illuminated photobioreactor: from submerged to attached cultivation. Algal Res 74:103169

Malla S, Sommer MOA (2014) A sustainable route to produce the scytonemin precursor using *Escherichia coli*. Green Chem 16(6):3255–3265

Matsui K, Nazifi E, Hirai Y, Wada N, Matsugo S, Sakamoto T (2012) The cyanobacterial UV-absorbing pigment scytonemin displays radical-scavenging activity. J Gen Appl Microbiol 58(2):137–144

McLafferty FW (1993) Interpretation of Mass Spectra, 4ed ed. University Science Books Sausalito, CaliforniaMilito A, Castellano I, Damiani E (2021) From sea to skin: is there a future for natural photoprotectants? Mar Drugs 19(7):379

Misonou T, Saitoh J, Oshiba S, Tokimoto Y, Maegawa M, Inoue Y, Hori H, Sakurai T (2003) UV-absorbing substance in the red alga *Porphyra yezoensis* (Bangiales, Rhodophyta) block thymine photodimer production. Mar Biotechnol 5:194–200

Moco S, Bino RJ, De Vos RCH, Vervoort J (2007) Metabolomics technologies and metabolite identification. Trends Anal Chem 26(9):855–866

Mogany T, Kumari S, Swalaha FM, Bux F (2022) In silico analysis of enzymes involved in mycosporine-like amino acids biosynthesis in *Euhalothece* sp.: structural and functional characterization. Algal Res 66:102806

Moisan TA, Mitchell BG (2001) UV absorption by mycosporine-like amino acids in *Phaeocystis Antarctica* Karsten induced by photosynthetically available radiation. Mar Biol 138:217–227

Nägeli C (1849) Gattungen einzelliger Algen, physiologisch und systematisch bearbeitet, Germany

Nakamura H, Kobayashi JI, Hirata Y (1981) Isolation and structure of a 330 nm UV-absorbing substance, asterina-330 from the starfish *Asterina pectinifera*. Chem Lett 10(10):1413–1414

Nakayama R, Tamura Y, Kikuzaki H, Nakatani N (1999) Antioxidant effect of the constituents of susabinori (*Porphyra yezoensis*). J Am Oil Chem Soc 76(5):649–653

Naurin S, Bennett J, Videau P, Philmus B, Soule T (2016) The response regulator Npun_F1278 is essential for scytonemin biosynthesis in the cyanobacterium *Nostoc punctiforme* ATCC 29133. J Phycol 52(4):564–571

Nicholson JK, Wilson ID (1989) High resolution proton magnetic resonance spectroscopy of biological fluids. Prog Nucl Magn Reson Spectrosc 21:449–501

Nicholson JK, Buckingham MJ, Sadler PJ (1983) High resolution ^{1}H NMR studies of vertebrate blood and plasma. Biochem J 211(3):605–615

Nishida Y, Miyabe Y, Kishimura H, Kumagai Y (2021) Monthly variation and ultraviolet stability of mycosporine-like amino acids from red alga dulse *Palmaria palmata* in Japan. Phycologia 1(2):119–128

Oyamada C, Kaneniwa M, Ebitani K, Murata M, Ishihara K (2008) Mycosporine-like amino acids extracted from scallop (*Patinopecten yessoensis*) ovaries: UV protection and growth stimulation activities on human cells. Mar Biotechnol 10:141–150

Pandey A, Pathak J, Singh DK, Ahmed H, Singh V, Kumar D, Sinha RP (2020) Photoprotective role of UV-screening pigment scytonemin against UV-B-induced damages in the heterocyst-forming cyanobacterium *Nostoc* sp. strain HKAR-2. Braz J Bot 43(1):67–80

Parailloux M, Godin S, Fernandes SC, Lobinski R (2020) Untargeted analysis for mycosporines and mycosporine-like amino acids by hydrophilic interaction liquid chromatography (HILIC)-electrospray orbitrap MS2/MS3. Antioxidants 9(12):1185

Pathak J, Rajneesh AH, Richa SRP (2017) Metabolomic profiling of cyanobacterial UV-protective compounds. Curr Metab 5(2):138–163

Pathak J, Ahmed H, Singh SP, Häder D-P, Sinha RP (2019) Genetic regulation of scytonemin and mycosporine-like amino acids (MAAs) biosynthesis in cyanobacteria. Plant Gene 17:100172

Pathak J, Singh PR, Sinha RP, Rastogi RP (2022) Evolution and distribution of cyanobacteria. In: Ecophysiology and biochemistry of cyanobacteria. Springer Nature, Singapore, pp 1–30

Pope MA, Spence E, Seralvo V, Gacesa R, Heidelberger S, Weston AJ, Dunlap WC, Shick JM, Long PF (2015) O-Methyltransferase is shared between the pentose phosphate and shikimate pathways and is essential for mycosporine-like amino acid biosynthesis in Anabaena variabilis ATCC 29413. Chem Bio Chem 16(2):320–327

Portwich A, Garcia-Pichel F (2003) Biosynthetic pathway of mycosporines (mycosporine-like amino acids) in the cyanobacterium *Chlorogloeopsis* sp. strain PCC 6912. Phycologia 42(4):384–392

Proteau PJ, Gerwick WH, Garcia-Pichel F, Castenholz RW (1993) The structure of scytonemin, an ultraviolet sunscreen pigment from the sheath of cyanobacteria. Exp Dermatol 49:825–829

Przybylski M, Glocker MO (1996) Electrospray mass spectrometry of bio-macromolecular complexes with non-covalent interactions new analytical perspectives for supra molecular chemistry and molecular recognition processes. Angewandte Chem 35(8):807–826

Raj S, Kuniyil AM, Sreenikethanam A, Gugulothu P, Jeyakumar RB, Bajhaiya AK (2021) Microalgae as a source of mycosporine-like amino acids (MAAs): advances and future prospects. Int J Environ Res Public Health 18(23):12402

Rastogi RP, Incharoensadki A (2013) UV radiation-induced accumulation of photoprotective compounds in the green alga *Tetraspora* sp. CU2551. Plant Physiol Biochem 70:7–13

Rastogi RP, Sinha RP, Incharoensakdi A (2013) Partial characterization, UV-induction and photoprotective function of sunscreen pigment, scytonemin from *Rivularia* sp. HKAR-4. Chemosphere 93(9):1874–1878

Rastogi RP, Sonani RR, Madamwar D (2015) Cyanobacterial sunscreen scytonemin: role in photoprotection and biomedical research. Appl Biochem Biotechnol 176:1551–1563

Rastogi RP, Sonani RR, Madamwar D, Incharoensakdi A (2016) Characterization and antioxidant functions of mycosporine-like amino acids in the cyanobacterium *Nostoc* sp. R76DM. Algal Res 16:110–118

Richa, Sinha RP (2015) Biochemical characterization of sunscreening mycosporine-like amino acids from two *Nostoc* species inhabiting diverse habitats. Protoplasma 252:199–208

Rosic NN (2012) Phylogenetic analysis of genes involved in mycosporine-like amino acid biosynthesis in symbiotic dinoflagellates. Appl Microbiol Biotechnol 94:29–37

Rosic NN (2021) Recent advances in the discovery of novel marine natural products and mycosporine-like amino acid UV-absorbing compounds. Appl Microbiol Biotechnol 105(19):7053–7067

Rosic N (2022) Genome mining as an alternative way for screening the marine organisms for their potential to produce UV-absorbing mycosporine-like amino acid. Mar Drugs 20(8):478

Rosic NN, Dove S (2011) Mycosporine-like amino acids from coral dinoflagellates. Appl Environ Microbiol 77(24):8478–8486

Rosic N, Climstein M, Boyle GM, Thanh Nguyen D, Feng Y (2023) Exploring mycosporine-like amino acid UV-absorbing natural products for a new generation of environmentally friendly sunscreens. Mar Drugs 21(4):253

Ručová D, Vilková M, Sovová S, Vargová Z, Kostecká Z, Frenák R, Routray D, Bačkor M (2023) Photoprotective and antioxidant properties of scytonemin isolated from Antarctic cyanobacterium *Nostoc commune* Vaucher ex Bornet & Flahault and its potential as sunscreen ingredient. J Appl Phycol 35:2839–2850

Salazar G, Paoli L, Alberti A, Huerta-Cepas J, Ruscheweyh HJ, Cuenca M, Wincker P (2019) Gene expression changes and community turnover differentially shape the global ocean metatranscriptome. Cell 179(5):1068–1083

Sancar A (2016) Mechanisms of DNA repair by photolyase and excision nuclease (Nobel lecture). Angew Chem Int Ed 55(30):8502

Schmid FX (2001) Biological macromolecules: UV-visible spectrophotometry. In: Encyclopedia of life sciences. Macmillan Publishers Ltd, Nature Publishing Group, UK

Schneider G, Figueroa FL, Vega J, Avilés A, Horta PA, Korbee N, Bonomi-Barufi J (2022) Effects of UV-visible radiation on growth, photosynthesis, pigment accumulation and UV-absorbing compounds in the red macroalga *Gracilaria cornea* (Gracilariales, Rhodophyta). Algal Res 64:102702

Schwarz D, Orf I, Kopka J, Hagemann M (2013) Recent applications of metabolomics toward cyanobacteria. Meta 3(1):72–100

Sekikawa I, Kubota C, Hiraoki T, Tsujino I (1986) Isolation and structure of a 357 nm UV absorbing substance, usujirene, from the red alga *Palmaria palmata* (L.) O Kuntze Jpn. J Phycol 34:185–188

Sen S, Mallick N (2022) Scytonemin: unravelling major progress and prospects. Algal Res 64:102–678

Sen A, Dewaker V, Debnath U, Jana K, Rath J, Joardar N, Sinha Babu SP (2023) *In silico* exploration and *in vitro* validation of the filarial thioredoxin reductase inhibitory activity of scytonemin and its derivatives. J Biomol Struct Dyn 21:1–13

Shick JM, Dunlap WC, Buettner R (2000) Ultraviolet (UV) protection in marine organisms II. Biosynthesis, accumulation, and sunscreening function of mycosporine-like amino acids. In: Yoshikawa T, Toyokuni S, Yamamoto Y, Naito Y (eds) Free radicals in chemistry, biology and medicine. London, UK, OICA International, pp 215–228

Silverstein RM, Webster FX, Kiemle DJ (2005) Spectrometric identification of organic compounds, 7th edn. Wiley, New York

Simeonov A, Michaelian K (2017) Properties of cyanobacterial UV-absorbing pigments suggest their evolution was driven by optimizing photon dissipation rather than photoprotection. arXiv preprint arXiv:1702.03588

Singh SP, Häder D-P, Sinha RP (2010a) Cyanobacteria and ultraviolet radiation (UVR) stress: mitigation strategies. Ageing Res Rev 9(2):79–90

Singh SP, Klisch M, Sinha RP, Häder D-P (2010b) Genome mining of mycosporine-like amino acid (MAA) synthesizing and non-synthesizing cyanobacteria: a bioinformatics study. Genomics 95(2):120–128

Singh DK, Pathak J, Pandey A, Rajneesh, Singh V, Sinha RP (2022) Purification, characterization and assessment of stability, reactive oxygen species scavenging and antioxidative potentials of mycosporine-like amino acids (MAAs) isolated from cyanobacteria. J Appl Phycol 34(6):3157–3175

Sinha RP, Ambasht NK, Sinha JP, Klisch M, Häder D-P (2003) UV-B-induced synthesis of mycosporine-like amino acids in three strains of *Nodularia* (cyanobacteria). J Photochem Photobiol B 71(1–3):51–58

Sinha RP, Singh SP, Häder D-P (2007) Database on mycosporines and mycosporine-like amino acids (MAAs) in fungi, cyanobacteria, macroalgae, phytoplankton and animals. J Photochem Photobiol B 89(1):29–35

Soule T, Stout V, Swingley WD, Meeks JC, Garcia-Pichel F (2007) Molecular genetics and genomic analysis of scytonemin biosynthesis in *Nostoc punctiforme* ATCC 29133. J Bacteriol 189(12):4465–4472

Soule T, Garcia-Pichel F, Stout V (2009a) Gene expression patterns associated with the biosynthesis of the sunscreen scytonemin in *Nostoc punctiforme* ATCC 29133 in response to UV-A radiation. J Bacteriol 191(14):4639–4646

Soule T, Palmer K, Gao Q, Potrafka RM, Stout V, Garcia-Pichel F (2009b) A comparative genomics approach to understanding the biosynthesis of the sunscreen scytonemin in cyanobacteria. BMC Genomics 10:336–345

Spence E, Dunlap WC, Shick JM, Long PF (2012) Redundant pathways of sunscreen biosynthesis in a cyanobacterium. Chembiochem 13(4):531–533

Squier AH, Airs RL, Hodgson DA, Keely BJ (2004) Atmospheric pressure chemical ionisation liquid chromatography/mass spectrometry of the ultraviolet screening pigment scytonemin: characteristic fragmentations. Rapid Commun Mass Spectrom 18(23):2934–2938

Stevenson CS, Capper EA, Roshak AK, Marquez B, Eichman C, Jackson JR, Mattern M, Gerwick WH, Jacobs RS, Marshall LA (2002) The identification and characterization of the marine natural product scytonemin as a novel antiproliferative pharmacophore. J Pharmacol Exp Ther 303(2):858–866

Stochaj WR, Dunlap WC, Shick JM (1994) Two new UV-absorbing mycosporine-like amino acids from the sea anemone *Anthopleura elegantissima* and the effects of zooxanthellae and spectral irradiance on chemical composition and content. Mar Biol 118:149–156

Takano S, Uemura D, Hirata Y (1978) Isolation and structure of a new amino acid, palythine, from the zoanthid *Palythoa tuberculosa*. Tetrahedron Lett 19(26):2299–2300

Takano S, Uemura D, Hirata Y (1979) Isolation and structure of a 334 nm UV absorbing substance, porphyra-334 from the red algae *Porphyra tenera* Kjellman. Chem Lett 8(4):419–420

Tamamizu K, Sakamoto T, Kurashige Y, Nozue S, Kumazaki S (2023) Scytonemin redox status in a filamentous cyanobacterium visualized by an excitation-laser-line-scanning spontaneous raman scattering spectral microscope. Spectrochim Acta A Mol Biomol Spectrosc 290:122258

Tamre E, Fournier GP (2022) Inferred ancestry of scytonemin biosynthesis proteins in cyanobacteria indicates a response to Paleoproterozoic oxygenation. Geobiology 20:764–775

Tang Y, Craven CB, Wawryk NJ, Qiu J, Li F, Li XF (2020) Advances in mass spectrometry-based omics analysis of trace organics in water. Trends Anal Chem 128:115918

Teai T, Raharivelomanana P, Bianchini JP, Faura R, Martin PMV, Cambon A (1997) Structure de deux nouvelles iminomycosporines isolees de *Pocillopora eydouxy*. Tetrahedron Lett 38(33):5799–5800

Tsujino I, Yabe K, Sekikawa I, Hamanaka N (1978) Isolation and structure of a mycosporine from the red alga *Chondrus yendoi*. Tetrahedron Lett 16:1401–1402

Uemura D, Katayama C, Wada A, Hirata Y (1980) Crystal and molecular structure of palythene possessing a novel 360 nm chromophore. Chem Lett 9(6):755–756

Varnali T, Edwards HG (2014) Reduced and oxidised scytonemin: theoretical protocol for raman spectroscopic identification of potential key biomolecules for astrobiology. Spectrochim Acta A Mol Biomol Spectrosc 117:72–77

Vega J, Schneider G, Moreira BR, Herrera C, Bonomi-Barufi J, Figueroa FL (2021) Mycosporine-like amino acids from red macroalgae: UV-photoprotectors with potential cosmeceutical applications. Appl Sci 11(11):5112

Volkmann M, Gorbushina AA, Kedar L, Oren A (2006) Structure of euhalothece-362, a novel red-shifted mycosporine-like amino acid, from a halophilic cyanobacterium (*Euhalothece* sp.). FEMS Microbiol Lett 258(1):50–54

Wada N, Sakamoto T, Matsugo S (2015) Mycosporine-like amino acids and their derivatives as natural antioxidants. Antioxidants 4(3):603–646

Weiss EL, Cape MR, Pan BJ, Vernet M, James CC, Smyth TJ, Ha SY, Iriarte JL, Mitchell BG (2022) The distribution of mycosporine-like amino acids in phytoplankton across a Southern Ocean transect. Front Mar Sci 9:1022957

Whitehead K, Hedges JI (2002) Analysis of mycosporine-like amino acids (MAAs) in aquatic plankton by liquid chromatography electrospray-ionization mass spectrometry. Mar Chem 80(1):27–39

Whitehead K, Karentz D, Hedges JI (2001) Mycosporine-like amino acids (MAAs) in phytoplankton, a herbivorous pteropod (*Limacina helicina*), and its pteropod predator (*Clione Antarctica*) in McMurdo Bay, Antarctica. Mar Biol 139:1013–1019

Wu JJ, Chalker BE, Rideout JA (1997) Two new UV-absorbing compounds from *Stylophora pistillata*: sulfate esters of mycosporine-like amino acids. Tetrahedron Lett 38(14):2525–2526

Xie C-L, Liu Q, Xia JM, Gao Y, Yang Q, Shao ZZ, Liu G, Yang X-W (2017) Anti-allergic compounds from the deep-sea-derived actinomycete *Nesterenkonia flava* MCCC 1 K00610. Mar Drugs 15(3):71

Yoshiki M, Tsuge K, Tsuruta Y, Yoshimura T, Koganemaru K, Sumi T, Matsui T, Matsumoto K (2009) Production of new antioxidant compound from mycosporine-like amino acid, porphyra-334 by heat treatment. Food Chem 113(4):1127–1132

Zhang L, Li L, Wu Q (2007) Protective effects of mycosporine-like amino acids of *Synechocystis* sp. PCC 6803 and their partial characterization. J Photochem Photobiol B Biology 86(3):240–245

Zhang A, Sun H, Wang P, Han Y, Wang X (2012) Modern analytical techniques in metabolomics analysis. Analyst 137(2):293–300

Zhang HW, Lv C, Zhang LJ, Guo X, Shen YW, Nagle DG, Zhou YD, Liu SH, Zhang WD, Luan X (2021) Application of omics-and multi-omics-based techniques for natural product target discovery. Biomed Pharmacother 141:111833

Metabolomics: An Approach to Improve Production of Cyanobacterial Natural Products

19

Varsha K. Singh, Sapana Jha, Palak Rana, Ashish P. Singh, Riya Tripathi, Neha Kumari, Donat-P. Häder, and Rajeshwar P. Sinha

Abstract

Cyanobacteria are oxygenic prokaryotes that show both immense biological and chemical diversity. These ancient organisms are present in almost every type of habitat on Earth. They are a source of numerous natural products (NPs) that have significant ecological functions and biotechnological and nutritional benefits. The study of cyanobacterial metabolites has benefited from developments in bioinformatics and analytical instruments used in metabolomic investigations. This chapter examines a few studies that demonstrate the application of both targeted and untargeted metabolomics to the study of cyanobacteria's NPs. To develop operative engineering solutions for growth enhancement, photosynthetic efficiency, and increased production, a deeper understanding of carbon/nitrogen metabolism, genetic regulation, route flux distribution, and integrative studies pertaining to the targeted biochemicals is needed. To present the state of knowledge regarding cyanobacterial metabolism, this chapter highlights the latest developments of genome-scale metabolic models (GSMMs), metabolomic analysis, and integrative modeling techniques.

Keywords

Biotechnological advancements · Metabolomics · Nutritional benefits · Photosynthetic efficiency · Secondary metabolites

V. K. Singh · S. Jha · P. Rana · A. P. Singh · R. Tripathi · N. Kumari · R. P. Sinha (✉)
Laboratory of Photobiology and Molecular Microbiology, Department of Botany, Institute of Science, Banaras Hindu University, Varanasi, India

D.-P. Häder
Emeritus from Friedrich-Alexander University, Department of Biology, Möhrendorf, Germany

M. Kesheri et al. (eds.), *Multi-Omics in Biomedical Sciences and Environmental Sustainability*, https://doi.org/10.1007/978-981-96-7067-3_19

Objectives

- Targeted and untargeted metabolomics can help researchers investigate how gene expression patterns relate to metabolites.
- Metabolomic approach can help us understand the significance of genetic and metabolic engineering in the enhanced synthesis of non-native and native biochemicals.
- Metabolomics can help identify the biomarkers for diseases, especially those related to diet and energy balance.
- Aims to state the knowledge regarding cyanobacterial metabolism and the importance of heterologous expression for cyanobacterial compounds.

19.1 Introduction

The demand for energy has grown and the depletion of natural resources has accelerated due to the fast-expanding human population and industrialization. The ecology has suffered greatly as a result of the overuse of natural resources, and a worldwide energy crisis is now imminent. Cyanobacteria are a rich source of unusual natural chemicals, but they also show enormous potential for the long-term, sustainable synthesis of chemical molecules. The field of cyanobacterial biotechnology has grown beyond its infancy. There is a growing number of businesses focusing cyanobacteria on a big scale for use as fertilizers, food additives, cosmetics, or as a platform to produce biopolymers, biofuels, and pigments. With just sunlight, CO_2, and minerals, cyanobacteria can produce organic compounds thanks to their autotrophic metabolism.

Green biotechnology production achieves the goal of creating a climate-neutral circular bioeconomy by becoming CO_2-neutral or even negative (Hitchcock et al. 2020). As a result, there are significant initiatives underway to broaden the product line and improve synthetic biology techniques for cyanobacteria. Consequently, there is a growing need to develop environmental-friendly and sustainable methods for creating sustainable chemical and fuel substitutes (Luan and Lu 2018). Here, a number of naturally occurring microorganisms as well as their genetically modified counterparts have proven useful in the production of diverse renewable and sustainable biochemicals (Fig. 19.1). But the majority of these systems rely on heterotrophs, like *Saccharomyces cerevisiae* and *Escherichia coli,* which need organic carbon sources (Jojima et al. 2010; Wang et al. 2019). In contrast, through photosynthesis, cyanobacteria use atmospheric carbon to build biomass (Singh et al. 2016). Additionally, because they may develop without freshwater-based substrate, marine cyanobacteria provide further sustainability benefits (Merlo et al. 2021).

The strains of *Synechococcus elongatus* UTEX 2973 and *Synechococcus elongatus* BDU 130192 have further enhanced the list of promising cyanobacterial strains suitable for biotechnological and industrial usage (Pathania et al. 2022). These strains not only show growth benefits, but also important features, including increased biomass concentration and increased resistance toward a range of abiotic

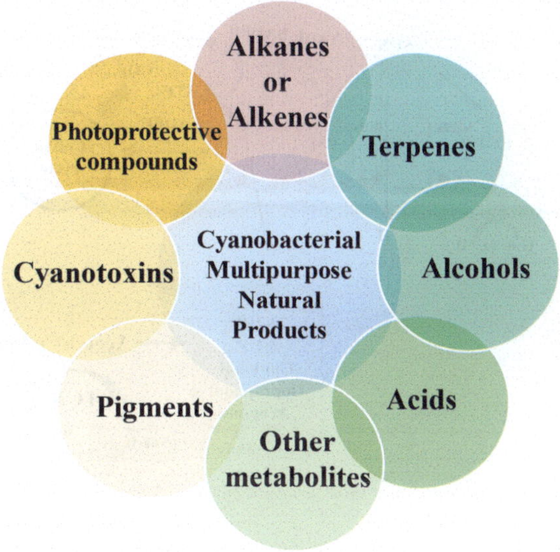

Fig. 19.1 List of a few most important value-added biochemicals metabolized by cyanobacterial cells

conditions. Cyanobacteria possess the metabolic processes essential for the conversion of solar energy into chemical energy and produce a range of energy-rich compounds. Given the advantages that cyanobacteria offer, numerous workers have created and successfully employed the tools that optimize the metabolic pathways that alter the native metabolic networks in an effort to reroute the photosynthetic carbon flux toward NPs such as alkanes, lipids, butanol, and sugars. (Knoot et al. 2018; Pattharaprachayakul et al. 2020).

19.2 Metabolic Flux Understanding in Cyanobacteria

Metabolic flux analysis (MFA) quantifies fluxes in metabolism within biochemical processes, facilitating the evaluation of the synthesis and utilization of metabolic rates inside an organism. One of the fundamental instruments of metabolic engineering, MFA has been established and extensively employed for assessing the distribution of carbon flux in significant metabolic pathways and thoroughly analyzing cell metabolism (Antoniewicz 2021). This is crucial to take into deliberation the primary pathways in cyanobacteria, which include the glycolysis and gluconeogenesis pathways, the tricarboxylic acid (TCA) pathway, the pentose phosphate pathway (PPP), and the Calvin-Benson-Bassham (CBB) cycle (Fig. 19.2).

Although carbon dioxide is transformed into organic molecules via these pathways and subsequently redistributed to fulfill metabolic requirements, its flow is not directly linked to the synthesis of readily measurable sugar sinks in plant tissues. It has been proven possible to analyze cyanobacteria using two distinct ^{13}C MFA techniques: isotopically nonstationary MFA (INST-MFA) and steady-state MFA (SS-MFA). In order to study central carbon metabolism in mixotrophic and

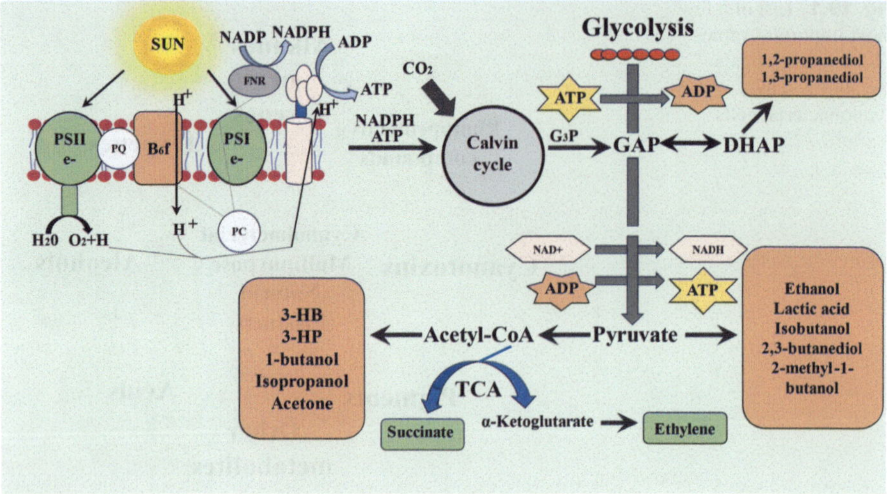

Fig. 19.2 Schematic representation of cyanobacterial photosynthetic metabolism and production of value-added biochemicals. The byproducts of photosynthesis and respiration (ATP, NADPH, and CO_2) generated by the photosynthetic and respiratory complexes present in the thylakoid membrane are utilized in the Calvin-Benson metabolism and glycolysis. The value-added biochemicals may be the result of native cyanobacterial metabolism or engineered metabolic pathways. (Modified from Knoot et al. 2018)

heterotrophic environments, steady-state MFA was initially used for *Synechocystis* sp. PCC 6803 (Yang et al. 2002). It was discovered that heterotrophic cells mainly produce energy by using the oxidative PPP (oxPPP), which metabolizes over 90% of incoming glucose to synthesize NADPH for respiration and growth. It is thought that cyanobacteria's special respiratory route, which links NADPH synthesis to oxidative phosphorylation, compensates for the absence of the entire cyanobacterial TCA cycle (Pelroy et al. 1972). NADPH and ATP are produced from photosynthetic light reactions that utilize the Calvin cycle flux under mixotrophic circumstances, which was roughly two times larger than glucose intake. Still, just one-third of the overall carbon intake was explained by flow through RuBisCO.

Several aspects restrict the study of cyanobacteria's metabolism. For instance, the cell's spatial distribution of metabolic processes such as light harvesting and carbon fixing. Furthermore, flux analysis under photoautotrophic conditions involves, in most cases, more computationally and experimentally intensive methods than steady-state MFA experiments applied to characterize heterotrophic metabolism, such as dynamic $^{13}CO_2$ labeling experiments and isotopically nonstationary MFA (INST-MFA) (Adebiyi et al. 2015; Angermayr et al. 2015). Systems metabolic engineering (SME) is a larger field that includes MFA and combines classical metabolic engineering with synthetic biology and systems biology (Lee et al. 2012; Choi et al. 2019). As biocatalysts, cyanobacteria have demonstrated considerable promise in directly converting CO_2 into valuable compounds like biofuels and medicinal substances.

MFA is an efficient way to determine metabolic fluxes in vivo. These mass balances can then be utilized to construct a unique flux solution from a restricted collection of observations. Flux data from MFA has been efficiently utilized to (i) determine metabolic bottlenecks that limit production rate, (ii) characterize unknown host organisms, and (iii) identify inefficient pathways that reduce product output (McAtee et al. 2015). Furthermore, it is critical to recognize and measure the most competitive branch points in secondary metabolism that direct flux away from the desirable secondary metabolites. Metabolic engineering can be used to identify the "wasteful fluxes" leading to undesirable side products and increase product yield in this way (Shih and Morgan 2020).

One of the main goals of MFA in cyanobacteria is to measure the fluxes that provide the important intermediate precursors for secondary metabolites. For instance, estimating the flux at the pyruvate and phosphoenolpyruvate (PEP) nodes is essential to calculating the quantity of fixed carbon going toward aromatic amino acids, alkaloids, and MAAs (shikimate) compared to the tricarboxylic acid cycle (TCA). The carbon directed toward the intermediates of the MEP pathway can also be determined by assessing the flux surrounding the pyruvate and glyceraldehyde-3-phosphate nodes.

To comprehend the flux redistribution in *Synechococcus* sp. PCC 7002, an astaxanthin-producing recombinant strain (Hasunuma et al. 2019) used ^{13}C labeling-based metabolomics. Through the quantitative representation of overlapping C/N metabolic networks that MFA offers, it is possible to connect phenome analysis and genome profiling to identify the pathways linked to bloom production.

19.2.1 GSMMs and Flux Balance Analysis

The term "gene-protein-reaction associations" refers to a large-scale stoichiometric model (GSM) that uses mathematical optimization to simulate mass balances of participating metabolites and stoichiometric coefficients to describe all metabolic pathways. These associations can be determined theoretically or experimentally (Broddrick et al. 2016; Qian et al. 2017; Gu et al. 2019). Based on the stoichiometric correlations among biochemical reactions within a metabolic network, GSMs employ linear programming to forecast the metabolic fluxes within the network in various scenarios (Gu et al. 2019). Genomics sequence models (GSMs) comprise thousands of processes and metabolites and are often built using metabolic pathway databases and genomic annotations. In addition to offering a thorough understanding of the metabolic network, GSMs can be applied to forecast how the network would behave in various environmental scenarios. GSMs, however, are unable to capture the metabolic pathways dynamics or the kinetics of specific events (Gu et al. 2019). Because of this, they may be less accurate in forecasting how cells will behave in non-steady-state situations or in response to outside disturbances. An exponential growth phase is the ideal time to simulate primary metabolism since GSMs operate under the presumption of a pseudo-steady state. Crucially, GSMs make it possible to use optimization methods like flux balance analysis (FBA). It makes use of linear programming to forecast metabolic flux values for the whole metabolic reaction network (He et al. 2015).

Constraint-based analysis (FBA) is a commonly used technique to investigate biochemical networks. It is mainly beneficial for genome-scale reconstructions and gap-filling processes that may be lacking in current GSMs. It would be easier to comprehensively assess secondary metabolite production capacity using metabolic models if there was more detailed information available on secondary metabolite biosynthesis reactions. By exposing the metabolic processes and enzymes that are essential for performing particular metabolic goals, FBA and MFA offer complementary methods to direct genome-scale engineering. To reroute the flow of metabolites toward a specific product or pathway, these methods can be utilized to discover metabolic pathways and enzymes that require modification. In addition to this knowledge, synthetic metabolic pathways that produce high-quality metabolites can be designed.

Pelroy et al. reported on the functioning of the PPP and the Calvin cycle during the growth of cyanobacterial cells under various trophic circumstances (Pelroy et al. 1972). The glycolytic pathway has been introduced to the network as the *Synechocystis* genome and includes genes with conserved structures for each enzymatic step of glycolysis. Cyanobacteria lack α-ketoglutarate dehydrogenase and have an incomplete TCA cycle (Pearce et al. 1969). In cyanobacteria, it has been shown that the enzymes isocitrate lyase and malate synthase are present and essential for the glyoxylate cycle to function (Pearce and Carr 1967). It was assumed that *Synechocystis* was a host of the glyoxylate pathway. It is believed that the genes for phosphoenolpyruvate synthetase and malic enzyme, which are present across the entire *Synechocystis* genome, are responsible for the first gluconeogenetic processes rather than phosphoenolpyruvate carboxykinase, which is not present in *Synechocystis* (Pearce and Carr 1967).

During photoautotrophic growth, cyanobacteria exhibit significantly lower activity levels of glycolysis, the PPP, and the TCA cycle as compared to model heterotrophs (Steinhauser et al. 2012). Nonstationary isotopic ^{13}C labelling assays show a rapid build-up of ^{13}C in the metabolites of the CBC and gluconeogenesis pathways, but the TCA cycle intermediates do not exhibit any detectable label accumulation. This suggests a delayed turnover of these metabolites (Xiong et al. 2015). This metabolic adaptability could be the secret to the continued success of metabolic engineering research. After glyceraldehyde 3-phosphate (GAP) is converted into pyruvate and dihydroxyacetone phosphate (DHAP) in one to five chemical steps, respectively, these compounds are produced close to the crucial carbon-fixation reactions. For alcohols in cyanobacteria, titers for compounds made from pyruvate, or DHAP, are typically among the highest reported (Shen and Liao 2012). In one of the earliest studies on cyanobacterial metabolic engineering, ethanol was produced by expressing aldehyde dehydrogenase and pyruvate decarboxylase. When NADH-consuming dehydrogenase was absent, the presence of transhydrogenase resulted in a growth deficiency in cells by neutralizing redox levels (Angermayr et al. 2015). A significant thermodynamic driving force for the synthesis routes is provided by the irreversible decarboxylative condensation that occurs as the first stage of biosynthesis. By further decarboxylation and reduction, the intermediate branched-chain amino acids (BCAAs) are converted to alcohols and aldehydes. After 2-acetolactate is decarboxylated to acetoin and then reduced to 2,3-butanediol (2,3-BD), 2,3-BD is produced (Oliver et al. 2013). Since acetyl-CoA and the TCA cycle receive less carbon during light periods, the product titers from metabolites derived from acetyl-CoA are often smaller than those obtained from pyruvate. Since photosynthesis can produce enough energy and reducing equivalents, it has been proposed that cyanobacteria mainly use these routes to produce carbon precursors for cellular components when exposed to light.

Driving enough carbon to acetyl-CoA is therefore a major constraint. One way is to induce dark fermentation of the glycogen reserves produced during periods of light to initiate glycolysis (Hirokawa et al. 2015). The production of target compounds has occasionally increased as a result of the deletion of the enzymes that catalyze the conversion of acetyl-CoA to acetate or storage polyhydroxy butyrate (PHB) (Wang et al. 2016). When growth conditions are fulfilled, the cyanobacterial TCA cycle is triggered, although it is not as active as it is in heterotrophic organisms. TCA cycle intervention has culminated in the synthesis of succinate and ethylene (Lan and Wei 2016). The cells secreted succinate into the liquid. The most common chemical ingredient produced worldwide is ethylene. The ethylene-forming enzyme, which is produced by a single gene, transforms α-ketoglutarate into ethylene, which is subsequently extracted straight from the culture perspective. The collaborating study in the TCA cycle is a challenge because amino acids and other biological components utilize α-ketoglutarate and other intermediates as their carbon skeletons. To maintain biomass production, the designed catabolic processes must be balanced with substrate regeneration (Lan and Wei 2016).

19.3 Cyanobacterial Strain Designing Approaches

It has been determined that significant genetic and metabolic engineering will be necessary to improve the synthesis of non-native and native biochemicals of interest in cyanobacteria. These problems call for a thorough comprehension of the regulation of cyanobacterial metabolism at various levels and the creation of an all-encompassing picture of that regulation. Furthermore, in order to effectively cultivate the recently discovered fast-growing strains as cyanobacterial biofactories, such information is needed. Therefore, it is critical to carefully identify and eliminate any waste cycles, metabolic and regulatory obstructions, or routes that are lowering the productivity of targeted molecules (Hagemann and Hess 2018). Synthetic biology techniques have been applied to control the targeted gene's expression in cyanobacterial strains that are ideal hosts for biotechnological and industrial purposes. These techniques include CRISPR-interference (CRISPRi) and Antisense Expression technologies (Table 19.1).

Table 19.1 List of technologies and algorithms implied in targeted production through strain-based designing

Techniques/algorithms	Strain	Descriptions
Randomly barcoded transposon insertion sequencing (RB-Tnseq)	*Synechococcus* sp. PCC 7942	Essential gene sets were identified.
RB-Tnseq	*Synechococcus* sp. PCC 7942	Diurnal growth genes were identified.
FBA algorithm	*Synechocystis* sp. PCC 6803	n-Butanol is obtained as a product.
Pooled CRISPR-interference	*Synechocystis* sp. PCC 6803	Increased production of L-lactate.
Antisense expression technology	*Synechococcus* sp. PCC 7942	Short-chain fatty acids production increased.
CRISPR-interference	*Synechocystis* sp. PCC 6803	Essential gene sets were identified.
Grad-seq	*Synechocystis* sp. PCC 6803	Ribonucleoprotein complexes were identified.
CRISPR/Cpf1	*Synechocystis* sp. PCC 6803	Gene modification (insertion and deletion).
FBA algorithm	*Synechocystis* sp. PCC 6803	Ethanol is obtained as a product.
FVA algorithm	*Synechocystis* sp. PCC 6803	Isoprene is obtained as a product.
Multiobjective hybrid machine learning	*Nostoc* sp. CCC-403	Phycobiliproteins (PBPs) are obtained as a product.
MCA algorithm	*Synechococcus* sp. PCC 7942	Limonene is obtained as a product.
MCA algorithm	*Synechocystis* sp. PCC 6803	Ethanol is obtained as a product.
FBA with MOMA	*Synechococcus elongatus* PCC 7942	1,3-Propanediol and glycerol were obtained as product.

Modified from Pathania et al. (2022)

19.3.1 GSMM

A computational method called GSMM makes use of mathematical modeling to forecast and improve microorganisms' metabolic pathways for the production of biofuel. In order to produce a full model of the metabolic network, GSMM entails building a network of metabolic reactions that take place within a microbe and integrating omics data, such as transcriptomic, proteomic, and genomes (O' Brien and Palsson 2015).

The first step toward rebuilding GSMM that aids in understanding cyanobacteria's basic metabolism relies on the genome sequences availability and precise annotation data (Broddrick et al. 2016). According to Gu et al. (2019), GSMMs are structural models based on stoichiometry that offer insights into metabolic networks without necessitating the knowledge of enzyme kinetic characteristics. Depending on the limitations and steady-state presumptions, these models can forecast intracellular flux patterns, growth rates, product production, and theoretical yield. To determine the cellular metabolic flux distributions, the GSMMs are frequently used in conjunction with other experimental observations, such as growth rate, O_2 production, and CO_2 and photon uptake rates. (Vu et al. 2012).

A number of automated tools including "FAME (Flux Analysis and Modeling Environment), Merlin (Metabolic Models Reconstruction using Genome-Scale Information), RAVEN (Reconstruction, Analysis, and Visualization of Metabolic Networks), ModelSEED, etc." have been developed to shorten the reconstruction time (Boele et al. 2012; Dias et al. 2015; Agren et al. 2013; Henry et al. 2010). By analyzing the GSMMs, techniques like RobustKnock, OptORF, OptKnock, OptForce, etc., have been developed to determine the gene modifications so as to target the desired product(s) (Tepper and Shlomi 2010; Kim and Reed 2010; Burgard et al. 2003; Ranganathan et al. 2010).

With the use of GSMM, it is possible to forecast metabolic fluxes, determine the best circumstances for targeted production, and pinpoint possible sites for metabolic engineering to increase yields and efficiency. For instance, Kanno et al. (2017) employed GSMM to determine metabolic engineering targets to enhance cyanobacteria's capacity for CO_2 fixation and chemical synthesis. Through research, they validated their theories and demonstrated that the anticipated genetic changes may indeed increase the production of the targeted products. Short-chain alkanes are a form of biofuel that are produced by cyanobacteria whose metabolic pathways have been engineered using GSMM (Choi and Lee 2013). The strains that emerged produced much more of the intended products. The fluxes or metabolite concentrations are computed by the kinetic or dynamic models using thermodynamics, enzyme kinetics, and other metabolic regulatory data (Kim et al. 2017). To address the overproduction of limonene and ethanol in cyanobacteria, kinetic modeling employing the Metabolic Control Analysis (MCA) approach has been utilized (Fell 1992).

19.3.2 Synthetic Biology Circuits

Genomes are made up of genes that work together to create biological functions. Scientists have created a set of standardized DNA parts, called BioBricks, that can be used to construct artificial genetic circuits with predetermined behavior. These circuits act like logical circuits, such as inverter, NAND, or NOR gates, and can be wired together using specific gene products. By combining these circuits, scientists can create new biological functions. This technology has been used to create light-driven switches, multicellular systems, and gene regulatory networks. Synthetic biology circuits are genetic circuits that can be designed and engineered to control gene expression in microorganisms. These circuits can be used to control the expression of genes involved in targeted production, allowing for precise control over the production process (Khalil and Collins 2010).

Synthetic biology circuits have been used in targeted production to enhance the metabolic pathways of microorganisms and increase their efficiency. Researchers applied synthetic biology techniques to engineer the cyanobacterium *Synechococcus elongatus* PCC 7942 to produce isobutyraldehyde, a precursor for advanced biofuels. They constructed a synthetic pathway consisting of multiple genes, including those encoding enzymes for isobutyraldehyde synthesis. Through the precise assembly and optimization of this synthetic pathway, they successfully achieved increased production of isobutyraldehyde in the engineered cyanobacterial strain (Ruffing and Jones 2012). Oliver et al. (2013) used synthetic biology techniques in a different study to increase free fatty acids (FFAs) production in the cyanobacterium *Synechococcus* sp. PCC 7002. To control FFA synthesis and enhance carbon flux toward it, they created and inserted artificial genetic circuits. The altered cyanobacterial strain's FFA production rose as a result of this synthetic biology strategy. Another study utilized a synthetic biology approach to improve hydrogen production in the cyanobacterium *Synechocystis* sp. PCC 6803. The researchers engineered a synthetic circuit by introducing hydrogenase genes under the control of a synthetic promoter. This circuit allowed for the controlled expression of hydrogenase enzymes, resulting in increased hydrogen production in the strain.

Synthetic biology has also made significant contributions to enhance desired production in algae (Fig. 19.3). To manufacture free fatty acids (FFAs), which can be used as building blocks for biofuels, scientists modified microalga *Chlamydomonas reinhardtii*. They created artificial genetic circuits to control the synthesis of FFA and enhanced the expression of vital lipid manufacturing enzymes. The altered algal strain's FFA production was successfully increased by this synthetic biology strategy.

These examples demonstrate the potential of synthetic biology circuits in improving the efficiency of microbial chemical production. By controlling the expression

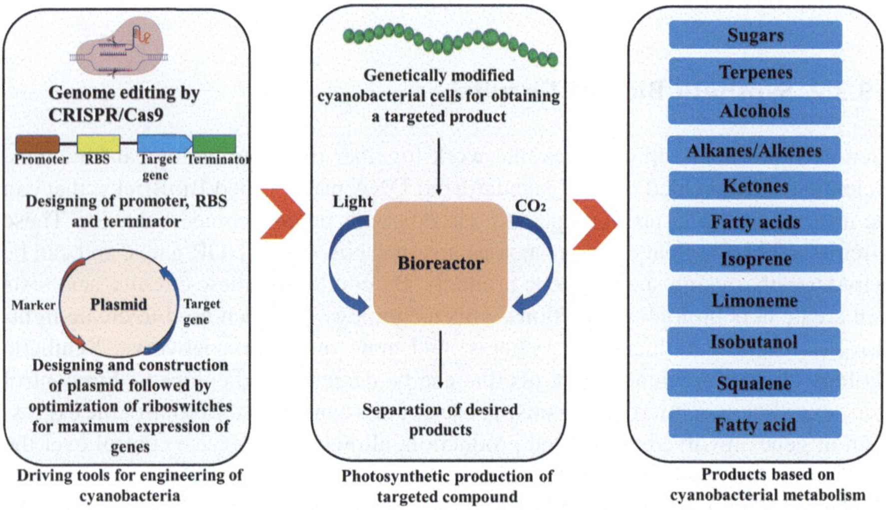

Fig. 19.3 Driving tools involved in the engineering of cyanobacteria to obtain value-added biochemicals. (Modified from Mishra et al. 2023)

of genes involved in metabolic pathways, synthetic biology circuits can optimize the production of NPs and reduce the cost of production.

19.3.3 Omics Technologies

Omics technologies (genomics, transcriptomics, proteomics, and metabolomics) have been increasingly used for the optimization of systems biology approaches in targeted production. These techniques enable the measurement and analysis of various molecular components and their interactions within cells, providing valuable insights into the cellular processes underlying targeted production (Sassetti et al. 2019).

The development of new sequencing technologies (NST) has enabled rapid and accurate analysis of genomic variations in microorganisms, including those that can produce biofuels. NSTs have allowed researchers to quickly and accurately identify and analyze the genomic makeup of microorganisms, leading to the discovery of new pathways and enzymes that can be used for targeted production (Jung et al. 2018).

Other omics technologies can also be used to understand the metabolic pathways involved in production and to identify potential targets for metabolic engineering. For instance, transcriptomics can be used to study gene expression patterns and identify key regulatory genes involved in production (Jin et al. 2016). Proteomics can be used to analyze protein expression levels and post-translational modifications, providing insights into the metabolic pathways and enzymes involved in synthesis (Sun et al. 2019; Nielsen et al. 2015).

The integration of omics data using computational and mathematical models can help to optimize the metabolic pathways and identify key targets for genetic engineering to enhance the production of desired NPs (Lerman et al. 2012). Moreover, the application of machine learning algorithms to omics data can help to predict the performance of different biofuel production strategies and identify the most promising approaches for further optimization (García-Sánchez et al. 2019). Overall, omics technologies have become powerful tools for the optimization of systems biology approaches in targeted production, providing valuable insights into the underlying cellular processes and enabling the identification of key targets for genetic engineering and process optimization.

19.4 Metabolomics Applied to Cyanobacterial NPs

A metabolome is the complete collection of all metabolites involved in the metabolic processes of an organism. It is generated via metabolomics, an analytical component of the "omic" that can be used for system analysis. Metabolomics is a branch of genetics, transcriptomics, and proteomics that studies metabolites (Fiehn 2001, 2002). The metabolome, which is defined by the genome's coding capacity, gets affected by alterations in the transcriptomics and proteomics, which further influence enzymatic and transport functions (Roessner et al. 2000; Desbrosses et al. 2005; Lei et al. 2011). Analogously with other "omics" technologies, metabolomics emerged from the significant advancements in metabolite analysis techniques. These advancements primarily resulted from the combination of new mass spectroscopy (MS) or nuclear magnetic resonance (NMR)-based compound identification/quantification tools with highly efficient separation techniques such as capillary electrophoresis (CE) and gas or liquid chromatography (GC or LC). These technologies made it possible to analyze complicated biological samples for metabolites in a thorough and consistent manner. In order to comprehend cyanobacterial NPs, we shall go over metabolomic techniques in this section. The photoautotrophic prokaryotes of the domain bacteria make up the phylum Cyanobacteria. An endosymbiotic process later transferred oxygenic photosynthesis—which gave rise to plastids—from ancient cyanobacteria approximately three billion years ago (Hohmann and Blankenship 2011). The analysis of metabolite targets focuses on a subset of metabolites that have common reactions or pathways (Fiehn et al. 2003; Dunn et al. 2005). Using a standard analytical method, metabolite profiling examines the number of a set of identified metabolites that can be processed simultaneously (Fig. 19.4) (Roessner et al. 2000; Fiehn 2001). The goal of metabolic fingerprinting is to measure the amount of metabolites inside the cells—which may include metabolites that are currently unknown—in a highly rapid, unbiased manner, or what is known as the endo-metabolome. In contrast, the exo-metabolome, or collection of metabolites expelled from cells, is examined by metabolic footprinting (Fiehn 2001; Allen et al. 2012). Lastly, the processes within an organism or cell are revealed by the flux analysis using stable isotopes like ^{13}C, ^{15}N, and ^{18}O or conventionally radioactive isotopes.

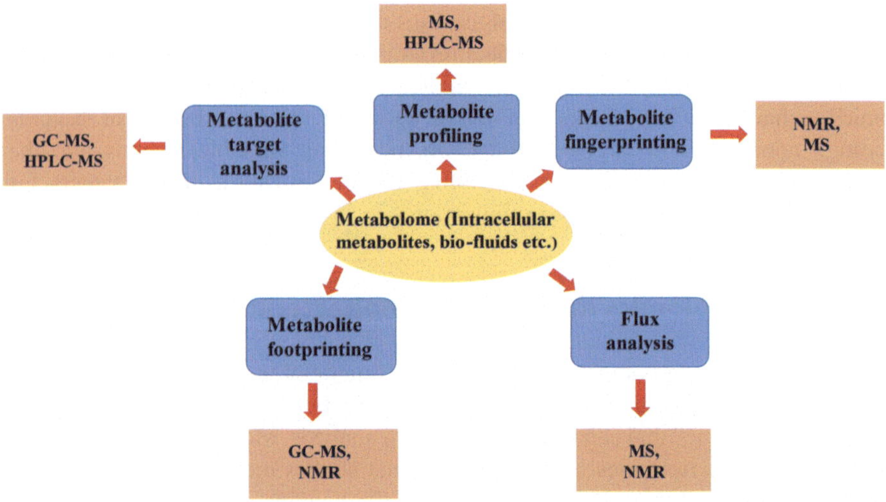

Fig. 19.4 Different techniques used for metabolomics profiling of cyanobacterial NPs

19.4.1 Methods for Cyanobacterial Metabolome Profiling Technologies

In biological systems, metabolomics is the measurement and identification of all metabolic substances in an unbiased and nontargeted manner, while metabolic profiling is the identification and measurement of a predetermined subset of metabolites that are typically associated with a particular metabolic pathway (Ellis and Goodacre 2012). Untargeted metabolomics is a broad and all-inclusive technique that measures every metabolite in a sample, even those with unknown targets. Targeted metabolomics, on the other hand, involves measuring a specific group of identified and biochemically annotated analytes. It utilizes numerous technologies for analysis, including methods that combine separation as well as detection of a specific substance. The cyanobacterial metabolome is currently analyzed using capillary electrophoresis (CE), liquid chromatography (LC), and gas chromatography (GC). Organic acids, phosphates, and nucleotides are examples of charged metabolites that are best separated from primary metabolism via CE (Takahashi et al. 2008). For GC and LC, the restriction to charged metabolites does not hold true. Since ultra-performance liquid chromatography (UPLC) produces better compound separation than traditional LC settings, it is commonly used (Narainsamy et al. 2013). Using reversed or normal stationary phases, polar chemicals were extracted from cyanobacteria using LC techniques (Bennette et al. 2011). Due to separation issues with low molecular mass primary metabolites, LC techniques for broad-coverage metabolome studies are typically restricted to the investigation of many primary metabolites, metabolites with secondary functions, or lipid complexes. GC-MS profiling is a commonly used technology that was among the first to be applied to cyanobacteria (Eisenhut et al. 2008). GC analysis is first used for substances that are volatile.

Chemical modifications of natural compounds into volatile products have expanded the scope of GC-MS in metabolomics. The profiling of primary metabolism, which includes the identification of neutral primary metabolites as well as secondary products, has been found to benefit greatly from this technology's ability to analyze an extensive variety of different chemical metabolites in a single analysis. Cyanobacteria produce a large number of secondary metabolites, each with a distinct function, in order to compete and thrive in a variety of environments. There are many complex secondary metabolites, including alkaloids, depsipeptides, polyketides, peptides, and photoprotective secondary compounds such as mycosporine-like amino acids (MAAs) and scytonemin (Thajuddin and Subramanian 2005; Sielaff et al. 2006; Spolaore et al. 2006; Sinha and Häder 2008; Rastogi and Sinha 2009; Pathak et al. 2015; Richa et al. 2015). It deals with the comprehensive information on the structure, synthesis, function in photoprotection, and several methods utilized for the metabolomic profiling of these cyanobacterial secondary metabolites such as scytonemin, and MAAs (Ellis and Goodacre 2012). In addition to aiding in the functional assignment of genes, metabolite profiling frequently results in the discovery of novel metabolic capacities. Techniques such as HPLC, ESI-MS, FTIR, and NMR are used to frequently discover novel metabolites or metabolites, revealing uncharacterized metabolic capabilities (Fig. 19.5) (Kind and Fiehn 2010; Bowen and Northern 2010; Baran et al. 2010). The process of liquid extraction is used to remove the target metabolites. An aqueous solvent is used for polar compounds like MAAs, an organic solvent is used for lipid-soluble compounds like scytonemin, or a mixture of solvents is used. The chemical characteristics of the solvent utilized determine the kind of metabolites that are extracted (Moco et al. 2007).

A typical method for analyzing secondary metabolites like MAAs, scytonemin, chlorophylls, and carotenoids is UV-VIS spectrophotometry. It quantifies how much ultraviolet radiation (UVR) is absorbed by a substance when it is in solution. Mass spectrometry offers precise mass measurements for molecular formula determination as well as sensitive detection and metabolite identification through mass spectrum interpretation and comparison (Dunn et al. 2005). Prior to mass detection, the majority of MS applications in metabolomics employ a separation technique, usually LC, GC, or CE. Applications of GC-MS based metabolomic techniques for the investigation of volatile and thermally stable polar as well as nonpolar metabolites have evolved quickly.

HPLC is a significant separation method that was once incorrectly labeled as high-pressure liquid chromatography. It makes use of small particle size stationary phases, typically with a diameter of 5–10 mm and a small range of particle sizes, and can resist pressures of up to 40 MPa (Wilson 2010). HPLC-MS is the gold standard analytical instrument for both qualitative and quantitative studies. Particularly for evaluating the functional groups found in metabolites, FTIR has become more widely used in biological systems. Since FTIR functions as a "fingerprint technique," neither of the samples has identical FTIR spectra in terms of peak numbers

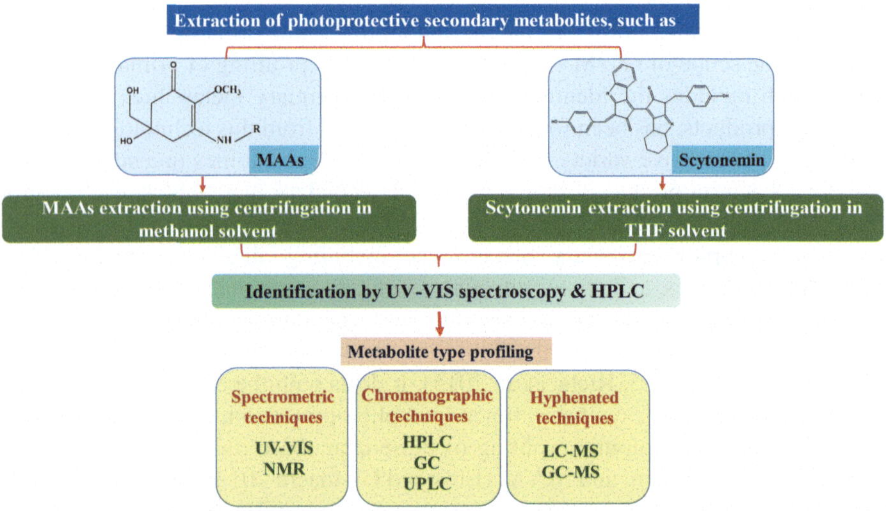

Fig. 19.5 A comprehensive and integrated technique for characterizing the metabolomic types of cyanobacterial compounds. (Modified from Pathak et al. 2017)

or the highest peak intensities (Pavia et al. 2001). NMR-based metabolomics is useful and well-suited since it may offer a "holistic view" of the metabolic substances in specific circumstances (Zhang et al. 2012). The NMR spectrum shows a signal with a certain frequency that corresponds to the chemical shift value, δ, which indicates the nuclear magnetic resonance of a molecule. NMR is regarded as one of the most selective methods for elucidating compounds because the NMR spectra of each molecule are distinct. The metabolic profiling technologies, such as MS or NMR, can help with metabolite identification, but a complete chemistry of a molecule is frequently only possible by combining metabolite information from multiple sources (Moco et al. 2007). Therefore, the most common analytical techniques for characterizing secondary metabolites like MAAs and scytonemin are UV-VIS spectroscopy, followed by HPLC, ESI-MS/MS, FTIR, and NMR.

Untargeted metabolomics, also known as nontarget analysis, is a qualitative technique that makes it easier to find metabolites and possibly novel compounds by examining the overall changes in the metabolome without focusing on any particular compounds. Metabolite target analysis focuses on examining the metabolites that perform common tasks or functions in certain specific processes. It has been difficult to develop specialized analytical tools due to the wide range of chemical variations found in bioactive compounds of cyanobacteria, including cyclic peptides, polyketides, alkaloids, and derivatives of amino acids that exhibit varying levels of toxicity. In the past, antibody-based methods or cell cultures were frequently used in bioassays for the analysis of cyanobacterial toxins. With the development of chromatography and NMR technologies, toxin classes could be specifically analyzed (Humpage et al. 2010). Metabolic fingerprinting is a quick, nontargeted, and highly efficient technique for determining the quantitative composition of the cellular metabolome, or metabolite spectrum. The process of separating low molecular mass compounds from cyanobacteria by CE, GC, or LC is typically combined with MS detection and identification in metabolic fingerprinting. Metabolome investigations of extracellular chemicals in growth media are commonly referred to as metabolic footprinting.

19.5 Heterologous Production of Cyanobacterial Compounds

19.5.1 Tools from Metabolic Engineering and Synthetic Biology

Various new technologies pertaining to synthetic biology and genetic engineering has been developed by which the identification of NPs from both innate and heterologous hosts has been altered (Smanski et al. 2016). The main focus for the creation of cyanobacterial NPs by heterologous means might be on the assembly and cloning of biosynthetic gene clusters (BGCs), the choice and enhancement of heterologous hosts, and the adjustment of BGC expression through transcription and translation.

19.5.2 Cloning and Assembly of BGCs

The process of assembly and cloning involves inserting a single targeted BGC into an appropriate vector. A reliable and extensively used method for genetically engineered cyanobacteria is the vector that integrates the required fragment of DNA into the neutral site I (NSI) by homologous recombination (Hitchcock et al. 2020). In cyanobacterial genomes, the NSI is a region that is neutral in function where the incorporation of heterologous DNA results in no appreciable phenotypic modification, at least under the pertinent growing circumstances (Ng et al. 2015). However, the development of novel vectors for cyanobacterial engineering has garnered attention recently, especially for heterologous expression by transferring cyanobacterial BGCs. Because of this, cyanobacterial genetic engineering has made extensive use of replicative plasmids based on the two main replication origins, RSF1010 and orp MB1 (Scherzinger et al. 1984; Rosano and Ceccarelli 2014; Kim et al. 2017; Vogel et al. 2017).

For cyanobacterial genetic engineering, shuttle vectors such as endogenous plasmid segments and *E. coli* plasmid replicons (pMB1) are frequently utilized. But the broad use of endogenous plasmids is limited by their frequent size, difficulty in further engineering, and species-specificity (Taton et al. 2014; Jin et al. 2018). RSF1010, which is a non-native replicon, has a wide host range (Bishé et al. 2019). Following the replication of pMB1 from *E. coli* and the likely replicon of plasmid pCC5.2, pSCB-YFP, a high copy number variant, was subsequently generated from the low copy number pPMQAK1 vector (Jin et al. 2018). Although *Synechocystis* sp. PCC6803 was genetically modified using this vector, its broad applicability in altering various cyanobacterial strains has not been confirmed (Jin et al. 2018). Recently, *Synechocystis* sp. PCC 6803 and *Synechococcus elongatus* UTEX 2973 were engineered using a flexible method named CyanoGate, which was based on the Plant Golden Gate MoClo kit (Vasudevan et al. 2019). The main methods for cloning and assembling BGCs of various sizes for heterologous expression are illustrated in Fig. 19.6. By using the traditional restriction digestion and ligation method, tiny BGCs may be easily cloned from native producers.

Prochlorosins from *Prochlorococcus* sp. MIT9313, hapalindoles obtained from *F. ambigua* UTEX 1903, MAAs isolated from *Cylindrospermum stagnale* PCC7417, and shinorine from *Fischerella* sp. PCC9339 were all cloned (Tang and Donk 2012; Katoch et al. 2016; Yang et al. 2018). Forming genetic libraries of natural producers utilizing vectors that have different capabilities for collecting information, like fosmids (about 40 kb), cosmids (about 45 kb), and bacterial artificial chromosomes (about 200 kb), is another common method for cloning and assembling BGCs. Southern blotting or polymerase chain reaction (PCR) is used to select clones containing the BGCs from the libraries. Sequencing the positive clones yields specific details about the suspected BGCs. Using this method, the BGCs of many cyanobacterial NPs have been characterized, including microviridin from *Microcystis aeruginosa* NIES843, hectochlorin from *Lyngbya majuscula*, and nostophycin from *Nostoc* sp. strain (Ramaswamy et al. 2007; Fewer et al. 2011). The expression of the BGCs of lyngbyatoxin from *Moorea producens*, microcystins from *M. aeruginosa*,

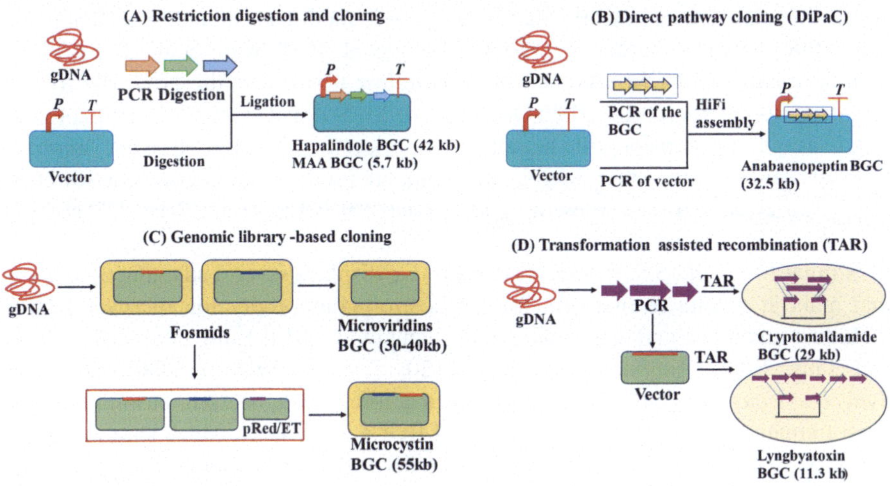

Fig. 19.6 Main methods for cyanobacterial BGC cloning and assembly in various sizes for heterologous expression. BGCs can either be constructed through PCR amplification and ligation in an appropriate vector (**a**) or directly with the vector (**b**). The clone containing a full BGC can be produced straight after the genomic library has been created. Alternatively, methods like Red-ET in *E. coli* (**c**) or TAR in yeast (**d**) can be used to recombine numerous clones to create a full BGC or genomic DNA. (Modified from Dhakal et al. 2021)

and patellamides from *Prochloron* are examples of how BGC-containing fosmids may be exploited directly for heterologous synthesis (Ongley et al. 2013; Liu et al. 2017).

Homologous recombination-based methods, like transformation-associated recombination (TAR) in *Saccharomyces cerevisiae* and lambda red (λ/Red) recombination in *E. coli*, can be highly effective for mobilizing a large fragment of DNA for expressing heterologous genes (Datsenko and Wanner 2000; Kouprina and Larionov 2016). For expression, the constructed BGCs need to be introduced into host cells. Natural transformation refers to the capacity of some cyanobacterial hosts to naturally introduce genetic elements (Onai et al. 2004).

19.5.3 Selection of Heterologous Hosts for Producing Cyanobacterial NPs

An appropriate host should be able to grow quickly and be easily manipulated genetically in the lab or in industrial settings. On the other hand, efficient expression of protein and the accessibility of precursors and cofactors for all stages required for biosynthesis are also necessary for the efficacious expression of a BGC (Zhang et al. 2019). A significant amount of work has gone into selecting and perfecting suitable heterologous hosts in order to effectively manufacture cyanobacterial NPs. *E. coli* is one of the most popular hosts for the heterologous generation of BGCs from several NP families, because of its quick growth, cost-effectiveness, ease of cultivation, and access to well-established genetic modification methods (Watanabe et al. 2006; Sanchez-Garcia et al. 2016; Zhang et al. 2018; Kaur et al. 2018). These include many RiPPs, MAAs, the NRP *Lyngbya* toxin, and the NRP-PK hybrid microcystins (Ziemert et al. 2008; Ongley et al. 2013; Liu et al. 2017). Dermatotoxin, lyngbyatoxin, the depsipeptide hapalosin, microcystin-LR (1), herbicide cyanobacterin, and microginin, which inhibits protease, are produced in *E. coli* (Ongley et al. 2013; Liu et al. 2017; D'Agostino and Gulder 2018; Liu et al. 2019; D'Agostino et al. 2022; Eusébio et al. 2022).

Similar to *E. coli*, *Saccharomyces cerevisiae* displays several advantageous traits that make it a suitable platform for the heterologous production of several NPs, including taxadiene and artemisinic acid (Ding et al. 2014; Kung et al. 2018; Bond and Tang 2019). *Anabaena* sp. PCC 7120, *Synechocystis* sp. PCC 6803, and *Synechococcus elongatus* PCC 7942 are the three cyanobacteria strains that are most frequently employed. Small or partial BGCs, such as the MAA shinorine, and polybrominated diphenyl ether, which was derived from a sponge, were successfully produced using *Synechocystis* and *S. elongatus* (Yang et al. 2018; Agarwal et al. 2017; Videau et al. 2020). It was revealed that a 21 kb BGC could produce the meroterpenoid tolypodiol and several of its analogues (Back et al. 2023). From a NRPS/PKS megasynthase BGC having a size of 28.7 kb, *S. elongatus* produces cryptomaldamide, which shows anti-infective property, and Columbamides are produced from 28.5 kb NRPS/PKS hybrid BGC (Taton et al. 2022; Back et al. 2023).

19.5.4 Transcriptional and Translational Tuning of BGC Expression

Optimizing the degree of biosynthetic gene expression is a well-established method for raising the targeted NPs' production in innate or heterologous hosts (Xu et al. 2020). At the transcriptional level, gene expression may be easily manipulated by utilizing various promoters. There are three primary categories of cyanobacterial promoters. The −10 consensus element (5-TATAAT-3) and the −35 element (5-TTGACA-3) are both present in type I promoters, but type II promoters only include the −10 element. Type III sigma (σ) factors control type III promoters in response to different stressors and stimuli; these promoters lack both consensus regions (Liu et al. 2017, 2019).

One tactic for adjusting the cyanobacterial BGCs' heterologous expression in a heterologous host might be sigma factor engineering. Constitutive and induced promoters are two methods for achieving transcriptional control of BGC expression in cyanobacterial hosts. In *Synechocystis* sp., two potent constitutive promoters are PrnpB and Pcpc560 (Englund et al. 2016; Wang et al. 2018). In *Anabaena* sp. PCC 7120, the Philmus group studied the transcriptional activity of many constitutive promoters of cyanobacterial BGCs (Videau et al. 2016).

The ability of an RBS to attract ribosomes for translation is another factor that determines how quickly proteins are produced from mRNA transcripts. In cyanobacteria, the RBS element can affect how proteins are expressed. For instance,

Synechococcus elongatus PCC 7942's output increased 13-fold when limonene synthase expression was tuned using a modified RBS (Wang et al. 2016; Englund et al. 2016; Liu and Pakrasi 2018). Bisabolene synthesis in *Synechocystis* sp. PCC 6803 was increased by applying specified RBS sequences that were calculated using the RBS calculator (Sebesta and Peebles 2020). In *Synechocystis* sp. PCC 6803, riboregulators have been discovered to be used for translation regulation (Abe et al. 2014; Sakai et al. 2015; Ueno et al. 2017). Biosynthetic gene codon optimization is a useful tactic to enhance heterologous expression. A synthetic version of the complete barbamide BGC was developed with codon optimization for heterologous expression in *S. venezuelae* (Kim et al. 2012). The codon-optimized and synthesized mpnD from *M. thermotolerans* SCSIO0065 and the tleC and tleD from *Streptomyces blastmyceticus* NBRC 12747 were also used for the best possible generation of teleocidin and pendolmycin in the cyanobacterium host *Anabaena* sp. PCC 7120 (Videau et al. 2020).

19.6 Conclusion

Metabolomics, the comprehensive study of metabolites within a biological system, offers a powerful approach to enhance the production of cyanobacterial NPs. Cyanobacteria, known for their diverse and valuable secondary metabolites, hold great potential in various industries, including pharmaceuticals, biofuels, and agriculture. By applying metabolomics, researchers can gain a detailed understanding of the metabolic pathways and regulatory networks involved in the biosynthesis of these NPs. Through metabolomics, it is possible to identify key metabolites and metabolic bottlenecks, enabling targeted genetic and environmental interventions to optimize production. This approach can also reveal the influence of external factors, such as nutrient availability and stress conditions, on metabolite profiles. Additionally, metabolomics facilitates the discovery of novel compounds and the engineering of cyanobacteria to produce higher yields of desired products or entirely new metabolites. Overall, integrating metabolomics into cyanobacterial research enhances our ability to harness and improve the production of valuable NPs, contributing to advancements in biotechnology and sustainable production methods.

19.7 Short Chapter Summary

This chapter reviews the findings that show how to use both targeted and untargeted metabolomics to investigate the natural products of cyanobacteria. Metabolomics makes it easier to find new molecules and to design cyanobacteria to create novel metabolites or enhanced quantities of desired products. Overall, this chapter summarizes the current understanding of cyanobacterial metabolism and emphasizes the most recent advancements in integrative modeling, genome-scale metabolic models, and metabolomic analysis.

Acknowledgements Varsha K. Singh (09/0013(12862)/2021-EMR-I) and Palak Rana (09/0013(16603)/2023-EMR-I) are thankful to the Council of Scientific & Industrial Research (CSIR), New Delhi, India, for the Senior Research Fellowship (SRF) and Junior Research Fellowship (JRF). Sapana Jha (No. R/Dev. /Sch./ UGC Non-NET Fello./2022-23/52561) is thankful to BHU for providing institutional fellowship. Ashish P. Singh (NTA Ref. No. 191620014505) and Riya Tripathi (NTA Ref. No. 231620041285) are thankful to the University Grants Commission (UGC), New Delhi, India, for the financial assistance in the form of Senior Research Fellowship (SRF) and Junior Research Fellowship (JRF). The incentive grant received from IoE (Scheme no. 6031), Banaras Hindu University, Varanasi, India, to Rajeshwar P. Sinha is highly acknowledged.

Author Contributions Chapter concept and design, Varsha K. Singh and Rajeshwar P. Sinha; chapter writing, Varsha K. Singh, Sapana Jha, Ashish P. Singh, Palak Rana, Riya Tripathi, and Neha Kumari; chapter editing and review, Donat-P. Häder and Rajeshwar P. Sinha.

References

Abe K, Miyake K, Nakamura M, Kojima K, Ferri S, Ikebukuro K, Sode K (2014) Engineering of a green-light inducible gene expression system in *Synechocystis* sp. PCC 6803. Microb Biotechnol 7(2):177–183

Adebiyi AO, Jazmin LJ, Young JD (2015) 13 C flux analysis of cyanobacterial metabolism. Photosynth Res 126:19–32

Agarwal V, Blanton JM, Podell S, Taton A, Schorn MA, Busch J, Lin Z, Schmidt EW, Jensen PR, Paul VJ, Biggs JS (2017) Metagenomic discovery of polybrominated diphenyl ether biosynthesis by marine sponges. Nat Chem Biol 13(5):537–543

Agren R, Liu L, Shoaie S, Vongsangnak W, Nookaew I, Nielsen J, Maranas CD (2013) The RAVEN toolbox and its use for generating a genome-scale metabolic model for *Penicillium chrysogenum*. PLoS Comput Biol 9(3):e1002980

Allen J, Davey HM, Broadhurst D, Heald JK, Rowland JJ, Oliver SG, Kell DB, Angermayr SA, Paszota M, Hellingwerf KJ (2012) Engineering a cyanobacterial cell factory for production of lactic acid. Appl Environ Microbiol 78(19):7098–7106

Angermayr SA, Rovira AG, Hellingwerf KJ (2015) Metabolic engineering of cyanobacteria for the synthesis of commodity products. Trends Biotechnol 33(6):352–361

Antoniewicz MR (2021) A guide to metabolic flux analysis in metabolic engineering: methods, tools and applications. Metab Eng 63:2–12

Back D, O'Donnell TJ, Axt KK, Gurr JR, Vanegas JM, Williams PG, Philmus B (2023) Identification, heterologous expression, and characterization of the *Tolypodiol* biosynthetic gene cluster through an integrated approach. ACS Chem Biol 18(8):1797–1807

Baran R, Bowen BP, Bouskill NJ, Brodie EL, Yannone SM, Northen TR (2010) Metabolite identification in *Synechococcus* sp. PCC 7002 using untargeted stable isotope assisted metabolite profiling. Anal Chem 82(21):9034–9042

Bennette NB, Eng JF, Dismukes GC (2011) An LC-MS-based chemical and analytical method for targeted metabolite quantification in the model cyanobacterium *Synechococcus* sp. PCC 7002. Anal Chem 83(10):3808–3816

Bishé B, Taton A, Golden JW (2019) Modification of RSF1010 based broad-host-range plasmids for improved conjugation and cyanobacterial bioprospecting. iScience 20:216–228

Boele J, Olivier BG, Teusink B (2012) FAME, the flux analysis and modeling environment. BMC Syst Biol 6:8

Bond CM, Tang Y (2019) Engineering *Saccharomyces cerevisiae* for production of simvastatin. Metab Eng 51:1–8

Bowen BP, Northen TR (2010) Dealing with the unknown: metabolomics and metabolite atlases. J Am Soc Mass Spectrom 21:1471–1476

Broddrick JT, Rubin BE, Welkie DG, Du N, Mih N, Diamond S, Palsson BO (2016) Unique attributes of cyanobacterial metabolism revealed by improved genome-scale metabolic modeling and essential gene analysis. Proc Natl Acad Sci USA 113(51):E8344–E8353

Burgard AP, Pharkya P, Maranas CD (2003) Optknock: a bilevel programming framework for identifying gene knockout strategies for microbial strain optimization. Biotechnol Bioeng 84(6):647–657

Choi YJ, Lee SY (2013) Microbial production of short-chain alkanes. Nature 502(7472):571–574

Choi KR, Jang WD, Yang D, Cho JS, Park D, Lee SY (2019) Systems metabolic engineering strategies: integrating systems and synthetic biology with metabolic engineering. Trends Biotechnol 37(8):817–837

D'Agostino PM, Gulder TA (2018) Direct pathway cloning combined with sequence-and ligation-independent cloning for fast biosynthetic gene cluster refactoring and heterologous expression. ACS Synth Biol 7(7):1702–1708

D'Agostino PM, Seel CJ, Ji X, Gulder T, Gulder TA (2022) Biosynthesis of cyanobacterin, a paradigm for furanolide core structure assembly. Nat Chem Biol 18(6):652–658

Datsenko KA, Wanner BL (2000) One-step inactivation of chromosomal genes in *Escherichia coli* K-12 using PCR products. Proc Natl Acad Sci USA 97:6640–6645

Desbrosses GG, Kopka J, Udvardi MK (2005) *Lotus japonicus* metabolic profiling. Development of gas chromatography-mass spectrometry resources for the study of plant-microbe interactions. Plant Physiol 137:1302–1318

Dhakal D, Chen M, Luesch H, Ding Y (2021) Heterologous production of cyanobacterial compounds. J Ind Microbol Biotechnol 48(3-4):kuab003

Dias O, Rocha M, Ferreira EC, Rocha I (2015) Reconstructing genome-scale metabolic models with merlin. Nucleic Acids Res 43:3899–3910

Ding MZ, Yan HF, Li LF, Zhai F, Shang LQ, Yin Z, Yuan YJ (2014) Biosynthesis of taxadiene in *Saccharomyces cerevisiae*: selection of geranylgeranyl diphosphate synthase directed by a computer-aided docking strategy. PLoS One 9:e109348

Dunn WB, Bailey NJC, Johnson HE (2005) Measuring the metabolome: current analytical technologies. Analyst 130:606–625

Eisenhut M, Ruth W, Haimovich M, Bauwe H, Kaplan A, Hagemann M (2008) The photorespiratory glycolate metabolism is essential for cyanobacteria and might have been conveyed endosymbiontically to plants. Proc Natl Acad Sci USA 105(44):17199–17204

Ellis DI, Goodacre R (2012) Metabolomics-assisted synthetic biology. Curr Opin Biotechnol 23:22–28

Englund E, Liang F, Lindberg P (2016) Evaluation of promoters and ribosome binding sites for biotechnological applications in the unicellular cyanobacterium *Synechocystis* sp. PCC 6803. Sci Rep 6:36640

Eusébio N, Castelo-Branco R, Sousa D, Preto M, D'Agostino P, Gulder TA, Leão PN (2022) Discovery and heterologous expression of microginins from *Microcystis aeruginosa* LEGE 91341. ACS Synth Biol 11(10):3493–3503

Fell DA (1992) Metabolic control analysis: a survey of its theoretical and experimental development. Biochem J 286(2):313–330

Fewer DP, Österholm J, Rouhiainen L, Jokela J, Wahlsten M, Sivonen K (2011) Nostophycin biosynthesis is directed by a hybrid polyketide synthase-non ribosomal peptide synthetase in the toxic cyanobacterium *Nostoc* sp. strain 152. Appl Environ Microbiol 77:8034–8040

Fiehn O (2001) Combining genomics, metabolome analysis, and biochemical modelling to understand metabolic networks. Compar Funct Genom 2:155–168

Fiehn O (2002) Metabolomics – the link between genotypes and phenotypes. Plant Mol Biol 48:155–171

Fiehn O, Kopka J, Dörmann P, Altmann T, Trethewey RN, Willmitzer L (2003) Metabolite footprinting. Nat Biotechnol 21:92–696

García-Sánchez CE, Hidalgo-López JE, Flores-Sánchez L, Morales-Martínez D (2019) Machine learning algorithms applied to omics data for predicting biofuel production. Appl Sci 9(15):3125

Gu C, Kim GB, Kim WJ, Kim HU, Lee SY (2019) Current status and applications of genome-scale metabolic models. Genome Biol 20:121

Hagemann M, Hess W (2018) Systems and synthetic biology for the biotechnological application of cyanobacteria. Curr Opin Biotechnol 49:94–95

Hasunuma T, Takaki A, Matsuda M, Kato Y, Vavricka CJ, Kondo A (2019) Single-stage astaxanthin production enhances the nonmevalonate pathway and photosynthetic central metabolism in *Synechococcus* sp. PCC 7002. ACS Synth Biol 8(12):2701–2709

He L, Wu SG, Wan N, Reding AC, Tan YJ (2015) Simulating cyanobacterial phenotypes by integrating flux balance analysis, kinetics, and a light distribution function. Microb Cell Factories 14:1–11

Henry CS, DeJongh M, Best AA, Frybarger PM, Linsay B, Stevens RL (2010) High-throughput generation, optimization and analysis of genome-scale metabolic models. Nat Biotechnol 28(9):977–982

Hirokawa Y, Suzuki I, Hanai T (2015) Optimization of isopropanol production by engineered cyanobacteria with a synthetic metabolic pathway. J Biosci Bioeng 119(5):585–590

Hitchcock A, Hunter CN, Canniffe DP (2020) Progress and challenges in engineering cyanobacteria as chassis for light-driven biotechnology. Microb Biotechnol 13(2):363–367

Hohmann-Marriott MF, Blankenship RE (2011) Evolution of photosynthesis. Annu Rev Plant Biol 62:515–548

Humpage AR, Magalhaes VF, Froscio SM (2010) Comparison of analytical tools and biological assays for detection of paralytic shellfish poisoning toxins. Anal Bioanal Chem 397:1655–1671

Jin H, Chen L, Wang J (2016) Synergistic action of a β-ketoacyl-acyl carrier protein synthase and a lipid-activated transcription factor for lipid production in photosynthetic microbes. Plant Biotechnol 14:1604–1613

Jin H, Wang Y, Idoine A, Bhaya D (2018) Construction of a shuttle vector using an endogenous plasmid from the cyanobacterium *Synechocystis* sp. PCC6803. Front Microbiol 9:1662

Jojima T, Omumasaba CA, Inui M, Yukawa H (2010) Sugar transporters in efficient utilization of mixed sugar substrates: current knowledge and outlook. Appl Microbiol Biotechnol 85(3):471–480

Jung S, Lee J, Park BC, Park C, Lee JH, Kim KH (2018) Genome-scale reconstruction and in silico analysis of ethanolamine utilization pathway in *Saccharomyces cerevisiae*. Sci Rep 8(1):1–11

Kanno M, Carroll AL, Atsumi S (2017) Global metabolic rewiring for improved CO_2 fixation and chemical production in cyanobacteria. Nat Commun 8(1):14724

Katoch M, Mazmouz R, Chau R, Pearson LA, Pickford R, Neilan BA (2016) Heterologous production of cyanobacterial mycosporine-like amino acids mycosporine-ornithine and mycosporine-lysine in *Escherichia coli*. Appl Environ Microbiol 82:6167–6173

Kaur J, Kumar A, Kaur J (2018) Strategies for optimization of heterologous protein expression in *E. coli*: roadblocks and reinforcements. Int J Biol Macromol 106:803–822

Khalil AS, Collins JJ (2010) Synthetic biology: applications come of age. Nat Rev Genet 11(5):367–379

Kim J, Reed JL (2010) OptORF: optimal metabolic and regulatory perturbations for metabolic engineering of microbial strains. BMC Syst Biol 4:53

Kim EJ, Lee JH, Choi H, Pereira AR, Ban YH, Yoo YJ, Kim E, Park JW, Sherman DH, Gerwick WH, Yoon YJ (2012) Heterologous production of 4-O-demethylbarbamide, a marine cyanobacterial natural product. Org Lett 14(23):5824–5827

Kim WJ, Lee SM, Um Y, Sim SJ, Woo HM (2017) Development of SyneBrick vectors as a synthetic biology platform for gene expression in *Synechococcus elongatus* PCC 7942. Front Plant Sci 8:293

Kind T, Fiehn O (2010) Advances in structure elucidation of small molecules using mass spectrometry. Bioanal Rev 2:23–60

Knoot CJ, Ungerer J, Wangikar PP, Pakrasi HB (2018) Cyanobacteria: promising biocatalysts for sustainable chemical production. J Biol Chem 293(14):5044–5052

Kouprina N, Larionov V (2016) Transformation-associated recombination (TAR) cloning for genomics studies and synthetic biology. Chromosoma 125:621–632

Kung SH, Lund S, Murarka A, McPhee D, Paddon CJ (2018) Approaches and recent developments for the commercial production of semi-synthetic artemisinin. Front Plant Sci 9:87

Lan EI, Wei CT (2016) Metabolic engineering of cyanobacteria for the photosynthetic production of succinate. Metab Eng 38:483–493

Lee JW, Na D, Park JM, Lee J, Choi S, Lee SY (2012) Systems metabolic engineering of microorganisms for natural and non-natural chemicals. Nat Chem Biol 8(6):536–546

Lei Z, Huhman DV, Summer LW (2011) Mass spectrometry strategies in metabolomics. J Biol Chem 286:25435–25442

Lerman JA, Hyduke DR, Latif H (2012) Portraying the expression landscapes of microbial genomes through the landscape viewer. Nucleic Acids Res 40(17):134

Liu D, Pakrasi HB (2018) Exploring native genetic elements as plug-in tools for synthetic biology in the cyanobacterium *Synechocystis* sp. PCC 6803. Microb Cell Factories 17:48

Liu T, Mazmouz R, Ongley SE, Chau R, Pickford R, Woodhouse JN, Neilan BA (2017) Directing the heterologous production of specific cyanobacterial toxin variants. ACS Chem Biol 12(8):2021–2029

Liu T, Mazmouz R, Pearson LA, Neilan BA (2019) Mutagenesis of the microcystin tailoring and transport proteins in a heterologous cyanotoxin expression system. ACS Synth Biol 8(5):1187–1194

Luan, Lu G, Xuefeng (2018) Tailoring cyanobacterial cell factory for improved industrial properties. Biotechnol Adv 36(2):430–442

McAtee AG, Jazmin LJ, Young JD (2015) Application of isotope labeling experiments and 13C flux analysis to enable rational pathway engineering. Curr Opin Biotechnol 36:50–56

Merlo S, Gabarrell Durany X, Pedroso Tonon A, Rossi S (2021) Marine microalgae contribution to sustainable development. Water 13:1373

Mishra S, Kumari N, Singh VK, Sinha RP (2023) Cyanobacterial biofuel: a platform for green energy. Adv Environ Eng Res 5(3):1–42

Moco S, Bino RJ, De Vos RCH, Vervoort J (2007) Metabolomics technologies and metabolite identification. Trends Anal Chem 26:855–866

Narainsamy K, Cassier-Chauvat C, Junot C, Chauvat F (2013) High performance analysis of the cyanobacterial metabolism via liquid chromatography coupled to a LTQ-Orbitrap mass spectrometer: evidence that glucose reprograms the whole carbon metabolism and triggers oxidative stress. Metabolomics 9:21–32

Ng AH, Berla BM, Pakrasi HB (2015) Fine-tuning of photoautotrophic protein production by combining promoters and neutral sites in the cyanobacterium *Synechocystis* sp. PCC6803. Appl Environ Microbiol 81:6857–6863

Nielsen J, Jewett MC, Cai Z (2015) Systems-level engineering of non-model microbes for the production of advanced biofuels and chemicals. Curr Opin Biotechnol 33:206–213

O'Brien EJ, Palsson BO (2015) Computing the functional proteome: recent progress and future prospects for genome-scale models. Curr Opin Biotechnol 34:125–134

Oliver JW, Machado IM, Yoneda H, Atsumi S (2013) Cyanobacterial conversion of carbon dioxide to 2, 3-butanediol. Proc Natl Acad Sci USA 110(4):1249–1254

Onai K, Morishita M, Kaneko T, Tabata S, Ishiura M (2004) Natural transformation of the thermophilic cyanobacterium *Thermosynechococcus elongatus* BP-1: a simple and efficient method for gene transfer. Mol Genet Genomics 271:50–59

Ongley SE, Bian X, Zhang Y, Chau R, Gerwick WH, Müller R, Neilan BA (2013) High-titer heterologous production in *E. coli* of lyngbyatoxin, a protein kinase C activator from an uncultured marine cyanobacterium. ACS Chem Biol 8:1888–1893

Pathak J, Rajneesh R, Kannaujiya VK, Sonker AS, Sinha RP (2015) Diverse functions and applications of novel and unique cyanobacterial sheath pigment, scytonemin. In: Sinha RP, Richa, Rastogi RP (eds) Biological sciences: innovations and dynamics. New India Publishing Agency, New Delhi, pp 237–262

Pathak J, Ahmed H, Sinha RP (2017) Metabolomic profiling of cyanobacterial UV-protective compounds. Curr Metab 5(2):138–163

Pathania R, Srivastava A, Srivastava S, Shukla P (2022) Metabolic systems biology and multi-omics of cyanobacteria: perspectives and future directions. Bioresour Technol 343:126007
Pattharaprachayakul N, Choi J, Incharoensakdi A, Woo HM (2020) Metabolic engineering and synthetic biology of cyanobacteria for carbon capture and utilization. Biotechnol Bioproc E 25(6):829–847
Pavia DL, Lampman GM, Kriz GS (2001) Introduction to spectroscopy, 3rd edn. Tomson Learning, Victoria
Pearce J, Carr NG (1967) The metabolism of acetate by the blue-green algae, *Anabaena* variabilis and *Anacystis nidulans*. J Gen Microbiol 49(2):301–313
Pearce J, Leach CK, Carr NG (1969) The incomplete tricarboxylic acid cycle in the blue-green alga *Anabaena variabilis*. J Gen Microbiol 55(3):371–378
Pelroy RA, Rippka R, Stanier RY (1972) Metabolism of glucose by unicellular blue-green algae. Arch Mikrobiol 87:303–322
Qian X, Kim MK, Kumaraswamy GK, Agarwal A, Lun DS, Dismukes GC (2017) Flux balance analysis of photoautotrophic metabolism: uncovering new biological details of subsystems involved in cyanobacterial photosynthesis. Biochim Biophys Acta Bioenerg 1858(4):276–287
Ramaswamy AV, Sorrels CM, Gerwick WH (2007) Cloning and biochemical characterization of the hectochlorin biosynthetic gene cluster from the marine cyanobacterium *Lyngbya majuscula*. J Nat Prod 70:1977–1986
Ranganathan S, Suthers PF, Maranas CD, Price ND (2010) OptForce: an optimization procedure for identifying all genetic manipulations leading to targeted overproductions. PLoS Comput Biol 6(4):e1000744
Rastogi RP, Sinha RP (2009) Biotechnological and industrial significance of cyanobacterial secondary metabolites. Biotechnol Adv 27(4):521–539
Richa, Sinha RP, Häder D-P (2015) Physiological aspects of UV-excitation of DNA. Top Curr Chem 356:203–248
Roessner U, Wagner C, Kopka J, Trethewey RN, Willmitzer L (2000) Simultaneous analysis of metabolites in potato by gas chromatography-mass spectrometry. Plant J 23:131–142
Rosano GL, Ceccarelli EA (2014) Recombinant protein expression in *Escherichia coli*: advances and challenges. Front Microbiol 5:172
Ruffing AM, Jones HD (2012) Physiological effects of free fatty acid production in genetically engineered *Synechococcus elongatus* PCC 7942. Biotechnol Bioeng 109(9):2190–2199
Sakai Y, Abe K, Nakashima S, Ellinger JJ, Ferri S, Sode K, Ikebukuro K (2015) Scaffold-fused riboregulators for enhanced gene activation in *Synechocystis* sp. PCC 6803. Microbiol Open 4:533–540
Sanchez-Garcia L, Martín L, Mangues R, Ferrer-Miralles N, Vázquez E, Villaverde A (2016) Recombinant pharmaceuticals from microbial cells: a 2015 update. Microb Cell Factories 15:33
Sassetti CM, Baldi F, Cocconcelli PS (2019) Omics technologies for the study of biofuels production processes. Front Microbiol 10:2317
Scherzinger E, Bagdasarian MM, Scholz P, Lurz R, Rückert B, Bagdasarian M (1984) Replication of the broad host range plasmid RSF1010: requirement for three plasmid-encoded proteins. Proc Natl Acad Sci USA 81:654–658
Sebesta J, Peebles CA (2020) Improving heterologous protein expression in *Synechocystis* sp. PCC 6803 for alpha bisabolene production. Metab Eng Commun 10:e00117
Shen CR, Liao JC (2012) Photosynthetic production of 2-methyl-1-butanol from CO_2 in cyanobacterium *Synechococcus elongatus* PCC7942 and characterization of the native acetohydroxyacid synthase. Ener Environ Sci 5(11):9574–9583
Shih ML, Morgan JA (2020) Metabolic flux analysis of secondary metabolism in plants. Metab Eng Commun 10:e00123
Sielaff H, Christiansen G, Schwecke T (2006) Natural products from cyanobacteria: exploiting a new source for drug discovery. IDrugs Investig Drugs J 9(2):119–127
Singh JS, Kumar A, Rai AN, Singh DP (2016) Cyanobacteria: a precious bioresource in agriculture, ecosystem, and environmental sustainability. Front Microbiol 7:529
Sinha RP, Häder D-P (2008) UV-protectants in cyanobacteria. Plant Sci 174(3):278–289

Smanski MJ, Zhou H, Claesen J, Shen B, Fischbach A, Voigt CA (2016) Synthetic biology to access and expand nature's chemical diversity. Nat Rev Microbiol 14:135–149

Spolaore P, Joannis-Cassan C, Duran E, Isambert A (2006) Commercial applications of microalgae. J Biosci Bioeng 101(2):87–96

Steinhauser D, Fernie AR, Araújo WL (2012) Unusual cyanobacterial TCA cycles: not broken just different. Trends Plant Sci 17(9):503–509

Sun H, Yan Y, Liu J (2019) Proteomics for understanding microbial metabolism of biofuels. Trends Biotechnol 37(3):312–324

Takahashi H, Uchimiya H, Hihara Y (2008) Difference in metabolite levels between photoautotrophic and photomixotrophic cultures of *Synechocystis* sp. PCC 6803 examined by capillary electrophoresis electrospray ionization mass spectrometry. J Exp Bot 59(11):3009–3018

Tang W, van der Donk WA (2012) Structural characterization of four prochlorosins: a novel class of lantipeptides produced by planktonic marine cyanobacteria. Biochemistry 51:4271–4279

Taton A, Unglaub F, Wright NE, Zeng WY, Paz-Yepes J, Brahamsha B, Palenik B, Peterson TC, Haerizadeh F, Golden SS, Golden JW (2014) Broad-host-range vector system for synthetic biology and biotechnology in cyanobacteria. Nucleic Acids Res 42:e136

Taton A, Rohrer S, Diaz B, Reher R, Caraballo Rodriguez AM, Pierce ML, Dorrestein PC, Gerwick L, Gerwick WH, Golden JW (2022) Heterologous expression in *Anabaena* of the columbamide pathway from the cyanobacterium *Moorena bouillonii* and production of new analogs. ACS Chem Biol 17(7):1910–2322

Tepper N, Shlomi T (2010) Predicting metabolic engineering knockout strategies for chemical production: accounting for competing pathways. Bioinformatics 26(4):536–543

Thajuddin N, Subramanian G (2005) Cyanobacterial biodiversity and potential applications in biotechnology. Curr Sci 89:47–57

Ueno K, Sakai Y, Shono C, Sakamoto I, Tsukakoshi K, Hihara Y, Sode K, Ikebukuro K (2017) Applying a riboregulator as a new chromosomal gene regulation tool for higher glycogen production in *Synechocystis* sp. PCC 6803. Appl Microbiol Biotechnol 101:8465–8474

Vasudevan R, Gale GA, Schiavon AA, Puzorjov A, Malin J, Gillespie MD, Vavitsas K, Zulkower V, Wang B, Howe CJ, Lea-Smith D (2019) CyanoGate: a Golden Gate modular cloning suite for engineering cyanobacteria based on the plant MoClo syntax. Plant Physiol 180:39–55

Videau P, Wells KN, Singh AJ, Gerwick WH, Philmus B (2016) Assessment of *Anabaena* sp. strain PCC 7120 as a heterologous expression host for cyanobacterial natural products: production of lyngbyatoxin A. ACS Synth Biol 5(9):978–988

Videau P, Wells KN, Singh AJ, Eiting J, Proteau PJ, Philmus B (2020) Expanding the NPs heterologous expression repertoire in the model cyanobacterium *Anabaena* sp. strain PCC 7120: production of pendolmycin and teleocidinB-4. ACS Synth Biol 9:63–75

Vogel AIM, Lale R, Hohmann-Marriott MF (2017) Stream lining recombination-mediated genetic engineering by validating three neutral integration sites in *Synechococcus* sp. PCC 7002. J Biol Eng 11:19

Vu TT, Stolyar SM, Pinchuk GE, Hill EA, Kucek LA, Brown RN, Lipton MS, Osterman A, Fredrickson JK, Konopka AE, Beliaev AS, Reed JL, Segre D (2012) Genome-scale modeling of light-driven reductant partitioning and carbon fluxes in diazotrophic unicellular cyanobacterium *Cyanothece* sp. ATCC 51142. PLoS Comput Biol 8(4):e1002460

Wang X, Liu W, Xin C, Zheng Y, Cheng Y, Sun S, Li R, Zhu XG, Dai SY, Rentzepis PM, Yuan JS (2016) Enhanced limonene production in cyanobacteria reveals photosynthesis limitations. Proc Natl Acad Sci USA 113:14225–14230

Wang B, Eckert C, Maness PC, Yu J (2018) A genetic tool box for modulating the expression of heterologous genes in the cyanobacterium *Synechocystis* sp. PCC 6803. ACS Synth Biol 7:276–286

Wang L, York SW, Ingram LO, Shanmugam KT (2019) Simultaneous fermentation of biomass-derived sugars to ethanol by a co-culture of an engineered *Escherichia coli* and *Saccharomyces cerevisiae*. Bioresour Technol 273:269–276

Watanabe K, Hotta K, Praseuth AP, Koketsu K, Migita A, Boddy CN, Wang CC, Oguri H, Oikawa H (2006) Total biosynthesis of antitumor nonribosomal peptides in *Escherichia coli*. Nat Chem Biol 2:423–428

Wilson K (2010) Chromatographic techniques. In: Principles and techniques of biochemistry and molecular biology, 7th edn. Cambridge, UK, Cambridge University Press, p 446

Xiong W, Morgan JA, Ungerer J, Wang B, Maness PC, Yu J (2015) The plasticity of cyanobacterial metabolism supports direct CO_2 conversion to ethylene. Nat Plants 1(5):1–7

Xu W, Klumbys E, Ang EL, Zhao H (2020) Emerging molecular biology tools and strategies for engineering natural product biosynthesis. Metabol Eng Commun 10:e00108

Yang C, Hua Q, Shimizu K (2002) Integration of the information from gene expression and metabolic fluxes for the analysis of the regulatory mechanisms in *Synechocystis*. Appl Microbiol Biotechnol 58:813–822

Yang G, Cozad MA, Holland DA, Zhang Y, Luesch H, Ding Y (2018) Photosynthetic production of sunscreen shinorine using an engineered cyanobacterium. ACS Synth Biol 7(2):664–671

Zhang A, Sun H, Wang P, Han Y, Wang X (2012) Modern analytical techniques in metabolomics analysis. Analyst 137:293–300

Zhang Y, Chen M, Bruner SD, Ding Y (2018) Heterologous production of microbial ribosomally synthesized and post translationally modified peptides. Front Microbiol 9:1801

Zhang JJ, Tang X, Moore BS (2019) Genetic platforms for heterologous expression of microbial NPs. Nat Prod Rep 36(9):1313–1332

Ziemert N, Ishida K, Liaimer A, Hertweck C, Dittmann E (2008) Ribosomal synthesis of tricyclic depsipeptides in bloom-forming cyanobacteria. Angew Chem 47(7756):7759

Importance of Omics in Microalgal Biofuel Production

20

Ashish P. Singh, Amit Gupta, Varsha K. Singh, Sapana Jha, Palak Rana, and Rajeshwar P. Sinha

Abstract

The oxygenic photoautotrophs, cyanobacteria, possess vital biochemical pathways in their metabolism that enable them to fix atmospheric CO_2 and synthesize a variety of metabolites. Nevertheless, the economic viability of cyanobacteria-based biofuels poses a barrier, prompting the development of various strategies to enhance the production performance of these microorganisms. The creation of bioengineering methods has made it possible to manipulate cyanobacterial metabolic pathways to produce a range of valuable bioproducts through photosynthetic processes. The efficient use of cyanobacteria as photosynthetic cell factories requires a thorough comprehension of their metabolism and how it interacts with other cellular processes. The application of systems and synthetic biology tools has produced a lot of data on various metabolic pathways. However, to create effective engineering strategies for additional growth, photosynthetic efficiency, and increased production of target biochemicals, a thorough understanding of their carbon/nitrogen metabolism, pathway flux distribution, genetic regulation, and integrative analyses is needed. The field of systems biology and genomics has led to the recognition of a new paradigm in systems metabolic engineering. Specifically, a systems-based approach essential for whole-cell inquiry and prediction is a reconstruction and modeling of the genome-scale metabolic network. To highlight our present understanding of cyanobacterial metabolism, we address recent developments in integrative modeling techniques and omics investigations (genomics, metabolomics, transcriptomics, and proteomics) in this chapter.

A. P. Singh · A. Gupta · V. K. Singh · S. Jha · P. Rana · R. P. Sinha (✉)
Laboratory of Photobiology and Molecular Microbiology, Department of Botany, Institute of Science, Banaras Hindu University, Varanasi, India

M. Kesheri et al. (eds.), *Multi-Omics in Biomedical Sciences and Environmental Sustainability*, https://doi.org/10.1007/978-981-96-7067-3_20

Keywords

Biofuels · Cyanobacteria · Lipid production · Omics · Systems biology

Objectives

- Applications of recent omics approaches may contribute to enhanced biofuel production from cyanobacteria and microalgae.
- Bio-based fuels may mitigate the alarming impact of greenhouse gas emissions and slow down global warming.
- Advanced gene editing and multi-omics analysis, like systems biology, synthetic biology, and metabolic engineering, will help us understand the recent advancements in the field of biotechnology.
- Aims to highlight the approaches for a sustainable and enhanced biofuel production.

20.1 Introduction

Biocompounds and cyanobacterial by-products can be used to provide alternative feedstock for the synthesis of biofuels, including ethanol, hydrogen, and biodiesel, among others. To mitigate the alarming impact of greenhouse gas emissions and slow down global warming, feedstocks for bio-based fuels are receiving increased attention. Cyanobacteria/Microalgae's rapid growth rate allows them to convert solar sunlight into fuels at higher photosynthetic efficiencies than terrestrial biofuel feedstocks (Arora et al. 2018). Consequently, studies aimed at creating model systems and technologies for the production of renewable biofuels focus on cyanobacteria (Singh et al. 2024). Several approaches are being used to produce biofuel on a large scale and sustainably, as well as to fully use biomass produced.

Large-scale biomass production places constraints on the area of biofuel research. Once this is accomplished, identifying, characterizing, and overproducing lipids will be the next problems. The last step is to find the most economical methods for separating lipids from microalgal cells (Hasunuma et al. 2016). Abiotic environmental stressors include pH, heat, light intensity, salinity, nutrition depletion, phytohormones, UV radiation, and others that may encourage the buildup of lipids inside cyanobacteria/microalgal cells (Salama et al. 2019; Satya et al. 2024). Large volumes of data may also be gathered using sophisticated methods to describe model organisms for effective biofuel production. The development of powerful genetic techniques (forward or reverse genetic approaches) and the identification of the proper genes responsible for bioenergy and pathways in cyanobacteria have made it possible to engineer cyanobacterial strains through either transgene expression or gene disruption (Beer et al. 2009).

Numerous researchers have used genomic analysis in conjunction with other methods to boost lipid production and accumulation. The overproduction of the

DGAT-2 gene increases lipid expression in *Phaeodactylum tricomutum* (Niu et al. 2013). Similar to this, overexpression of the *NoD12* gene in *Nannochloropsis oceanica* results in increased lipid synthesis (Kaye et al. 2015). Through the study of RNA transcripts and different genetic modifications of cyanobacteria/microalgae for the effective generation of biofuel, transcriptomics offers a comprehensive understanding of the function of genes. According to transcriptome research, FAME rose by 50% in *Chlorella vulgaris* when nitrogen was depleted (Guarnieri et al. 2011). Using a metabolomic method on *Phaeodactylum tricornutum*, Popko et al. (2016) discovered that betaine lipids are a major source of triglyceride production and that sedoheptulose buildup occurs under conditions of nitrogen shortage. Omics methods are mostly diagnostic, providing information on cellular homeostasis strategies and enzymatic control. Comprehending the expression of genes implicated in lipid production is necessary for genomics techniques. It will be critical to comprehend the causes of the excess lipid production. The proteomics techniques need the enzymes involved in the lipid biosynthesis process. A thorough grasp of the biochemical, biophysical, and structural aspects of enzyme characterization is necessary. Ultimately, it is necessary to create the technology for separating lipids from biomass.

20.2 Understanding Cyanobacteria by Multi-omics

Advanced gene editing and multi-omics analysis were essential to biotechnology's quick growth, which gave rise to new words like systems biology and metabolic engineering. The goal of synthetic biology is to find standard protocols to quickly and creatively use biology. It has been possible to extract the functional protein from several species by using genomic sequencing and analysis. Moreover, the inducibility of promoters might be predicted by transcriptome analysis in conjunction with genome sequencing. Through genetic circuit design and genome editing to reroute the metabolic engineering and carbon flux seek to build a robust host for high-level synthesis of the value-added chemical. Under specific circumstances, proteomics and metabolomics are also involved. A complicated biological system is described by system biology using two omics analyses (Choi et al. 2019). Below is a summary of the current contributions of omics to the study of cyanobacteria and microalgae biofuel production (Fig. 20.1).

20.2.1 Genomics Analysis

The direct gene annotation was made possible by high gene density, or fewer introns, and the majority of genomics analysis efforts have been focused on cyanobacteria and *Nanochloropsis* sp. (Lockhart and Winzeler 2000). According to recent genome-wide comparative research conducted on 41 cyanobacterial genomes, T-terminal codons are primarily found in the genome regardless of GC concentration, and there is no bias in the codon use of the genes (Prabha et al. 2017). In

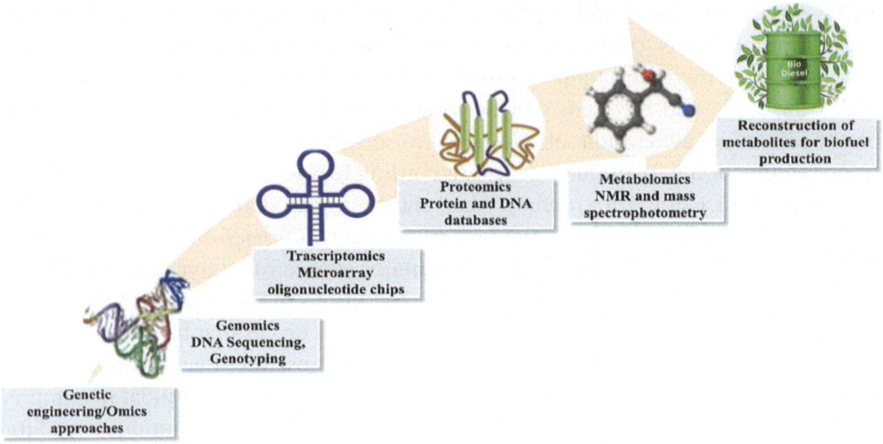

Fig. 20.1 Schematic of systemic omics approach for strain development and reconstruction of metabolites for biofuel production

addition to the comparison of DNA aspects, the prediction of proteomes and metabolomics was aided by the genome-wide comparative study. Since transcription factor regulates the phenotype of cyanobacteria and microalgae, it is a necessary protein. Hu et al. (2014) conducted a thorough examination of the regulatory pathways involved in TAG production by identifying the binding site (TFBS) and transcription factor (TF) in the microalgae *Nannochlorpsis*.

In addition to using TF as the target, the biochemical process was also demonstrated by the prediction of protein–protein interaction (PPI) on the basis of genome-based comparison. For instance, PPI prediction in *Synechocystis* sp. PCC6803 was effectively used to study photosynthesis and DNA repair. Furthermore, the role of the protein ssl3451 in photosynthesis, ssl0822 (sigma factor), and sll1252 (organic ion transmembrane transporter) have all been satisfactorily characterized (Lv et al. 2015). Numerous naturally occurring synthesis pathways for secondary metabolites have been demonstrated to be supported by cyanobacteria through comparative genomic research.

These pathways give rise to intermediate or final products that hold promise for a range of therapeutic applications, including the prevention of cancer, the preservation of multiple drugs, the inhibition of enzymes, and antifungal, antibacterial, anti-inflammatory, and antiviral properties (Dittmann et al. 2015). When the molecule was still unknown, Leao et al. (2017) used the comparative genomics technique to determine the capability of the cyanobacterial genus *Moorea*, demonstrating the unique cluster of genes for new secondary metabolites. The initial step is to identify the differences using genomics analysis; meanwhile, experiment-based characterization would expeditiously advance synthetic biology.

Based on a comparative genome analysis between the model cyanobacterium *Synechocystis elongatus* PCC 7942 and *Synechocystis elongatus* UTEX 2973, two successful examples have been reported. It was found that the similar gene set differs by at most 55 single-nucleotide polymorphisms (SNPs), 7.5 kb deletions, and 188 kb inversions, resulting in a vast difference in phenotype (Yu et al. 2015). After conducting a thorough mutational investigation on *S. elongatus* UTEX 2973, Ungerer et al. (2018) discovered three genes—*ppnK*, *atpA*, and *rpaA*—with SNPs causing fast growth. *S. elongatus* PCC 7942's cell growth was improved after the allele gene was replaced, and the mutant strain's doubling time was 2.25 h (Ungerer et al. 2018). Additionally, Li et al. (2018) carried out a thorough gene information analysis and discovered that the *pilN* gene, which has an SNP that causes early termination of translation, may be the primary cause of the naturally non-transformable character. Li et al. (2018) reported that *S. elongatus* UTEX 2973 was able to regain its natural transformability using complementation of the allelic gene from *S. elongatus* PCC 7942, with a good transformation efficiency. Thus, sufficient genetic data expedites the cyanobacteria's use in biorefineries.

20.2.2 Transcriptomics

The primary focus of transcriptomics is the nature, composition, and roles of transcripts that are generated by a cell in certain circumstances. Genomic data for cyanobacteria may yield sufficient knowledge for their design. However, to gain whole genome information, transcriptomics and genomics data must be combined due to the intricate intron and exon structure found in a cyanobacterium (Gao et al. 2014). Furthermore, transcription start site (TSS) and stimuli-response characterization may be achieved by transcriptome coupling with genetic data. Liu et al. (2018) reported a noteworthy occurrence of the mechanisms involved in the light response, which is frequently employed in synthetic biology to manipulate cell response by light exposure via transcriptome analysis. According to Tan et al. (2018), TSS analysis allowed for the precise in-length targeted RNA design, particularly in CRISPR technology, to achieve high on-target efficiency. It also allowed for the prediction of RBS position. In addition to providing insights into the genome, the comparative transcriptome was a crucial tool for identifying exceptionally suitable mutant strains. The green microalgae *Dunaliella tertiolecta* had a higher content of lipids following the random insertion of plasmid pGreenI1000, which caused photosynthesis pathways upregulation and enhanced inositol phosphate metabolism (Yao et al. 2015a, b).

In addition to random insertion, certain cyanobacteria/microalgae exposed to atmospheric and room temperature plasma (ARTP) produced a mutant *Chlamydomonas reinhardii* that produced 5.2 times more H_2 than the natural type. *C. reinhardii* provided a recent successful example, showing about 5.2 times more H_2 production than the wild type. Additionally, transcriptome comparison analysis showed that enhanced H_2 generation is a result of the overexpression of several photosynthesis-related genes, including ATPase, photosystem I and II, electron transporter, and cytochrome b_6/f complex (Ban et al. 2019). Adaptive laboratory evolution (ALE) served as the foundation for the alternative strategy. Two strains of

Chlorella underwent mutations that allowed them to acquire high flu gas and high phenol tolerance (Zhou et al. 2017; Cheng et al. 2019).

Comparative transcriptome analysis of a strain of *Chlorella* with high phenol tolerance revealed upregulation of the antioxidant enzyme, photosynthesis pathway, and carotenoid biosynthesis, whereas flu-gas-tolerant *Chlorella* demonstrated upregulation of extracellular sulfur transporter, oxidative phosphorylation, and photosynthesis (Cheng et al. 2019). At 5% CO_2, the biomass of a *Nannochloropsis* sp. that lacks carbonic anhydrase is increased. A further instance was conducted on a *Nannochloropsis* sp. that has overexpressed NobZIP1, exhibiting a significant increase in lipid synthesis. A comparative transcriptome study revealed that NobZIP1 overexpression resulted in the downregulation of UDP-glucose dehydrogenase (UGDH) and the upregulation of the four genes of the TGA biosynthesis pathway. Remarkably, UGDH silencing changed the cell wall composition and caused lipids to be secreted (Wei et al. 2019; Li et al. 2019).

20.2.3 Proteomics

Proteomics is the methodical analysis of all proteins produced by cells or organisms under certain physiological conditions to determine the identity and amount of their protein sequences (D'Agostino et al. 2016; Armengaud 2023). Yang et al. (2014) documented the use of proteomics in confirming predicted genes, correlating start and stop codon locations, and identifying new proteins absent from the genome annotation (Yang et al. 2014). Furthermore, in the bioengineering of microalgae and cyanobacteria, many effective uses of proteome analysis have been documented. These applications assessed changes at the translational level, which offered important insights for the fields of synthetic biology and metabolic engineering in these organisms. A comparative proteomics analysis was conducted on a salinity-tolerant *C. reinhardtii*. The results demonstrated that enhanced protein abundance in stress and defense, iron absorption and metabolism, membrane transport and trafficking, and protein degradation all contributed to high salinity tolerance. Proteomics revealed, notably, that this study's hypothesized salinity-specific post-translational modification process (PTM) altered several housekeeping proteins (Sithtisarn et al. 2017).

The lactate toxicity was first thought to be the cause of significant growth retardation seen in *Synchocystis* sp. PCC 6803, a lactate-producing organism. However, another explanation was put forth using proteomics, which involves an imbalance in the redox pair caused by dehydrogenation with different up-regulated proteins (Borirak et al. 2015). By integrating transcriptomes, genomes, and proteomics, a thorough knowledge might be attained, particularly regarding the distinction between transcription and translation (Carrieri et al. 2017). The notion that distinct folding patterns in transcriptomes and proteomics indicate distinct translational mechanisms in a CA-deficient *Nannochloropsis* sp. has been discussed recently (Wei et al. 2019).

20.2.4 Metabolomics

The study of all the metabolites in an organism, both quantitatively and qualitatively, is called metabolomics. Various chromatographic techniques, such as liquid and gas chromatography, are used to measure the known metabolite concentrations following separation. Using adaptive laboratory evolution (ALE), *Crypthecodinium cohnii*, a type of green algae, was shown to be glucose-tolerant. Comparative metabolomics revealed that the glucose tolerance was a result of the cells producing more glutamate, glycerol, succinate, and malonate, which could shield them from substrate inhibition, and less xylose and fructose, which would put them under osmotic stress (Li et al. 2017). Nuclear magnetic resonance (NMR) was utilized to discover unknown metabolites in addition to the recognized metabolites. A new columbamide A was discovered as a secondary metabolite in the unusual and common cyanobacteria *Moorea* using metabolomics in conjunction with nuclear magnetic resonance (Kleigrewe et al. 2015).

Moreover, isotope analysis using mass spectrometry can validate the distribution of metabolic fluxes. In order to boost CO_2 fixation, increase the generation of 2,3-butadoil, and improve glucose utilization, Kanno et al. (2017) designed *Synechococcus elongatus* PCC7942. They showed the role of RuBP in CO_2 fixation and synthesis of different chemicals in cyanobacteria using metabolomics studies in conjunction with the C13 isotope supply (Kanno et al. 2017). *Synechocystis* sp. PCC6803 (Wang et al. 2016) and *Synechococcus* sp. PCC7002 (Kopka et al. 2017) has been the subject of a strategy that combines proteomics, genomics, and metabolomics to understand their cellular processes. Recently, a prolific ethanol-producing *Synechocystis* sp. PCC6803 was recently built using in silico modeling of the ethanol production process by utilizing trans-omics data (metabolomics and genomes combined) (Nishiguchi et al. 2019).

20.3 Genetic Engineering and Omics Approaches for Cyanobacterial Biofuel Production

Biotechnological treatments can enhance the biomass buildup and nutrient removal efficiency of cyanobacteria and microalgae. The chloroplast, nuclear, and mitochondrial genomes are the bases for these modifications (Sarwan et al. 2024; Anand and Padmanabhan 2024). According to Gimpel et al. (2015), most of the enzymes involved in secondary product production are encoded by the nuclear genome. Targeting a specific gene modification through transformation and selection techniques is an essential phase in the genetic engineering of cyanobacteria. *Agrobacterium* construction, particle bombardment, electroporation technique, and agitation with glass beads are the four methods used in transformation. Genome editing and gene interference are the most effective gene-targeting techniques (Bashir et al. 2016).

Recent advances in genome editing include the utilization of zinc-finger nucleases (ZFN), transcription activator-like effector nucleases (TALEN), and CRISPR-Cas9 for gene modification. The use of gene-interfering tools in genetic modification, including microRNA (miRNA), and silencing RNA (siRNA), to activate or repress specific gene expression, can aid in the attainment of desired cyanobacterial phenotypes (Ng et al. 2017). In wastewater, molecular phylogeny paired with omics technologies to identify genotypic and phenotypic features is superior to increasing biomass production and the pace at which nutrients are removed (Chamberlin et al. 2018).

An understanding of metabolic processes with dynamic shifts in protein synthesis routes, metabolic, and bioremediation pathways, and the impact of various environmental factors on organism's metabolism are all provided by "omics" (Fig. 20.2). A thorough comprehension of these changes can lead to the development of more effective environmental protection and energy production (Patel et al. 2015). The multipurpose value of cyanobacteria biomass culture is enhanced by the incorporation of omics in feedstock, biofuel, and industrial biotechnology. Microalgal sequence data, not only used to identify distinct genes and their order but also facilitate genome comparisons, is included in genomics (Guarnieri and Pienkos 2015). Another crucial molecular technique in algomics is transcriptomics, which studies gene structure, expression, and other genomic regulators (Anand et al. 2019) (Fig. 20.2).

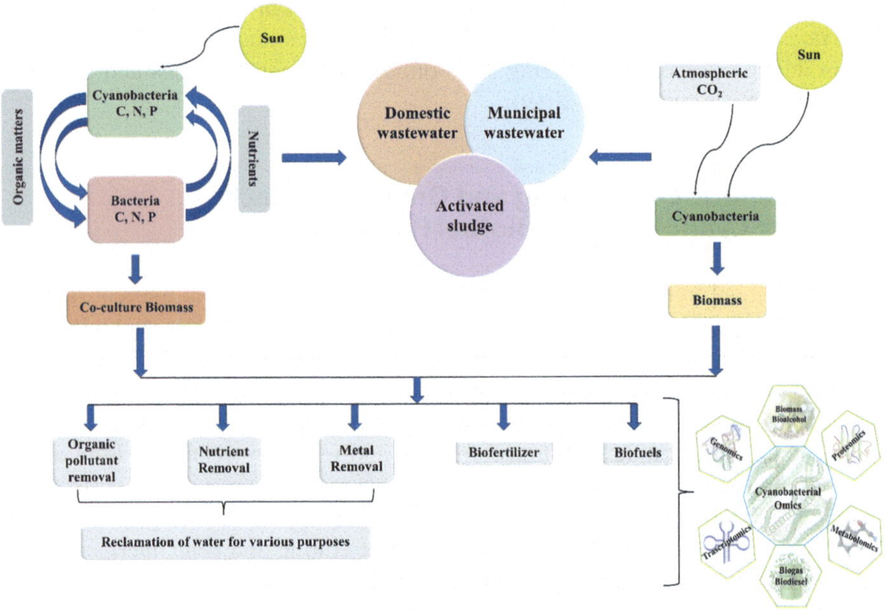

Fig. 20.2 Mechanism of cyanobacterial association with bacteria in wastewater treatment and different applications of the emerging cyanobacterial omics approaches

Proteomic analysis is another facet of omics that offers rich insights into the genetic makeup that is crucial for identifying the proteins involved in protein trafficking systems, protein profiling, and metabolic processes (Perera et al. 2019). Analyzing the metabolites in microalgal cells both qualitatively and quantitatively to enhance the synthesis of targeted products is known as metabolomics, which is another important field. Increasing the flux rate of pathways, decreasing the number of harmful substances in the environment, and eliminating competing pathways are the main tactics needed for metabolic optimization (Kato et al. 2022). Additionally, for photosynthetic species, a fourth method of enhancing carbon absorption is proposed. Genetic techniques can significantly improve biomass accumulation to raise the possibility of producing biofuels.

20.3.1 Biofuel

The biological approach to producing ethanol has garnered a lot of research interest in recent years. In the past, a two-step biosynthetic process was employed, which involved gathering plant-based feedstocks and then microbially fermenting them to produce fuels (Stephanopoulos 2007). Interest in using photosynthetic prokaryotes for the straight transformation of carbon dioxide into fuels has increased since it appears that this indirect biosynthetic strategy is inefficient in turning recalcitrant biomass into biofuels (Joshi and Mishra 2022). Even now, native cyanobacterial strains are not able to make significant amounts of ethanol by nature. Consequently, it is important to increase these strains' biosynthetic effectiveness to reach a commercially viable level (Lau et al. 2015; Kumar et al. 2021). It is now possible to effectively rewire microbial cell pathways to achieve higher production through the use of synthetic biology-based techniques, such as replacing natural enzymes with effective substitutes, optimizing and engineering enzymes, modifying photosynthesis and carbon fixation, installing novel pathways, and tailoring regulators or cofactors (Liang et al. 2018; Ruffing et al. 2022).

Deng and Coleman (1999) created a new way to increase the production of ethanol by inserting the genes *adh* and *pdc*, which encode alcohol dehydrogenase II and pyruvate decarboxylase, respectively, into a *Synechococcus* 7942 that was derived from the bacterium *Zymomonas mobilis*. This was done for the first time, and the results demonstrated that the modified cyanobacterial strain produced a large quantity of ethanol. Later, a genetically modified strain of *Synechocystis* 6803 was created by Gao et al. (2012) with a significantly higher ethanol production efficiency of 5.5 g/L. The methods involved inactivating the poly-β-hydroxybutyrate biosynthesis pathway, overexpressing endogenous alcohol dehydrogenase, and introducing an extra copy of pyruvate decarboxylase. Additionally, research was done on the roles of different growing environments, including metal ions, tap water, and anoxic air, in the production of ethanol. Given their easy-to-cultivate nature and higher production efficiency of ethanol, cyanobacteria appear to be a viable platform for the bioconversion of solar energy and CO_2 into a variety of bioproducts with added value.

20.3.2 Green Hydrogen Production

When filamentous cyanobacteria are grown in an environment deficient in nitrogen, hydrogen gas is released as a consequence of nitrogen fixation. It is regarded as one of the most favored alternatives to conventional energy sources based on fossil fuels because of its high efficiency, environmental friendliness, and renewability. Microorganisms that are very promising for producing hydrogen are cyanobacteria. Cyanobacterial strains are a more viable and economically justifiable means of producing hydrogen than traditional chemical and photoelectrical approaches (Dutta et al. 2005). In many photosynthetic cyanobacterial strains, hydrogen production has been studied, and at least 14 cyanobacterial species are found to create hydrogen in a variety of growing conditions (Pinto et al. 2002).

Two different hydrogenase types are found in cyanobacterial strains: reversible hydrogenases, which can be involved in both the intake and synthesis of hydrogen and oxygen oxidizing uptake hydrogenases (Tamagnini et al. 2002). The hydrogenase's high oxygen sensitivity and the propensity of (NiFe) thermodynamically favor hydrogen absorption and limit the efficiency of some cyanobacterial hydrogen production (Ghirardi et al. 2007) (Table 20.1). By obstructing or deactivating pathways that compete with hydrogenases for reductant consumption, elevated hydrogen production can be achieved. When the *ldhA* gene was used to inactivate NADH in *Synechococcus* 7002, the result was a five-fold increase in hydrogen production and a significant improvement in the $NADH/NAD^+$ ratio (McNeely et al. 2010).

Similar to this, Ducat et al. (2011) reported that exogenous hydrogenases from *Clostridium acetobutylicum* were heterologously expressed in *Synechococcus* 7942, leading to an enhanced hydrogen titer. It has been demonstrated that engineering cyanobacteria to reroute glycogen catabolism by using the pentose phosphate pathway (PPP) significantly improves intracellular NADPH buildup, which in turn increases hydrogen generation. Compared to the wild-type strain, Kumaraswamy et al. (2013) observed a 2.3–3-fold rise in hydrogen levels in the engineered *Synechococcus* 7002 strain, which had the gap1 gene inactivated and overexpressed NAD^+-dependent glyceraldehyde-3-phosphate ($GAPDH^{-1}$). In a more recent study, Maswanna et al. (2018) used an entrapment technique to immobilize *Tetraspora* CU2551 cells in an alginate matrix to examine the cells' decreased exposure to O_2.

Parameters of 4.0% alginate concentration, 0.125 mg dry cell/mL alginate, and 2.8–3.35 mm bead size, the culture medium's hydrogen production increased dramatically to 0.3 mL H_2/mL medium. Fascinatingly, it was discovered that this titer was six times more than cell suspension, 2–10 times greater than green algae, and 10–50 times greater than other cyanobacteria. From this, it can be concluded that immobilized cells on supports represent a viable strategy for increasing photosynthetic organisms' ability to produce photobiological hydrogen.

Table 20.1 List of biotechnologically created recombinant cyanobacterial strains used for the generation of biofuels

S. No.	Cyanobacterial strain	Culture conditions	Biofuel precursor	Engineered status/involved genes	Product concentration	References
1.	*Cyanothece* sp. ATCC 51142	ASP-2 medium	Hydrogen	Disruption of uptake hydrogenase gene	400 mmol H_2/mg Chl	Masukawa et al. (2002)
2.	*S. elongatus* 7942	BG-11 Medium (modified)	Iso-butanol	Overexpression of RuBISCO	450 mg/L	Atsumi et al. (2009)
3.	*Synechocystis* sp. PCC 6803	BG-11 medium	Ethanol	PpsbA2 promoter-expressed *pdc-adh* gene set	5.5 g/L	Dexter and Fu (2009)
4.	*Synechocystis* sp. PCC 6803	Nitrogen limiting medium	Hydrogen	nitrate reductase (*InarB*), nitrite reductase (*InirA*)	186 nmol/mg Chl *a*	Baebprasert et al. (2011)
5.	*S. elongatus* 7942	BG-11 Medium (modified)	1-butanol	substitution of bifunctional aldehyde/AdhE2 with separate Bldh and YqhD	30 mg/L	Lan and Liao (2012)
6.	*S. elongatus* 7942	BG-11 medium	2,3 butanediol	Integrated *alsS*, *alsD*, and *adh*	2.4 g/L	Oliver et al. (2013)
7.	*Synechocystis* sp. PCC 6803	BG-11 medium	Isobutanol	*Ptac* expression of the Ehrlich pathway's *kivD* and *adhA* gene set	114 mg/L	Varman et al. (2013)
8.	*Synechococcus* sp. PCC 7002	Agar plates of medium A $^+$	Fatty acids	*fadD*, *tesA*, and *rbcLS*	130 mg/L	Ruffing (2014)
9.	*Synechocystis* sp. PCC 6803	BG-11 medium	Fatty alcohol for biodiesel	fatty acyl-CoA reductase gene	761 μg/g dcw	Qi et al. (2013)
10.	*Synechococcus* sp. PCC 7002	CO_2-enriched seawater medium	Ethanol	PDC and *ADH* gene	0.25% (v/v)	Kopka et al. (2017)

20.3.3 Biogas Production

One of the most promising clean fuels are ethylene, and hydrogen, when they burn, release just water and generate a considerable amount of heat per mole. Furthermore, burning microalgae directly might also be a source of clean energy (Choi et al. 2019). The ability to produce hydrogen is found in *Scenedesmus obliquus*, *Chlorella vulgaris*, and *C. reinhardii*; however, only the latter produces a high amount of H_2 through genetic engineering of hydrogen-producing enzymes (Khetkorn et al. 2017), with cyanobacteria producing much less than that. Three different kinds of hydrogen metabolism-related enzymes have been identified in the cyanobacterium *Anabaena* sp. PCC 7120: a nitrogenase, an uptake hydrogenase (Hup), and a reversible bidirectional hydrogenase (Hox). The rate at which H_2 is produced by the △hupL and △hupL/△hoxH mutants was 3–6 times higher than in the wild-type cells (Masukawa et al. 2002) (Table 20.2).

Baebprasert et al. (2011) reported that a high rate of hydrogen generation was detected in *Synechocystis* sp. PCC 6803 following the redirection of the electron supply from the nitrate absorption route by a mutation in △narB:△- nirA. HydA activity may sustain light-dependent hydrogen generation in *S. elongatus* PCC7942 in anoxic conditions; this resulted in a maximal production rate of 2.8 μmol/h/mg Chl *a* (Ducat et al. 2011) (Table 20.2). Rare reports suggested that genetically modified eukaryotic microalgae might be able to produce hydrogen, but the effective gene regulation encouraging hydrogen generation was supported by the cyanobacteria's unique genomic information and remarkable transformation efficiency (Khetkorn et al. 2017).

The *xylA* and *xylB* genes from *E. coli* have recently been used to investigate genetic ethylene-producing cyanobacteria, which was necessary to control xylose utilization (Lee et al. 2015). The rate of ethylene synthesis in the presence of xylose was significantly increased by introducing both *xylAB* and the xylose transporter

Table 20.2 Production of various biogases reported in different engineered cyanobacteria

Products	Species	Genes target	Genes/pathways involved	References
Hydrogen	*Anabaena* sp. 7120	*hup*L	*hup*L and *hup*H	Masukawa et al. (2002)
	Synechocystis sp. PCC 6803	*nar*B and *nir*A	*nar*B and *nir*A	Baebprasert et al. (2011)
	S. elongatus PCC7942	*hyd*A	*gyd*A	Ducat et al. (2011)
Ethylene	*Synechocystis* sp. PCC 6803	–	Expressing ethylene-forming enzyme (EFE)	Wang et al. (2019)
	Synechocystis sp. PCC 6803	*xyl*AB and *glg*C	*xyl*AB and *glg*C	Lee et al. (2015)
	Synechocystis sp. PCC 6803	*kgt*P	*kgt*P	Zhu et al. (2015)
	S. elongatus PCC 7942	efe	efe overexpression from *P. syringae* pv	Carbonell et al. (2019)

xylFGH into the ethylene-producing strain *Synechocystis* sp. PCC 6803. However, *Synechocystis* sp. PCC 6803 was able to produce more ethylene after introducing a codon-optimized ethylene-forming enzyme (EFE) from *Pseudomonas syringae* (Zhu et al. 2015).

20.3.4 Ethanol and Butanol

Regarding the bioethanol generated by cyanobacteria that have been modified, *S. elongates* PCC7942 co-expresses *ictB*, *ecaA*, and *acsAB*. The cyanobacteria may create up to 7.2 g/L of ethanol when co-fermented with *Z. mobilis* (Chow et al. 2015). After adopting a similar idea to *Synechococcus* sp. PCC7002, ethanol productivity was effectively provided through the integration of the pdc-slr1192 in route on a neutral site NS0027 (Wang et al. 2019) (Table 20.3). Among the several biofuel goals, n-butanol has drawn a lot of attention due to its applicability as a chemical feedstock in the existing infrastructure. However, because the metabolic route for synthesizing n-butanol originates from stringent anaerobes and differs greatly from that of cyanobacteria, engineering cyanobacteria to make n-butanol has proven to be difficult. Since *S. elongatus* PCC 7942 enzymes with NADPH-dependent alcohol dehydrogenase (YqhD) are linked to polyhedral bodies and are crucial for the degradation of 1,2-propanediol by *S. enterica* (PduP) enzymes, competitive ethanol production would take place following the introduction of these enzymes (Lan and Liao 2012).

Table 20.3 Production of various biofuels: ethanol, butanol, and isobutanol reported in different engineered cyanobacteria

Products	Species	Genes target	Strategy	References
Ethanol	*S. elongatus* PCC 7942	ictB, ecaA, and acsAB	Overexpression of ictB, ecaA, and acsAB	Chow et al. (2015)
	Synechocystis sp. PCC 6803	Pdc and adh	Overexpression of adh and Pdc	Gao et al. (2012)
	Synechocystis sp. PCC 6803	zwf	*zwf* overexpression	Choi and Park (2016)
	Synechocystis sp. PCC 7002	Pdc and adh	Overexpression of adh and Pdc	Wang et al. (2019)
Butanol	*S. elongatus* PCC 7942	ter, nphT7, bldH, yqhD, phaJ, and phaB	Overexpression of ter, bldH, nphT7, phaJ, yqhD, and phaB	Lan and Liao (2012)
	S. elongatus PCC 7942	Ter, pduP, CAAase, nphT7, yqhD, phaJ, and phaB	Overexpression of ter, pduP, phaB, nphT7, yqhD, phaJ, and CAAase	Fathima et al. (2018)
Isobutanol	*Synechocystis* sp. PCC 6803	kivD and adhA	Overexpression of kivD and adhA	Varman et al. (2013)
	Synechocystis sp. PCC 6803	Mutant kivDS286T, *adhA*, *ilvC*, and *ilvD*	Overexpression of mutant $kivD^{S286T}$, adhA, and ilvCD	Miao et al. (2018)

Furthermore, the build-up of acetyl-CoA indicated that the n-butanol pathway's rate-limiting step was Acetyl-CoA carboxylase (ACCase). According to Fathima et al. (2018), the strain that was produced showed potential for use in the future in the photosynthetic microbial-based synthesis of n-butanol (Table 20.3). Isobutanol and butanol have several notable characteristics. Isobutanol is a next-generation biofuel made from pyruvate that requires fewer chemical steps to go from pyruvate to various products than n-butanol. To produce isobutanol in *Synechocystis* sp. PCC 6803, it was necessary to overexpress two heterologous Ehrlich pathway genes. Additionally, a plasmid containing the *Lactococcus lactis* genes *kivD* and *adhA* under the control of an IPTG-inducible promoter, Ptac, had to be created (Varman et al. 2013). Due to its toxicity to cells and potential for photochemical degradation by hydroxyl radicals during the culture process, researchers used oleyl alcohol as a solvent to remove isobutanol in situ. In mixotrophic culture settings, this led to a final net concentration of isobutanol of 298 mg/L. Nonetheless, when co-expressed in a single operon and a convergent-oriented operon downstream of another gene, *kivD*'s expression level was considerably impacted. Thus, the co-expression of *IlvC*, *IlvD* gene, and the expression of the ADH encoded by codon-optimized slr1192 were shown to be viable strategies to further increase isobutanol synthesis (Miao et al. 2018).

20.3.5 Lipid Production

The wide definition of microalgal lipids includes fatty acids, cholesterol, triglycerides, and phospholipids. Phospholipids and fatty acids are more desirable and profitable. The cyanobacteria's metabolic pathway for phospholipid and fatty acid synthesis included the elongation cycle and the two-step start reaction of *fabD* and *fadH,* generating β-ketoacetyl-ACP after acetyl-CoA (acc) was transformed to malonyl-CoA by acetyl-CoA carboxylase (accABCD). An order of reactions would occur in the elongation cycle, including reduction, dehydration, and elongation by the condensation of extra malonyl-ACP molecules (*FabB* and *FabF*). The mechanism that 2,4-dienoyl-CoA reductase (fadH) catalyzes is said to be the rate-limiting step of fatty acid production by *Synechococcus* sp. PCC 7002 (Kuo and Khosla 2014).

Recently, Santos-Merino et al. (2018) reported that the introduction of delta (12)-fatty acid desaturase (*desA* and *desB*) into *Synechococcus* sp. PCC 7002, deletion of *fadD*, encoding acyl-CoA synthetase that competes for the flux from the β-oxidation to elongation cycle, and overexpression of β-ketoacyl-ACP synthase II (*fabF*) all increased omega-3 fatty acid levels (Santos-Merino et al. 2018). *Synechocystis* sp. PCC 6803 was engineered by Eungrasamee et al. (2019) to overexpress acyl-ACP synthetase, which allows for the recycling of free fatty acids. The engineered strain produced 41.4 mg/L/day and 34.5% of total lipid content in DCW.

For instance, to improve the synthesis of shorter fatty acids—which are highly attractive for use in biofuel applications—the exogenous Acyl-ACP thioesterase gene was introduced into *Phaeodactylum tricornutum* (Radakovits et al. 2011). Without affecting biomass, *P. tricornutum* was genetically modified by antisense silencing of *pdk*, a crucial gene that participates in lipid metabolism (Ma et al. 2014) (Table 20.4). The pyruvate to acetyl-CoA oxidative decarboxylation is catalyzed by the complex of pyruvate dehydrogenase and is inhibited by pyruvate dehydrogenase kinase (PDK). In order to increase the lipid content of the *Thalassiosira pseudonana* without harming its development, a multifunctional lipase/phospholipase/acyltransferase was shut down (Trentacoste et al. 2013). During exponential development, the transgenic cultures showed a 2.4–3.3-fold increase in lipid content; following 40 h of silicon deprivation, the strains showed a 4.1–fold increase relative to the wild type. *Chlorella vulgaris* was able to increase the amount of C16:0, saturated fatty acid (SFA), and accumulate more than 30% of the total lipid content without harming its development by introducing a defective ω-3 fatty acid desaturase (*fad3*) (Lau et al. 2017).

The phosphoenolpyruvate carboxylase (PEPC) gene, which is associated with the tricarboxylic acid (TCA) cycle, was downregulated in the first report using the CRISPRi system in cell wall-deprived *C. reinhardtii* CC400 (Kao and Ng 2017). This strategy has successfully shifted the carbon flux from the protein synthesis TCA cycle to the fatty acid generation. According to the results, the productivity and lipid content of the transformants were increased to 28.5% DCW and 34.9 mg/L/day, respectively. The capacity of the industrial alga *Nannochloropsis* spp. to accumulate large amounts of polyunsaturated fatty acids (PUFAs) and triacylglycerol (TAG) attracted attention (Ma et al. 2014) (Table 20.4).

However, under conditions of nutrient deprivation, the microalgae *Nannochloropsis gaditana* could maximize their production of lipids; however, this frequently resulted in a low growth rate (Wang et al. 2016). ZnCys, a lipid regulator in *N. gaditana*, was discovered by a reverse genetics pipeline built on the CRISPR-Cas9 system (Ajjawi et al. 2017). While lipid output rose under semicontinuous growing conditions without compromising the capacities of growth, under nutrient-replete conditions, lipid accumulation in transformants improved to 40–55% (Ajjawi et al. 2017).

20.4 Biofuel Production with Wastewater Treatment

According to Park and Lee (2016), the potential for cyanobacteria/microalgae to produce biofuel ranges from 5.6 TOE ha^{-1} y^{-1} to 106.8 TOE ha^{-1} y^{-1}. Using wastewater as a growing medium has the potential to increase biomass productivity and the concentration of biocomponents, which may then be collected for the generation of biofuels (biodiesel, biogas, bioethanol, and biohydrogen) (Fig. 20.3) (Kadri et al. 2023). The procedures of converting microalgae and cyanobacteria biomass into biofuels include the following: biodiesel (biomass transesterification); biogas (anaerobic digestion); bioethanol (fermentation); and biohydrogen (dark fermentation or photosynthetic activities) (Nasir et al. 2023). According to Nguyen et al. (2019), *C. vulgaris* culture found in seafood wastewater effluent displayed an enhanced lipid content of 32.15% for the manufacture of biodiesel. According to Arora et al. (2020), *Chlorella sorokiniana* grown in raw urban wastewater

Table 20.4 Production of various lipids reported in different engineered cyanobacteria

Species	Genes targeted	Strategy	Product	References
Nostoc punctiforme PCC 73,102	Fatty Alka(e)nes	overexpression of *Aar*, *Adc*, and *Npun_F5141*	DCW (12.9%)	Wang et al. (2020)
Synechocystis sp. PCC 6803	acyl-ACP synthetase gene (*aas*)	*aas* overexpression	Lipid	Eungrasamee et al. (2019)
S. elongatus PCC 7942	*fabF*, *fadD*, *desAB*	Deletion of *desAB* and overexpression of *desAB* and *fadF*	omega-3 fatty acid	Santos-Merino et al. (2018)
Synechocystis sp. PCC 6803	*tesA*, *slr1609*	Addition and removal of *tesA* and *slr1609* genes, respectively	Free fatty acids	Liu et al. (2011)
Synechocystis sp. PCC 6803	*TesA*, *fap*	The strain's Aas mutant with truncated *TesA* and *fap* gene.	Fatty Alka(e)nes	Yunus et al. (2018)
Synechocystis sp. PCC 6803	–	Growing copies of Farex pressing cassettes	Fatty alcohols	Qi et al. (2013)
Synechococcus sp. PCC 7002	*slr1609/fadD*, *tesA*	Removal of the *tesA* and *slr1609/fadD* genes	Free fatty acids	Ruffing (2014)
Synechocystis sp. PCC 6803	CAR	Heterologous expression of CAR	Fatty alcohols	Yunus and Jones (2018)
Synechocystis sp. PCC 6803	*far, At3g11980*	Overexpression of two copies of *far* and *At3g11980*	Fatty alcohols	Chen et al. (2020)
Synechocystis sp. PCC 6803	*far*	*far* of Jojoba	Fatty alcohols	Tan et al. (2011)
S. elongatus PCC 7942	Phospho-ketolase pathway	heterologous phosphorketolase pathway	Fatty acid ethyl esters	Lee et al. (2017)

demonstrated 97% fertilizer removal and 1.13 g L^{-1} biomass productivity. To produce biodiesel, attached microalgae-based lipid underwent biofixation of CO_2 to reveal 97.7% (by lipid dry weight) of fatty acid methyl esters (FAMEs), mostly composed of C_{16} and C_{18}, which were effective for burning (Rosli et al. 2019). *Desmodesmus* sp. PW1, a kind of microalgae, can effectively extract nutrients like N and P from piggery effluent.

For all diluted samples, N and P removal rates of up to 90% and 70%, respectively, were observed; however, for undiluted piggery wastewater, the N content dropped dramatically over eight days of cultivation, from 296.7 to 4.9 mg L^{-1}, indicating a 98.3% N removal rate. Hexadecanoic acid, oleic acid, and linoleic acid made up the majority of PW1's fatty acid makeup (Chen et al. 2020). Moreover, Eladel et al. (2019) found that growing in wastewater notably decreased polyunsaturated fatty acids by 36.0% in comparison to Bold Basal Medium (BBM), favoring monounsaturated fatty acids. These investigations have demonstrated the role of microalgal biomass in the generation of lipids (biodiesel) and wastewater treatment. Furthermore, the production of alcohol, biogas, and biohydrogen is correlated with

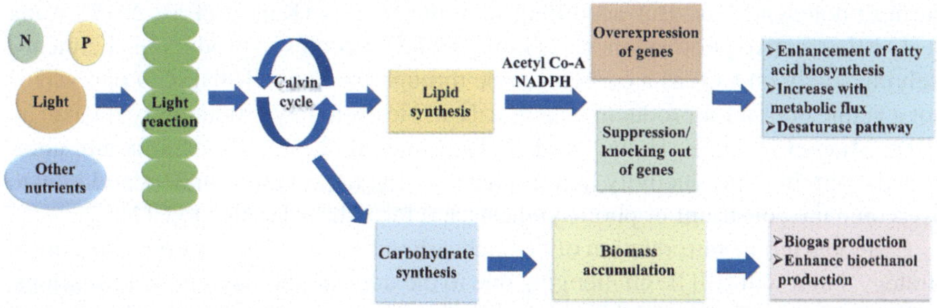

Fig. 20.3 Biotechnological procedures to improve wastewater nutrient absorption and omics strategies to further increase biomass production using wastewater. (Modified from El-Sheekh et al. 2022)

microalgae that have a high carbohydrate content. When grown in municipal wastewater, *N. gaditana* produced 94.3 mg of bioethanol per gram of biomass (Onay 2018) (Fig. 20.3). *Chlamydomonas* sp. QWY37, grown in swine wastewater, produced 944 mg L^{-1} d^{-1} of carbohydrates and removed 81% of COD, 96% of TN, and 100% of TP. The microalgal carbohydrate content yielded 61 g L^{-1} of bioethanol (Qu et al. 2020). Because of the high carbohydrate content, *Chlorella* sp. grown on granular sludge demonstrated biohydrogen productivity of 54.0 mL H_2 g^{-1} VS (Giang et al. 2019).

20.5 Conclusion

There are still many obstacles in the path of cyanobacterial and microalgal genetic engineering. For example, the progress of genetic engineering in microalgae, like *Chlorella*, has been hampered by the absence of a sufficient genomic resource center (Yang et al. 2016). The ineffectiveness of transformation techniques is another barrier, which might be brought on by the cell wall's resilience and stiffness (Tanwar et al. 2018). Scale-up and industrial applications require a tremendous amount of research attention from an economic perspective. Although cyanobacteria could convert CO_2 into different compounds, the concentration was significantly lower than the appropriate value in the contemporary commercial setting (Yazdani and Gonzalez 2008). However, cell growth decreased as a result of the glycogen synthesis disruption, which also decreased the utilization of glycogen and increased targeted chemical production by uplifting the carbon flux (Gordon et al. 2016).

In the future, it will be crucial to preserve or enhance the balance between the production of more desirable chemicals and the utilization of glycogen. Can cyano bacteria and super-producer strains be assembled into a consortium to fix CO_2 and achieve high-level tailored chemistry simultaneously? Cyanobacteria have historically been common hosts when it comes to engineering simultaneous CO_2 fixation and chemical synthesis. According to multi-omics research, any modified cyanobacterium with exceptional characteristics increases the expression of proteins during photosynthesis. But according to some reports (Kanno et al. 2017; Wang et al. 2016), native photosynthesis is limited when it comes to producing chemicals. Relying solely on CO_2 as a carbon source through the native RuBISCO pathway is not a viable option for producing chemicals and leads to low productivity (Gao et al. 2012; Miao et al. 2018; Huang et al. 2016; Lin et al. 2017). To manufacture compounds with high productivity autotrophically, it is important to understand how to overcome the constraint of photosynthesis and raise the tolerance to CO_2.

Cyanobacteria's introduction of distinct carbon fixation pathways may be a novel strategy, but it will still be challenging due to the unclear photosynthetic limitations. To find the mechanism, additional multi-omics analysis is required, which might lead to an exponential acceleration in the development of cyanobacterial synthetic biology. Therefore, the task will be to develop a super-producer strain using a multi-omics technique that can simultaneously mitigate CO_2 and metabolize intracellular glycogen to produce the desired chemical. The paucity of genetic data makes it difficult to apply the knowledge base from cyanobacterial and microalgal genetic engineering.

Double homologous recombination, for instance, is a popular method for cyanobacteria engineering. Based on homologous recombination, the same idea might be used for eukaryotic microalgae, such as *Chlamydomonas*, where the *rbcS* intron was successfully constructed. Therefore, research on cyanobacteria may also hasten the development of eukaryotic microalgae synthetic biology. Despite these circumstances, work on using cyanobacteria and microalgae for genetic engineering is well underway. A growing number of recently created genetic tools and techniques, such as pretreatment cell-wall removal to increase transformation efficiency, have been published to enhance the expression and transformation system. The content of some bio-compounds in microalgae may also be greatly influenced by cultivation conditions; optimal productivity could be achieved with the right combination of genetic manipulation and culture processes.

However, the T-DNA integration, together with other transformation techniques, proved to be a versatile and effective platform for the future genetic engineering of microalgae. Here, genome sequences and transcriptomics data have been used to provide genome sequences and gene annotations for the design of genome editing, such as homologous arm selection or sgRNA design. Productivity may rise by transformation into other fast-growing microalgae or cyanobacteria, although this is dependent on an effective genome editing technique. Pathway design is made possible and logical by multi-omics. Furthermore, the successful production of transgenic microalgae and cyanobacteria is dependent on the use of an accurate, efficient, and multisite genome editing technique. In the future, it will be possible to engineer microalgae or cyanobacteria to produce large quantities of specific chemicals while maintaining the same growth rate by combining multi-omics with sophisticated gene editing tools.

20.6 Short Chapter Summary

This chapter explores the use of cyanobacteria and microalgae as alternative feedstocks for biofuel production, addressing challenges and opportunities in this field. It highlights the potential of these microorganisms to convert solar energy into biofuels more efficiently than terrestrial crops, leading to research focused on developing large-scale, sustainable biofuel production methods. The chapter emphasizes the use of multi-omics approaches (genomics, transcriptomics, proteomics, and metabolomics) coupled with genetic engineering to enhance biofuel production in cyanobacteria and microalgae. Genetic engineering techniques are employed to improve lipid accumulation and biofuel production, particularly for compounds like ethanol, hydrogen, and biodiesel. The chapter also addresses obstacles in cyanobacterial and microalgal genetic engineering, including transformation inefficiencies, limited genomic resources, and the need to balance production with cell growth. Overall, combining multi-omics, advanced gene-editing, and metabolic engineering will accelerate the development of microalgal and cyanobacterial strains that effectively contribute to sustainable biofuel production and environmental bioremediation.

Acknowledgments Ashish P. Singh (NTA Ref. No. 191620014505) is grateful to the University Grants Commission (UGC), New Delhi, India, for the financial support in the form of a Senior Research Fellowship. Amit Gupta (09/013(0912)/2019-EMR-I) and Varsha K. Singh (09/0013(12862)/2021-EMR-1) are thankful to the Council of Scientific and Industrial Research, New Delhi, for the financial support in the form of Senior Research Fellowship. Sapana Jha (No. R/Dev. /Sch./ UGC Non-NET Fello./2022-23/52561) is thankful to BHU for providing institutional fellowship. Palak Rana is thankful to the Council of Scientific and Industrial Research, New Delhi, for the financial support in the form of a Junior Research Fellowship. Rajeshwar P. Sinha would like to express gratitude for the incentive grant received from IoE (Scheme no. 6031), Banaras Hindu University, Varanasi, India.

Authors Contributions Chapter concept and design, Ashish P. Singh and Rajeshwar P. Sinha; chapter writing, Ashish P. Singh, Amit Gupta, Varsha K. Singh, Sapana Jha, and Palak Rana; chapter editing and review, Rajeshwar P. Sinha.

References

Ajjawi I, Verruto J, Aqui M, Soriaga LB, Coppersmith J, Kwok K, Peach L, Orchard E, Kalb R, Xu W, Carlson TJ (2017) Lipid production in *Nannochloropsis gaditana* is doubled by decreasing expression of a single transcriptional regulator. Nat Biotechnol 35(7):647–652

Anand S, Padmanabhan P (2024) Omics approach for enhanced microalgae biomass production with the improved concentration of desired biomolecules. In: Bharadvaja N, Kumar L, Pandit S, Banerjee S, Anand R (eds) Recent trends and developments in algal biofuels and biorefinery. Springer Nature Switzerland, Cham, pp 367–381

Anand V, Kashyap M, Samadhiya K, Kiran B (2019) Strategies to unlock lipid production improvement in algae. Int J Environ Sci Technol 16:1829–1838

Armengaud J (2023) Metaproteomics to understand how microbiota function: the crystal ball predicts a promising future. Environ Microbiol 25(1):115–125

Arora N, Dubey D, Sharma M, Patel A, Guleria A, Pruthi PA, Kumar D, Pruthi V, Poluri KM (2018) NMR-based metabolomic approach to elucidate the differential cellular responses during mitigation of arsenic (III, V) in a green microalga. ACS Omega 3(9):11847–11856

Arora N, Jaiswal KK, Kumar V, Vlaskin MS, Nanda M, Pruthi V, Chauhan PK (2020) Small-scale phyco-mitigation of raw urban wastewater integrated with biodiesel production and its utilization for aquaculture. Bioresour Technol 297:122489

Atsumi S, Higashide W, Liao JC (2009) Direct photosynthetic recycling of carbon dioxide to isobutyraldehyde. Nat Biotechnol 27(12):1177–1180

Baebprasert W, Jantaro S, Khetkorn W, Lindblad P, Incharoensakdi A (2011) Increased H_2 production in the cyanobacterium *Synechocystis* sp. strain PCC 6803 by redirecting the electron supply via genetic engineering of the nitrate assimilation pathway. Metab Eng 13(5):610–616

Ban S, Lin W, Luo Z, Luo J (2019) Improving hydrogen production of *Chlamydomonas reinhardtii* by reducing chlorophyll content via atmospheric and room temperature plasma. Bioresour Technol 275:425–429

Bashir KMI, Kim MS, Stahl U, Cho MG (2016) Microalgae engineering toolbox: selectable and screenable markers. Biotechnol Bioprocess Eng 21:224–235

Beer LL, Boyd ES, Peters JW, Posewitz MC (2009) Engineering algae for biohydrogen and biofuel production. Curr Opin Biotechnol 20(3):264–271

Borirak O, De Koning LJ, Van Der Woude AD, Hoefsloot HC, Dekker HL, Roseboom W, De Koster CG, Hellingwerf KJ (2015) Quantitative proteomics analysis of an ethanol-and a lactate-producing mutant strain of *Synechocystis* sp. PCC6803. Biotechnol Biofuels 8:1–4

Carbonell V, Vuorio E, Aro EM, Kallio P (2019) Enhanced stable production of ethylene in photosynthetic cyanobacterium *Synechococcus elongatus* PCC 7942. World J Microbiol Biotechnol 35:77

Carrieri D, Lombardi T, Paddock T, Cano M, Goodney GA, Nag A, Old W, Maness PC, Seibert M, Ghirardi M, Yu J (2017) Transcriptome and proteome analysis of nitrogen starvation responses in *Synechocystis* 6803 ΔglgC, a mutant incapable of glycogen storage. Algal Res 21:64–75

Chamberlin J, Harrison K, Zhang W (2018) Impact of nutrient availability on tertiary wastewater treatment by *Chlorella vulgaris*. Water Environ Res 90(11):2008–2016

Chen CY, Kuo EW, Nagarajan D, Ho SH, Dong CD, Lee DJ, Chang JS (2020) Cultivating *Chlorella sorokiniana* AK-1 with swine wastewater for simultaneous wastewater treatment and algal biomass production. Bioresour Technol 302:122814

Cheng D, Li X, Yuan Y, Yang C, Tang T, Zhao Q, Sun Y (2019) Adaptive evolution and carbon dioxide fixation of *Chlorella* sp. in simulated flue gas. Sci Total Environ 650:2931–2938

Choi YN, Park JM (2016) Enhancing biomass and ethanol production by increasing NADPH production in *Synechocystis* sp. PCC 6803. Bioresour Technol 213:54–57

Choi KR, Jang WD, Yang D, Cho JS, Park D, Lee SY (2019) Systems metabolic engineering strategies: integrating systems and synthetic biology with metabolic engineering. Trends Biotechnol 37(8):817–837

Chow TJ, Su HY, Tsai TY, Chou HH, Lee TM, Chang JS (2015) Using recombinant cyanobacterium (*Synechococcus elongatus*) with increased carbohydrate productivity as feedstock for bioethanol production via separate hydrolysis and fermentation process. Bioresour Technol 184:33–41

D'Agostino PM, Woodhouse JN, Makower AK, Yeung AC, Ongley SE, Micallef ML, Moffitt MC, Neilan BA (2016) Advances in genomics, transcriptomics and proteomics of toxin-producing cyanobacteria. Environ Microbiol Rep 8(1):3–13

Deng MD, Coleman JR (1999) Ethanol synthesis by genetic engineering in cyanobacteria. Appl Environ Microbiol 65(2):523–528

Dexter J, Fu P (2009) Metabolic engineering of cyanobacteria for ethanol production. Energy Environ Sci 2(8):857–864

Dittmann E, Gugger M, Sivonen K, Fewer DP (2015) Natural product biosynthetic diversity and comparative genomics of the cyanobacteria. Trends Microbiol 23(10):642–652

Ducat DC, Sachdeva G, Silver PA (2011) Rewiring hydrogenase-dependent redox circuits in cyanobacteria. Proc Natl Acad Sci USA 108(10):3941–3946

Dutta D, De D, Chaudhuri S, Bhattacharya SK (2005) Hydrogen production by cyanobacteria. Microb Cell Factories 4:36

Eladel H, Abomohra AE, Battah M, Mohmmed S, Radwan A, Abdelrahim H (2019) Evaluation of *Chlorella sorokiniana* isolated from local municipal wastewater for dual application in nutrient removal and biodiesel production. Bioprocess Biosyst Eng 42:425–433

El-Sheekh M, El-Dalatony MM, Thakur N, Zheng Y, Salama ES (2022) Role of microalgae and cyanobacteria in wastewater treatment: genetic engineering and omics approaches. IJEST 1:1–22

Eungrasamee K, Miao R, Incharoensakdi A, Lindblad P, Jantaro S (2019) Improved lipid production via fatty acid biosynthesis and free fatty acid recycling in engineered *Synechocystis* sp. PCC 6803. Biotechnol Biofuels 12:8

Fathima AM, Chuang D, Laviña WA, Liao J, Putri SP, Fukusaki E (2018) Iterative cycle of widely targeted metabolic profiling for the improvement of 1-butanol titer and productivity in *Synechococcus elongatus*. Biotechnol Biofuels 11:188

Gao Z, Zhao H, Li Z, Tan X, Lu X (2012) Photosynthetic production of ethanol from carbon dioxide in genetically engineered cyanobacteria. Energy Environ Sci 5(12):9857–9865

Gao C, Wang Y, Shen Y, Yan D, He X, Dai J, Wu Q (2014) Oil accumulation mechanisms of the oleaginous microalga *Chlorella protothecoides* revealed through its genome, transcriptomes, and proteomes. BMC Genomics 15:582

Ghirardi ML, Posewitz MC, Maness PC, Dubini A, Yu J, Seibert M (2007) Hydrogenases and hydrogen photoproduction in oxygenic photosynthetic organisms. Annu Rev Plant Biol 58(1):71–91

Giang TT, Lunprom S, Liao Q, Reungsang A, Salakkam A (2019) Improvement of hydrogen production from *Chlorella* sp. biomass by acid-thermal pretreatment. Peer J 7:6637

Gimpel JA, Henriquez V, Mayfeld SP (2015) In metabolic engineering of eukaryotic microalgae: potential and challenges come with great diversity. Front Microbiol 6:1376

Gordon GC, Korosh TC, Cameron JC, Markley AL, Begemann MB, Pfleger BF (2016) CRISPR interference as a titratable, trans-acting regulatory tool for metabolic engineering in the cyanobacterium *Synechococcus* sp. strain PCC 7002. Metab Eng 38:170–179

Guarnieri MT, Pienkos PT (2015) Algal omics: unlocking bioproduct diversity in algae cell factories. Photosynth Res 123:255–263

Guarnieri MT, Nag A, Smolinski SL, Darzins A, Seibert M, Pienkos PT (2011) Examination of triacylglycerol biosynthetic pathways via *de novo* transcriptomic and proteomic analyses in an unsequenced microalga. PLoS One 6(10):25851

Hasunuma T, Matsuda M, Kondo A (2016) Improved sugar-free succinate production by *Synechocystis* sp. PCC 6803 following identification of the limiting steps in glycogen catabolism. Metab Eng Commun 3:130–141

Hu J, Wang D, Li J, Jing G, Ning K, Xu J (2014) Genome-wide identification of transcription factors and transcription-factor binding sites in oleaginous microalgae *Nannochloropsis*. Sci Rep 4(1):5454

Huang CH, Shen CR, Li H, Sung LY, Wu MY, Hu YC (2016) CRISPR interference (CRISPRi) for gene regulation and succinate production in cyanobacterium *S. elongatus* PCC 7942. Microb Cell Factories 15:196

Joshi S, Mishra S (2022) Recent advances in biofuel production through metabolic engineering. Bioresour Technol 352:127037

Kadri MS, Singhania RR, Haldar D, Patel AK, Bhatia SK, Saratale G, Parameswaran B, Chang JS (2023) Advances in algomics technology: application in wastewater treatment and biofuel production. Bioresour Technol 387:129636

Kanno M, Carroll AL, Atsumi S (2017) Global metabolic rewiring for improved CO_2 fixation and chemical production in cyanobacteria. Nat Commun 8(1):14724

Kao PH, Ng IS (2017) CRISPRi mediated phosphoenolpyruvate carboxylase regulation to enhance the production of lipid in *Chlamydomonas reinhardtii*. Bioresour Technol 245:1527–1537

Kato Y, Inabe K, Hidese R, Kondo A, Hasunuma T (2022) Metabolomics-based engineering for biofuel and bio-based chemical production in microalgae and cyanobacteria: a review. Bioresour Technol 344:126196

Kaye Y, Grundman O, Leu S, Zarka A, Zorin B, Didi-Cohen S, Khozin-Goldberg I, Boussiba S (2015) Metabolic engineering toward enhanced LC-PUFA biosynthesis *in Nannochloropsis oceanica*: overexpression of endogenous Δ12 desaturase driven by stress-inducible promoter leads to enhanced deposition of polyunsaturated fatty acids in TAG. Algal Res 11:387–398

Khetkorn W, Rastogi RP, Incharoensakdi A, Lindblad P, Madamwar D, Pandey A, Larroche C (2017) Microalgal hydrogen production–a review. Bioresour Technol 243:1194–1206

Kleigrewe K, Almaliti J, Tian IY, Kinnel RB, Korobeynikov A, Monroe EA, Duggan BM, Di Marzo V, Sherman DH, Dorrestein PC, Gerwick L (2015) Combining mass spectrometric metabolic profiling with genomic analysis: a powerful approach for discovering natural products from cyanobacteria. J Nat Prod 78(7):1671–1682

Kopka J, Schmidt S, Dethloff F, Pade N, Berendt S, Schottkowski M, Martin N, Dühring U, Kuchmina E, Enke H, Kramer D (2017) Systems analysis of ethanol production in the genetically engineered cyanobacterium *Synechococcus* sp. PCC 7002. Biotechnol Biofuels 10:1–21

Kumar S, Singh NK, Kumar A, Shukla P (2021) Next generation biofuel production in the omics era: potential and prospects. In: Kumar A, Kumar R, Shukla P, Pandey MK (eds) Omics Technologies for Sustainable Agriculture and Global Food Security (Vol II). Springer, pp 293–311

Kumaraswamy GK, Guerra T, Qian X, Zhang S, Bryant DA, Dismukes GC (2013) Reprogramming the glycolytic pathway for increased hydrogen production in cyanobacteria: metabolic engineering of NAD^+-dependent GAPDH. Energy Environ Sci 6(12):3722–3731

Kuo J, Khosla C (2014) The initiation ketosynthase (FabH) is the sole rate-limiting enzyme of the fatty acid synthase of *Synechococcus* sp. PCC 7002. Metab Eng 22:53–59

Lan EI, Liao JC (2012) ATP drives direct photosynthetic production of 1-butanol in cyanobacteria. Proc Natl Acad Sci USA 109(16):6018–6023

Lau NS, Matsui M, Abdullah AA (2015) Cyanobacteria: photoautotrophic microbial factories for the sustainable synthesis of industrial products. Biomed Res Int 1:754934

Lau CC, Loh SH, Aziz A, San Cha T (2017) Effects of disrupted omega-3 desaturase gene construct on fatty acid composition and expression of four fatty acid biosynthetic genes in transgenic *Chlorella vulgaris*. Algal Res 26:143–152

Leao T, Castelão G, Korobeynikov A, Monroe EA, Podell S, Glukhov E, Allen EE, Gerwick WH, Gerwick L (2017) Comparative genomics uncovers the prolific and distinctive metabolic potential of the cyanobacterial genus *Moorea*. Proc Natl Acad Sci USA 114(12):3198–3203

Lee TC, Xiong W, Paddock T, Carrieri D, Chang F, Chiu HF, Ungerer J, Hank Juo SH, Maness PC, Yu J (2015) Engineered xylose utilization enhances bioproducts productivity in the cyanobacterium *Synechocystis* sp. PCC 6803. Metab Eng 30:179–189

Lee HJ, Choi J, Lee SM, Um Y, Sim SJ, Kim Y, Woo HM (2017) Photosynthetic CO_2 conversion to fatty acid ethyl esters (FAEEs) using engineered cyanobacteria. J Agric Food Chem 65(6):1087–1092

Li X, Pei G, Liu L, Chen L, Zhang W (2017) Metabolomic analysis and lipid accumulation in a glucose tolerant *Crypthecodinium cohnii* strain obtained by adaptive laboratory evolution. Bioresour Technol 235:87–95

Li S, Sun T, Xu C, Chen L, Zhang W (2018) Development and optimization of genetic toolboxes for a fast-growing cyanobacterium *Synechococcus elongatus* UTEX 2973. Metab Eng 48:163–174

Li DW, Balamurugan S, Yang YF, Zheng JW, Huang D, Zou LG, Yang WD, Liu JS, Guan Y, Li HY (2019) Transcriptional regulation of microalgae for concurrent lipid overproduction and secretion. Sci Adv 5(1):3795

Liang F, Englund E, Lindberg P, Lindblad P (2018) Engineered cyanobacteria with enhanced growth show increased ethanol production and higher biofuel to biomass ratio. Metab Eng 46:51–59

Lin PC, Saha R, Zhang F, Pakrasi HB (2017) Metabolic engineering of the pentose phosphate pathway for enhanced limonene production in the cyanobacterium *Synechocystis* sp. PCC 6803. Sci Rep 7(1):17503

Liu X, Sheng J, Curtiss R III (2011) Fatty acid production in genetically modified cyanobacteria. Proc Natl Acad Sci 108(17):689–6904

Liu Z, Zhang J, Jin J, Geng Z, Qi Q, Liang Q (2018) Programming bacteria with light—sensors and applications in synthetic biology. Front Microbiol 9:2692

Lockhart DJ, Winzeler EA (2000) Genomics, gene expression and DNA arrays. Nature 405(6788):827–836

Lv Q, Ma W, Liu H, Li J, Wang H, Lu F, Zhou C, Shi T (2015) Genome-wide protein-protein interactions and protein function exploration in cyanobacteria. Sci Rep 5(1):15519

Ma YH, Wang X, Niu YF, Yang ZK, Zhang MH, Wang ZM, Yang WD, Liu JS, Li HY (2014) Antisense knockdown of pyruvate dehydrogenase kinase promotes the neutral lipid accumulation in the diatom *Phaeodactylum tricornutum*. Microb Cell Factories 13:100

Masukawa H, Mochimaru M, Sakurai H (2002) Disruption of the uptake hydrogenase gene, but not of the bidirectional hydrogenase gene, leads to enhanced photobiological hydrogen production by the nitrogen-fixing cyanobacterium *Anabaena* sp. PCC 7120. Appl Microbiol Biotechnol 58:618–624

Maswanna T, Phunpruch S, Lindblad P, Maneeruttanarungroj C (2018) Enhanced hydrogen production by optimization of immobilized cells of the green alga *Tetraspora* sp. CU2551 grown under anaerobic condition. Biomass Bioenergy 111:88–95

McNeely K, Xu Y, Bennette N, Bryant DA, Dismukes GC (2010) Redirecting reductant flux into hydrogen production via metabolic engineering of fermentative carbon metabolism in a cyanobacterium. Appl Environ Microbiol 76(15):5032–5038

Miao R, Xie H, Lindblad P (2018) Enhancement of photosynthetic isobutanol production in engineered cells of *Synechocystis* PCC 6803. Biotechnol Biofuels 11:1–9

Nasir ASM, Mohamed B, Li LY (2023) Comparative life cycle assessment of co-pyrolysing sewage sludge and wastewater-grown microalgae for biofuel production. Resour Conserv Recycl 190:106780

Ng IS, Tan SI, Kao PH, Chang YK, Chang J-S (2017) Recent developments on genetic engineering of microalgae for biofuels and bio-based chemicals. Biotechnol J 12(10):1600644

Nguyen TD, Nguyen DH, Lim JW, Chang CK, Leong HY, Tran TN, Vu TB, Nguyen TT, Show PL (2019) Investigation of the relationship between bacteria growth and lipid production cultivating of microalgae *Chlorella vulgaris* in seafood wastewater. Energies 12(12):2282

Nishiguchi H, Hiasa N, Uebayashi K, Liao J, Shimizu H, Matsuda F (2019) Transomics data-driven, ensemble kinetic modelling for system-level understanding and engineering of the cyanobacteria central metabolism. Metab Eng 52:273–283

Niu YF, Zhang MH, Li DW, Yang WD, Liu JS, Bai WB, Li HY (2013) Improvement of neutral lipid and polyunsaturated fatty acid biosynthesis by overexpressing a type 2 diacylglycerol acyltransferase in marine diatom *Phaeodactylum tricornutum*. Mar Drugs 11(11):4558–4569

Oliver JWK, Machado IMP, Yoneda H, Atsumi S (2013) Cyanobacterial conversion of carbon dioxide to 2,3-butanediol. Proc Natl Acad Sci USA 110(4):1249–1254

Onay M (2018) Bioethanol production from *Nannochloropsis gaditana* in municipal wastewater. Energy Procedia 153:253–257

Park H, Lee CG (2016) Theoretical calculations on the feasibility of microalgal biofuels: utilization of marine resources could help realizing the potential of microalgae. Biotechnol J 11(11):1461–1470

Patel AK, Huang EL, Low-Decarie E, Lefsrud MG (2015) Comparative shotgun proteomic analysis of wastewater-cultured microalgae: nitrogen sensing and carbon fixation for growth and nutrient removal in *Chlamydomonas reinhardtii*. J Proteome Res 14(8):3051–3067

Perera IA, Abinandan S, Subashchandrabose SR, Venkateswarlu K, Naidu R, Megharaj M (2019) Advances in the technologies for studying consortia of bacteria and cyanobacteria/ microalgae in wastewaters. Crit Rev Biotechnol 39(5):709–731

Pinto FA, Troshina O, Lindblad P (2002) A brief look at three decades of research on cyanobacterial hydrogen evolution. Int J Hydrog Energy 27(11–12):1209–1215

Popko J, Herrfurth C, Feussner K, Ischebeck T, Iven T, Haslam R, Hamilton M, Sayanova O, Napier J, Khozin-Goldberg I, Feussner I (2016) Metabolome analysis reveals betaine lipids as

major source for triglyceride formation, and the accumulation of sedoheptulose during nitrogen-starvation of *Phaeodactylum tricornutum*. PLoS One 11(10):0164673
Prabha R, Singh DP, Sinha S, Ahmad K, Rai A (2017) Genome-wide comparative analysis of codon usage bias and codon context patterns among cyanobacterial genomes. Mar Genomics 32:31–39
Qi F, Yao L, Tan X, Lu X (2013) Construction, characterization and application of molecular tools for metabolic engineering of *Synechocystis* sp. Biotechnol Lett 35:1655–1661
Qu W, Loke Show P, Hasunuma T, Ho S-H (2020) Optimizing real swine wastewater treatment efficiency and carbohydrate productivity of newly microalga *Chlamydomonas* sp. QWY37 used for cell-displayed bioethanol production. Bioresour Technol 305:123072
Radakovits R, Eduafo PM, Posewitz MC (2011) Genetic engineering of fatty acid chain length in *Phaeodactylum tricornutum*. Metab Eng 13(1):89–95
Rosli SS, Lim JW, Jumbri K, Lam MK, Uemura Y, Ho CD, Tan WN, Cheng CK (2019) Modeling to enhance attached microalgal biomass growth onto fluidized beds packed in nutrients-rich wastewater whilst simultaneously biofixing CO_2 into lipid for biodiesel production. Energy Convers Manag 185:1–10
Ruffing AM (2014) Improved free fatty acid production in cyanobacteria with *Synechococcus* sp. PCC 7002 as host. Front Bioeng Biotechnol 2:17
Ruffing AM, Davis RW, Lane TW (2022) Advances in engineering algae for biofuel production. Curr Opin Biotechnol 78:102830
Salama ES, Govindwar SP, Khandare RV, Roh HS, Jeon BH, Li X (2019) Can omics approaches improve microalgal biofuels under abiotic stress? Trends Plant Sci 24(7):611–624
Santos-Merino M, Garcillán-Barcia MP, De La Cruz F (2018) Engineering the fatty acid synthesis pathway in *Synechococcus elongatus* PCC 7942 improves omega-3 fatty acid production. Biotechnol Biofuels 11:239
Sarwan J, Nair MV, Sharma S, Uddin N, Bose KJC (2024) Omics technology approaches for the generation of biofuels. In: Shah MP, Kaur P (eds) Biomass energy for sustainable development. CRC Press, pp 165–191
Satya S, Sharma S, Choudhary G, Kaushik G (2024) Advances in environmental microbiology: a multi-omic perspective. In: Kesheri M, Kanchan S, Salisbury TB, Sinha RP (eds) Microbial omics in environment and health. Springer, Singapore, pp 175–204
Singh VK, Jha S, Rana P, Soni R, Lalnunpuii R, Singh PK, Sinha RP, Singh G (2024) Cyanobacteria as a biocatalyst for sustainable production of biofuels and chemicals. Energies 17(2):408
Sithtisarn S, Yokthongwattana K, Mahong B, Roytrakul S, Paemanee A, Phaonakrop N, Yokthongwattana C (2017) Comparative proteomic analysis of *Chlamydomonas reinhardtii* control and a salinity-tolerant strain revealed a differential protein expression pattern. Planta 246:843–856
Stephanopoulos G (2007) Challenges in engineering microbes for biofuels production. Science 315(5813):801–804
Tamagnini P, Axelsson R, Lindberg P, Oxelfelt F, Wünschiers R, Lindblad P (2002) Hydrogenases and hydrogen metabolism of cyanobacteria. Microbiol Mol Biol Rev 66(1):1–20
Tan X, Yao L, Gao Q, Wang W, Qi F, Lu X (2011) Photosynthesis driven conversion of carbon dioxide to fatty alcohols and hydrocarbons in cyanobacteria. Metab Eng 13(2):169–176
Tan X, Hou S, Song K, Georg J, Klähn S, Lu X, Hess WR (2018) The primary transcriptome of the fast-growing cyanobacterium *Synechococcus elongatus* UTEX 2973. Biotechnol Biofuels 11:218
Tanwar A, Sharma S, Kumar S (2018) Targeted genome editing in algae using CRISPR/Cas9. Indian J Plant Physiol 23:653–669
Trentacoste EM, Shrestha RP, Smith SR, Glé C, Hartmann AC, Hildebrand M, Gerwick WH (2013) Metabolic engineering of lipid catabolism increases microalgal lipid accumulation without compromising growth. Proc Natl Acad Sci USA 110(49):19748–19753
Ungerer J, Wendt KE, Hendry JI, Maranas CD, Pakrasi HB (2018) Comparative genomics reveals the molecular determinants of rapid growth of the cyanobacterium *Synechococcus elongatus* UTEX 2973. Proc Natl Acad Sci USA 115(50):11761–11770

Varman AM, Xiao Y, Pakrasi HB, Tang YJ (2013) Metabolic engineering of *Synechocystis* sp. strain PCC 6803 for isobutanol production. Appl Environ Microbiol 79(3):908–914

Wang Y, Chen L, Zhang W (2016) Proteomic and metabolomic analyses reveal metabolic responses to 3-hydroxypropionic acid synthesized internally in cyanobacterium *Synechocystis* sp. PCC 6803. Biotechnol Biofuels 9:1–5

Wang M, Luan G, Lu X (2019) Systematic identification of a neutral site on chromosome of *Synechococcus* sp. PCC7002, a promising photosynthetic chassis strain. J Biotechnol 295:37–40

Wang L, Chen L, Yang S, Tan X (2020) Photosynthetic conversion of carbon dioxide to oleochemicals by cyanobacteria: recent advances and future perspectives. Front Microbiol 11:634

Wei L, Shen C, El Hajjami M, You W, Wang Q, Zhang P, Ji Y, Hu H, Hu Q, Poetsch A, Xu J (2019) Knockdown of carbonate anhydrase elevates *Nannochloropsis* productivity at high CO_2 level. Metab Eng 54:96–108

Yang MK, Yang YH, Chen Z, Zhang J, Lin Y, Wang Y, Xiong Q, Li T, Ge F, Bryant DA, Zhao JD (2014) Proteogenomic analysis and global discovery of posttranslational modifications in prokaryotes. Proc Natl Acad Sci USA 111(52):5633–5642

Yang B, Liu J, Jiang Y, Chen F (2016) *Chlorella* species as hosts for genetic engineering and expression of heterologous proteins: Progress, challenge and perspective. Biotechnol J 11(10):1244–1261

Yao L, Cengic I, Anfelt J, Hudson EP (2015a) Multiple gene repression in cyanobacteria using CRISPRi. ACS Synth Biol 5(3):207–212

Yao L, Tan TW, Ng YK, Ban KHK, Shen H, Lin H, Lee YK (2015b) RNA-Seq transcriptomic analysis with Bag2D software identifies key pathways enhancing lipid yield in a high lipid-producing mutant of the non-model green alga *Dunaliella tertiolecta*. Biotechnol Biofuels 8:191

Yazdani SS, Gonzalez R (2008) Engineering *Escherichia coli* for the efficient conversion of glycerol to ethanol and co-products. Metab Eng 10(6):340–351

Yu J, Liberton M, Cliften PF, Head RD, Jacobs JM, Smith RD, Koppenaal DW, Brand JJ, Pakrasi HB (2015) *Synechococcus elongatus* UTEX 2973, a fast-growing cyanobacterial chassis for biosynthesis using light and CO_2. Sci Rep 5(1):8132

Yunus IS, Jones PR (2018) Photosynthesis-dependent biosynthesis of medium chain length fatty acids and alcohols. Metab Eng 49:59–68

Yunus IS, Wichmann J, Wordenweber R, Lauersen KJ, Kruse O, Jones PR (2018) Synthetic metabolic pathways for photobiological conversion of CO_2 into hydrocarbon fuel. Metab Eng 49:201–211

Zhou L, Cheng D, Wang L, Gao J, Zhao Q, Wei W, Sun Y (2017) Comparative transcriptomic analysis reveals phenol tolerance mechanism of evolved *Chlorella* strain. Bioresour Technol 227:266–272

Zhu T, Xie X, Li Z, Tan X, Lu X (2015) Enhancing photosynthetic production of ethylene in genetically engineered *Synechocystis* sp. PCC 6803. Green Chem 17(1):421–434

Lichen Symbiosis: A Systems Biology Perspective

21

Palak Rana, Varsha K. Singh, Sapana Jha, Riya Tripathi, Ashish P. Singh, Amit Gupta, and Rajeshwar P. Sinha

Abstract

Lichens are symbionts that inhabit a wide range of ecological conditions on Earth, ranging from hot and humid tropical rainforests to cold alpines and polar ecosystems. In this symbiotic association, a fungus and a photosynthetic partner, either green algae or cyanobacteria, live in a distinctive symbiotic relationship and thrive well by mutually exchanging food, nutrients, and shelter between fungus and phototroph. In contrast to other fungal symbiotic association, lichens form noticeable and easily identifiable macroscopic vegetative thalli that display a range of diverse colors and different morphologies. Although this symbiotic association has successfully been studied for many years, very little information is available about the physiological and molecular mechanisms operating in this stable symbiotic association. This chapter will firstly give a brief account of the evolution that this symbiotic association has undergone. Moreover, it addresses the metabolic pathways and signaling and recognition mechanisms between the photobiont and the mycobiont. This chapter provides an overview of the steps that have been made thus far to apply the function of omics approaches, including transcriptomics, metabolomics, genomics, and so on, in understanding the regulation, signaling, and metabolism of lichen symbiosis. Thus, a stronger comprehension of this association will result from this.

Keywords

Lichen · Metabolic pathways · Mycobiont · Omics · Photobiont

P. Rana · V. K. Singh · S. Jha · R. Tripathi · A. P. Singh · A. Gupta · R. P. Sinha (✉)
Laboratory of Photobiology and Molecular Microbiology, Department of Botany, Institute of Science, Banaras Hindu University, Varanasi, India

M. Kesheri et al. (eds.), *Multi-Omics in Biomedical Sciences and Environmental Sustainability*, https://doi.org/10.1007/978-981-96-7067-3_21

Objectives

- Omics approach will help us understand the functional diversity within the lichen flora.
- Aims to decipher the mechanism of signaling between the mycobiont and photobiont.
- Highlights the metabolism and its regulation in the lichen symbiosis.
- Summarizes the existing information about the evolution and molecular mechanisms underlying the signaling that occurs in the lichen.

21.1 Introduction

Symbiosis is a cooperative behavior displayed by two separate organisms that, rather than competing for food, shelter, resources, and space, exist in the same niche, benefiting each other and forming a permanent arrangement. Scientists have been debating the evolutionary implications of symbiosis ever since it was first identified as a phenomenon. The first life form to prompt such lines of inquiry was the lichen (Spribille et al. 2022). Lichens form a remarkably stable and ecologically important symbiotic relationship in which algae and fungi from two distinct kingdoms live together. Algae and fungi are organisms from separate niches with diverse biology that come together to form one organism, known as the lichen, which is morphologically and physiologically distinct and possesses very unique properties different from its contributing organisms. The fungal companion supplies nutrients from the substrate, acts as a shelter, and helps to screen out harmful light radiation, thus protecting the photobiont. Algal partners, on the other hand, use photon energy to reduce the carbon dioxide in the atmosphere, producing organic molecules. It also converts air nitrogen to the assimilatory form via nitrogen fixation, thus supplying sustenance for the entire lichen thallus. Fungi comprise most of the visible section of this thallus and provide the shape, size, structure, and color that lichens have. They produce a variety of secondary metabolites belonging to depsides, depsidones, depsones, chromones, xanthones, and dibenzofurans derivatives (Bhattacharyya et al. 2016).

These secondary metabolites aid in the protection of this symbiotic relationship from various stresses and is advantageous in flourishing the lichen to varying environmental conditions. Thus, the lichen association is beneficial for both partners. The fungal partner effectively uptakes the mineral nutrients, protects and benefits the algal partners. The latter also allows the former to grow in circumstances where they cannot succeed alone (Dar et al. 2022). Lichen associations are a typical example for successful and lasting interactions between microorganisms and have been investigated for many years. However, there are considerable gaps in our understanding of the molecular mechanisms behind the lichen symbiosis. How this association originated, flourished over time, and evolved into the present-day lichen, which can thrive not only in favorable conditions but also in adverse environmental conditions such as polar tropics, deserts, and artic regions, is a very interesting topic for current research (Honegger 1998; Aubert et al. 2007).

Some paleontological evidence exists that can explain the genesis and evolution of this flora. Although many studies have been undertaken and will take place to learn about the evolutionary importance of lichen evolution, omics approaches such as genomics have made these studies much more comprehensive. However, little is known about the signaling processes that occur between the photobiont and mycobiont partners of this association. Is symbiosis based on biological slavery or cooperation? If so, how does it remain stable? What is the reciprocity that stabilizes it? How do these partners come together? What stimulates them to do so and how do they operate this process so seamlessly is still an area of interest for various researchers. The use of omics has made it easy to address all these questions. This chapter deals with summarizing the existing information about the evolution and molecular mechanisms underlying the signaling that occurs in the lichen. It also emphasizes the use of contemporary omics approaches to improve the traditional knowledge of this connection. The advent of omics techniques such as genomics, transcriptomics, and metabolomics, as well as current molecular methodologies, has unravelled various questions about how these interactions work. These techniques have helped to better understand the metabolic pathways and mechanisms underlying the symbiosis of this relationship (Nazem-Bokaee et al. 2021).

21.2 Evolutionary History of Lichen Symbiosis

Fossil evidence indicates that lichens are ancient species that have existed for more than 400 million years. Although the precise origins of lichens are unknown, they are believed to have originated from free-living fungi that partnered with algae or cyanobacteria and are considered to have evolved along separate lines in a number of fungal lineages, giving rise to the diversified lichen flora that exists today (Schoch et al. 2009; Gueidan et al. 2008). Lichens have long and often been proposed as a precursor vegetation, maybe dating back to the Neoproterozoic, which could have supported carbon cycle feedbacks and created soil before the formation of tracheophytes.

Lichens are common pioneers of bare substrates in current systems. Although significant improvements have been achieved in elucidating the evolution of lichens, there is currently no synthetic viewpoint that connects the histories of fungi and algae participating in lichen formation, on the level of macro-evolution (Eriksson 2005). In the Early Devonian (415 Myr), lichens were among the earliest organisms to successfully adapt to life on land (Honegger et al. 2013). Many thalloid fossils from the Devonian and Silurian periods are known, but the fossil deposits are insufficient to differentiate between lichens, extinct algal experiments, microbial mats, and land plant gametophytes. The key evolutionary innovation of lichens is their symbiotic relationship. Fungal members offer protection and framework, whereas algal or cyanobacterial members carry out photosynthesis, which supplies nutrients to the association. Lichens are able to flourish in conditions where neither pair could live by themselves because of their mutualistic relationship. Approximately 7% of the surface of our planet is covered with lichens, which also alter many elements of current biocrust-dominated ecosystems. Through photosynthesis and nitrogen-fixation by lichen symbionts, carbon and nitrogen are pulled into the ecosystem, facilitating the growth and development of lichen thalli and the transfer of nutrients upon breakdown or consumption. They sustain a variety of intra thalline bacterial and eukaryotic communities, which increase ecological complexity (Fluteau and Sepulchre 2021). They also provide a home and food supply for a variety of soil invertebrates. To comprehend the lichens' participation and roles in various geochemical cycles and climate change across geological periods, as well as their rise in the process of organic evolution, a thorough knowledge of the sequence of evolutionary lineage, stability, and timeline of this ancient popular symbiosis is essential (Eriksson 2005).

Fossil records, which resembled lichens, dated to the Neoproterozoic era have been rejected or deemed unclear due to their lack of characteristic traits of lichens. Nevertheless, their presence supports an early origin of lichens. There are a few charcoalified Silurian-Devonian fragments with stratified thalli that are undoubtedly similar to lichens. These fragments are surely known to be newer than vascular plants that are related to contemporary lycopsids (Nelsen et al. 2020). Therefore, the existence of pretracheophyte lichens is not well supported by fossil data. Additionally, the experimental reproduction of taphonomic processes shows that important features of contemporary lichen thalli have been destroyed, indicating that the discovery of breakthrough fossils may not be feasible. A study of phylogenetics can be used as another method to compute how old the lichen associations are. Recently, researchers have been seen to rely on a nonquantitative method in unrevealing the origin of lichenization in fungi, considering the obstacles in finding irrefutable proof of lichens in the fossil record. These evidences affirm that lichens evolved from ascomycetes in the Ordovician period, most probably alongside or earlier than tracheophytes. Molecular phylogenetics research has been found as another useful method to estimate the age of lichen associations, given the difficulties in finding indisputable fossilized proofs of lichens. Instead, recent research has focused on a nonquantitative method to distinguish the evolutionary history of lichenization in fungi, and it has suggested that lichenization originated from a single ascomycete not later than the Ordovician, either contemporaneous with or before tracheophytes (Lutzoni et al. 2001). Therefore, it is theoretically possible that lichens were accountable for Ordovician environmental effects that preceded tracheophytes.

In these early systems, the lack of lichens and nonvascular plants is thought to have resulted in soil ecosystems, which were shallow, dry, and not very stable, with less nutrient input and productivity as well as more straightforward food webs. Enhanced dust gathering and the stability require higher surficial texture, which was probably first supplied by terrestrial plants, but not by the lichens. Lichens' delayed evolution indicates that early land plants colonized comparatively basic terrestrial environments that relied on microbes. As a result, lichens did not contribute any prior system alteration or competitive interactions (Dighton and White 2017). This further demonstrates that the path of succession in present ecosystems does not have to follow their evolution, as it differs from the pattern of succession in current dryland ecosystems, where the early colonists included lichens that gave the soil its stability. This stability was favorable for later colonization by bryophytes. Many researchers believe that lichens have risen in evolutionary progression after vascular plants, and resultantly their contribution to modifying the environmental condition during the Neoproterozoic and early Paleozoic seems negligible. With high accuracy, one can say that ascomycete-lichen-forming fungi have evolved after tracheophytes. Phylogenetic studies revealed that Pezizomycotina, which is considered a lichen-forming clade, has earlier origins than the tracheophytes. Some of the lichen-forming fungi, which have been found limited to just five of Agaricomycetes lineages, originated from Basidiomycota, which predates tracheophytes in their origin. In order to accept the existence of lichen-forming fungi, we will have to invoke the evolutionary history of the process of lichen-formation in clades that lack any presently existing, lichenized descendants other than Pezizomycotina and Agaricomycetes (Spribille et al. 2022).

21.3 Omics: A Tool for Understanding Symbiosis

The molecular mechanism of signaling between the mycobiont and photobiont in the lichen symbiosis is poorly understood. Systems biology studies living systems by combining sophisticated high data-generating technologies like “omics” with computational tools like multiscale or constraint-based modeling to better understand the interdependencies between system components and the underlying complexity of the system (Kitano 2002). The study of lichens has historically been approached from the perspectives of ecology, evolution, and physiology; however, the use of systems biology has recently reignited interest in this field. This multidisciplinary method explores the complex relationships and regulatory networks inside lichens by combining high-throughput molecular data with computational studies. Through the analysis of genomes, transcriptomes, proteomes, and metabolomes, scientists hope to elucidate the molecular pathways that underlie lichen symbiosis systematically. In the case of the legume-rhizobium symbiosis, for example, the application of systems biology and genomics approaches has improved the knowledge of the molecular mechanisms operating behind this symbiotic interaction and brought us one step closer to developing more effective symbiotic N_2 fixation processes that will increase agricultural crop yields. It has not yet been possible to analyze the lichen symbiosis using a similar systems biology methodology (diCenzo et al. 2019). Studies of the fungal and algal partners at the genomic level have uncovered genetic adaptations that support their symbiotic living. The dynamics of gene expression in lichen formation, stress response, and nutrition exchange are clarified by transcriptomic research. The chemical variety of lichens and their unique secondary metabolites, which support their ecological roles and possible biotechnological uses, are revealed by metabolomic investigations.

The use of metabolic network modeling and signaling, along with "omics" techniques, may offer predictive methods for learning about lichen signaling and metabolic pathways. Due to the comparatively sluggish growth of mycobionts, research on the genesis and operation of lichen symbiosis has lagged in applying traditional experimental microbiology methodologies and coculture investigations. Even though some new studies (Joneson et al. 2011; Armaleo et al. 2019; Kono et al. 2020) have focused on the early stages of lichenization, there is still a lot to learn about how fungal and algal cells interact at the molecular level during lichenization, as well as how the symbionts' metabolism and signaling work. The application of modern systems biology techniques could help resolve prolonged concerns underlying this symbiont. Previously, a lichen was thought to have a single fungal and algal genotype, which aligns with its location in the plant kingdom, rather than the fungus, which is well-known for forming colonies with heterogeneous genotypes. Various investigations comparing the anatomy of lichens and fungi, which were sufficient to question the notion that lichens are a member of the plant kingdom, failed to satisfy the researchers for over a century. The genomics research and use of molecular techniques such as DNA sequencing for this approach could eventually place lichens in the fungal lineage and transfer their position from the plant kingdom to fungi. Recent research has shown that lichens, which were previously thought to be single, discrete lichen individuals, are in fact heterogeneous organisms made up of several different fungal genotypes that live seamlessly together (Allen and Lendemer 2022).

Moreover, we now know that lichens are composed of more than just the photobiont and mycobiont communities; stable, specialized, and structurally linked bacterial communities are also acknowledged as partners in the lichen symbiosis. Nevertheless, unlike the well-established roles of the fungal and algal partners, the exact involvement of these partners has remained mostly unknown. Present-day omics methodologies offer perfect instruments for investigating the functional diversity present in the lichen microbiome. These methods also provide thorough baseline data for additional experimental research. For example, an integrated genomics approach provided valuable insights into the nitrogen fixation methods used by the photobiont (cyanobacteria) in the lichen. The utilization of meta-omics and amplicon sequencing has enabled recognition of the diverse bacterial microbiota found in lichens. In one such study, it was found that the lichen flora *Alphaproteobacteria* appears to be the predominant species (Aschenbrenner et al. 2014; Grube et al. 2015; Hodkinson et al. 2014; Bertrand et al. 2018). Thus, Meta-omics techniques are useful tools for uncovering the diversity of purposes served by the lichen microbiome. Cernava et al. (2017), in their study on *Lobaria holobiome,* explored the structural composition and functional roles of different bacterial partners by integration of transcriptome data with genomic and proteomic information. Their omics approaches aided in the discovery of various unidentified active bacterial partners in the lichen symbiosis and advanced the investigation of the holobiomes in lichens. A complementary microscopic approach (FISH/CLSM imaging) supports this strategy.

By employing omics techniques, including metagenomes, metaproteomes, and *16S rRNA* gene amplicon libraries, the cyanobacteria demonstrated a 30% abundance. In addition, a high abundance of five bacterial orders was discovered: Sphingomonadales, Rhodospirillales, Myxococcales, Chthoniobacterales, and Sphingobacteriales. The transcriptome data supported the functional potential of lichen bacteria and gave expression data for the bacterial orders, in conjunction with metagenomic and metaproteomic data. The microbiota of lichens is complex and functionally varied. Some members provide food (feeders) and other members dispense protection and/or the ability to regulate growth in the hosts, i.e., protectors. Participants of the known functional guild perform both these functions within the *Lobaria holobiome*, with their primary functions being those of probiotics and detoxifiers (Cernava et al. 2017). Transcriptomics data revealed that Chthoniobacterales are involved in the production of several vitamins, the breakdown of aromatic compounds, and the provision of defense against oxidative stress and fluoroquinolones. Rhodospirillales participate in the transfer of potassium and ammonia, having assimilated around 94% and 96% of each, respectively. Sphingobacteriales were mostly involved in the metabolism of iron and sulfur. The principal agents responsible for the breakdown of n-phenylalkanoic acid were Chthoniobacterales. Rhodospirillales were implicated in the breakdown of xylenols, cresols, and phenylpropanoids; on the other hand, Myxococcales break down chloroaromatic compounds (Fig. 21.1).

Metagenomic and metatranscriptomic investigations have revealed diverse approaches employed by related microorganisms to endure in challenging environments. Extracellular reactive oxygen species (ROS), e.g., hydrogen peroxide, superoxide, and hydroxyl radicals, are known to be generated at large rates by lichens

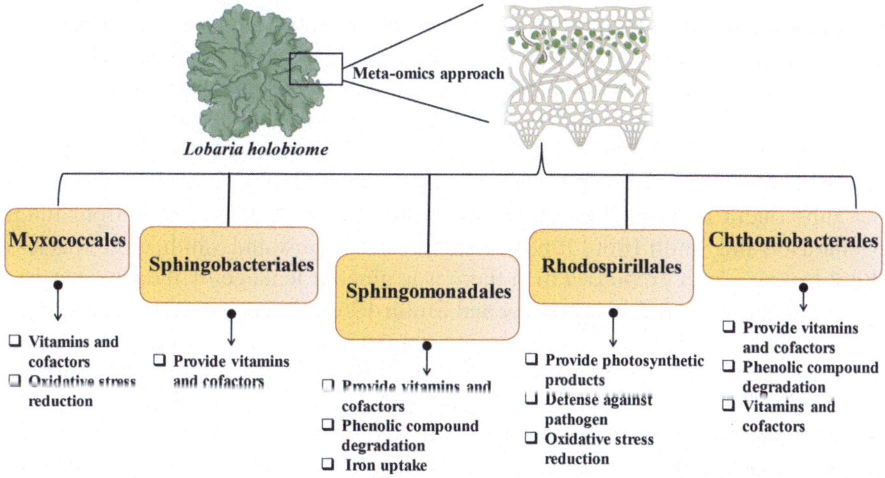

Fig. 21.1 Functional diversity of various bacterial orders (Sphingomonadales, Rhodospirillales, Myxococcales, Chthoniobacterales, and Sphingobacteriales) within the thallus of the lichen *Lobaria holobiome*. (Modified from Cernava et al. 2017)

following desiccation-rehydration episodes. The fungi (that have become part of the lichen) primarily lacking the crystals of the probable antibacterial lichen chemicals on their surface may benefit from extracellular ROS production in their defence against bacterial and fungal infections (Beckett et al. 2003, 2005). This theory is also favored by metagenomics data, which shows that different bacterial species have similar correlative responses to oxidative stress, typically, production of catalase enzymes and low-molar-mass antioxidant molecules like glutathione. The active portion of the metatranscriptome was also discovered to have these roles. Further examination sheds light on the functional roles played by the aforementioned conglomeration of bacteria. Later, it was contextualized within an ecological framework. Diverse meta-omics strategies were used to overcome the shortcomings of the individual approaches (Brakni et al. 2018). Whenever researchers want to investigate the structural formation of the community of microbes in never-before-tested settings, cost-effective amplicon sequencing studies are frequently chosen. On the other hand, in comparison to metagenomic and metatranscriptomic datasets, they provide less thorough information. When the benefits of different NGS techniques are combined, the overall accuracy is typically higher (Cernava et al. 2017).

21.4 Metabolic Pathways Regulation in Lichen Symbiosis

Lichens are remarkable examples of symbiotic organisms that emerge from a mutualistic relationship between an alga (phycobiont) or cyanobacteria and a fungus. In this type of symbiotic relationship, the fungus provides refuge and nutrients to the algae, whereas, if the photobiont is a cyanobacterium, it supplies organic carbon captured by photosynthesis from atmospheric carbon dioxide in addition to nitrogen (Nash 1996). The metabolic pathway regulation of this symbiosis involves numerous mechanisms and guarantees the life and appropriate operation of both partners (Table 21.1). An outline of the regulation of these pathways is provided below.

Carbohydrates like glucose or ribitol (sugar alcohol) are produced by the photobiont during photosynthesis and are then transported to the fungal partner. According to Richardson and Smith, ribitol is the primary form in which carbon assimilated by photosynthetic processes travels from the algae to the fungus in *Xanthoria aureola*. The subsequent accumulation of mannitol indicates that it originates from fungi (Richardson and Smith 1966, 1968). As observed by Drew and Smith, there was less ribitol leakage from *Trebouxia* than there was glucose leakage by the *Nostoc* symbiont of *Peltigera polydactyla* (Drew and Smith 1966). In exchange, the photobiont receives vital nutrients and a suitable environment from the mycobiont. Signaling molecules and transport proteins control this exchange thoroughly. Through the process of photosynthesis, ATP is produced, and atmospheric carbon is converted to carbohydrates by the photobiont. Cyanolichens translocate glucose-based carbohydrates to their fungal partner, whereas chlorolichens convert carbohydrates to polyols, which are then transmitted to their corresponding mycobionts. Such carbohydrate translocation powers the metabolism of mycobionts and supplies nitrogen to the photobiont (if it is a chlorolichen).

Table 21.1 Various metabolic pathway operating in the photobiont and mycobiont members of lichen and their end product

Metabolic pathway	Enzyme	Organism	Function	Products	References
Carbohydrate metabolism	RuBisCO	Algal partner	Catalyzes the first step of carbon fixation	3-Phosphoglycerate	Hill (1976)
Nitrogen metabolism	Nitrate reductase	Algal partner	Reduces nitrate to nitrite	Nitrite	Millbank and Kershaw (1973)
	Nitrite reductase	Algal partner	Reduces nitrite to ammonium	Ammonium	Millbank and Kershaw (1973)
Secondary metabolism	Phenylalanine ammonia lyase	Fungal partner	Catalyzes the conversion of phenylalanine to trans-cinnamate	Trans-cinnamate	Calcott et al. (2018)
	Polyketide synthase	Fungal partner	Catalyzes the biosynthesis of plyketides	Polyketides	Calcott et al. (2018)

According to latest research, lichen symbionts may have coupled metabolisms in which photobionts provide oxygen (along with carbohydrates) in exchange for carbon dioxide from their fungal partner (Schwartzman 2010; Hom and Murray 2014; Ten Veldhuis et al. 2020). It is thought that the identification of symbionts is mostly dependent on the lichen compounds that mycobionts (primarily) produce and the photobionts that produce phytohormones and/or carbohydrates. Over time, a sequence of events that usually occur in the stroma region of the chloroplast or in the cytoplasmic inclusions of cyanobacterial cells fix CO_2, variously referred to as the Calvin, C-3, or reductive pentose cycle. Free-living cyanobacteria and algae convert carbon dioxide to carbohydrates; lichens appear to also use the Calvin cycle for photosynthesis (Hill 1976; Brown and Kershaw 1986). For the conversion of 3-phosphoglycerate into 3-phosphoglyceraldehyde, the Calvin cycle needs NADPH and ATP. Additionally, ATP is needed to convert ribulose-5-phosphate into ribulose-1,5-bisphosphate. The one point in the cycle when carbon dioxide is taken up is mediated by the intriguing bifunctional enzyme ribulosebisphosphate carboxylase/oxygenase (RuBisCO), which carboxylates ribulose-1,5-bisphosphate in the presence of light. RuBisCO is found in lichens (Vicente and Legaz 1988) and also in other autotrophs, which may not be surprising. It is a massive molecule that is made up of eight large and eight small subunits in both cyanobacteria and chloroplasts (Tabita 1987).

In cyanobacteria, transcription of the small and large subunits occurs simultaneously (Tabita 1987). However, in green algae, transcription of the subunits must occur independently, as the big subunit gene is located in the chloroplast while the small subunit gene is located in the cell nucleus. Lichens lack RuBisCO and fix carbon via the reductive noncyclic acetyl-CoA pathway or the reductive citric acid cycle. However, alternative mechanisms to ingest CO_2 have not been found in lichens. Important enzymes other than RuBisCO are needed in each scenario (Strauss and Fuchs, 1993). Prokaryotes have been shown to employ 3-hydroxypropionate as an intermediary and acetyl-CoA carboxylase and propionyl-CoA carboxylase as absorbing enzymes in an alternate manner of autotrophic assimilation (Eisenreich et al. 1993). Fungal pathogens most likely absorb nutrients from higher plant cells through diffusion across the host cell membrane. The solute transport along a concentration gradient appears to be sustained by high levels of food reserves in the host and the rapid conversion of the transported molecules into fungal-specific compounds in the presence of pathogens (Smith et al. 1969; Hall et al. 1992).

Light and the availability of nutrients are two examples of environmental elements that might affect the rate of metabolism. An essential step in lichens containing cyanobacterial photobionts is nitrogen fixation. Only the lichens with *Nostoc* and *Calothrix*, among the eight genera of cyanobacteria identified as phycobionts have been demonstrated to fix nitrogen (Ahmadjian 1967; Duncan 1970). The nitrogen-fixing bacteria in the cephalodia of *Peltigera aphthosa* were assumed to be identical to legume root nodules (Goebel 1926). However, Goebel (1926) believed that nitrogen fixation was carried out by *Nostoc* in the cephalodia. Henriksson (1958), reported that, although the nature of the compounds was unclear, an isolated strain of *Nostoc* secreted molecules that can promote the growth of the fungal partner.

Lichen fungi often lack biotin and thiamine, and the phycobiont's production of such compounds plays an important role in their healthy growth. The fungus companion in lichens uses intricate biochemical pathways to produce the majority of the secondary compounds. Most of the secondary metabolites found in lichens that have been isolated so far are derived from fungi. These are phenolic compounds that are mostly generated from polymalonates; metabolites have been documented, with the majority being phenolic polyketides found only in lichens (Culberson 1969; Herz et al. 1984; Nguyen et al. 2013). These are made from a small set of monoaromatic subunits generated by polyketide synthase, two (or sometimes more) of which are linked together by ether, ester, or carbon–carbon bonds. Polycyclic aromatic chemicals, such as xanthones, anthra/naptha-quinones, and chromones, are also produced by many lichens from longer polyketide chains (Elix and Stocker-Wörgötter 2008; Nguyen et al. 2013). Over 1000 lichen secondary phenylalanines are precursors to biosynthesize a variety of unusual aromatic compounds with an oxalane-carbonyl group, such as pulvinic acid (Goga et al. 2020). Terpenes and carotenoids are among the other natural compounds from lichens that have been documented to be derived from mevalonate. In addition, they create several xanthones, including lichexanthone-like metabolites such as depsides and depsidones (Fig. 21.2).

The majority of research on lichen metabolism has been concentrated on figuring out how symbionts exchange essential nutrients (Lines et al. 1989; Kono et al. 2020; Ten Veldhuis et al. 2020) as well as discovering secondary metabolites and the mechanisms by which they are produced (Nazem-Bokaee et al. 2021). Studies on the metabolic interplay in lichens were initiated in the 1960s by examining the assimilation and transport of carbohydrates among the symbionts of *Peltigera polydactyla* (Smith and Drew 1965; Drew and Smith 1967) and a number of related observations on many other lichens (Smith et al. 1969). The primary objective of those investigations was to determine the type of carbon that was transferred between lichen symbionts. The majority of lichen metabolism research has focused on studying the entire carbon and nitrogen metabolism in lichens, particularly in relation to the fixation of nitrogen by cyanolichens, carbon sinks, overall carbon fixation, and lichen growth (Honegger et al. 1993; Dahlman et al. 2004; Nash 2008). Although this has not been fully realized to study the lichen metabolism at a system-level, Eisenreich et al. (2011) suggested that combining "omics" methods with isotope labeling experiments—often referred to as "fluxomics"—could improve our knowledge of pathways involved in lichen metabolism.

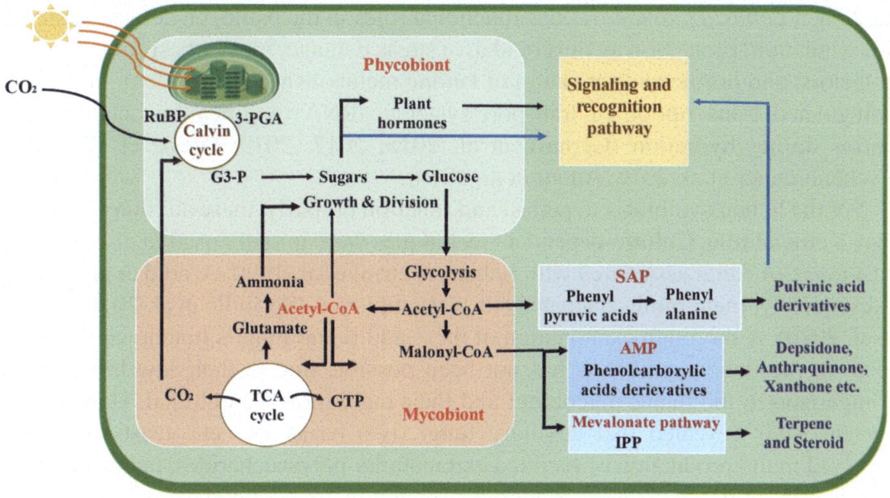

Fig. 21.2 The intricate regulation of metabolic pathways in the lichen symbiosis, with an emphasis on interactions between the photobiont and the mycobiont. It encompasses important metabolic functions such as carbon metabolism, nitrogen assimilation, and secondary metabolite synthesis, as shown in the image. It depicts the symbiotic exchange of nutrients, such as nitrogenous compounds and carbohydrates, between the algal and fungal partners (RuBp Ribulose 1,5-bisphosphate, 3-PGA 3- Phosphoglyceric acid, G-3P Glyceraldehyde 3 phosphate, TCA Tricarboxylic acid cycle, SAP Shikimic acid pathway, AMP Acetyl malonyl pathway, IPP Isopentenyl pyrophosphate, GTP Guanosine triphosphate). (Modified from Nazem-Bokaee et al. 2021)

A comprehensive knowledge of the primary symbionts and other partners involved in lichen metabolism at the molecular level is still absent, despite the fact that it is widely acknowledged that lichen symbionts exchange carbohydrates and macromolecules. Owing to this deficiency, important parts of metabolism have remained unknown. These features include the influence of the components of the cell wall and biosynthesis on symbiotic proliferation and exchange of metabolites, or the primary aspects of carbon metabolism with regard to lichen compartmentalization. According to studies conducted to date on a variety of terrestrial and marine lichens, Rhizobiales have been identified as a prominent bacterial order in the microbiome (Grube et al. 2009; Hodkinson and Lutzoni 2009; Hodkinson et al. 2012; Erlacher et al. 2015). Particularly, it was discovered that Sphingomonadales and Bacteroidetes are prevalent in cyanolichens, while Rhodospirillales are abundant in chlorolichens (Hodkinson et al. 2012; Graham et al. 2018; West et al. 2018).

The richness and makeup of the bacterial population in association with lichens are thought to be influenced by multiple causes. As cyanobacterial photobionts, some additional bacteria were believed to be capable of fixing freely available nitrogen and could be crucial as a supply of nitrogen for the lichen symbiosis (Hodkinson and Lutzoni 2009). Furthermore, a variety of bacteria that would not typically flourish in such nitrogen-starved conditions have been found to be present in cyanobacterial lichens, which are frequently found in such habitats (Hodkinson et al. 2012). In addition to nitrogen fixation, studies using meta-omics (e.g., metagenomics, metatranscriptomics, and meta-proteomics) have shown that the microbial community of the lichen *Lobaria pulmonaria* has functional roles in the pathogen defense system, detoxification, protection against oxidative stress damage, biosynthesis of vitamins, cofactors, and hormones, activation of ketone metabolism during dehydration, and upregulated transcription of transport systems, tRNA modification, and various porins during hydration (Cernava et al. 2015, 2017, 2019; Grube et al. 2015; Aschenbrenner et al. 2016; Auður et al. 2016).

For the lichen symbiosis to persist and function properly, these auxiliary bacteria play a crucial role. Culture-dependent techniques were initially used to discover the vast range of fungi associated with lichens (Petrini et al. 1990; Arnold et al. 2009); subsequently, meta-omics data analysis was employed (Spribille et al. 2016; Smith et al. 2020). A thorough examination of these additional fungi's functional involvement in the lichen symbiosis has not been possible due to their low biomass in comparison to the major mycobiont and their inability to be cultured. However, a current study revealed that auxiliary fungi (two basidiomycete yeasts) may be involved in the production of secreted extracellular polysaccharides, lichen nutrient acquisition, and secondary metabolite production based on the analysis of the metagenome of the lichen *Alectoria sarmentosa* (Tagirdzhanova et al. 2021). As a result, they probably have a significant impact on how the lichen symbiosis continues to exist and function.

21.5 Omics Approach for Comprehending the Signaling in Lichens

The term "symbiosis" was initially used by Anton de Bary, a German mycologist to characterize the state in which dissimilar species coexist, thus providing Schwendener's concept that lichens are composed of two distinct living beings. (Oulhen et al. 2016). Lichens represent a unique association among the various symbiotic relationships that occur in nature, formed specifically between fungi and either green algae or cyanobacteria (Sønstebø and Rohrlack 2011). Numerous kinds of interconnected species, usually described as the "holobiont," have mutualistic-antagonistic connections that are necessary for the stability of this symbiotic association (Douglas and Werren 2016). The stability of the environmental conditions, partner availability, and the interactive intimacy of the symbionts all influence this complicated connection (Rafferty et al. 2015; Chomicki and Renner 2017). Developing universal signaling models that initiate lichen symbiosis is a difficult task. First, no signaling molecule found in any of the lichens studied has been revealed to have any known or proposed importance. Secondly, the signaling pathways associated with the compounds that may have recognition roles are still unclear. Thirdly, according to Schoch et al. (2009), lichens may have independently evolved within several fungal lineages, indicating that the specific species of interest may influence the nature of these communication pathways. However, several studies have started to identify the molecular features that support partner signaling and identification at different stages of lichenization due to advancements in genetic and analytical methods (Athukorala and Piercey-Normore 2015).

21.5.1 Lectin-Ligand Signaling in Lichens

Glycoproteins, called lectins, are found in every aspect of life and are ubiquitous. Additionally, lectins from both cyano- and chlorolichens have been identified and isolated. Due to the presence of diverse sites for binding carbohydrates, lectins have the ability to bind cells, acting as receptors and potentially promoting more interfacial communication. The mycobiont synthesizes lectins with glycosidic moieties that contain different mergers of groups of carbohydrates, which can bond with particular ligands from the photobiont. In this nontraditional receptor-ligand system, mycobiont lectins function as receptors for ligands coupled to photobionts. Figure 21.3 shows a potential mechanism for the mycobiont to recognize and attract photobionts based on in-depth research on the lichens *Evernia prunastri* and *Xanthoria parietina* (Millanes et al. 2004). The urease-like recognition-related protein (RR1) was identified in the lichen *Cladonia rangiferina* and is thought to function as a ligand on the cell wall of the suitable photobiont of this lichen relationship.

All or some cyanolichens may possess the lectin recognition and signaling mechanism summarized for chlorolichens in Fig. 21.3 (Diaz et al. 2016). In addition, the ATPase of a photobiont and a contractile protein are involved in the process, which

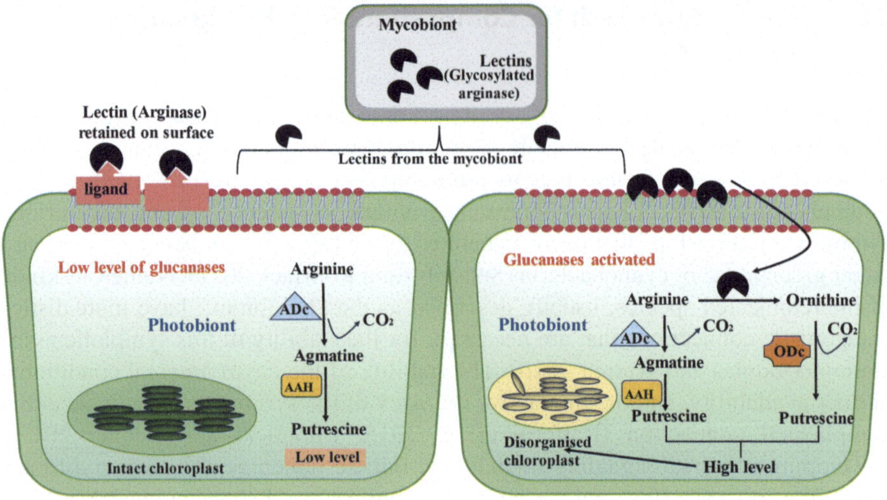

Fig. 21.3 Lectin-ligand signaling model in lichens represents the signaling between the mycobiont and photobiont. The lectins, such as arginase from the mycobiont, come and interact with the ligand α-1,4-glycosalated urease present on the photobiont. The compactable photobiont (on the left-hand side of figure) cell containing this ligand interacts with arginase and induces desirable changes, whereas the photobiont that is incompatible (on the right side) fails to do so (ADc arginine decarboxylase, AAH agmatine amidino hydrolase, ODc ornithine decarboxylase). (Modified from Singh and Walia 2014)

sets up a sequence of contraction-relaxation phases that lead to photobiont emigration in the direction of the mycobiont lectin (Diaz et al. 2011). Some lichens' mycobionts are considered to be dependent on lectin-ligand recognition mechanisms not only to establish the first photobiont contact but also to support the growth and replication of newly formed algal cells inside the lichen thallus (Diaz et al. 2016). *Asterochloris glomerata*, an algal partner of *Cladonia grayi*, and its transcriptome analysis, conducted by Armaleo et al. (2019), investigated the differential gene expression which are part of the symbiosis and signaling.

21.5.2 Exudates Signaling in Lichens

Drew and Smith initially proposed the release of carbohydrates and their translocation from the photobiont to the mycobiont of lichens in the mid-1960s. They traced a radioactive isotope to determine the amount of labeled carbon [^{14}C] in sodium bicarbonate at [^{14}C] fixed glucose by the cyanobacterial symbiont (*Nostoc*) of *Peltigera polydactyla* in comparison to its free-living and cultivated forms. The amounts of carbohydrates released by the photobionts either drastically decrease or disappear when symbiotic interaction is not established (Armaleo et al. 2019). The exudate signaling model is based on the mobility and release of ribitol. It is mostly based on separate investigations that have shown ribitol release in *Gyalolechia bracteate* cultures (Meessen et al. 2013) and *Cladonia grayi* (Joneson et al. 2011). It is hypothesized that a carbon-nitrogen exchange may have been the initial catalyst for symbiotic association, despite the fact that the specific composition of the released chemicals in this exudate-signaling scenario remains somewhat unclear. Coculturing model fungi *Saccharomyces cerevisiae*, *Aspergillus nidulans*, or *Neurospora crassa* with the alga *Chlamydomonas reinhardtii* may facilitate mutualistic connections by allowing carbon and nitrogen exchanges under specific growth conditions (Hom and Murray 2014). Additionally, their results suggest that the photobiont may be able to reabsorb carbon produced during mycobiont respiration (as CO_2), allowing for efficient carbon recycling within the lichen symbiosis (Schwartzman 2010). Therefore, the initial stage of symbiotic interaction in lichens may be triggered by the requirement for nutrition exchange between species. The next section will cover signaling network modeling as one method to produce insights into the possible roles that particular exudate components could have in the gross signaling path via the believed "exudates signaling" mechanism.

21.5.3 Signaling Network Modeling

The links between the many "signals" and "receptors" that comprise a signaling network are established by the signal transduction mechanisms that control the network. Any or all of these signals and receptors could be proteins or other biomolecules, inorganic chemicals like phosphates, organic substances (e.g., ATP), or enzymes like kinases. Predictive and descriptive subtypes are common divisions in signaling network models. While predictive models can estimate system behaviors under novel perturbations and capture the kinetics of the signaling pathway, or the rates at which reactions occur, descriptive models are typically more straightforward and offer a qualitative overview of the signaling pathway structure (Antebi et al. 2017). There are currently no published models for the signaling networks of lichens, and their modeling faces many of the same difficulties as other symbiotic systems. There are currently greater possibilities for building lichen signaling models. For instance, by measuring the amounts of lectin and urease in lichen cultures at pre- and post-contact stages, one may use the proteomics approach to ascertain the relative abundances of these proteins. The proteome profile of these cultures can reveal more details about the initiation of the recognition process, as well as whether or not additional specific proteins are present during the pertinent stages of lichenization. These proteins may be correlated with levels of lectin or urease activity. The development of a signaling pathway model may examine the relationship between lectin production and putrescine biosynthesis in the inhibition of cell wall disruption by suitable photobionts (Nazem-Bokaee et al. 2021).

21.6 Conclusion

Lichens are a flora of great evolutionary importance and are remarkable models for studying the symbiotic association in nature, providing deeper knowledge about the interspecific knowledge of symbiotic interactions. In spite of the fact that lichens have evolved very early, still very few studies have been done on lichens because, firstly, they are the organization of two different individuals into the same organism, and therefore it is difficult to study the molecularity of these floras. Secondly, they grow slowly and cannot be cultured in the laboratory. Therefore, lichens are still a mystery. From the viewpoint of biology-omics, studies on lichen symbiosis have revealed the complex biological processes that underlie this collaboration, including proteomics, metabolomics, transcriptomics, and genomics. The genetic basis for the reciprocal dependence between photosynthetic partners and fungi has been clarified by genomic studies that have demonstrated their genetic compatibility and adaptation. Furthermore, continuous modifications in gene expression that enhance nutrition exchange, stress tolerance, and cellular communication within the lichen thallus have been found by transcriptomic analysis. While metabolomics has discovered special secondary metabolites that are essential for defence, communication, and environmental adaptability, proteomics has helped to clarify the functional proteins that contribute to lichen stability and resilience.

A thorough comprehension of the symbiotic interactions has been made possible by integrating different omics techniques, and this understanding has potential applications in biotechnology, agriculture, and environmental sciences. The knowledge on metabolism, signaling mechanisms, and the function of omics approaches in constructing the mechanism of their symbiosis has been gathered, and this chapter has made progress in this regard. The chapter includes a detailed description of lichen metabolism as well as an examination of the signaling processes that lichens use. Additionally, it has been discussed how modern omics might help in our understanding of the molecular mechanism underlying the interaction between mycobionts and photobiota. Even with these developments, there are still a lot of unanswered questions concerning the development of lichen partners, the management of symbiosis, and the ecological effects of their interactions. Subsequent investigations utilizing sophisticated omics technologies in conjunction with bioinformatics and systems biology will persist in deciphering the intricacies of lichen symbiosis. This research has the potential to advance our knowledge of symbiotic relationships, advance more comprehensive theories of ecology and evolution, and stimulate creative thinking in the application of such mutualistic relationships to sustainable solutions across a range of domains. To sum up, the biology-omics approach to lichen symbiosis has greatly expanded our understanding of these complex relationships. It is important to note that lichens are important models for researching mutualistic symbiosis, and further exploration in this field has enormous potential to reveal novel biological principles and useful applications.

21.7 Short Chapter Summary

Lichens, a symbiotic relationship between a mycobiont and a photobiont, acts in a unique manner, and interpreting this symbiotic interaction still fascinates the scientific community. Omics techniques have proven to be an effective tool for understanding the symbiosis mechanism and the evolution of the lichen flora. Metabolomics, transcriptomics, and other methodologies have helped to understand the functional diversity within the flora, as well as the signaling and metabolic processes that regulate lichen symbiosis. Therefore, omics techniques are a great tool for gaining a deeper knowledge of this flora and deciphering the mechanism behind their symbiosis.

Acknowledgments Palak Rana (09/0013(16603)/2023-EMR-I) and Varsha K. Singh (09/0013 (12862)/2021-EMR-1) are thankful to the Council of Scientific & Industrial Research (CSIR), New Delhi, India, for the Junior Research Fellowship (JRF) and Senior Research Fellowship (SRF). Ashish P. Singh (NTA Ref. No. 191620014505) and Riya Tripathi (NTA Ref. No. 231620041285) are thankful to the University Grants Commission (UGC), New Delhi, India, for the Junior Research Fellowship (JRF) and Senior Research Fellowship (SRF). Sapana Jha (No. R/ Dev. /Sch./ UGC Non-NET Fello. /2022-23/52561) is thankful to BHU for providing institutional fellowship. The incentive grant received from IoE (Scheme no. 6031), Banaras Hindu University, Varanasi, India, to Rajeshwar P. Sinha is highly acknowledged.

Author Contributions Study concept and design, Palak Rana and Rajeshwar P. Sinha; chapter writing, Palak Rana, Varsha K. Singh, Sapana Jha, Riya Tripathi, Ashish P. Singh, and Amit Gupta; chapter editing and review, Rajeshwar P. Sinha.

References

Ahmadjian V (1967) A guide to the algae occurring as lichen symbionts, isolation, culture, cultural physiology and identification. Phycologia 6(2–3):127–160

Allen JL, Lendemer JC (2022) A call to reconceptualize lichen symbioses. Trends Ecol Evol 37(7):582–589

Antebi YE, Nandagopal N, Elowitz MB (2017) An operational view of intercellular signalling pathways. Curr Opin Syst Biol 1:16–24

Armaleo D, Muller O, Lutzoni F, Andresson OS, Blanc G, Bode HB (2019) The lichen symbiosis re-viewed through the genomes of *Cladonia grayi* and its algal partner *Asterochloris glomerata*. BMC Genomics 20:1–33

Arnold AE, Miadlikowska J, Higgins KL, Sarvate SD, Gugger P, Way A, Lutzoni F (2009) A phylogenetic estimation of trophic transition networks for ascomycetous fungi: are lichens cradles of symbiotrophic fungal diversification? Syst Biol 58(3):283–297

Aschenbrenner IA, Cardinale M, Berg G, Grube M (2014) Microbial cargo: do bacteria on symbiotic propagules reinforce the microbiome of lichens? Environ Microbiol 16(12):3743–3752

Aschenbrenner IA, Cernava T, Berg G, Grube M (2016) Understanding microbial multi-species symbioses. Front Microbiol 7:180

Athukorala SNP, Piercey-Normore MD (2015) Recognition- and defense-related gene expression at 3 resynthesis stages in lichen symbionts. Can J Microbiol 61(1):1–12

Aubert S, Juge C, Boisson AM, Gout E, Bligny R (2007) Metabolic processes sustaining the reviviscence of lichen *Xanthoria elegans* (Link) in high mountain environments. Planta 226:1287–1297

Auður SM, Andrésson ÓS, Vilhelmsson O (2016) Nutrient scavenging activity and antagonistic factors of non-photobiont lichen-associated bacteria: a review. World J Microbiol Biotechnol 32:1–11
Beckett RP, Minibayeva FV, Vylegzhanina NN, Tolpysheva T (2003) High rates of extracellular superoxide production by lichens in the suborder Peltigerineae correlate with indices of high metabolic activity. Plant Cell Environ 26(11):1827–1837
Beckett RP, Minibayeva FV, Laufer Z (2005) Extracellular reactive oxygen species production by lichens. Lichenologist 37(5):397–407
Bertrand RL, Abdel-Hameed M, Sorensen JL (2018) Lichen biosynthetic gene clusters part II: homology mapping suggests a functional diversity. J Nat Prod 81(4):732–748
Bhattacharyya S, Deep PR, Singh S, Nayak B (2016) Lichen secondary metabolites and its biological activity. Am J Pharm Tech Res 6(6):1–7
Brakni R, Ali Ahmed M, Burger P, Schwing A, Michel G, Pomares C, Michel T (2018) UHPLC-HRMS/MS based profiling of Algerian lichens and their antimicrobial activities. Chem Biodivers 15(4):e1800031
Brown D, Kershaw KA (1986) Seasonal changes in the kinetic parameters of a photosynthetic fructose-1, 6-bisphosphatase isolated from *Peltigera rufescens*. Plant Physiol 82(2):457–461
Calcott MJ, Ackerley DF, Knight A, Keyzers RA, Owen JG (2018) Secondary metabolism in the lichen symbiosis. Chem Soc Rev 47(5):1730–1760
Cernava T, Müller H, Aschenbrenner IA, Grube M, Berg G (2015) Analyzing the antagonistic potential of the lichen microbiome against pathogens by bridging metagenomic with culture studies. Front Microbiol 6:620
Cernava T, Erlacher A, Aschenbrenner IA, Krug L, Lassek C, Riedel K, Berg G (2017) Deciphering functional diversification within the lichen microbiota by meta-omics. Microbiome 5:1–13
Cernava T, Aschenbrenner IA, Soh J, Sensen CW, Grube M, Berg G (2019) Plasticity of a holobiont: desiccation induces fasting-like metabolism within the lichen microbiota. ISME J 13(2):547–556
Chomicki G, Renner SS (2017) Partner abundance controls mutualism stability and the pace of morphological change over geologic time. Proc Natl Acad Sci USA 114(15):3951–3956
Culberson CF (1969) Chemical and botanical guide to lichen products. University of North Carolina Press, North Carolina
Dahlman L, Persson J, Palmqvist K, Näsholm T (2004) Organic and inorganic nitrogen uptake in lichens. Planta 219:459–467
Dar TUH, Dar SA, Islam SU, Mangral ZA, Dar R, Singh BP, Haque S (2022) Lichens as a repository of bioactive compounds: an open window for green therapy against diverse cancers. In: Theresa V (ed) Seminars in cancer biology, vol 86. Academic Press, United States, pp 1120–1137
Diaz EM, Vicente-Manzanares M, Sacristan M, Vicente C, Legaz ME (2011) Fungal lectin of *Peltigera canina* induces chemotropism of compatible *Nostoc* cells by constriction-relaxation pulses of cyanobiont cytoskeleton. Plant Signal Behav 6(10):1525–1536
Diaz E, Sánchez Elordi E, Santiago R, Vicente Córdoba C, Legaz González ME (2016) Algal-fungal mutualism: cell recognition and maintenance of the symbiotic status of lichens. J Vet Med Re 3:1052
diCenzo GC, Zamani M, Checcucci A, Fondi M, Griffitts JS, Finan TM (2019) Multidisciplinary approaches for studying rhizobium-legume symbioses. Can J Microbiol 65(1):1–33
Dighton J, White JF (2017) The fungal community: its organization and role in the ecosystem. CRC Press, Florida
Douglas AE, Werren JH (2016) Holes in the hologenome: why host-microbe symbioses are not holobionts. MBio 7(2):10–1128
Drew EA, Smith DC (1966) The physiology of the symbiosis in *Peltigera polydactyla* (Neck.). Hoffm Lichenologist 3(2):197–201
Drew EA, Smith DC (1967) Studies in the physiology of lichens: VIII. Movement of glucose from alga to fungus during photosynthesis in the thallus of *Peltigera polydactyla*. New Phytol 66(3):389–400

Duncan UK (1970) Introduction to British Lichens. T. Buncle & Co. Ltd, Arbroath
Eisenreich W, Strauss G, Werz U, Fuchs G, Bacher A (1993) Retrobiosynthetic analysis of carbon fixation in the phototrophic eubacterium *Chloroflexus aurantiacus*. Eur J Biochem 215(3):619–632
Eisenreich W, Knispel N, Beck A (2011) Advanced methods for the study of the chemistry and the metabolism of lichens. Phytochem Rev 10:445–456
Elix JA, Stocker-Wörgötter E (2008) Biochemistry and secondary metabolites. In: Nash TH III (ed) Lichen biology. Cambridge University Press, Cambridge, UK, pp 104–133
Eriksson OE (2005) Origin and evolution of Ascomycota—the protolichenes hypothesis. Sven Mykol Tidskr 26:30–33
Erlacher A, Cernava T, Cardinale M, Soh J, Sensen CW, Grube M, Berg G (2015) Rhizobiales as functional and endosymbiontic members in the lichen symbiosis of *Lobaria pulmonaria* L. Front Microbiol 6:53
Fluteau F, Sepulchre P (2021) Climate evolution on the geological timescale and the role of paleogeographic changes. In: Ramstein G, Landais A, Bouttes N, Sepulchre P, Govin A (eds) Paleoclimatology: Frontiers in earth sciences. Springer Cham, Cham, pp 255–269
Goebel K (1926) Ein Beitrag zur Biologie der Flechten. Ann Jard Bot Buitenzorg 36(1)
Goga M, Elečko J, Marcinčinová M, Ručová D, Bačkorová M, Bačkor M (2020) Lichen metabolites: an overview of some secondary metabolites and their biological potential. In: Jean-Michel M, Kishan GR (eds) Co-evolution of secondary metabolites. Springer Nature, London, pp 175–209
Graham LE, Trest MT, Will-Wolf S, Miicke NS, Atonio LM, Piotrowski MJ, Knack JJ (2018) Microscopic and metagenomic analyses of *Peltigera ponojensis* (Peltigerales, Ascomycota). Int J Plant Sci 179(3):241–255
Grube M, Cardinale M, de Castro Jr JV, Müller H, Berg G (2009) Species-specific structural and functional diversity of bacterial communities in lichen symbioses. ISME J 3(9):1105–1115
Grube M, Cernava T, Soh J, Fuchs S, Aschenbrenner I, Lassek C, Berg G (2015) Exploring functional contexts of symbiotic sustain within lichen-associated bacteria by comparative omics. ISME J 9(2):412–424
Gueidan C, Villasenor CR, De Hoog GS, Gorbushina AA, Untereiner WA, Lutzoni F (2008) A rock-inhabiting ancestor for mutualistic and pathogen-rich fungal lineages. Stud Mycol 61(1):111–119
Hall JL, Aked J, Gregory AJ, Storr T (1992) Carbon metabolism and transport in a biotrophic fungal association. In: Pollock CJ, Farrar JF, Gordon AJ (eds) Carbon partitioning within and between organisms. Bios Scientific Publishers, Oxford, UK, pp 181–198
Henriksson E (1958) Studies on the physiology of the lichen *Collema* II: a preliminary report on the isolated fungal partner with special regard to its behaviour when growing together with the symbiotic alga. Sven Bot Tidskr 52:391–396
Herz W, Grisebach H, Kirby GW, Elix JA, Whitton AA, Sargent MV (1984) Recent progress in the chemistry of lichen substances. Springer Vienna, pp 103–234
Hill DJ (1976) Physiology of lichen symbiosis. In: Lichenology: progress and problems; proceedings of an international symposium
Hodkinson BP, Lutzoni F (2009) A microbiotic survey of lichen-associated bacteria reveals a new lineage from the Rhizobiales. Symbiosis 49:163–180
Hodkinson BP, Gottel NR, Schadt CW, Lutzoni F (2012) Photoautotrophic symbiont and geography are major factors affecting highly structured and diverse bacterial communities in the lichen microbiome. Environ Microbiol 14(1):147–161
Hodkinson BP, Allen JL, Forrest LL, Goffinet B, Sérusiaux E, Andrésson ÓS, Miao V, Bellenger JP, Lutzoni F (2014) Lichen-symbiotic cyanobacteria associated with *Peltigera* have an alternative vanadium-dependent nitrogen fixation system. Eur J Phycol 49(1):11–19
Hom EF, Murray AW (2014) Niche engineering demonstrates a latent capacity for fungal-algal mutualism. Science 345(6192):94–98
Honegger R (1998) The lichen symbiosis—what is so spectacular about it? Lichenologist 30(3):193–212

Honegger R, Kutasi V, Ruffner HP (1993) Polyol patterns in eleven species of aposymbiotically cultured lichen mycobionts. Mycol Res 97(1):35–39

Honegger R, Edwards D, Axe L (2013) The earliest records of internally stratified cyanobacterial and algal lichens from the lower Devonian of the Welsh borderland. New Phytol 197(1):264–275

Joneson S, Armaleo D, Lutzoni F (2011) Fungal and algal gene expression in early developmental stages of lichen-symbiosis. Mycologia 103(2):291–306

Kitano H (2002) Systems biology: a brief overview. Science 295(5560):1662–1664

Kono M, Kon Y, Ohmura Y, Satta Y, Terai Y (2020) In vitro resynthesis of lichenization reveals the genetic background of symbiosis-specific fungal algal interaction in *Usnea hakonensis*. BMC Genomics 21:671

Lines CEM, Ratcliffe RG, Rees TAV, Southon TE (1989) A ^{13}C NMR study of photosynthate transport and metabolism in the lichen *Xanthoria calcicole* Oxner. New Phytol 111(3):447–456

Lutzoni F, Pagel M, Reeb V (2001) Major fungal lineages are derived from lichen symbiotic ancestors. Nature 411(6840):937–940

Meessen J, Eppenstein S, Ott S (2013) Recognition mechanisms during the pre-contact state of lichens: II. Influence of algal exudates and ribitol on the response of the mycobiont of *Fulgensia bracteate*. Symbiosis 59:131–143

Millanes AM, Fontaniella B, Garcia ML, Solas MT, Vicente C, Legaz ME (2004) Cytochemical location of urease in the cell wall of two different lichen phycobionts. Tissue Cell 36(6):373–377

Millbank JW, Kershaw (1973) Nitrogen metabolism. In: Ahmadjian V, Hale ME (eds) The Lichens. Academic Press, Londan, pp 289–307

Nash TH (1996) Lichen biology. Cambridge University Press, Cambridge, UK

Nash TH III (2008) Nitrogen, its metabolism and potential contribution to ecosystems. In: Lichen Biology, pp 216–233

Nazem-Bokaee H, Hom EF, Warden AC, Mathews S, Gueidan C (2021) Towards a systems biology approach to understanding the lichen symbiosis: opportunities and challenges of implementing network modelling. Front Microbiol 12:667864

Nelsen MP, Lücking R, Boyce CK, Lumbsch HT, Ree RH (2020) No support for the emergence of lichens prior to the evolution of vascular plants. Geobiology 18(1):3–13

Nguyen KH, Chollet-Krugler M, Gouault N, Tomasi S (2013) UV-protectant metabolites from lichens and their symbiotic partners. Nat Prod Rep 30(12):1490–1508

Oulhen N, Schulz BJ, Carrier TJ (2016) English translation of Heinrich Anton de Bary's 1878 speech, 'Die Erscheinung der Symbiose' ('De la symbiose'). Symbiosis 69(3):131–139

Petrini O, Hake U, Dreyfuss MM (1990) An analysis of fungal communities isolated from fruticose lichens. Mycologia 82(4):444–451

Rafferty NE, Caradonna PJ, Bronstein JL (2015) Phenological shifts and the fate of mutualisms. Oikos 124(1):14–21

Richardson DHS, Smith DC (1966) The physiology of the symbiosis in *Xanthoria aureola* (Ach.) Erichs. Lichenologist 3(2):202–206

Richardson DHS, Smith DC (1968) Lichen physiology: IX. Carbohydrate movement from the *Trebouxia* symbiont of *Xanthoria aureola* to the fungus. New Phytol 67(1):61–68

Schoch CL, Sung GH, López-Giráldez F, Townsend JP, Miadlikowska J, Hofstetter V, Spatafora JW (2009) The ascomycota tree of life: a phylum-wide phylogeny clarifies the origin and evolution of fundamental reproductive and ecological traits. Syst Biol 58(2):224–239

Schwartzman DW (2010) Was the origin of the lichen symbiosis triggered by declining atmospheric carbon dioxide levels? Biol Lichens Symbios 105:191–196

Singh RS, Walia AK (2014) Characteristics of lichen lectins and their role in symbiosis. Symbiosis 62:123–134

Smith DC, Drew EA (1965) Studies in the physiology of lichens: V. Translocation from the algal layer to the medulla in *Peltigera polydactyla*. New Phytol 64(2):195–200

Smith D, Muscatine L, Lewis D (1969) Carbohydrate movement from autotrophs to heterotrophs in parasitic and mutualistic symbiosis. Biol Rev 44:17–90

Smith HB, Dal Grande F, Muggia L, Keuler R, Divakar PK, Grewe F, Leavitt SD (2020) Metagenomic data reveal diverse fungal and algal communities associated with the lichen symbiosis. Symbiosis 82(1):133–147

Sønstebø JH, Rohrlack T (2011) Possible implications of Chytrid parasitism for population subdivision in freshwater cyanobacteria of the genus *Planktothrix*. Appl Environ Microbiol 77(4):1344–1351

Spribille T, Tuovinen V, Resl P, Vanderpool D, Wolinski H, Aime MC, McCutcheon JP (2016) Basidiomycete yeasts in the cortex of ascomycete macrolichens. Science 353(6298):488–492

Spribille T, Resl P, Stanton DE, Tagirdzhanova G (2022) Evolutionary biology of lichen symbioses. New Phytol 234(5):1566–1582

Strauss G, Fuchs G (1993) Enzymes of a novel autotrophic CO_2 fixation pathway in the phototrophic bacterium *Chloroflexus aurantiacus,* the 3-hydroxypropionate cycle. Eur J Biochem 215:633–643

Tabita FR (1987) Carbon dioxide fixation and its regulation in cyanobacteria. In: Fay P, Van Baalen C (eds) The cyanobacteria. Elsevier, Amsterdam, pp 95–117

Tagirdzhanova G, Saary P, Tingley JP, Díaz-Escandón D, Abbott DW, Finn RD, Spribille T (2021) Predicted input of uncultured fungal symbionts to a lichen symbiosis from metagenome-assembled genomes. Genome Biol Evol 13(4):evab047

Ten Veldhuis MC, Ananyev G, Dismukes GC (2020) Symbiosis extended: exchange of photosynthetic O_2 and fungal-respired CO_2 mutually power metabolism of lichen symbionts. Photosynth Res 143(3):287–299

Vicente C, Legaz ME (1988) Lichen enzymology. In: Galun M (ed) Handbook of lichenology. CRC Press, Florida, pp 239–281

West NJ, Parrot D, Fayet C, Grube M, Tomasi S, Suzuki MT (2018) Marine cyanolichens from different littoral zones are associated with distinct bacterial communities. PeerJ 6:e5208

The manufacturer's authorised representative in the EU is Springer Nature Customer Service Centre GmbH, Europaplatz 3, 69115 Heidelberg, Germany. If you have any concerns regarding our products, please contact ProductSafety@springernature.com

Printed and bound by CPI Group (UK) Ltd, Croydon, CR0 4YY
01/10/2026
02225922-0017